Roberto de Franchis
Editor

Portal Hypertension VII

Proceedings of the 7th Baveno Consensus Workshop: Personalized Care in Portal Hypertension

Springer

Editor
Roberto de Franchis
Gastroenterology Unit
University of Milan
Milan, Milano, Italy

ISBN 978-3-031-08551-2 ISBN 978-3-031-08552-9 (eBook)
https://doi.org/10.1007/978-3-031-08552-9

This Springer imprint is published by the registered company Springer Nature Switzerland AG
The registered company address is: Gewerbestrasse 11, 6330 Cham, Switzerland

Baveno VII was dedicated to the memory of four Baveno legends, colleagues who have recently passed away and gave fundamental contributions to the success of the Baveno endeavor.

Tom Boyer
Andrew K. Burroughs,
Roberto J. Groszmann
Luigi Pagliaro

Preface

Baveno VII was a sequel of the Baveno I-VI Workshops, which were held every 5 years from 1990 to 2015. The aims of these meetings were to develop definitions of key events in portal hypertension, to review the existing evidence on the natural history, the diagnosis, and the therapeutic modalities of portal hypertension, and to issue evidence-based recommendations for the conduct of clinical trials and the management of patients. All these meetings were successful and produced consensus recommendations mainly centered on the management of varices and variceal hemorrhage. Globally, the consensus reports of the Baveno I-VI workshops received more than 4300 citations in the medical literature.

To continue and expand the work of the previous meetings, a Baveno VII workshop was planned for March 20–21, 2020. However, the COVID 19 pandemic and the consequent lockdown forced the organizers to postpone the workshop until the end of October 2021 and to change the format from face to face to a virtual meeting. Despite these limitations, many of the experts responsible for the major achievements over the last years in the field of portal hypertension and its complications participated in the workshop. Many of them had attended the previous meetings.

In 2016, shortly after Baveno VI and inspired by the spirit of the Baveno meetings, the Baveno Cooperation was launched with the aim to create a network of collaborating experts in portal hypertension aiming to establish a continuous, high-quality research agenda. In 2019, the European Association for the Study of the Liver (EASL) endorsed the Baveno Cooperation as an official EASL consortium.

Patients with cirrhosis transition through different prognostic stages, the main ones being the compensated and decompensated stages. The hallmark of transition from the compensated to the decompensated stage is the development of complications such as ascites, variceal hemorrhage, and overt hepatic encephalopathy. At the Baveno VI conference, the concept of compensated advanced chronic liver disease (cACLD) was introduced, based on noninvasive tests that would predict the development of complications of cirrhosis. Among patients with compensated cirrhosis or cACLD, at least two different stages have been identified, those with and without clinically significant portal hypertension (CSPH). The various disease stages differ in outcomes, and therefore patients in different stages have different diagnostic and therapeutic needs. Accordingly, the Baveno VII workshop was entitled "Personalized Care for Portal Hypertension." The structure of the workshop was similar to that of the previous ones—there were four sessions with a total of nine parts, each dealing with a key topic: the relevance and indications for measuring the hepatic venous pressure gradient (HVPG) as a gold standard, the use of noninvasive tools for the diagnosis of cACLD and CSPH, the impact of etiological and of non-etiological therapies in the course of cirrhosis, the prevention of the first episode of decompensation, the management of the acute bleeding episode, the prevention of further decompensation, as well as the diagnosis and management of splanchnic vein thrombosis and other vascular disorders of the liver. For each of these topics, a thorough review of the medical literature was made, and a series of consensus statements/recommendations were discussed and agreed upon. Between sessions, there were seven more lectures, held by world experts. The topics of the lectures were: New concept of risk stratification, Clinical stages and ordinal outcomes in portal hypertension, Lifestyle, and genetic modifiers of progression of cACLD, PELS in progression. Can they regress? Fibrogenesis and regression of fibrosis, Angiogenesis, and progression of ACLD, and Drugs to modify liver fibrosis progression and regression.

These proceedings follow closely the structure of the workshop. The consensus statements that were agreed upon in each session are reported at the end of the pertinent chapters. Whenever applicable, the level of existing evidence was evaluated, and the recommendations were ranked according to the GRADE System which ranks the scientific evidence from A (high) to D (very low). The strength of the recommendations was graded 1 (strong) and 2 (weak). Each recommendation is labeled as "new," "changed," or "unchanged" in comparison with the Baveno VI consensus statements.

We wish to warmly thank the friends who accepted to give the lectures and to serve as chairpersons and panelists of the sessions, and who helped us by working hard during the past 3 years in the preparation of the workshop and the chapters of this book.

In addition, we are grateful to the following organizations who endorsed and supported Baveno VII: (a) International Scientific Societies: EASL (European Association for the Study of the Liver). (b) National scientific societies: AASLD (American Association for the Study of Liver Disease); AEEH (Spanish Association for the Study of the Liver); AFEF (French Association for the Study of the Liver); AIGO (Italian Association of Hospital Gastroenterologists and Endoscopists); AISF (Italian Association for the Study of the Liver); CIBERehd (Spanish network of biomedical investigation in liver and digestive diseases); ÖGGH (Austrian Society for Gastroenterology and Hepatology); SASL (Swiss Association for the Study of the Liver); SIGE (Italian Society of Gastroenterology). (c) International Research groups: Decision, Galaxy, Microb-Predict.

We are indebted to the EASL and Meta-Fusion for the use of the EASL Platform, to Cyriac Couvas, and to the Meta-Fusion personnel for the technical assistance during the workshop.

We also wish to thank Annamaria Sorresso, Denise Santi, and the staff of ADB Eventi e Congressi, who managed brilliantly the organization of the workshop.

Finally, we wish to thank all the companies who sponsored the workshop, Catherine Mazars of Springer for her encouragement and cooperation in this project, and Springer for the timely and excellent production of this volume.

Department of Clinical and Biomedical Sciences
University of Milan
Milan, Italy

Roberto de Franchis

Thomas D. Boyer, MD (1943–2018)—A Tribute

Tom Boyer was a master hepatologist, accomplished scientist, inspiring mentor, and a friend of the Baveno community.

His passion for the liver arose when he was a Liver Disease fellow (1974–1976) at the University of Southern California with the legendary Telfer "Pete" Reynolds, where he honed his skills in clinical hepatology. He then moved to San Francisco for GI fellowship at the University of California in San Francisco, where he honed his skills in basic sciences in the lab of Rudi Schmidt. He stayed on as faculty at UCSF and quickly rose through the ranks becoming Professor of Medicine and Chief of GI at the VA San Francisco in 1981. In 1990, he moved to Emory where he was Professor of Medicine and Adjunct Professor of Biochemistry. His final professional home was in Tucson, Arizona, where he was Chairman of Medicine from 2008 to 2014 before transitioning to Director of the Liver Research Unit. In recognition to his contributions to hepatology, the University of Arizona honored him by renaming this unit as the "Thomas D Boyer Liver Institute."

Tom loved patient care and was a great clinician, but he was also a physician scientist that not only performed laboratory research focused on the biology of oxidant stress and drug detoxification but also clinical research. In fact, his true passion was in patient-oriented research in the areas of cirrhosis, portal hypertension, and the complications of cirrhosis where he performed top-notch research, oftentimes as the leader of multicenter collaborative trials.

Tom's extraordinary breadth of talent and interest made him the perfect co-editor of his famous hepatology textbook, *Zakim and Boyer's Hepatology* which is now in its seventh edition. He was also a real leader and was past President of both the AASLD (2002) and the International Association for the Study of the Liver (2008–2010) having received the AASLD Distinguished Service Award in 2012 for his contributions to the association.

Tom Boyer was an active participant of the Baveno consensus conferences from 2000 to 2015. He would always stand in the back of the room and make incredibly insightful comments that would lead to modification of the recommendations. We will miss his candor and common sense.

At this Baveno VII conference, his absence was clearly felt because, as one can glean from his vast publication list, he would have had something to contribute to each of the panels. On measurement of hepatic venous pressure gradient (HVPG–panel 1), Tom was one of the first to publish on the performance of direct transhepatic portal pressure measurements, and he wrote about changing the practice of clinical hepatology by measuring HVPG. Tom participated in multicenter studies on etiological and non-etiological therapies for cirrhosis of different etiologies (panels 3 and 4). He performed many studies regarding the transjugular intrahepatic portosystemic shunt (TIPS, discussed in panels 5,6,7,8, and 9) and, in fact, together with an interventional radiologist, Ziv Haskal, he authored two iterations of the AASLD guideline on TIPS. Two of his seminal randomized control trials were the second North-American placebo-controlled trial of terlipressin for hepatorenal syndrome (panel 7) and the multicenter trial of distal splenorenal shunt vs. TIPS for variceal bleeding (panel 6).

Remarkably, Tom Boyer never lost perspective of the most important objective of research and the key focus of Baveno conferences. He expressed this perfectly well in the title of an editorial regarding clinical trials for variceal bleeding in which he concluded that, at the end, the winner was the patient. Let us not forget this. Tom Boyer's memory will live on in the Baveno community.

New Haven, CT, USA
West Haven, CT, USA

Guadalupe Garcia-Tsao

Memories of Prof Andrew K. Burroughs, MD (1953–2014)

A founding father of the Baveno Group, an amazing clinician, a true European, and a friend to so many. Just some of the many tributes written and spoken over the years in recognition of Prof Andrew Kenneth Burroughs (A.K.B.).

It is hard to believe that it was in 2014 that he died—it often feels much more recent than that and particularly at meetings such as Baveno. So what I am not going to do is describe Andy's clinical and scientific achievements—these are legion and well described both in paper and online.

What I want to do is describe what it was like working, and sharing an office, with Andy, and to do that I have to slightly describe myself.

I arrived at the Royal Free in 1993, a very young and not-very-bright trainee. I was informed, and rapidly realized, that I was working under someone quite exceptional, who delivered on those two pivotal leadership elements. He was both aspirational and inspirational.

First of all, he made Hepatology fun. He had an energy and vitality that was infectious. It would also be fair to say that he couldn't understand when other people didn't have that same emotion when confronted by yet another textbook case (usually with an extra twist!) of a rare disease.

Andy was not a morning person, preferring to work late in the evening, discussing papers, and going through abstracts with the large group of visiting fellows that were drawn to him—he was very much a true representative of the Dame Sheila

Sherlock legacy, sharing knowledge and experience both in and out of Hepatology to colleagues from around the world.

When I was his registrar, it was those evenings when we would go and see ward referrals—at a time when the hospital was quieter, and I selfishly would have him all to myself, and would bombard him with questions—no doubt driving him mad, but he never showed it.

Andy was generous with his time for everyone, and one of his proudest achievements was that no visiting fellow left without their name on a paper, and his pastoral care extended beyond just paper writing. At International meetings, he would introduce his junior colleagues—including me—to the good and the great within their field, be it portal hypertension, transplantation, or any of the other myriad conditions in which he not only had an interest but was regarded as an authority on.

Andy was an interesting mixture of contrasts. His office was a complete tip-paper, old meeting bags, journals, etc.—all apparently strewn in random order such that you were lucky if you could find a seat.

Once when he was on holiday, I took the opportunity of throwing out ten meeting bags containing abstract books, etc.—all at least 5 years old. Of course, he noticed—I kind of never doubted that he would—he wasn't particularly impressed!

But to visit him at home, especially at the weekend, you would be greeted by the archetypal Italian bar owner—unshaven in vest and trousers—properly relaxed, usually with some amazing Italian cheese on the kitchen table and the inevitable offer of a glass of wine.

And then the contrast continued, when he showed you his immaculately annotated stamp collection beautifully laid out in a small, but surprisingly tidy room—in such stark contrast to his office!

I look back and over the years have recognized that Andy was quite visionary. He had long talked about transplantation for ETOH-related hepatitis, recognized early the role of advanced nurse practitioners in the delivery of health care, and while not always the first to embrace new technology, also saw the digitization of health care as being inevitable.

He was also a champion for the disenfranchised and the downtrodden. This was often apparent at the transplant listing meeting, and his interaction with patients and their families. Many may not know but he oversaw the publication, by the Royal College of Physicians, of a collection of letters from transplant recipients to their donor's families and it makes for powerful reading.

For me, when Andy was alive, every day was a school day, and it is true to say that in his death, he continued to educate me and my colleagues. It is clear that none of us are irreplaceable, and that saving things up for retirement is a mistake, as the future is unpredictable. If there are things you wish to do, especially with your family, do them now—as ultimately it is they who will miss you the most when you have gone. Many of us saw ourselves as part of Andy's extended family, and yes, we miss him.

London, UK David Patch

Tribute to Roberto J. Groszmann

Prof Roberto J. Groszmann

Buenos Aires, 1939; New Haven (CT) 2021

Roberto Groszmann, the father of the scientific approach to portal hypertension died in early 2021, which was a devastating news for all the portal hypertension community that he so much contributed to enhance, shine, and accomplish tremendous advances. The faculty and organizers of the Baveno VII meeting wanted to pay a special tribute to his figure in this special occasion in recognition and gratitude for his continued dedication to the study of the mechanisms and therapy of portal hypertension for over 40 years.

Roberto was born in Buenos Aires, Argentina, as he always very proudly reminded, and grew up scientifically in the USA, where he went in 1965 just after his marriage with Aida for postgraduate training in Chicago, followed by a postdoctoral fellowship in Washington DC, where he started his studies on hemodynamics of portal hypertension at the Jay Cohn laboratory. After a period of 4 years as a Scientist back in Argentina, he returned to the USA in 1975 due to the political unrest and civil violence that preceded a military coup. He settled down as an Assistant Professor at Yale University, where he continued to work for the rest of his life. He retired in 2009, but continued to cooperate in many scientific, editorial, and teaching activities as an Emeritus Professor.

His contributions to portal hypertension have been immense. He was the first to conduct truly translational research in this field, and his lab at the West Haven VA has been the light-fare illuminating the horizon for all those interested in portal hypertension over the world, many of whom had the privilege of training with him and enjoying his mentorship and continued advice for the rest of their careers, including many of today's KOL worldwide.

Roberto's contributions have been fantastic. Just to mention a few, he was the first to characterize the hyperkinetic splanchnic circulation of patients with cirrhosis and portal hypertension, shown by decreased mesenteric, portal, and hepatic transit times, almost 50 years ago. He then developed experimental rat models where he studied the mechanism of this hyperkinetic circulation and its role in aggravating portal hypertension, which provided the rationale for using vasoactive drugs and non-selective beta-blockers for its treatment. Afterward, he focused on the study of the molecular mechanisms of the splanchnic vasodilatation, including seminal studies on the role of nitric oxide and of bacterial translocation, together with studies on isolated perfusion of mesenteric beds and of the portal-collateral circulation that allowed a better understanding of the complexities of the many mechanisms involved in the regulation of splanchnic and liver blood flow in chronic liver disease. On the turning of the century, he was deeply involved in studying the mechanisms of the increased hepatic vascular tone and of endothelial dysfunction in cirrhosis, through a series of brilliant translational experiments that pointed towards new targets for therapy. Clinically, these years were rich in cooperation, mainly through the Timolol study, a 10-year international effort funded by the NIH that put together a formidable group: the Yale team (Roberto and Lupe Garcia-Tsao), Boston (Norman Grace), Royal Free Hospital (London) with Andy Burroughs and David Patch, and the Barcelona Hepatic Hemodynamic Laboratory (with Juan Carlos Garcia-Pagàn and me). This cooperation, besides its scientific contributions such as introducing the concept of "clinically significant portal hypertension," was key for many other aspects, including to frame the germ of would later crystalize, thanks to the enthusiasm and hard work of Roberto de Franchis as the Baveno meetings. Roberto Groszmann was deeply involved in all of them, and we all will remember forever the vivid, endless discussions during the meetings, that started at breakfast and didn't end until late at night, after having dinner in an informal ambience where the young trainees and fellows could sit together with him and be seduced by the magic of portal hypertension and its mysteries, many of them becoming the ones that have now the responsibility of continuing his path carrying over new innovative research ultimately aiming for a better care for the patients suffering from portal hypertension.

Finally, I would like to comment that Roberto Groszmann was not only a great scientist, doctor, and teacher, but a man of many qualities, that enjoyed being with his family, his dog, and his life-long friends, many of them acquired at or around his work and who become part of his extended family. Roberto had a warm personality, was a man who enlightened one's life, inspired one's work, paved the path of one's career, provided the best advice on many aspects of life, and became one's most valued friend. The Baveno community will never forget him. He will be sorely missed.

Bern, Switzerland
Barcelona, Spain

Jaime Bosch

Tribute to Prof Luigi Pagliaro (1931–2020)

Professor Luigi Pagliaro was a brilliant man who dedicated his life to patient care, clinical research, teaching, and mentoring his fellows.

Graduated in Medicine in 1954, he initially focused his clinical research on viral hepatitis and in 1968 was among the first to demonstrate the histological evolution of viral hepatitis in cirrhosis.

He was an outstanding clinical researcher mentoring a number of fellows and creating a research group well known in clinical hepatology since the early 1980s. His activity in clinical research produced 243 published articles in major medical journals.

In 1970, he was a founding member of AISF (Italian Association for the Study of the Liver). Since 1973, he was head of a clinical Department of Medicine in the V. Cervello Hospital in Palermo which soon became one of the most important hepatological centers in Italy, with well-recognized clinical reputation and research activity particularly in viral hepatitis, cirrhosis, and portal hypertension. In 1978, he was charged as the president of the EASL annual meeting in Padua. He was full professor of Medicine in the University of Palermo for 30 years, until 2003 when he retired becoming Emeritus Professor of Medicine.

He was a pioneer in teaching and spreading of clinical methodology and evidence-based medicine (EBM) in the academic and the medical community in Italy. With a seminal meeting for hepatologists and statisticians held in Erice in

1986, he was the first to introduce the use of meta-analysis in hepatology. However, he also recognized the limits of EBM in individual patients out of the bounds of available evidence and emphasized the importance of soundly built clinical expertise and patient preferences.

However, beyond his critical scientific thinking, he will be remembered for his dedication to clinical medicine and for his extraordinary capability in taking care of patients as human beings. His humanity and humility were cornerstones of his personality. When he entered in the rooms to visit patients in the ward, he was always smiling, greeting patients with bowed head, and leaving the step to students and women. He was used to personally put his chair next to the patient and soon started with him a relationship made of complicity and loving care. His door was always open to patients and relatives without any limitation. His friendly and loving approach to patients became a paradigm for hundreds of doctors who worked with him and had the chance to be introduced to his own path, to clinical medicine and methodology. He focused his career on remodeling the field of Internal Medicine and Hepatology to better serve his patients.

He did not fail to be next to everyone in his staff, students, and cooperators supporting those who needed help and enthusiastically encouraging any new clinical research project. All those who worked closely with him and those with whom he collaborated in Italy and around the world will remember him as a person of profound vision and deep principles.

Words cannot capture the legacy he leaves behind, after decades of devotion to teaching, training, and research.

Passing away on September 7, 2020, he left his wife Enza and two children, Antonio and Laura, who were his strong support all along his career.

Darkness was like a dream. And people from my past came to visit me (Ko Un).

Palermo, Italy Gennaro D'Amico

Contents

Contributors

Juan G. Abraldes, MD, MMSc Liver Unit, Division of Gastroenterology, University of Alberta, Edmonton, AB, Canada

Laia Aceituno, MD Liver Unit, Hospital Universitari Vall d'Hebron, Vall d'Hebron Institute of Research (VHIR), Vall d'Hebron Barcelona Hospital Campus, Universitat Autònoma de Barcelona, Barcelona, Spain

Centro para la Investigación Biomédica en Red de Enfermedades Hepáticas y Digestivas (CIBERehd), Madrid, Spain

Agustín Albillos, MD Department of Gastroenterology and Hepatology, Hospital Ramón y Cajal, Instituto Ramón y Cajal de Investigación Sanitaria (IRYCIS), Centro de Investigación Biomédica Red de Enfermedades Hepáticas y Digestivas (CIBEREHD), Universidad de Alcalá, Madrid, Spain

Wayne W. H. Bai, MBChB Waikato District Health Board, University of Auckland, Auckland, New Zealand

Anna Baiges, MD Barcelona Hepatic Hemodynamic Laboratory, Liver Unit, Hospital Clínic, IDIBAPS, CIBERehd, Health care provider of the European Reference Network on Rare Liver Disorders (ERN Liver), Barcelona, Spain

Jasmohan S. Bajaj, MD Division of Gastroenterology, Hepatology and Nutrition, Virginia Commonwealth University and McGuire VA Medical Center, Richmond, VA, USA

Rafael Bañares, MD, PhD Faculty of Medicine, Complutense University of Madrid, Madrid, Spain

Digestive Disease Service, Hospital General Universitario Gregorio Marañón, Facultad de Medicina, Universidad Complutense, Madrid, Spain

CIBERehd, ISCiii, Madrid, Spain

Ana Barreira, MD Liver Unit, Hospital Universitari Vall d'Hebron, Vall d'Hebron Institute of Research (VHIR), Vall d'Hebron Barcelona Hospital Campus, Universitat Autònoma de Barcelona, Barcelona, Spain

Universitat Autònoma de Barcelona, Barcelona, Spain

Centro para la Investigación Biomédica en Red de Enfermedades Hepáticas y Digestivas (CIBERehd), Madrid, Spain

Marta Barrufet, MD Interventional Radiology Unit, CDI, Hospital Clinic, Barcelona, Spain

Department of Radiology, Hospital Clínic de Barcelona, Barcelona, Spain

Chiara Becchetti, MD Department of Visceral Surgery and Medicine, Inselspital, Bern University Hospital, University of Bern, Bern, Switzerland

Hepatology, Department for BioMedical Research, Visceral Surgery and Medicine, University of Bern, Bern, Switzerland

Lina Benajiba, MD Centre d'Investigations Cliniques, INSERM, CIC1427, Université Paris Cité, Hôpital Saint-Louis, AP-HP, Paris, France

Patricia Bermudez, MD Department of Radiology, Hospital Clínic de Barcelona, Barcelona, Spain

Interventional Radiology Unit, CDI, Hospital Clinic, Barcelona, Spain

Annalisa Berzigotti, MD, PhD Department of Visceral Surgery and Medicine, Inselspital, Bern University Hospital, University of Bern, Bern, Switzerland

Jaime Bosch, MD University Clinic for Visceral Surgery and Medicine, Inselspital-Bern, University Hospital of Bern, University of Bern, Bern, Switzerland

Department of Biomedical Research, University of Bern, Bern, Switzerland

IDIBAPS, Centro de Investigación Biomédica en Red de Enfermedades Hepáticas y Digestivas (CIBERehd), Barcelona, Spain

Department of Hepatology, University Clinic for Visceral Surgery and Medicine, Inselspital, Bern University Hospital, University of Bern, Bern, Switzerland

Charlotte Bouzbib, MD Groupement Hospitalier APHP-Sorbonne Université, Service d'Hépato-Gastro-Entérologie, Hôpital Pitié-Salpêtrière, Paris, France

Sorbonne Université, INSERM, Centre de Recherche Saint-Antoine (CRSA), Institute of Cardiometabolism and Nutrition (ICAN), Paris, France

Brain-Liver Pitié-Salpêtrière Study Group (BLIPS), APHP-Sorbonne Université, Paris, France

Christophe Bureau, MD Service d'Hépato-Gastroentérologie, CHU Toulouse cedex et Université Paul Sabatier, Toulouse, France

Seth M. Buryska Division of Gastroenterology and Hepatology, Mayo Clinic, Rochester, MN, USA

Vincenza Calvaruso, MD Section of Gastroenterology and Hepatology, Department of Health Promotion, Mother and Child Care, Internal Medicine and Medical Specialties (PROMISE), University of Palermo, Palermo, Italy

Gastroenterology and Hepatology Unit, Department of Health Promotion Sciences, Maternal and Infant Care, Internal Medicine and Medical Specialities (PROMISE), University of Palermo, Palermo, Italy

Andrés Cardenas GI & Liver Transplant Unit, Institut de Malalties Digestives I Metaboliques, Hospital Clínic Barcelona, University of Barcelona, Barcelona, Spain

Gastroenterology, Hepatology and Endoscopy, University of Barcelona, Barcelona, Spain

Institut d'Investigacions Biomèdiques August Pi-Sunyer (IDIBAPS), Barcelona, Spain

Ciber de Enfermedades Hepáticas y Digestivas (CIBERehd), Barcelona, Spain

Antonio Colecchia, MD Department of Gastroenterology, Verona University Hospital, Verona, Italy

Gennaro D'Amico, MD Gastroenterology Unit, Ospedale V Cervello, Palermo, Italy

Clinica La Maddalena, Palermo, Italy

Alessandra Dell'Era, MD Department of Biomedical and Clinical Sciences, University of Milan, Milan, Italy

Gastroenterology and Endoscopy Unit, ASST Fatebenfratelli Sacco, Milan, Italy

Angels Escorsell Digestive Diseases Unit, FGS Hospital de la Santa Creu i Sant Pau, Barcelona, Spain

CiberEHD, Madrid, Spain

Jonathan A. Fallowfield, MD Translational Liver Research, Centre for Inflammation Research, The Queen's Medical Research Institute, University of Edinburgh, Edinburgh, UK

Hector Ferral, MD Department of Radiology, Section of Interventional Radiology, Louisiana State University, New Orleans, LA, USA

NorthShore University HealthSystem, Evanston, IL, USA

Roberto de Franchis, MD, AGAF Department of Clinical and Biomedical Sciences, University of Milan, Milan, Italy

Sven Francque, MD Department of Gastroenterology and Hepatology, Antwerp University Hospital, Edegem, Belgium

Translational Science in Inflammation and Immunology, Faculty of Medicine and Health Sciences, University of Antwerp, Antwerp, Belgium

Laboratory of Experimental Medicine and Paediatrics, Faculty of Medicine and Health Science, University of Antwerp, Wilrijk, Belgium

Ron C. Gaba, MD, MS Division of Interventional Radiology, Department of Radiology, College of Medicine, University of Illinois at Chicago, Chicago, IL, USA

Juan Carlos Garcia-Pagàn, MD, PhD Barcelona Hepatic Hemodynamic Laboratory, Liver Unit, Hospital Clínic, Institut de Investigacions Biomèdiques August Pi i Sunyer (IDIBAPS), University of Barcelona, Barcelona, Spain

CIBEREHD (Centro de Investigación Biomédica en Red Enfermedades Hepáticas y Digestivas), Barcelona, Spain

Health Care Provider of the European Reference Network on Rare Liver Disorders (ERN-Liver), ERN-Liver, Barcelona, Spain

Guadalupe Garcia-Tsao, MD, FAASLD Digestive Diseases Section, School of Medicine, Yale University, New Haven, CT, USA

Digestive Diseases Section, Yale University, New Haven, CT, USA

Digestive Diseases Section, VA-CT Healthcare System, West Haven, CT, USA

Joan Genescà, MD Liver Unit, Hospital Universitari Vall d'Hebron, Vall d'Hebron Institute of Research (VHIR), Vall d'Hebron Barcelona Hospital Campus, Barcelona, Spain

Universitat Autònoma de Barcelona, Barcelona, Spain

CIBERehd-Instituto de Salud Carlos III, Barcelona, Spain

Andrea De Gottardi, MD Department of Gastroenterology and Hepatology, Ente Ospedaliero Cantonale, Lugano, Switzerland

Faculty of Biomedical Sciences, Università della Svizzera Italiana, Lugano, Switzerland

Jordi Gracia-Sancho, MD Liver Vascular Biology, IDIBAPS Research Institute, Hospital Clínic Barcelona, CIBEREHD, Barcelona, Spain

Hepatology, Department for Biomedical Research, University of Berne and Inselspital, Bern, Switzerland

Antonio Guerrero, MD Servicio de Gastroenterología, y Hepatología, Hospital Universitario Ramón y Cajal, Instituto Ramón y Cajal de Investigación Sanitaria (IRYCIS), CIBERehd, Universidad de Alcalá, Madrid, Spain

Guohong Han, MD Department of Liver Diseases and Digestive Interventional Radiology, National Clinical Research Center for Digestive Diseases and Xijing Hospital of Digestive Diseases, Fourth Military Medical University, Xi'an, China

Department of Liver Diseases and Interventional Radiology, Digestive Diseases Hospital, Xi'an International Medical Center Hospital, Northwest University, Xi'an, China

Ziv J. Haskal, MD Department of Radiology and Medical Imaging/Interventional Radiology, University of Virginia School of Medicine, Richmond, VA, USA

Virginia Hernandez-Gea, MD Barcelona Hepatic Hemodynamic Laboratory, Liver Unit, Hospital Clinic-IDIBAPS, Health Care Provider of the European Reference Network on Rare Liver Disorders (ERN-Liver), Barcelona, Spain

Centro de Investigación Biomédica Red de Enfermedades Hepáticas y Digestivas (CIBERehd), Barcelona, Spain

Luis Ibáñez-Samaniego, MD Digestive Diseases Department, Hospital General Universitario Gregorio Marañón, Facultad de Medicina, Universidad Complutense, Madrid, Spain

Centro de Investigación Biomédica en Red (CIBERehd), ISCiii, Madrid, Spain

Jidong Jia, MD Beijing Key Laboratory of Translational Medicine on Liver Cirrhosis, Beijing, China

National Clinical Research Center of Digestive Diseases, Beijing, China

Liver Research Center, Beijing Friendship Hospital, Capital Medical University, Beijing, China

Kosuke Kaji Department of Gastroenterology, Nara Medical University, Kashihara, Nara, Japan

Jean-Jacques Kiladjian, MD Centre d'Investigations Cliniques, INSERM, CIC1427, Université Paris Cité, Hôpital Saint-Louis, AP-HP, Paris, France

Aleksander Krag, MD Centre for Liver Research, Department of Gastroenterology and Hepatology, Odense University Hospital, Odense, Denmark

Institute of Clinical Research, University of Southern Denmark, Odense, Denmark

Wim Laleman, MD PhD Department of Gastroenterology and Hepatology, Section of Liver and Biliopancreatic Disorders, University Hospitals Leuven, Leuven, Belgium

Department of Gastroenterology and Hepatology, University Hospital Gasthuisberg, KU Leuven, Leuven, Belgium

Hélène Larrue Servie d'Hépatologie Hôpital Rangueil CHU Toulouse cedex et Université Paul Sabatier, Toulouse, France

Sabela Lens, MD Liver Unit, Hospital Clínic de Barcelona, Universidad de Barcelona, IDIBAPS, Barcelona, Spain

Centro de Investigación Biomédica Red de enfermedades hepáticas y digestivas (CIBERehd), Madrid, Spain

Xuefeng Luo, MD Department of Gastroenterology and Hepatology, Sichuan University-University of Oxford Huaxi Joint Centre for Gastrointestinal Cancer, West China Hospital, Sichuan University, Chengdu, China

Department of Gastroenterology and Hepatology, West China Hospital, Sichuan University, Chengdu, China

Yong Lv, MD Department of Liver Diseases and Digestive Interventional Radiology, National Clinical Research Centre for Digestive Diseases and Xijing Hospital of Digestive Diseases, Fourth Military Medical University, Xi'an, China

Military Medical Innovation Center, Fourth Military Medical University, Xi'an, China

Mattias Mandorfer Vienna Hepatic Hemodynamic Lab, Division of Gastroenterology and Hepatology, Department of Internal Medicine III, Medical University of Vienna, Vienna, Austria

Yuly P. Mendoza Hepatology, University Clinic for Visceral Surgery and Medicine, Inselspital Bern University Hospital, University of Bern, Bern, Switzerland

Hepatology, Department for BioMedical Research, University Clinic for Visceral Surgery and Medicine, University of Bern, Bern, Switzerland

Graduate School for Health Sciences (GHS), University of Bern, Bern, Switzerland

Schalk Van der Merwe, MD Department of Gastroenterology and Hepatology, Section of Liver and Biliopancreatic Disorders, University Hospitals Leuven, KU Leuven, Leuven, Belgium

Vincenzo La Mura, MD Fondazione IRCCS Ca' Granda Ospedale Maggiore Policlinico, Angelo Bianchi Bonomi Hemophilia and Thrombosis Center, Milan, Italy

Università degli Studi di Milano, Department of Pathophysiology and Transplantation, Milan, Italy

Fondazione IRCCS Ca' Granda, Ospedale Maggiore Policlinico, Dipartimento di Scienze Biomediche per la Salute and CRC AM e A Migliavacca per lo Studio delle Malattie del Fegato, University of Milan, Milan, Italy

Sarwa Darwish Murad, MD, PhD Department of Gastroenterology and Hepatology, Erasmus MC University Medical Center Rotterdam, Rotterdam, The Netherlands

Tadashi Namisaki, MD Department of Gastroenterology, Nara Medical University, Kashihara, Nara, Japan

Pol Olivas, MD Barcelona Hepatic Hemodynamic Laboratory, Liver Unit, Hospital Clínic-IDIBAPS, Health Care Provider of the European Reference Network on Rare Liver Disorders (ERN-Liver), CIBERehd, Barcelona, Spain

Centro de Investigación Biomédica Red de Enfermedades Hepáticas y Digestivas (CIBERehd), Madrid, Spain

Naaventhan Palaniyappan, MD NIHR Nottingham Biomedical Research Centre, Nottingham University Hospitals NHS Trust, University of Nottingham, Nottingham, UK

Valerie Paradis, MD Department of d'Anatomie Pathologique, Hôpital Beaujon, Clichy, France

Pathology Unit, Hôpital Beaujon, Assistance Publique-Hôpitaux de Paris, Clichy, France

David Patch, MD Royal Free Hospital, London, UK

Valeria Perez-Campuzano, MD Barcelona Hepatic Hemodynamic Laboratory, Liver Unit, Hospital Clínic, Institut de Investigacions Biomèdiques August Pi i Sunyer (IDIBAPS), University of Barcelona, Barcelona, Spain

CIBEREHD (Centro de Investigación Biomédica en Red Enfermedades Hepáticas y Digestivas), Barcelona, Spain

Salvatore Piano, MD Unit of Internal Medicine and Hepatology, Department of Medicine—DIMED, University of Padova, Padua, Italy

Massimo Pinzani, MD University College London, Institute for Liver and Digestive Health, Division of Medicine, Royal Free Hospital, London, UK

Aurelie Plessier, MD Université de Paris cité, AP-HP, Hôpital Beaujon, Service d'Hépatologie, DMU DIGEST, Centre de Référence des Maladies Vasculaires du Foie, FILFOIE, Valdig an EASL consortium, ERN RARE-LIVER, Centre de recherche sur l'inflammation, Inserm, UMR 1149, Paris, France

Mónica Pons, MD Liver Unit, Hospital Universitari Vall d'Hebron, Vall d'Hebron Research Institute (VHIR), Vall d'Hebron Barcelona Hospital Campus, Universitat Autònoma de Barcelona, Barcelona, Spain

Centro para la Investigación Biomédica en Red de Enfermedades Hepáticas y Digestivas (CIBEREHD), Madrid, Spain

Massimo Primignani, MD First Gastroenterology Unit, IRCCS Ca' Granda Policlinico Foundation, Milan, Italy

Bogdan Procopet, MD Gastroenterology Department, 3rd Medical Clinic, "Iuliu Hatieganu" University of Medicine and Pharmacy, Cluji-Napoca, Romania

Gastroenterology and Hepatology Department, Regional Institute of Gastroenterology and Hepatology "Octavian Fodor",, Cluji-Napoca, Romania

Pierre Emmanuel Rautou, MD Service d'Hépatologie, and Inserm UMR1149, Hôpital Beaujon, Clichy, France

Thomas Reiberger, MD Division of Gastroenterology and Hepatology, Department of Medicine III, Medical University of Vienna, Vienna, Austria

Cristina Ripoll First Department of Internal Medicine, Martin Luther University Halle-Wittenberg, Halle (Saale), Germany

Internal Medicine IV, Jena University Hospital Friedrich-Schiller University, Jena, Germany

Kyle E. Robinson, MD Division of Gastroenterology and Hepatology, Mayo Clinic, Rochester, MN, USA

Susana G. Rodrigues, MD Hepatology, University Clinic for Visceral Surgery and Medicine, Inselspital Bern, University of Bern, Bern, Switzerland

Hepatology, Department for BioMedical Research, Visceral Surgery and Medicine, University of Bern, Bern, Switzerland

Marika Rudler Groupement Hospitalier APHP-Sorbonne Université, Service d'Hépato-Gastro-Entérologie, Hôpital Pitié-Salpêtrière, Paris, France

Unité de Soins Intensifs d'Hépato-Gastro-Entérologie, Groupement Hospitalier APHP-Sorbonne Université, Paris, France

Sorbonne Université, INSERM, Centre de Recherche Saint-Antoine (CRSA), Institute of Cardiometabolism and Nutrition (ICAN), Paris, France

Brain-Liver Pitié-Salpêtrière Study Group (BLIPS), APHP-Sorbonne Université, Paris, France

Barcelona Hepatic Hemodynamic Unit, Liver Unit, Hospital Clinic, University of Barcelona, Barcelona, Spain

Marvin Ryou, MD Division of Gastroenterology, Hepatology and Endoscopy, Brigham and Women's Hospital, Harvard Medical School, Boston, MA, USA

Shiv K. Sarin, MD Institute of Liver and Biliary Sciences, New Delhi, India

Filippo Schepis, MD Hepatic Hemodynamic Laboratory, Division of Gastroenterology, AOU of Modena, Modena, Italy

University of Modena and Reggio Emilia, Modena, Italy

Marco Senzolo, MD Gastroenterology and Multivisceral Transplant Unit, Department of Surgery, Oncology and Gastroenterology, Padua University Hospital, Padua, Italy

Vijay Shah, MD Division of Gastroenterology and Hepatology, Mayo Clinic, Rochester, MN, USA

Akash Shukla, MD Department of Gastroenterology, Seth GS Medical College & KEM Hospital, Mumbai, India

Puneeta Tandon, MD, FRCP(C), MSc (Epi) Division of Gastroenterology (Liver Unit), University of Alberta, Edmonton, AB, Canada

Luis Tellez, MD Servicio de Gastroenterología y Hepatología, Hospital Universitario Ramón y Cajal, Instituto Ramón y Cajal de Investigación Sanitaria (IRYCIS), Centro de Investigación Biomédica Red de Enfermedades Hepáticas y Digestivas (CIBEREHD), Universidad de Alcalá, Madrid, Spain

Dominique Thabut, MD Groupement Hospitalier APHP-Sorbonne Université, Service d'Hépato-Gastro-Entérologie, Hôpital Pitié-Salpêtrière, Paris, France

Sorbonne Université, INSERM, Centre de Recherche Saint-Antoine (CRSA), Institute of Cardiometabolism and Nutrition (ICAN), Paris, France

Brain-Liver Pitié-Salpêtrière Study Group (BLIPS), APHP-Sorbonne Université, Paris, France

Hôpital de la Pitié-Salpétrière, APHP, UPMC, Sorbonne University, Paris, France

Maja Thiele, MD, PhD FLASH Center for Liver Research, Department of Gastroenterology and Hepatology, Odense University Hospital, Odense, Denmark

Department of Clinical Research, University of Southern Denmark, Odense, Denmark

Ferran Torres Medical Statistics Core Facility, IDIBAPS, Hospital Clinic, Barcelona, Spain

Biostatistics Unit, Faculty of Medicine, Universitat Autònoma de Barcelona, Barcelona, Spain

Jonel Trebicka, MD Translational Hepatology, Department of Internal Medicine I, University Clinic Frankfurt, Frankfurt, Germany

Department of Internal Medicine B, University of Münster, Münster, Germany

European Foundation for the Study of Chronic Liver Failure—EF Clif, Barcelona, Spain

Diraj Tripathi, MD Liver Unit, University Hospitals, Birmingham NHS Foundation Trust, Birmingham, UK

Institute of Immunology and Immunotherapy, University of Birmingham, Birmingham, UK

Emmanuouil A. Tsochatzis, MD UCL Institute for Liver and Digestive Health and Sheila Sherlock Liver Unit, Royal Free Hospital and UCL, London, UK

Laura Turco, MD Division of Internal Medicine for the Treatment of Severe Organ Failure, IRCCS Azienda Ospedaliero-Universitaria di Bologna, Bologna, Italy

Fanny Turon, MD Barcelona Hepatic Hemodynamic Laboratory, Liver Unit, Hospital Clínic, IDIBAPS, CIBERehd, Health care provider of the European Reference Network on Rare Liver Disorders (ERN Liver), Barcelona, Spain

Dominique-Charles Valla, MD Service d'hépatologie Hôpital Beaujon, Clichy, France

Emma Vanderschueren, MD Department of Gastroenterology and Hepatology, Section of Liver and Biliopancreatic Disorders, University Hospitals Leuven, KU Leuven, Leuven, Belgium

Marina Vilaseca, PhD Laboratory for Cellular Epigenomics, Riken, Yokohama, Kanagawa, Japan

Càndid Villanueva Gastrointestinal Bleeding Unit, Department of Gastroenterology, Hospital de Sant Pau, Biomedical Research Institute Sant Pau (IIB Sant Pau), Barcelona, Spain

Universitat Autònoma de Barcelona, Barcelona, Spain

Centro de Investigación Biomédica en Red de Enfermedades Hepáticas y Digestivas (CIBERehd), Barcelona, Spain

Élise Vuille-Lessard, MD Department of Visceral Surgery and Medicine, Inselspital, Bern University Hospital, University of Bern, Bern, Switzerland

Sarah Wang, MD Division of Gastroenterology (Liver Unit), University of Alberta, Edmonton, AB, Canada

Ian Wanless, MD Department of Pathology, Dalhousie University, Halifax, NS, Canada

Li Yang, MD Department of Gastroenterology and Hepatology, Sichuan University-University of Oxford Huaxi Joint Centre for Gastrointestinal Cancer, West China Hospital, Sichuan University, Chengdu, Sichuan, China

Hitoshi Yoshiji, MD Third Department of Internal Medicine, Department of Gastroenterology, Endocrinology and Metabolism, Nara Medical University, Nara, Japan

Alberto Zanetto, MD Gastroenterology and Multivisceral, Transplant Unit, Department of Surgery, Oncology, and Gastroenterology, Padua University Hospital, Padua, Italy

Hepatic Hemodynamic Laboratory, Division of Gastroenterology, AOU of Modena and University of Modena and Reggio Emilia, Modena, Italy

Part I

Introductory Lectures

Introduction: Baveno I to Baveno VII … and Beyond

1

Roberto de Franchis

Since 1986, ten international consensus meetings on portal hypertension have been held. After the first organized in Groningen, the Netherlands, by Andrew Burroughs [1], the other nine took place in Baveno in 1990 (Baveno I) [2] and 1995 (Baveno II) [3, 4], in Milan in 1992 [5], in Reston, USA in 1996 [6], in Stresa in 2000 (Baveno III) [7, 8], again in Baveno in 2005 (Baveno IV) [9, 10], in Atlanta, USA in 2007 [11] and again in Stresa in 2010 (Baveno V) [12, 13], and Baveno in 2015 (Baveno VI) [14, 15]. This is the 11th meeting of this kind, the seventh with the name of Baveno.

Baveno I to VI

Attendance at the Baveno Workshops

The attendance to the Baveno workshops was 205 in Baveno I, 252 in Baveno II, 385 in Baveno III, 485 in Baveno IV, 314 in Baveno V, 243 in Baveno VI, and 509 in Baveno VII. The proportion of international participants rose steadily from 19% in Baveno I to 87% in Baveno VII. The countries represented rose from 18 in Baveno I to 50 in Baveno V and VII.

Publications Derived from the Baveno Workshops

Reports of the Baveno workshops have been published in the *Journal of Hepatology* in 1992 [2] (Baveno I), in 1996 [3] (Baveno II), in 2000 [7] (Baveno III), in 2005 [9] (Baveno IV), in 2010 [12] (Baveno V), and 2015 (Baveno VI) [14]. Starting from

R. de Franchis (✉)
Department of Biomedical and Clinical Sciences, University of Milan, Milan, Italy
e-mail: Roberto.defranchis@unimi.it

R. de Franchis (ed.), *Portal Hypertension VII*,
https://doi.org/10.1007/978-3-031-08552-9_1

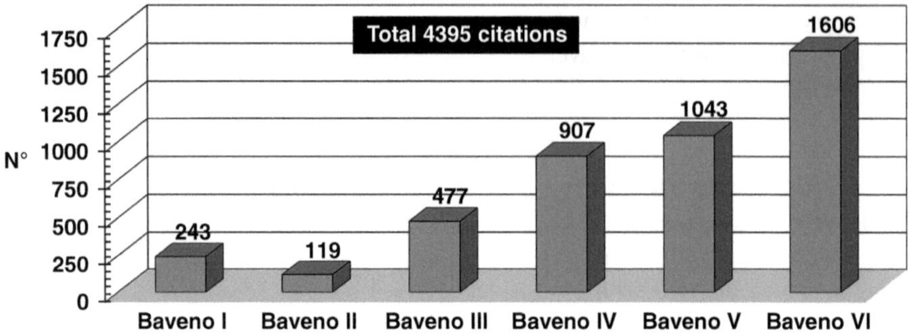

Fig. 1.1 Citations of the Baveno reports I to VI (data from Scopus 22.10.2021)

Baveno II, the proceeding books of the workshops were published by Blackwell Science in 1996 [4] and 2001 [8], by Blackwell Publications in 2006 [10], by Wiley-Blackwell in 2011 [13], and by Springer in 2016 [15].

Impact of the Baveno Consensus on the Medical Literature

Figure 1.1 shows the number of citations of the Baveno I–VI consensus reports in the medical literature between January 1993 and October 22nd, 2021. Overall, the reports had 4395 citations. The number of citations more than doubled between 2015 and 2021.

The Baveno workshop's proceeding books were also highly successful. As an example, the Baveno VI consensus proceedings book: "Portal Hypertension VI: expanding consensus in portal hypertension" received more than 40,000-chapter downloads from 2016 to 2021.

Validation of the Baveno Definitions and Recommendations

The validity of the definitions and recommendations proposed at each Baveno workshop underwent verification in the following years. In this respect, Baveno VI was no exception. As an example, at Baveno VI new criteria for ruling out the presence of high-risk varices (HRV) in patients with compensated advanced chronic liver disease (cACLD) by noninvasive means were proposed (the so-called Baveno VI criteria). These criteria stated that cACLD patients with a liver stiffness (measured by transient elastography) of <20 kPa and a platelet count of $>150{,}000/\text{mm}^3$ had *a* < 5% risk of having HRV, and thus could avoid undergoing screening endoscopy. If proven true, these criteria would have important clinical implications, since avoiding endoscopy in patients fulfilling the Baveno VI criteria would significantly decrease the burden of endoscopy units. In the years between 2016 and 2021, more than 30 studies were carried out to test the validity of the Baveno VI criteria. A metanalysis published in this book [16] showed that the cumulative negative predictive value of the Baveno VI criteria to exclude the presence of HRV was 0.99, and that the proportion of saved endoscopies ranged between 8% and 60% in the different studies.

Beyond Baveno VII, the Baveno Cooperation

Inspired by the friendly atmosphere of the Baveno meetings, the Baveno Cooperation was launched in 2016 with the aim of creating a permanent network of research groups involved in the organization of the Baveno workshops, to take advantage of the many relationships, personal links, and synergies created by working together over the years. Its goal is to promote the progress of research in the pathophysiology, diagnosis, noninvasive evaluation, and management of portal hypertension. In 2019, the Baveno Cooperation has been officially endorsed as an EASL Research Group. Cooperative studies carried out within the framework of the Baveno Cooperation have been presented for the first time at Baveno VII and will be published under the Baveno Cooperation umbrella as a quality mark. Along this line, several items of the research agenda presented at Baveno VII will be the topic of future Baveno Cooperation studies. We are convinced that the Baveno Cooperation will be the ideal framework to continue in the future the friendly collaboration that has always been the hallmark of the Baveno enterprise.

The Baveno I–VII Workshops Were a Concerted Effort of the Following

Speakers and Chairpersons

Argentina: J Vorobioff; **Austria**: G Krejs, M Mandorfer, M Peck, T Reiberger; **Belgium**: S Francque, W Laleman, F Nevens; **Canada**: J Abraldes, J Heathcote, S Ling, N Marcon, G Pomier Layrargues, P Tandon, I Wanless; **China**: G Han, J Jia, X Luo; **Denmark**: U Becker, F Bendtsen, E Christensen, C Gluud, A Krag, S Møller, TIA Sørensen; M Thiele; **Egypt**: G Shiha; **France**: L Benajiba, B Bernard-Chabert, C Bureau, P Calès, L Castéra, JJ Kiladjian, D Lebrec, R Moreau, V Paradis, JP Pascal, A Plessier, P E Rautou, M Rudler, C Silvain, D Thabut, D Valla, JP Vinel; **Germany**: K Binmøller, W Fleig, G Richter, C Ripoll, M Rössle, T Sauerbruch, M Schepke, D Schuppan, M Staritz, J Trebicka, A Zipprich; **Great Britain**: AK Burroughs, E Elias, J Fallowfield, P Hayes, J O'Beirne D Patch, D Tripathi, E Tsochatzis, D Westaby; **India**: YC Chawla, A Kumar, SK Sarin, A Shukla; **Israel**: I Gralnek; **Italy**: E Ancona, M Angelico, G Balducci, G Barosi, G Battaglia, M Bolognesi, L Bolondi, L Cestari, GC Caletti, V Calvaruso, F Cosentino, G D'Amico, R de Franchis, A Dell'Era, A Gatta, G Gerunda, V La Mura, A Liberati, A, Maffei Faccioli, PM Mannucci, C Merkel, M Merli, G Minoli, A Morabito, L Pagliaro, A Peracchia, S Piano, M Pinzani, M Primignani, O Riggio, P Rossi, C Sabbà, D Sacerdoti, F Salerno, F Schepis, M Senzolo, GP Spina, F Tinè, A Tripodi, L Turco,

V Ziparo, M Zoli; **Japan**: I Yoshiji; **Norway**: L Aabakken; **Pakistan**: S Abid; **Portugal**: P Alexandrino; **Romania**: B Procopet; **Spain**: A Albillos, S Augustin, A Baiges, R Bañares, M Barrufet, A Cardenas, A Escorsell, JC Garcia-Pagàn, J Genescà, P Ginés, J Gracia, V Hernandez-Gea, S Lens, M Navasa, J Piqué, R Planas,, J Rodès, L Tellez, F Turon, C Villanueva; **Switzerland**: A Berzigotti, J Bosch, A de Gottardi, AS Gomes Rodrigues, Hadengue, P Gertsch, C Sieber, R Wiest; **Sweden**: C Søderlund; **Taiwan**: FY Lee, HC Lin, J H Lo; **The Netherlands**: H Janssen, F Leebeek; SD Murad, H van Buuren; **U.S.A.**: J Bajaj, A Blei, T Boyer, N Chalasani, M Fallon, H Ferral, R Gaba, G Garcia-Tsao, N Grace, R Groszmann, JM Henderson, Y Iwakiri, P Kamath, WR Kim, D Kravetz, L Laine, B Mittman, A Sanyal, V Shah, B Shneider, J Talwalkar, G van Stiegmann.

Organization: S Covre, A M Sorresso, D Santi, G Sabattini, ADB Eventi e Congressi.

References

1. Burroughs AK, editor. Methodology and review of clinical trials in portal hypertension, excerpta medical congress service No. 763. New York: Oxford; 1987.
2. de Franchis R, Pascal JP, Ancona E, et al. Definitions, methodology and therapeutic strategies in portal hypertension. J Hepatol. 1992;15:256–61.
3. de Franchis R. Developing consensus in portal hypertension. J Hepatol. 1996;25:390–4.
4. de Franchis R, editor. Portal hypertension II. Proceedings of the Second Baveno international consensus workshop on definitions, methodology and therapeutic strategies. Oxford: Blackwell Science; 1996.
5. Spina GP, Arcidiacono R, Bosch J, et al. Gastric endoscopic features in portal hypertension: final report of a consensus conference. J Hepatol. 1994;21:461–7.
6. Grace ND, Groszmann R, Garcia-Tsao G, et al. Portal hypertension and variceal bleeding: an AASLD single topic symposium. Hepatology. 1998;28:868–80.
7. de Franchis R. Updating consensus in portal hypertension: report of the Baveno III consensus workshop on definitions, methodology and therapeutic strategies in portal hypertension. J Hepatol. 2000;33:846–52.
8. de Franchis R, editor. Portal hypertension III. Proceedings of the Third Baveno international consensus workshop on definitions, methodology and therapeutic strategies. Oxford: Blackwell Science; 2001.
9. de Franchis R. Evolving consensus in portal hypertension—report of the Baveno IV consensus workshop on methodology of diagnosis and therapy in portal hypertension. J Hepatol. 2005;43:167–76.
10. de Franchis R, editor. Portal hypertension IV. Proceedings of the Fourth Baveno international consensus workshop on definitions, methodology and therapeutic strategies. Oxford: Blackwell Publications; 2006.
11. Garcia-Tsao G, Bosch J, Groszmann R. Portal hypertension and variceal bleeding, unresolved issues. Summary of an American Association for the study of liver disease and European Association for the study of the liver single-topic conference. Hepatology. 2008;47:1764–72.
12. de Franchis R, Baveno V Faculty. Revising consensus in portal hypertension: report of the Baveno V consensus workshop on methodology of diagnosis and therapy in portal hypertension. J Hepatol. 2010;53:762–8.
13. de Franchis R, editor. Portal hypertension V. Proceedings of the Fifth Baveno international consensus workshop on definitions, methodology and therapeutic strategies. Oxford: Wiley-Blackwell; 2011.

14. de Franchis R, on behalf of the Baveno VI Faculty. Expanding consensus in portal hypertension: report of the Baveno VI consensus workshop: stratifying risk and individualizing care for portal hypertension. J Hepatol. 2015;63:743–52.
15. de Franchis R, editor. Portal hypertension VI. Proceedings of the VIth Baveno consensus workshop: stratifying risk and individualizing care. New York: Springer; 2016.
16. Bai W, Abraldes JG. Varices and screening endoscopy. In: de Franchis R, editor. Portal Hypertension VII. Proceedings of the VIIth Baveno consensus workshop: personalized care for portal hypertension. New York: Springer; 2022.

New Concepts in Risk Stratification

2

Juan Gonzalez Abraldes

Introduction

Risk prediction models are tools that utilize several factors in combination to produce the estimates of the risk of outcomes *for an individual* [1]. Ideally, risk prediction models should provide useful information to guide decision-making, including making treatment decisions [2]. It is important to note that research dealing with risk prediction is different from research aimed at understanding causal associations, even if it shares many analytical tools with causal inference research, such as regression modeling [3]. Indeed, the variables used to predict an outcome do not need to be causally associated with that outcome.

This chapter deals with some relevant concepts related to risk prediction models that are relevant to advance in the personalization of care in portal hypertension.

Risk Prediction and Probabilistic Thinking

To understand the constraints of what we can expect from risk prediction models, it is essential to acknowledge that in the clinical context most events are nondeterministic [4], but something that occurs at the end of a long chain of random processes [5, 6]. For example, a patient with compensated cirrhosis might have progressed to the point of having a high risk of decompensation (i.e., having a portal pressure gradient of 24 mmHg), and yet remain compensated for a longer period of time. On the contrary, a patient at a theoretical low risk might experience a random event (i.e., an infection) that triggers decompensation. Therefore, there is inherent randomness in whether patients develop events or not so; at best, we only can capture tendencies with our risk prediction models.

J. G. Abraldes (✉)
Division of Gastroenterology, University of Alberta, Edmonton, AB, Canada
e-mail: juan.g.abraldes@ualberta.ca

R. de Franchis (ed.), *Portal Hypertension VII*,
https://doi.org/10.1007/978-3-031-08552-9_2

Diagnosis as a Risk Prediction Problem

Establishing a diagnosis in a patient is most useful when that diagnosis conveys prognostic information and, even more, if it has an impact on management [1]. For example, diagnosing the presence of clinically significant portal hypertension (CSPH) not only implies that the patient is at risk of decompensation, but also that the patient would benefit from being treated with beta-blockers [7, 8]., Since the gold standard for diagnosing CSPH (the hepatic vein catheterism [9, 10]) is difficult to use in clinical practice, we need simple tools to predict the presence of CSPH [11].

When using these diagnostic tools, traditional metrics of diagnostic performance are not useful for decision-making [12]. For example, sensitivity is the probability of getting a positive test given the presence of the disease [13], which is never the relevant clinical question. It implies that the test can be classified as positive or negative and is a backward probability since the flow of information goes in the wrong direction (from the presence of the disease/condition to the test result). The relevant clinical question would be answered by providing a forward probability: in a given patient, given certain values of a diagnostic test/diagnostic model, what would be the probability that the patient has the target condition. This was the approach followed in the ANTICIPATE study [11], in which different risk prediction models were provided to predict CSPH. Figure 2.1 shows a model based on liver stiffness by transient elastography (LSM by TE) and platelet count, subsequently validated in Pons et al [14] According to this model, for example, a patient with an LSM by TE of 24 kPa and a platelet count of 135 × 10^9 would have a risk of having a CSPH of ~65%.

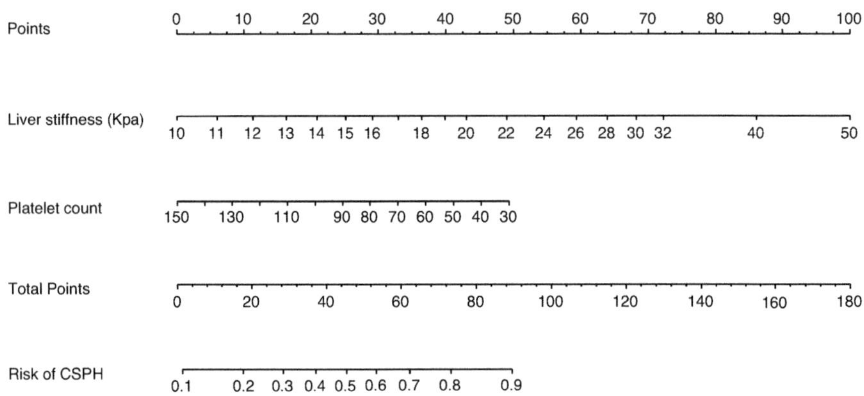

Fig. 2.1 Example of a risk prediction model (represented as a nomogram) for clinically significant portal hypertension (CSPH), based on liver stiffness measured by transient elastography and platelet count. This model is based on the ANTICIPATE study [11] and was subsequently validated (for alcohol-related and hepatitis C-related cirrhosis) in an independent sample in the study by Pons et al [14] To calculate the probability of CSPH trace, a vertical line from each of the predictors' axis to the first line ("points"). Add the total points and trace a vertical line from the "total points" axis to the risk axis to calculate the risk of CSPH

Decision Thresholds and Risk Stratification

One of the advantages of this approach is that the probabilities given by the model carry their own error measurements. Indeed, in a patient with a probability of CSPH of 65%, we can estimate that, if we do not treat the patient with beta-blockers, the chances of this being an error would be 65%, whereas if we decide to treat that patient with beta-blockers the chances of it being an error would be 35%. Knowledge of these error rates allows a sound utility analysis of the benefits/harms of implementing an intervention at a certain risk level and is the basis for establishing risk thresholds. Defining these thresholds and subsequently risk strata, therefore, is independent from the model building process [15], and this definition would depend on the intervention/decision that is considered. If the utilities of the intervention change, the decision thresholds would likely change accordingly.

Another important concept in risk stratification is that patients within a stratum are heterogeneous. An example of this is provided in Fig. 2.2. In the case shown in the figure, by defining a stratum with patients above the risk threshold for CSPH of 0.65, we end up with a group of patients with an overall risk of CSPH of 0.88. This value would be different if the distribution of patients within the strata changes (for

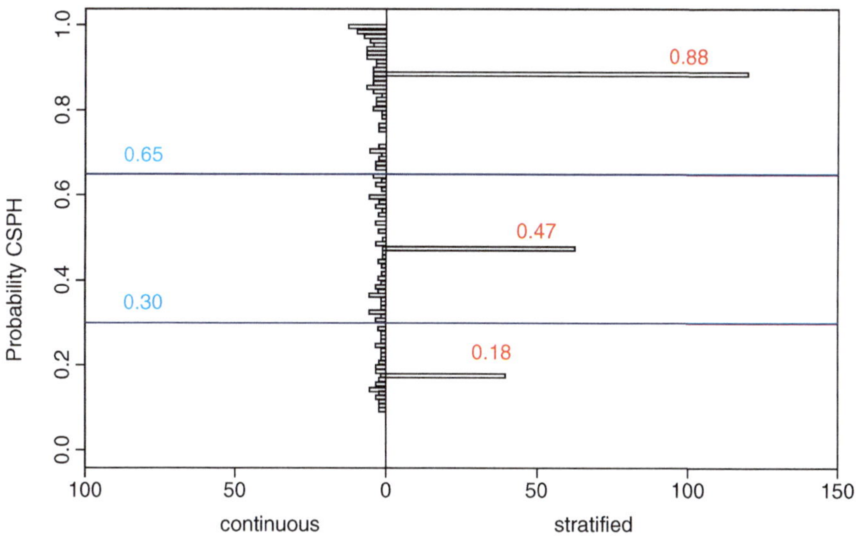

Fig. 2.2 Illustration of the effect of the heterogeneity within a stratum and the calculated risk within that stratum. The histogram on the left shows the distribution of 200 patients according to their risk of CSPH. The histogram on the right shows the distribution of risks after stratifying the sample into three groups. First group is composed of patients with risk between 0% and 30%. Second group between 30% and 65%. Third group between 65% and 100%. This graph shows that even if the last group includes all patients with a risk over 65%, the estimated risk of that strata is 88%. If we were only presented with the data on the right histogram, the reader would not be able to grasp the variability in risk within that group. Furthermore, the overall risk of that strata would vary if we had a different sample of patients enriched with lower or higher risks of CSPH

example, if more patients with more advanced risk are evaluated, the risk of this group would be higher). Since we evaluate one patient at a time, it makes little sense that the risk that we assign to a particular patient depends on the "group risk" to which this patient belongs. Both the physician and the patient would be interested to know what the predicted risk (in a continuous scale) in the patient context is, whether it is above or below the risk threshold for a given decision (but also how much above or below), so that the patient and physician can make an informed decision.

Risk Prediction and the Interpretation of Treatment Effects in Randomized Controlled Trials

When reporting the results of randomized controlled trials (RCTs), we usually provide a global metric of relative efficacy, for example, a hazard ratio (HR) or an odds ratio (OR) with their 95% confidence intervals (CI), and the absolute reduction in the proportion of events in the treatment vs. the control arm. For example, in the recently published PREDESCI RCT [7], we can read in the abstract that the use of beta-blockers in patients with compensated cirrhosis and CSPH reduced the incidence of decompensation with an HR of 0.51 (95% CI 0.26–0.97), and that the primary endpoint occurred in 16% in the BB group vs. 27% in the placebo group, which would be an absolute risk reduction (ARR) of 11% over the follow-up of the trial. Additional ARRs could be inferred from the cumulative incidence plots at 1, 2, 3, and 4 years. Indeed, at 4 years, the ARR would be quite impressive ~16%. The magnitude of treatment effect measured on the absolute scale is most relevant for treatment decisions [16], and it is well established that patients have a better understanding of the risks and harms of an intervention when these are expressed in absolute numbers [17].

The number needed to treat (NNT) which is a frequent metric used in evidence-based medicine and represents that the inverse of the ARR would be 9 patients [9]. It would follow that 9 patients would be needed to be treated for a median of 37 months (the median follow-up of the trial) for preventing one decompensating event. The use of NNTs has several issues [18]. It conveys the notion that only 1 out of 9 patients benefitted from the intervention, when in a parallel trial it is impossible to know if all patients had a small benefit or 1 patient had a large benefit and the other 8 did not [19, 20]. The other issue with the NNT is that it would only apply to patients with a baseline risk of decompensation of around 27% since the ARR applies to a baseline risk of 27%. Assuming a constant effect of beta-blockers on the relative scale, for every level of risk (i.e., that the HR would be a constant 0.5 independently of what is the baseline risk of the event), the ARR would be different for every level of baseline risk.

Figure 2.3a plots the predicted absolute risk reduction given the baseline risk of being event-free and the HR. For example, with an HR like the one reported in the PREDESCI trial (0.5), in a patient with a baseline risk of 15% of decompensation, the ARR would be 7% and with a baseline risk of 10% the ARR would be

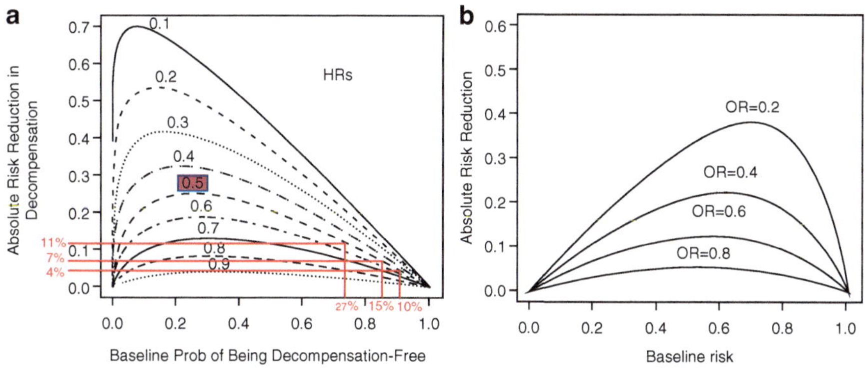

Fig. 2.3 Nomogram to estimate the ARR based on the baseline risks of outcomes (or of being outcome-free) for different hazard ratios (**a**) or odd ratios (**b**). In (**a**), we use the HR observed in the PREDESCI [7] study (0.5) to show three potential values of the absolute risk reduction according to three potential baseline risks of decompensation

4%. If follows that the NNTs would be, respectively, 14 and 25. It is also important to note that the global ARR and NNT do not even apply to a theoretical "average patient with CSPH" since, like in every other RCT, the sample of patients included in the PREDESCI randomized trial was not a random sample of the population of patients with CSPH. Figure 2.3b shows a similar nomogram to estimate the ARR according to the baseline risk of the event but in this case for different ORs.

The recent PATH (Predictive Approaches to Treatment effect Heterogeneity) Statement extensively addresses this issue [21], presenting two approaches for addressing the heterogeneity of treatment effect, both in the relative and the absolute scales. In one approach, an external well-calibrated model is applied to assess the absolute risk reduction by the new treatment at different levels of risk. In the second approach, data from the RCT (ideally from an individual patient meta-analysis of several large, randomized trials) is used to develop a risk prediction model, introducing a term for the treatment assignment and its interactions with baseline covariates. Presentation of the treatment effects across full levels of baseline risk (rather than in a few strata) could lead to more informed treatment decisions, especially when there is a narrow balance between benefits and harms. The PATH statement also deals with subgroup analysis, proposing again a risk modeling approach in which all relevant variables are modeled together, rather than the traditional one at a time subgroup analysis.

Conclusion

In summary, wider use of well-calibrated risk prediction models in the context of diagnosis, prognosis, and the interpretation of RCTs might advance the goal of providing more personalized care to patients with portal hypertension.

References

1. Hemingway H, Croft P, Perel P, et al. Prognosis research strategy (PROGRESS) 1: a framework for researching clinical outcomes. BMJ. 2013;346:e5595.
2. Steyerberg EW, Moons KGM, van der Windt DA, et al. Prognosis research strategy (PROGRESS) 3: prognostic model research. PLoS Med. 2013;10(2):e1001381.
3. Hernán MA, Hsu J, Healy B. A second chance to get causal inference right: a classification of data science tasks. Chance. 2019;32(1):42–9.
4. Harrell F. Classification vs. prediction. Stat Think. 2017; https://www.fharrell.com/post/classification/
5. Gelman A. A counterexample to the potential-outcomes model for causal inference. https://statmodeling.stat.columbia.edu/2021/07/26/a-counterexample-to-the-potential-outcomes-model-for-causal-inference/
6. Smith GD. Epidemiology, epigenetics and the 'gloomy prospect': embracing randomness in population health research and practice. Int J Epidemiol. 2011;40(3):537–62.
7. Villanueva C, Albillos A, Genescà J, et al. β blockers to prevent decompensation of cirrhosis in patients with clinically significant portal hypertension (PREDESCI): a randomised, double-blind, placebo-controlled, multicentre trial. Lancet. 2019;393(10181):1597–608.
8. Garcia-Tsao G, Abraldes JG. Nonselective beta-blockers in compensated cirrhosis: preventing variceal hemorrhage or preventing decompensation? Gastroenterology. 2021;161(3):770–3.
9. Bosch J, Abraldes JG, Berzigotti A, García-Pagan JC. The clinical use of HVPG measurements in chronic liver disease. Nat Rev Gastroenterol Hepatol. 2009;6(10):573–82.
10. Veldhuijzen van Zanten D, Buganza E, Abraldes JG. The role of hepatic venous pressure gradient in the management of cirrhosis. Clin Liver Dis. 2021;25(2):327–43.
11. Abraldes JG, Bureau C, Stefanescu H, et al. Noninvasive tools and risk of clinically significant portal hypertension and varices in compensated cirrhosis: the "anticipate" study. Hepatology. 2016;64(6):2173–84.
12. Harrell F. Clinicians' misunderstanding of probabilities makes them like backwards probabilities such as sensitivity, specificity, and type I error. Stat Think. 2017; https://www.fharrell.com/post/backwards-probs/
13. Guggenmoos-Holzmann I, van Houwelingen HC. The (in)validity of sensitivity and specificity. Stat Med. 2000;19(13):1783–92.
14. Pons M, Augustin S, Scheiner B, et al. Noninvasive diagnosis of portal hypertension in patients with compensated advanced chronic liver disease. Am J Gastroenterol. 2021;116(4):723–32.
15. Wynants L, van Smeden M, McLernon DJ, et al. Three myths about risk thresholds for prediction models. BMC Med. 2019;17(1):192.
16. Lesko CR, Henderson NC, Varadhan R. Considerations when assessing heterogeneity of treatment effect in patient-centered outcomes research. J Clin Epidemiol. 2018;100:22–31.
17. Fagerlin A, Zikmund-Fisher BJ, Ubel PA. Helping patients decide: ten steps to better risk communication. J Natl Cancer Inst. 2011;103(19):1436–43.
18. Mayo. S. Senn: Personal perils: are numbers needed to treat misleading us as to the scope for personalised medicine? (Guest Post). Error Stat Philos. 2018; https://errorstatistics.com/2018/07/11/s-senn-personal-perils-are-numbers-needed-to-treat-misleading-us-as-to-the-scope-for-personalised-medicine-guest-post/
19. Senn S. Mastering variation: variance components and personalised medicine. Stat Med. 2016;35(7):966–77.
20. Senn S. Statistical pitfalls of personalized medicine. Nature. 2018;563(7733):619–21.
21. Kent DM, Paulus JK, van Klaveren D, et al. The predictive approaches to treatment effect heterogeneity (PATH) statement. Ann Intern Med. 2020;172(1):35–45.

Clinical Stages and Ordinal Outcomes in Portal Hypertension

3

Gennaro D'Amico

Background

The concept of clinical stages of cirrhosis has been introduced to identify well-recognized conditions linked to clearly different expected disease progression and mortality rates [1]. Based on the impressive difference in survival expectations, two major clinical stages are widely used in clinical practice and research for patient characterization and risk stratification: compensated and decompensated cirrhosis. The median survival of compensated cirrhosis is in the order of 12 years [2] and it decreases to 2–4 years [3, 4] in decompensated cirrhosis. Based on the progressive increase in mortality, substages have been proposed either in compensated or in decompensated cirrhosis, according to the presence of clinically significant portal hypertension (defined as a hepatic vein portal pressure gradient, HVPG, ≥ 10 mmHg) or esophageal varices in compensated cirrhosis, and to the type and number of decompensating events in decompensated cirrhosis [5]. Importantly, the proposed stages of cirrhosis have been qualified as "clinical" because they refer to different clinical conditions which may present without a predictable sequence as it would be expected in a pathophysiological severity staging.

The increasing risk of mortality across the proposed stages of cirrhosis makes them conceptually similar to ordinal outcomes, which are by definition a rank-ordered sequence of a given outcome according to severity. In this chapter, the concept of clinical stages of cirrhosis, their similarity to ordinal outcomes, and the applicability of ordinal outcomes in cirrhosis are reviewed.

G. D'Amico (✉)
Gatroenterology Unit, Ospedale V Cervello, Palermo, Italy

Clinica La Maddalena, Palermo, Italy

R. de Franchis (ed.), *Portal Hypertension VII*,
https://doi.org/10.1007/978-3-031-08552-9_3

Definition and Incidence of Decompensation

Decompensated cirrhosis has been traditionally defined by the presence or history of one or more of four major complications of cirrhosis: ascites, bleeding, encephalopathy, and jaundice [2, 6, 7]. The definition of compensated cirrhosis is somehow the complement to this or the absence of any of these complications. However, although there is a large consensus on the clinical relevance of classifying cirrhosis as compensated or decompensated, there is also appreciable inconsistency in the definition of decompensation as reported in the medical literature; also, a debate is growing on the appropriateness of including jaundice in the definition, based on its weigh in cholestatic disease and in the setting of acute liver damage which may fully recover without leaving consequences.

In a systematic review of predictors of decompensation in cirrhosis [8], the definition of decompensation in 104 cohorts including 152,320 patients with cirrhosis, from different etiology, included in 91 studies published before 2019 was based on the following criteria (Table 3.1 and Fig. 3.1): the development of ascites, bleeding or encephalopathy in 42 cohorts; the development of ascites, bleeding, encephalopathy, or jaundice in 31; any combination of ascites, bleeding, encephalopathy, or jaundice with hepatocellular carcinoma in 13; any combination of ascites, bleeding, encephalopathy, or jaundice with an increase of Child–Pugh score of at least two points in 9; and any combination of ascites, bleeding, encephalopathy or jaundice with an increase in prothrombin time or development of varices or need of diuretics in 9. Of note, jaundice was included in the definition of decompensation in only 1

Table 3.1 Definition of decompensation in 104 cohorts including 152,320 patients with cirrhosis, from different etiology, included in 91 studies published before 2019[a]

	Etiology						
Definition	Alcohol	HBV	HCV	NAFLD	Biliary	Unselected	Total
Abe[b]	2	5	14	3	7	11	42
Abej[c]	1	6	11	0	1	12	31
Plus hcc[d]	0	2	6	2	1	2	13
Plus cps[e]	0	4	2	2	0	1	9
Plus oth[f]	2	2	2	1	1	1	9
Total	5	19	35	8	10	27	104

[a] Numbers are the number of cohorts per each type of etiology. *HBV* Hepatitis B virus, *HCV* Hepatitis C virus, *NAFLD* non-alcoholic fatty liver disease, *Biliary* primary biliary cholangitis and primary sclerosing cholangitis, *Unselected* cohorts including patients with cirrhosis from different etiologies without separate analysis

[b] *abe* ascites, bleeding, encephalopathy

[c] *abej* ascites, bleeding, encephalopathy, jaundice

[d] *plus hcc* combination of any of ascites, bleeding, encephalopathy, or jaundice, with hepatocellular carcinoma (hcc)

[e] *plus cps* combination of any of ascites, bleeding, encephalopathy, or jaundice, with increase of two points in Child–Pugh score (cps)

[f] *plus other* combination of any of ascites, bleeding, encephalopathy, or jaundice, with an increase in prothrombin time, or development of varices or need of diuretics

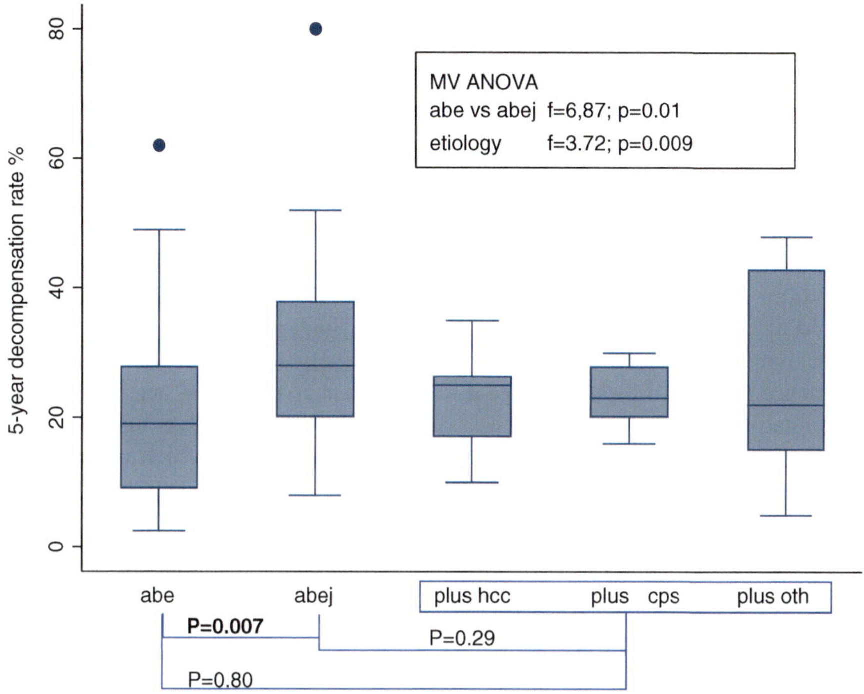

Fig. 3.1 Box plots of 5-year incidence rate of decompensation according to etiology of cirrhosis. Data from a systematic review of 91 studies of predictors of decompensation [8], reporting on 104 cohorts including 152,320 participants, from different etiologies. The incidence rate was significantly higher in cohorts where decompensation definition included also jaundice together with ascites, bleeding, and encephalopathy, compared to those where jaundice was not included. The difference holds significant even when adjusted by the etiology of liver disease in multivariable analysis. abeascites, bleeding, encephalopathy; abej ascites, bleeding, encephalopathy, jaundice; plus hcc: combination of any of ascites, bleeding, encephalopathy or jaundice, with hepatocellular carcinoma (hcc); plus cps: combination of any of ascites, bleeding, encephalopathy or jaundice, with an increase of two points in Child–Pugh score (cps); plus other: combination of any of ascites, bleeding, encephalopathy, or jaundice, with an increase in prothrombin time, or development of varices or need of diuretics. MV multivariable, ANOVA analysis of variance

of 11 cohorts with cholestatic liver disease: the bilirubin cutoff to define decompensation was set at 11 mg/dL in this cohort [9]. The incidence of decompensation was significantly different across different etiologies and was significantly higher in cohorts where the definition included jaundice together with ascites, bleeding, and encephalopathy, compared to cohorts where it was not included. The increase in decompensation incidence when including jaundice in the definition of decompensation holds significant even when adjusted by etiology. These findings suggest that in non-cholestatic diseases, the definition of decompensation of cirrhosis should be based on the development of any ascites, bleeding, encephalopathy, and jaundice and that excluding jaundice may impact either in clinical practice or in clinical research. However, the role of jaundice in the context of acute liver damage occurring in patients with compensated cirrhosis requires further study.

Other areas of uncertainty in the definition of decompensation include the role of minimal ascites (only detectable on ultrasounds) or minimal hepatic encephalopathy, which per Baveno VII consensus [10] are currently considered only predictors of worse outcomes in compensated cirrhosis.

In the above quoted systematic review of predictors of decompensation [8], the median 5-year decompensation rate was 23% (ranging from 2.5% to 80%) across all the definitions and etiologies. To investigate which is the first decompensating event to occur, in a multicenter European Latino-American study including a total of 2296 patients still unpublished, a cohort of 1007 patients with compensated cirrhosis has been prospectively followed. The cumulative incidence of each of ascites, variceal bleeding, encephalopathy, and jaundice or any combination of them has been assessed by a competing risks analysis. The first and most frequent decompensating event was ascites which occurred in almost 50% of patients in 10 years, followed by any combination of two or more events in 15%, bleeding alone in 11%, jaundice alone in 7.5%, and encephalopathy alone in 5%.

Clinical Stages of Cirrhosis

Substages have been proposed either in compensated or in decompensated cirrhosis [1, 5, 11]. In compensated cirrhosis, two stages have been defined based on the presence/absence of esophageal varices and their related risk of death and disease progression. However, since approximately 50% of patients without varices [12] still have mild portal hypertension (MPH) (HVPG > 5 mmHg and < 10 mmHg), while CSPH (HVPG $\geq$ 10 mmHg) is already present in the remaining, and the risk of disease progression is significantly higher in presence of CSPH, it has been proposed that patients without esophageal varices are substaged according to the presence of CSPH, while patients with varices have CSPH by definition. Importantly, liver stiffness measurement alone or in combination with platelet count may allow to rule out CSPH with specificity in the order of 0.90 (details in Chap. 9 in this book). Five-year death risk has been reported in the order of 1% and 10%, respectively, in patients without or with esophagogastric varices and compensated cirrhosis. Survival without CSPH and without varices is not yet well defined although it is expected to be significantly longer than with CSPH. Therefore, three stages with increasing mortality risk/severity have been proposed in compensated cirrhosis: stage 1, no varices and no CSPH; stage 2, no varices with CSPH; stage 3, gastroesophageal varices.

In decompensated cirrhosis, three further stages have been proposed according to the type and number of events marking decompensation: 5-year mortality for patients transitioning to decompensated cirrhosis through variceal bleeding without other decompensating events is approximately 20%; for those who have experienced a single nonbleeding event, it is in the order of 30%; and for those who have had any combination of two or more events, it is >80%. The stages of decompensated cirrhosis are therefore defined as follows: stage 4, variceal bleeding alone; stage 5, any single nonbleeding event; and stage 6, any combination of ≥2 events.

Ordinal Outcomes

An ordinal outcome is a rank-ordered sequence of a given indicator for which the difference between levels is not known. The indicator is the outcome itself. The rank-ordered sequence is represented by a sequence of grades of the outcome, from the lowest to the highest, and the rank order is set according to increasing (or decreasing) severity. An example of a largely used ordinal outcome is the Rankin scale for neurologic disability [13]. In this scale, scores range from 0 to 6, with 0 indicating no symptoms, 1 no clinically significant disability, 2 slight disability, 3 moderate disability, 4 moderately severe disability, 5 severe disability, and 6 death.

In randomized clinical trials (RCTs), ordinal outcomes may offer some advantages over the traditionally used binary outcomes. One of such advantages lies in stopping rules. In RCTs using conventional binary outcomes, patients experiencing the outcome event are withdrawn from the trial and treated thereafter with the best available treatment. Using ordinal outcomes implies that a patient receiving the experimental treatment may still be maintained in the trial across several grades of the outcome until reaching the one for which the stopping rule has been set. As an example, in an RCT of treatment to reduce portal pressure, the development of esophageal varices would be considered a failure. However, stopping the trial at this point would hamper to assess the treatment efficacy to prevent other events such as ascites, even while failing to prevent the development of varices. This has been observed in an RCT that showed that non-selective beta-blockers (NSBB) prevent decompensation in compensated cirrhosis but not varices [14], while in an older RCT [15], treatment withdrawal at the time of development of varices hampered the assessment of treatment effect on other events, such as ascites.

In fact, ordinal outcomes allow to assess the effects of different therapies (e.g., active vs. placebo) across a relevant range of events with increasing severity if ethically consistent stopping rules may be set [6]. This concept may be applied to cirrhosis by using as an example the proposed clinical stages, which are ranked according to increasing mortality thus responding to the basic criteria of ordinal outcomes. Because each of these stages has a different predominant pathophysiological mechanism [12, 16, 17], determining the effect of therapy at a given stage would also be the key in elucidating its most likely mechanism of action. A proposed ordinal outcome for compensated cirrhosis is shown in Fig. 3.2. In this proposal, grade 1 of the ordinal outcome (corresponding to the least severe disease phenotype) corresponds to compensated cirrhosis without gastroesophageal

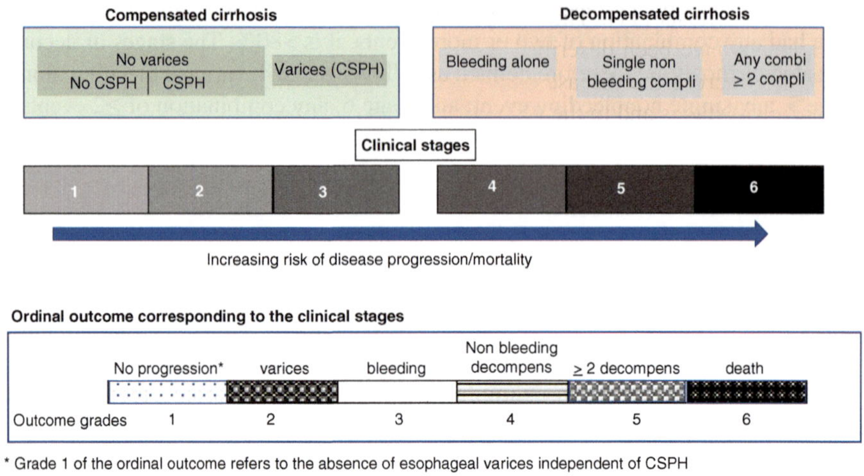

Fig. 3.2 Clinical stages of cirrhosis and a proposed ordinal outcome. Compensated and decompensated cirrhosis are represented in the upper part of the cartoon in boxes including the description of the proposed clinical stages. The clinical stages are ranked from 1 to 6 according to increased risk of disease progression or death (middle part of the cartoon). A proposed six grades ordinal outcome corresponding to the clinical stages of the disease is depicted in the lowest part of the cartoon. CSPH clinically significant portal hypertension. Note that grade 1 of the outcome corresponds to the absence of gastroesophageal varices independent of CSPH because the risk of disease progression or death in compensated cirrhosis without CSPH is not yet clearly defined, albeit considered nil

varices, independent of CSPH because the risk of disease progression/mortality in this disease stage without CSPH is not yet defined.

A second potential advantage of using ordinal outcomes is that they allow (require) more granularity in the description of the disease course both in RCTs and in observational studies. In this regard, it is worth to note that, based on these considerations, a recent consensus of experts regarding clinical trial design in portal hypertension [18] proposed more granularity in the definition of outcomes to allow for the assessment of treatment effect on several events and recommended the use of ordinal outcomes and relevant statistical approaches whenever appropriate [19].

A third relevant advantage of ordinal outcomes is that they require smaller sample sizes than binary outcomes [19–22], because of the usually larger number of events accounted for compared with binary outcomes which would only constitute a part of the spectrum of [18, 23].

Examples of Application of an Ordinal Outcome in Compensated Cirrhosis

To explore the applicability of ordinal outcomes in cirrhosis, we used the 5-year outcome of a cohort of 202 consecutive patients with compensated cirrhosis without gastroesophageal varices prospectively followed in a study of the clinical course of cirrhosis [11]. The study methodology and baseline characteristics of participants have been described in detail elsewhere [11]. The clinical course of this cohort across the disease stages and the final 5-year distribution of patients according to stages and to the proposed ordinal outcome is reported in Fig. 3.3. As shown in the figure, the ordinal outcome summarizes the final condition of participants with much more granularity compared to a binary outcome on say mortality or decompensation (respectively grade 6 or grades 3 to 6 of the proposed ordinal outcome).

The same outcome may be used to assess the disease progression over time. As an example, the baseline condition, 24- and 60-month outcome of the same cohort above reported is shown in Fig. 3.4. At 24 months, a total of 42 new events had been observed, and the patient distribution across the grades of the outcome was the following: no progression (grade 1) 160, esophageal varices (grade 2) 25, variceal bleeding (grade 3) 1, first nonbleeding decompensation (grade 4) 10, two or more decompensating events (grade 5) 1, and death (grade 6) 5. After 60 months of follow-up, a total of 73 events had occurred and the final patient distribution according to the assessed ordinal outcome was the following: 129 (grade 1), 43 (grade 2), 2 (grade 3), 7 (grade 4), 5 (grade 5), and 16 (grade 6). The figure shows the patient distribution according to the ordinal outcome on a percent scale, providing an immediate vision of the disease progression over time.

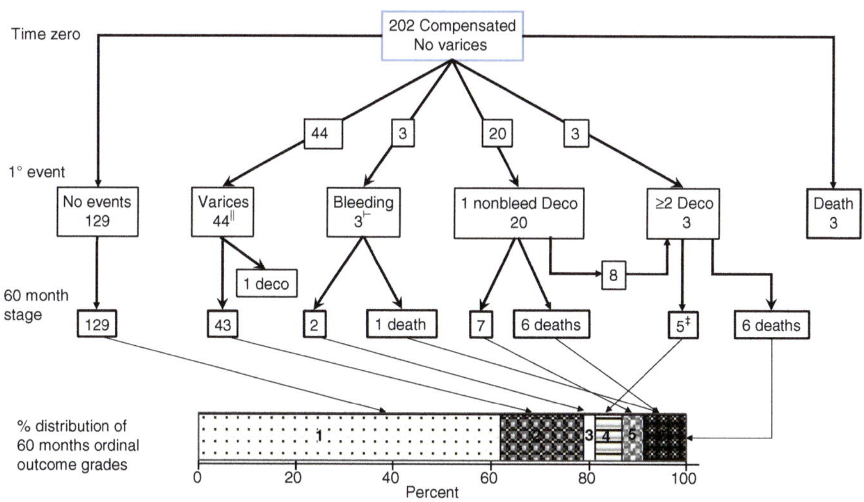

Fig. 3.3 Application of the proposed 60-month ordinal outcome for compensated cirrhosis. The clinical course of a cohort of 202 participants with compensated cirrhosis without esophageal varices at time zero is shown. All the patients were in stage 1 at baseline. The first event was the development of varices in 44 participants, portal hypertensive bleeding in 3, a first nonbleeding decompensating event in 20, two or more decompensating events at the same time in 3, and death (from non-liver-related causes) in 3. Following the first event, several new transitions across stages and 13 deaths were observed. These new transitions resulted in the 60-month ordinal outcome shown in the lower part of the figure and representing the participant distribution across the clinical stages of the disease

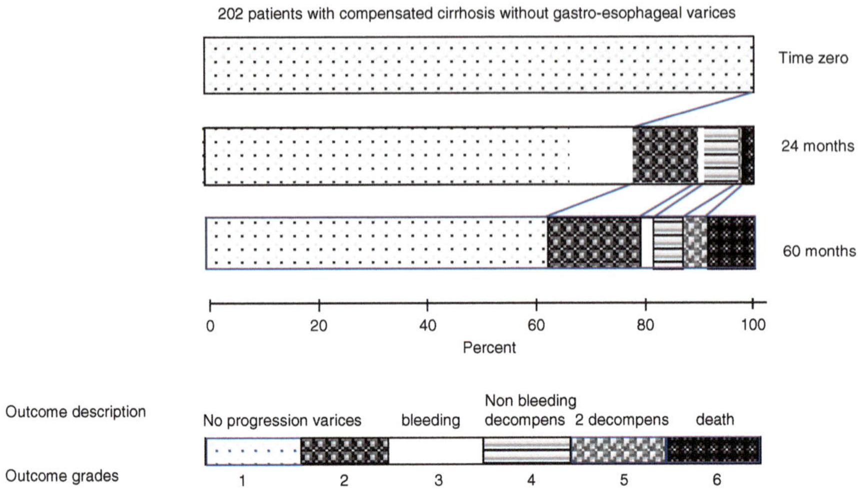

Fig. 3.4 60-month disease progression according to an ordinal outcome. 202 patients with compensated cirrhosis without gastroesophageal varices at time zero. After 24 months, a disease progression occurred in 42 patients, leaving 79% of patients free of any progression (stage 1) and the remaining 21% distributed across the other grades of the outcome. At 60 months, the proportion of patients still free of disease progression was 63%, while 37 had experienced new events. The proportion of patients at each outcome grade progressively increased from baseline to the end of follow-up

Ordinal Outcomes for Randomized Clinical Trials in Cirrhosis

To assess the applicability of ordinal outcomes in cirrhosis, a simulation of an RCT for the prevention of decompensation in the above-presented cohort of patients with cirrhosis is presented. We divided the cohort into two groups of patients similar for all the baseline characteristics but for one which would simulate the treatment tested in the RCT. The grouping variable is the platelet's count and the cutoff for the two groups is 150×10^9/L, with the experimental treatment group being those with PLT >150 and the control group being those with PLT ≤150. To make the simulation realistic, the two groups' 24-month outcome is reported and compared likewise in an RCT. Statistical methods for comparing ordinal outcomes in different groups of patients have been described elsewhere [19, 20] and a tutorial for handling such analyses is also available [24]. Specific software is included in most statistical packages under the heading of "ordinal logistic analysis."

In this simulation, disease progression was defined as the development of esophageal varices, decompensation, or death. To explore the advantages of ordinal outcomes, we compared the two groups' outcomes at 24 months, using (a) the ordinal outcome as described above and (b) a binary outcome corresponding to the ordinal outcome dichotomization at the level of variceal bleeding (no progression or development of varices vs. any decompensation or death).

There were 101 patients with PLT ≤150 × 10^9/L (control group) and 101 with PLT >150 (treatment group). A graphical representation of the ordinal outcome in the two study groups is shown in Fig. 3.5.

To compare the ordinal outcome of the two groups, a basic analysis using a Chi-square test for the 2 by 6 Table (5 degrees of freedom) was first performed and showed no significant differences. In contrast, the proportional odds cumulative logistic model showed a significant difference between the two groups according to the ordinal outcome: the odds ratio for disease progression was 0.48 (0.24–0.96) for patients with platelet count >150 × 10^9/L compared to those with values ≤150 × 10^9/L (Fig. 3.5 and Table 3.2). By contrast, when using the binary composite outcome (no disease progression vs. decompensation or death), the difference between the two groups was not significant either with the binary logistic analysis or with the Cox model (Table 3.2). The lack of statistical significance of the difference between the two groups may be at least in part explained by the marked difference in the total number of outcome events accounted for by the two different types of outcomes: 42 with the ordinal vs. 17 for the binary outcome. It is also important to note that in this simulated trial, a participant experiencing variceal bleeding would be maintained in the study until the end of the study if the comparison was based on an ordinal outcome with ethically consistent stopping rules. By contrast, bleeding would represent a study endpoint in the case of a binary composite endpoint. Therefore, whether the experimental treatment had the potential of preventing further disease progression could only be assessed by using an ordinal outcome and not with a binary one.

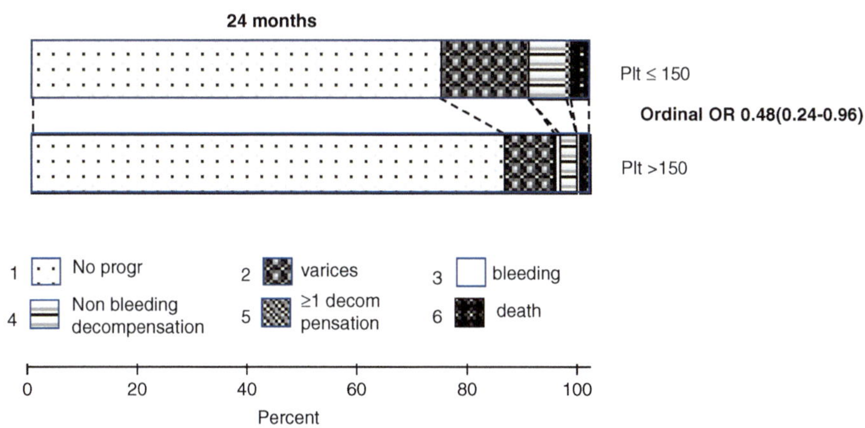

Fig. 3.5 Ordinal outcome at 24 months for patients with a platelet count of ≤ and, respectively, >150 × 10^9/L. The ordinal outcome was significantly worse in patients with lower platelet count. OR denotes odds ratio. Grades refer to the corresponding grade of the ordinal outcome: no progression, grade 1; development of varices, grade 2; bleeding, grade 3; nonbleeding decompensation grade 4; any second decompensation grade 5; death, grade 6

Sample Size Estimation with Ordinal Compared to Binary Outcomes

Methods for sample size calculation using ordinal outcomes are available [24, 25] and it has been consistently shown that in RCTs the needed sample size is lower with ordinal than with binary outcomes [19, 20]. In fact, by dichotomizing an ordinal outcome, part of the outcome events will be considered as "nonevent" [20], thus reducing the total number of events available for the relevant comparison and by reducing the information given by the ordinal outcome (Table 3.2). To illustrate this, a sample size estimation is presented hypothesizing an RCT like the one emulated in the previous paragraphs: the number of expected events is the same as above reported. The sample size estimation is performed either for the ordinal outcome (6 grades from no disease progression to death) or for a binary composite outcome resulting from the dichotomization of the ordinal outcome at two different grades: (a) no progression vs. any progression (including the development of varices, decompensation, or death) or grade 1 vs. grades 2 to 6 of the ordinal outcome; (b)

Table 3.2 24-month disease progression[a] comparison of two groups of patients with compensated cirrhosis and without esophageal varices according to whether they had a platelet count of ≤150 or > 150 × 10^9/L

	Group (PLT × 10^9/L)			Group (PLT × 10^9/L)	
Ordinal outcome[b]	≤150 (*N* = 101)	>150 (*N* = 101)	Dichotomized outcome[c]	≤150 (*N* = 101)	>150 (*N* = 101)
1 = no prog	74	86	No decompensation	90	95
2 = Varices	16	9			
3 = GI bleeding	0	1	Decompensation Or death	11	6
4 = single nonbleeding decompensating event	7	3			
5 = ≥2 decompensating events	1	0			
6 = death	3	2			
Total	27	15			
Total events	42		Total events	17	
OR, ordinal Logistic[d]	0.48 (0.24–0.96)		OR, binary Logistic[e]	0.51 (0.18–1.45)	
			HR, cox model[f]	0.54 (0.20–1.45)	

OR odds ratio, *HR* hazards ratio

[a]Disease progression was defined according to an ordinal outcome in the left part of the table and according to a binary composite outcome corresponding to the dichotomization of the ordinal outcome at grade 3 in the right part

[b]The grades of the ordinal outcome correspond to the clinical stages of cirrhosis and encompass the whole disease course

[c]The binary composite outcome consists of no decompensation, including grades 1 and 2 of the ordinal outcome vs. decompensation or death, including grades 3 to 6 of the ordinal outcome

[d]Ordinal logistic is the proportional ordinal logistic model for ordinal outcomes with underlying proportional OR across the outcome grades (details in [23])

[e]Binary logistics is to the traditional logistic model for binary outcomes

[f]Cox model is the proportional hazards cox model for binary time-dependent outcomes

no progression or development of esophageal varices vs. any decompensating event or death (including bleeding or any decompensating event or death) or grades 1–2 vs. 3–6 of the ordinal outcome (Table 3.3); decompensating events for the binary outcome are bleeding, ascites, hepatic encephalopathy, or jaundice. The sample size estimation is performed under the hypothesis that the experimental treatment would reduce the baseline risk of disease progression by 50% after 24 months. The risk of events in the control group (the baseline underlying risk) is therefore 42/202 or 0.21 with the ordinal outcome and with the binary outcome obtained by dichotomizing the ordinal outcome at grade 2, but it is only 17/202 or 0.08 with the binary outcome corresponding to dichotomization of the ordinal outcome at grade 3 (Table 3.3).

With these baseline risks, the required total sample size for this RCT to show a 50% risk reduction would be 354 patients with the ordinal outcome while it would be 385 with the binary outcome for any disease progression (ordinal outcome dichotomized at grade 2) and 1115 with the binary outcome for decompensation or death (ordinal outcome dichotomized at grade 3) (Table 3.3). Note that in the case of dichotomization at grade 2 (varices), the advantage of using the ordinal outcome would only be of adding the severity of the outcome to the efficacy assessment. In fact, if the development of varices was a stopping rule, it would not be possible to assess the treatment effect after variceal development.

The sample size estimates for the ordinal outcome and for the two binary composite endpoints above considered are based on the log-rank method for the binary outcomes and on the method for ordered categorical variables (ordinal outcome) [25]. Details of calculations have been reported elsewhere [23].

Table 3.3 Sample size required for a superiority RCT with an ordinal outcome or with two binary outcomes obtained by dichotomization of the ordinal outcome at two different levels

	24-month expected outcome based on previous experience with 202 patients with compensated cirrhosis		
		Level of dichotomization for two binary outcomes	
Grades of the ordinal outcome	Ordinal outcome	Varices (grade 2)	Bleeding (grade 3)
1 = no progression	160	160	185
2 = varices	25	42	
3 = bleeding	2		17
4 = single nonbleeding decompensating event	13		
5 = ≥2 decompensating events	1		
6 = death	1		
Total expected events	42	42	17
Sample size calculation			
Baseline risk	0.21	0.21	0.08
Risk seduction hypothesis (24 months)	0.5	0.5	0.5
Total sample size required[a]	354	385	1115

[a] Detail of sample size calculation is reported in [23]

Using Ordinal Outcomes in Portal Hypertension

The example of ordinal outcome above shown has been developed for studies of disease progression, in patients with compensated cirrhosis. Clearly, this outcome does not apply to any research aim. As an example, in an RCT for the prevention of further decompensation in patients with ascites, a possible ordinal outcome could include the need for large volume paracentesis, the development of spontaneous bacterial peritonitis, acute kidney injury, and of course other non-ascites-dependent complications like portal hypertensive bleeding, encephalopathy, and jaundice. In fact, in a cohort of 513 patients with ascites alone as their first decompensating event, after 24 months of follow-up, 256 had not experienced any other event, while the remaining, after a series of crossing events among several other complications were in the following conditions: 26 had experienced one of the possible ascites-related complications; 59 had experienced one or more of bleeding, encephalopathy, or jaundice; 20 had any combination of ascites related or not related complications; and 152 had died. The risk of death significantly increased across the first three of these conditions. Therefore, a possible ordinal outcome for patients with ascites and free of other complications could be the following: grade 1, no progression; grade 2, ascites-related complications; grade 3, non-ascites-related complications; grade 4, any combination of ascites or non-ascites-related complications; and grade 5, death. The sample size required to show a 0.5 relative risk reduction would be 95 with this ordinal outcome, 170 by dichotomizing it at grade 2 (no progression vs. progression), and 340 if it was dichotomized at grade 5 (alive vs. dead).

This example shows that when appropriate, ordinal outcomes may be applicable in different segments of the clinical course of cirrhosis. The optimal outcome should be set according to the specific study aim and grades of the outcome should be ranked according to increasing severity.

Conclusions

Compensated and decompensated cirrhosis are the major stages of cirrhosis. Decompensated cirrhosis has been usually defined as the presence or history of any variceal bleeding, ascites, hepatic encephalopathy, or jaundice although the role of jaundice in the definition of decompensation is currently under debate. Six substages have been defined, 3 in the compensated and 3 in the decompensated disease. The six substages are not defined on a pathophysiological basis but on their clinical characteristics and different mortality risk, and consequently, they do not occur in a predictable sequence. Because of the progressive increase in their mortality risk, the proposed stages of cirrhosis fulfill the criteria of ordinal outcomes. Based on the clinical stages of cirrhosis, an ordinal outcome for cirrhosis progression has been proposed encompassing the following grades: 1, no disease progression; 2, esophageal varices; 3, variceal bleeding; 4, any single nonbleeding decompensating event; 5, any association of two or more decompensating events; and 6, death. Applicability and potential advantages of this ordinal outcome for prognosis or treatment research in compensated cirrhosis have been shown. Similarly, ordinal outcomes may be set according to specific aims and settings as appropriate in different segments of the clinical course of cirrhosis when a possible outcome indicator may be rank-ordered according to its severity.

References

1. D'Amico G, Garcia-Tsao G, Pagliaro L. Natural history and prognostic indicators of survival in cirrhosis: a systematic review of 118 studies. J Hepatol. 2006;44:217–31.
2. Ginés P, Quintero E, Arroyo V, Terés J, Bruguera M, Rimola A, Caballería J, Rodés J, Rozman C. Compensated cirrhosis. Natural history and prognostic factors. Hepatology. 1987;7:122–8.
3. D'Amico G, Morabito A, Pagliaro L, Marubini E. Survival and prognostic indicators in compensated and decompensated cirrhosis. Dig Dis Sci. 1986;31:468–75.
4. Planas R, Montoliu S, Ballestè B, Rivera M, Mireia M, Masnou H, et al. Natural history of patients hospitalized for management of cirrhotic ascites. Clin Gastroenterol Hepatol. 2006;4:1385–94.
5. D'Amico G, Morabito A, D'Amico M, Pasta L, Malizia G, Rebora P, Valsecchi MG. Clinical states of cirrhosis and competing risks. J Hepatol. 2018;68(3):563–76.
6. Saunders JB, Walters JRF, Davies P, Paton A. A 20-year prospective study of cirrhosis. Brit Med J. 1981;282:263–36.
7. Tsochatzis EA, Bosch J, Burroughs AK. Liver cirrhosis. Lancet. 2014;383(9930):1749–61.
8. D'Amico G, Perricone G. Prediction of decompensation in patients with compensated cirrhosis: does etiology matter? Curr Hepatol Rep. 2019;18:144–56.
9. Papatheodoridis GV, Hadziyannis ES, Deutsch M, Hadziyannis SJ. Ursodeoxycholic acid for primary biliary cirrhosis: final results of a 12-year, prospective, randomized, controlled trial. Am J Gastroenterol. 2002;97(8):2063–70.
10. De Franchis R, Bosch J, Garcia-Tsao G, Reiberger T, Ripoll C, Baveno VII Faculty. Renewing consensus in portal hypertension. Report of the Baveno VII consensus workshop: personalized care in portal hypertension. J Hepatol. 2022;76(4):959–74.
11. D'Amico G, Pasta L, Morabito A, D'Amico M, Caltagirone M, Malizia G, et al. Competing risks and prognostic stages in cirrhosis: a 25-year inception cohort study of 494 patients. Aliment Pharmacol Ther. 2014;39:1180–19.
12. Villanueva C, Albillos A, Jenescà J, Abraldes J, Calleca JL, Aracil C, et al. Development of hyperdinamic circulation and response to β-blockers in compensated cirrhosis with portal hypertension. Hepatology. 2016;63:197–206.
13. Broderick JP, Adeoye O, Jordan Elm J. Evolution of the modified Rankin scale and its use in future stroke trials. Stroke. 2017;48:2007–12.
14. Villanueva C, Albillos A, Genescà J, Garcia-Pagan JC, Calleja JL, Aracil C, et al. β blockers to prevent decompensation of cirrhosis in patients with clinically significant portal hypertension (PREDESCI): a randomised, double-blind, placebo-controlled, multicentre trial. Lancet. 2019;393:1597–608.
15. Groszmann RJ, Garcia-Tsao G, Bosch J, Grace ND, Burroughs AK, Planas R, et al. Beta-blockers to prevent gastroesophageal varices in patients with cirrhosis. N Engl J Med. 2005;353:2254–61.
16. Turco L, Garcia-Tsao G, Magnani I, Bianchini M, Costetti M, Caporali C, Colopi S, et al. Cardiopulmonary hemodynamics and C-reactive protein as prognostic indicators in compensated and decompensated cirrhosis. J Hepatol. 2018;68:949–58.
17. Bernardi M, Moreau R, Angeli P, Schnabl B, Arroyo V. Mechanisms of decompensation and organ failure in cirrhosis: from peripheral arterial vasodilatation to systemic inflammation hypothesis. J Hepatol. 2015;63:1272–84.

18. Abraldes JG, Trebicka J, Chalasani N, D'Amico G, Rockey D, Sha V, Bosch J, Garcia-Tsao G. Prioritization of therapeutic targets and trial design in cirrhotic portal hypertension. Hepatology. 2019;69:1287–99.
19. Scott SC, Goldberg MS, Mayo NE. Statistical assessment of ordinal outcomes in comparative studies. J Clin Epidemiol. 1997;50(1):45–55.
20. Murray GD, Barer D, Choi S, Fernandes H, Gregson B, Lees K, et al. Design and analysis of phase III trials with ordered outcome scales: the concept of the sliding dichotomy. J Neurotrauma. 2005;22:511–7.
21. Roozenbeek B, Lingsma HF, Perel P, Edwards P, Roberts I, Murray GD, Maas AIR, et al. The added value of ordinal analysis in clinical trials: an example in traumatic brain injury. Crit Care. 2011;15:R127.
22. Agresti A. Analysis of ordinal categorical data (Wiley series in probability and statistics). Hoboken, NJ: Wiley; 2010.
23. D'Amico G, Abraldes JG, Rebora P, Valsecchi MG, Garcia-Tsao G. Ordinal outcomes are superior to binary outcomes for designing and evaluating clinical trials in compensated cirrhosis. Hepatology. 2020;72(3):1029–42.
24. Harrell FE Jr, Margolis PA, Gove S, Mason KE, Mulholland EK, Lehmann D. Development of a clinical prediction model for an ordinal outcome: the World Health Organization multicentre study of clinical signs and etiological agents of pneumonia, sepsis and meningitis in young infants. Stat Med. 1998;17:909–44.
25. Whitehead J. Sample size calculations for ordered categorical-data. Stat Med. 1993;12:2257–71.

Lifestyle and Genetic Modifiers of Liver Disease Progression

4

Mattias Mandorfer and Annalisa Berzigotti

Lifestyle Modifiers of Liver Disease Progression

Lifestyle is broadly defined as the way an individual lives, and includes habits, attitudes, and tastes. The main components of lifestyle include nutrition, physical activity/exercise, and behaviors modulating the response to stressors and mental health.

Unhealthy Lifestyle

A large body of evidence shows that some "unhealthy" lifestyle factors (alcohol intake, obesity, malnutrition, lack of physical activity, and cigarette smoke) influence the likelihood of suffering from liver disease, favor the progression of liver disease to cirrhosis and hepatic decompensation, further decompensation, and death, and reduce the likelihood of liver disease regression after removal of a main cause of liver injury (e.g., HCV infection). The burden of "unhealthy" lifestyle is difficult to estimate, but there has been an attempt to quantify its impact on the risk of liver-related mortality. In subjects included in a nationwide, prospective cohort study, subjects were considered at low risk if having all the following characteristics: never smoked or past moderate smoking, no or moderate alcohol use, BMI between 18.5 and 24.9, weekly physical activity, and diet meeting at least 40 points on a scale for healthy diet criteria. Multivariate-adjusted hazard ratios (HRs) for five

M. Mandorfer
Vienna Hepatic Hemodynamic Lab, Division of Gastroenterology and Hepatology, Department of Internal Medicine III, Medical University of Vienna, Vienna, Austria
e-mail: mattias.mandorfer@meduniwien.ac.at

A. Berzigotti (✉)
Department of Visceral Surgery and Medicine, Inselspital, Bern University Hospital, University of Bern, Bern, Switzerland
e-mail: annalisa.berzigotti@insel.ch

R. de Franchis (ed.), *Portal Hypertension VII*,
https://doi.org/10.1007/978-3-031-08552-9_4

vs. zero modifiable risk factors were 3.59 [95% confidence interval (95%CI): 1.50–7.42]) for incident hepatocellular carcinoma (HCC) and 4.27 (95%CI: 56–98) for cirrhosis-related mortality [1].

Alcohol

In addition to being a leading cause of chronic liver disease (CLD) worldwide [2, 3], alcohol intake is a well-recognized risk factor for the progression of liver fibrosis in patients with CLD of other etiologies [4]. In patients with cirrhosis, alcohol intake briskly increases portal pressure and porto-collateral blood flow [5, 6]. Importantly, alcohol and obesity synergistically interact [7], leading to a much higher risk of cirrhosis and hepatocellular carcinoma than each of these two individual risk factors [8, 9]. Ongoing alcohol intake is associated with higher mortality after a first decompensation episode [10] and after an episode of alcoholic hepatitis [11, 12]. On the other hand, remaining abstinent from alcohol improves survival by over 20% at 5 years in patients with alcoholic cirrhosis, and reduces the risk of decompensation by improving portal hypertension (PH) [13]. Alcohol abstinence also allows recovery of gut integrity, as proven by an improved diversity of the gut microbiota and by a reduced permeability [14]. Complete abstinence from alcohol should be recommended in patients with cACLD, irrespective of the underlying etiology.

Obesity

Obesity has been reported in 20%–40% of patients with compensated cirrhosis in the recently reported series [15, 16]. In addition, the incidence of cirrhosis due to NASH, which is almost invariably associated with obesity, is increasing worldwide. Obesity is also no longer uncommon in patients with decompensated cirrhosis, and has been reported in 12%–25% of patients included in the recent series [17–19].

Obesity is a cofactor of the progression of liver fibrosis, and increases the risk of liver-related events, so hepatologists should become familiar with management of it. In the HALT-C trial, histological progression to cirrhosis or clinical progression to decompensation increased by 14% per each BMI quartile [15]. In addition, body weight gain >5% at 1 year of inclusion added a further 35% of risk of progression of liver disease.

In patients with compensated cirrhosis and PH [defined by hepatic venous pressure gradient (HVPG) >5 mmHg] and no varices on endoscopy included in a randomized controlled trial on timolol vs. placebo to prevent the onset of esophageal varices, BMI as a continuous variable was associated with an increase in the risk of decompensation, independent of HVPG and albumin [16]. Risk of decompensation in patients with obesity was approximately tripled vs. patients with a normal BMI [16]. Interestingly, not only obesity per se, but also body composition markers of dysfunctional adipose tissue, such as higher subcutaneous fat density, predicted first decompensation in patients with compensated cirrhosis [20].

In addition, obesity may promote other conditions that contribute to clinical deterioration in patients with cirrhosis. In one study in patients awaiting liver transplantation, obesity was an independent predictor of onset of portal vein thrombosis (hazard ratio 13.1) [21]. Moreover, the risk of developing hepatocellular carcinoma increases in patients with obesity; however, a large part of the effect of the association might be explained by the coexistence of diabetes mellitus [22].

Finally, obesity reduces the likelihood of histological regression of cirrhosis in patients in whom the main cause of liver injury has been removed (e.g., chronic hepatitis B on long-term viral suppression) [23].

From the above-described data, obesity emerges as an important but potentially modifiable risk factor in patients with compensated advanced chronic liver disease (cACLD). Three studies demonstrated that intentional weight loss of at least 5% is able to improve liver fibrosis on histology in the long term and to impact favorably on the HVPG [15, 24, 25]. A weight loss of at least 5% was achieved in over 50% of cases, except in diabetic patients who achieved lower success rates. In the SportDiet study, a proof-of-concept study looking at changes in HVPG after 16 weeks of intensive lifestyle changes (moderate caloric restriction and supervised exercise) in compensated cirrhosis with PH, a weight loss >10% was associated with a statistically significant and clinically meaningful reduction of HVPG [25].

In decompensated cirrhosis, data are less abundant. Overall, obesity increases the risk of serious bacterial infection requiring ICU admission [26], and severe obesity is associated with increased risk of acute-on-chronic liver failure [27] and represents a challenge if liver transplantation is indicated due to higher risk of complications [28, 29]. In one paper, obesity was associated with a lower risk of death during the ICU stay [30], but in another one, obesity was a risk factor for death after admission for septic shock [31]. Therefore, whether an "obesity paradox" in critically ill patients with cirrhosis exists, remains to be confirmed.

Malnutrition, Sarcopenia, and Frailty

Protein-energy malnutrition (PEM) is a common complication of cirrhosis present in almost all patients with decompensated cirrhosis, which contributes to skeletal muscle mass and function loss (sarcopenia) and to frailty, which can be defined as decreased physiologic reserve and increased susceptibility to health stressors. While defining malnutrition in cirrhosis can be challenging, sarcopenia can be assessed by several different methods, ranging from simple tools (hand-grip strength test; mid-arm circumference measurement) to imaging methods [32]. Skeletal muscle index assessed on a single-slice CT scan at L3-L4 is considered accurate and reproducible in cirrhosis. Irrespective of the method used for assessment, sarcopenia and frailty have been consistently associated with increased mortality in patients with decompensated cirrhosis, independent of liver function [33, 34]. In compensated cirrhosis, there are less data, which are however fully in line with what observed in decompensated cirrhosis. Old studies suggested that malnourished patients (low hand-grip strength or low mid-arm circumference) [35, 36] have higher risks of decompensation and bacterial infections. In a recent large monocentric study, liver frailty index held predictive value for decompensation in compensated patients and for mortality in decompensated patients [37]. Nutritional assessment and supplementation should be part of the clinical routine in cirrhosis, together with measures aimed at improving physical activity [34, 38].

Cigarette Smoke

Cigarette smoke is pro-fibrogenic in the liver [39, 40], and increases the risk of liver cirrhosis independent of alcohol intake [41]. Importantly, cigarette smoke is a major risk factor for the development of HCC in patients with alcoholic and viral cirrhosis [42–44], and the risk increases in a dose-dependent manner in the general population [45]. In patients with alcoholic cirrhosis, cigarette smoke also synergistically increases the risk of oral cavity, throat, and esophagus cancer [46]. Vice versa, smoking cessation was associated with a reduction in HCC risk [45], and this lifestyle change should be recommended in patients with cirrhosis.

Healthy Lifestyle—Protective Factors

Physical Activity and Exercise

Sedentary lifestyle is very frequent in the general population, and even more in patients with cirrhosis, who according to the available data spend on average 76% of waking hours in sedentary state [47]. Physical inactivity is a risk factor for the onset of non-alcoholic fatty liver disease (NAFLD), and in later stages of CLD it contributes to sarcopenia and physical deconditioning [48]. Exercise decreases intrahepatic fat content independent of weight loss in NAFLD and improves aerobic capacity (oxygen consumption) and has multiple beneficial effects on the cardiovascular system, lung function, endothelial function, mental health, and eventually on quality of life [48]. In murine models of liver disease and in large epidemiological studies, exercise reduces the risk of HCC in a dose-dependent manner [49].

In patients with cirrhosis, 10 randomized controlled trials using 8–16 weeks of supervised or unsupervised exercise as intervention have been published so far. The studies are heterogenous in design and include mostly compensated patients (Child A and B), but in this population no safety issue has been detected. The results showed homogenous positive effects on muscle mass, functional status, HVPG, and quality of life. Rehabilitation including exercise tailored to patients' status in decompensated cirrhosis is currently being implemented in several centers.

Increasing physical activity and exercising should be considered a cornerstone of lifestyle management for CLD.

Coffee Consumption and Mediterranean Diet

In a large study based on participants of the UK Biobank study including 384,818 coffee drinkers and 109,767 non-coffee drinkers, all types of coffee were found to be protective against the development of CLD and cirrhosis [50], confirming previous smaller reports. Coffee drinkers had lower adjusted HRs of CLD (HR: 0.79, 95%: CI 0.72–0.86), CLD or steatosis (HR: 0.80, 95%CI: 0.75–0.86), death from CLD (HR: 0.51, 95%CI: 0.39–0.67), and HCC (HR: 0.80, 95%CI: 0.54–1.19). These results have been confirmed using systematic review and meta-analysis approaches [51, 52].

The benefits of Mediterranean diet (rich in olive oil, vegetables, and fruit) on cardiovascular risk and on other chronic diseases have been established by several large studies [53–57]. In patients with cirrhosis, one study reported a lower risk of hospitalization due to liver-related events in patients adhering to Mediterranean diet vs. Western diet. In this study, patients with cirrhosis on Mediterranean diet showed a higher gut microbiome diversity, which may explain the better clinical outcomes [58].

Genetic Modifiers of Progression of cACLD

There is an extensive body of evidence indicating that genetic factors impact the susceptibility for CLD and the progression to advanced chronic liver disease (ACLD), particularly in NAFLD [59], alcohol-related liver disease (ALD) [60], and hepatitis C [61]. Moreover, some genetic factors have also been shown to promote disease progression beyond this point. However, with the exception of rare monogenetic liver diseases, individual variants explain only a small proportion of the variation in CLD (e.g., NAFLD) [62] prevalence/severity, highlighting the polygenic nature of CLD. Polygenic risk scores have been recently shown to improve (vs. simple blood tests for liver fibrosis) the prediction of liver-related events (incident cirrhosis, HCC, or liver transplantation) in the general population as well as in subjects at risk for NAFLD [63], and profoundly modify the likelihood of cirrhosis in subjects with heavy alcohol consumption [60]. However, even when combining multiple disease-modifying variants, only a minor proportion of variance is explained [62], which could be attributed to the limited knowledge on the genetics underlying CLD, or more plausibly, to the overwhelming contribution of lifestyle factors. Moreover, genetic factors are not modifiable (at present). Despite this, genetic studies offer unique possibilities: First, they circumvent the chicken or the egg causality dilemma, which commonly impedes the interpretation of studies on ACLD. Second, the genetic factors are constant over time; this is in contrast to other patient characteristics (including, but not limited to lifestyle factors) and conventional biomarkers that are applied for risk stratification and treatment individualization.

Genetic factors that have been studied in context of ACLD can be divided into two categories:

1. Those, that are primarily modifiers of liver metabolism, and thus, metabolic-associated fatty liver disease. However, the same variants may also impact the course of other etiologies of CLD. Examples are *patatin-like phospholipase domain-containing protein 3* (*PNPLA3*) *rs738409 C > G* and *17β-hydroxysteroid dehydrogenase type 13* (*HSD17B13*) *rs72613567 T > TA*. Both of them affect hepatic lipid metabolism and are common variants that are associated with only moderate changes in risk for ACLD development due to MAFLD [59, 60]. *Serpin peptidase inhibitor clade A member 1* (*SERPINA1 rs28929474 G > A*), which encodes the alpha-1 antitrypsin deficiency (A1AD) Z-allele, is less common. While homozygosity for the Z-allele results in A1AD-related liver disease (i.e., a rare, monogenetic CLD with incomplete penetrance due to a toxic

gain-of-function), heterozygosity (affecting approximately 2% of Europeans [64] with considerable regional differences) is accompanied by profoundly increased risks of ACLD [65].

2. In addition, there are several genetic factors that orchestrate etiology-independent pathophysiological mechanisms that may contribute to the progression of liver disease/PH. Besides inherited thrombophilia promoting liver fibrosis progression [66–69], variants in *nucleotide-binding oligomerization domain-containing protein 2* (*NOD2*; encoding an intracellular pathogen recognition receptor) and nuclear receptor subfamily 1 group H member 4 (*NR1H4*; encoding the Farnesoid X receptor) have been shown to modify the course of ACLD.

Individual Genetic Variants

PNPA3 and HSD17B13

PNPLA3 encodes a lipase with activity toward triglycerides localized on the surface of lipid droplets [70]. The loss-of-function *rs738409 C > G*-variant not only increases hepatic triglyceride content upon accumulation of the mutant protein on the surface of lipid droplets [70] but also potentiates the proinflammatory and -fibrogenic features of HSC [71]. It has been linked to NAFLD/NASH and disease severity [72, 73] and ALD-induced cirrhosis [74–77]. Moreover, it was associated with hepatic steatosis/liver fibrosis in HCV-monoinfected [78] and HIV/HCV-coinfected patients [79], although its impact in viral hepatitis was less consistent, as compared to MAFLD. Importantly, in the context of CLD, *PNPLA3* is the most thoroughly investigated variant and there are also studies suggesting a role as a disease-modifying variant in established ACLD. In cACLD patients with NAFLD, harboring the *G*-allele doubled the risk of hepatic decompensation [80]. Moreover, the *PNPLA3 G/G*-genotype doubled the mortality risk of patients with PH due to MAFLD, even after the development of CSPH (defined by HVPG ≥10 mmHg) [81]. This is in line with an earlier study of Friedrich et al. [82], which investigated the effect of *PNPLA3* in patients listed for liver transplantation and reported increased risks of (further) hepatic decompensation and mortality. An analysis based on data of the Steroids or Pentoxifylline for Alcoholic Hepatitis (STOPAH) trial [83], which included patients with and without ACLD, revealed an association between *PNPLA3* genotype and disease severity as well as mortality; of note, the latter effect was limited to patients who were abstinent. Although this may be interpreted as evidence for *PNPLA3* genotype impacting the course of ACLD after eradication/suppression of the primary etiological factor, *PNPLA3* and other variants did not impact the dynamics of liver disease in ACLD patients who achieved HCV cure [84].

HSD17B13 rs72613567 T > TA has been shown to decrease the susceptibility for ALD/NAFLD as well as the risk of cirrhosis in these etiologies; in patients undergoing bariatric surgery, it decreased the odds of NASH and liver fibrosis [85–87]. While patients with ACLD of viral or MAFLD etiology harboring the protective variant had less pronounced liver disease at baseline, *HSD17B13* did not impact progression of ACLD in a longitudinal analysis [88]. Accordingly, the importance of this variant in the context of ACLD remains to be established.

SERPINA1/Alpha-1 Antitrypsin Deficiency

SERPINA1 rs28929474 G > A (i.e., the Z-allele) is less prevalent, as compared to the previously mentioned variants; however, heterozygosity for this allele is an even stronger (odds ratios of 6 to 7) risk factor for cirrhosis of non-alcoholic and alcoholic etiology [65] and also CSPH [89]. Approaching the resulting MZ genotype from the other direction (i.e., assessing subjects without previously known CLD who have been diagnosed by genetic testing) indicates that lifestyle factors (obesity and diabetes mellitus) are key modulators of the risk of liver fibrosis in subjects heterozygous for this strong risk allele [90]. This supports the notion that genetic, metabolic, and environmental factors contribute to MAFLD—a concept that may also be extrapolated to CLD in general [91]. Finally, there is also some evidence suggesting that the MZ genotype may confer a worse prognosis in patients with established cirrhosis [92].

NOD2

NOD2 senses muramyl dipeptide (MDP) in the cytosol and activates NF-κB signaling. Loss-of-function variants adversely impact gut barrier function and bacterial translocation, leading to increases in bacterial DNA and interleukin 6 in the systemic circulation, as well as increased risks of spontaneous bacterial peritonitis (SBP) and bacterial infections in general [93–99]. These observations have led to the hypothesis that NOD2 variants may guide the use of norfloxacin in patients with cirrhosis and ascites, but without a history of SBP. In the "Impact of NOD2 genotype-guided antibiotic prevention on survival in patients with liver Cirrhosis and Ascites (INCA)" trial, patients harboring a NOD2 variant are randomized to receive norfloxacin or placebo. Survival and SBP/bacterial infections/hospitalization within 1 year are being evaluated as primary and secondary endpoints, respectively. Accordingly, NOD2 provides an example for the potential clinical application of genetics for risk stratification, i.e., the identification of patients who may be at a particularly high risk of a specific outcome, thereby increasing the absolute risk reduction and decreasing the number need to treat/sample size for a clinical trial.

Of note, there are also other genetic variants in innate immunity receptors, which may be of relevance in patients with acute decompensation and bacterial infection [100].

NR1H4/FXR

Farnesoid X receptor (i.e., a bile acid receptor) [101] signaling has important implications for the gut-liver-axis and is a promising target in the treatment of PH [102]. ACLD patients harboring the *NR1H4 rs35724 rs35724 G > C* allele had a decreased risk of first hepatic decompensation and mortality. These findings may be explained by the gain-of-function that is conferred by this variant, which was accompanied by increased hepatic FXR expression [103]. In the latter study, it was also found to be linked to decreased odds of steatosis, steatohepatitis, and liver fibrosis in patient who underwent liver biopsy for the suspicion of NASH. These findings add to the body of evidence supporting the detailed evaluation of FXR agonists in the context of PH. The implications of this variant on the efficacy of FXR agonists have yet to be evaluated in clinical trials. Of note, this variant may not only inform about risk in an individual patient (see *NOD2*), but also modify the effectiveness of pharmacological intervention, i.e., the relative risk reduction.

Conclusions and Outlook

The importance of genetic variants in patients who have already progressed to ACLD is less well established, as compared to their impact on the progression to ACLD. Although the contribution of genetic factors seems to be considerably lower, as compared to lifestyle factors, genetics may provide information on the risk of specific complications of ACLD and mortality in an individual, which may guide the use of pathophysiology-oriented interventions (e.g., norfloxacin). Moreover, improvements in the understanding of genetics may lead to novel treatments for (A) CLD in the future, as *HSD17B13* (e.g., ARO-HSD; NCT04202354) and *SERPINA1* (e.g., ARO-AAT; NCT03946449) are druggable by small interfering RNA (siRNA). Finally, the impact of *NR1H4* variants on the effectiveness of FXR agonists (e.g., obeticholic acid and numerous other compounds [101]) warrants further study.

The interaction between genetic background and lifestyle on the risk of decompensation in the specific setting of ACLD has not been studied, and is a topic for future research.

References

1. Simon TG, Kim MN, Chong D, Fuchs C, Meyerhardt J, Giovannucci E, Stampfer M, Zhang X, Chan A. The impact of healthy lifestyle on the incidence of hepatocellular carcinoma and cirrhosis-related mortality among U.S. adults. Hepatology. 2019;70(Supp 1):11A.
2. Ventura-Cots M, Ballester-Ferre MP, Ravi S, Bataller R. Public health policies and alcohol-related liver disease. JHEP Rep. 2019;1(5):403–13.
3. Blachier M, Leleu H, Peck-Radosavljevic M, Valla DC, Roudot-Thoraval F. The burden of liver disease in Europe: a review of available epidemiological data. J Hepatol. 2013;58(3):593–608.
4. Monto A, Patel K, Bostrom A, Pianko S, Pockros P, McHutchison JG, et al. Risks of a range of alcohol intake on hepatitis C-related fibrosis. Hepatology. 2004;39(3):826–34.
5. Luca A, Garcia-Pagan JC, Bosch J, Feu F, Caballeria J, Groszmann RJ, et al. Effects of ethanol consumption on hepatic hemodynamics in patients with alcoholic cirrhosis. Gastroenterology. 1997;112(4):1284–9.
6. Spahr L, Goossens N, Furrer F, Dupuis M, Vijgen S, Elkrief L, et al. A return to harmful alcohol consumption impacts on portal hemodynamic changes following alcoholic hepatitis. Eur J Gastroenterol Hepatol. 2018;30(8):967–74.
7. Diehl AM. Obesity and alcoholic liver disease. Alcohol. 2004;34(1):81–7.
8. Loomba R, Yang HI, Su J, Brenner D, Barrett-Connor E, Iloeje U, et al. Synergism between obesity and alcohol in increasing the risk of hepatocellular carcinoma: a prospective cohort study. Am J Epidemiol. 2013;177(4):333–42.
9. Mahli A, Hellerbrand C. Alcohol and obesity: a dangerous Association for Fatty Liver Disease. Dig Dis. 2016;34(Suppl 1):32–9.
10. Powell WJ Jr, Klatskin G. Duration of survival in patients with Laennec's cirrhosis. Influence of alcohol withdrawal, and possible effects of recent changes in general management of the disease. Am J Med. 1968;44(3):406–20.

11. Louvet A, Labreuche J, Artru F, Bouthors A, Rolland B, Saffers P, et al. Main drivers of outcome differ between short term and long term in severe alcoholic hepatitis: a prospective study. Hepatology. 2017;66(5):1464–73.
12. Altamirano J, Lopez-Pelayo H, Michelena J, Jones PD, Ortega L, Gines P, et al. Alcohol abstinence in patients surviving an episode of alcoholic hepatitis: prediction and impact on long-term survival. Hepatology. 2017;66(6):1842–53.
13. Mann RE, Smart RG, Govoni R. The epidemiology of alcoholic liver disease. Alcohol Res Health. 2003;27(3):209–19.
14. Tilg H, Cani PD, Mayer EA. Gut microbiome and liver diseases. Gut. 2016;65(12):2035–44.
15. Everhart JE, Lok AS, Kim HY, Morgan TR, Lindsay KL, Chung RT, et al. Weight-related effects on disease progression in the hepatitis C antiviral long-term treatment against cirrhosis trial. Gastroenterology. 2009;137(2):549–57.
16. Berzigotti A, Garcia-Tsao G, Bosch J, Grace ND, Burroughs AK, Morillas R, et al. Obesity is an independent risk factor for clinical decompensation in patients with cirrhosis. Hepatology. 2011;54(2):555–61.
17. Carias S, Castellanos AL, Vilchez V, Nair R, Dela Cruz AC, Watkins J, et al. Nonalcoholic steatohepatitis is strongly associated with sarcopenic obesity in patients with cirrhosis undergoing liver transplant evaluation. J Gastroenterol Hepatol. 2016;31(3):628–33.
18. Choudhary NS, Saigal S, Saraf N, Mohanka R, Rastogi A, Goja S, et al. Sarcopenic obesity with metabolic syndrome: a newly recognized entity following living donor liver transplantation. Clin Transpl. 2015;29(3):211–5.
19. Montano-Loza AJ, Angulo P, Meza-Junco J, Prado CM, Sawyer MB, Beaumont C, et al. Sarcopenic obesity and myosteatosis are associated with higher mortality in patients with cirrhosis. J Cachexia Sarcopenia Muscle. 2016;7(2):126–35.
20. Tapper EB, Zhang P, Garg R, Nault T, Leary K, Krishnamurthy V, et al. Body composition predicts mortality and decompensation in compensated cirrhosis patients: a prospective cohort study. JHEP Rep. 2020;2(1):100061.
21. Ayala R, Grande S, Bustelos R, Ribera C, Garcia-Sesma A, Jimenez C, et al. Obesity is an independent risk factor for pre-transplant portal vein thrombosis in liver recipients. BMC Gastroenterol. 2012;12:114.
22. Tseng CH. Metformin and risk of hepatocellular carcinoma in patients with type 2 diabetes. Liver Int. 2018;38(11):2018–27.
23. Marcellin P, Gane E, Buti M, Afdhal N, Sievert W, Jacobson IM, et al. Regression of cirrhosis during treatment with tenofovir disoproxil fumarate for chronic hepatitis B: a 5-year open-label follow-up study. Lancet. 2013;381(9865):468–75.
24. Macias-Rodriguez RU, Ilarraza-Lomeli H, Ruiz-Margain A, Ponce-de-Leon-Rosales S, Vargas-Vorackova F, Garcia-Flores O, et al. Changes in hepatic venous pressure gradient induced by physical exercise in cirrhosis: results of a pilot randomized open clinical trial. Clin Transl Gastroenterol. 2016;7(7):e180.
25. Berzigotti A, Albillos A, Villanueva C, Genesca J, Ardevol A, Augustin S, et al. Effects of an intensive lifestyle intervention program on portal hypertension in patients with cirrhosis and obesity: the SportDiet study. Hepatology. 2017;65(4):1293–305.
26. Sundaram V, Kaung A, Rajaram A, Lu SC, Tran TT, Nissen NN, et al. Obesity is independently associated with infection in hospitalised patients with end-stage liver disease. Aliment Pharmacol Ther. 2015;42(11–12):1271–80.
27. Sundaram V, Jalan R, Ahn JC, Charlton MR, Goldberg DS, Karvellas CJ, et al. Class III obesity is a risk factor for the development of acute-on-chronic liver failure in patients with decompensated cirrhosis. J Hepatol. 2018;69(3):617–25.
28. LaMattina JC, Foley DP, Fernandez LA, Pirsch JD, Musat AI, D'Alessandro AM, et al. Complications associated with liver transplantation in the obese recipient. Clin Transpl. 2012;26(6):910–8.
29. Terjimanian MN, Harbaugh CM, Hussain A, Olugbade KO Jr, Waits SA, Wang SC, et al. Abdominal adiposity, body composition and survival after liver transplantation. Clin Transpl. 2016;30(3):289–94.

30. Choi C, Lennon RJ, Choi DH, Serafim LP, Allen AM, Kamath PS, et al. Relationship between body mass index and survival among critically ill patients with cirrhosis. J Intensive Care Med. 2022;37(6):817–24.
31. Kok B, Karvellas CJ, Abraldes JG, Jalan R, Sundaram V, Gurka D, et al. The impact of obesity in cirrhotic patients with septic shock: a retrospective cohort study. Liver Int. 2018;38(7):1230–41.
32. Tandon P, Raman M, Mourtzakis M, Merli M. A practical approach to nutritional screening and assessment in cirrhosis. Hepatology. 2017;65(3):1044–57.
33. Sinclair M, Gow PJ, Grossmann M, Angus PW. Review article: sarcopenia in cirrhosis—aetiology, implications and potential therapeutic interventions. Aliment Pharmacol Ther. 2016;43(7):765–77.
34. Lai JC, Tandon P, Bernal W, Tapper EB, Ekong U, Dasarathy S, et al. Malnutrition, frailty, and sarcopenia in patients with cirrhosis: 2021 practice guidance by the American Association for the Study of Liver Diseases. Hepatology. 2021;74(3):1611–44.
35. Alvares-da-Silva MR, Reverbel da Silveira T. Comparison between handgrip strength, subjective global assessment, and prognostic nutritional index in assessing malnutrition and predicting clinical outcome in cirrhotic outpatients. Nutrition. 2005;21(2):113–7.
36. Alberino F, Gatta A, Amodio P, Merkel C, Di Pascoli L, Boffo G, et al. Nutrition and survival in patients with liver cirrhosis. Nutrition. 2001;17(6):445–50.
37. Wang S, Whitlock R, Xu C, Taneja S, Singh S, Abraldes JG, et al. Frailty is associated with increased risk of cirrhosis disease progression and death. Hepatology. 2022;75(3):600–9.
38. European Association for the Study of the Liver. EASL clinical practice guidelines on nutrition in chronic liver disease. J Hepatol. 2019;70(1):172–93.
39. Altamirano J, Bataller R. Cigarette smoking and chronic liver diseases. Gut. 2010;59(9):1159–62.
40. Azzalini L, Ferrer E, Ramalho LN, Moreno M, Dominguez M, Colmenero J, et al. Cigarette smoking exacerbates nonalcoholic fatty liver disease in obese rats. Hepatology. 2010;51(5):1567–76.
41. Dam MK, Flensborg-Madsen T, Eliasen M, Becker U, Tolstrup JS. Smoking and risk of liver cirrhosis: a population-based cohort study. Scand J Gastroenterol. 2013;48(5):585–91.
42. Marrero JA, Fontana RJ, Fu S, Conjeevaram HS, Su GL, Lok AS. Alcohol, tobacco and obesity are synergistic risk factors for hepatocellular carcinoma. J Hepatol. 2005;42(2):218–24.
43. Koh WP, Robien K, Wang R, Govindarajan S, Yuan JM, Yu MC. Smoking as an independent risk factor for hepatocellular carcinoma: the Singapore Chinese health study. Br J Cancer. 2011;105(9):1430–5.
44. Pessione F, Ramond MJ, Njapoum C, Duchatelle V, Degott C, Erlinger S, et al. Cigarette smoking and hepatic lesions in patients with chronic hepatitis C. Hepatology. 2001;34(1):121–5.
45. Petrick JL, Campbell PT, Koshiol J, Thistle JE, Andreotti G, Beane-Freeman LE, et al. Tobacco, alcohol use and risk of hepatocellular carcinoma and intrahepatic cholangiocarcinoma: the liver cancer pooling project. Br J Cancer. 2018;118(7):1005–12.
46. Pelucchi C, Gallus S, Garavello W, Bosetti C, La Vecchia C. Alcohol and tobacco use, and cancer risk for upper aerodigestive tract and liver. Eur J Cancer Prev. 2008;17(4):340–4.
47. Dunn MA, Josbeno DA, Schmotzer AR, Tevar AD, DiMartini AF, Landsittel DP, et al. The gap between clinically assessed physical performance and objective physical activity in liver transplant candidates. Liver Transpl. 2016;22(10):1324–32.
48. Tandon P, Ismond KP, Riess K, Duarte-Rojo A, Al-Judaibi B, Dunn MA, et al. Exercise in cirrhosis: translating evidence and experience to practice. J Hepatol. 2018;69(5):1164–77.
49. Baumeister SE, Leitzmann MF, Linseisen J, Schlesinger S. Physical activity and the risk of liver cancer: a systematic review and meta-analysis of prospective studies and a bias analysis. J Natl Cancer Inst. 2019;111(11):1142–51.
50. Kennedy OJ, Fallowfield JA, Poole R, Hayes PC, Parkes J, Roderick PJ. All coffee types decrease the risk of adverse clinical outcomes in chronic liver disease: a UK biobank study. BMC Public Health. 2021;21(1):970.

51. Kennedy OJ, Roderick P, Buchanan R, Fallowfield JA, Hayes PC, Parkes J. Systematic review with meta-analysis: coffee consumption and the risk of cirrhosis. Aliment Pharmacol Ther. 2016;43(5):562–74.
52. Kennedy OJ, Roderick P, Buchanan R, Fallowfield JA, Hayes PC, Parkes J. Coffee, including caffeinated and decaffeinated coffee, and the risk of hepatocellular carcinoma: a systematic review and dose-response meta-analysis. BMJ Open. 2017;7(5):e013739.
53. Grosso G, Marventano S, Yang J, Micek A, Pajak A, Scalfi L, et al. A comprehensive meta-analysis on evidence of Mediterranean diet and cardiovascular disease: are individual components equal? Crit Rev Food Sci Nutr. 2017;57(15):3218–32.
54. Schwingshackl L, Missbach B, Konig J, Hoffmann G. Adherence to a Mediterranean diet and risk of diabetes: a systematic review and meta-analysis. Public Health Nutr. 2015;18(7):1292–9.
55. Schwingshackl L, Schwedhelm C, Galbete C, Hoffmann G. Adherence to mediterranean diet and risk of cancer: an updated systematic review and meta-analysis. Nutrients. 2017;9(10):1063.
56. Sofi F, Abbate R, Gensini GF, Casini A. Accruing evidence on benefits of adherence to the Mediterranean diet on health: an updated systematic review and meta-analysis. Am J Clin Nutr. 2010;92(5):1189–96.
57. Sofi F, Macchi C, Abbate R, Gensini GF, Casini A. Mediterranean diet and health status: an updated meta-analysis and a proposal for a literature-based adherence score. Public Health Nutr. 2014;17(12):2769–82.
58. Bajaj JS, Idilman R, Mabudian L, Hood M, Fagan A, Turan D, et al. Diet affects gut microbiota and modulates hospitalization risk differentially in an international cirrhosis cohort. Hepatology. 2018;68(1):234–47.
59. Trepo E, Valenti L. Update on NAFLD genetics: from new variants to the clinic. J Hepatol. 2020;72(6):1196–209.
60. Whitfield JB, Schwantes-An T-H, Darlay R, Aithal GP, Atkinson SR, Bataller R, et al. A genetic risk score and diabetes predict development of alcohol-related cirrhosis in drinkers. J Hepatol. 2022;76(2):275–82.
61. Heim MH, Bochud PY, George J. Host—hepatitis C viral interactions: the role of genetics. J Hepatol. 2016;65(1 Suppl):S22–32.
62. Paternostro R, Staufer K, Traussnigg S, Stattermayer AF, Halilbasic E, Keritam O, et al. Combined effects of PNPLA3, TM6SF2 and HSD17B13 variants on severity of biopsy-proven non-alcoholic fatty liver disease. Hepatol Int. 2021;15(4):922–33.
63. De Vincentis A, Tavaglione F, Jamialahmadi O, Picardi A, Antonelli Incalzi R, Valenti L, et al. A polygenic risk score to refine risk stratification and prediction for severe liver disease by clinical fibrosis scores. Clin Gastroenterol Hepatol. 2022;20(3):658–73.
64. Martinez-Gonzalez C, Blanco I, Diego I, Bueno P, Miravitlles M. Estimated prevalence and number of PiMZ genotypes of Alpha-1 antitrypsin in seventy-four countries worldwide. Int J Chron Obstruct Pulmon Dis. 2021;16:2617–30.
65. Strnad P, Buch S, Hamesch K, Fischer J, Rosendahl J, Schmelz R, et al. Heterozygous carriage of the alpha1-antitrypsin pi*Z variant increases the risk to develop liver cirrhosis. Gut. 2019;68(6):1099–107.
66. Wright M, Goldin R, Hellier S, Knapp S, Frodsham A, Hennig B, et al. Factor V Leiden polymorphism and the rate of fibrosis development in chronic hepatitis C virus infection. Gut. 2003;52(8):1206–10.
67. Poujol-Robert A, Boelle PY, Wendum D, Poupon R, Robert A. Association between ABO blood group and fibrosis severity in chronic hepatitis C infection. Dig Dis Sci. 2006;51(9):1633–6.
68. Maharshak N, Halfon P, Deutsch V, Peretz H, Berliner S, Fishman S, et al. Increased fibrosis progression rates in hepatitis C patients carrying the prothrombin G20210A mutation. World J Gastroenterol. 2011;17(45):5007–13.

69. Plompen EP, Darwish Murad S, Hansen BE, Loth DW, Schouten JN, Taimr P, et al. Prothrombotic genetic risk factors are associated with an increased risk of liver fibrosis in the general population: the Rotterdam study. J Hepatol. 2015;63(6):1459–65.
70. Meroni M, Longo M, Rametta R, Dongiovanni P. Genetic and epigenetic modifiers of alcoholic liver disease. Int J Mol Sci. 2018;19(12):3857.
71. Bruschi FV, Claudel T, Tardelli M, Caligiuri A, Stulnig TM, Marra F, et al. The PNPLA3 I148M variant modulates the fibrogenic phenotype of human hepatic stellate cells. Hepatology. 2017;65(6):1875–90.
72. Romeo S, Kozlitina J, Xing C, Pertsemlidis A, Cox D, Pennacchio LA, et al. Genetic variation in PNPLA3 confers susceptibility to nonalcoholic fatty liver disease. Nat Genet. 2008;40(12):1461–5.
73. Stattermayer AF, Traussnigg S, Aigner E, Kienbacher C, Huber-Schonauer U, Steindl-Munda P, et al. Low hepatic copper content and PNPLA3 polymorphism in non-alcoholic fatty liver disease in patients without metabolic syndrome. J Trace Elem Med Biol. 2017;39:100–7.
74. Stickel F, Buch S, Lau K, Meyer zu Schwabedissen H, berg T, Ridinger M, et al. Genetic variation in the PNPLA3 gene is associated with alcoholic liver injury in caucasians. Hepatology. 2011;53(1):86–95.
75. Trepo E, Gustot T, Degre D, Lemmers A, Verset L, Demetter P, et al. Common polymorphism in the PNPLA3/adiponutrin gene confers higher risk of cirrhosis and liver damage in alcoholic liver disease. J Hepatol. 2011;55(4):906–12.
76. Salameh H, Raff E, Erwin A, Seth D, Nischalke HD, Falleti E, et al. PNPLA3 gene polymorphism is associated with predisposition to and severity of alcoholic liver disease. Am J Gastroenterol. 2015;110(6):846–56.
77. Buch S, Stickel F, Trepo E, Way M, Herrmann A, Nischalke HD, et al. A genome-wide association study confirms PNPLA3 and identifies TM6SF2 and MBOAT7 as risk loci for alcohol-related cirrhosis. Nat Genet. 2015;47(12):1443–8.
78. Stattermayer AF, Scherzer T, Beinhardt S, Rutter K, Hofer H, Ferenci P. Review article: genetic factors that modify the outcome of viral hepatitis. Aliment Pharmacol Ther. 2014;39(10):1059–70.
79. Mandorfer M, Payer BA, Schwabl P, Steiner S, Ferlitsch A, Aichelburg MC, et al. Revisiting liver disease progression in HIV/HCV-coinfected patients: the influence of vitamin D, insulin resistance, immune status, IL28B and PNPLA3. Liver Int. 2015;35(3):876–85.
80. Grimaudo S, Pipitone RM, Pennisi G, Celsa C, Camma C, Di Marco V, et al. Association between PNPLA3 rs738409 C>G variant and liver-related outcomes in patients with nonalcoholic fatty liver disease. Clin Gastroenterol Hepatol. 2020;18(4):935–44. e3
81. Mandorfer M, Scheiner B, Stattermayer AF, Schwabl P, Paternostro R, Bauer D, et al. Impact of patatin-like phospholipase domain containing 3 rs738409 G/G genotype on hepatic decompensation and mortality in patients with portal hypertension. Aliment Pharmacol Ther. 2018;48(4):451–9.
82. Friedrich K, Wannhoff A, Kattner S, Brune M, Hov JR, Weiss KH, et al. PNPLA3 in end-stage liver disease: alcohol consumption, hepatocellular carcinoma development, and transplantation-free survival. J Gastroenterol Hepatol. 2014;29(7):1477–84.
83. Atkinson SR, Way MJ, McQuillin A, Morgan MY, Thursz MR. Homozygosity for rs738409:G in PNPLA3 is associated with increased mortality following an episode of severe alcoholic hepatitis. J Hepatol. 2017;67(1):120–7.
84. Semmler G, Binter T, Kozbial K, Schwabl P, Chromy D, Bauer D, et al. Influence of genetic variants on disease regression and outcomes in HCV-related advanced chronic liver disease after SVR. J Pers Med. 2021;11(4):281.
85. Abul-Husn NS, Cheng X, Li AH, Xin Y, Schurmann C, Stevis P, et al. A protein-truncating HSD17B13 variant and protection from chronic liver disease. N Engl J Med. 2018;378(12):1096–106.
86. Pirola CJ, Garaycoechea M, Flichman D, Arrese M, San Martino J, Gazzi C, et al. Splice variant rs72613567 prevents worst histologic outcomes in patients with nonalcoholic fatty liver disease. J Lipid Res. 2019;60(1):176–85.

87. Ma Y, Belyaeva OV, Brown PM, Fujita K, Valles K, Karki S, et al. HSD17B13 is a hepatic retinol dehydrogenase associated with histological features of non-alcoholic fatty liver disease. Hepatology. 2019;69(4):1504–19.
88. Scheiner B, Stättermayer AF, Schwabl P, Bucsics T, Paternostro R, Bauer D, et al. Impact of HSD17B13 rs72613567 genotype on hepatic decompensation and mortality in patients with portal hypertension. Liver Int. 2020;40(2):393–404.
89. Mandorfer M, Bucsics T, Hutya V, Schmid-Scherzer K, Schaefer B, Zoller H, et al. Liver disease in adults with alpha1-antitrypsin deficiency. United European Gastroenterol J. 2018;6(5):710–8.
90. Schneider CV, Hamesch K, Gross A, Mandorfer M, Moeller LS, Pereira V, et al. Liver Phenotypes of European Adults Heterozygous or Homozygous for Pi*Z Variant of AAT (Pi*MZ vs Pi*ZZ genotype) and Noncarriers. Gastroenterology. 2020;159(2):534–48.e11.
91. Eslam M, Sanyal AJ, George J, International Consensus P. MAFLD: a consensus-driven proposed nomenclature for metabolic associated fatty liver disease. Gastroenterology. 2020;158(7):1999–2014. e1
92. Schaefer B, Mandorfer M, Viveiros A, Finkenstedt A, Ferenci P, Schneeberger S, et al. Heterozygosity for the alpha-1-antitrypsin Z allele in cirrhosis is associated with more advanced disease. Liver Transpl. 2018;24(6):744–51.
93. Appenrodt B, Grunhage F, Gentemann MG, Thyssen L, Sauerbruch T, Lammert F. Nucleotide-binding oligomerization domain containing 2 (NOD2) variants are genetic risk factors for death and spontaneous bacterial peritonitis in liver cirrhosis. Hepatology. 2010;51(4):1327–33.
94. Bruns T, Peter J, Reuken PA, Grabe DH, Schuldes SR, Brenmoehl J, et al. NOD2 gene variants are a risk factor for culture-positive spontaneous bacterial peritonitis and monomicrobial bacterascites in cirrhosis. Liver Int. 2012;32(2):223–30.
95. Reiberger T, Ferlitsch A, Payer BA, Mandorfer M, Heinisch BB, Hayden H, et al. Non-selective betablocker therapy decreases intestinal permeability and serum levels of LBP and IL-6 in patients with cirrhosis. J Hepatol. 2013;58(5):911–21.
96. Lutz P, Kramer B, Kaczmarek DJ, Hubner MP, Langhans B, Appenrodt B, et al. A variant in the nuclear dot protein 52kDa gene increases the risk for spontaneous bacterial peritonitis in patients with alcoholic liver cirrhosis. Dig Liver Dis. 2016;48(1):62–8.
97. Harputluoglu MM, Dertli R, Otlu B, Demirel U, Yener O, Bilgic Y, et al. Nucleotide-binding Oligomerization domain-containing protein 2 variants in patients with spontaneous bacterial peritonitis. Dig Dis Sci. 2016;61(6):1545–52.
98. Dinya T, Tornai T, Vitalis Z, Tornai I, Balogh B, Tornai D, et al. Functional polymorphisms of innate immunity receptors are not risk factors for the non-SBP type bacterial infections in cirrhosis. Liver Int. 2018;38(7):1242–52.
99. Reichert MC, Ripoll C, Casper M, Greinert R, Vandieken E, Grünhage F, et al. Common NOD2 risk variants as major susceptibility factors for bacterial infections in compensated cirrhosis. Clin Transl Gastroenterol. 2019;10(1):e00002.
100. Schaapman JJ, Amoros A, van der Reijden JJ, Laleman W, Zeuzem S, Banares R, et al. Genetic variants of innate immunity receptors are associated with mortality in cirrhotic patients with bacterial infection. Liver Int. 2020;40(3):646–53.
101. Simbrunner B, Trauner M, Reiberger T. Therapeutic aspects of bile acid signalling in the gut-liver axis. Aliment Pharmacol Ther. 2021;54(10):1243–62.
102. Simbrunner B, Mandorfer M, Trauner M, Reiberger T. Gut-liver axis signaling in portal hypertension. World J Gastroenterol. 2019;25(39):5897–917.
103. Grimaudo S, Dongiovanni P, Pihlajamaki J, Eslam M, Yki-Jarvinen H, Pipitone RM, et al. NR1H4 rs35724 G>C variant modulates liver damage in nonalcoholic fatty liver disease. Liver Int. 2021;41(11):2712–9.

Part II

HVPG as a Gold Standard

HVPG as a Gold Standard: Accuracy Is Essential

5

Juan Carlos Garcia-Pagàn, Filippo Schepis, Ron C. Gaba, Alberto Zanetto, Valeria Perez-Campuzano, Ziv J. Haskal, and Hector Ferral

J. C. Garcia-Pagàn (✉)
Barcelona Hepatic Hemodynamic Laboratory, Liver Unit, Hospital Clínic, Institut de Investigacions Biomèdiques August Pi i Sunyer (IDIBAPS), University of Barcelona, Barcelona, Spain

CIBEREHD (Centro de Investigación Biomédica en Red Enfermedades Hepáticas y Digestivas), Barcelona, Spain

Health Care Provider of the European Reference Network on Rare Liver Disorders (ERN-Liver), ERN-Liver, Barcelona, Spain
e-mail: jcgarcia@clinic.cat

V. Perez-Campuzano
Barcelona Hepatic Hemodynamic Laboratory, Liver Unit, Hospital Clínic, Institut de Investigacions Biomèdiques August Pi i Sunyer (IDIBAPS), University of Barcelona, Barcelona, Spain

CIBEREHD (Centro de Investigación Biomédica en Red Enfermedades Hepáticas y Digestivas), Barcelona, Spain
e-mail: vperez2@clinic.cat

F. Schepis
Hepatic Hemodynamic Laboratory, Division of Gastroenterology, AOU of Modena and University of Modena and Reggio Emilia, Modena, Italy
e-mail: fschepis@unimore.it

R. C. Gaba
Division of Interventional Radiology, Department of Radiology, University of Illinois at Chicago, Chicago, IL, USA
e-mail: rgaba@UIC.EDU

A. Zanetto
Hepatic Hemodynamic Laboratory, Division of Gastroenterology, AOU of Modena and University of Modena and Reggio Emilia, Modena, Italy

Gastroenterology and Multivisceral Transplant Unit, Department of Surgery, Oncology, and Gastroenterology, Padova University Hospital, Padova, Italy
e-mail: alberto.zanetto@unipd.it

Z. J. Haskal
Department of Radiology and Medical Imaging/Interventional Radiology, University of Virginia School of Medicine, Richmond, VA, USA

H. Ferral
Department of Radiology, Section of Interventional Radiology, Louisiana State University, New Orleans, LA, USA

NorthShore University HealthSystem, Evanston, IL, USA

R. de Franchis (ed.), *Portal Hypertension VII*,
https://doi.org/10.1007/978-3-031-08552-9_5

The progression of portal hypertension in patients with chronic liver disease (CLD) is directly related to the risk of complications such as ascites and variceal hemorrhage. Equally, the reduction of portal pressure (PP) reduces the risk of these events. PP measurements provide both diagnostic and prognostic information. The only validated method for assessing liver hemodynamics is the wedged hepatic venous pressure (WHVP), which indirectly reflects the actual portal vein pressure. Temporary catheter occlusion of a hepatic vein allows measurement of the pressure head of hepatic sinusoidal blood, which reflects the direct pressure of the portal venous system. The correlation between WHVP and portal vein pressure is high as long as presinusoidal portal vein occlusion is absent. The hepatic venous pressure gradient (HVPG) is calculated by subtracting the free hepatic vein pressure (FHVP) from the WHVP. This gradient is the gold standard for the diagnosis of clinically significant PH. It is also used to determine prognosis and response to treatments.

Accurate calculation of hepatic hemodynamics requires carefully standardized techniques. However, the U.S. and European surveys have suggested significant variability in the way these pressures are measured, potentially leading to variabilities and inaccuracies that could affect the validity of the results. The main objective of this section is to emphasize the importance of a standardized and reproducible technique to calculate liver venous pressures [1].

Procedure Technique

HVPG procedures should be performed in a state-of-the-art catheterization laboratory with standard pressure measurement equipment that can record digital or printed paper tracings. This facilitates assessment of venous waveforms, decreases inter- and intra-observer measurement variability, and allows external centralized review of data. Without waveform and pressure recordings, recording numerical measures directly from a monitor screen makes stable assessments of WHVP more difficult, prevents waveform inspection, reduces accuracy, and makes data inaccessible to other investigators, which is particularly important in research trials. For these reasons, isolated numerical recording of transient "on-screen" pressure values data should be avoided.

Measurement procedures are often performed under conscious (moderate) sedation to assure patient comfort using standard sedatives such as midazolam and fentanyl (or meperidine and propofol). Increased doses of these medications result in more profound depression of consciousness and may affect hemodynamic parameters. Two studies evaluated the effect of intravenous sedation, confirming that doses exceeding 0.02 mg/kg of midazolam reduced the accuracy of pressure data. Waveform changes also result from larger doses of sedatives as respiration becomes irregular and more profound [2, 3]. Importantly, low dose midazolam (0.02 mg/kg) did not modify the HVPG and currently is the only acceptable sedation for HVPG measurement [2].

Real-time ultrasound guidance should be used to assure safe jugular or femoral venous access. Using fluoroscopic guidance, the right or middle hepatic veins (HV) are typically catheterized, followed by measurements of free and wedged hepatic vein pressures. These pressure measurements can be obtained using an end-hole angiographic catheter that is manipulated as distally as possible into the hepatic vein to achieve a "wedged" position. Preferably, an end-hole, compliant balloon occlusion catheter can be positioned within a more central portion of the HV. The occlusion balloon catheter allows a larger area of liver parenchyma peripheral to the point of temporary venous occlusion to be evaluated. Studies comparing angiographic catheters with compliant occlusion balloons demonstrate the superiority of the balloon approach, yielding more accurate and precise measurements of WHVP and, thus a better reflection of the direct PP. Based on these studies, compliant balloon occlusion catheters are recommended [4, 5] Of note, European and U.S. surveys performed for the Baveno VII conference revealed that approximately 40% of responding operators still use an end-hole angiographic catheter to measure the WHVP; this indicates areas for improvement and standardization.

Finally, with the occlusion balloon inflated within the HV, a small amount of contrast (5–10 mL) should be injected to confirm the complete occlusion of hepatic outflow [4–9] and exclude hepatic venovenous communications which can artifactually lead to underestimation of the actual PP [9, 10, 11]. If hepatic vein to hepatic vein connections are venographically demonstrated during balloon occlusion, they should be reported, and other sites of measurement should be sought.

Pressure Measurements and Data Recording

The pressure transducer should be placed at the level of the right atrium, i.e., at the midaxillary line. The transducer should record "zero" when open to air [9]. Monitor scales should be set at "central venous" pressure settings rather than "arterial" as their lower pressure ranges (up to 50–60 mmHg) are more suitable for detecting small changes in pressures (0.5 mmHg). In contrast, arterial pressure settings (up to 200 mmHg) are difficult to interpret [9]. Slow speed (up to 7.5 mm/s) permanent recording of pressure tracings is recommended [12].

Technical Aspects

The WHVP is recorded once the occlusion balloon is in its optimal occluded position. Recording should last at least 90 s to allow the pressure to plateau to its maximum level. WHVP should be measured in triplicate to reduce inconsistencies [9]. Once the wedge pressures have been measured, the balloon is deflated, and the catheter is withdrawn to a position within 2–3 cm from the junction of the HV to the inferior vena cava (IVC) to measure the free hepatic venous pressure (FHVP) [1]; this may also be measured prior to balloon inflation. IVC pressure should be measured as an internal control. The ideal site for the calculation of the pressure gradient has been a matter of debate. Some operators have advocated the use of pressures from the IVC or right atrium as an alternative to the FVHP. Results from a comparative study by La Mura show that the HVPG has a significantly superior clinical prognostic value than the wedge-hepatic to atrial pressure gradient. Therefore, WHVP and FHVP must be used to calculate HVPG [6]. If a pressure gradient greater than 2 mmHg is found between FHV and IVC, contrast should be injected to rule out HV stenosis [13]; still, FHVP must be used for gradient calculation. Right atrial pressure is measured to rule out a post-sinusoidal component.

Diagnosis of Clinically Significant Portal Hypertension (CSPH) and Prediction of Main Outcomes in Patients with Different Etiologies of Cirrhosis

The Timolol study [14], a RCT comparing non-selective beta-blockers (NSBBs) to placebo for preventing the development of varices in patients with viral and alcoholic cirrhosis, identified an HVPG ≥10 mmHg as a high-risk marker for development of esophageal varices. A nested analysis of this study by Ripoll et al. found that an HVPG ≥10 mmHg identified patients at higher risk of decompensation, defined by the development of ascites, variceal hemorrhage (VH), or hepatic encephalopathy (HE) [15]. Robic et al. reported a 2-year prospective study of 100 patients with alcohol or viral CLD (65 with cirrhosis), wherein an HVPG ≥10 mmHg predicted the first event of decompensation (ascites, HE, or VH) [16]. These studies showed that an HVPG ≥10 mmHg identified compensated patients at risk for decompensation. Therefore, such patients must be considered as having CSPH [17].

In patients with CPSH, it is likely that the risk of decompensation increases in parallel with the severity of PH. A retrospective, single-center study of 86 patients with compensated cirrhosis not treated with NSBBs (54 viral, 11 alcohol, and 21 multifactorial/others) reported the incidence of the first decompensation to be significantly higher in patients with a baseline HVPG ≥16 mmHg compared with those with pressures <16 mmHg (35.1% vs. 11.5%, $p = 0.02$) [18]. A retrospective study of 741 patients with compensated cirrhosis of both viral and non-viral etiologies stratified them by HVPG: 6 to <12 mmHg; 12 to <20 mmHg; and ≥20 mmHg. All patients with an HVPG ≥12 mmHg were treated with carvedilol. An HVPG ≥12 mmHg was independently predictive of decompensation. Moreover, an HVPG ≥20 mmHg yielded a twofold higher risk of decompensation compared with an HVPG between 12 and <20 mmHg, and a 4.5-fold higher risk compared with an HVPG between 6 and <12 mmHg [19].

Non-alcoholic steatohepatitis (NASH) is an increasing cause of CLD. In a study of 258 patients with compensated NASH (95% Child-Pugh A), 19% experienced liver-related complications that were mainly associated with baseline HVPGs of ≥10 mmHg. Indeed, only 8% of those patients with an HVPG <10 mmHg developed decompensation, which led the authors to hypothesize that PH in NASH patients may partly depend upon a presinusoidal component unassessed by HVPG [20]. Furthermore, a recent study comparing direct PP measurement with WHVP showed that WHVP underestimated PP in NASH patients compared with other etiologies [21]. Finally, a large retrospective cross-sectional multicenter study, assessing the association between HVPG values and clinical signs of PH in patients with advanced NAFLD (aNAFLD) showed that aNAFLD patients had a higher prevalence of portal hypertension-related decompensation at any value of HVPG as compared to aHCV patients; 9% of those patients with an HVPG <10 mmHg had decompensation, mainly with ascites [22].

Two independent cohort studies comparing patients with primary biliary cholangitis (PBC), alcohol and viral etiologies, showed a poor correlation between directly measured PP and WHVP in PBC [23, 24]. In a study by Navasa et al., five patients had esophageal varices despite HVPG <6 mmHg; this indirectly indicated the presence of a presinusoidal component of PH in PBC [25]. Porto-sinusoidal vascular liver disorder (PSVD) is another condition with a clear presinusoidal component wherein WHVP underestimates PP. In these patients, an HVPG <10 mmHg is frequently found despite the presence of severe complications of PH [26].

In conclusion, an HVPG ≥10 mmHg defines the presence of CPSH in patients with alcohol, viral, and NASH-related compensated cirrhosis. In patients with PBC and PSVD, HVPG is unreliable in defining the presence and severity of PH.

Variceal Hemorrhage

Variceal hemorrhage requires values of HVPG ≥12 mmHg. Conversely, patients whose HVPG is reduced to <12 mmHg are protected from PH-related bleeding [27–31] (Table 5.1). There is a consensus that higher values of HVPG are correlated with worse outcomes. One early study, performed when modern endoscopic treatments of VH were unavailable, demonstrated that an HVPG ≥16 mmHg measured within 48 h of hospitalization was strongly correlated with continued bleeding or

Table 5.1 Diagnostic and prognostic values of HVPG in patients with cirrhosis

Single HVPG measurement
≥10 mmHg: Defines "clinically significant portal hypertension" for the increased risk of developing varices, clinical decompensation (variceal hemorrhage, ascites, and hepatic encephalopathy) and HCC
≥10 mmHg: Increased risk of decompensation after hepatic resection for HCC
≥12 mmHg: Increased risk of rupture of varices
≥16 mmHg: Increased risk of death
≥20 mmHg: Treatment failure, early rebleeding, and mortality in variceal hemorrhage
≥16 mmHg and ≥ 20 mmHg: High and very high risk of death after non-hepatic surgery
Repeat HVPG measurement
Reduction to <12 mmHg: Abolition of risk of first variceal hemorrhage and recurrent hemorrhage
Reduction of ≥10% from baseline: Reduced risk first episode of variceal hemorrhage or other decompensating events
Reduction of ≥20% from baseline: Reduced risk of recurrent variceal hemorrhage, ascites, and mortality
Reduction of ≥10% from baseline after acute intravenous propranolol
Administration: Reduced risk of first variceal bleeding, rebleeding, and mortality

HCC hepatocellular carcinoma, *HVPG* hepatic vein pressure gradient

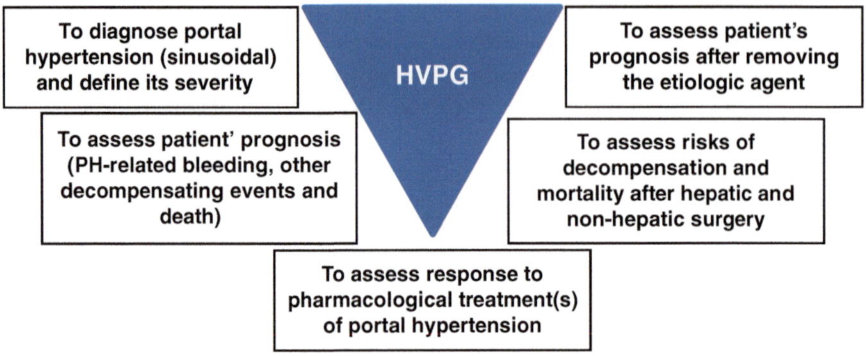

Fig. 5.1 Potential applications of HVPG in clinical practice

early rebleeding [32]. Multiple prospective cohorts, including a seminal randomized controlled trial, confirmed the association between increasing HVPG and a lower probability of hemorrhage control, and consistently found that an HVPG ≥20 mmHg strongly predicted failure, early rebleeding [33–38], and higher mortality [37, 38]. In these studies, a majority of patients were alcoholics or had viral-related cirrhosis. The correlation between HVPG and late rebleeding is less clear [36, 39]. Thus, the assessment of HVPG provides useful information to assess the risk of variceal bleeding or rebleeding in patients with viral and/or alcoholic cirrhosis (Fig. 5.1).

Hepatocellular Carcinoma (HCC)

Three studies evaluated the correlation between HVPG and risk of developing HCC [40–42], the largest by Ripoll et al. ($n = 213$) [41]. The authors found that HVPG was an independent predictor of HCC development, and that patients with CSPH had a sixfold higher risk of HCC compared with those with less severe PH [41]. Two other smaller studies confirmed that HVPG is an independent predictor of HCC in cirrhosis [40, 42].

Survival

The correlation between HVPG and survival in cirrhosis has been extensively studied. Most studies converge upon a strong and independent association between increased HVPG and the risk of death, especially in compensated patients (Table 5.1). In a 42-month prospective study of 81 patients with biopsy-proven alcoholic cirrhosis, Tage-Jansen et al. confirmed that death increased in parallel with the severity of baseline HVPG (<12 vs. 12–20 vs. ≥20 mmHg) [43]. Another study confirmed the association of CSPH with increased risk of mortality, independent of Model for End-Stage Liver Disease (MELD) scores [44]. However, the risk thresholds were inconsistent among studies: ≥10 mmHg in 1 retrospective study [45]; ≥15 mmHg in 1 prospective [46]; ≥16 mmHg in 5 studies (3 retrospective and 2 prospective) [13, 18, 30, 39, 47]; ≥18 mmHg in 1 retrospective [48]; and ≥20 mmHg in 2 studies (1 retrospective and 1 prospective) [35, 42]. On the other hand, two studies with a relatively smaller sample size did not find that HVPG per se provided prognostic significant information for survival [31, 49]. These results likely reflect the marked heterogeneity of patient demographics, ranges of HVPG values, use of NSBBs, management of cirrhosis, and durations of follow-up. In conclusion, HVPG is an independent predictor of mortality in patients with cirrhosis, particularly in compensated ones. Despite marked heterogeneity among studies, a value of HVPG ≥16 mmHg appears to be the best threshold for identifying patients at higher risk for death.

Assessment of HVPG in Patients Receiving NSBBs for Prevention of Variceal Hemorrhage and Decompensation

There is strong evidence that lowering PH by NSBBs reduces the risk of the first episode of VH [29, 50–52]. The first RCT, by Groszmann et al., randomized 51 patients with esophageal varices to propranolol vs. 51 patients treated with placebo [29]. All patients who bled during follow-up had an HVPG >12 mmHg compared with none among patients with an HVPG <12 mmHg (Table 5.1). Merkel et al. reported similar findings in a prospective study of 49 cirrhotic patients (alcohol and HCV-related) with varices at risk who were started on NSBBs +/− isosorbide mononitrate (ISMN) [50]. Response to NSBBs was defined as a decrease of >20% from the baseline value. The cumulative probability of VH was higher in poor vs. good responders. Remarkably, no patient who achieved an HVPG <12 mmHg experienced VH during a 5-year follow-up. These findings were confirmed in a second 71 patient cohort treated with NSBBs + ISMN [52]. Merkel's target thresholds were modified by a larger study by Villanueva et al., in which reduction of HVPG by ≥10% from baseline showed a significantly higher prognostic value compared with the previously proposed ≥20% [53]. The same study also demonstrated that acute response to NSBBs, as assessed by an HVPG measurement before vs. after intravenous infusion of propranolol, provided useful information for the long-term management of cirrhotic patients at risk of VH. Chronic response to NSBBs was also associated with a significantly lower risk of ascites. This further emphasizes that the assessment of hemodynamic response after starting NSBBs is useful for distinguishing among groups at different risks of decompensation during follow-up [53]. This was corroborated by a prospective cohort study of patients with compensated, mostly HCV-related, cirrhosis with HVPGs >12 mmHg; NSBBs were correlated with reduced risk of ascites and increased survival during 5-year follow-up [54].

HVPG has value in secondary prophylaxis of VH. This was first evaluated in a prospective cohort of 69 bleeding cirrhotic patients in whom HVPG was reassessed 3 months after the start of NSBBs [55]. Rebleeding was significantly less in patients with a HVPG reduction of ≥20%. Similarly, another multicenter cohort with 8-year follow-up showed that the cumulative probability of freedom from rebleeding was significantly higher in responders vs. non-responders. The cumulative probability of freedom from ascites, spontaneous bacterial peritonitis, HE, and overall patient survival was also higher in the responders [56]. A third study confirmed these results [57].

La Mura et al. investigated the utility of HVPG to improve prognostic stratification in 424 patients receiving secondary prophylaxis [58]. By combining clinical data such as ascites and/or HE plus severity of PH (HVPG ≥16 mmHg), they identified two groups of patients at significantly different risks of recurrent VH and mortality. The "Low" risk group included patients without ascites or HE, and patients with VH plus ascites or HE but an HVPG <16 mmHg. The "High" risk group included patients with ascites and/or HE and an HVPG ≥16 mmHg unresponsive to NSBBs. If confirmed by future prospective studies, this schema would further reinforce the utility of HVPG to identify patients at risk despite first-line secondary prophylaxis with ligation plus NSBBs, i.e., ones in whom TIPS may be considered. Further studies are also needed to explore whether an HVPG-guided strategy for secondary prophylaxis of VH may reduce the risk of rebleeding and improve survival as was indicated by a proof-of-concept seminal RCT, one in which, however, the standard of care treatment for PH was not applied [59].

A recent multicenter RCT of compensated patients with a high risk of decompensation (HVPG ≥10 mmHg) demonstrated that NSBBs patients had significantly better survival free of decompensation, particularly ascites, compared with the placebo group [60]. A post-hoc analysis showed that an HVPG reduction ≥10% correlated with a higher chance survival without decompensation.

In summary, in patients with viral and alcohol-related cirrhosis, a NSBB-driven decrease in HVPG significantly reduced the risks of variceal bleeding and other decompensating events. For primary prophylaxis of VH, an HVPG <12 mmHg or a decrease by 10% from the baseline value is clinically significant. For secondary prophylaxis, achieving an HVPG <12 mmHg or decreasing it by 20% from baseline protects patients from recurrent VH. For prevention of ascites, a decrease in HVPG of at least 10% from baseline is clinically relevant and reduces decompensation and liver-related death.

HVPG Predicts Risk of Decompensation and Mortality after Hepatic and Non-hepatic Surgery

Patients with Cirrhosis and HCC: Candidates for Hepatic Resection

Patients with cirrhosis, CSPH, and HCC who undergo hepatic surgery are at increased risk of postoperative decompensation and mortality [61–70]. A prospective series of 46 consecutive Child-Pugh A patients without clinical signs of PH and potentially resectable HCC reported a postoperative 1-year rate of ascites in 0% of patients without CSPH, compared with 30% in patients with HVPG from 10.5 to 12.5 mmHg [67]. There remains a need for defining a good-risk subset of CSPH patients for hepatic resection. While HVPG can stratify risks of postoperative decompensation, approximately 25% of patients with CSPH may nonetheless experience a normal postoperative course [65]. In addition, in these patients, a laparoscopic approach may mitigate the risks due to CSPH [62, 66, 69, 70]. In a retrospective report of 79 patients with CSPH, laparoscopic resection was the only independent predictor of a "best" outcome [61]. A prospective study comparing 10 laparoscopic resection patients with HVPG ≥10 mmHg with six patients who underwent open surgery found that rates of postoperative ascites and death were significantly higher in the open surgery group [70]. Reduction in postoperative risk in laparoscopic patients with CSPH has been observed in two other studies [64, 66].

In summary, the presence of CSPH, evaluated by HVPG measurement, is independently associated with increased risks of post-surgical decompensation and death. However, further longitudinal studies, which should consider the amount of resected liver and the application of minimally invasive approaches, are needed.

Patients with Cirrhosis Who Undergo Extrahepatic Surgery

In a prospective multicenter study, Reverter et al. described the utility of HVPG to predict outcomes of non-hepatic elective surgery in 140 patients with cirrhosis; 116 (83%) had CPSH [71]. The variables independently associated with outcome were ASA class, high-risk surgery, and HVPG. An HVPG >16 mmHg (HR >2.5) was associated with significant increase in mortality. Death was particularly high (44%) in patients with HVPG values ≥20 mmHg [71]. Further studies on whether the use of TIPS *prior* to surgery may help to improve survival in this setting are awaited.

PPG in the Setting of Tips

PPG Measurement

Abundant evidence supports the critical relationship between HVPG/PPG and the development of PH complications, and their recurrence after TIPS [27–31, 72–74]. PPG should always be measured before and after TIPS creation. When measuring PPG, the impact of sedation and measurement timing on hepatic hemodynamics should be considered. In 2014, Reverter et al. reported a prospective study examining the impact of sedation on hepatic hemodynamics in 44 patients undergoing HVPG and PPG measurement during TIPS under deep sedation [3]. The investigators reported that deep sedation added substantial variability and uncertainty to HVPG and PPG measurements. In 2017, Silva-Junior et al. retrospectively investigated the effect of timing on PPG measurement [75] in 155 TIPS patients. PPG was measured immediately post-TIPS, at least 24 h post-TIPS in stable, non-sedated patients (early PPG), and 1-month post-TIPS (late PPG). The immediate PPG differed from early PPG during general anesthesia (8.5 vs. 10 mmHg, $P = 0.015$), and deep sedation (12 vs. 10.5 mmHg, $P < 0.001$). There was no difference between early PPG and late PPG values (8.5 vs. 8 mmHg, $P > 0.05$). Thus, the immediate post-TIPS PPG may be influenced by various procedural factors and may not represent long-term PPG. PPG measurements in non-sedated hemodynamically stable patients without vasoactive agents or recent volume expansion may better reflect durable post-TIPS PPG values. Therefore, studies seeking correlations between post-TIPS PPG values and clinical outcomes should measure PPG accordingly.

Anatomic Location for PPG Measurement

La Mura et al. demonstrated, in 99 TIPS patients, that the post-TIPS porto-atrial gradient was a mean of 2.5 mmHg higher than the porto-caval gradient [6]. In considering a target gradient of 12 mmHg, 20% of the porto-caval gradients were <12 mmHg but had a porto-atrial gradient >12 mmHg; without needed perspective this could have prompted further TIPS dilation. Moreover, in the 1998 study by Casado et al., post-TIPS clinical outcomes were correlated with portal to caval gradients [72]. Notably, an unpublished survey of North American Interventional Radiologists (SIR Connect, September 2021) demonstrated the predominant use of right atrial pressure for post-TIPS PPG measurement (67% of 61 respondents), indicating a broad use of right atrial pressure to calculate the PP gradient. This could explain why a significant number of published studies have used the right atrial pressure for PPG calculation. Although these studies have supported the clinical effectiveness of TIPS while employing right atrial pressure, this does not mean that right atrial pressure is equivalent to IVC [76, 77]. In consideration of these data, anatomic locations for post-TIPS PPG measurement also should include the main portal vein and the IVC at the shunt outflow.

Optimal PPG Threshold for Portal Hypertensive Bleeding/Ascites

In a study of 122 TIPS patients, Casado et al. correlated clinical events to hemodynamic findings, reporting that all patients with rebleeding had a PPG (portal to caval) >12 mmHg [72]. In 2001, Rössle et al. reported a longitudinal study of 225 TIPS patients with variceal bleeding, wherein 80% of rebleeding occurred with PPGs similar to or greater than the baseline PPG, while only one (0.4%) and three (1.3%) patients rebled with a PPG <12 mmHg or PPG reduction by >50% [73]. In a 2007 retrospective observational cohort study of 118 TIPS patients, Biecker et al. found that patients with an initial PPG reduction >60% rarely suffered from rebleeding [74]. On these bases, it is recommended that in patients with variceal bleeding undergoing TIPS, reduction of absolute PPG to <12 mmHg is associated with near complete protection from portal hypertensive bleeding and is the preferred target for TIPS hemodynamic success. Relative reduction of PPG by at least 50% from the pre-TIPS baseline may be also useful but further studies are needed.

The optimal PPG threshold for ascites has been studied in several investigations. The 1998 study by Casado et al. found that all patients ($n = 26$) who developed ascites after TIPS had a PPG >12 mmHg [72]. In 2003, Sanyal et al. published a multicenter, prospective clinical trial of 109 ascites patients randomized to medical therapy or medical therapy + TIPS and found no relationship between PP reduction (mean final PPG = 8.3 mmHg) and ascites recurrence [78]. In 2004, Nair et al. reported a retrospective observational cohort study of 28 patients who underwent TIPS for ascites (mean final PSG = 8.6 mmHg) and did not identify post-TIPS PPG as an independent predictor of response [79]. In 2007, Salerno et al. presented a meta-analysis of four RCTs of TIPS versus paracentesis for ascites, reporting recurrent ascites in 42% of TIPS patients with a mean final PPG of 11.4 mmHg versus 89% of paracentesis patients ($P < 0.0001$) [80]. In 2014, Parvinian et al. published a retrospective single center study of 80 ascites patients treated with TIPS (mean final PPG = 6.8 mmHg) and reported an ascites response rate of ~80%, but uncovered no optimal PPG threshold associated with clinical response (response rate for 8, 10, and 12 mmHg thresholds = 79%, 79%, 78%, $P = 0.965$) [81]. In conclusion, the optimal PPG decrease to control medically refractory ascites remains unclear. Further investigation correlating TIPS hemodynamic outcomes and ascites clinical response is necessary.

PPG Thresholds in Overshunting Adverse Events

Excessive reduction of PPG by TIPS is associated with a higher risk of overshunting-related adverse events, such as HE [82–84]. Although interventions to address overshunting-related adverse events (e.g., TIPS reduction) have been studied, the approaches, PPG modifications, and clinical outcomes still vary [85–95]. As such, the optimal PPG target or degree of PPG elevation for interventions to address overshunting-related adverse events (e.g., TIPS reduction) is unknown. Further investigation to define the relationship between PPG and the resolution of overshunting adverse events is necessary.

References

1. van Zanten DV, Buganza E, Abraldes JG. The role of hepatic venous pressure gradient in the management of cirrhosis. Clin Liver Dis. 2021;25(2):327–43.
2. Steinlauf AF, Garcia-Tsao G, Zakko MF, Dickey K, Gupta T, Groszmann RJ. Low-dose midazolam sedation: an option for patients undergoing serial hepatic venous pressure measurements. Hepatology. 1999;29(4):1070–3.
3. Reverter E, Blasi A, Abraldes JG, Martínez-Palli G, Seijo S, Turon F, et al. Impact of deep sedation on the accuracy of hepatic and portal venous pressure measurements in patients with cirrhosis. Liver Int. 2014;34(1):16–25.
4. Zipprich A, Winkler M, Seufferlein T, Dollinger MM. Comparison of balloon vs. straight catheter for the measurement of portal hypertension. Aliment Pharmacol Ther. 2010;32(11–12):1351–6.
5. Maleux G, Willems E, Fieuws S, Heye S, Vaninbroukx J, Laleman W, et al. Prospective study comparing different indirect methods to measure portal pressure. J Vasc Interv Radiol. 2011;22(11):1553–8.
6. la Mura V, Abraldes JGJG, Berzigotti A, Erice E, Flores-Arroyo A, Garcia-Pagan JC, et al. Right atrial pressure is not adequate to calculate portal pressure gradient in cirrhosis: a clinical-hemodynamic correlation study. Hepatology. 2010;51(6):2108–16.

7. la Mura V, Garcia-Guix M, Berzigotti A, Abraldes JG, García-Pagán JC, Villanueva C, et al. A new prognostic algorithm based on stage of cirrhosis and HVPG to improve risk-stratification after variceal bleeding. Hepatology. 2020;72(4):1353–65.
8. Bosch J, García-Pagán JC. Calculating hepatic venous pressure gradient: feel free to stay free. J Vasc Interv Radiol. 2016;27(8):1138–9.
9. Groszmann RJ, Wongcharatrawee S. The hepatic venous pressure gradient: anything worth doing should be done right. Hepatology. 2004;39(2):280–2.
10. Bai W, Al-Karaghouli M, Stach J, Sung S, Matheson GJ, Abraldes JG. Test–retest reliability and consistency of HVPG and impact on trial design: a study in 289 patients from 20 randomized controlled trials. Hepatology. 2021;74(6):3301–15.
11. Ma J, Gong X, Luo J, Gu J, Yan Z, Zhang W, et al. Impact of intrahepatic Venovenous shunt on hepatic venous pressure gradient measurement. J Vasc Interv Radiol. 2020;31(12):2081–8.
12. Tandon P, Ripoll C, Assis D, Wongcharatrawee S, Groszmann RJ, Garcia-Tsao G. The interpretation of hepatic venous pressure gradient tracings—excellent interobserver agreement unrelated to experience. Liver Int. 2016;36(8):1160–6.
13. Silva-Junior G, Baiges A, Turon F, Torres F, Hernández-Gea V, Bosch J, et al. The prognostic value of hepatic venous pressure gradient in patients with cirrhosis is highly dependent on the accuracy of the technique. Hepatology. 2015;62(5):1584–92.
14. Groszmann RJRJ, Garcia-Tsao G, Bosch J, Grace NDND, Burroughs AKAK, Planas R, et al. Beta-blockers to prevent gastroesophageal varices in patients with cirrhosis. N Engl J Med. 2005;353(21):2254–61.
15. Ripoll C, Groszmann R, Garcia-Tsao G, Grace N, Burroughs A, Planas R, et al. Hepatic venous pressure gradient predicts clinical decompensation in patients with compensated cirrhosis. Gastroenterology. 2007;133(2):481–8.
16. Robic MA, Procopet B, Metivier S, Peron JM, Selves J, Vinel JP, et al. Liver stiffness accurately predicts portal hypertension related complications in patients with chronic liver disease: a prospective study. J Hepatol. 2011;55(5):1017–24.
17. D'Amico G, Garcia-Tsao G, Cales P, Escorsell A, Nevens F, Cestari R, et al. Session 2—diagnosis of portal hypertension: how and when. In: Portal Hypertension III: Proceedings of the Third Baverno International Consensus Workshop on Definitions, Methodology and Therapeutic Strategies, vol. 31. Oxford: Blackwell Science Ltd; 2008. p. 36–64.
18. Berzigotti A, Rossi V, Tiani C, Pierpaoli L, Zappoli P, Riili A, et al. Prognostic value of a single HVPG measurement and Doppler-ultrasound evaluation in patients with cirrhosis and portal hypertension. J Gastroenterol. 2011;46(5):687–95.
19. Jindal A, Bhardwaj A, Kumar G, Sarin SK. Clinical decompensation and outcomes in patients with compensated cirrhosis and a hepatic venous pressure gradient ≥20 mm hg. Am J Gastroenterol. 2020;115(10):1624–33.
20. Sanyal AJ, Harrison SA, Ratziu V, Abdelmalek MF, Diehl AM, Caldwell S, et al. The natural history of advanced fibrosis due to nonalcoholic steatohepatitis: data from the simtuzumab trials. Hepatology. 2019;70(6):1913–27.
21. Ferrusquía-Acosta J, Bassegoda O, Turco L, Reverter E, Pellone M, Bianchini M, et al. Agreement between wedged hepatic venous pressure and portal pressure in non-alcoholic steatohepatitis-related cirrhosis. J Hepatol. 2021;74(4):811–8.
22. Bassegoda O, Olivas P, Turco L, Mandorfer M, Serra-Burriel M, Tellez L, et al. Decompensation in advanced non-alcoholic fatty liver disease may occur at lower hepatic venous pressure gradient levels that in patients with viral disease. Clin Gastroenterol Hepatol. 2021; https://doi.org/10.1016/j.cgh.2021.10.023.
23. Pomier-Layrargues G, Kusielewicz D, Willems B, Villeneuve JP, Marleau D, Cote J, et al. Presinusoidal portal hypertension in non-alcoholic cirrhosis. Hepatology. 1985;5(3):415–8.
24. Boyer TD, Triger DR, Horisawa M, Redeker AG, Reynolds TB. Direct transhepatic measurement of portal vein pressure using a thin needle. Comparison with wedged hepatic vein pressure. Gastroenterology. 1977;72(4 Pt 1):584–9.
25. Navasa M, Pares A, Bruguera M, Caballeria J, Bosch J, Rodes J. Portal hypertension in primary biliary cirrhosis. Relationship with histological features. J Hepatol. 1987;5(3):292–8.

26. Seijo S, Reverter E, Miquel R, Berzigotti A, Abraldes JG, Bosch J, et al. Role of hepatic vein catheterisation and transient elastography in the diagnosis of idiopathic portal hypertension. Dig Liver Dis. 2012;44(10):855–60.
27. Conn HO, Grace ND, Bosch J, Groszmann RJ, Rodes J, Wright SC, et al. Propranolol in the prevention of the first hemorrhage from esophagogastric varices: a multicenter, randomized clinical trial. The Boston-New Haven-Barcelona portal hypertension study group. Hepatology. 1991;13(5):902–12.
28. Garcia-Tsao G, Groszmann RJ, Fisher RL, Conn HO, Atterbury CE, Glickman M. Portal pressure, presence of gastroesophageal varices and variceal bleeding. Hepatology. 1985;5(3):419–24.
29. Groszmann RJ, Bosch J, Grace ND, Conn HO, Garcia-Tsao G, Navasa M, et al. Hemodynamic events in a prospective randomized trial of propranolol versus placebo in the prevention of a first variceal hemorrhage [see comments]. Gastroenterology. 1990;99(5):1401–7.
30. Stanley AJ, Robinson I, Forrest EH, Jones AL, Hayes PC. Haemodynamic parameters predicting variceal haemorrhage and survival in alcoholic cirrhosis. QJM. 1998;91(1):19–25.
31. Urbain D, Muls V, Makhoul E, Jeghers O, Thys O, Ham HR. Prognostic significance of hepatic venous pressure gradient in medically treated alcoholic cirrhosis: comparison to aminopyrine breath test. Am J Gastroenterol. 1993;88(6):856–9.
32. Ready JB, Robertson AD, Goff JS, Rector WG Jr. Assessment of the risk of bleeding from esophageal varices by continuous monitoring of portal pressure. Gastroenterology. 1991;100(5 Pt 1):1403–10.
33. Abraldes JG, Villanueva C, Bañares R, Aracil C, Catalina MV, Garci A-Pagán JC, et al. Hepatic venous pressure gradient and prognosis in patients with acute variceal bleeding treated with pharmacologic and endoscopic therapy. J Hepatol. 2008;48(2):229–36.
34. Ardevol A, Alvarado-Tapias E, Garcia-Guix M, Brujats A, Gonzalez L, Hernández-Gea V, et al. Early rebleeding increases mortality of variecal bleeders on secondary prophylaxis with β-blockers and ligation. Dig Liver Dis. 2020;52:1017–25.
35. Moitinho E, Escorsell A, Bandi JC, Salmerón JM, García-Pagán JC, Rodés J, Bosch J. Prognostic value of early measurements of portal pressure in acute variceal bleeding. Gastroenterology. 1999;117(3):626–31.
36. Liu C, Liu Y, Shao R, Wang S, Wang G, Wang L, et al. The predictive value of baseline hepatic venous pressure gradient for variceal rebleeding in cirrhotic patients receiving secondary prevention. Ann Transl Med. 2020;8(4):91.
37. Monescillo A, Martinez-Lagares F, Ruiz-del-Arbol L, Sierra A, Guevara C, Jimenez E, et al. Influence of portal hypertension and its early decompression by TIPS placement on the outcome of variceal bleeding. Hepatology. 2004;40(4):793–801.
38. Villanueva C, Ortiz J, Minana J, Soriano G, Sabat M, Boadas J, et al. Somatostatin treatment and risk stratification by continuous portal pressure monitoring during acute variceal bleeding. Gastroenterology. 2001;121(1):110–7.
39. Patch D, Armonis A, Sabin C, Christopoulou K, Greenslade L, McCormick A, et al. Single portal pressure measurement predicts survival in cirrhotic patients with recent bleeding. Gut. 1999;44(2):264–9.
40. Kim MY, Baik SK, Yea CJ, Lee IY, Kim HJ, Park KW, et al. Hepatic venous pressure gradient can predict the development of hepatocellular carcinoma and hyponatremia in decompensated alcoholic cirrhosis. Eur J Gastroenterol Hepatol. 2009;21(11):1241–6.
41. Ripoll C, Groszmann RJRJ, Garcia-Tsao G, Bosch J, Grace N, Burroughs A, et al. Hepatic venous pressure gradient predicts development of hepatocellular carcinoma independently of severity of cirrhosis. J Hepatol. 2009;50(5):923–8.
42. Suk KT, Kim EJ, Kim DJ, Kim HS, Bang CS, Park TY, et al. Prognostic significance of hemodynamic and clinical stages in the prediction of hepatocellular carcinoma. J Clin Gastroenterol. 2017;51(3):285–93.
43. Tage-Jensen U, Henriksen JH, Christensen E, Widding A, Ring-Larsen H, Christensen NJ. Plasma catecholamine level and portal venous pressure as guides to prognosis in patients with cirrhosis. J Hepatol. 1988;6(3):350–8.

44. Ripoll C, Banares R, Rincon D, Catalina M-V, Iacono OL, Salcedo M, et al. Influence of hepatic venous pressure gradient on the prediction of survival of patients with cirrhosis in the MELD Era. Hepatology. 2005;42(4):793–801.
45. Zipprich A, Garcia-Tsao G, Rogowski S, Fleig WE, Seufferlein T, Dollinger MM. Prognostic indicators of survival in patients with compensated and decompensated cirrhosis. Liver Int. 2012;32:1407–14.
46. Gluud C, Henriksen JH, Nielsen G. Prognostic indicators in alcoholic cirrhotic men. Hepatology. 1988;8(2):222–7.
47. Merkel C, Bolognesi M, Bellon S, Zuin R, Noventa F, Finucci G, et al. Prognostic usefulness of hepatic vein catheterization in patients with cirrhosis and esophageal varices. Gastroenterology. 1992;102(3):973–9.
48. Suk KT, Kim CH, Park SH, Sung HT, Choi JY, Han KH, et al. Comparison of hepatic venous pressure gradient and two models of end-stage liver disease for predicting the survival in patients with decompensated liver cirrhosis. J Clin Gastroenterol. 2012;46(10):880–6.
49. Moller S, Bendtsen F, Christensen E, Henriksen JH. Prognostic variables in patients with cirrhosis and oesophageal varices without prior bleeding. J Hepatol. 1994;21(6):940–6.
50. Merkel C, Bolognesi M, Sacerdoti D, Bombonato G, Bellini B, Bighin R, et al. The hemodynamic response to medical treatment of portal hypertension as a predictor of clinical effectiveness in the primary prophylaxis of variceal bleeding in cirrhosis. Hepatology. 2000;32(5):930–4.
51. Sharma P, Kumar A, Sharma BC, Sarin SK. Early identification of haemodynamic response to pharmacotherapy is essential for primary prophylaxis of variceal bleeding in patients with "high-risk" varices. Aliment Pharmacol Ther. 2009;30(1):48–60.
52. Turnes J, Garcia-Pagan JCJC, Abraldes JGJG, Hernandez-Guerra M, Dell'Era A, Bosch J, et al. Pharmacological reduction of portal pressure and long-term risk of first variceal bleeding in patients with cirrhosis. Am J Gastroenterol. 2006;101:506–12.
53. Villanueva C, Aracil C, Colomo A, Hernandez-Gea V, Lopez-Balaguer JM, Varez-Urturi C, et al. Acute hemodynamic response to beta-blockers and prediction of long-term outcome in primary prophylaxis of variceal bleeding. Gastroenterology. 2009;137(1):119–28.
54. Hernández-Gea V, Aracil C, Colomo A, Garupera I, Poca M, Torras X, et al. Development of ascites in compensated cirrhosis with severe portal hypertension treated with β-blockers. Am J Gastroenterol. 2012;107(3):418–27.
55. Feu F, Garcia-Pagan JC, Bosch J, Luca A, Teres J, Escorsell A, et al. Relation between portal pressure response to pharmacotherapy and risk of recurrent variceal haemorrhage in patients with cirrhosis. Lancet. 1995;346(8982):1056–9.
56. Abraldes JGGJG, Tarantino I, Turnes J, Garcia-pagan JCJC, Rodes J, Bosch J, et al. Hemodynamic response to pharmacological treatment of portal hypertension and long-term prognosis of cirrhosis. Hepatology. 2003;37(4):902–8.
57. Villanueva C, Lopez-Balaguer JM, Aracil C, Kolle L, Gonzalez B, Minana J, et al. Maintenance of hemodynamic response to treatment for portal hypertension and influence on complications of cirrhosis. J Hepatol. 2004;40(5):757–65.
58. la Mura V, Garcia-Guix M, Berzigotti A, Abraldes JG, García-Pagán JC, Villanueva C, et al. A prognostic strategy based on stage of cirrhosis and HVPG to improve risk stratification after Variceal bleeding. Hepatology. 2020;72(4):1353–65.
59. Villanueva C, Graupera I, Aracil C, Alvarado E, Miñana J, Puente Á, et al. A randomized trial to assess whether portal pressure guided therapy to prevent variceal rebleeding improves survival in cirrhosis. Hepatology. 2017;65(5):1693–707.
60. Villanueva C, Albillos A, Genescà J, Garcia-pagan JC, Calleja JL, Aracil C, et al. β blockers to prevent decompensation of cirrhosis in patients with clinically significant portal hypertension (PREDESCI): a randomised, double-blind, placebo-controlled, multicentre trial. Lancet. 2019;6736(18):1–13.
61. Azoulay D, Ramos E, Casellas-Robert M, Salloum C, Lladó L, Nadler R, et al. Liver resection for hepatocellular carcinoma in patients with clinically significant portal hypertension. JHEP Rep. 2021;3(1):100190.

62. Boleslawski E, Petrovai G, Truant S, Dharancy S, Duhamel A, Salleron J, et al. Hepatic venous pressure gradient in the assessment of portal hypertension before liver resection in patients with cirrhosis. Br J Surg. 2012;17:855–63.
63. Bruix J, Castells A, Bosch J, Feu F, Fuster J, Garcia-Pagan JCC, et al. Surgical resection of hepatocellular carcinoma in cirrhotic patients: prognostic value of preoperative portal pressure. Gastroenterology. 1996;111(4):1018–22.
64. Casellas-Robert M, Lim C, Lopez-Ben S, Lladó L, Salloum C, Codina-Font J, et al. Laparoscopic liver resection for hepatocellular carcinoma in Child–Pugh A patients with and without portal hypertension: a multicentre study. World J Surg. 2020;44(11):3915–22.
65. Cucchetti A, Cescon M, Golfieri R, Piscaglia F, Renzulli M, Neri F, et al. Hepatic venous pressure gradient in the preoperative assessment of patients with resectable hepatocellular carcinoma. J Hepatol. 2016;64(1):79–86.
66. Lim C, Osseis M, Lahat E, Doussot A, Sotirov D, Hemery F, et al. Safety of laparoscopic hepatectomy in patients with hepatocellular carcinoma and portal hypertension: interim analysis of an open prospective study. Surg Endosc. 2018;33(3):811–20.
67. Llop E, Berzigotti A, Reig M, Erice E, Reverter E, Seijo S, et al. Assessment of portal hypertension by transient elastography in patients with compensated cirrhosis and potentially resectable liver tumors. J Hepatol. 2012;56(1):103–8.
68. Llovet JM, Fuster J, Bruix J. Intention-to-treat analysis of surgical treatment for early hepatocellular carcinoma: resection versus transplantation. Hepatology. 1999;30(6):1434–40.
69. Molina V, Sampson-Dávila J, Ferrer J, Fondevila C, Díaz del Gobbo R, Calatayud D, et al. Benefits of laparoscopic liver resection in patients with hepatocellular carcinoma and portal hypertension: a case-matched study. Surg Endosc. 2017;32(5):2345–54.
70. Truant S, Bouras AF, Hebbar M, Boleslawski E, Fromont G, Dharancy S, et al. Laparoscopic resection vs. open liver resection for peripheral hepatocellular carcinoma in patients with chronic liver disease: a case-matched study. Surg Endosc. 2011;25(11):3668–77.
71. Reverter E, Cirera I, Albillos A, Debernardi-Venon W, Abraldes JG, Llop E, et al. The prognostic role of hepatic venous pressure gradient in cirrhotic patients undergoing elective extrahepatic surgery. J Hepatol. 2019;71(5):942–50.
72. Casado M, Bosch J, Garcia-Pagan JCC, Bru C, Banares R, Bandi JCC, et al. Clinical events after transjugular intrahepatic portosystemic shunt: correlation with hemodynamic findings. Gastroenterology. 1998;114(6):129.
73. Ro M, Siegerstetter V, Olschewski M, Ochs A, Berger E, et al. How much reduction in portal pressure is necessary to prevent variceal rebleeding? A longitudinal study in 225 patients with transjugular intrahepatic portosystemic shunts. Am J Gastroenterol. 2001;96(12):3379–83.
74. Biecker E, Roth F, Heller J, Schild HH, Sauerbruch T, Schepke M. Prognostic role of the initial portal pressure gradient reduction after TIPS in patients with cirrhosis. Eur J Gastroenterol Hepatol. 2007;19(10):846–52.
75. Silva-Junior G, Turon F, Baiges A, Cerda E, García-Criado Á, Blasi A, et al. Timing affects measurement of portal pressure gradient after placement of transjugular intrahepatic portosystemic shunts in patients with portal hypertension. Gastroenterology. 2017;152(6):1358–65.
76. Charon JP, Alaeddin FH, Pimpalwar SA, Fay DM, Olliff SP, Jackson RW, et al. Results of a retrospective multicenter trial of the Viatorr expanded polytetrafluoroethylene-covered stent-graft for transjugular intrahepatic portosystemic shunt creation. J Vasc Interv Radiol. 2004;15(11):1219–30.
77. Lakhoo J, Bui JT, Zivin SP, Lokken RP, Minocha J, Ray CE, et al. Root cause analysis of rebleeding events following transjugular intrahepatic portosystemic shunt creation for variceal hemorrhage. J Vasc Interv Radiol. 2015;26(10):1444–53.
78. Sanyal AJ, Genning C, Reddy KR, Wong F, Kowdley KV, Benner K, et al. The North American study for the treatment of refractory ascites. Gastroenterology. 2003;124(3):634–41.
79. Nair S, Singh R, Yoselewitz M. Correlation between portal/hepatic vein gradient and response to transjugular intrahepatic portosystemic shunt creation in refractory ascites. J Vasc Interv Radiol. 2004;15(12):1431–4.

80. Salerno F, Cammà C, Enea M, Rössle M, Wong F. Transjugular intrahepatic portosystemic shunt for refractory ascites: a meta-analysis of individual patient data. Gastroenterology. 2007;133:825–34.
81. Parvinian A, Bui JT, Knuttinen MG, Minocha J, Gaba RC. Transjugular intrahepatic portosystemic shunt for the treatment of medically refractory ascites. Diagn Interv Radiol. 2014;20(1):58.
82. Chung HH, Razavi MK, Sze DY, Frisoli JK, Kee ST, Dake MD, et al. Portosystemic pressure gradient during transjugular intrahepatic portosystemic shunt with Viatorr stent graft: what is the critical low threshold to avoid medically uncontrolled low pressure gradient related complications? J Gastroenterol Hepatol. 2008;23(1):95–101.
83. Schepis F, Vizzutti F, Garcia-Tsao G, Marzocchi G, Rega L, de Maria N, et al. Under-dilated TIPS associate with efficacy and reduced encephalopathy in a prospective, non-randomized study of patients with cirrhosis. Clin Gastroenterol Hepatol. 2018;16(7):1153–1162.e7.
84. Riggio O, Merlli M, Pedretti G, Servi R, Meddi P, Lionetti R, et al. Hepatic encephalopathy after transjugular intrahepatic portosystemic shunt. Incidence and risk factors. Dig Dis Sci. 1996;41(3):578–84.
85. Kochar N, Tripathi D, Ireland H, Redhead DN, Hayes PC. Transjugular intrahepatic portosystemic stent shunt (TIPSS) modification in the management of post-TIPSS refractory hepatic encephalopathy. Gut. 2006;55(11):1617–23.
86. Jacquier A, Vidal V, Monnet O, Varoquaux A, Gaubert J-Y, Champsaur P, et al. A modified procedure for transjugular intrahepatic portosystemic shunt flow reduction. J Vasc Interv Radiol. 2006;17(8):1359–63.
87. Maleux G, Verslype C, Heye S, Wilms G, Marchal G, Nevens F. Endovascular shunt reduction in the management of transjugular portosystemic shunt-induced hepatic encephalopathy: preliminary experience with reduction stents and stent-grafts. Am J Roentgenol. 2007;188(3):659–64.
88. Sze DY, Hwang GL, Kao JS, Frisoli JK, STW K, Razavi MK, et al. Bidirectionally adjustable TIPS reduction by parallel stent and stent-graft deployment. J Vasc Interv Radiol. 2008;19(11):1653–8.
89. Cookson DT, Zaman Z, Gordon-Smith J, Ireland HM, Hayes PC. Management of transjugular intrahepatic portosystemic shunt (TIPS)-associated refractory hepatic encephalopathy by shunt reduction using the parallel technique: outcomes of a retrospective case series. Cardiovasc Intervent Radiol. 2011;34(1):92–9.
90. Blue RC, Lo GC, Kim E, Patel RS, Scott Nowakowski F, Lookstein RA, et al. Transjugular intrahepatic portosystemic shunt flow reduction with adjustable polytetrafluoroethylene-covered balloon-expandable stents using the "sheath control" technique. Cardiovasc Intervent Radiol. 2016;39(6):935–9.
91. Joseph AS, Sandhu B, Khalil A, Lopera J. Transjugular portosystemic shunt reductions: a retrospective single-center experience. J Vasc Interv Radiol. 2019;30(6):876–84.
92. Sarwar A, Esparaz AM, Chakrala N, Mangano M, Ganguli S, Malik R, et al. Efficacy of TIPS reduction for refractory hepatic encephalopathy, right heart failure, and liver dysfunction. Am J Roentgenol. 2021;216(5):1267–72.
93. Fanelli F, Salvatori FM, Rabuffi P, Boatta E, Riggio O, Lucatelli P, et al. Management of refractory hepatic encephalopathy after insertion of TIPS: long-term results of shunt reduction with hourglass-shaped balloon-expandable stent-graft. Am J Roentgenol. 2009;193(6):1696–702.
94. de Keyzer B, Nevens F, Laenen A, Heye S, Laleman W, Verslype C, et al. Percutaneous shunt reduction for the management of TIPS-induced acute liver decompensation: a follow-up study. Ann Hepatol. 2016;15(6):911–7.
95. Schindler P, Seifert L, Masthoff M, Riegel A, Köhler M, Wilms C, et al. TIPS modification in the management of shunt-induced hepatic encephalopathy: analysis of predictive factors and outcome with shunt modification. J Clin Med. 2020;9(2):567.

HVPG as a Gold Standard: Consensus Statements of Panel 1

6

Hector Ferral, Juan Carlos Garcia-Pagàn, Filippo Schepis, Ron C. Gaba, Alberto Zanetto, Valeria Perez-Campuzano, and Ziv J. Haskal

H. Ferral
NorthShore University HealthSystem, Evanston, IL, USA

Department of Radiology, Section of Interventional Radiology, Louisiana State University, New Orleans, USA

J. C. Garcia-Pagàn (✉)
Barcelona Hepatic Hemodynamic Laboratory, Liver Unit, Hospital Clínic, Institut de Investigacions Biomèdiques, August Pi i Sunyer (IDIBAPS), University of Barcelona, Barcelona, Spain

CIBEREHD (Centro de Investigación Biomédica en Red Enfermedades Hepáticas y Digestivas), Barcelona, Spain
e-mail: jcgarcia@clinic.cat

F. Schepis
Hepatic Hemodynamic Laboratory, Division of Gastroenterology, AOU of Modena and University of Modena and Reggio Emilia, Modena, Italy
e-mail: fschepis@unimore.it

R. C. Gaba
Department of Radiology, College of Medicine, University of Illinois at Chicago, Chicago, IL, USA
e-mail: rgaba@UIC.EDU

A. Zanetto
Gastroenterology and Multivisceral, Transplant Unit, Department of Surgery, Oncology, and Gastroenterology, Padua University Hospital, Padua, Italy
e-mail: alberto.zanetto@unipd.it

V. Perez-Campuzano
Barcelona Hepatic Hemodynamic Laboratory, Liver Unit, Hospital Clínic, Institut de Investigacions Biomèdiques, August Pi i Sunyer (IDIBAPS), University of Barcelona, Barcelona, Spain

CIBEREHD (Centro de Investigación Biomédica en Red Enfermedades Hepáticas y Digestivas), Barcelona, Spain

Health Care Provider of the European Reference Network on Rare Liver Disorders (ERN-Liver), Barcelona, Spain
e-mail: vperez2@clinic.cat

Z. J. Haskal
Department of Radiology and Medical Imaging/Interventional Radiology, University of Virginia School of Medicine, Richmond, VA, USA

R. de Franchis (ed.), *Portal Hypertension VII*,
https://doi.org/10.1007/978-3-031-08552-9_6

Descriptions of HVPG Measurement

1.1. The use of an end-hole, compliant balloon occlusion catheter reduces the random error of wedged hepatic vein pressure (WHVP) measurements and is preferred over the use of a conventional straight catheter (A.1). (New)
1.2. A small volume of contrast medium should be injected when the occlusion balloon is inflated to confirm a satisfactory occluded position and to exclude the presence of hepatic venous-to-venous communications (A.1). (New)
1.3. Hepatic venous-to-venous communications may result in underestimation of the WHVP and must be reported (A.1). (New)
1.4. Deep sedation during liver hemodynamic measurement may cause inaccurate HVPG values (B.1). If light sedation is required, low-dose midazolam (0.02 mg/kg) does not modify the HVPG and is acceptable (B.1). (New)
1.5. Slow-speed (up to 7.5 mm/s) permanent tracings of pressures, recorded either on paper or electronically, are recommended. Digital, on-screen readings are much less accurate and should not be used (A.1). (New)
1.6. To properly reflect the portal venous pressure, WHVP requires a stabilization time. Recording of WHVP requires a minimum of 1 min, with particular attention to stability during the last 20–30 s. WHVP should be recorded in triplicate (D.1). (New)
1.7. The wedged to free hepatic vein pressure gradient has a superior clinical prognostic value than wedged to right atrial pressure gradient and should be used as the standard reference (B.1). Right atrial pressure can be measured to rule out a post-hepatic component of portal hypertension (B.1). (New)
1.8. Free hepatic vein pressure (FHVP) must be measured in the hepatic vein (HV) within 2–3 cm from the HV-Inferior vena cava (IVC) confluence. IVC pressure should be measured as an internal control, at the level of the hepatic vein ostium. If the FHVP is more than 2 mmHg above IVC pressures, the presence of a hepatic vein outflow obstruction should be ruled out with the injection of a small amount of contrast medium (A.1). (New)

Diagnosis of CSPH in Patients with Cirrhosis

1.9. Hepatic venous pressure gradient (HVPG) measurement values >5 mmHg indicate sinusoidal portal hypertension (A.1). (Unchanged)
1.10. In patients with viral and alcohol-related cirrhosis, HVPG measurement is the gold-standard method to assess the presence of "clinically significant portal hypertension" (CSPH), which is defined as an HVPG ≥10 mmHg (A.1). (Changed)
1.11. In patients with primary biliary cholangitis, there may be an additional pre-sinusoidal component of portal hypertension that cannot be assessed by HVPG (B.1). As such, in these patients, HVPG may underestimate the prevalence and severity of PH (B.1). (New)

1.12. In patients with NASH cirrhosis, although an HVPG ≥10 mmHg remains strongly associated with the presence of clinical signs of portal hypertension, these signs can also be present in a small proportion of patients with HVPG values <10 mmHg (C.2). (New)

1.13 In patients with chronic liver disease and clinical signs of portal hypertension (gastroesophageal varices, ascites, and portosystemic collateral vessels) but with HVPG <10 mmHg, Porto-sinusoidal vascular disorder (PSVD) must be ruled out (B.1). (New)

1.14 In alcohol-related or viral cirrhosis, a decrease in HVPG in response to NSBB is associated with a significant reduction in the risk of variceal bleeding or of other decompensating events (A.1). (Changed)

Inclusion of HVPG Assessment in Trial Design

1.15 HVPG measurements should be encouraged in clinical trials investigating novel therapies but are not essential if portal hypertension-associated endpoints are well defined (B.1). (Unchanged)

1.16 In viral, alcohol-related, and reasonably in NASH cirrhosis, HVPG response assessment is recommended as a surrogate endpoint in phase II clinical trials where a low rate of events is expected (D.2). (Changed)

1.17 Test–retest reliability of HVPG measurement is excellent but influenced by the stage of liver disease (lower in decompensated patients) and its etiology (higher in alcoholic patients). This should be taken into consideration in designing clinical trials based on HVPG assessment (C.1).

Assessment of Surgical Risks

1.18 The presence of CSPH, determined either by HVPG ≥10 mmHg or by clinical manifestations of portal hypertension, is associated with higher risk of decompensation and mortality in patients with cirrhosis undergoing liver resection for HCC (A.1). (New)

1.19. In candidates for non-hepatic abdominal surgery, an HVPG ≥16 mmHg is associated with an increased risk of short-term mortality after surgery (C.1). (New)

Portal Pressure Gradient (PPG) in the Setting of TIPS

1.20. PPG should be measured before and after TIPS insertion (A.1). (New)

1.21. Anatomic locations for post-TIPS PPG measurement should include the main portal vein and the IVC (at the shunt outflow) (B.1). (New)

1.22 The immediate post-TIPS PPG may be influenced by various factors, such as general anesthesia, use of vasoactive agents, and hemodynamic instability, and therefore immediate post-TIPS PPG may not represent long-term PPG (B.1). PPG measurements in hemodynamically stable, non-sedated patients better reflect post-TIPS PPG values and are recommended (B.1). (New)

1.23 In patients with variceal bleeding undergoing TIPS, reduction of absolute PPG to <12 mmHg is associated with near complete protection from portal hypertensive bleeding and is the preferred target for TIPS hemodynamic success (A.1). Relative reduction of PPG, by at least 50% from the pre-TIPS baseline, may also be useful (B.2). (New)

1.24 PPG remeasurement is indicated if there is clinical or Doppler ultrasonographic suspicion of TIPS dysfunction to evaluate the need for TIPS revision (B.1). (New)

Research Agenda

- The usefulness, safety, and accuracy of direct portal pressure measurement by the endoscopic US require further evaluation.
- The prognostic role of HVPG and the definition of specific cutoffs in patients with NASH cirrhosis requires further investigation.
- The utility of HVPG-guided therapy needs further confirmation in randomized clinical trials.
- The prognostic role of HVPG in patients undergoing extra-hepatic surgery needs further investigation in prospective cohorts that should compare HVPG with noninvasive tests.
- To examine the test–retest HVPG reliability at an individual level and factors that determine variability.
- Further investigation evaluating portocaval-versus-portoatrial measured PPG and clinical outcomes after TIPS (e.g., rebleeding) is warranted.
- The optimal PPG decrease to control medically recurrent/refractory ascites is unclear. Further investigation correlating TIPS hemodynamic outcomes and ascites' clinical response is necessary.
- The optimal PPG increase (in the context of TIPS reduction) needed to ameliorate adverse events related to over-shunting should be investigated.

Part III

Noninvasive Tools for cACLD and Portal Hypertension

7 Results of the Baveno VII Questionnaire on the Use of Noninvasive Tools for cACLD and Portal Hypertension

Annalisa Berzigotti, Jonathan A. Fallowfield, Juan G. Abraldes, Maja Thiele, and Joan Genescà

Introduction

The Baveno VI consensus workshop set a major change in comparison with the previous recommendations, by proposing simple criteria based on noninvasive tests (NITs) to identify patients (a) in an advanced stage of chronic liver disease while

A. Berzigotti (✉)
Department of Visceral Surgery and Medicine, Inselspital, Bern University Hospital, University of Bern, Bern, Switzerland
e-mail: annalisa.berzigotti@insel.ch

J. A. Fallowfield
University of Edinburgh Centre for Inflammation Research, The Queen's Medical Research Institute, Edinburgh BioQuarter, Edinburgh, UK
e-mail: Jonathan.Fallowfield@ed.ac.uk

J. G. Abraldes
Liver Unit, Division of Gastroenterology, University of Alberta, Edmonton, Canada
e-mail: Juan.g.abraldes@ualberta.ca

M. Thiele
FLASH Center for Liver Research, Department of Gastroenterology and Hepatology, Odense University Hospital, Odense, Denmark

Department of Clinical Research, University of Southern Denmark, Odense, Denmark
e-mail: maja.thiele@rsyd.dk

J. Genescà
Liver Unit, Hospital Universitari Vall d'Hebron, Vall d'Hebron Research Institute (VHIR), Vall d'Hebron Barcelona Hospital Campus, Barcelona, Spain

Universitat Autònoma de Barcelona, Barcelona, Spain

Centro para la Investigación Biomédica en Red de Enfermedades Hepáticas y Digestivas (CIBERehd), Madrid, Spain
e-mail: jgenesca@vhebron.net

R. de Franchis (ed.), *Portal Hypertension VII*,
https://doi.org/10.1007/978-3-031-08552-9_7

still being compensated (cACLD), (b) at risk of having clinically significant portal hypertension (CSPH), and (c) at low risk of having varices needing treatment. Among these new NIT-based assessments, the so-called "Baveno VI criteria" (Liver stiffness measurement—LSM < 20 kPa + Platelet count >150 × 10^9/L), to spare unnecessary upper gastrointestinal endoscopy have been validated by a large number of studies in the last 5 years, bringing NITs well within the clinical routine in this important setting.

Given the success of the Baveno VI criteria, our panel aimed to better understanding the opinion of the experts in the field of portal hypertension (PH) on the current practice and potential future use of invasive and noninvasive methods in the following aspects:

1. Concept and cutoffs of cACLD.
2. Use of NITs to diagnose CSPH.
3. Triage of patients for screening endoscopy.
4. Use of NITs to monitor cACLD.
5. Use of spleen stiffness.
6. Emerging NITs in the above-mentioned settings.

A questionnaire was sent to all Baveno expert faculty ($n = 64$). The questionnaire was completely filled-in by 42 (66%) and its main results are presented in the following paragraphs.

Concept and Cutoffs of cACLD

Do you think that the cACLD concept is clinically useful and do you use it in clinical practice?

- 85% of the faculty answered positively to this question.

Do you think that the two liver stiffness measurement cutoffs (using transient elastography) rule out (10 kPa) and rule in (15 kPa) cACLD are accurate?

- 76% of the faculty answered positively to this question.

Do you think that the elastography cutoffs for cACLD are etiology-dependent?

- 83% of the faculty answered positively to this question.

Summarizing the above-mentioned results, a majority of the faculty found it useful to apply the concept of cACLD based on NITs and accepted the previously proposed criteria as sufficiently accurate. However, it was felt that the criteria need to be applied in an etiology-based fashion. As it will be observed in the following chapters, proposed changes in the definition and scope of cACLD will also be challenged by the need of gathering more information about etiology-based differences in this concept.

Use of NITs to Diagnose CSPH

Do you use LSM by TE (Transient elastography) to estimate the presence/absence of CSPH in your clinical practice? If Yes, what for?

- 71% of the faculty answered positively to the question, and 80% of the respondents answered that they use the parameter systematically for general prognosis.
- 43% of the respondents use this NIT also to assess the risk of decompensation pre-surgery.
- 30% to assess the risk of decompensation to initiate beta-blocker preventive therapy.

To diagnose CSPH using TE, would you favor the use of a single cutoff (above confirms/below discards CSPH) or two different thresholds (one to rule in, one to rule out, grey zone in between)? In the same line, and regarding etiologies and CSPH diagnosis, which option do you think would be more useful for clinical practice?

- 73% of the faculty answered that they would favor the use of two cutoffs, one to rule in and one to rule out CSPH, accepting a grey zone in between.
- As for the use of common or etiology-specific cutoffs, 69% stated that they would favor the use of a cutoff/set of cutoffs (rule in/out) a little less accurate but applicable to all etiologies.

How would you use LSM to diagnose CSPH in patients with cACLD?

The two possible options were between a single-point LSM or repeated LSM, and to interpret any change above/below cutoffs as a change in the CSPH status at any given time point.

- 71% of the faculty was in favor of using LSM in a dynamic way, with controls over the follow-up.

Taken together, these answers provide expert support of using LSM alone or combined with other NITs to detect patients at high risk of CSPH for different purposes, mainly for prognostic evaluation of cACLD patients. Also, a dual cutoff approach applicable to all etiologies if possible and considering follow-up changes would be the preferred option. In the following chapters, new evidence regarding the estimation of CSPH by LSM, and the utility of follow-up changes will be presented, refining and improving the noninvasive criteria developed at Baveno VI for evaluating CSPH.

Triage of Patients for Screening Endoscopy

What noninvasive criteria do you use to triage patients with compensated cirrhosis in need for endoscopy (multiple choices possible)

- 52% of the respondents use the Baveno VI criteria.
- 16% use the expanded Baveno VI criteria.
- Additional 23% use other NIT-based criteria (LSPS 9%, "other" 14%).
- Only 31% of the faculty continue doing endoscopy in all patients.

Do you think different thresholds for noninvasive tests should be used for different etiologies in the setting of identifying patients who can skip endoscopy?

- 54% of the respondents answered that etiology-specific cutoffs are not needed.
- 46% answered that they would be needed.

Do you think Baveno VI criteria (or other criteria) for detecting varices needing treatment are still useful after the results of the PREDESCI study (showing that the use of beta-blockers in patients with CSPH but no varices or with small varices might prevent decompensation)?

- 76% of the faculty answered that the Baveno VI criteria will be still useful.

If you have access to different liver elastography techniques other than VCTE (pSWE or 2D-SWE), do you use the same thresholds of liver stiffness to indicate endoscopy i.e., *20 kPa for Baveno VI criteria, or 25 kPa for extended Baveno VI criteria?*

- 50% of the faculty has no access to other ultrasound elastography techniques; among the other half of respondents, a majority (66%) do not use the same thresholds.

These answers show the high uptake of the Baveno VI criteria, though not universal. Expanded criteria, alternative models, and other elastography techniques are seldom used. The majority did not support the use of etiology-specific criteria. Indeed, these results are in keeping with the updated Baveno VII recommendations, based on a systematic review of the literature, showing (a) the reliability of Baveno VI criteria, (b) the high rate of false negatives with expanded Baveno VI criteria, and (c) the homogeneous performance of Baveno VI criteria across etiologies.

Use of NITs to Monitor cACLD

Do you use liver stiffness for monitoring patients with cACLD? If yes, for which indication?

- 85% of the faculty answered positively. Indications were multiple and are summarized in Table 7.1. A majority of reasons for monitoring were to decide whether to initiate screening endoscopy, for monitoring disease progression or regression, either as part of the natural history or after etiology-specific treatment. Only two respondents used LSM to evaluate hemodynamic response to NSBB.

If you use liver stiffness to monitor patients with cACLD, please indicate here the interval you use

- 62% of the faculty answered that they use LSM at yearly intervals.
- 19% answered that they use it at 6-month intervals. None of those who used LSM for monitoring repeated the measurement in intervals exceeding 12 months.

What do you consider a clinically relevant change in liver stiffness in cACLD patients with index LSM ≥ 10 kPa measured by FibroScan?

- 40% of the faculty considers that a reduction or increase of at least 20% of the baseline value should be considered clinically relevant.
- 26% of the faculty does not use a percentage, but an absolute number, and our question did not address what absolute numbers were considered.
- A minority (22%) found other relative changes to be clinically relevant,
- and 12% answered "other" criteria.

Do you repeat index liver stiffness measurement to increase the true-positive rate for diagnosis of advanced fibrosis (e.g., *if LSM ≥ 8 kPa)?*

- 55% of the faculty repeats the LSM in this situation.

Do you regard LSM as a prognostic marker in cACLD?

- 83% of the faculty regards LSM as a prognostic marker in cACLD.

Table 7.1 Do you use liver stiffness for monitoring patients with cACLD? If yes, for which indication?

Increase above 20 kPa using FibroScan in cACLD patients with platelet count ≥150 triggers variceal surveillance	69%
For estimating fibrosis progression or regression rate	57%
Antiviral treatment effect on chronic HCV	54%
Effect of alcohol rehabilitation in ALD	47%
Antiviral treatment effect on chronic HBV	45%
Effect of weight loss or pharmaceutical treatment in NAFLD	42%
Treatment effect in autoimmune hepatitis	38%
If increasing LSM then more frequent visits	35%
If stable or declining LSM, then less frequent outpatient visits	26%
Assess the efficacy of TIPS	11%
Assess hemodynamic response to NSBB	4%

- 74% indicated that they regard LSM as a prognostic marker of liver-related events/decompensation or time to decompensation.
- 10% used it for the prediction of liver-related death, while only one respondent answered HCC and one answered all-cause mortality.

If you are using FibroScan for prognostication in cACLD, how do you use cutoffs?

- 62% of the faculty uses global (not etiology specific) cutoffs,
- while 21% uses etiology-specific cutoffs.

Do you regard LSM as a prognostic marker in decompensated liver disease?

- 57% of the faculty does not regard LSM as a prognostic marker in decompensated patients. The commonest reason for using LSM in decompensated cirrhosis patients was the time to next liver-related event (38%) and liver-related mortality (26%).

Use of Spleen Stiffness

Do you use spleen stiffness measurement to assess portal hypertension in your patients with cACLD?

- 73% of the faculty does not use spleen stiffness measurement (SSM). Among the respondents who use SSM, the main use is to identify or rule-out clinically significant portal hypertension.

Which elastography method do you use to measure spleen stiffness

- The use included vibration-controlled transient elastography (FibroScan) with standard or spleen-specific probe, pSWE, 2D-SWE, and MRE, without a prevalence of any of the mentioned techniques.

Emerging NITs in the Above-Mentioned Settings

Which noninvasive imaging methods (excluding ultrasound elastography) could play a future role for assessment of PH in routine clinical practice? Multiple answers are possible.

The answers provided are summarized in Table 7.2. Percentages indicate the % of respondents (42) who indicated the given method.

What context of use would you prioritize for a new noninvasive biomarker in PH?

The answers provided are summarized in Table 7.3. Percentages indicate the % of respondents (42) who selected the given indication.

Table 7.2 Which noninvasive imaging methods (excluding ultrasound elastography) could play a future role for assessment of PH in routine clinical practice?

Multiparametric MRI (i.e., combined structural changes and dynamic blood flow)	52%
Ultrasound/Doppler ultrasound/CEUS-based methods	52%
MR elastography	42%
MR-based flow measurements	26%
Multiphase contrast-enhanced CT	9%

Table 7.3 What context of use would you prioritize for a new noninvasive biomarker in PH?

Dynamic monitoring of portal hypertension over time	76%
To predict the presence of CSPH or ruling-out high-risk varices	76%
Assessment of treatment efficacy of NSBB (i.e., baseline and on-treatment)	73%
Prognostication of long-term clinical outcomes	71%
Early phase (proof-of-concept) clinical trials of new therapies for portal hypertension	64%

Table 7.4 Regarding other noninvasive tests for PH, which of the following (if any) do you consider as potentially useful to predict CSPH, varices or outcomes?

Indocyanine Green 15-min retention test	9%
13C-methacetin breath test	19%
ELF test	19%
Von Willebrand factor antigen	19%
Soluble CD163	2%
Pro-C5	0%
Combination tests or other	33%
None	38%

If an MRI-based method reliably tracked HVPG, would you use it to assess hemodynamic response to NSBB (i.e., *baseline and on-treatment)?*

- 61% of the faculty answered that they might consider the use of MRI-based methods, depending on cost/availability, etc.
- 28% of the faculty responded that they would use an MRI-based method regardless,
- while 9% would not.

Regarding other noninvasive tests for PH, which of the following (if any) do you consider as potentially useful to predict CSPH, varices or outcomes?

The answers provided are summarized in Table 7.4. Percentages indicate the % of respondents (42) who selected the given method.

Taken together, these responses highlight an ongoing unmet need for additional NITs across different contexts of use. Priority areas for novel NITs include dynamic monitoring of PH over time and assessment of treatment efficacy—clinical indications for which performing repeated HVPG is impractical and where measurement of liver stiffness using TE or MRE does not accurately track portal pressure. Additionally, experts indicated general enthusiasm for most modalities, especially MRI and CEUS-based assessments that are discussed in detail in the specific chapter of this book. There was relatively weak support, at present, for the use of serum-based NITs and liver clearance tests although studies are ongoing and further validation may emerge in the coming years.

Compensated Advanced Chronic Liver Disease (cACLD)

8

Mònica Pons, Ana Barreira, and Joan Genescà

From Baveno VI to Baveno VII

Important new concepts were introduced by Baveno VI consensus guidelines in 2015 regarding the extended use of noninvasive tests, especially transient elastography (TE), for the management of patients in the advanced stages of chronic liver disease (CLD) [1]. The new term "compensated advanced chronic liver disease" (cACLD) was introduced; indications about diagnosing clinically significant portal hypertension (CSPH) were provided, and new noninvasive criteria for avoiding screening endoscopies for varices were developed.

The reasons for the introduction of the cACLD term were several but originated in a progressive change in clinical practice when managing CLD patients, due to the appearance of liver stiffness measurement (LSM) by elastography, and more specifically transient elastography (TE) [2]. The widespread use of TE allowed the early detection of CLD patients with advanced disease at risk of developing CSPH, and consequently, liver-related events during follow-up. This fact has been paralleled by a progressive reduction in the use of liver biopsy for staging purposes in CLD. Therefore, the term cACLD was an attempt to denominate a new clinical scenario derived from the extensive use of TE as an important staging method for CLD, also reflecting that, in the absence of a liver biopsy, it was not possible to distinguish between severe fibrosis and cirrhosis. Because of the clinical implications of this new entity, it was advised that suspected cACLD patients should be referred to a liver disease specialist.

M. Pons · A. Barreira · J. Genescà (✉)
Liver Unit, Hospital Universitari Vall d'Hebron, Vall d'Hebron Institute of Research (VHIR), Vall d'Hebron Barcelona Hospital Campus, Barcelona, Spain

Universitat Autònoma de Barcelona, Barcelona, Spain

Centro para la Investigación Biomédica en Red de Enfermedades Hepáticas y Digestivas (CIBERehd), Madrid, Spain
e-mail: abarreira@vhebron.net; jgenesca@vhebron.net

R. de Franchis (ed.), *Portal Hypertension VII*,
https://doi.org/10.1007/978-3-031-08552-9_8

Based on a very pragmatical and empirical approach, at the Baveno VI conference, it was decided to use a dual TE cutoff approach to maximize the selection of two groups of patients with very different risks of developing CSPH, and consequently liver-related outcomes. Obviously, the proposed cutoffs were derived from the extensive literature reflecting the relationship between different TE cutoffs and liver fibrosis stages in different etiologies of CLD. Consequently, LSM <10 kPa was proposed as a safe cutoff for excluding (ruling out) cACLD, selecting a population of CLD patients with a low prevalence of severe fibrosis/cirrhosis and portal hypertension, and a low risk of developing CSPH and liver-related events, while LSM >15 kPa was highly suggestive (ruling in) of cACLD, selecting CLD patients with a high prevalence of severe fibrosis/cirrhosis and portal hypertension, and at risk of developing CSPH and liver-related events.

Since the publication of the guidelines in 2015, many studies have used the term cACLD; until September 2021, there have been 79 publications indexed in PubMed with the term "cACLD" or "compensated advanced chronic liver disease," 27 of them published in 2020. In a significant proportion, the main topic of the studies where the term was used was the validation of noninvasive criteria to avoid screening endoscopy for varices.

As explained in a previous chapter, in the present consensus workshop, a panelist's survey showed that most (85%) experts believe that the cACLD concept is clinically useful and most of them use it in their clinical practice. Likewise, 76% believe that the cutoff points established in Baveno VI (10 kPa and 15 kPa) for ruling-out and ruling-in are accurate. Despite this, 83% consider that the cutoffs are dependent on the etiology of liver disease.

From Histological Staging to a Noninvasive Clinical Staging of Chronic Liver Disease

Liver histology has been the reference tool for staging CLD through the identification of the degree of liver fibrosis. Liver fibrosis correlates with patient outcomes represented by the development of liver-related events during follow-up. However, the limitations of liver biopsy are widely known (complications, invasiveness, sampling error, etc.); additional drawbacks are the low reliability of liver biopsy evaluation [2], inadequacy for repeated measures, and lack of linearity and granularity in the information provided. On top of that, modern personalized medicine is moving to the use of biomarkers, patients do not like liver biopsies, and their voice and patient centrality are to be increasingly considered. Finally, other current issues like the Covid-19 pandemic will push to find alternatives to invasive procedures.

LSM by elastography possesses many of the desired theoretical properties of a good biomarker and almost none of the limitations and drawbacks of liver biopsy. One of the conceptual problems in hepatology is that we tend to refer and validate every new tool to what we consider a "gold standard," and in that case liver biopsy. However, pretending that a new biomarker is a perfect diagnostic tool of an imperfect standard is probably unrealistic.

Considering the dual TE cutoff approach to cACLD, it would be expected that the ruled-in population (LSM >15 kPa) would be enriched with patients with a high prevalence (positive predictive value—PPV) of severe fibrosis/cirrhosis, portal hypertension, and at higher risk of liver outcomes. The other way around would be expected with the population selected by the ruling out criteria (LSM <10 kPa) showing a low prevalence (1-negative-predicted value, NPV) of these features. Between these two populations, we will obviously have a grey zone with patients showing intermediate values of these clinical features.

In order to be clinically sound and scientifically valid, the two proposed cutoffs should be able to identify these high and low-risk populations. Regarding liver histology, it would be desirable that the prevalence of severe fibrosis/cirrhosis in each group reached 90% and 10%, respectively, with intermediate values in the grey zone. However, considering the imperfections of liver biopsy as a reference tool, prevalences of 80% and 20% might be considered acceptable. Far more important than that, these three subgroups of patients should present very different risks of clinical events during follow-up, indicating that subgrouping of CLD patients based on LSM values holds prognostic implications independently of not selecting populations with a 100% and 0% prevalence of severe fibrosis/cirrhosis.

Other relevant issues to be considered concerning the cACLD LSM cutoffs are in relation to practical issues of daily clinical practice; practitioners tend to use what is simple and readily applicable. In that sense, and provided that the scientific value is not lost, the use of the same cutoffs for different etiologies and values with numbers easy to remember are helpful assets.

Finally, changes in the proposed cACLD cutoffs should be balanced against losing important clinical relevance in the identification of the LSM subgroups.

Excluding cACLD (LSM <10 kPa)

Several studies have intended to evaluate the cACLD cutoffs performance by analyzing their ability to detect and exclude severe fibrosis/cirrhosis or portal hypertension. These studies allow us to assess the prevalences of these clinical indicators in the populations selected by the two cACLD cutoffs in the different etiologies of CLD (Table 8.1) [3–9].

As shown in Table 8.1, the prevalence of severe fibrosis/cirrhosis is low in patients with LSM <10 kPa, around 10% in most studies, but ranging between 4% and 20% depending on the etiology of CLD [3–8]. The highest prevalences were observed in hepatitis B patients (16.3%) [5] and obese non-alcoholic fatty liver disease (NAFLD) patients (27%) from the study by Wong et al. [3] using the XL probe of the TE. As seen, other studies with more patients, observed lower prevalences of severe fibrosis/cirrhosis in NAFLD patients with LSM <10 kPa. In the case of NAFLD, it is plausible that a selection bias might exist to obtain histology from patients who are candidates for clinical trials, not adequately reflecting the general population with NAFLD.

Table 8.1 Presence of compensated advanced chronic liver disease (cACLD) clinical features (Fibrosis F3–F4 or hepatic venous pressure gradient-HVPG >5 mmHg) in the three subgroups defined by liver stiffness cutoffs

Study	Etiology	No. of patients	Patient selection	XL probe	cACLD feature	cACLD subgroups		
						<10 kPa	10–15 kPa	>15 kPa
Wong et al. [3]	Non-obese NAFLD	231	All patients	No	F3-F4	21/158 (13.3%)	–	22/24 (91.4%)
	Obese[a] NAFLD	194	=	Yes	=	35/129 (27.1%)	–	25/29 (86.2%)
Piccinni et al. [4]	Mixed[b]	111	≥10 kPa	Yes	=	–	27/47 (58%)	43/64 (67%)
Papatheodoridi et al. [5]	Mixed	5483	All patients	No	=	431/3606 (12.0%)	469/891 (52.6%)	817/986 (82.9%)
	HCV	2864	=	=	=	243/1966 (12.4%)	265/456 (58.1%)	371/442 (83.9%)
	HBV	704	=	=	=	85/522 (16.3%)	59/103 (57.3%)	67/79 (84.8%)
	Alcohol	932	=	=	=	46/515 (8.9%)	52/118 (44.1%)	253/299 (84.6%)
	NAFLD	983	=	=	=	57/602 (9.5%)	93/214 (43.5%)	125/167 (74.9%)
	Non-obese Patients	3530	=	=	=	310/2496 (12.4%)	288/496 (58.1%)	473/538 (87.9%)
	Obese[a] Patients	1056	=	=	=	72/560 (12.9%)	105/242 (43.4%)	189/254 (74.4%)
Ji et al. [6]	MAFLD[c]	220	=	No	=	5/124 (4.0%)	9/57 (15.8%)	22/39 (56.4%)
	Non-obese Patients	174	=	=	=	5/110 (4.6%)	7/35 (20.0%)	16/29 (55.2%)
	Obese[a] Patients	46	=	=	=	0/14 (0%)	2/22 (9.1%)	6/10 (60.0%)
Zhou et al. [7]	NAFLD	830	=	Yes	=	45/582 (7.7%)	74/161 (46.0%)	62/87 (71.3%)
	Non-obese NAFLD	433	=	=	=	30/358 (8.4%)	31/54 (57.4%)	16/21 (76.2%)
	Obese[d] NAFLD	397	=	=	=	15/224 (7.4%)	43/107 (40.2%)	46/66 (69.7%)
Rivera et al. [8]	NAFLD	501	All patients	Yes	=	27/218 (12.4%)	63/161 (39.1%)	91/122 (74.6%)
	Non-obese NAFLD	164	=	=	=	10/86 (11.6%)	22/42 (52.4%)	32/36 (88.9%)
	Obese[a] NAFLD	332	=	=	=	17/131 (13.0%)	40/116 (34.5%)	59/85(69.4%)

(continued)

Table 8.1 (continued)

Study	Etiology	No. of patients	Patient selection	XL probe	cACLD feature	cACLD subgroups		
						<10 kPa	10–15 kPa	>15 kPa
Pons et al. [9]	Mixed	836	≥10 kPa	No	HVPG >5 mmHg	–	130/211 (61.6%)	564/625 (90.2%)
	HCV	358	=	=	=	–	69/90 (76.7%)	253/268 (94.4%)
	HBV	27	=	=	=	–	6/8 (75%)	19/19 (100%)
	Alcohol	203	=	=	=	–	20/24 (83.3%)	176/179 (98.3%)
	NAFLD[e]	248	=	Yes	=	–	35/89 (39.3%)	116/159 (73%)
	Non-obese NAFLD	101	=	=	=	–	14/35 (40%)	55/66 (83.3%)
	Obese[a] NAFLD	133	=	=	=	–	19/53 (35.8%)	54/85 (63.5%)

[a] BMI ≥30 kg/m^2
[b] 36% obese, 64% metabolic component
[c] 129 patients had coexisting chronic liver disease (mainly chronic hepatitis B)
[d] BMI ≥28 kg/m^2
[e] 68.5% with XL probe availability
NAFLD non-alcoholic fatty liver disease, *HCV* hepatitis C virus, *HVB* hepatitis B virus, *MAFLD* metabolic-associated fatty liver disease

In Table 8.2, several studies that have evaluated different liver-related outcomes in the populations selected by the <10 kPa cutoff or similar ruling-out cACLD are described [10–21]. As shown, liver-related events are low in patients with LSM <10 kPa, independently of the etiology; in most studies, cumulative incidence rates at 3 years or event rates are around or below 1%. Data from a collaborative study with 2638 NAFLD patients from France, Hong Kong, Canada, and Spain indicate that the cumulative rate of any liver-related event in 1820 patients with LSM <10 kPa during 3 years of follow-up was 0.1% (unpublished data) (Fig. 8.1). In addition, in the specific population of 365 patients with LSM between ≥8 and <10 kPa, the liver event rate was 0.

Some authors have proposed to lower the cutoff for ruling out cACLD to LSM <7–8 kPa with the aim of increasing the sensitivity to exclude severe fibrosis/cirrhosis, minimizing the false-negative rates [5]. As largely explained, we do not support the concept that cACLD and the TE values that define it should become perfect diagnostic tools for excluding or diagnosing severe fibrosis/cirrhosis, but rather clinically useful rules to categorize CLD patients. By lowering the 10 kPa cutoff, the grey zone increases with patients who are at very low risk of events, and consequently, referrals to hepatologists will increase with such patients. In addition, the transition from normal LSM or absence of fibrosis to possible severe fibrosis will almost disappear.

Despite not being labeled as cACLD, patients with abnormal TE values but below <10 kPa should be monitored for changes indicating progression to cACLD. Since the risk of liver events in these patients is very low within a 3-year time period, reassessment in 2–3 years seems a reasonable strategy.

Table 8.2 Liver-related events during follow-up in different studies evaluating patients with chronic liver disease selected by a liver stiffness value below 10 kPa or similar values

Study	Etiology	Patients (*n*)	Liver event	Follow-up (months)	LSM cutoff	Event rate
Masuzaki et al. [10]	HCV	866	HCC	36 (mean)	≤10 kPa	CI: 0.4% (3 years) ER: 2/511 (0.4%)
Fung et al. [11]	HBV	528	LRD + HCC	35 (median)	<10 kPa	CI: 0 (3 years) ER: 0/445
Vergniol et al. [12]	HCV	1457	OS	47.3 (median)	≤9.5 kPa	OS: 96% (5 years)
Jung et al. [13]	HBV	1130	HCC	30.7 (median)	≤8 kPa	CI: 1.58% (3 years)
Coperchot et al. [14]	PBC	150	LRE	28 (mean)	≤9.6 kPa	ER: 1/113 (0.8%)
Klibansky et al. [15]	Mixed	400	LRE	28 (median)	<10.5 kPa	ER: 3/224 (1.3%)
Pang et al. [16]	Mixed	2052	LRE	15.6 (median)	<10 kPa	CI: 3.9% (3 years)
Coperchot et al. [17]	PSC	168	LRE	48 (mean)	≤9.9 kPa	ER: 6/112 (5%) OS: 97% (3 years)
Tatsumi et al. [18]	HCV	470	HCC	23 (median)	≤12 kPa	CI: 0 (2 years) ER: 1/363 (0.3%)
Shili-Masmoudi et al. [19]	NAFLD	2245	LRE	27 (median)	≤12 kPa	CI: 0.2% (3 years) OS: 96.5% (3 years)
Rasmussen et al. [20]	ALD	443	LRE[a]	49 (median)	<10 kPa	CI: 1.1% (3 years) ER: 9/303 (3%)
Grgurevic et al. [21]	T2D-78% NAFLD	454	LRE	25 (median)	<9.6 kPa	ER: 0

[a] Including alcoholic hepatitis

HCV hepatitis C virus, *HBV* hepatitis B virus, *PBC* primary biliary cholangitis, *PSC* primary sclerosing cholangitis, *NAFLD* non-alcoholic fatty liver disease, *ALD* alcoholic liver disease, *HCC* hepatocellular carcinoma, *CI* cumulative incidence, *ER* event rate, *LRD* liver-related mortality, *OS* overall survival, *LRE* liver-related events

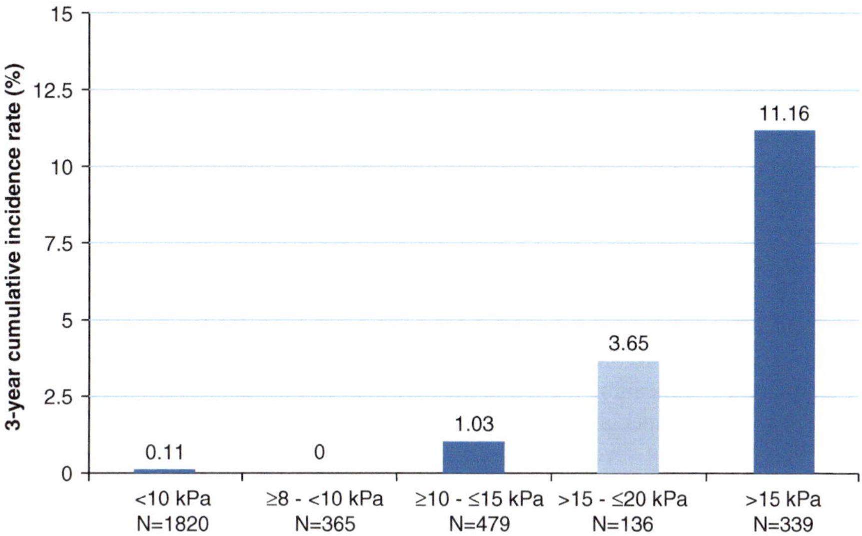

Fig. 8.1 Liver-related events (3-year cumulative incidence rate) in a cohort of 2638 patients with non-alcoholic fatty liver disease distributed in subgroups defined by different liver stiffness cutoffs, including the values that define compensated advanced chronic liver disease (cACLD)

Detecting Highly Suggestive cACLD (LSM >15 kPa)

Similar to what was observed with the ruling out criterium for cACLD, the ruling in criterium for cACLD of >15 kPa selects a population of patients with a prevalence of advanced fibrosis/cirrhosis higher than 80% across the different etiologies of CLD (Table 8.1). In addition, the prevalence of portal hypertension, as defined by a hepatic venous pressure gradient ≥5 mmHg, was higher than 90% in most patients evaluated in one study [9] (Table 8.1). NAFLD patients, especially obese patients, present lower prevalences of advanced fibrosis/cirrhosis or portal hypertension in most studies [5–7, 9]. Although not completely clear, it is possible that obesity might interfere with LSM measurements reducing the rate of NAFLD patients with severe fibrosis/cirrhosis independently of the probe (M or XL) used [22].

There are few studies that have explored the outcomes in the specific population of patients with LS >15 kPa. Rasmussen et al. [20] demonstrated in a cohort of patients with early alcohol-related liver disease that patients with LSM >15 kPa have a higher risk of presenting liver-related events (54% at 4 years of follow-up) compared to those with LSM between 10 and 15 kPa (intermediate risk-21% of events) and those with LSM <10 kPa (3% of events-Table 8.2). Many other studies have not explored specifically the cutoff of 15 kPa but have demonstrated that the prognosis worsens substantially when LSM increases and specially in the range from 15 to 25 kPa; this will be extensively discussed in the next chapter (Outcome and Prognosis). The risk of developing hepatocellular carcinoma was also higher in those patients with LSM between 15.1 and 20 kPa (cumulative incidence of 19% at 3 years) and intermediate in those with LSM 10.1–15 kPa (11.7% at 3 years) as compared to patients with LSM ≤10 kPa (Table 8.2) in a cohort of patients with chronic hepatitis C [10]. Again, data from the collaborative study from France, Hong Kong, Canada, and Spain indicate that the cumulative rate of any liver-related event in 339 patients with LSM >15 kPa during 3 years of follow-up was 11% (unpublished data) (Fig. 8.1).

It has been proposed to decrease the cutoff for ruling in cACLD to >12 kPa with the aim of decreasing a too high specificity of the >15 kPa cutoff to rule in severe fibrosis/cirrhosis, minimizing the false-negative rates [5]. By lowering the 15 kPa cutoff, the grey zone decreases, but the population of cACLD increases at the expense of a significant reduction in the percentage of patients with severe fibrosis/cirrhosis. This is especially dramatic for NAFLD patients, since in some series the prevalence (PPV) of severe fibrosis/cirrhosis might be decreased to less than 50%. More importantly, the cACLD population would increase at the expense of patients with a minimal rate of liver-related events.

Suggestive cACLD (Grey Zone/LSM ≥10 to <15 kPa)

As expected, patients classified in the grey or intermediate zone (LSM ≥10 to <15 kPa) present intermediate prevalences of the clinical features of cACLD (Table 8.1). The presence of severe fibrosis/cirrhosis is around 50% in many cohorts and portal hypertension (HVPG >5 mmHg) can be detected in 75%–80% of patients. Again, patients with NAFLD show lower prevalences of these clinical features.

In terms of liver-related outcomes, in the few studies mentioned above that have specifically evaluated patients in the suggestive cACLD group, intermediate incidences of liver-related events are demonstrated. Again, data from the collaborative study from France, Hong Kong, Canada, and Spain indicate that the cumulative rate of any liver-related event in 479 patients in the grey zone (LSM ≥10 to <15 kPa) during 3 years of follow-up was 1% (unpublished data) (Fig. 8.1). This is 10 times higher than patients with LSM <10 kPa and 10 times lower than patients with LSM >15 kPa. In addition, in the specific subpopulation of 136 patients with LSM between >15 and 20 kPa, the liver event rate increased to 3.6%.

Baveno VI suggested that in the grey zone, invasive procedures such as liver biopsy demonstrating at least severe fibrosis, endoscopy confirming the presence of varices or an HVPG confirming the presence of portal hypertension must be performed to confirm cACLD [1]. Endoscopy should only be indicated if Baveno VI criteria for screening endoscopy are met and performing a liver biopsy or HVPG to these patients is not routinely carried out in many centers. These procedures should probably be individualized considering the risk-benefit of the intervention. What is clear is that monitoring for progression is required.

Other Elastography Techniques

Acoustic radiation force impulse (ARFI) techniques, as TE, use shear-wave elastography (SWE) for the noninvasive assessment of liver fibrosis. ARFI techniques can be divided into point shear wave elastography (pSWE) and multidimensional shear wave elastography (2D-SWE and 3D-SWE) [23, 24]. Although ARFI techniques have been available for almost 10 years and they provide some technological advances compared to TE, their use in daily clinical practice has been rather modest. One of the main limitations of their use is that they use different proprietary algorithms to determine velocity of the shear wave and hence liver stiffness. As a consequence, the cutoffs for fibrosis staging vary across different systems from different vendors. However, in the recent years, the Quantitative Imaging Biomarker Alliance (QIBA) committee of the Radiologic Society of North America has contributed to diminish this variability by developing standardized phantoms that vendors use to harmonize their measurements [25, 26].

Both pSWE and 2D-SWE have been demonstrated to have high accuracy (similar to TE) for fibrosis staging. Moreover, 2D-SWE performed with the Aixplorer machine was shown to have a good concordance with TE. Casinotto, et al. [27] indicated that 2D-SWE values were slightly higher compared to TE in low percentiles and lower in high percentiles, being the best concordance in values between 7 and 9 kPa. Best accuracy cutoff values of 2D-SWE for identifying TE values <10 kPa and >15 kPa were 10 kPa and 14 kPa, respectively.

According to the previous evidence and to classify cACLD patients, the Society of Radiologists in Ultrasound [25] have proposed a vendor-neutral "rule of four" (5, 9, 13, 17 kPa) for the ARFI techniques for viral etiologies and NAFLD, being the cutoff for ruling out <9 kPa (<1.7 m/s) (in the absence of other known clinical signs) and >13 kPa (>2.1 m/s) the cutoff for ruling in cACLD. According to the consensus, those patients with >17 kPa (>2.4 m/s) values are suggestive of having CSPH, but additional tests may be required. The authors have also suggested that cACLD cutoffs may be lower in some NAFLD patients and follow-up or additional tests are needed for those patients with liver stiffness values between 7 and 9 kPa [25].

Magnetic resonance elastography (MRE) has demonstrated a good accuracy for fibrosis staging in the main etiologies of chronic liver disease, especially NAFLD; however, its cost and lack of extended availability have limited its use in clinical practice [28, 29].

Summary

The concept of cACLD with the dual LSM cutoffs identifies two very clinically different populations of CLD patients with high and low prevalence of severe fibrosis/cirrhosis, portal hypertension, and most importantly, liver-related outcomes during follow-up.

A question remains open regards to the use of etiology-dependent cACLD thresholds. This was already an issue when using the cACLD definition regarding the presence of severe fibrosis/cirrhosis and it was also indicated in the panelist's questionnaire. However, to be able to provide this information, more data on the rate of liver events in the different etiologies is needed, especially in the LSM range of 10–15 kPa. In addition, it would also be helpful to have a clear definition of what it is considered low or high event rate (and at what time frame), what are the different types of liver-related events to consider, and what are the implications of changing the thresholds.

References

1. de Franchis R, Baveno VI Faculty. Expanding consensus in portal hypertension: report of the Baveno VI consensus workshop: stratifying risk and individualizing care for portal hypertension. J Hepatol. 2015;63:743–52.
2. Davison B, Harrison S, Cotter G, Alkhouri N, Sanyal A, Edwards C, et al. Suboptimal reliability of liver biopsy evaluation has implications for randomized clinical trials. J Hepatol. 2020;73:1322–32.
3. Wong VWS, Irles M, Wong GLH, Shili S, Chan AWH, Merrouche W, et al. Unified interpretation of liver stiffness measurement by M and XL probes in non-alcoholic fatty liver disease. Gut. 2019;68:2057–64.
4. Piccinni R, Rodrigues SG, Montani M, Murgia G, Delgado MG, Casu S, et al. Controlled attenuation parameter reflects steatosis in compensated advanced chronic liver disease. Liver Int. 2020;40:1151–8.
5. Papatheodoridi M, Hiriart JB, Lupsor-Platon M, Bronte F, Boursier J, Elshaarawy O, et al. Refining the Baveno VI elastography criteria for the definition of compensated advanced chronic liver disease. J Hepatol. 2021;74:1109–16.
6. Ji Y, Li J, He Q, Zhou L, Chen J. Baveno VI elastography criteria for ruling in cACLD works well in MAFLD patients. J Hepatol. 2021;75:1243–5.
7. Zhou Y, Gao F, Liu W, Wong GL, Mahadeva S, Raihan Nik Mustapha N, et al. Screening for compensated advanced chronic liver disease using refined Baveno VI elastography cutoffs in Asian patients with nonalcoholic fatty liver disease. Aliment Pharmacol Ther. 2021;54:470–80.
8. Rivera J, Calleja J, Aller R, Gonzalo-Benito H, Arias-Loste M, Ireuzubieta P, et al. Prevalence estimation of significant fibrosis due to non-alcoholic steatohepatitis combining transient elastography and histology. J Hepatol. 2021;75:S255.
9. Pons M, Augustin S, Scheiner B, Guillaume M, Rosselli M, Rodrigues SG, et al. Noninvasive diagnosis of portal hypertension in patients with compensated advanced chronic liver disease. Am J Gastroenterol. 2021;116:723–32.
10. Masuzaki R, Tateishi R, Yoshida H, Goto E, Sato T, Ohki T, et al. Prospective risk assessment for hepatocellular carcinoma development in patients with chronic hepatitis C by transient elastography. Hepatology. 2009;49:1954–61.
11. Fung J, Lai CL, Seto WK, Wong DKH, Yuen MF. Prognostic significance of liver stiffness for hepatocellular carcinoma and mortality in HBeAg-negative chronic hepatitis B. J Viral Hepat. 2011;18:738–44.
12. Vergniol J, Foucher J, Terrebonne E, Bernard P, Le Bail B, Merrouche W, et al. Noninvasive tests for fibrosis and liver stiffness predict 5-year outcomes of patients with chronic hepatitis C. Gastroenterology. 2011;140:1970–1979.e3.
13. Jung KS, Kim SU, Ahn SH, Park YN, Kim DY, Park JY, et al. Risk assessment of hepatitis B virus-related hepatocellular carcinoma development using liver stiffness measurement (FibroScan). Hepatology. 2011;53:885–94.
14. Corpechot C, Carrat F, Poujol-Robert A, Gaouar F, Wendum D, Chazouillères O, et al. Noninvasive elastography-based assessment of liver fibrosis progression and prognosis in primary biliary cirrhosis. Hepatology. 2012;56:198–208.

15. Klibansky DA, Mehta SH, Curry M, Nasser I, Challies T, Afdhal NH. Transient elastography for predicting clinical outcomes in patients with chronic liver disease. J Viral Hepat. 2012;19:e184–93.
16. Pang JXQ, Zimmer S, Niu S, Crotty P, Tracey J, Pradhan F, et al. Liver stiffness by transient elastography predicts liver-related complications and mortality in patients with chronic liver disease. PLoS One. 2014;22:e95776.
17. Corpechot C, Gaouar F, Naggar AEI, Kemgang A, Wendum D, Poupon R, et al. Baseline values and changes in liver stiffness measured by transient elastography are associated with severity of fibrosis and outcomes of patients with primary sclerosing cholangitis. Gastroenterology. 2014;146:970–979.e6.
18. Tatsumi A, Maekawa S, Sato M, Komatsu N, Miura M, Amemiya F, et al. Liver stiffness measurement for risk assessment of hepatocellular carcinoma. Hepatol Res. 2015;45:523–32.
19. Shili-Masmoudi S, Wong GLH, Hiriart JB, Liu K, Chermak F, Shu SST, et al. Liver stiffness measurement predicts long-term survival and complications in non-alcoholic fatty liver disease. Liver Int. 2020;40:581–9.
20. Rasmussen DN, Thiele M, Johansen S, Kjærgaard M, Lindvig KP, Israelsen M, et al. Prognostic performance of seven biomarkers compared to liver biopsy in early alcohol-related liver disease. J Hepatol. 2021;75(5):1017–25. https://doi.org/10.1016/j.jhep.2021.05.037.
21. Grgurevic I, Salkic N, Mustapic S, Bokun T, Podrug K, Marusic S, et al. Liver and nonliver-related outcomes at 2 years are not influenced by the results of the FIB-4 test and liver Elastography in a real-life cohort of patients with type 2 diabetes. Can J Gastroenterol Hepatol. 2021;2021:5582813.
22. Petta S, Wai-Sun Wong V, Bugianesi E, Fracanzani AL, Cammà C, Hiriart JB, et al. Impact of obesity and alanine aminotransferase levels on the diagnostic accuracy for advanced liver fibrosis of noninvasive tools in patients with nonalcoholic fatty liver disease. Am J Gastroenterol. 2019;114:916–28.
23. Dietrich C, Bamber J, Berzigotti A, Bota S, Cantisani V, Castera L, et al. EFSUMB guidelines and recommendations on the clinical use of liver ultrasound Elastography, update 2017 (long version). Ultraschall Med. 2017;38:e16–47.
24. Ferraioli G, Wong V, Castera L, Berzigotti A, Sporea I, Dietrich C, et al. Liver ultrasound Elastography: an update to the World Federation for Ultrasound in medicine and biology guidelines and recommendations. Ultrasound Med Biol. 2018;44:2419–40.
25. Barr RG, Wilson SR, Rubens D, Garcia-Tsao G, Ferraioli G. Update to the Society of Radiologists in ultrasound liver elastography consensus statement. Radiology. 2020;296:263–74.
26. Palmeri M, Nightingale K, Fielding S, Rouze N, Deng Y, Lynch T, et al. RSNA QIBA ultrasound shear wave speed phase II phantom study in viscoelastic media. IEEE Int Ultrason Symp. 2015;2015:1–4.
27. Cassinotto C, Lapuyade B, Guiu B, Marraud des Grottes H, Piron L, Merrouche W, et al. Agreement between 2-dimensional shear wave and transient elastography values for diagnosis of advanced chronic liver disease. Clin Gastroenterol Hepatol. 2020;18:2971–2979.e3.
28. European Association for the Study of the Liver. EASL Clinical practice guidelines on non-invasive tests for evaluation of liver disease severity and prognosis—2021 update. J Hepatol. 2021;75:659–89.
29. Singh S, Venkatesh S, Wang Z, Miller F, Motosugi U, Low R, et al. Diagnostic performance of magnetic resonance elastography in staging liver fibrosis: a systematic review and meta-analysis of individual participant data. Clin Gastroenterol Hepatol. 2015;13:440–451.e6.

Noninvasive Detection of Clinically Significant Portal Hypertension with Liver Elastography

9

Mònica Pons, Laia Aceituno, and Joan Genescà

From Baveno VI to Baveno VII

One of the hallmarks in the natural history of cACLD or cirrhotic patients is the presence of clinically significant portal hypertension (CSPH), determined by a hepatic venous pressure gradient (HVPG) of ≥10 mmHg, indicating a starting point for developing varices and clinical decompensation. The importance of detecting CSPH has gained relevance since the publication of the PREDESCI study demonstrating that β-blockers decrease liver-related events in cirrhotic patients with CSPH, independently of the presence of varices [1]. As a consequence of that, the interest for finding a noninvasive approach to detect patients at high risk of CSPH has been growing.

Among different noninvasive approaches, liver stiffness measurement (LSM) by transient elastography (TE) has emerged as the more assessed and reliable method to detect patients with CSPH [2]. The addition of other parameters, such as platelet count or spleen diameter, was suggested in some studies [3, 4], but the exact roles of these were not clearly defined. Since most studies indicated that LSM values from 20 to 25 kPa were useful to rule in CSPH [5, 6], the Baveno VI consensus indicated that LSM by TE ≥20–25 kPa alone or combined with platelets and spleen size were sufficient to rule in CSPH in virus-related cACLD patients [7]. No recommendations regarding ruling out were made at that time.

Since Baveno VI, several publications have allowed refinement of the proposed recommendations and demonstrated that most of the assumptions made were true.

M. Pons · L. Aceituno · J. Genescà (✉)
Liver Unit, Hospital Universitari Vall d'Hebron, Vall d'Hebron Institute of Research (VHIR), Vall d'Hebron Barcelona Hospital Campus, Universitat Autònoma de Barcelona, Barcelona, Spain

Centro para la Investigación Biomédica en Red de Enfermedades Hepáticas y Digestivas (CIBEREHD), Madrid, Spain
e-mail: laceituno@vhebron.net; jgenesca@vhebron.net

R. de Franchis (ed.), *Portal Hypertension VII*,
https://doi.org/10.1007/978-3-031-08552-9_9

Also, the panelist's survey showed that most of the panelists (71%) use TE to estimate the presence/absence of CSPH in clinical practice. Although most of them used it for general prognosis (80%), it is noteworthy that almost one-third used TE for assessing the risk of decompensation to initiate β-blocker preventive therapy. A majority of the panelists (73%) consider the two cutoff approach better than a single cutoff approach and the use of the same cutoffs for all etiologies (69%), although this could imply a little less accuracy.

Ruling in CSPH

After Baveno VI, two publications have contributed to tune and improve the assessment and stratification of cACLD patients based on their risk of CSPH. The ANTICIPATE study provided risk prediction models for CSPH using LSM alone or LSM plus platelet count in a cACLD population mainly composed of viral and alcohol etiologies [8]. These models were subsequently validated in a different cohort with a similar composition [9]. In this last study, classification rules for CSPH were also provided.

As already determined in Baveno VI, the ruling in LSM cutoff for CSPH ranges from 20 to 25 kPa. In the largest multicenter study published with more than 800 cACLD patients [9], the accumulated prevalence of CSPH were 83.5% for LSM ≥20 kPa, 91% for LSM ≥25 kPa, and 93.7% for LSM ≥30 kPa. LSM ≥ 25 kPa was chosen as the optimal cutoff for ruling in CSPH with *a* > 90% prevalence or positive predictive value (PPV) and specificity. This LSM cutoff was adequate for CSPH in viral and alcoholic cACLD, but not for non-alcoholic steatohepatitis (NASH) cACLD patients (Fig. 9.1). The 25 kPa cutoff has already been validated in a recent publication with 76 cACLD patients and a PPV of 87% [10].

However, the number of cACLD patients with CSPH below the proposed cutoff of 25 kPa is still high (40% to 75%) [9]. The detection of these patients can be improved with the use of platelet counts by calculating the risk predicted by the

	STAGES OF CHRONIC LIVER DISEASE						
	No cirrhosis		Early compensated cirrhosis	Late compensated cirrhosis			Decompensated cirrhosis
	CLD	Early cACLD		Late cACLD			dACLD
Liver fibrosis	F1-F2	F3	F4	F4			F4
HVPG (mm Hg)	<5	5 - < 10		≥10			≥10
Portal hypertension	No	Mild		CSPH			CSPH
Liver stiffness (kPa)	<10	10 - < 25		15 - < 20	20 - < 25	≥25	Unneeded
Platelet count (K/mm^3)	Any	Normal		< 110	< 150	Any	Any
Preventing decompensation	No	No		Yes			No

cACLD: compensated advanced chronic liver disease; HVPG: hepatic venous pressure gradient

Fig. 9.1 Identification of patients with chronic liver disease (CLD) at risk of clinically significant portal hypertension (CSPH) and preventive measures to avoid decompensation in the different stages of the disease by using noninvasive tests including liver stiffness. *cACLD* compensated advanced chronic liver disease, *HVPG* hepatic venous pressure gradient

ANTICIPATE model with LSM plus platelets for CSPH. As shown in Fig. 9.1, patients with LSM between 20 and 25 kPa and platelets below 150,000/mm^3 present an average minimum predicted risk of CSPH of 60% and the same applies to patients with LSM between 15 and 20 kPa and platelets below 110,000/mm^3. These patients can be considered as at high risk of CSPH and might be candidates for the prevention of decompensation. A similar recommendation regarding the use of LSM and platelet count for identifying patients at risk of CSPH has already been published [11].

Patients with NASH cACLD showed different interactions with LSM and the prediction of CSPH compared to other etiologies. Obesity (BMI $\geq$ 30 kg/m^2) seems to play an important role in changing the association between LSM and portal pressure. This was demonstrated in the study by Pons et al. [9], in which for a given LSM value or a combination of LSM plus platelet count values, the HVPG was lower for higher BMIs. The ANTICIPATE model overpredicted CSPH in NASH patients, and a new model (ANTICIPATE–NASH model) to predict CSPH was constructed on the basis of LSM, BMI, and platelet counts; this predictive model with the corresponding nomogram can be used to assess the individual risk of CSPH in NASH cACLD patients. Also, the ruling in cutoff of LSM $\geq$25 kPa only reached a 90% prevalence of PPV in non-obese NASH patients but had less PPV for CSPH in obese NASH patients. In Table 9.1, examples of NASH patients with a minimum predicted risk of CSPH of 60% with different combinations of BMI, LSM, and platelets are shown in order to facilitate a rapid identification for the clinician of possible candidates for the prevention of decompensation.

Table 9.1 Platelet count needed for different combinations of liver stiffness measurement (LSM) and body mass index (BMI) values to obtain a clinically significant portal hypertension risk prediction of 60% in patients with non-alcoholic steatohepatitis (NASH) using the risk prediction of the NASH–ANTICIPATE model

LSM (kPa)	BMI range (kg/m^2)	Platelet count ($\times 10^9$/L)
15	20	80
	25	75
	30	60
	35	45
20	20	135
	25	120
	30	100
	35	90
25	20	160
	25	150
	30	140
	35	125
30	20	190
	25	180
	30	165
	35	155

Ruling Out CSPH

Excluding CSPH in cACLD patients has been a difficult task. This is related, in large part, to the high prevalence (>50%) of CSPH in most series evaluating noninvasive predictors of CSPH. By using LSM alone, several cutoffs have been evaluated, including the first one proposed of 13.6 kPa [12]. Using the combination of LSM and platelet count seems to work better in ruling out, and the combination of LSM <25 kPa and platelets >150,000 selected a subgroup of cACLD patients from the original ANTICIPATE cohort with a risk of CSPH of 17% [13]. Data from the large multicenter study already mentioned indicated that it was also not possible to identify a single LSM cutoff with a high NPV to exclude CSPH [9]. However, adding platelet count $\geq$150,000/mm^3 to an LSM $\leq$15 kPa cutoff could exclude CSPH with a negative predicted value (NPV) and sensitivity of >90% in most etiologies, including hepatitis C virus, alcohol, and NASH cACLD. Patients with hepatitis B virus cACLD were underrepresented and could not be evaluated. This combined ruling out criteria has recently been validated in a recent publication with 76 cACLD patients with an NPV and sensitivity of 100% [10].

Other Elastographic Techniques

As recently compiled in a review [2], pSWE and 2D-SWE have been demonstrated to have a significant correlation with HVPG and good accuracy for detecting CSPH; however, the variability of the devices and published cutoffs for diagnosing CSPH makes it hard to recommend their use for noninvasive diagnosis of CSPH. This variability is mainly due to the heterogeneity of the selected population in the different studies.

In an individual patient meta-analysis including 328 patients from five different studies [14], only 27% of them with cACLD concluded that liver stiffness <14 kPa by 2D-SWE performed with the Aixplorer device may be used to rule out CSPH.

Jansen et al. [15, 16], in a study including 109 patients (43 patients Child–Pugh B/C), developed an algorithm including liver stiffness and spleen stiffness measurements by 2D-SWE (Aixplorer device), where CSPH was safely ruled out and ruled in; however, the accuracy of the algorithm was demonstrated to be insufficient for its clinical practice use [17].

In conclusion, further studies are needed with cACLD patients, the target population for diagnosing CSPH, to validate the utility of other elastography techniques for ruling in and ruling out CSPH.

Summary

In summary, with the use of simple and readily available noninvasive parameters, the risk of CSPH can be reasonably assessed in the main etiologies of cACLD, facilitating the identification of potential candidates for prophylaxis of decompensation. In patients with virus and/or alcohol-related cACLD and non-obese (BMI <30 kg/m^2) NASH cACLD, an LSM value ≥25 kPa is sufficient to rule in CSPH. In patients with virus and/or alcohol-related cACLD with LSM values <25 kPa and NASH cACLD, the ANTICIPATE and ANTICIPATE–NASH models could be, respectively, used to predict the risk of CSPH. In cACLD patients of viral, alcohol, and NASH etiologies, a value of LSM ≤15 kPa plus a platelet count ≥150,000/mm^3 can be used to rule out CSPH.

References

1. Villanueva C, Albillos A, Genescà J, Garcia-Pagan JC, Calleja JL, Aracil C, et al. β blockers to prevent decompensation of cirrhosis in patients with clinically significant portal hypertension (PREDESCI): a randomised, double-blind, placebo-controlled, multicentre trial. Lancet. 2019;393:1597–608. https://doi.org/10.1016/S0140-6736(18)31875-0.
2. Vuille-Lessard É, Rodrigues SG, Berzigotti A. Noninvasive detection of clinically significant portal hypertension in compensated advanced chronic liver disease. Clin Liver Dis. 2021;25:253–89. https://doi.org/10.1016/J.CLD.2021.01.005.
3. Berzigotti A, Seijo S, Arena U, Abraldes JG, Vizzutti F, García–Pagán JC, et al. Elastography, spleen size, and platelet count identify portal hypertension in patients with compensated cirrhosis. Gastroenterology. 2013;144:102–111.e1. https://doi.org/10.1053/J.GASTRO.2012.10.001.
4. Kim BK, Han KH, Park JY, Ahn SH, Kim JK, Paik YH, et al. A liver stiffness measurement-based, noninvasive prediction model for high-risk esophageal varices in B-viral liver cirrhosis. Am J Gastroenterol. 2010;105:1382–90. https://doi.org/10.1038/AJG.2009.750.
5. You M, Kim K, Pyo J, Huh J, Kim H, Lee S, et al. A meta-analysis for the diagnostic performance of transient elastography for clinically significant portal hypertension. Ultrasound Med Biol. 2017;43:59–68. https://doi.org/10.1016/J.ULTRASMEDBIO.2016.07.025.
6. Song J, Ma Z, Huang J, Liu S, Luo Y, Lu Q, et al. Comparison of three cut-offs to diagnose clinically significant portal hypertension by liver stiffness in chronic viral liver diseases: a meta-analysis. Eur Radiol. 2018;28:5221–30. https://doi.org/10.1007/S00330-018-5478-Z.
7. de Franchis R, Baveno VI Faculty. Expanding consensus in portal hypertension: report of the Baveno VI consensus workshop: stratifying risk and individualizing care for portal hypertension. J Hepatol. 2015;63:743–52. https://doi.org/10.1016/j.jhep.2015.05.022.
8. Abraldes JG, Bureau C, Stefanescu H, Augustin S, Ney M, Blasco H, et al. Noninvasive tools and risk of clinically significant portal hypertension and varices in compensated cirrhosis: the "Anticipate" study. Hepatology. 2016;64:2173–84. https://doi.org/10.1002/hep.28824.
9. Pons M, Augustin S, Scheiner B, Guillaume M, Rosselli M, Rodrigues SG, et al. Noninvasive diagnosis of portal hypertension in patients with compensated advanced chronic liver disease. Am J Gastroenterol. 2021;116:723–32. https://doi.org/10.14309/ajg.0000000000000994.
10. Podrug K, Trkulja V, Zelenika M, Bokun T, Madir A, Kanizaj TF, O'Beirne J, Grgurevic I. Validation of the new diagnostic criteria for clinically significant portal hypertension by platelets and elastography. Dig Dis Sci. 2022;67(7):3327–32. https://doi.org/10.1007/s10620-021-07277-8.
11. Garcia-Tsao G, Abraldes JG. Nonselective beta-blockers in compensated cirrhosis: preventing variceal hemorrhage or preventing decompensation? Gastroenterology. 2021;161:770–3. https://doi.org/10.1053/J.GASTRO.2021.04.077.
12. Vizzutti F, Arena U, Romanelli RG, Rega L, Foschi M, Colagrande S, et al. Liver stiffness measurement predicts severe portal hypertension in patients with HCV-related cirrhosis. Hepatology. 2007;45:1290–7. https://doi.org/10.1002/HEP.21665.
13. Augustin S, Pons M, Santos B, Ventura M, Genescà J. Identifying compensated advanced chronic liver disease: when (not) to start screening for varices and clinically significant portal hypertension. In: de Franchis R, editor. Portal Hypertens VI. Springer International Publishing; 2016. p. 39–49. https://doi.org/10.1007/978-3-319-23018-4_5.

14. Thiele M, Hugger MB, Kim Y, Rautou P-E, Elkrief L, Jansen C, et al. 2D shear wave liver elastography by Aixplorer to detect portal hypertension in cirrhosis: an individual patient data meta-analysis. Liver Int. 2020;40:1435–46. https://doi.org/10.1111/LIV.14439.
15. Jansen C, Bogs C, Verlinden W, Thiele M, Möller P, Görtzen J, et al. Shear-wave elastography of the liver and spleen identifies clinically significant portal hypertension: a prospective multicentre study. Liver Int. 2017;37:396–405. https://doi.org/10.1111/LIV.13243.
16. Jansen C, Bogs C, Verlinden W, Thiele M, Möller P, Görtzen J, et al. Algorithm to rule out clinically significant portal hypertension combining shear-wave elastography of liver and spleen: a prospective multicentre study. Gut. 2016;65:1057–8. https://doi.org/10.1136/GUTJNL-2016-311536.
17. Elkrief L, Ronot M, Andrade F, Burgio MD, Issoufaly T, Zappa M, et al. Non-invasive evaluation of portal hypertension using shear-wave elastography: analysis of two algorithms combining liver and spleen stiffness in 191 patients with cirrhosis. Aliment Pharmacol Ther. 2018;47:621–30. https://doi.org/10.1111/APT.14488.

Varices and Screening Endoscopy

10

Wayne W. H. Bai and Juan G. Abraldes

The Baveno VI Criteria

The first five Baveno consensus conferences recommended performing endoscopy on every patient with a diagnosis of cirrhosis. For the first time, Baveno VI conference introduced a ***two-step strategy*** for the screening of esophageal varices in patients with compensated advanced chronic liver disease (cACLD), establishing what became known as the Baveno VI criteria [1]. These resulted from the combination of a liver stiffness measurement by transient elastography (LSM by TE) of less than 20 kPa, together with a platelet count over 150×10^9/L. The risk of high-risk varices (HRV) when fulfilling both criteria was considered low enough to circumvent the performance of an endoscopy. In those patients outside Baveno VI criteria, an endoscopy should be performed. These criteria have been widely adopted in practice and recommended by subsequent guidelines for the management of a chronic liver disease [2]. Furthermore, since the Baveno VI conference, the criteria have undergone an unprecedented level of validation, with 28 fully published manuscripts up to March 2021 testing its performance.

W. W. H. Bai
Waikato District Health Board, University of Auckland, Hamilton, New Zealand

J. G. Abraldes (✉)
Liver Unit, Division of Gastroenterology, University of Alberta, Edmonton, AB, Canada
e-mail: juan.g.abraldes@ualberta.ca

R. de Franchis (ed.), *Portal Hypertension VII*,
https://doi.org/10.1007/978-3-031-08552-9_10

The basic tenet of the proposal at Baveno VI was that *"for high risk varices (HRV: medium-large varices or small with red signs) the acceptable risk of missing varices should be near 0 or 5% at the most")* [3]. This put the emphasis on not missing an opportunity of doing primary prophylaxis of bleeding (either with beta-blockers or variceal ligation), and considered much less harmful the performance of an unneeded endoscopy since the invasiveness of endoscopy is low. Indeed, setting that threshold at the 5% level means that we would accept to do an endoscopy with up to 95% chance of a false-positive result, and up to 5% chance of a false-negative result, or what would be equivalent to the value of not missing a case of HRV (a false negative) was 19 (95/5) times greater than the value of doing an unneeded endoscopy (false positive).

The subsequent ANTICIPATE study showed that the pointwise estimate of a combination of LSM by TE of 20 kPa and a platelet count of 150 × 10^9/L corresponds to the predicted risk of HRV of ~5% [4]. The Baveno VI criteria, therefore, set the maximum allowed risk of missing varices at the 5% mark (Fig. 10.1). Provided that the ANTICIPATE model is well calibrated, the negative predicted value (NPV) of the criteria would tend to be >95%, since all patients with negative criteria would have a theoretical risk of HRV below 5%. This seems very reasonable in a two-step strategy where, in the first step, a high NPV is favored.

The interpretation of the 5% threshold raised some questions [5], including whether this should correspond to a sensitivity of >95% for the criteria or a NPV

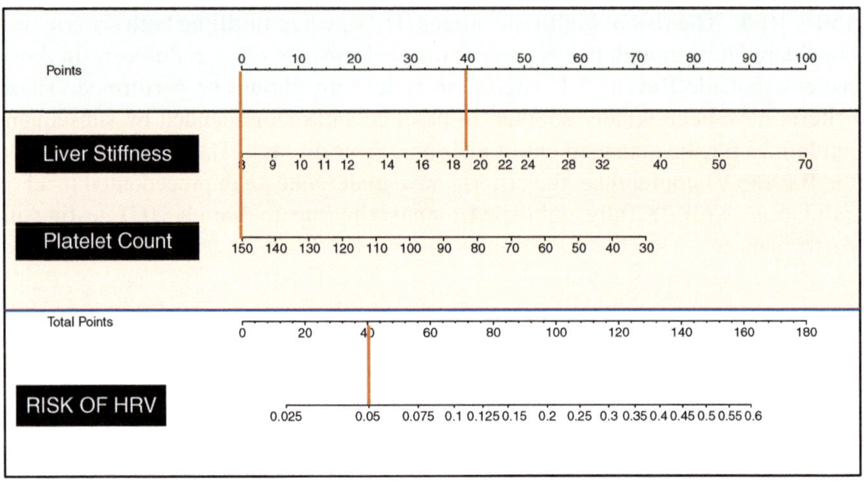

Baveno VI criteria: LSM by TE < 20 kPa OR Platelet count >150
Maximum risk of HRV: 5%

Fig. 10.1 ANTICIPATE nomogram for predicting HRV based on LSM by TE and platelet count. A LSM of just below 20 kPa, together with a platelet count of 150 × 10^9/L yields a probability of HRV of 5% [4]

of >95%. It is important to emphasize that sensitivity is a backward probability. For example, the sensitivity of the Baveno VI criteria would be the probability of having positive criteria provided the patient has HRV. Therefore, sensitivity never reflects a clinical question, but might be relevant in the early development of a diagnostic test, especially when the study is based on a case-control study. The NPV better reflects the clinical question addressed here: what would be the probability of not having HRV provided that the patient has values within the Baveno VI criteria.

Evolving recommendations for the management of compensated cirrhosis, based on the result of the PREDESCI study [6] might result in the use of beta-blockers in patients with clinically significant portal hypertension (CSPH), regardless of the presence of varices. This shift in paradigm, if widely adopted, will decrease the relevance of Baveno VI criteria for the assessment of varices, since patients already on beta-blockers do not require endoscopy. However, a relevant proportion of patients would have either contraindications or intolerance to beta-blockers, and these patients would require endoscopic assessment unless they are within the Baveno VI criteria.

Validation of Baveno VI Criteria

We performed a systematic search of fully published studies up to March 2021 assessing the performance of Baveno VI criteria. The search strategy is reported in detail in supplementary data 11.1. We identified 28 studies, of which the main characteristics are reported in Table 10.1.

We performed a univariate quantitative meta-analysis of proportions to pool NPVs since, as discussed above, this is the metric we consider the benchmark for validation of the performance of the criteria. The forest plot is shown in Fig. 10.2, with further methodological details in the figure legend. The pooled NPV was 0.99 (95% CI 0.99–1.00), with no significant heterogeneity. The proportion of saved endoscopies ranged from 8% to 60%. The interpretation of the proportion of saved endoscopies must be taken with caution, since they are highly dependent on the spectrum of diseases assessed in individual studies. Results of the meta-analysis of sensitivity are provided in supplementary data 11.2 (pooled sensitivity 0.99; 95% CI 0.98–0.99). We did not use bivariate models (which take into account the covariance of sensitivity and specificity) in our meta-analysis for two reasons. First, bivariate models require continuity correction, which adds a 0.5 to cells with a zero value. Since the number of zero cells was high (a number of studies had a sensitivity of 100% or an NPV of 100%), adding that 0.5 would artificially bring down sensitivity and NPV. Second, as detailed above, Baveno VI is meant to be used in the first step of the approach to diagnosing varices, so the main goal is a high NPV, and positive predictive value and specificity have much less relevance.

Table 10.1 Main characteristics of the included validation studies

Study	Inclusion criteria	Patients (*n*)	Age, Mean (yr.) (SD)	Male %	BMI, Mean (SD)	Child Pugh A, %. (Score, SD)	Time interval (month)	Etiology						Prevalence of HRV's, %	Prevalence of any varices, %
								Viral (all), %	HCV,%	HBV,%	Alcohol %	NAFLD/ NASH, %	Other, %		
Augustin et al., [7]	LSM ≥ 10. Compensated cirrhosis	925	59.4 (11.2)	55.4	27.1	95.7%	3–12 months	N/A	62.8	N/A	N/A	N/A	N/A	9.9	24.9
Bae et al., [8]	LSM ≥ 10 kPa Compensated cirrhosis	282	54.0[a] [6]	67.1	25.1 (3.7)	100.0	6 months	N/A	12.0	60.6	13.1	14.3	5.9	19.5	44.0
Bellan et al., [9]	Hep C cirrhosis LSM >6 kPa	160	65[a]	56.9	26[a]	90.5 [b]MELD 8	N/A	100.0	100.0	0	0	0	0	10.6	35.6
Cales et al., [10]	Cirrhosis	287	55.4 (10.7)	72.1	27.2 (5.6)	60.3% Score 6.7	3 months	25.8	N/A	N/A	64.5	5.6	4.2	17.4	44.2
Calvaruso et al., [5]	Hep C cirrhosis. LSM ≥ 12 kPa or Stage 4 fibrosis or evidence of GEV	1381	65.9	59.8	26.0	88.6	1 year	100.0	100.0	0	0	0	0	9.4	49.2
Colecchia et al., Prospective cohort [11]	LSM ≥ 10 kPa	115	58	67.8	26.4	86.9% [b]MELD 7.0	3 months	N/A	40.9	4.4	16.5	16.5	21.7	13	52.2
Colecchia et al, Retrospective cohort [11]	LSM ≥ 10 kPa	498	60 [10]	58.4	25.9	70.3% [b]MELD 8.0	3 months	N/A	85.1	5.9	7	N/a	N/A	20.1	50.6

Duan et al., Beijing cohort [12]	LSM ≥ 10 kPa Compensated cirrhosis	104	52[a]	54.8	23.2[a]	80.8%	3 months	34.6	0	34.6	10.6	N/A	50.0	18.3	51.9
Gaete et al., [13]	Compensated cirrhosis. Child-Pugh A	300	61	59.7	N/A	[b]MELD 8	6 months	N/A	10.7	0.7	4.3	67.3	17.0	18.0	N/A
Galizzi et al., 2021 [14]	NASH Compensated F3/4 fibrosis	21	61	19.0	31.4	100%	12 months	0	0	0	0	100.0	0	14.3	28.6
Jangouk et al., (US) [15]	LSM ≥ 10 kPa Compensated cirrhosis	161	62	99.4	29	92.4% [b]MELD 9	3 months	N/A	73.3	N/A	13.0	10.6	3.1	8.7	34.1
Jangouk et al., (IT) [15]	LSM ≥ 10 kPa Compensated cirrhosis	101	63	72.3	25	91.1% [5]	2 months	N/A	66.4	N/A	11.8	3.0	18.8	16.8	52.4
Kew et al., [16]	LSM ≥ 10 kPa	352	61	61.1	N/A	100%	12 months	42.3	16.8	25.6	8.8	32.1	16.8	15.9	N/A
Kotwal et al., Development cohort [17]	LSM ≥ 10 kPa Compensated cirrhosis	372	59.4	59.4	N/A	100%	12 months	92.2	87.1	5.1	32.0	11.0	0	9.1	27.2
Kotwal et al., Validation cohort [17]	LSM ≥ 10 kPa Compensated cirrhosis	200	61	49.0	N/A	100%	12 months	23.0	23.0	0	24.0	48.5	4.5	5.5	25.0
Lee et al., [18]	LSM ≥ 10 kPa Compensated cirrhosis	1218	56.0 (11.5)	63.9	N/A	76.6% [b]MELD 6.8 (1.1)	6 months	N/A	12.1	39.7	29.2	19.0	19.0	20.4	N/A
Matsui et al., [19]	cACLD	384	64.4	53.6	24.2	100% cACLD	12 months	45.8	39.3	6.5	9.4	38.8	6.0	2.9	14.8

(continued)

Table 10.1 (continued)

Study	Inclusion criteria	Patients (*n*)	Age, Mean (yr.) (SD)	Male %	BMI, Mean (SD)	Child Pugh A, %. (Score, SD)	Time interval (month)	Etiology						Prevalence of HRV's, %	Prevalence of any varices, %
								Viral (all), %	HCV,%	HBV,%	Alcohol %	NAFLD/ NASH, %	Other, %		
Maurice et al., [20]	LSM ≥ 10 kPa cACLD	310	58	67.4	N/A	89% MELD 7	12 months	N/A	54.5	7.7	12.9	13.5	11.3	4.8	23.2
Moctezuma-Velazquez et al., **PBC** group [21]	LSM ≥ 10 kPa PBC and PSC only	147	59.1 (11.5)	14.0	N/A	[b]MELD 8.2 (3.0)	12 months	0.0	0.0	0.0	0.0	0.0	100.0	14	33
Moctezuma-Velazquez et al., **PSC** group [21]	LSM ≥ 10 kPa PBC and PSC only	80	44.8 (16.9)	68.0	N/A	[b]MELD 9.6 (5.3)	12 months	0.0	0.0	0.0	0.0	0.0	100.0	12	36
Nawalerspanya et al., [22]	Age ≥ 18 yo Compensated cirrhosis	128	57.4 (11.3)	60.2	23.5 (2.1)	100%	6 months	N/A	37.5	32.8	4.7	5.5	19.5	7.8	N/A
Petta et al., [23]	Cirrhosis LSM >11.5 (M) LSM > 11.0 (XL)	790	62 [10]	55.0	32.6 (6.7)	100%	6 months	0.0	0.0	0.0	0.0	100.0	0.0	11.5	31.3
Protopapas et al., [24]	cACLD LSM >12.	107	63.7 (12.1)	60.7	N/A	5.7 (0.2)	6 months	45.8	N/A	N/A	24.2	N/A	31.8	20.6	47.7
Sharma et al., [25]	cACLD LSM ≥10 kPa	895	41.4	71.3	N/A	100% cACLD	3 months	55.3	19.1	36.2	19.1	21.9	3.1	29.5	56.0
Silva et al., [26]	LSM >12.5 Compensated cirrhosis	97	54.3 (11.2)	76.3	N/A	100%	12 months	0.0	78.4	3.1	8.2	N/A	10.3	14.4	44.3

Sousa et al., [27]	cACLD	104	57.0	69.2	N/A	N/A	12 months	0.0	79.8	3.8	11.5	N/A	4.8	8.6	25.0
Stanislas et al., [28]	cACLD LSM ≥ 11 kPa	60	48.8	75.0	22.4	66.7	N/A	100.0	0	100.0	0	0	0	26.7	40.0
Stefanescu et al., [29]	cACLD	185	59[a]	35.0	26.0[a]	MELD 9.2	6 months	66.9	59.6	7.3	30.4	N/A	2.7	23.2	N/A
Thabut et al., [30]	cACLD	891	53.9[a]	67.5	25.5[b]	100	1 year	0.0	81.0	16.6	N/A	N/A	N/A	8.1	24.7
Tosetti et al., [31]	cACLD LSM ≥ 10 kPa	442	60[a]	64.2	25.4[b]	100%	1 year	79.4	68.8	10.6	5.9	14.7	0	31.2	7.0
Wang et al., [32]	HBV cirrhosis	341	48[a]	82.4	23.4[a]	94.7%	<2 days	100.0	0	100.0	0	0	0	20.5	61.9
Wong et al., [33]	cACLD Compensated cirrhosis	267	58 [11]	72.3	24.8 (3.7)	100	2 weeks	0.0	7.7	77.7	N/A	7.3	7.3	13.9	N/A

[a] Median
[b] MELD score

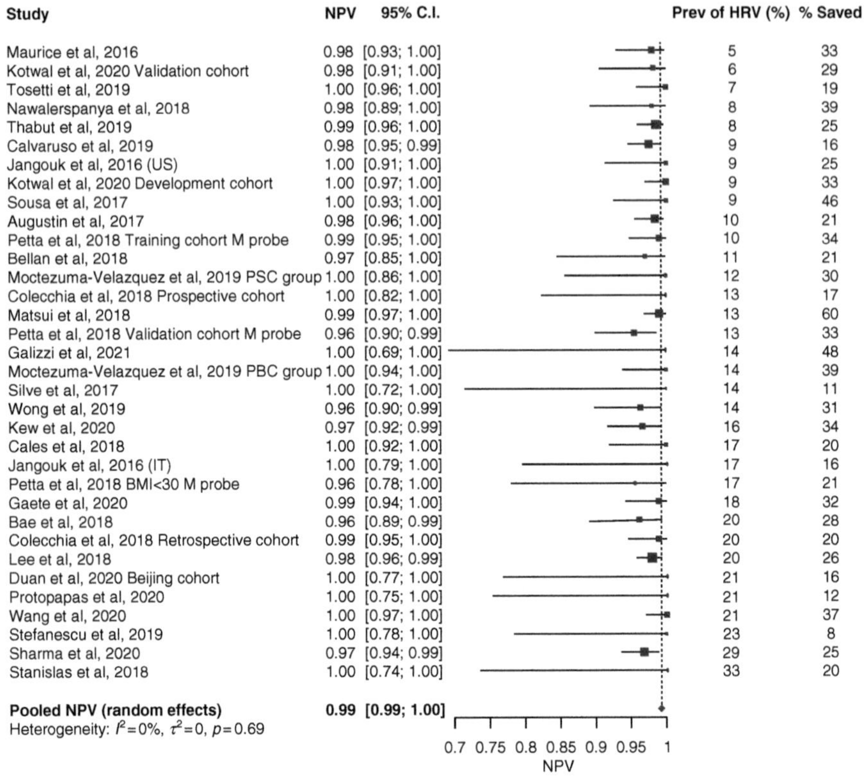

Fig. 10.2 Forest plot showing NPVs of the Baveno VI validation studies. Studies were ordered by prevalence. Prevalence and proportion of saved endoscopies are shown in the right columns. Meta-analysis was performed after double arcsine transformation of the proportions, pooled with random effects. *Note*: Calés et al. 2018 study data was extracted from the following systematic review, since more complete data was obtained from the authors [34]

Impact of Etiology on the Performance of Baveno VI Criteria

Even if there was no heterogeneity in the NPVs of Baveno VI across studies, to confirm that Baveno VI criteria perform well across etiologies, we conducted a subgroup meta-analysis. Only 12 of the 28 studies performed etiology-specific analysis. Figure 10.3 demonstrates the forest plot of the studies. There were no significant differences in NPVs according to etiology subgroups.

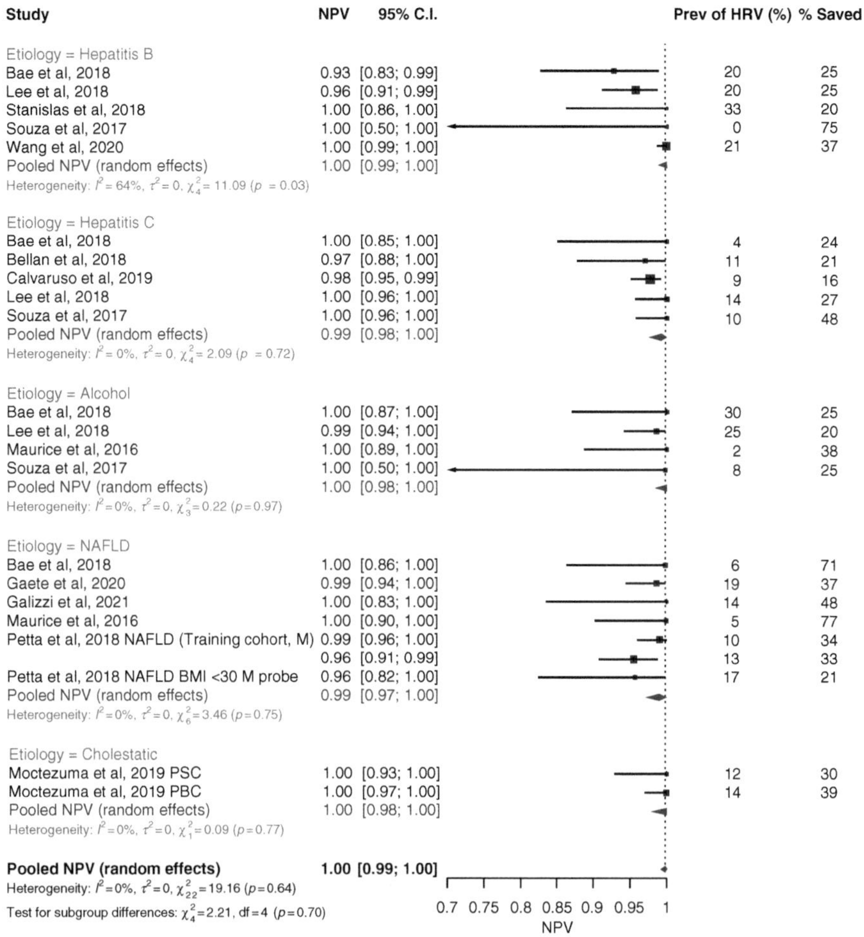

Fig. 10.3 Etiology-specific pooled negative predictive values of Baveno VI criteria

Can Baveno VI Criteria Be Expanded?

The Baveno VI criteria is undoubtedly a validated tool to select for low-risk cACLD patients who can safely avoid a surveillance gastroscopy. However, it has been suggested that the number of saved endoscopies is low. As stated above, this is an unsound metric to compare different studies since it largely depends on how early in the natural history of cACLD these criteria are applied.

In an attempt to increase the proportion of saved endoscopies, the Expanded Baveno VI criteria were proposed after the Baveno VI conference, in which the

LSM by TE threshold was increased to 25 kPa, and the platelet threshold decreased to 110 × 10^9/L [7]. A systematic search identified 16 studies assessing the expanded Baveno VI criteria, and results are shown in Fig. 10.4. Pooled NPV was 0.97 (95% CI 0.95–0.98). However, distinct from Baveno VI criteria, performance of Expanded Baveno VI showed significant heterogeneity ($p < 0.0001$). Results of the meta-analysis of sensitivities are provided in Supplementary data 11.3 (pooled sensitivity 0.90; 95% CI 0.87–0.93).

To address the sources of heterogeneity, we first evaluated whether etiology was associated with different performance of the expanded criteria. Eight studies showed etiology-specific data. Subgroup meta-analysis did not show any differences in performance across etiologies (Fig. 10.5).

We then evaluated the impact of the prevalence of HRV on the performance of Expanded Baveno VI criteria. The group of patients within the Expanded Baveno VI criteria comprises those who are within Baveno VI, and those beyond Baveno VI. The latter is the group that either shows a LSM of 20–25 kPa or a platelet count between 110 and 150. The pointwise risk of HRV of a LSM of 25 and a platelet count of 110 according to the ANTICIPATE model is ~12% [4]. Therefore, patients beyond Baveno VI but within Expanded Baveno VI would have a predicted risk of HRV between 5% and 12%. The prevalence of varices in patients within the Expanded Baveno VI criteria would depend largely on the distribution of the patients

Study	NPV	95% C.I.	Prev of HRV (%)	% Saved
Kotwal et al, 2020 Validation cohort	0.99	[0.94; 1.00]	6	49
Tosetti et al, 2019	1.00	[0.98; 1.00]	7	44
Nawalerspanya et al, 2018	0.97	[0.91; 1.00]	8	59
Calvaruso et al, 2019	0.97	[0.96; 0.99]	9	36
Kotwal et al, 2020 Development cohort	0.99	[0.96; 1.00]	9	53
Augustin et al, 2017	0.98	[0.96; 0.99]	10	40
Petta et al, 2018 Training cohort M probe	0.96	[0.92; 0.98]	10	58
Moctezuma-Velazquez et al, 2019 PSC group	0.97	[0.85; 1.00]	12	45
Petta et al, 2018 Validation cohort M probe	0.96	[0.92; 0.98]	13	54
Galizzi et al, 2021	0.93	[0.68; 1.00]	14	71
Moctezuma-Velazquez et al, 2019 PBC group	0.94	[0.87; 0.98]	14	59
Cales et al, 2018	0.99	[0.93; 1.00]	17	37
Petta et al, 2018 BMI <30 M probe	0.87	[0.75; 0.95]	17	49
Gaete et al, 2020	0.97	[0.93; 0.99]	18	52
Bae et al, 2018	0.93	[0.88; 0.97]	20	52
Lee et al, 2018	0.94	[0.92; 0.96]	20	44
Duan et al, 2021 Beijing cohort	0.95	[0.76; 1.00]	21	23
Protopapas et al, 2020	0.97	[0.83; 1.00]	21	28
Wang et al, 2020	0.96	[0.93; 0.98]	21	61
Stefanescu et al, 2019	0.89	[0.75; 0.96]	23	24
Sharma et al, 2020	0.93	[0.90; 0.95]	29	44
Pooled NPV (random effects)	**0.97**	**[0.95; 0.98]**		

Heterogeneity: $I^2 = 71\%$, $\tau^2 = 0.0038$, $p < 0.01$

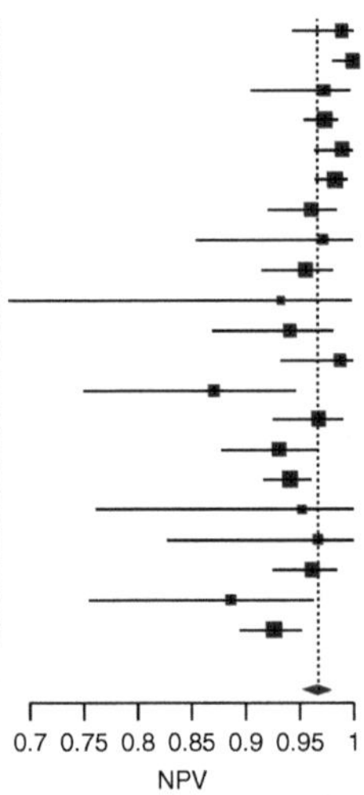

Fig. 10.4 Forest plot showing NPVs of Expanded Baveno VI validation studies. Studies were ordered by prevalence. Methodology to pool the NPVs was similar to that shown in Fig. 10.2

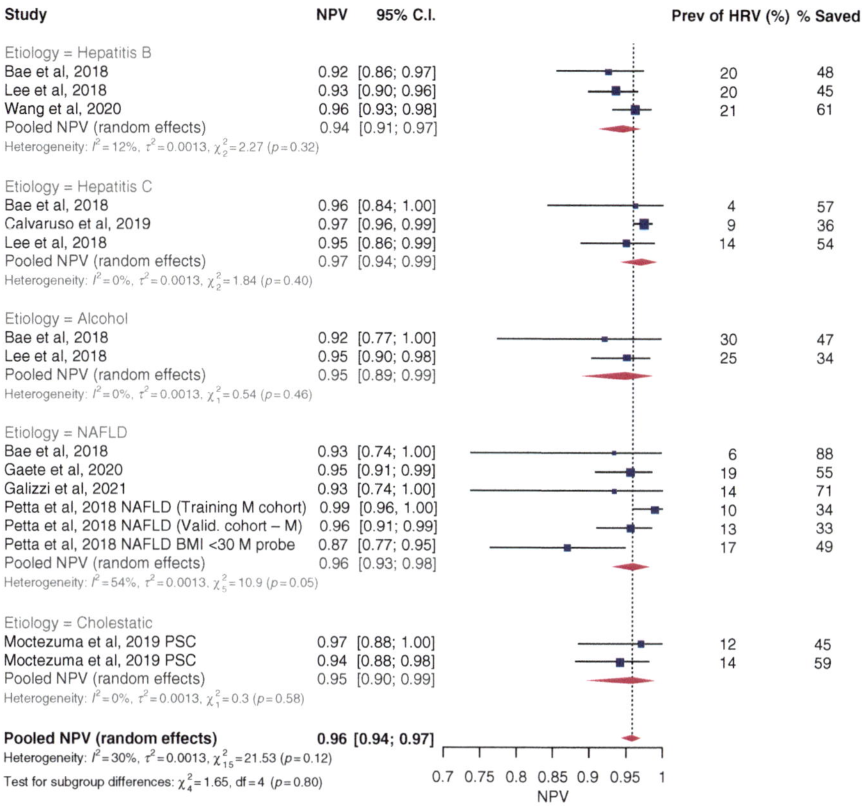

Fig. 10.5 Stratified meta-analysis of NPVs of Expanded Baveno VI criteria according to etiology. There were no significant subgroup differences across different etiologies

in those two groups (within Baveno VI and beyond Baveno VI). We therefore predicted that in series with higher prevalence of HRV, that would predictably have a higher number of patients beyond Baveno VI, the NPV of Expanded Baveno VI would decrease.

To assess this hypothesis, we performed a meta-regression analysis of NPV on the prevalence of HRV. There was a strong association between NPV and prevalence of HRV (Fig. 10.6a), with prevalence explaining 77% of the heterogeneity in NPVs observed across studies. There was no significant association between prevalence of HRV and the NPV of original Baveno VI (Fig. 10.6b), which is likely explained by the fact that all patients within Baveno VI have a theoretical <5% risk of HRV [4].

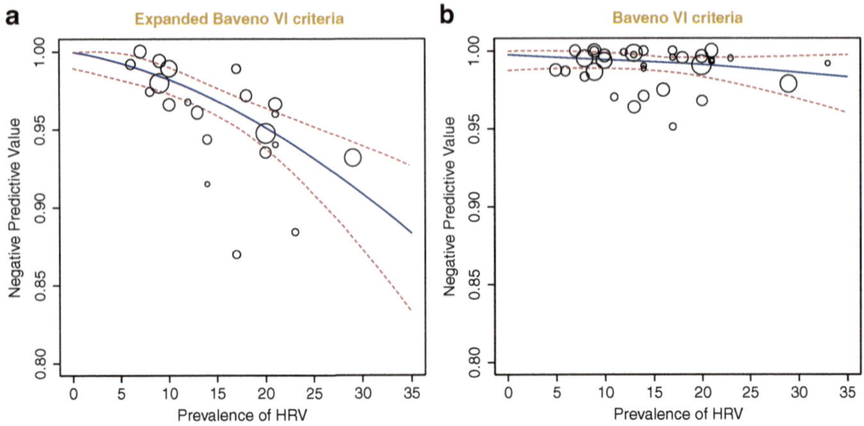

Fig. 10.6 Meta-regression assessing the association between prevalence of HRV and NPV of the Expanded Baveno VI criteria (**a**) and Baveno VI criteria (**b**)

Other Elastography Methods and Baveno VI Criteria

Point shear wave elastography (pSWE) and 2D-SWE have witnessed increased use in the last few years [35]. The main unsolved issue with these methods is the multiplicity of devices with proprietary algorithms that lead to differences in the quantification of the speed of shear wave, and consequently provide values of liver stiffness that are not identical [35–39]. Therefore, the same liver stiffness thresholds defined for TE cannot be directly applied to pSWE or 2D-SWE [40].

Methods Beyond Baveno VI and Expanded Baveno VI Criteria

Several other models have been proposed for the noninvasive prediction of HRV, including the use of spleen stiffness, spleen diameter or blood-based tests, with only limited or no external validation. Several of these models are reviewed in specific chapters of the book.

Summary and Conclusions

The Baveno VI criteria have been extensively validated as a decision rule for not doing an endoscopy in patients with compensated cirrhosis. The pooled NPV in series with a wide range of prevalences of HRVs (from 5% to 33%) is 99% (95% CI 99%–100%). The expanded Baveno VI cannot be recommended at the present time in any etiology.

References

1. de Franchis R. Revising consensus in portal hypertension: report of the Baveno V consensus workshop on methodology of diagnosis and therapy in portal hypertension. J Hepatol. 2010;53(4):762–8.
2. Garcia-Tsao G, Abraldes JG, Berzigotti A, Bosch J. Portal hypertensive bleeding in cirrhosis: risk stratification, diagnosis, and management: 2016 practice guidance by the American Association for the Study of Liver Diseases. Hepatology. 2017;65(1):310–35.
3. Augustin S, Pons M, Genesca J. Validating the Baveno VI recommendations for screening varices. J Hepatol. 2017;66(2):459–60.
4. Abraldes JG, Bureau C, Stefanescu H, Augustin S, Ney M, Blasco H, et al. Noninvasive tools and risk of clinically significant portal hypertension and varices in compensated cirrhosis: the 'Anticipate' study. Hepatology. 2016;64(6):2173–84.
5. Calvaruso V, Cacciola I, Licata A, Madonia S, Benigno R, Petta S, et al. Is transient elastography needed for noninvasive assessment of high-risk varices? The REAL experience. Am J Gastroenterol. 2019;114(8):1275–82.
6. Villanueva C, Albillos A, Genescà J, Garcia-Pagan JC, Calleja JL, Aracil C, et al. β blockers to prevent decompensation of cirrhosis in patients with clinically significant portal hypertension (PREDESCI): a randomised, double-blind, placebo-controlled, multicentre trial. Lancet. 2019;393(10181):1597–608.
7. Augustin S, Pons M, Maurice JB, Bureau C, Stefanescu H, Ney M, et al. Expanding the Baveno VI criteria for the screening of varices in patients with compensated advanced chronic liver disease. Hepatology. 2017;66(6):1980–8.
8. Bae J, Sinn DH, Kang W, Gwak G-Y, Choi MS, Paik Y-H, et al. Validation of the Baveno VI and the expanded Baveno VI criteria to identify patients who could avoid screening endoscopy. Liver Int. 2018;38(8):1442–8.
9. Bellan M, Sainaghi PP, Minh MT, Minisini R, Molinari L, Baldrighi M, et al. Gas6 as a predictor of esophageal varices in patients affected by hepatitis C virus related-chronic liver disease. Biomark Med. 2018;12(1):27–34.
10. Calès P, Sacher-Huvelin S, Valla D, Bureau C, Olivier A, Oberti F, et al. Large oesophageal varice screening by a sequential algorithm using a cirrhosis blood test and optionally capsule endoscopy. Liver Int. 2018;38(1):84–93.
11. Colecchia A, Ravaioli F, Marasco G, Colli A, Dajti E, Di Biase AR, et al. A combined model based on spleen stiffness measurement and Baveno VI criteria to rule out high-risk varices in advanced chronic liver disease. J Hepatol. 2018;69(2):308–17.
12. Duan Z, Li L, Li J, Zhou S. Validation of the combined model based on platelet count and albumin to rule out high-risk varices in liver cirrhosis. Biomed Res Int. 2020;2020:5783748.
13. Gaete MI, Díaz LA, Arenas A, González K, Cattaneo M, Fuster F, et al. Baveno VI and Expanded Baveno VI criteria successfully predicts the absence of high-risk gastro-oesophageal varices in a Chilean cohort. Liver Int. 2020;40(6):1427–34.
14. Galizzi HO, Couto CA, Taranto DOL, Araújo SIO, Vilela EG. Accuracy of non-invasive methods/models for predicting esophageal varices in patients with compensated advanced chronic liver disease secondary to nonalcoholic fatty liver disease. Ann Hepatol. 2021;20:100229.
15. Jangouk P, Turco L, De Oliveira A, Schepis F, Villa E, Garcia-Tsao G. Validating, deconstructing and refining Baveno criteria for ruling out high-risk varices in patients with compensated cirrhosis. Liver Int. 2017;37(8):1177–83.
16. Kew GS, Chen ZJ, Yip AW, Huang YWC, Tan LY, Dan YY, et al. Identifying patients with cirrhosis who might avoid screening endoscopy based on serum albumin and bilirubin and platelet counts. Clin Gastroenterol Hepatol. 2021;19(1):199–201.e2.
17. Kotwal V, Mbachi C, Wang Y, Attar B, Randhawa T, Flores E, et al. A novel score to predict esophageal varices in patients with compensated advanced chronic liver disease. Dig Dis Sci. 2021;66(6):2084–91.

18. Lee HA, Kim SU, Seo YS, Lee Y-S, Kang SH, Jung YK, et al. Prediction of the varices needing treatment with non-invasive tests in patients with compensated advanced chronic liver disease. Liver Int. 2019;39(6):1071–9.
19. Matsui N, Imajo K, Yoneda M, Kessoku T, Honda Y, Ogawa Y, et al. Magnetic resonance elastography increases usefulness and safety of non-invasive screening for esophageal varices. J Gastroenterol Hepatol. 2018;33(12):2022–8.
20. Maurice JB, Brodkin E, Arnold F, Navaratnam A, Paine H, Khawar S, et al. Validation of the Baveno VI criteria to identify low risk cirrhotic patients not requiring endoscopic surveillance for varices. J Hepatol. 2016;65(5):899–905.
21. Moctezuma-Velazquez C, Saffioti F, Tasayco-Huaman S, Casu S, Mason A, Roccarina D, et al. Non-invasive prediction of high-risk varices in patients with primary biliary cholangitis and primary sclerosing cholangitis. Am J Gastroenterol. 2019;114(3):446–52.
22. Nawalerspanya S, Sripongpun P, Chamroonkul N, Kongkamol C, Piratvisuth T. Validation of original, expanded Baveno VI, and stepwise & platelet-MELD criteria to rule out varices needing treatment in compensated cirrhosis from various etiologies. Ann Hepatol. 2020;19(2):209–13.
23. Petta S, Sebastiani G, Bugianesi E, Viganò M, Wong VW-S, Berzigotti A, et al. Non-invasive prediction of esophageal varices by stiffness and platelet in non-alcoholic fatty liver disease cirrhosis. J Hepatol. 2018;69(4):878–85.
24. Protopapas AA, Mylopoulou T, Papadopoulos VP, Vogiatzi K, Goulis I, Mimidis K. Validating and expanding the Baveno VI criteria for esophageal varices in patients with advanced liver disease: a multicenter study. Ann Gastroenterol. 2020;33(1):87–94.
25. Sharma S, Agarwal S, Gunjan D, Kaushal K, Anand A, Saraya A. Deciding among noninvasive tools for predicting varices needing treatment in chronic liver disease: an analysis of Asian cohort. Am J Gastroenterol. 2020;115(10):1650–6.
26. Silva MJ, Bernardes C, Pinto J, Loureiro R, Duarte P, Mendes M, et al. Baveno VI recommendation on avoidance of screening endoscopy in cirrhotic patients: are we there yet? GE Port J Gastroenterol. 2017;24(2):79–83.
27. Sousa M, Fernandes S, Proença L, Silva AP, Leite S, Silva J, et al. The Baveno VI criteria for predicting esophageal varices: validation in real life practice. Rev Esp Enferm Dig. 2017;109(10):704–7.
28. Stanislas DA, Constant A, Wilson NNA, Dimitri KH, Demba B, Anzouan-Kacou K, et al. Liver transient elastography combined to platelet count (Baveno VI) predict high esophageal varices in Black African patient with compensated hepatitis B related cirrhosis. Open J Gastroenterol. 2018;8(5):192–200.
29. Stefanescu H, Marasco G, Calès P, Fraquelli M, Rosselli M, Ganne-Carriè N, et al. A novel spleen-dedicated stiffness measurement by FibroScan® improves the screening of high-risk oesophageal varices. Liver Int. 2020;40(1):175–85.
30. Thabut D, Bureau C, Layese R, Bourcier V, Hammouche M, Cagnot C, et al. Validation of Baveno VI criteria for screening and surveillance of esophageal varices in patients with compensated cirrhosis and a sustained response to antiviral therapy. Gastroenterology. 2019;156(4):997–1009.e5.
31. Tosetti G, Primignani M, La Mura V, D'Ambrosio R, Degasperi E, Mezzina N, et al. Evaluation of three 'beyond Baveno VI' criteria to safely spare endoscopies in compensated advanced chronic liver disease. Dig Liver Dis. 2019;51(8):1135–40.
32. Wang H, Wen B, Chang X, Wu Q, Wen W, Zhou F, et al. Baveno VI criteria and spleen stiffness measurement rule out high-risk varices in virally suppressed HBV-related cirrhosis. J Hepatol. 2021;74(3):584–92.
33. Wong GL-H, Liang LY, Kwok R, Hui AJ, Tse Y-K, Chan HL-Y, et al. Low risk of variceal bleeding in patients with cirrhosis after variceal screening stratified by liver/spleen stiffness. Hepatology. 2019 Sep;70(3):971–81.

34. Stafylidou M, Paschos P, Katsoula A, Malandris K, Ioakim K, Bekiari E, et al. Performance of Baveno VI and expanded Baveno VI criteria for excluding high-risk varices in patients with chronic liver diseases: a systematic review and meta-analysis. Clin Gastroenterol Hepatol. 2019;17(9):1744–1755.e11.
35. Piscaglia F, Salvatore V, Mulazzani L, Cantisani V, Schiavone C. Ultrasound shear wave elastography for liver disease. A critical appraisal of the many actors on the stage. Ultraschall Med. 2016;37(1):1–5.
36. Cassinotto C, Lapuyade B, Guiu B, Marraud des Grottes H, Piron L, Merrouche W, et al. Agreement between 2-dimensional shear wave and transient elastography values for diagnosis of advanced chronic liver disease. Clin Gastroenterol Hepatol. 2020;18(13):2971–2979.e3.
37. Ronot M, Ferraioli G, Müller H-P, Friedrich-Rust M, Filice C, Vilgrain V, et al. Comparison of liver stiffness measurements by a 2D-shear wave technique and transient elastography: results from a European prospective multi-centre study. Eur Radiol. 2021;31(3):1578–87.
38. Conti F, Serra C, Vukotic R, Felicani C, Mazzotta E, Gitto S, et al. Assessment of liver fibrosis with elastography point quantification vs other noninvasive methods. Clin Gastroenterol Hepatol. 2019;17(3):510–517.e3.
39. Ferraioli G, De Silvestri A, Lissandrin R, Maiocchi L, Tinelli C, Filice C, et al. Evaluation of inter-system variability in liver stiffness measurements. Ultraschall Med. 2019;40(1):64–75.
40. Kang SH, Baik SK, Kim MY. Application of Baveno criteria and modified Baveno criteria with shear-wave elastography in compensated advanced chronic liver disease. J Korean Med Sci. 2020;35(30):e249.

Liver Elastography for Prognostication and Monitoring Patients With Compensated Advanced Chronic Liver Disease

11

Maja Thiele

Introduction

Patients with chronic liver disease (CLD) worry about their health: Will their condition deteriorate to symptomatic complaints, decompensation, and life-threatening disease? Will they ever experience improvement? [1] A diagnosis of compensated advanced chronic liver disease (cACLD) worries patients due to the risk of developing symptoms of decompensation that affects daily living, leads to frequent hospital visits, the need for pharmacological therapy, invasive interventions, and worsening in the mental and physical aspects of health-related quality of life. These aspects of chronic liver disease hold more clinical relevance than the diagnosis itself [2].

Baveno VI established the use of liver stiffness measurements (LSM) by transient elastography (TE) to stratify patients with CLD according to their probability of having cACLD, with 10 kPa as the rule out cutoff, and 15 kPa for ruling in cACLD. Baveno VII marks a shift from diagnosis to prognosis, thereby focusing directly on the quality and length of patients' lives. The change from a diagnostic to a prognostic focus is possible due to evidence from meta-analyses and high-quality prospective cohorts, showing the prognostic accuracy of liver stiffness in patients with CLD [3–12]. Most evidence concerns the major liver disease etiologies (HVC, HBV, NAFLD, ALD), but there is also evidence of a comparable prognostic accuracy of TE in more rare CLD etiologies such as primary biliary cholangitis and primary sclerosing cholangitis [13, 14].

M. Thiele (✉)
FLASH Center for Liver Research, Department of Gastroenterology and Hepatology, Odense University Hospital, Odense, Denmark

Department for Clinical Medicine, University of Southern Denmark, Odense, Denmark
e-mail: Maja.thiele@rsyd.dk

R. de Franchis (ed.), *Portal Hypertension VII*,
https://doi.org/10.1007/978-3-031-08552-9_11

Other elastography techniques than TE (point and two-dimensional shear-wave elastography, pSWE, 2D-SWE) also possess prognostic ability, but the generalizability of those studies is limited by heterogeneity in elastography techniques, cutoffs, and study populations [7, 15, 16]. Further, the pSWE and 2D-SWE elastography systems offered by several manufacturers are not comparable due to differences in both software and hardware [17]. Similarly, it is important to stress that LSM values by TE are not comparable to pSWE or 2D-SWE values [18]. It is therefore not currently possible to make recommendations regarding prognostication with elastography equipment other than transient elastography.

Liver Stiffness by Transient Elastography as a Prognostic Tool

Transient elastography provides a continuous measure of liver stiffness, with increasing liver stiffness indicating higher risk of decompensation and mortality. The dose–response relationship between liver stiffness and outcomes is however not linear, as indicated by two meta-analyses [3, 4]. Both studies find that the relative risk of liver-related events and all-cause mortality increases substantially in patients with LSM above 10 kPa, whereas the slope wanes off after 25 kPa, marking the point where other factors become more important than liver stiffness for progression of portal hypertension and liver dysfunction.

The generalizability of the available meta-analyses is limited by the fact that they were generated from a majority of studies on chronic viral hepatitis: In the most recent, 46% of studies investigated HCV, 32% HBV, while 22% of publications studied a mixed population [3]. Further, not all included patients have cACLD; many have LSM < 10 kPa, others are decompensated at the time of inclusion.

Fortunately, several recent, high-quality single-etiology studies in NAFLD, ALD, or HCV confirm the good prognostic accuracy of baseline LSM by TE to predict decompensation and mortality, all-cause or liver-related [6–9, 11, 12, 19–23]. The cutoffs reported in the various studies converge on roughly four particular points of LSM: 10, 15, 20, and 25 kPa (see Table 11.1). This leads to the "*rule of five,*" as an easy-to-use rule of thumb for the assessment of the relative risk of decompensation or liver-related mortality in a patient with chronic liver disease. The risk of decompensation within 2–5 years is negligible if LSM is below 10 kPa, after which the relative risk increases in steps of 5 kPa.

It is only possible to make generalizations across liver disease etiologies regarding the relative risks of decompensation and death. This is due to large differences in the incidence of decompensation and death between individual disease etiologies. For example, reports in alcohol-related liver disease indicate an 8–10-times higher rate of liver-related mortality than in NAFLD [19, 24].

In CLD patients with decompensation, there are more accurate prognostic scores than LSM, typically the model for end-stage liver disease [25]. Liver stiffness therefore has no current role in patients with decompensated CLD, except for addressing decompensation (see Chap. 47). While some studies indicate a benefit of combining MELD and LSM for prediction of further decompensation, this concept needs to be validated [15, 26].

Table 11.1 Table of evidence for individual studies assessing the prognostic ability of liver stiffness measurements with transient elastography to predict risk of liver-related events, decompensation, and/or mortality

Study	Etiology	N	Events/endpoint	Follow-up	Prognostic intervals			
					<10 kPa	10–15 kPa	15–20 kPa	≥20 kPa
Rasmussen 2021[a] [7]	ALD	462	Total events: 87 LRE	Median 4.1 years (IQR 31–70 months)	Cumulative incidence: 0.2% PY 3 years: 1.1% Event rate: 11/304 (4%)	Cumulative incidence: 2.0% PY 3 years: 10.2% Event rate: 9/42 (21%)	Cumulative incidence: 8.0% PY. 3 years: 24% Event rate: 5/15 (33%)	Cumulative incidence: 10% PY. 3 years 36% Event rate: 48/81 (59%)
Genesca 2021[a]	NAFLD	2638	Total events: 45 LRE	Median 2.2 years (IQR 1.8–2.9)	Cumulative incidence: 0.04% PY. 3 years: 0.11% Event rate: 2/1820 (0.1%)	Cumulative incidence: 0.3% PY. 3 years: 1.03% Event rate: 4/479 (0.8%)	Cumulative incidence (≥15 kPa): 4.2% PY. 3 years: 11.16% Event rate: 8/136 (5.9%)	Event rate: 31/203 (15.3%)
Decraecker 2021 [19]	NAFLD ($n = 1698$), ALD ($n = 1667$)	3365	Total events: 563 deaths (510 ALD, 53 NAFLD)	Median 4.5 years (IQR 2.5–7.2)	Cumulative incidence[b] at 3 years: 3%	Cumulative incidence[b] at 3 years: 5%	Cumulative incidence[b] at 3 years: 15%	Cumulative incidence[b] at 3 years: 28%

Study	Etiology	*N*	Events/endpoint	Follow-up	Other cutoffs		
Liu 2021 [12]	cACLD	661	Decompensation, total events not stated	Median 3.4 years (IQR 2.3–5.1)	<u>10–19.9 kPa</u> sHR 9.8 if platelet count <150 × 10^9/L	<u>20–25 kPa</u> sHR 16.8	<u>≥25 kPa</u> sHR 38.0
Mendoza 2021 [11]	NAFLD with cACLD	233	Total events: 14 LRE	Median 1.4 years	<u>10–21 kPa</u> Events 3/147 (2%)	<u>>21 kPa</u> Events 11/86 (13%)	

(continued)

Table 11.1 (continued)

Study	Etiology	*N*	Events/endpoint	Follow-up	Other cutoffs		
Petta 2021 [8]	NAFLD with cACLD	1039	Total events: 71 decompensations	Median 2.9 years (IQR 1.6–5.3)	10–20.9 kPa Event rate: 2% decompensations	≥21 kPa Event rate: 14% decompensations. Cumulative incidence at	
Poynard 2014 [23]	HCV	3031	Total events: 104 deaths	Median 5.2 years	<9.5 kPa 39/2344 (1.7%) event rate	9.5–20 kPa 29/486 (6.0%) event rate	≥20 kPa 36/201 (18%) event rate
Robic 2011 [22]	Mixed (65% cirrhosis, 38% ALD)	100	Total events: 18 decompensations	Mean 1.3 years ±0.8	<21.1 kPa: Event rate: 0/57 (0%)	≥21.1 kPa Event rate: 18/43 (53%)	
Shili-Masmoudi 2020 [9]	NAFLD	2245	Total events: 55 deaths, 3 LTX, 21 LRE (decomp. And HCC)	Median 2.3 years (IQR 25–38 months)	≤12 kPa Cumulative incidences: Death at 1-3-5 years: 0.5%-2.9%-3.4% LRE at 1-3-5 years: 0.2%-0.2%-0.3%	>12 kPa Cumulative incidences: Death at 1-3-5 years: 2.0%-9.1%-13.8% LRE at 1-3-5 years: 2.1%-2.5%-10.2%	

Table of evidence for individual studies assessing the prognostic ability of liver stiffness measurements with transient elastography to predict risk of liver-related events, decompensation, and/or mortality. Focusing only on studies which report cutoffs close to 10, 15, 20, and/or 25 kPa

ALD alcohol-related liver disease, *cACLD* compensated advanced chronic liver disease, *HCC* hepatocellular carcinoma, *HCV* hepatitis C, *LRE* liver-related events, *LTX* liver transplantation, *NAFLD* non-alcoholic fatty liver disease, *PY* person-years

[a] Unpublished data, analyses done for Baveno VII conference

[b] Based on KM-curves

How to Monitor Patients with Chronic Liver Disease Using Liver Stiffness

There are three clinical scenarios where monitoring patients with chronic liver disease using LSM by TE is of relevance: (A) In patients with elevated liver stiffness, but baseline LSM below 10 kPa threshold for ruling out cACLD. (B) In patients with baseline LSM by TE ≥10 kPa, to control for false positives. (C) In the management of cACLD patients, where LSM is monitored to guide decision-making during outpatient care.

Given the very low rate of decompensation in patients with LSM < 10 kPa from studies with follow-up periods spanning 2 to 5 years, it is probably safe to monitor patients with LSM 7–9.9 kPa every 3 to 5 years [20]. However, management should be on a case-by-case basis. In a mixed etiology cohort study of CLD patients, time-dependent ROC curves showed that the optimal predictive performance of LSM by TE lasted 2–3 years in patients with LSM < 6.7 kPa; compared to 1 year in patients with LSM 6.7–17.6 kPa [27]. A long time interval, or no follow-up in older people, seems relevant in patients at low risk of liver fibrosis progression, whereas patients with several risk factors for progression and LSM close to 10 kPa should probably be monitored more closely.

Due to the risk of false positives, an elevated index LSM should be repeated in a fasting state when feasible [28, 29]. Two consecutively elevated measurements increase sensitivity in both ALD and NAFLD [30, 31]. As the sensitivity of diagnostic tests is always lower in low-prevalence populations, this is particularly important in case LSM is used in primary care or the general population, for referral pathways [32]. If there are reasons to doubt the validity of the index LSM, investigators may also consider a confirmatory test with a blood-based biomarker (Table 11.2). This is in accordance with guidelines on noninvasive tests [33]. In head-to-head comparisons, though, LSM has a better positive-predictive value than both FIB-4, the ELF test, FibroTest or similar serum tests [34, 35].

Patients with cACLD may be monitored using annual LSM measurements, if the longitudinal measurements have implications for patient management, using 12 months as a feasible and preferred interval. Of seven studies evaluating longitudinal LSM by TE, three use annual TE, three repeat after 3 years, and one after 6–12 months (see Table 11.3). In addition, almost two-thirds of the Baveno VII faculty prefer annual monitoring over other time intervals (see Chap. 7).

Table 11.2 Suggested blood-based biomarkers and their cutoffs which can be used complementary to index LSM

	ALD		NAFLD		Viral	
Diagnosis of ≥F3	Sens	Spec	Sens	Spec	Sens	Spec
ELF ≥9.8	89%	77%	65%	86%	60%	91%
FibroTest ≥0.58	66%	89%	–	–	67%	88%
FibroTest ≥0.48	75%	86%	37%	90%	–	–
FIB-4 ≥ 2.67	70%	89%	30%	94%	–	–

Suggested blood-based biomarkers and their cutoffs which can be used complementary to index LSM. Selected based on diagnostic studies using biopsy-controlled advanced fibrosis as outcome [34, 36–40]. The cutoffs also show prognostic accuracy [5, 6, 11, 38, 41, 42]

Table 11.3 Studies that have investigated the prognostic value of longitudinal LSM

Study	Etiology	N	Endpoint	Follow-up	Changes in LSM as prognostic indicator
Wang 2014 [21]	93% viral	220	Portal hypertension progression	Median 37 months LSM every 6–12 months	Baseline LSM < 17 kPa and no worsening: 11/149 (7%) events Baseline LSM < 17 kPa but worsening: 2/12 (17%) events Baseline LSM ≥ 17 kPa regardless of LSM during FU: 17/59 (29%) events
Vergniol 2014 [44]	HCV	1025	Death or LTX	Median 38 months. LSM after 3 years	<7 kPa or 7–14 kPa without worsening: Very low cumulative risk (4 years <5%). 7–14 kPa and worsening, or ≥ 14 kPa and improvement: Moderate cumulative risk (4 years 20%) ≥14 kPa and increase: High cumulative risk (4 years 50%)
Kamaraj 2018 [45]	NAFLD	90	Liver-related events	Median 37 months LSM after 1 year	All four events happened in patients with LSM ≥ 15 kPa at baseline and no improvement
Pons 2019 [10]	HCV after DAA, baseline LSM ≥ 10 kPa	572	Portal hypertension-related events	Median 2.9 years LSM after 1 year	All seven patients with portal hypertension related events had LSM > 20 kPa at baseline and 4/5 (80%) did not improve ≥20% from baseline during FU

Table 11.3 (continued)

Study	Etiology	N	Endpoint	Follow-up	Changes in LSM as prognostic indicator
Semmler 2021 [43]	HCV after DAA	276	12 with hepatic decompensation, 5 liver-related deaths	Median 37 months	Baseline LSM cutoff of 25 kPa for predicting decompensation. Patients without decompensation decreased on average 21% in LSM, versus a 22% increase in patients with decompensation. LSM at follow-up ≤12.4 kPa: No decompensations LSM at follow-up 12.4–25.3 kPa: 2.6% 3-year cumulative risk of decompensation LSM at follow-up ≥25.3 kPa: 17.4% 3-year cumulative risk of decompensation
Rasmussen 2021[a]	ALD	219	Liver-related events	Median 49 months LSM after 3.1 years (IQR 2.1–4.1)	If LSM < 10 kPa at follow-up, regardless of baseline: 1/167 (0.6%) events. Baseline LSM < 10 kPa and worsening to LSM ≥ 10: 1/10 (10%) events Baseline LSM ≥ 10 kPa, but improvement ≥20% and LSM < 20 kPa at follow-up; or decrease to LSM <10 kPa: 3/178 (1.7%) LRE. Baseline LSM ≥ 10 kPa and no response: 7/39 (18%) LRE

(continued)

Table 11.3 (continued)

Study	Etiology	N	Endpoint	Follow-up	Changes in LSM as prognostic indicator
Petta 2021 [8]	NAFLD	533 with cACLD	Decompensation, HCC and liver-related death	Median 35 months (19–63). LSM after 1 year	Baseline and delta-LSM both predicted liver decompensation. Delta-LSM also predicted all-cause mortality. Improvement (>20% reduction in LSM): 3.8% decompensation event rate (0% if baseline LSM < 21 kPa) Stable (between 20% reduction and 20% increase in LSM): 6.2% decompensation event rate (3.2% if baseline LSM < 21 kPa). Impairment (>20% increase in LSM): 14.4% decompensation rate (10% if LSM < 21 kPa at baseline)

ALD alcohol-related liver disease, *cACLD* compensated advanced chronic liver disease, *DAA* direct acting antivirals, *HCC* hepatocellular carcinoma, *HCV* hepatitis C, *LRE* liver-related events, *LSM* liver stiffness measurement by TE, *LTX* liver transplantation, *NAFLD* non-alcoholic fatty liver disease

[a] Unpublished data, analyses done for Baveno VII conference based on data from [7]

The Clinical Relevance of Changes in Liver Stiffness in Patients with Chronic Liver Disease

There is a widespread availability of pharmaceutical, dietary, or psychosocial interventions that can attenuate or reverse liver disease progression: antivirals for chronic hepatitis, weight loss for NAFLD, and alcohol rehabilitation for ALD are the most common. Combined with a widespread availability of LSM, it has become a pressing need to map the prognostic relevance of longitudinal changes in liver stiffness. So far, six studies have investigated the prognostic value of longitudinal LSM (Table 11.3). Four studies investigated LSM in chronic viral hepatitis (two including

HCV patients before and after DAA), and another two are in NAFLD patients following the natural history of disease. A seventh set of analyses, on ALD, were conducted for the Baveno VII conference, but have not yet been published.

Some overall trends can be deduced from the six published monitoring studies. First, $a \geq 20\%$ change in LSM seems to be clinically relevant: two studies used it as a predefined endpoint, while a third study found an average 22% increase in LSM in HCV patients who decompensated during follow-up, while patients free of decompensation showed a 21% decrease in LSM [8, 10, 43]. Second, highly elevated liver stiffnesses at follow-up, above 17–25 kPa, result in a substantial risk of decompensation or death regardless of whether LSM improved or worsened from baseline. Consequently, a CLD patient with a LSM decrease of 20% or more is at very low risk of LRE, if the LSM at follow-up is below approximately 20 kPa. If the follow-up LSM in cACLD patients improves to below 10 kPa, the prognostic evidence shown first in this chapter indicates a negligible risk of decompensation, regardless of the proportional change.

An effective intervention to reverse disease progression in cACLD patients should result in a very low risk of liver-related events or liver-related mortality. Such a significant improvement in LSM may consequently be defined as a reduction ≥20% and LSM < 20 kPa, or any improvement to LSM < 10 kPa.

For Baveno VII, we tested this definition in a cohort of 219 ALD patients without decompensation at baseline, and repeated LSM measurements after 1–4 years (see Table 11.3). Of the patients with cACLD at baseline, 1.7% (3/178) experienced an event if LSM improved ≥20% and LSM was <20 kPa at follow-up, or if LSM decreased to <10 kPa. In comparison, 10% (1/10) of patients with baseline LSM < 10 kPa but worsening to LSM ≥ 10 experienced events, and so did 18% (7/39) of patients with cACLD and no substantial improvement in LSM. The suggested definition of a clinically relevant response in LSM is probably a conservative estimate, especially for NAFLD patients achieving weight decrease and HCV patients after DAA. Higher LSM values during follow-up monitoring may be acceptable for these patient groups.

An important exception to LSM as a monitoring tool is for changes in HVPG after non-selective beta-blockers or Carvedilol. Changes in LSM do not correlate with HVPG after NSBB, nor with a clinically significant response to NSBB [46, 47]. LSM can therefore not be recommended for evaluation of changes in portal pressure after NSBB treatment.

References

1. Younossi ZM, Boparai N, Price LL, Kiwi ML, McCormick M, Guyatt G. Health-related quality of life in chronic liver disease: the impact of type and severity of disease. Am J Gastroenterol. 2001;96(7):2199–205.
2. Rowe IA. Too much medicine: overdiagnosis and overtreatment of non-alcoholic fatty liver disease. Lancet Gastroenterol Hepatol. 2018;3(1):66–72.
3. Shen Y, Wu SD, Wu L, Wang SQ, Chen Y, Liu LL, et al. The prognostic role of liver stiffness in patients with chronic liver disease: a systematic review and dose-response meta-analysis. Hepatol Int. 2019;13(5):560–72.

4. Wang J, Li J, Zhou Q, Zhang D, Bi Q, Wu Y, et al. Liver stiffness measurement predicted liver-related events and all-cause mortality: a systematic review and nonlinear dose–response meta-analysis. Hepatol Commun. 2018;2(4):467–76.
5. Rhodes FA, Trembling P, Panovska-Griffiths J, Tanwar S, Westbrook RH, Rodger A, et al. Systematic review: investigating the prognostic performance of four non-invasive tests in alcohol-related liver disease. J Gastroenterol Hepatol. 2021;36(6):1435–49.
6. Boursier J, Vergniol J, Guillet A, Hiriart J-B, Lannes A, Le Bail B, et al. Diagnostic accuracy and prognostic significance of blood fibrosis tests and liver stiffness measurement by FibroScan in non-alcoholic fatty liver disease. J Hepatol. 2016;65(3):570–8.
7. Rasmussen DN, Thiele M, Johansen S, Kjærgaard M, Lindvig KP, Israelsen M, et al. Prognostic performance of 7 biomarkers compared to liver biopsy in early alcohol-related liver disease. J Hepatol. 2021;75(5):1017–25.
8. Petta S, Sebastiani G, Viganò M, Ampuero J, Wai-Sun Wong V, Boursier J, et al. Monitoring occurrence of liver-related events and survival by transient elastography in patients with nonalcoholic fatty liver disease and compensated advanced chronic liver disease. Clin Gastroenterol Hepatol. 2021;19(4):806–15.e5.
9. Shili-Masmoudi S, Wong GL, Hiriart JB, Liu K, Chermak F, Shu SS, et al. Liver stiffness measurement predicts long-term survival and complications in non-alcoholic fatty liver disease. Liver Int. 2020;40(3):581–9.
10. Pons M, Rodriguez-Tajes S, Esteban JI, Marino Z, Vargas V, Lens S, et al. Non-invasive prediction of liver related events in HCV compensated advanced chronic liver disease patients after oral antivirals. J Hepatol. 2020;72(3):472–80.
11. Mendoza YP, Shengir M, Bosch J, Sebastiani G, Berzigotti A. FIB-4 improves LSM-based prediction of complications in overweight or obese patients with compensated advanced chronic liver disease. Clin Gastroenterol Hepatol. 2021; https://doi.org/10.1016/j.cgh.2021.03.007.
12. Liu Y, Liu C, Li J, Kim TH, Enomoto H, Qi X. Risk stratification of decompensation using liver stiffness and platelet counts in compensated advanced chronic liver disease. J Hepatol. 2022;76(1):248–50.
13. Corpechot C, Carrat F, Poujol-Robert A, Gaouar F, Wendum D, Chazouilleres O, et al. Noninvasive elastography-based assessment of liver fibrosis progression and prognosis in primary biliary cirrhosis. Hepatology. 2012;56(1):198–208.
14. Corpechot C, Gaouar F, El Naggar A, Kemgang A, Wendum D, Poupon R, et al. Baseline values and changes in liver stiffness measured by transient elastography are associated with severity of fibrosis and outcomes of patients with primary sclerosing cholangitis. Gastroenterology. 2014;146(4):970–9.e6.
15. Trebicka J, Gu W, de Ledinghen V, Aubé C, Krag A, Praktiknjo M, et al. Two-dimensional shear wave elastography predicts survival in advanced chronic liver disease. Gut. 2022;71(2):402–14.
16. Hernandez Sampere L, Vermehren J, Mücke VT, Graf C, Peiffer KH, Dultz G, et al. Point shear-wave elastography using acoustic radiation force impulse imaging for the prediction of liver-related events in patients with chronic viral hepatitis. Hepatol Commun. 2021;5(1):112–21.
17. Piscaglia F, Salvatore V, Mulazzani L, Cantisani V, Schiavone C. Ultrasound shear wave elastography for liver disease. A critical appraisal of the many actors on the stage. Ultraschall Med. 2016;37(01):1–5.
18. Cassinotto C, Lapuyade B, Guiu B, Marraud des Grottes H, Piron L, Merrouche W, et al. Agreement between 2-dimensional shear wave and transient elastography values for diagnosis of advanced chronic liver disease. Clin Gastroenterol Hepatol. 2020;18(13):2971–9.e3.
19. Decraecker M, Dutartre D, Hiriart J-B, Irles-Depé M, Marraud des Grottes H, Chermak F, et al. Long-term prognosis of patients with alcohol-related liver disease or non-alcoholic fatty liver disease according to metabolic syndrome or alcohol use. Liver Int. 2022;42(2):350–62.
20. Vergniol J, Foucher J, Terrebonne E, Bernard PH, le Bail B, Merrouche W, et al. Noninvasive tests for fibrosis and liver stiffness predict 5-year outcomes of patients with chronic hepatitis C. Gastroenterology. 2011;140(7):1970–9.

21. Wang JH, Chuah SK, Lu SN, Hung CH, Kuo CM, Tai WC, et al. Baseline and serial liver stiffness measurement in prediction of portal hypertension progression for patients with compensated cirrhosis. Liver Int. 2014;34(9):1340–8.
22. Robic MA, Procopet B, Métivier S, Péron JM, Selves J, Vinel JP, et al. Liver stiffness accurately predicts portal hypertension related complications in patients with chronic liver disease: a prospective study. J Hepatol. 2011;55(5):1017–24.
23. Poynard T, Vergniol J, Ngo Y, Foucher J, Munteanu M, Merrouche W, et al. Staging chronic hepatitis C in seven categories using fibrosis biomarker (FibroTest™) and transient elastography (FibroScan®). J Hepatol. 2014;60(4):706–14.
24. Kim D, Li AA, Gadiparthi C, Khan MA, Cholankeril G, Glenn JS, et al. Changing trends in etiology-based annual mortality from chronic liver disease, from 2007 through 2016. Gastroenterology. 2018;155(4):1154–63.
25. Cho EJ, Kim MY, Lee JH, Lee IY, Lim YL, Choi DH, et al. Diagnostic and prognostic values of noninvasive predictors of portal hypertension in patients with alcoholic cirrhosis. PLoS One. 2015;10(7):e0133935.
26. Klibansky DA, Mehta SH, Curry M, Nasser I, Challies T, Afdhal NH. Transient elastography for predicting clinical outcomes in patients with chronic liver disease. J Viral Hepat. 2012;19(2):184–93.
27. Bertrais S, Boursier J, Ducancelle A, Oberti F, Fouchard-Hubert I, Moal V, et al. Prognostic durability of liver fibrosis tests and improvement in predictive performance for mortality by combining tests. J Gastroenterol Hepatol. 2017;32(6):1240–9.
28. Nascimbeni F, Lebray P, Fedchuk L, Oliveira CP, Alvares-da-Silva MR, Varault A, et al. Significant variations in elastometry measurements made within short-term in patients with chronic liver diseases. Clin Gastroenterol Hepatol. 2015;13(4):763–71.
29. Kjaergaard M, Thiele M, Jansen C, Staehr Madsen B, Gortzen J, Strassburg C, et al. High risk of misinterpreting liver and spleen stiffness using 2D shear-wave and transient elastography after a moderate or high calorie meal. PLoS One. 2017;12(4):e0173992.
30. Chow JC, Wong GL, Chan AW, Shu SS, Chan CK, Leung JK, et al. Repeating measurements by transient elastography in non-alcoholic fatty liver disease patients with high liver stiffness. J Gastroenterol Hepatol. 2019;34(1):241–8.
31. Legros L, Bardou-Jacquet E, Turlin B, Michalak S, Hamonic S, Le Gruyer A, et al. Transient elastography accurately screens for compensated advanced chronic liver disease in patients with ongoing or recent alcohol withdrawal. Clin Gastroenterol Hepatol. 2021; https://doi.org/10.1016/j.cgh.2021.02.021.
32. Usher-Smith JA, Sharp SJ, Griffin SJ. The spectrum effect in tests for risk prediction, screening, and diagnosis. BMJ. 2016;353:i3139.
33. Berzigotti A, Tsochatzis E, Boursier J, Castera L, Cazzagon N, Friedrich-Rust M, et al. EASL clinical practice guidelines on non-invasive tests for evaluation of liver disease severity and prognosis—2021 update. J Hepatol. 2021;75(3):659–89.
34. Thiele M, Madsen BS, Hansen JF, Detlefsen S, Antonsen S, Krag A. Accuracy of the enhanced liver fibrosis test vs fibrotest, elastography and indirect markers in detection of advanced fibrosis in patients with alcoholic liver disease. Gastroenterology. 2018;154(5):1369–79.
35. Anstee QM, Lawitz EJ, Alkhouri N, Wong VW, Romero-Gomez M, Okanoue T, et al. Noninvasive tests accurately identify advanced fibrosis due to NASH: baseline data from the STELLAR trials. Hepatology. 2019;70(5):1521–30.
36. Vali Y, Lee J, Boursier J, Spijker R, Löffler J, Verheij J, et al. Enhanced liver fibrosis test for the non-invasive diagnosis of fibrosis in patients with NAFLD: a systematic review and meta-analysis. J Hepatol. 2020;73(2):252–62.
37. Vali Y, Lee J, Boursier J, Spijker R, Verheij J, Brosnan MJ, et al. FibroTest for evaluating fibrosis in non-alcoholic fatty liver disease patients: a systematic review and meta-analysis. J Clin Med. 2021;10(11):2415.
38. Connoley D, Patel PJ, Hogan B, Tanwar S, Rhodes F, Parkes J, et al. The enhanced liver fibrosis test maintains its diagnostic and prognostic performance in alcohol-related liver disease: a cohort study. BMC Gastroenterol. 2021;21(1):268.

39. Mózes FE, Lee JA, Selvaraj EA, Jayaswal ANA, Trauner M, Boursier J, et al. Diagnostic accuracy of non-invasive tests for advanced fibrosis in patients with NAFLD: an individual patient data meta-analysis. Gut. 2022;71(5):1006–19.
40. Leroy V, Hilleret MN, Sturm N, Trocme C, Renversez JC, Faure P, et al. Prospective comparison of six non-invasive scores for the diagnosis of liver fibrosis in chronic hepatitis C. J Hepatol. 2007;46(5):775–82.
41. Hagström H, Talbäck M, Andreasson A, Walldius G, Hammar N. Ability of noninvasive scoring systems to identify individuals in the population at risk for severe liver disease. Gastroenterology. 2020;158(1):200–14.
42. Irvine KM, Wockner LF, Shanker M, Fagan KJ, Horsfall LU, Fletcher LM, et al. The enhanced liver fibrosis score is associated with clinical outcomes and disease progression in patients with chronic liver disease. Liver Int. 2016;36(3):370–7.
43. Semmler G, Binter T, Kozbial K, Schwabl P, Hametner-Schreil S, Zanetto A, et al. Noninvasive risk stratification after HCV eradication in patients with advanced chronic liver disease. Hepatology. 2021;73(4):1275–89.
44. Vergniol J, Boursier J, Coutzac C, Bertrais S, Foucher J, Angel C, et al. Evolution of noninvasive tests of liver fibrosis is associated with prognosis in patients with chronic hepatitis C. Hepatology. 2014;60(1):65–76.
45. Kamarajah SK, Chan W-K, Nik Mustapha NR, Mahadeva S. Repeated liver stiffness measurement compared with paired liver biopsy in patients with non-alcoholic fatty liver disease. Hepatol Int. 2018;12(1):44–55.
46. Kim HY, So YH, Kim W, Ahn DW, Jung YJ, Woo H, et al. Non-invasive response prediction in prophylactic carvedilol therapy for cirrhotic patients with esophageal varices. J Hepatol. 2019;70(3):412–22.
47. Marasco G, Dajti E, Ravaioli F, Alemanni LV, Capuano F, Gjini K, et al. Spleen stiffness measurement for assessing the response to β-blockers therapy for high-risk esophageal varices patients. Hepatol Int. 2020;14(5):850–7.

12 Spleen Stiffness

Antonio Colecchia, Élise Vuille-Lessard, and Annalisa Berzigotti

Rationale for the Use of Spleen Stiffness for Portal Hypertension

Splenomegaly is a hallmark of portal hypertension. Once portal pressure increases, whatever the cause, passive congestion of the spleen occurs, leading to its increase in size and stiffness. In addition, splanchnic arterial vasodilation leads to increased splenic arterial flow, further aggravating this phenomenon. From a microscopical point of view, splenic lymphoid tissue activation, angiogenesis, and fibrogenesis occur. Altogether, this leads to an increase in the stiffness of the organ [1].

Using spleen stiffness measurement (SSM) as a marker of portal hypertension in cACLD potentially overcomes two of the main limitations of liver stiffness measurement (LSM), since SSM a) is devoid of the confounding effect of liver congestion, inflammation, infiltration, or cholestasis and b) takes into account the flow-related component of portal hypertension, not mirrored by LSM [2].

After the initial papers published by Stefanescu et al. [3] and Colecchia et al. [4] showing that, using transient elastography (standard 50 Hz probe, FibroScan, Echosens, France), spleen stiffness measurement (SSM) correlates with the size of esophageal varices and with HVPG, there has been an increasing interest in the use of this novel parameter in patients with cACLD. Up to now, about 50 studies presented data on SSM measured either by ultrasound elastography (transient elastography, TE; point shear wave elastography, pSWE; 2D shear wave elastography, 2D-SWE) or by magnetic resonance elastography (MRE) as a marker of portal

A. Colecchia
Department of Gastroenterology, Verona University Hospital, Verona, Italy
e-mail: antonio.colecchia@aovr.veneto.it

É. Vuille-Lessard · A. Berzigotti (✉)
Department of Visceral Surgery and Medicine, Inselspital, Bern University Hospital, University of Bern, Bern, Switzerland
e-mail: elise.vuille-lessard@insel.ch; Annalisa.berzigotti@insel.ch

R. de Franchis (ed.), *Portal Hypertension VII*,
https://doi.org/10.1007/978-3-031-08552-9_12

Table 12.1 Studies reporting on SSM according to the measurement technique. Small studies with less than 30 cases were not included

Study	Year	Method used	N included and etiology	Failure rate	Endpoint	AUROC for the selected endpoint	Chosen cutoff for the selected endpoint	Sensitivity	Specificity
SSM by TE									
Stefanescu et al. [3]	2011	TE	174, mixed	14.4%	EV	0.781	46.4 kPa	83.6%	71.4%
Colecchia et al. [5]	2012	TE	113, HCV, compensated	11.5%	CSPH	0.966	40.0 kPa (rule out)	98.5%	74.3%
							52.8 kPa (rule in)	76.9%	97.1%
					EV	0.941	41.3 kPa (rule out)	98.1%	66.0%
							55.0 kPa (rule in)	71.7%	95.7%
Sharma et al. [32]	2013	TE	200, mixed	13%	EV	0.898	40.8 kPa	94%	76%
Calvaruso et al. [10]	2013	TE (modified range)	112, HCV, compensated	14.3%	EV	0.701	50.0 kPa	65%	61%
					LEV	0.820	54.0 kPa	80%	70%
Zykus et al. [33]	2015	TE	107, mixed, most compensated	7.5%	CSPH	0.846	47.6 kPa	77.3%	79.2%
Stefanescu et al. [34]	2015	TE	136, mixed	N/A	HRV	0.742	53 kPa	89%	54%
Wong et al. [35]	2016	TE	176, HBV	15.9%	EV	0.685	21.4 kPa (rule out)	90.3%	43.4%
							50.5 kPa (rule in)	45.2%	90.3%

Colecchia et al. [8]	2018	TE	498 (derivation cohort 258, 85% HCV; internal validation cohort 240, 40% HCV); external validation cohort 115, mixed	26 (4.5%)	HRV	0.847	46.0 (rule out)	97.8%	43.8%
Arribas Anta et al. [36]	2019	TE	66, mixed	9.1%	EV	0.800	48 kPa	87%	69%
Stefanescu et al. [9]	2020	TE (spleen-dedicated, 100 Hz)	260, mixed	7.5% (vs. 24% for 50 Hz)	CSPH	0.811	34.15 kPa	N/A	N/A
					EV	0.728	33.3 kPa (rule out)	90.3%	33.7%
							70 kPa (rule in)	29.1%	90.5%
					HRV	0.756	41.3 kPa (rule out)	91.3%	40.8%
							79.9 kPa (rule in)	26.1%	90.1%
Wang et al. [11]	2021	TE	341, HBV cirrhosis with viral suppression	4.1%	HRV	N/A	46 kPa	95.7%	65.3%
SSM by pSWE									
Rifai et al. [37]	2011	pSWE (VTQ)	100, mixed	22%	CSPH	0.680	3.29 m/s	47%	73%
Bota et al. [38]	2012	pSWE (VTQ)	145, mixed	2.1%	LEV	0.578	2.55 m/s	96.7%	21.0%
Ye et al. [39]	2012	pSWE (VTQ)	204, HBV	N/A	EV LEV	0.830 0.839	3.16 m/s 3.39 m/s	84.1% 78.9%	81% 78.3%

(continued)

Table 12.1 (continued)

Study	Year	Method used	N included and etiology	Failure rate	Endpoint	AUROC for the selected endpoint	Chosen cutoff for the selected endpoint	Sensitivity	Specificity
Vermehren et al. [40]	2012	pSWE (VTQ)	166, mixed	0%	LEV	0.580	3.04 m/s	90%	25%
Takuma et al. [41]	2013	pSWE (VTQ)	340, mixed	4.5%	EV HRV	0.937 (viral) 0.923 (others) 0.930 (all)	3.18 m/s 3.24 m/s 3.30 m/s	98.9% 97.7% 98.9%	59.9% 65.2% 62.9%
Rizzo et al. [42]	2014	pSWE (VTQ)	54, HCV	N/A	EV	0.959	3.10 m/s 2.32 m/s	96.4%	88.5%
Attia et al. [43]	2015	pSWE (VTQ)	78, mixed, some decompensated, 90% CSPH, 76% EV	0%	CSPH	0.968		96%	89%
Kim et al. [44]	2015	pSWE (VTQ)	132, mixed	4.5%	EV LEV	0.785 0.786	3.16 m/s 3.40 m/s	87.0% 78.9%	60.4% 63.0%
Park et al. [45]	2016	pSWE (ElastPQ)	366, viral and alcohol	24%	EV	0.859	29.9 kPa	85.1 kPa	79.1 kPa
Takuma et al. [46]	2016	pSWE (VTQ)	62, mixed, most compensated	3.2%	CSPH HVPG ≥12 EV LEV	0.943 0.963 0.937 0.955	3.10 m/s 3.15 m/s 3.36 m/s 3.51 m/s	97.1% 96.6% 95.8% 93.8%	57.7% 61.3% 77.8% 84.1%

Fierbinteanu-Braticevici et al. [47]	2019	pSWE (VTQ)	135, mixed	0%	EV HRV	0.776 0.972	2.5 m/s (rule out) 3.5 m/s (rule in) 3.2 m/s (rule out) 3.8 m/s (rule in)	92% 47% 97% 55%	22% 96% 69% 98%
Peagu et al. [48]	2019	pSWE (VTQ)	178, viral	N/A	EV LEV	0.872 0.969	2.89 m/s 3.30 m/s	91.4% 96.4%	67.7% 88.5%
Darweesh et al. [49]	2019	pSWE (VTQ)	200, HCV	1%	EV	0.760	3.25 m/s	85%	58%
Giuffrè et al. [50]	2020	pSWE (ElastPQ)	210, mixed, compensated	4.5%	EV	0.95	31 kPa (rule out) 69 kPa (rule in)	100% 14%	60% 100%
SSM by 2D-SWE									
Elkrief et al. [51]	2015	2D-SWE (SSI) TE	79, mixed, most decompensated, 89% CSPH, 69% child-Pugh B-C	3% 58%	CSPH	0.640	34.7 kPa	40%	100%
					LEV CSPH LEV	0.580 0.630 0.650	32.3 kPa 56.3 kPa 73.5 kPa	48% 73% 54%	71% 67% 78%
Procopet et al. [7]	2015	2D-SWE (SSI)	55, mixed, most compensated	34%	CSPH	0.725	22.7 kPa (rule out)	90%	N/A
							40 kPa (rule in)	N/A	90%

(continued)

Table 12.1 (continued)

Study	Year	Method used	N included and etiology	Failure rate	Endpoint	AUROC for the selected endpoint	Chosen cutoff for the selected endpoint	Sensitivity	Specificity
Cassinotto et al. [12]	2015	2D-SWE (SSI)	401, mixed, some decompensated	29.2%	EV	0.80	N/A	N/A	N/A
					HRV	0.78 (all)	N/A	N/A	N/A
						0.75 (compensated)	25.6 kPa (with NPV >90%)	94%	36%
Grgurevic et al. [19]	2015	2D-SWE (SSI)	126, mixed	29.4%	EV	0.790	30.3 kPa	79.6%	75.8%
Jansen et al. [52]	2017	2D-SWE (SSI)	158, mixed, some decompensated	18.8%	CSPH	0.840	26.3 kPa 21.7 kPa (rule out) 35.6 kPa (rule in)	79.7% 91.9% 51.4%	84.2% 50% 92%
Zhu et al. [53]	2019	2D-SWE (SSI)	104, HBV, most compensated	24.6%	CSPH	0.810	23.2 kPa (rule out)	>90%	N/A
							34.2 kPa (rule in)	N/A	>90%
Karagiannakis et al. [14]	2019	2D-SWE (SSI)	64, mixed, compensated	9.8%	HRV	0.792 (all) 0.854 (excluding cholestatic LD)	33.7 kPa (rule out) 35.8 kPa (rule out)	91.7% 88.9%	60.0% 72.4%
Cho et al. [54]	2020	2D-SWE	274, mixed, compensated	N/R	HRV	0.844	≤27.3 kPa (rule out)	98.1%	35.9%
SSM by MRE									
Danielsen et al. [55]	2021	2D-MRE	52, mixed etiologies, some decompensated	Not reported	HVPG HVPG ≥12	Correlation 0.94 0.810 (0.64–0.97)	10.5 kPa	80%	79%

hypertension or varices using HVPG measurement or endoscopy as gold standards. The studies with a larger sample size are summarized in Table 12.1.

A recent systematic review and meta-analysis of 32 studies using any of the above-mentioned techniques in 3952 patients concluded that spleen stiffness had a summary area under the ROC curve (sAUROC) of over 0.90, with a sensitivity of 0.85 and specificity of 0.86 for detecting CSPH. As for high-risk varices (HRV), the sAUROC was 0.83 with a sensitivity of 0.87 and specificity of 0.66. The performance of SSM was superior in Asian subjects, who had a lower body mass index.

SSM Using Transient Elastography

Ten large ($n > 100$) studies on SSM using TE have been published so far. Over 80% of study patients had a viral etiology of liver disease (untreated HCV or HBV, or HBV on viral suppression). SSM was measured using the standard liver probe with 50 Hz frequency in all studies except one. Reproducibility has been proven excellent [5–7]. Due to technical requirements not being met in small spleens, SSM had a high failure rate up to 15%–27%, which constitutes a major limitation of the method. When ultrasound was used to locate the spleen, applicability improved significantly [5, 8]. Similarly, the failure rate of SSM using a novel, spleen-dedicated probe with 100 Hz frequency improved to 7.5% [9].

Since the spleen is stiffer than the liver, with normal values up to 21 kPa, a ceiling effect at 75 kPa was occurring with the standard probe in patients with ACLD, as proven by the use of a modified software able to provide a range up to 150 kPa [10]. The novel spleen-dedicated probe provides values up to 100 kPa [9].

In the published studies using HVPG as a gold standard, SSM correlated with the HVPG with a similar or even closer correlation coefficient than LSM. The best cutoff value to rule out and rule in CSPH has not yet been set. From the analysis of the existing data, mainly in patients with cACLD due to HBV or HCV and using the standard 50 Hz probe, it seems that SSM < 21–30 kPa can rule out CSPH with a sensitivity >90%, while SSM above 50 kPa could rule in CSPH with a specificity >90%. Validation in other etiologies and large prospective series is needed.

As for ruling out and ruling in HRV, the available data suggest that SSM below 40 kPa (standard probe) rules out HRV with a sensitivity >90% (Tables 12.1 and 12.2). In two independent studies which proposed [8] or applied [11] a slightly higher SSM cutoff value (46 kPa), SSM alone or used in combination with the Baveno VI criteria increased the rate of spared endoscopies in comparison to the Baveno criteria, while maintaining the rate of missed varices requiring treatment below 5% (Table 12.2). In the only study published so far, using a spleen specific 100 Hz TE probe allowed improving the results obtained by the standard 50 Hz probe in terms of spared endoscopies [9].

Table 12.2 Performance of SSM combined to the Baveno VI criteria or with LSM alone

Study	Year	Method used	N included and etiology	N (%) HRV	Chosen SSM cutoff to spare endoscopy	% Spared endoscopies and missed HRV using Baveno VI	% spared endoscopies and missed HRV using SSM	% spared endoscopies and missed HRV using Baveno VI + SSM
Wong et al. [56]	2018	TE-randomized open label-controlled trial	548 (274 per arm), 85% viral hepatitis (>>HBV)	11 (4%) in the NITs arm, 5.8% in the standard of care arm	41.3 kPa + LSM < 12.5 kPa	N/A	N/A	N/A LSSM (LSM + SSM) strategy spared 41.8% endoscopies
Stefanescu et al. [9]	2020	TE (standard 50 Hz) TE (spleen-dedicated, 100 Hz)	260, mixed	69 (26.5%)	40.1 kPa 41.3 kPa	8.1%; 0 8.1%; 0	18.4%; 4.7% 30.8%; 4.7%	26.5%; 4.7 38.1%; 4.7
Colecchia et al. [8]	2018	TE	Derivation cohort 258, 85% HCV Internal validation cohort 240, 40% HCV External validation cohort 115	54 (20.9%) 46 (19%) 28 (13%)	46 kPa	21.7%; 2.2% 16.5%; 0	35.8%; 2.2% 30.4%; 0	43.8%; 4.3% 37.4%; 0
Wang et al. [11]	2021	TE	341, HBV cirrhosis with viral suppression	70 (20.5%)	46 kPa	37.0%; 0	52.1%; 0	61.6%; 4.3%
Cho et al. [54]	2020	2D-SWE	274, mixed, compensated	54 (19.7%)	27.3 kPa	18.6% (LSM <16 kPa + Plt > 150 G/L); 0	28.8%; 1.9%	36.1%; 1.9%

SSM Using Other Ultrasound Elastography Methods

The applicability of pSWE and 2D-SWE is affected by similar factors, including the absence of splenomegaly, obesity, movements caused by heart beating and ascites [12]. Even though several studies are available with both methods (Table 12.1), there is a considerable heterogeneity in the type of included patients, and several studies included decompensated ACLD patients.

With these limitations, the analysis of the data suggests that using pSWE (Virtual Touch Siemens; pSWE by other devices has too limited data) SSM values <2.5 m/s could be used to rule out CSPH and HRV, while values >3.5 m/s might suggest EV (see Table 12.1). In a prospective study using pSWE (Virtual Touch Siemens) in patients with cACLD mostly due to HBV, SSM predicted variceal bleeding with an AUROC of 0.911 [13]. The best cutoff value discriminating patients developing variceal bleeding from those who did not (with an incidence of 7.3% over 32 months of follow-up) was 3.48 m/s.

As for 2D-SWE (Supersonic Imagine; 2D-SWE by other devices has too limited data), values of SSM < 21–25 kPa could be used to rule out CSPH (cutoff value closer to TE), while values <35 kPa could be used to rule out HRV (see Table 12.1). Karagiannakis et al. [14] showed that SSM by this method might help sparing a larger proportion of endoscopies than the Baveno VI criteria, without missing more HRV.

SSM Using Magnetic Resonance Elastography

SSM by MRE has been evaluated in eight studies, most of which included a very small number of patients. Data regarding the prediction of varices and HRV are in line with those provided by ultrasound elastography methods, but a direct comparison of the accuracy of these methods is not possible yet [15]. Availability and cost limit the routine use of MRE to measure SSM in cACLD.

SSM for the Prediction of Liver-Related Events, Mortality, and Response to Therapy

SSM predicted the first clinical decompensation and mortality in five studies [4, 16–19] and predicted HCC recurrence in one study [20]. The best cutoff value predicting decompensation using TE was 54 kPa. In patients with HCV cirrhosis experiencing sustained virological response, SSM decreases significantly [21, 22], and SSM was an independent predictor of liver-related events (decompensation [23] and HCC [24]). Two studies (one using pSWE [25] and one with TE [26]) showed that SSM might predict the hemodynamic response to NSBB in patients started on primary prophylaxis. SSM decreases after TIPS, suggesting that it parallels the decrease in portal pressure [27–31].

Summary

The data summarized in this chapter show that SSM can be considered as a marker of portal hypertension and should be included as a complementary noninvasive test in the armamentarium of hepatologists to assess CSPH and varices in addition to the Baveno VI criteria. In patients with cACLD due to viral causes, SSM used in combination with the Baveno VI criteria seems to allow to safely expand the rate of spared endoscopies. However, SSM applicability remains an issue, and evidence is not strong enough to recommend cutoff values to rule out/rule in varices requiring treatment by techniques other than TE. In addition, data in patients with cACLD due to non-viral causes are scarce, and it is still difficult to draw solid conclusions in this context. Furthermore, whether the use of the novel TE spleen-specific probe allows better risk stratification remains to be ascertained in future studies.

References

1. Mejias M, Garcia-Pras E, Gallego J, Mendez R, Bosch J, Fernandez M. Relevance of the mTOR signaling pathway in the pathophysiology of splenomegaly in rats with chronic portal hypertension. J Hepatol. 2010;52:529–39.
2. Tseng Y, Li F, Wang J, Chen S, Jiang W, Shen X, et al. Spleen and liver stiffness for noninvasive assessment of portal hypertension in cirrhotic patients with large esophageal varices. J Clin Ultrasound. 2018;46:442–9.
3. Stefanescu H, Grigorescu M, Lupsor M, Procopet B, Maniu A, Badea R. Spleen stiffness measurement using Fibroscan for the noninvasive assessment of esophageal varices in liver cirrhosis patients. J Gastroenterol Hepatol. 2011;26:164–70.
4. Colecchia A, Colli A, Casazza G, Mandolesi D, Schiumerini R, Reggiani LB, et al. Spleen stiffness measurement can predict clinical complications in compensated HCV-related cirrhosis: a prospective study. J Hepatol. 2014;60:1158–64.
5. Colecchia A, Montrone L, Scaioli E, Bacchi-Reggiani ML, Colli A, Casazza G, et al. Measurement of spleen stiffness to evaluate portal hypertension and the presence of esophageal varices in patients with HCV-related cirrhosis. Gastroenterology. 2012;143:646–54.
6. Balakrishnan M, Souza F, Muñoz C, Augustin S, Loo N, Deng Y, et al. Liver and spleen stiffness measurements by point shear wave elastography via acoustic radiation force impulse: intraobserver and interobserver variability and predictors of variability in a US population. J Ultrasound Med. 2016;35:2373–80.
7. Procopet B, Berzigotti A, Abraldes JG, Turon F, Hernandez-Gea V, García-Pagán JC, et al. Real-time shear-wave elastography: applicability, reliability and accuracy for clinically significant portal hypertension. J Hepatol. 2015;62:1068–75.
8. Colecchia A, Ravaioli F, Marasco G, Colli A, Dajti E, Di Biase AR, et al. A combined model based on spleen stiffness measurement and Baveno VI criteria to rule out high-risk varices in advanced chronic liver disease. J Hepatol. 2018;69:308–17.
9. Stefanescu H, Marasco G, Cales P, Fraquelli M, Rosselli M, Ganne-Carrie N, et al. A novel spleen-dedicated stiffness measurement by FibroScan(R) improves the screening of high-risk oesophageal varices. Liver Int. 2020;40:175–85.
10. Calvaruso V, Bronte F, Conte E, Simone F, Craxì A, Di Marco V. Modified spleen stiffness measurement by transient elastography is associated with presence of large oesophageal varices in patients with compensated hepatitis C virus cirrhosis. J Viral Hepat. 2013;20:867–74.
11. Wang H, Wen B, Chang X, Wu Q, Wen W, Zhou F, et al. Baveno VI criteria and spleen stiffness measurement rule out high-risk varices in virally suppressed HBV-related cirrhosis. J Hepatol. 2021;74:584–92.

12. Cassinotto C, Charrie A, Mouries A, Lapuyade B, Hiriart JB, Vergniol J, et al. Liver and spleen elastography using supersonic shear imaging for the non-invasive diagnosis of cirrhosis severity and oesophageal varices. Dig Liver Dis. 2015;47:695–701.
13. Takuma Y, Nouso K, Morimoto Y, Tomokuni J, Sahara A, Takabatake H, et al. Prediction of oesophageal variceal bleeding by measuring spleen stiffness in patients with liver cirrhosis. Gut. 2016;65:354–5.
14. Karagiannakis DS, Voulgaris T, Koureta E, Chloupi E, Papatheodoridis GV, Vlachogiannakos J. Role of spleen stiffness measurement by 2D-shear wave elastography in ruling out the presence of high-risk varices in cirrhotic patients. Dig Dis Sci. 2019;64:2653–60.
15. Singh R, Wilson MP, Katlariwala P, Murad MH, McInnes MDF, Low G. Accuracy of liver and spleen stiffness on magnetic resonance elastography for detecting portal hypertension: a systematic review and meta-analysis. Eur J Gastroenterol Hepatol. 2021;32(2):237–45.
16. Meister P, Dechêne A, Büchter M, Kälsch J, Gerken G, Canbay A, et al. Spleen stiffness differentiates between acute and chronic liver damage and predicts hepatic decompensation. J Clin Gastroenterol. 2019;53(6):457–63.
17. Takuma Y, Morimoto Y, Takabatake H, Toshikuni N, Tomokuni J, Sahara A, et al. Measurement of spleen stiffness with acoustic radiation force impulse imaging predicts mortality and hepatic decompensation in patients with liver cirrhosis. Clin Gastroenterol Hepatol. 2017;15:1782–1790.e1784.
18. Zhang Y, Mao DF, Zhang MW, Fan XX. Clinical value of liver and spleen shear wave velocity in predicting the prognosis of patients with portal hypertension. World J Gastroenterol. 2017;23:8044–52.
19. Grgurević I, Bokun T, Mustapić S, Trkulja V, Heinzl R, Banić M, et al. Real-time two-dimensional shear wave ultrasound elastography of the liver is a reliable predictor of clinical outcomes and the presence of esophageal varices in patients with compensated liver cirrhosis. Croat Med J. 2015;56:470–81.
20. Marasco G, Colecchia A, Colli A, Ravaioli F, Casazza G, Bacchi Reggiani ML, et al. Role of liver and spleen stiffness in predicting the recurrence of hepatocellular carcinoma after resection. J Hepatol. 2019;70:440–8.
21. Ravaioli F, Colecchia A, Dajti E, Marasco G, Alemanni LV, Tame M, et al. Spleen stiffness mirrors changes in portal hypertension after successful interferon-free therapy in chronic-hepatitis C virus patients. World J Hepatol. 2018;10:731–42.
22. Pons M, Santos B, Simon-Talero M, Ventura-Cots M, Riveiro-Barciela M, Esteban R, et al. Rapid liver and spleen stiffness improvement in compensated advanced chronic liver disease patients treated with oral antivirals. Ther Adv Gastroenterol. 2017;10:619–29.
23. Dajti E, Ravaioli F, Colecchia A, Marasco G, Bacchi Reggiani ML, Colli A, et al. Spleen stiffness measurements predict the risk of hepatic decompensation after direct-acting antivirals in HCV cirrhotic patients. Ultraschall Med. 2020; https://doi.org/10.1055/a-1205-0367.
24. Dajti E, Marasco G, Ravaioli F, Colecchia L, Ferrarese A, Festi D, et al. Risk of hepatocellular carcinoma after HCV eradication: determining the role of portal hypertension by measuring spleen stiffness. JHEP Rep. 2021;3:100289.
25. Kim HY, So YH, Kim W, Ahn DW, Jung YJ, Woo H, et al. Non-invasive response prediction in prophylactic carvedilol therapy for cirrhotic patients with esophageal varices. J Hepatol. 2019;70:412–22.
26. Marasco G, Dajti E, Ravaioli F, Alemanni LV, Capuano F, Gjini K, et al. Spleen stiffness measurement for assessing the response to beta-blockers therapy for high-risk esophageal varices patients. Hepatol Int. 2020;14:850–7.
27. Ran HT, Ye XP, Zheng YY, Zhang DZ, Wang ZG, Chen J, et al. Spleen stiffness and splenoportal venous flow: assessment before and after transjugular intrahepatic portosystemic shunt placement. J Ultrasound Med. 2013;32:221–8.
28. Gao J, Zheng X, Zheng YY, Zuo GQ, Ran HT, Auh YH, et al. Shear wave elastography of the spleen for monitoring transjugular intrahepatic portosystemic shunt function: a pilot study. J Ultrasound Med. 2016;35:951–8.

29. De Santis A, Nardelli S, Bassanelli C, Lupo M, Iegri C, Di Ciesco CA, et al. Modification of splenic stiffness on acoustic radiation force impulse parallels the variation of portal pressure induced by transjugular intrahepatic portosystemic shunt. J Gastroenterol Hepatol. 2018;33:704–9.
30. Buechter M, Manka P, Theysohn JM, Reinboldt M, Canbay A, Kahraman A. Spleen stiffness is positively correlated with HVPG and decreases significantly after TIPS implantation. Dig Liver Dis. 2018;50:54–60.
31. Attia D, Rodt T, Marquardt S, Hinrichs J, Meyer BC, Gebel M, et al. Shear wave elastography prior to transjugular intrahepatic portosystemic shunt may predict the decrease in hepatic vein pressure gradient. Abdom Radiol (NY). 2019;44:1127–34.
32. Sharma P, Kirnake V, Tyagi P, Bansal N, Singla V, Kumar A, et al. Spleen stiffness in patients with cirrhosis in predicting esophageal varices. Am J Gastroenterol. 2013;108:1101–7.
33. Zykus R, Jonaitis L, Petrenkienė V, Pranculis A, Kupčinskas L. Liver and spleen transient elastography predicts portal hypertension in patients with chronic liver disease: a prospective cohort study. BMC Gastroenterol. 2015;15:183.
34. Stefanescu H, Radu C, Procopet B, Lupsor-Platon M, Habic A, Tantau M, et al. Non-invasive ménage à trois for the prediction of high-risk varices: stepwise algorithm using lok score, liver and spleen stiffness. Liver Int. 2015;35:317–25.
35. Wong GL, Kwok R, Chan HL, Tang SP, Lee E, Lam TC, et al. Measuring spleen stiffness to predict varices in chronic hepatitis B cirrhotic patients with or without receiving non-selective beta-blockers. J Dig Dis. 2016;17:538–46.
36. Arribas Anta J, Garcia Gonzalez M, Torres Guerrero ME, Garrido Gomez E, Rodriguez de Santiago E, Lopez Duran S, et al. Prediction of the presence of esophageal varices using spleen stiffness measurement by transient elastography in cirrhotic patients. Acta Gastroenterol Belg. 2018;81:496–501.
37. Rifai K, Cornberg J, Bahr M, Mederacke I, Potthoff A, Wedemeyer H, et al. ARFI elastography of the spleen is inferior to liver elastography for the detection of portal hypertension. Ultraschall Med. 2011;32(Suppl 2):E24–30.
38. Bota S, Sporea I, Sirli R, Focsa M, Popescu A, Danila M, et al. Can ARFI elastography predict the presence of significant esophageal varices in newly diagnosed cirrhotic patients? Ann Hepatol. 2012;11:519–25.
39. Ye XP, Ran HT, Cheng J, Zhu YF, Zhang DZ, Zhang P, et al. Liver and spleen stiffness measured by acoustic radiation force impulse elastography for noninvasive assessment of liver fibrosis and esophageal varices in patients with chronic hepatitis B. J Ultrasound Med. 2012;31:1245–53.
40. Vermehren J, Polta A, Zimmermann O, Herrmann E, Poynard T, Hofmann WP, et al. Comparison of acoustic radiation force impulse imaging with transient elastography for the detection of complications in patients with cirrhosis. Liver Int. 2012;32:852–8.
41. Takuma Y, Nouso K, Morimoto Y, Tomokuni J, Sahara A, Toshikuni N, et al. Measurement of spleen stiffness by acoustic radiation force impulse imaging identifies cirrhotic patients with esophageal varices. Gastroenterology. 2013;144:92–101. e102
42. Rizzo L, Attanasio M, Pinzone MR, Berretta M, Malaguarnera M, Morra A, et al. A new sampling method for spleen stiffness measurement based on quantitative acoustic radiation force impulse elastography for noninvasive assessment of esophageal varices in newly diagnosed HCV-related cirrhosis. Biomed Res Int. 2014;2014:365982.
43. Attia D, Schoenemeier B, Rodt T, Negm AA, Lenzen H, Lankisch TO, et al. Evaluation of liver and spleen stiffness with acoustic radiation force impulse quantification elastography for diagnosing clinically significant portal hypertension. Ultraschall Med. 2015;36:603–10.
44. Kim HY, Jin EH, Kim W, Lee JY, Woo H, Oh S, et al. The role of spleen stiffness in determining the severity and bleeding risk of esophageal varices in cirrhotic patients. Medicine. 2015;94:e1031.
45. Park J, Kwon H, Cho J, Oh J, Lee S, Han S, et al. Is the spleen stiffness value acquired using acoustic radiation force impulse (ARFI) technology predictive of the presence of esophageal varices in patients with cirrhosis of various etiologies? Med Ultrason. 2016;18:11–7.

46. Takuma Y, Nouso K, Morimoto Y, Tomokuni J, Sahara A, Takabatake H, et al. Portal hypertension in patients with liver cirrhosis: diagnostic accuracy of spleen stiffness. Radiology. 2016;279:609–19.
47. Fierbinteanu-Braticevici C, Tribus L, Peagu R, Petrisor A, Baicus C, Cretoiu D, et al. Spleen stiffness as predictor of esophageal varices in cirrhosis of different etiologies. Sci Rep. 2019;9:16190.
48. Peagu R, Sararu R, Necula A, Moldoveanu A, Petrisor A, Fierbinteanu-Braticevici C. The role of spleen stiffness using ARFI in predicting esophageal varices in patients with hepatitis B and C virus-related cirrhosis. Rom J Intern Med. 2019;57:334–40.
49. Darweesh SK, Yosry A, Salah M, Zayed N, Khairy A, Awad A, et al. Acoustic radiation forced impulse-based splenic prediction model using data mining for the noninvasive prediction of esophageal varices in hepatitis C virus advanced fibrosis. Eur J Gastroenterol Hepatol. 2019;31:1533–9.
50. Giuffre M, Macor D, Masutti F, Abazia C, Tine F, Bedogni G, et al. Spleen Stiffness Probability Index (SSPI): a simple and accurate method to detect esophageal varices in patients with compensated liver cirrhosis. Ann Hepatol. 2020;19:53–61.
51. Elkrief L, Rautou PE, Ronot M, Lambert S, Dioguardi Burgio M, Francoz C, et al. Prospective comparison of spleen and liver stiffness by using shear-wave and transient elastography for detection of portal hypertension in cirrhosis. Radiology. 2015;275:589–98.
52. Jansen C, Bogs C, Verlinden W, Thiele M, Möller P, Görtzen J, et al. Shear-wave elastography of the liver and spleen identifies clinically significant portal hypertension: a prospective multicentre study. Liver Int. 2017;37:396–405.
53. Zhu YL, Ding H, Fu TT, Peng SY, Chen SY, Luo JJ, et al. Portal hypertension in hepatitis B-related cirrhosis: diagnostic accuracy of liver and spleen stiffness by 2-D shear-wave elastography. Hepatol Res. 2019;49(5):540–9.
54. Cho YS, Kim Y, Sohn JH. Application of supersonic shear imaging to the Baveno VI criteria and a combination model with spleen stiffness measurement to rule out high-risk varices in compensated advanced chronic liver disease. Ultraschall Med. 2020; https://doi.org/10.1055/a-1168-6271.
55. Danielsen KV, Hove JD, Nabilou P, Yin M, Chen J, Zhao M, et al. Using MR elastography to assess portal hypertension and response to beta-blockers in patients with cirrhosis. Liver Int. 2021;41:2149–58.
56. Wong GLH, Kwok R, Hui AJ, Tse YK, Ho KT, Lo AOS, et al. A new screening strategy for varices by liver and spleen stiffness measurement (LSSM) in cirrhotic patients: a randomized trial. Liver Int. 2018;38:636–44.

Emerging Non-invasive Markers: Imaging, Blood, and Liver Clearance Tests

13

Naaventhan Palaniyappan and Jonathan A. Fallowfield

Emerging Non-invasive Methods as a Surrogate for HVPG Measurement

Imaging Markers

Contrast-Enhanced Ultrasound

Contrast-enhanced ultrasound (CEUS) involves the intravenous administration of minute, gas-filled microbubbles that enhance the intravascular signal. The hepatic vein arrival time (HVAT) of the microbubble agent has been studied as a diagnostic tool for detecting advanced fibrosis and cirrhosis [1]. The peak enhancement time of the microbubble, which is defined as the interval time from the contrast onset in the splenic artery to the time to reach maximum intensity in the splenic vein, was shown to correlate with HVPG [2].

Several characteristics of the contrast agents have been studied in relation to PH. The subharmonic signal from the US contrast agents reflected pressure changes in the ambient fluid. This was the basis of the subharmonic-aided pressure estimation technique (SHAPE) [3]. The SHAPE gradient between the portal vein and the hepatic vein was initially shown to correlate with HVPG in a pilot study of 45 patients [4] and validated in a prospective cohort of 125 patients [5]. Assessment of the hepatic vascular network using computer-based analysis of the videos generated from CEUS yielded the 'hepatic vascular connectome' [6]. Patients with cirrhosis

N. Palaniyappan
NIHR Nottingham Biomedical Research Centre, Nottingham University Hospitals NHS Trust, University of Nottingham, Nottingham, UK
e-mail: naaventhan.palaniyappan@nottingham.ac.uk

J. A. Fallowfield (✉)
University of Edinburgh Centre for Inflammation Research, The Queen's Medical Research Institute, Edinburgh BioQuarter, Edinburgh, UK
e-mail: Jonathan.Fallowfield@ed.ac.uk

R. de Franchis (ed.), *Portal Hypertension VII*,
https://doi.org/10.1007/978-3-031-08552-9_13

had a lower clustering coefficient of the hepatic vascular connectome compared to healthy controls, and the clustering coefficient correlated with HVPG in 15 patients [6]. This method has been evaluated further in a multicentre study, but the results have not yet been published in detail. Although an excellent correlation with HVPG was reported in the initial results, the automated software was only able to provide portal pressure estimation in 56% of the patients studied [7]. The poor technical success rate may limit the generalisability of this technique.

Computed Tomography

The ratio of liver to spleen volume measured on contrast-enhanced computed tomography (CT) images was shown to correlate with HVPG [8]. These CT scans were obtained preoperatively in patients being evaluated for hepatocellular carcinoma (HCC) resection, and the CT-based model was developed to stratify patients for resection surgery and liver transplantation. However, the correlation was poor in patients at extremes of the HVPG spectrum (i.e., normal, or very high) [9, 10].

Advances in the imaging-based 3D modelling with computational fluid analysis allow non-invasive assessment of intravascular blood flow and pressure, and these techniques are used in cardiology for the assessment of coronary arteries [11]. Qi et al. recently used triple-phase CT angiography of the liver to develop a 'virtual HVPG' (vHVPG) [12]. The vHVPG was calculated using a mathematical model that included the portal vein velocity measured using Doppler US. There was a moderate positive correlation between vHVPG and invasive HVPG ($R = 0.61$), with an AUROC of 0.88 for diagnosing CSPH. Notably, interpretation of the vHVPG was time-consuming (~2.5 h/case) and the study included small numbers of patients without CSPH.

The development of a radiomics signature involves machine learning to extract high-dimensional quantitative features from radiological images. This was explored as a method to evaluate PH [13]. Regions of interest were drawn on portal venous phase CT images of the liver and spleen, and 20,648 radiomics features were retrieved. This was reduced to seven features from the liver and four features from the spleen, which were included in a regression model to develop radiomics-based HVPG (rHVPG). The rHVPG had an AUROC of 0.85 for diagnosing CSPH and was validated in four external cohorts. The direct correlation between rHVPG and HVPG was not reported.

These CT-based methods involve ionising radiation, which limits utility especially for repeated measurements. The intravenous contrast for CT is also recognised to be nephrotoxic and can cause contrast-induced acute kidney injury [14].

Magnetic Resonance Imaging

Haemodynamic Measures

Phase contrast (PC)-MRI is a non-invasive technique to measure flow in a blood vessel without the use of intravenous contrast. PC-MRI measures blood flow with high accuracy, confirmed by phantom models [15, 16] and by in vivo studies with direct measurement in deep canine arteries and veins [17].

Table 13.1 Summary of MRI parameters evaluated for the assessment of portal hypertension

MRI parameters	Study	Correlation with HVPG	Notes
Vessel flow (PC-MRI)			
• Azygous vein flow	Gouya et al. 2016 [18]	$r^2 = 0.77$	$n = 69$; HVPG range 3 to 25 mmHg
	Palaniyappan et al. 2016 [19]	$r = 0.66$	$n = 30$; mean HVPG 9.8 ± 6.1 mmHg
• Superior mesenteric artery velocity	Palaniyappan et al. 2016 [19]	$r = 0.53$	
• Splenic artery velocity	Palaniyappan et al. 2016 [19]	$r = 0.58$	
• Caval subtraction hepatic artery fraction	Chouhan et al. 2017 [20]	$r = 0.78$	$n = 12$; mean HVPG 12.3 ± 1.6 mmHg
Tissue perfusion (arterial spin labelling (ASL)			
• Liver tissue perfusion	Palaniyappan et al. 2016 [19]	$r = 0.38$	Correlation was absent in HVPG >10 mmHg
• Arrival time		$r = -0.47$	
Structural measures			
• Liver T1	Palaniyappan et al. 2016 [19]	$r = 0.84$	
• Spleen T1	Palaniyappan et al. 2016 [19]	$r = 0.40$	Correlation was absent in HVPG >10 mmHg
	Levick et al. 2019 [25]	$r = 0.69$	$n = 19$, median HVPG 9.0 (IQR 4.0–14.0)
Dynamic contrast-enhanced MRI (DCE-MRI)			
• Liver distribution volume (DV)	Wagner et al. 2018 [31]	$r = 0.49$	$n = 34$; 12 patients <5 mmHg, 13 patients 5–10 mmHg, 9 patients ≥10 mmHg
• Liver time to peak (TTP)		$r = 0.52$	
• Liver upslope		$r = -0.57$	

PC-MRI measured azygous vein flow that has been shown to correlate with grade of oesophageal varices [16] and HVPG [18, 19] (Table 13.1). However, the correlation was absent in patients with CSPH [19]. The relationship between hepatic inflow and portal pressure is less well established. Portal venous flow does not correlate with HVPG [18–20], but hepatic arterial fraction of the hepatic inflow was shown to correlate with HVPG in 12 patients [20]. The hepatic artery flow in this study was obtained indirectly by subtracting the portal vein flow from the total hepatic blood flow (the difference between infra- and suprahepatic vena cava). PC-MRI measured flow in the splanchnic circulation (splenic artery and superior mesenteric artery) correlated significantly with HVPG [19].

Arterial spin labelling (ASL)-MRI is a non-invasive technique that quantifies tissue perfusion by using magnetically labelled arterial blood water protons as an endogenous tracer [21]. The liver perfusion and tissue arrival time measured using ASL-MRI have been shown to correlate with HVPG [19].

Structural and Architectural Changes

Native (non-contrast) longitudinal relaxation time (T1) can detect pathologically important processes in tissues [22] and is an established composite marker of liver inflammation and fibrosis [23, 24]. Furthermore, liver T1 correlated with HVPG, and this relationship was maintained in patients with CSPH [19]. The T1 measurement was respiratory-triggered and multi-slice, and therefore a large volume of the liver could be sampled in a reasonable timeframe. Interestingly, the distribution of liver T1 values was shown to increase with worsening of PH, reflecting the increasing heterogeneity of T1 values across the liver volume. This underscores the sampling variability associated with liver biopsy (and potentially transient elastography).

A subsequent study reported an association between splenic T1 and HVPG but failed to show a correlation with liver T1 [25]. The methodology used to measure T1 could potentially explain this difference. In this study, a modified look-locker inversion recovery (MOLLI) T1 mapping method was used which requires breath-holding for every image slice acquired. The T1 measurements were obtained from regions of interest drawn on the liver and spleen and therefore could be susceptible to sampling variability due to the heterogeneity of T1 values across the organ. In addition, it has been shown that the hepatic fat content can be large enough to cause substantial MOLLI T1 alterations [26].

An MRI-based scoring system of the features of PH was studied as a surrogate measure of HVPG [27]. The PH score included the number of variceal sites, volume of ascites and maximum splenic diameter, with scores between 0 and 3 in each domain yielding a total score of 0 to 9 for each patient. The PH score correlated with HVPG and the AUROC for detection of PH and CSPH was 0.78 and 0.83, respectively.

Dynamic Contrast-Enhanced MRI (DCE-MRI)

Perfusion-weighted MRI can be performed by measuring the signal intensity in the tissue of interest after injection of contrast against time. Initial assumption of a linear relationship between the signal intensity and the concentration of gadolinium in the liver using a single-input single-compartment model is simplistic and inaccurate, as it does not take into account the separate portal venous and hepatic arterial contributions [28]. However, subsequent analysis of signal intensity over the portal vein, aorta and liver parenchyma against time can be fitted to a dual-input single-compartment model [29].

Decreased portal fraction, total liver perfusion, increased arterial fraction as well as increased mean transit time (MTT) were related to severity of PH [30]. Significant correlation of DCE-MRI parameters including contrast time-to-peak, liver distribution volume and liver upslope correlated with HVPG [31].

4D flow MRI mapping with gadolinium-based contrast allows 3D vascular coverage and is a promising technique for comprehensive haemodynamic analyses.

Increased 4D flow parameters in the splanchnic circulation (splenic artery peak velocity, superior mesenteric vein, and splenic vein flow) were related to severity of PH. However, HVPG was not measured in this study and the composite PH score [27] was used as a surrogate measure of portal pressure.

The potential adverse events associated with the use of contrast agents limit the use of DCE-MRI in patients with chronic liver disease. Gadolinium-based contrast agents are reported to cause nephrogenic systemic fibrosis in patients with renal failure [32].

Combination MRI Markers

The combination of liver T1 and splenic artery velocity correlated with HVPG [19]. These MR markers reflect the underlying pathophysiological changes (structural and haemodynamic) in the development and progression of PH. Moreover, the linear model provided good prediction of HVPG across the spectrum of HVPG values from normal to CSPH and performed better than liver T1 or splenic artery velocity alone.

Magnetic Resonance Elastography (MRE)

Magnetic resonance elastography (MRE) is an alternative to the ultrasonography-based method to evaluate liver and spleen stiffness. Recent studies have validated the use of MRE for evaluating liver fibrosis [33]. MRE has the theoretical advantage over ultrasonography-based elastography methods of evaluating stiffness over a larger area of liver, hence reducing sampling variability. Using MRE, the shear modulus can be assessed using either a 2D or 3D technique. MRE requires special hardware and software which could limit its widespread use.

In a recent meta-analysis, the diagnostic accuracy of spleen stiffness was higher than liver stiffness using MRE. The AUROC for detection of CSPH was 0.88 and 0.92 for liver and spleen stiffness, respectively [34]. The correlation of MRE-measured liver and spleen stiffness with HVPG is summarised in Table 13.2.

Table 13.2 Summary of the relationship between MRE measured liver and spleen stiffness with portal hypertension

Parameter	Sample size, *n*	Study	Correlation with HVPG (Correlation coefficient, *r*)
Liver stiffness	34	Wagner et al. 2018 [31]	0.486
	36	Ronot et al. 2014 [35]	0.44
	15	Gharib et al. 2017 [36]	0.64
	52	Danielsen et al. 2021 [37]	0.96
Spleen stiffness	34	Wagner et al. 2018 [31]	0.099 (NS)
	36	Ronot et al. 2014 [35]	0.57
	52	Danielsen et al. 2021 [37]	0.97

NS not significant

Serum Markers

Simple liver fibrosis scores were developed using combinations of routine blood tests as indirect markers of liver scarring. Although these markers only correlate moderately with HVPG, some (e.g., Lok score) can diagnose CSPH and the presence of varices (Table 13.3). Thrombocytopenia is an important indication of PH and many of the simple marker panels contain platelet count. The Enhanced Liver Fibrosis (ELF) test is derived from direct markers related to hepatic extracellular matrix turnover and has been extensively validated for the non-invasive assessment of liver fibrosis [38]. The direct and indirect markers of fibrosis perform well in identifying advanced liver fibrosis and early stages of PH when it is largely driven by increased intrahepatic vascular resistance due to structural changes. However, these markers are unlikely to reflect the haemodynamic changes that occur with severe PH. The ALBI score was originally devised as a measure of liver function in patients with hepatocellular carcinoma (HCC) [39]. There was a weak positive correlation between ALBI score and HVPG ($r = 0.307$, $P < 0.001$) [40]. The ALBI score has also been shown to predict patients at risk of decompensation [41].

sCD163, a scavenger receptor expressed on macrophages, is a specific marker of macrophage activation and is related to the severity of cirrhosis and PH [42]. Combination of ELF and sCD163 had a superior diagnostic accuracy in identifying CSPH compared to each component individually (AUROC of 0.82, 0.88 and 0.90 for sCD163, ELF and combination, respectively) [43]. Von Willebrand factor (vWF) is related to endothelial dysfunction and circulating levels of vWF correlated with HVPG [44]. vWF was also reported to be related to bacterial translocation and inflammation and associated with clinical outcomes independent of HVPG [45]. The VITRO score is calculated as the ratio of vWF to thrombocytes, and the diagnostic accuracy of the VITRO score for detecting cirrhosis [46] and CSPH [47] was superior to vWF alone.

Indocyanine green (ICG) is administered intravenously and nearly exclusively extracted by the hepatic parenchyma and rapidly excreted in bile. Therefore, ICG clearance, which is quantitatively assessed by spectrophotometry, reflects both hepatic function and hepatic blood flow. The ICG 15-minute retention test (ICG-R15) is performed on peripheral blood samples following a bolus injection of ICG and has been shown to be linearly correlated with HVPG ($r = 0.57 - 0.78$) [50, 74, 75]. The ICG-R15 can also be assessed in vivo using pulse dye densitometry finger probes, but this has not yet been correlated with HVPG.

The HepQuant SHUNT test quantifies hepatic function by simultaneously measuring flow-dependent clearance of cholate from both portal and systemic circulations. In a small study of 20 patients, the SHUNT test was shown to correlate with HVPG [76]. The derived disease severity index (DSI) has been shown to predict decompensation independent of MELD [77]. The 13C-methacetin breath test is another potential method to assess hepatic function and, in a study of 155 patients with NASH-related cirrhosis, detected CSPH with an AUROC of 0.83 [78].

These promising results with different liver 'clearance' tests need further validation in large multicentre studies.

Table 13.3 Summary of Serum Marker Tests in Evaluating Portal Hypertension; Correlation with Hepatic Venous Pressure Gradient (HVPG), Diagnostic Accuracy in Estimating Clinically Significant Portal Hypertension (CSPH), and High-Risk Varices (HRV)/Varices Needing Treatment (VNT) Liver Clearance Tests

Blood-based biomarkers		Sample size, *n*	Correlation with HVPG (Correlation coefficient, r)	Diagnosis of CSPH (AUROC)	Diagnosis of HRV/ VNT (AUROC)	Notes
Indirect fibrosis markers						
ALBI *(albumin, bilirubin)*	Hsieh et al. [40]	242	0.31			Retrospective, predominantly viral hepatitis (81%)
Lok index *(platelet, AST, ALT, INR)*	Zhou et al. [48]	132			0.81	Retrospective, CHB cirrhosis patients who did not meet Baveno VI criteria
	Hsieh et al. [40]	242	0.30			Retrospective, predominantly viral hepatitis (81%)
	Cho et al. [49]	219		0.76 (cut-off 0.8)	0.65 (cut-off 1.5)	Retrospective, alcohol-related cirrhosis
	Lisotti et al. [50]	96		0.83		Prospective, mixed etiology
	Sebastiani et al. [51]	510			0.70 (cut-off 1.5)	Retrospective, patients with cirrhosis and gastroscopy
	Wang et al. [52]	238		0.74 (cut-off 1.3)		Retrospective, mixed etiology
	Hassan et al. [53]	65			0.72 (cut-off 0.7)	Prospective, CHC cirrhosis
	Stefanescu et al. [54]	231			0.73 (cut-off 0.796)	Prospective, biopsy proven cirrhosis secondary to alcohol and CHC
	Farid et al. [55]	277			0.72	Prospective, CHC (Egypt)
	Alam et al. [56]	153			0.6 (cut-off 0.62)	Prospective, CHC cirrhosis (Pakistan)
Forns' index *(platelets, GGT, age, cholesterol)*	Cho et al. [49]	219		0.64 (cut-off 8.9)	0.52 (cut-off 9.1)	Retrospective, alcohol-related cirrhosis
	Sebastiani et al. [51]	510			0.66 (cut-off 8.8)	Retrospective, patients with cirrhosis and gastroscopy
	Wang et al. [52]	238		0.66 (cut-off 11.05)		Retrospective, mixed etiology
	Siregar et al. [57]	51			0.72 (cut-off 7.92)	Retrospective, predominantly CHB and CHC cirrhosis
	Hassan et al. [53]	65			0.73 (cut-off 6.9)	Prospective, CHC cirrhosis
	Stefanescu et al. [54]	231			0.65 (cut-off 8.54)	Prospective, biopsy proven cirrhosis secondary to alcohol and CHC

Table 13.3 (continued)

Blood-based biomarkers		Sample size, *n*	Correlation with HVPG (Correlation coefficient, r)	Diagnosis of CSPH (AUROC)	Diagnosis of HRV/VNT (AUROC)	Notes
AST-to-ALT ratio (AAR)	Lisotti et al. [50]	96		0.71		Prospective, mixed etiology
	Sebastiani et al. [51]	510			0.64 (cut-off 1.1)	Retrospective, patients with cirrhosis and gastroscopy
	Wang et al. [52]	238		0.57 (cut-off 1.59)		Retrospective, mixed etiology
	Farid et al. [55]	277			0.58	CHC (Egypt), prospective
Fibrosis-4 (FIB-4) score *(platelet count, AST, ALT, age)*	Hsieh et al. [40]	242	0.27			Retrospective, predominantly viral hepatitis (81%)
	Zhou et al. [48]	132			0.59 (NS)	Retrospective, CHB cirrhosis patients who did not meet Baveno VI criteria
	Cho et al. [49]	219		0.65 (cut-off 4.1)	0.56 (cut-off 2.6)	Retrospective, alcohol-related cirrhosis
	Lisotti et al. [50]	96		0.766		Prospective, mixed etiology
	Sebastiani et al. [51]	510			0.63 (cut-off 4.3)	Retrospective, patients with cirrhosis and gastroscopy
	Wang et al. [52]	238		0.69 (cut-off 2.72)		Retrospective, mixed etiology
	Hassan et al. [53]	65			0.76 (cut-off 3.3)	Prospective, CHC cirrhosis
	Stefanescu et al. [54]	231			0.63 (cut-off 6.75)	Prospective, biopsy proven cirrhosis secondary to alcohol and CHC
	Farid et al. [55]	277			0.7	CHC (Egypt), prospective
	Alam et al. [56]	153			0.6 (cut-off 3.07)	Prospective, CHC cirrhosis (Pakistan)

Table 13.3 (continued)

Blood-based biomarkers		Sample size, *n*	Correlation with HVPG (Correlation coefficient, r)	Diagnosis of CSPH (AUROC)	Diagnosis of HRV/VNT (AUROC)	Notes
AST to platelet ratio index (APRI)	Hsieh et al. [40]	242	0.24			Retrospective, predominantly viral hepatitis (81%)
	Zhou et al. [48]	132			0.59 (NS)	Retrospective, CHB cirrhosis patients who did not meet Baveno VI criteria
	Cho et al. [49]	219		0.64 (cut-off 1.0)	0.42 (cut-off 1.2)	Retrospective, alcohol-related cirrhosis
	Lisotti et al. [50]	96		0.74		Prospective, mixed etiology
	Sebastiani et al. [51]	510			0.57 (cut-off 1.5)	Retrospective, patients with cirrhosis and gastroscopy
	Hametner et al. [58]	236		0.62 (cut-off 1.74)		Retrospective, mixed etiology
	Wang et al. [52]	238		0.74 (cut-off 0.73)		Retrospective, mixed etiology
	Stefanescu et al. [42]	231			0.54 (cut-off 2.2)	Prospective, biopsy proven cirrhosis secondary to alcohol and CHC
	Farid et al. [55]	277			0.63	CHC (Egypt), prospective
	Alam et al. [56]	153			0.6 (cut-off 1.03)	Prospective, CHC cirrhosis (Pakistan)
Cirrhosis discriminant score (CDS) *(platelet count, ALT/AST ratio, INR)*	Hsieh et al. [40]	242	0.26			Retrospective, predominantly viral hepatitis (81%)
	Alam et al. [56]	153			0.6 (cut-off 6.5)	Prospective, CHC cirrhosis (Pakistan)
Goteborg university cirrhosis index (GUCI) *(AST, INR)*	Hsieh et al. [40]	242	0.21			Retrospective, predominantly viral hepatitis (81%)
	Farid et al. [55]	277			0.66	CHC (Egypt), prospective
	Alam et al. [56]	153			0.6 (cut-off 1.02)	Prospective, CHC cirrhosis (Pakistan)

(continued)

Table 13.3 (continued)

Blood-based biomarkers		Sample size, *n*	Correlation with HVPG (Correlation coefficient, r)	Diagnosis of CSPH (AUROC)	Diagnosis of HRV/VNT (AUROC)	Notes
FibroIndex *(platelet count, AST, gamma globulin)*	Sebastiani et al. [51]	510			0.65 (cut-off 2.5)	Retrospective, patients with cirrhosis and gastroscopy
Kings score *(age, AST, INR, platelet count)*	Wang et al. [52]	238		0.76 (cut-off 23.47)		Retrospective, mixed etiology
	Alam et al. [40]	153			0.6 (cut-off 20)	Prospective, CHC cirrhosis (Pakistan)
P2/MS (platelet count $[10^9/L])^2/$ (monocyte fraction [%] × segmented neutrophil fraction [%]) *(platelet count, monocyte fraction, segmented neutrophil fraction)*	Cho et al. [49]	219		0.67 (cut-off 60.2)	0.47 (cut-off 69.4)	Retrospective, alcohol-related cirrhosis
Direct fibrosis markers						
ELF	Hametner et al. [58]	236		0.68 (cut-off 11.4)		Retrospective, mixed etiology
	Palaniyappan et al. [19]	30	0.758			Prospective, mixed etiology
	Sandahl et al. [43]	80		0.88		Prospective, mixed etiology
	Mauro et al. [59]	112	0.671		0.884 (cut-off 10.83)	HCV infected OLT recipients achieving SVR
	Frankova et al. [60]	109	0.349			Liver transplant candidates, mixed etiology
	Ishida et al. [61]	127			0.48 (cut-off 11.75)	Retrospective (Japan), mixed etiology
	Simbrunner et al. [62]	201	0.443	0.833 (cut-off 10.5)	0.552	Prospective, mixed etiology

Table 13.3 (continued)

Blood-based biomarkers		Sample size, *n*	Correlation with HVPG (Correlation coefficient, r)	Diagnosis of CSPH (AUROC)	Diagnosis of HRV/VNT (AUROC)	Notes
HA	Palaniyappan et al. [19]	30	0.752			Prospective, mixed etiology
	Sandahl et al. [43]	80		0.86		Prospective, mixed etiology
	Frankova et al. [60]	109	0.288			Liver transplant candidates, mixed etiology
	Ishida et al. [61]	127			0.50 (cut-off 110.63)	Retrospective (Japan), mixed etiology
	Simbrunner et al. [62]	201	0.419	0.828 (cut-off 71.4)		Prospective, mixed etiology
TIMP1	Busk et al. [63]	84	0.40			Retrospective, alcohol
	Palaniyappan et al. [19]	30	0.512			Prospective, mixed etiology
	Sandahl et al. [43]	80		0.85		Prospective, mixed etiology
	Frankova et al. [60]	109	0.434			Liver transplant candidates, mixed etiology
	Ishida et al. [61]	127			0.48 (cut-off 379.9)	Retrospective (Japan), mixed etiology
	Simbrunner et al. [62]	201	0.368	0.722 (cut-off 281.4)		Prospective, mixed etiology
PIIINP	Palaniyappan et al. [19]	30	0.607			Prospective, mixed etiology
	Sandahl et al. [43]	80		0.74		Prospective, mixed etiology
	Frankova et al. [60]	109	0.271			Liver transplant candidates, mixed etiology
	Ishida et al. [61]	127			0.48 (cut-off 0.60)	Retrospective (Japan), mixed etiology
	Simbrunner et al. [62]	201	0.332	0.748 (cut-off 16.9)		Prospective, mixed etiology
FibroTest *(α2-macro-globulin, haptoglobin, apolipoprotein A1, bilirubin, GGT, age, gender)*	Thabut et al. [64]	130	0.58			Prospective, mixed etiology
	Thabut et al. [65]	99			0.77	Retrospective
Procollagen type V (pro-C5)	Leeming et al. [66]	94	0.33	0.73 (cut-off 330)		Retrospective, alcohol cirrhosis (90%)

(continued)

Table 13.3 (continued)

Blood-based biomarkers		Sample size, *n*	Correlation with HVPG (Correlation coefficient, r)	Diagnosis of CSPH (AUROC)	Diagnosis of HRV/VNT (AUROC)	Notes
Pro-peptide of type III collagen (pro-C3)	Jansen et al. [67]	58	0.354			Retrospective, HIV/HCV co-infection
Osteopontin	Bruha et al. [68]	154	0.25	0.763 (cut-off 80 ng/mL)		Retrospective
	Frankova et al. [60]	109	0.514			Liver transplant candidates, mixed etiology
Markers of inflammation						
Soluble CD163 (sCD163)	Holland-Fisher et al. [69]	36	0.49 [portal venous pressure gradient (PVPG) measured during TIPS]			Prospective, mixed etiology
	Grønbæk et al. [42]	81	$R^2 = 0.90$ (hyperbolic model, Michaelis-Menten function)	0.83 (cut-off 3.95)		Prospective, mixed etiology
	Sandahl et al. [43]	80		0.82		Prospective, mixed etiology
Combination of sCD163 and ELF	Sandahl et al. [43]	80		0.91		Prospective, mixed etiology
Markers of endothelial dysfunction						
Von Willebrand factor (vWF)	La Mura et al. [70]	42	0.47			Prospective, mixed etiology
	Ferlitsch et al. [71]	286	0.687	0.884 (cut-off 241)		Prospective, mixed etiology
	Horvatits et al. [72]	61	0.43			Prospective, mixed etiology
	Wu et al. [73]	60	0.696	0.885 (cut-off 1510.5)	0.83 (cut-off 1990)	Retrospective, cirrhosis due to chronic hepatitis B
	Hametner et al. [58]	236		0.79 (cut-off 226)		Retrospective, mixed etiology
	Mandorfer et al. [45]	225	0.333			Retrospective, mixed etiology
VITRO test (vWF/thrombocyte ratio)	Hametner et al. [58]	236		0.86 (cut-off 1.58)		Retrospective, mixed etiology

Non-invasive Methods for Assessment of HVPG Response

The non-invasive assessment of haemodynamic response following treatment for PH has been considered an unmet need in hepatology. Historically, a reduction in heart rate (HR) was assessed as a proxy for therapeutic response to non-selective beta-blockers (NSBB). However, changes in heart rate do not correlate with the changes in HVPG [79, 80]. Nevertheless, the absolute benefit of identifying HVPG 'responders' is not fully established. Indeed, in the PREDESCI study, NSBB treatment without using portal pressure response in the follow-up to guide therapy improved decompensation-free survival. There are two important attributes for a non-invasive test to reliably monitor HVPG response. Firstly, the test should correlate with HVPG across a broad spectrum of HVPG values. As discussed previously, most non-invasive methodologies have been developed and validated as a binary predictor of CSPH and/or presence of varices, but the data on correlation with HVPG as a continuous variable are limited. Secondly, the inherent variability of the measurement should be small enough to detect the relatively modest changes in HVPG that may occur with pharmacological treatments. The haemodynamic response is defined by 10%–20% changes in HVPG from baseline which could correspond to absolute pressure changes as small as 2–4 mmHg. It follows that any non-invasive test with significant variability will lack sufficient sensitivity to detect the minor differences in HVPG. Notwithstanding, in the context of clinical trials, the within-individual variance of HVPG itself is a potential confounder in evaluating the haemodynamic response to interventions, especially in decompensated patients [81].

Doppler US-based assessment of blood flow showed some promise in detecting HVPG response following treatment with terlipressin [82] and propranolol [83]. However, the initial results have not been reproduced [84]. This is likely due to the technical variation associated with Doppler US which limits the ability to reliably detect changes in HVPG. PC-MRI is another potential method to non-invasively evaluate blood flow alterations. In a small feasibility study, a reduction in cardiac output as measured by PC-MRI flow in the abdominal aorta was reported following NSBB, but there were no statistically significant changes in flow in the other vessels analysed [85]. In this study, there were also no contemporaneous HVPG measurements.

Spleen stiffness measured using MRE significantly decreased following the administration of intravenous NSBB, but no change was observed in liver stiffness [37]. However, the changes in spleen stiffness were not related to HVPG response.

In addition, non-imaging-based markers of HVPG response have been evaluated. The expression of specific vasoactive proteins of Ras homolog family member A (RhoA) and Rho-kinase (ROCK) pathway in the gastric mucosa correlated with acute haemodynamic response following intravenous propranolol [86]. The haemodynamic responders showed lower expression of beta-arrestin2 (βArr2) in antral biopsies. This is not strictly a non-invasive test as the tissue samples are obtained by upper gastrointestinal endoscopy. Alternatively, using a serum metabolomic profiling approach, the combination of two metabolites (phosphatidylcholine and eicosadienoic acid) also identified acute HVPG responders to intravenous propranolol with an AUROC of 0.801 [87].

Conclusion

Although HVPG is an invasive and highly specialised method for the diagnosis of PH and assessment of treatment response, it has an important role in specific clinical circumstances and in interventional trials. A number of non-invasive tests (other than ultrasound elastography) have been shown to correlate with HVPG and perform well for the diagnosis of CSPH [59], but further validation in larger cohorts of patients with diverse etiologies is generally required. Variability and reproducibility will remain a challenge for development of suitable PH monitoring biomarkers.

References

1. Kim G, Shim KY, Baik SK. Diagnostic accuracy of hepatic vein arrival time performed with contrast-enhanced ultrasonography for cirrhosis: a systematic review and meta-analysis. Gut Liver. 2017;11(1):93–101.
2. Shimada T, et al. Impact of splenic circulation: non-invasive microbubble-based assessment of portal hemodynamics. Eur Radiol. 2015;25(3):812–20.
3. Halldorsdottir VG, et al. Subharmonic contrast microbubble signals for noninvasive pressure estimation under static and dynamic flow conditions. Ultrason Imaging. 2011;33(3):153–64.
4. Eisenbrey JR, et al. Chronic liver disease: noninvasive subharmonic aided pressure estimation of hepatic venous pressure gradient. Radiology. 2013;268(2):581–8.
5. Gupta I, et al. Diagnosing portal hypertension with noninvasive subharmonic pressure estimates from a US contrast agent. Radiology. 2021;298(1):104–11.
6. Amat-Roldan I, et al. Assessment of hepatic vascular network connectivity with automated graph analysis of dynamic contrast-enhanced US to evaluate portal hypertension in patients with cirrhosis: a pilot study. Radiology. 2015;277(1):268–76.
7. Berzigotti A, et al. Non-invasive measurement of HVPG using graph analysis of dynamic contrast-enhanced ultrasound: the CLEVER study. J Hepatol. 2018;68:S76–7.
8. Iranmanesh P, et al. Accurate computed tomography-based portal pressure assessment in patients with hepatocellular carcinoma. J Hepatol. 2014;60(5):969–74.
9. Qi X, et al. Virtual portal pressure from anatomic CT angiography. J Hepatol. 2014;61(1):180–1.
10. Qi X, et al. Insufficient accuracy of computed tomography-based portal pressure assessment in hepatitis B virus-related cirrhosis: an analysis of data from CHESS-1601 trial. J Hepatol. 2018;68(1):210–1.
11. Lu MT, et al. Noninvasive FFR derived from coronary CT angiography management and outcomes in the PROMISE trial. JACC Cardiovasc Imaging. 2017;10(11):1350–8.
12. Qi XL, et al. Virtual hepatic venous pressure gradient with CT angiography (CHESS 1601): a prospective multicenter study for the noninvasive diagnosis of portal hypertension. Radiology. 2019;290(2):370–7.
13. Liu FQ, et al. Development and validation of a radiomics signature for clinically significant portal hypertension in cirrhosis (CHESS1701): a prospective multicenter study. EBioMedicine. 2018;36:151–8.
14. Andreucci M, et al. Update on the renal toxicity of iodinated contrast drugs used in clinical medicine. Drug Healthc Patient Saf. 2017;9:25–37.
15. Burkart DJ, et al. Volumetric flow-rates in the portal venous system—measurement with cine phase-contrast Mr imaging. Am J Roentgenol. 1993;160(5):1113–8.
16. Gouya H, et al. Chronic liver disease: systemic and splanchnic venous flow mapping with optimized cine phase-contrast MR imaging validated in a phantom model and prospectively evaluated in patients. Radiology. 2011;261(1):144–55.
17. Pelc LR, et al. Arterial and venous-blood flow—noninvasive quantitation with Mr imaging. Radiology. 1992;185(3):809–12.

18. Gouya H, et al. Portal hypertension in patients with cirrhosis: indirect assessment of hepatic venous pressure gradient by measuring azygos flow with 2D-cine phase-contrast magnetic resonance imaging. Eur Radiol. 2016;26(7):1981–90.
19. Palaniyappan N, et al. Non-invasive assessment of portal hypertension using quantitative magnetic resonance imaging. J Hepatol. 2016;65(6):1131–9.
20. Chouhan MD, et al. Caval subtraction 2D phase-contrast MRI to measure Total liver and hepatic arterial blood flow proof-of-principle, correlation with portal hypertension severity and validation in patients with chronic liver disease. Investig Radiol. 2017;52(3):170–6.
21. Odudu A, et al. Arterial spin labelling MRI to measure renal perfusion: a systematic review and statement paper. Nephrol Dial Transplant. 2018;33(Suppl 2):ii15–21.
22. Moon JC, et al. Myocardial T1 mapping and extracellular volume quantification: a Society for Cardiovascular Magnetic Resonance (SCMR) and CMR working Group of the European Society of cardiology consensus statement. J Cardiovasc Magn Reson. 2013;15:92.
23. Banerjee R, et al. Multiparametric magnetic resonance for the non-invasive diagnosis of liver disease. J Hepatol. 2014;60(1):69–77.
24. Hoad CL, et al. A study of T1 relaxation time as a measure of liver fibrosis and the influence of confounding histological factors. NMR Biomed. 2015;28(6):706–14.
25. Levick C, et al. Non-invasive assessment of portal hypertension by multi-parametric magnetic resonance imaging of the spleen: a proof of concept study. PLoS One. 2019;14(8):e0221066.
26. Mozes FE, et al. Influence of fat on liver T1 measurements using modified look-locker inversion recovery (MOLLI) methods at 3T. J Magn Reson Imaging. 2016;44(1):105–11.
27. Kihira S, et al. Non-invasive prediction of portal pressures using CT and MRI in chronic liver disease. Abdom Radiol (NY). 2016;41(1):42–9.
28. Pandharipande PV, et al. Perfusion imaging of the liver: current challenges and future goals. Radiology. 2005;234(3):661–73.
29. Materne R, et al. Assessment of hepatic perfusion parameters with dynamic MRI. Magn Reson Med. 2002;47(1):135–42.
30. Annet L, et al. Hepatic flow parameters measured with MR imaging and Doppler US: correlations with degree of cirrhosis and portal hypertension. Radiology. 2003;229(2):409–14.
31. Wagner M, et al. Noninvasive prediction of portal pressure with MR Elastography and DCE-MRI of the liver and spleen: preliminary results. J Magn Reson Imaging. 2018;48(4):1091–103.
32. Thomsen HS, Marckmann P, Logager VB. Update on nephrogenic systemic fibrosis. Magn Reson Imaging Clin N Am. 2008;16(4):551–60.
33. Selvaraj EA, et al. Diagnostic accuracy of elastography and magnetic resonance imaging in patients with NAFLD: a systematic review and meta-analysis. J Hepatol. 2021;75(4):770–85.
34. Singh R, et al. Accuracy of liver and spleen stiffness on magnetic resonance elastography for detecting portal hypertension: a systematic review and meta-analysis. Eur J Gastroenterol Hepatol. 2021;32(2):237–45.
35. Ronot M, et al. Assessment of portal hypertension and high-risk oesophageal varices with liver and spleen three-dimensional multifrequency MR elastography in liver cirrhosis. Eur Radiol. 2014;24(6):1394–402.
36. Gharib AM, et al. Magnetic resonance Elastography shear wave velocity correlates with liver fibrosis and hepatic venous pressure gradient in adults with advanced liver disease. Biomed Res Int. 2017;2017:2067479.
37. Danielsen KV, et al. Using MR elastography to assess portal hypertension and response to beta-blockers in patients with cirrhosis. Liver Int. 2021;41(9):2149–58.
38. Guha IN, et al. Noninvasive markers of fibrosis in nonalcoholic fatty liver disease: validating the European liver fibrosis panel and exploring simple markers. Hepatology. 2008;47(2):455–60.
39. Johnson PJ, et al. Assessment of liver function in patients with hepatocellular carcinoma: a new evidence-based approach-the ALBI grade. J Clin Oncol. 2015;33(6):550–8.
40. Hsieh YC, et al. Correlation and prognostic accuracy between noninvasive liver fibrosismarkers and portal pressure in cirrhosis: role of ALBI score. PLoS One. 2018;13(12):e0208903.
41. Guha IN, et al. Validation of a model for identification of patients with compensated cirrhosis at high risk of decompensation. Clin Gastroenterol Hepatol. 2019;17(11):2330–2338.e1.
42. Grønbaek H, et al. Soluble CD163, a marker of Kupffer cell activation, is related to portal hypertension in patients with liver cirrhosis. Aliment Pharmacol Ther. 2012;36(2):173–80.

43. Sandahl TD, et al. The macrophage activation marker sCD163 combined with markers of the enhanced liver fibrosis (ELF) score predicts clinically significant portal hypertension in patients with cirrhosis. Aliment Pharmacol Ther. 2016;43(11):1222–31.
44. Zou Z, et al. von Willebrand factor as a biomarker of clinically significant portal hypertension and severe portal hypertension: a systematic review and meta-analysis. BMJ Open. 2019;9(8):e025656.
45. Mandorfer M, et al. Von Willebrand factor indicates bacterial translocation, inflammation, and procoagulant imbalance and predicts complications independently of portal hypertension severity. Aliment Pharmacol Ther. 2018;47(7):980–8.
46. Maieron A, et al. Von Willebrand factor as a new marker for non-invasive assessment of liver fibrosis and cirrhosis in patients with chronic hepatitis C. Aliment Pharmacol Ther. 2014;39(3):331–8.
47. Mandorfer M, et al. Changes in hepatic venous pressure gradient predict hepatic Decompensation in patients who achieved sustained Virologic response to interferon-free therapy. Hepatology. 2020;71(3):1023–36.
48. Zhou H, et al. Liver stiffness and serum markers for excluding high-risk varices in patients who do not meet Baveno VI criteria. World J Gastroenterol. 2019;25(35):5323–33.
49. Cho EJ, et al. Diagnostic and prognostic values of noninvasive predictors of portal hypertension in patients with alcoholic cirrhosis. PLoS One. 2015;10(7):e0133935.
50. Lisotti A, et al. Indocyanine green retention test as a noninvasive marker of portal hypertension and esophageal Varices in compensated liver cirrhosis. Hepatology. 2014;59(2):643–50.
51. Sebastiani G, et al. Prediction of oesophageal varices in hepatic cirrhosis by simple serum non-invasive markers: results of a multicenter, large-scale study. J Hepatol. 2010;53(4):630–8.
52. Wang L, et al. Diagnostic efficacy of noninvasive liver fibrosis indexes in predicting portal hypertension in patients with cirrhosis. PLoS One. 2017;12(8):e0182969.
53. Hassan EM, et al. Can transient elastography, Fib-4, Forns index, and Lok score predict esophageal varices in HCV-related cirrhotic patients? Gastroenterol Hepatol. 2014;37(2):58–65.
54. Stefanescu H, et al. A new and simple algorithm for the noninvasive assessment of esophageal varices in cirrhotic patients using serum fibrosis markers and transient elastography. J Gastrointestin Liver Dis. 2011;20(1):57–64.
55. Farid K, et al. Development and evaluation of a novel score for prediction of large oesophageal varices in patients with hepatitis c virus-induced liver cirrhosis. Br J Biomed Sci. 2017;74(3):138–43.
56. Alam L, Saeed F. Non-invasive tools to assess the risk of varices-needing-treatment in cirrhosis secondary to HCV. PAFMJ. 2021;71:12–7.
57. Siregar RA, Dairi LB, Siregar GA. Forns index as a useful noninvasive predictor of esophageal varices in liver cirrhosis. Universa Medicina. 2016;35(3):199–205.
58. Hametner S, et al. The VITRO score (Von Willebrand factor antigen/thrombocyte ratio) as a new marker for clinically significant portal hypertension in comparison to other non-invasive parameters of fibrosis including ELF test. PLoS One. 2016;11(2):e0149230.
59. Mauro E, et al. Portal pressure and liver stiffness measurements in the prediction of fibrosis regression after sustained virological response in recurrent hepatitis C. Hepatology. 2018;67(5):1683–94.
60. Frankova S, et al. Liver stiffness measured by two-dimensional shear-wave elastography predicts hepatic vein pressure gradient at high values in liver transplant candidates with advanced liver cirrhosis. PLoS One. 2021;16(1):e0244934.
61. Ishida K, et al. Accuracy of fibrosis-4 index in identification of patients with cirrhosis who could potentially avoid variceal screening endoscopy. J Clin Med. 2020;9(11):3510.
62. Simbrunner B, et al. Non-invasive detection of portal hypertension by enhanced liver fibrosis score in patients with different aetiologies of advanced chronic liver disease. Liver Int. 2020;40(7):1713–24.
63. Busk TM, et al. TIMP-1 in patients with cirrhosis: relation to liver dysfunction, portal hypertension, and hemodynamic changes. Scand J Gastroenterol. 2014;49(9):1103–10.

64. Thabut D, et al. Relationship between the Fibrotest and portal hypertension in patients with liver disease. Aliment Pharmacol Ther. 2007;26(3):359–68.
65. Thabut D, et al. Non-invasive diagnosis of large oesophageal varices with FibroTest in patients with cirrhosis: a preliminary retrospective study. Liver Int. 2006;26(3):271–8.
66. Leeming DJ, et al. Pro-C5, a marker of true type V collagen formation and fibrillation, correlates with portal hypertension in patients with alcoholic cirrhosis. Scand J Gastroenterol. 2015;50(5):584–92.
67. Jansen C, et al. PRO-C3-levels in patients with HIV/HCV-co-infection reflect fibrosis stage and degree of portal hypertension. PLoS One. 2014;9(9):e108544.
68. Bruha R, et al. Osteopontin: a non-invasive parameter of portal hypertension and prognostic marker of cirrhosis. World J Gastroenterol. 2016;22(12):3441–50.
69. Holland-Fischer P, et al. Kupffer cells are activated in cirrhotic portal hypertension and not normalised by TIPS. Gut. 2011;60(10):1389–93.
70. La Mura V, et al. Von Willebrand factor levels predict clinical outcome in patients with cirrhosis and portal hypertension. Gut. 2011;60(8):1133–8.
71. Ferlitsch M, et al. von Willebrand factor as new noninvasive predictor of portal hypertension, decompensation and mortality in patients with liver cirrhosis. Hepatology. 2012;56(4):1439–47.
72. Horvatits T, et al. von Willebrand factor antigen for detection of hepatopulmonary syndrome in patients with cirrhosis. J Hepatol. 2014;61(3):544–9.
73. Wu H, et al. von Willebrand factor as a novel noninvasive predictor of portal hypertension and esophageal varices in hepatitis B patients with cirrhosis. Scand J Gastroenterol. 2015;50(9):1160–9.
74. Lisotti A, et al. Relationship between indocyanine green retention test, decompensation and survival in patients with child-Pugh a cirrhosis and portal hypertension. Liver Int. 2016;36(9):1313–21.
75. Møller S, et al. Indocyanine green retention test in cirrhosis and portal hypertension: accuracy and relation to severity of disease. J Gastroenterol Hepatol. 2019;34(6):1093–9.
76. Helmke SM, et al. Portal-systemic Shunt fraction measured by the HepQuant-SHUNT test correlates with the hepatic venous pressure gradient (HVPG). Hepatology. 2017;66:348A.
77. Fallahzadeh MA, et al. Predicting clinical decompensation in patients with cirrhosis using the Hepquant-SHUNT test. Aliment Pharmacol Ther. 2021;53(8):928–38.
78. Ilan Y, et al. 13C-Methacetin breath test accurately assesses clinically significant portal hypertension in patients with NASH cirrhosis. J Hepatol. 2017;66(1):S100.
79. Garcia-Tsao G, et al. Short-term effects of propranolol on portal venous pressure. Hepatology. 1986;6(1):101–6.
80. Heebøll S, et al. Propranolol treatment of portal hypertension in cirrhosis patients is better the higher the untreated pressure: a single-Centre prospective experience. Scand J Gastroenterol. 2013;48(8):969–73.
81. Bai W, et al. Test-retest reliability and consistency of HVPG and impact on trial design: a study in 289 patients from 20 randomized controlled trials. Hepatology. 2021;74(6):3301–15.
82. Baik SK, et al. Recent variceal bleeding: Doppler US hepatic vein waveform in assessment of severity of portal hypertension and vasoactive drug response. Radiology. 2006;240(2):574–80.
83. Kim MY, et al. Damping index of Doppler hepatic vein waveform to assess the severity of portal hypertension and response to propranolol in liver cirrhosis: a prospective nonrandomized study. Liver Int. 2007;27(8):1103–10.
84. Choi YJ, et al. Comparison of Doppler ultrasonography and the hepatic venous pressure gradient in assessing portal hypertension in liver cirrhosis. J Gastroenterol Hepatol. 2003;18(4):424–9.
85. McDonald N, et al. Assessment of Haemodynamic response to nonselective Beta-blockers in portal hypertension by phase-contrast magnetic resonance angiography. Biomed Res Int, vol. 2017; 2017. p. 9281450.
86. Trebicka J, et al. Assessment of response to beta-blockers by expression of beta Arr2 and RhoA/ROCK2 in antrum mucosa in cirrhotic patients. J Hepatol. 2016;64(6):1265–73.
87. Reverter E, et al. Metabolomics discloses potential biomarkers to predict the acute HVPG response to propranolol in patients with cirrhosis. Liver Int. 2019;39(4):705–13.

Noninvasive Surrogates for cACLD, CSPH, Varices: Consensus Statements of Panel 2

14

Annalisa Berzigotti, Joan Genescà, Juan G. Abraldes, Jonathan A. Fallowfield, and Maja Thiele

Definition of Compensated Advanced Chronic Liver Disease (cACLD)

2.1 The use of elastography in clinical practice has allowed the early identification of patients with untreated/active chronic liver disease (CLD) at risk of having clinically significant portal hypertension (CSPH) and consequently, at risk of developing decompensation and liver-related death (A;1). (Changed)

A. Berzigotti
Department of Visceral Surgery and Medicine, Inselspital, Berne University Hospital, Bern, Switzerland
e-mail: annalisa.berzigotti@insel.ch

J. Genescà (✉)
Liver Unit, Hospital Universitari Vall d'Hebron, Universitat Autònoma de Barcelona, Vall d'Hebron Research Institute (VHIR), Vall d'Hebron Barcelona Hospital Campus, Barcelona, Spain

CIBERehd, Instituto de Salud Carlos III, Madrid, Spain
e-mail: jgenesca@vhebron.net

J. G. Abraldes
Liver Unit, Division of Gastroenterology, University of Alberta, Edmonton, AB, Canada
e-mail: Juan.g.abraldes@ualberta.ca

J. A. Fallowfield
University of Edinburgh Centre for Inflammation Research, The Queen's Medical Research Institute, Edinburgh BioQuarter, Edinburgh, UK
e-mail: Jonathan.Fallowfield@ed.ac.uk

M. Thiele
FLASH Center for Liver Research, Department of Gastroenterology and Hepatology, Odense University Hospital, Odense, Denmark

Department of Clinical Research, University of Southern Denmark, Odense, Denmark
e-mail: maja.thiele@rsyd.dk

R. de Franchis (ed.), *Portal Hypertension VII*,
https://doi.org/10.1007/978-3-031-08552-9_14

2.2 The term "compensated advanced chronic liver disease (cACLD)" had been proposed to reflect the continuum of severe fibrosis and cirrhosis in patients with ongoing CLD. A pragmatic definition of cACLD based on liver stiffness measurement (LSM) is aimed at stratifying the risk of CSPH and decompensation at the point of care, irrespective of histological stage or the ability of LSM to identify these stages. (B;1). (Changed)

2.3 Currently, both terms "cACLD" and "compensated cirrhosis" are acceptable, but not equal (B;1). (Changed)

Criteria to Identify cACLD

2.4 LSM values by transient elastography (TE) <10 kPa in the absence of other known clinical/imaging signs rule out cACLD; values between 10 and 15 kPa are suggestive of cACLD; values >15 kPa are highly suggestive of cACLD (B;1). (Changed)

2.5 CLD patients with LSM < 10 kPa by TE have a negligible 3-year risk (≤1%) of decompensation and liver-related death (A;1) (New)

2.6 Patients with cACLD should be referred to a liver disease specialist for further work up (B;1). (Changed)

2.7 Invasive methods (liver biopsy, hepatic venous pressure gradient—HVPG) can be used for further work up in an individualized manner at referral centers (B;1). (Changed)

Outcome and Prognosis

2.8 LSM (irrespective of the technique used for its measurement) holds prognostic information in cACLD, both at index investigation and during follow-up (A;1). (New)

2.9 A rule of five for LSM by TE (10-15-20-25 kPa) should be used to denote progressively higher relative risks of decompensation and liver-related death independently of the etiology of CLD (B;1) (New)

How to Monitor?

2.10 Patients with LSM values 7–10 kPa and ongoing liver injury should be monitored on a case-by-case basis for changes indicating progression to cACLD (C;2). (New)

2.11 TE could have false-positive results; therefore, index LSM ≥10 kPa should be repeated in fasting conditions as soon as feasible or complemented with an established serum marker of fibrosis (FIB-4 ≥ 2.67, ELF test ≥9.8, FibroTest ≥0.58 for ALD/viral, FibroTest≥0.48 for NAFLD) (B;2). (New)

2.12 In cACLD patients, LSM could be repeated every 12 months to monitor changes (B;2).

2.13 A clinically significant decrease in LSM, which is associated with a substantially reduced risk of decompensation and liver-related death, can be defined as a decrease in LSM of ≥20% associated with LSM < 20 kPa or any decrease to a LSM <10 kPa (C;2). (New)

Diagnosis of CSPH in Patients with cACLD

2.14 Although the concept of CSPH is HVPG-driven, noninvasive tests are sufficiently accurate for estimating CSPH in clinical practice (A;1) (New)

2.15 LSM by TE ≤15 kPa plus platelet count ≥150 × 10^9/L rules out CSPH (sensitivity and negative predictive value >90%) in cACLD patients (B;2). (New)

2.16 In patients with virus and/or alcohol-related cACLD and non-obese (BMI <30 kg/m^2) NASH cACLD, a LSM value by TE ≥25 kPa is sufficient to rule in CSPH (specificity and positive-predictive value >90%), defining the group of patients at risk of having endoscopic signs of PH and at higher risk of decompensation (B;1). (Changed)

2.17 In patients with virus and/or alcohol-related and non-obese NASH cACLD with LSM values <25 kPa, the ANTICIPATE model can be used to predict the risk of CSPH. Based on this model, patients with LSM values between 20 and 25 kPa and platelet count <150 × 10^9/L or LSM values between 15 and 20 kPa and platelet count <110 × 10^9/L have a CSPH risk of at least 60% (B;2). (New)

2.18 In patients with NASH cACLD, the ANTICIPATE-NASH model (including LSM, platelet count, and BMI) may be used to predict the risk of CSPH, but further validation is needed (C;2). (New)

Varices and Screening Endoscopy in Patients That cannot Be Treated With NSSB

2.19 Patients with compensated cirrhosis not candidates for initiating NSBB (contraindication/intolerance) for the prevention of decompensation should undergo an endoscopy for variceal screening if LSM by TE ≥20 kPa or platelet count ≤150 × 10^9/L (A;1). (New)

2.20 Patients avoiding screening endoscopy can be followed up by yearly repetition of TE and platelet count. If LSM increases (≥20 kPa) or platelet count declines (≤150 × 10^9/L), these patients should undergo screening endoscopy (D;1). (Unchanged) (Fig. 14.1)

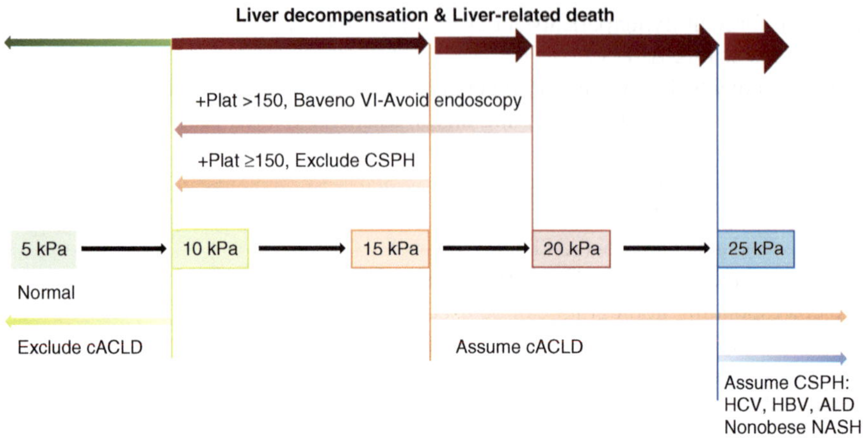

Fig. 14.1 Liver decompensation and liver-related death

Spleen Stiffness

2.21 SSM using TE can be used in cACLD due to viral hepatitis (untreated HCV; untreated and treated HBV) to rule out and rule in CSPH (SSM <21 kPa and SSM >50 kPa, respectively). Validation of the best cutoff using a 100 Hz-specific TE probe, as well as using pSWE and 2D-SWE is needed (B;2). (New)

2.22 In patients not candidates for initiating NSBB (contraindication/intolerance) for the prevention of decompensation and in whom endoscopy would be required according to the Baveno VI criteria (LSM by TE ≥20 kPa or platelet count ≤150 × 10^9/L), SSM ≤40 kPa by TE can be used to identify subjects at low probability of high-risk varices, in whom endoscopy can be avoided. (C;2). (New)

Research Agenda

- Define risk of decompensation associated with different LSM cutoffs in different etiologies of cACLD.
- Validation and refinement of noninvasive tools for CSPH in NASH patients.
- Evaluate the diagnostic value of LSM for CSPH in etiologies other than viral/alcohol/NASH.
- Establish whether gender and age require specific calibration of NITs for CSPH.
- Validation of circulating biomarkers for prediction of decompensation in all etiologies.
- Validation of LSM thresholds for CSPH, high-risk varices, and decompensation obtained from devices other that TE.

- Validation of what constitutes a clinically significant improvement or worsening of LSM in all etiologies.
- Validation of SSM in non-viral etiologies.
- Evaluation of emerging methods to diagnose CSPH and determine response to NSBB, such as contrast-enhanced ultrasound-based methods (SHAPE), magnetic resonance imaging methods, combination of elastography, novel imaging methods and tests addressing liver function.

Part IV

New Scenarios 1: Introductory Lectures—Progression and Regression of Cirrhosis

Progression and Regression of Cirrhosis: The Histologic Perspective

15

Ian R. Wanless

Introduction

Currently published opinion on the pathogenesis of cirrhosis was developed prior to recent observations in hepatic pathology that demonstrate that (1) hepatic vein obstruction causes a series of events that lead to tissue necrosis and collapse that remodels to create the features of cirrhosis and (2) removal of hepatocytes leads to tissue collapse and a marked condensation of structural collagen that is easily mistaken for fibrosis [1, 2]. From these two observations, our understanding of the pathogenesis of cirrhosis and the nature of hepatic fibrosis can be revised. The purpose of this chapter is to discuss these revisions as *the vascular theory of cirrhosis* and a *new classification of hepatic fibrosis*, illustrated by the four main histologic elements in cirrhosis (Figs. 15.1, 15.2, 15.3, 15.4).

I. R. Wanless (✉)
Department of Pathology, Dalhousie University, Halifax, NS, Canada

R. de Franchis (ed.), *Portal Hypertension VII*,
https://doi.org/10.1007/978-3-031-08552-9_15

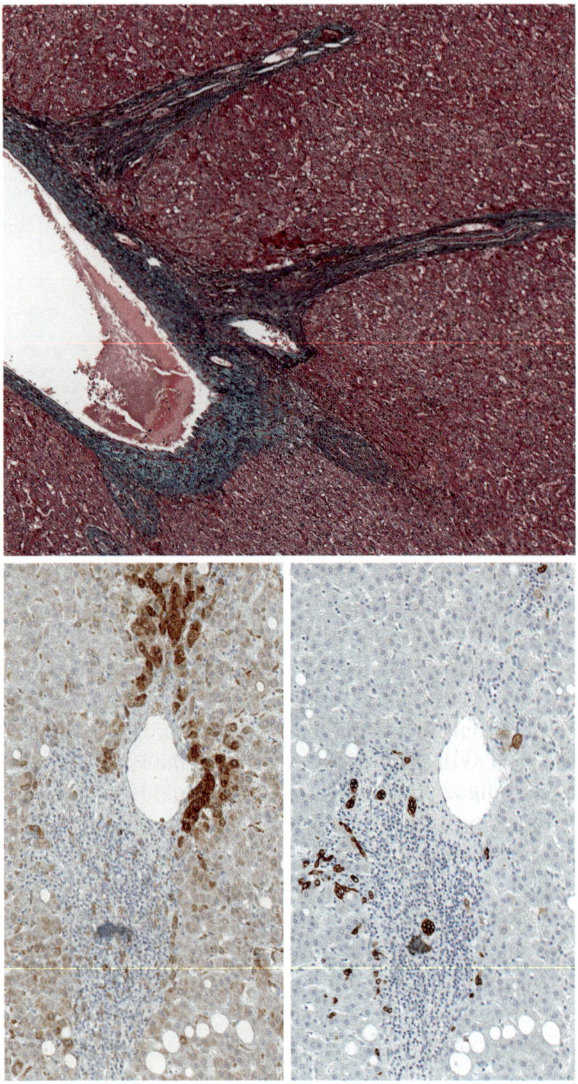

Fig. 15.1 Parenchymal extinction lesions (PELs) in early chronic liver disease. (Top) Four separate portal tracts are adherent to a large hepatic vein (0.7 mm diameter). The parenchyma between the portal tracts and vein is not visible because of collapse. (Elastic-trichrome stain). (Bottom) A small PEL involving one portal tract and adjacent hepatic vein (0.1 mm diameter). Glutamine synthetase stain (left) identifies the hepatic vein and the K7 stain (right) identifies ducts in the portal tract

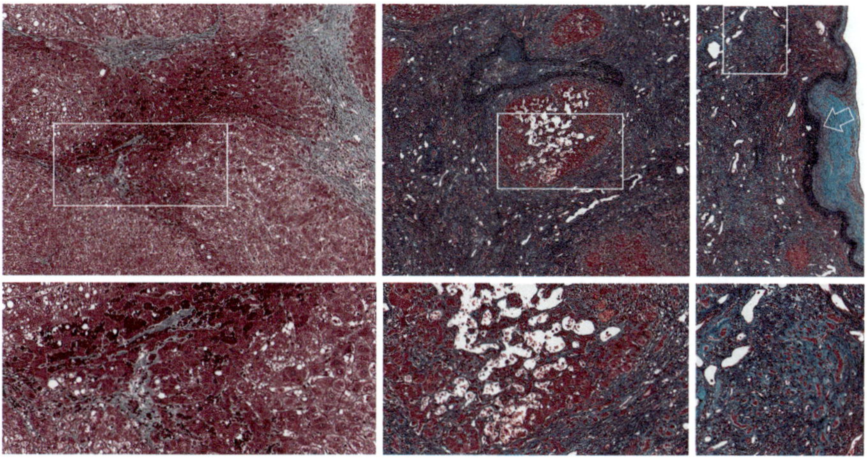

Fig. 15.2 Histologic elements in cirrhosis (Elastic-trichrome stain). (Left) A PEL is seen as focal sinusoidal dilatation and hepatocellular atrophy with minimal collapse at this early congestive phase. This lesion is superimposed on an older PEL (inset) showing approximation of a portal tract and a small obliterated hepatic vein (0.02 mm diameter). The two larger portal tracts have portal vein obliteration and stromal fibrosis. (Center) This liver has more advanced congestion and parenchymal extinction with almost total loss of hepatocytes and approximation of numerous portal tracts in the septa. The largest hepatic vein (0.35 mm diameter) is obstructed by fibrosis. The nodule (inset) shows atrophy and dropout with sinusoidal dilatation that presages further extinction and collapse. (Right) Capsular fibrous thickening is comprised of layers of collagen and elastic tissue. Beneath the capsule, parenchyma is extinct with numerous portal tracts that have collapsed together. Some arteries from portal tracts are penetrating the elastic layer (arrow). The area in the box, enlarged below, shows a cluster of ducts with fibrous thickening of the periductal tissue

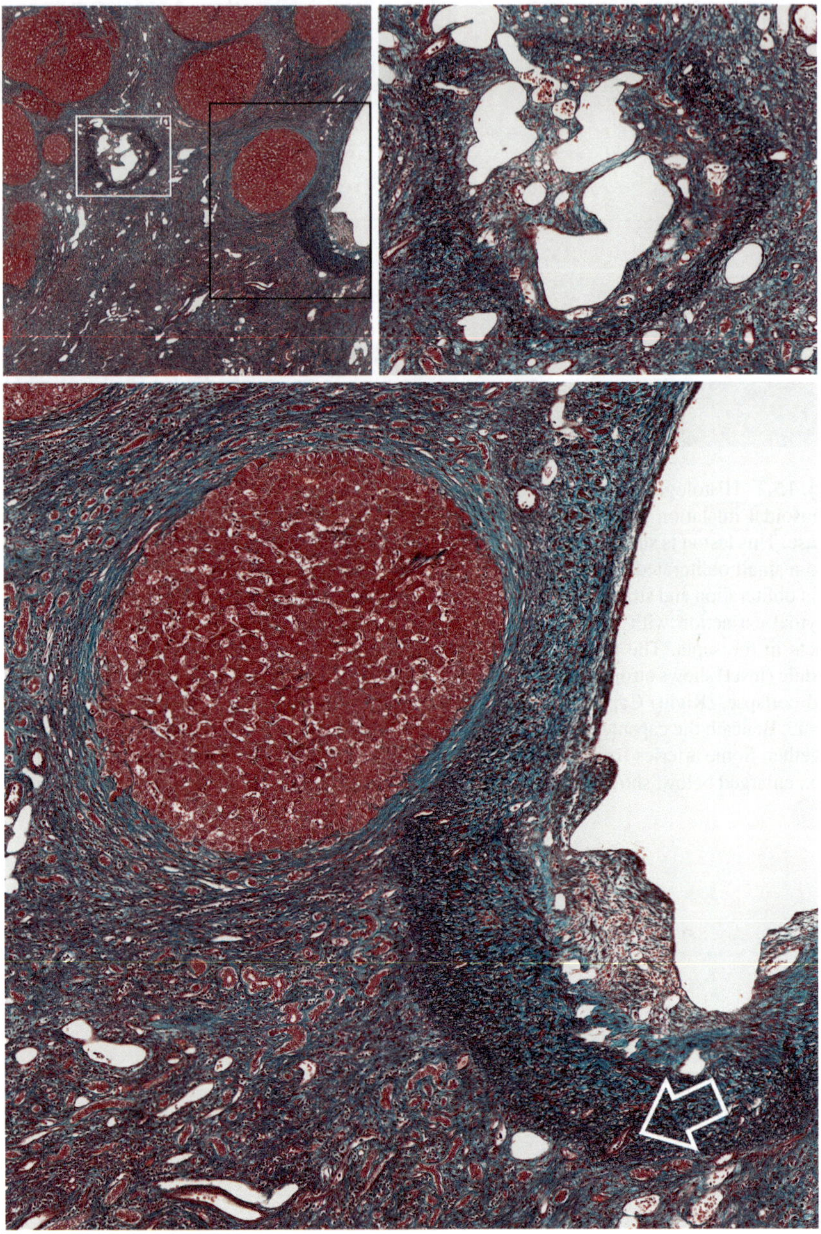

Fig. 15.3 Histologic elements in cirrhosis, showing patterns of fibrosis. (Elastic-trichrome stain). (Top left) Low magnification photo showing nodules separated by both thin and very thick septa. Hepatic veins show partial obstruction with intimal fibrosis (boxes, with higher magnification to right and bottom). (Top right) Hepatic vein (0.6 mm diameter) enlarged to show complex intimal thickening containing multiple channels, some of which are fed by arteries that have migrated from the adjacent portal tracts. (Bottom) The largest hepatic vein (1.0 mm diameter) has similar intimal changes and numerous arteries in the wall (arrow). There is a thin ring of dense collagen in the interstitium surrounding the nodule

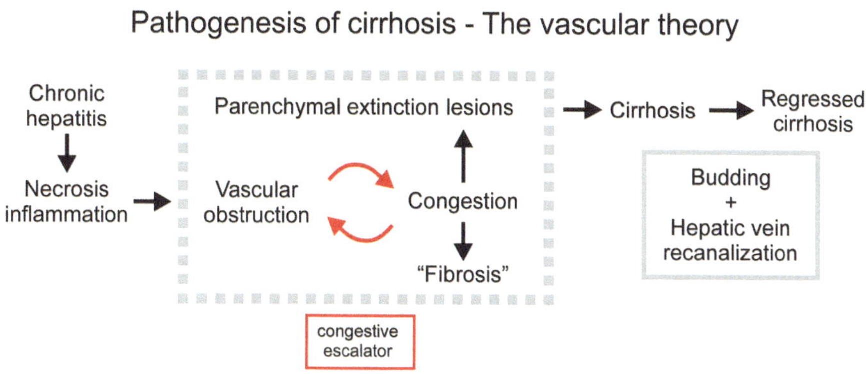

Fig. 15.4 Pathogenesis of cirrhosis. The previously described histologic elements are listed in the box with a dotted outline. The arrows indicate the sequence of events that result in cirrhosis and repair. The red arrows indicate a positive feedback loop that drives the extension of lesions to involve larger veins and their dependent parenchyma

Pathology of Cirrhosis—The Main Histologic Elements

Parenchymal extinction is the process by which tissue is lost. ***Parenchymal extinction lesions*** (PELs) are the histologic evidence of this process. PELs have a natural history that begins with hepatocellular injury (seen as ballooning, apoptosis, necrosis, and cell dropout) and vascular injury (seen as endothelial loss and edema or hemorrhage), followed by tissue collapse, recognized by approximation of hepatic veins and portal tracts (Figs. 15.1, 15.2, 15.3). Arteriovenous shunts often develop as the PELs collapse.

The concept of parenchymal extinction (and PELs) is useful because it defines and identifies both the process and the histologic evidence of significant (i.e., potentially progressive) disease. PELs are physiologically significant because vascular disruption prolongs the repair, allowing time for multiple small PELs to accumulate and form septa. Recognition of this extended natural history provides histologic evidence of the progression from stage 1 to cirrhosis and allows confirmation of the pathogenesis of this progression (see Fig. 15.5 and more detail in published videos) [1]. The number and size of PELs offer quantifiable evidence of the stage of disease progression.

Vascular Obstruction: Obstruction of medium and large hepatic veins is easily seen in cirrhosis, usually with marked intimal fibrosis (Figs. 15.2 and 15.3) [1, 3–5]. Smaller hepatic veins are obstructed early in the course of the disease, although the delicate walls are often hidden by collapse (Fig. 15.1) [1, 6]. Portal vein obstruction also occurs.

Congestion: Severe congestion is seen as sinusoidal engorgement, hemorrhage, and interstitial edema. Mild congestion is less obvious, usually seen as hepatocytes with atrophy and darker cytoplasm because of decreased glycogen and lipid contents (Fig. 15.2).

Fibrosis is defined as an increase in tissue matrix, especially collagen, often measured semi-quantitatively by histological examination (Figs. 15.2 and 15.3) or digital image analysis, aided by special stains or second harmonic microscopy [7].

Pathogenesis of Cirrhosis—The Vascular Theory

These histologic elements are physiologically related to each other, as indicated by the arrows in Fig. 15.4. Vascular obstruction is central to this theory because an imbalance of arterial inflow and venous outflow capacity leads to congestion with transmural pressure gradients that injure sinusoidal and venous endothelium, followed by secondary ischemia, parenchymal extinction, and exudative fibrosis [1]. Initiators of this process may be any cause of hepatocellular injury, most commonly NASH, alcohol, and chronic hepatitis [6, 8].

Anatomical progression of disease occurs as PELs increase in number and merge to form septa (Fig. 15.5). As more hepatocytes are lost, the septa become wider, and the liver gets smaller. While parenchymal extinction advances, the hepatic veins undergo progressive obstruction.

Mechanisms of Progression: Progression of the disease is driven by the primary injury (e.g., hepatitis) followed by vascular obstruction and congestion that operate in a positive feedback loop (the congestive escalator) (Fig. 15.4). The congestive

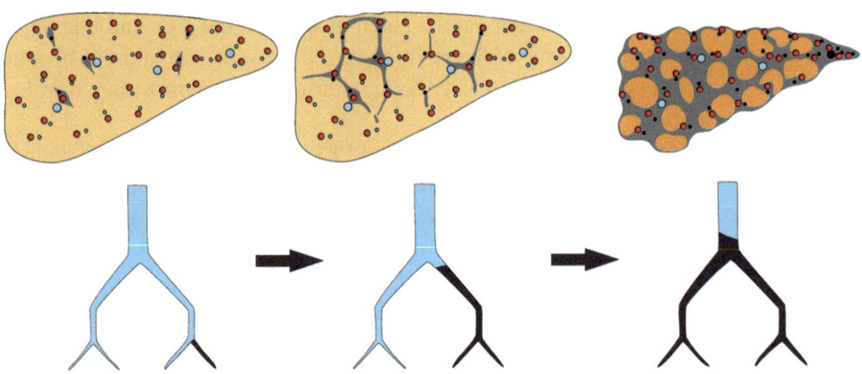

Fig. 15.5 Anatomic features during the progression of chronic hepatitis. These cartoons depict progression from early stage to severe cirrhosis. The dark grey regions represent parenchymal extinction lesions (PELs). Portal tracts are represented by red dots as the sites of arterial pressure. Hepatic veins are represented by blue and black dots in patent and obstructed states, respectively. The distribution of obstruction in the hepatic vein tree is represented below. PELs begin small and aggregate into larger PELs when adjacent structures are damaged, usually from the extension of vascular injury. Aggregates of PELs are the collapsed tissue that comprises the septa in cirrhotic livers. The liver gets smaller by net loss of hepatocytes and collapse of the original infrastructure. In most chronic liver diseases, obstruction begins in the small hepatic veins and progresses to include the larger veins. Thus, the concept of parenchymal extinction can explain how septa can be thin or thick, containing few or many portal tracks and hepatic veins, depending on the size and number of adjacent extinction events

escalator has at least three components: (1) edematous and fibrous thickening of the vessel walls, (2) expansion of the interstitium, and (3) arteriovenous shunts. The first causes further hepatic vein obstruction and therefore worsening of congestion; the second causes external vascular compression and a regional compartment syndrome effect. The third component, arteriovenous shunts, develops during PEL formation (see video in [1]). These shunts bring arterial pressure to the interstitium, largely within septa and hepatic vein walls (Fig. 15.3).

In summary, the primary injury is important during active primary disease and abates after remission or successful therapy. This primary injury is important because it initiates endothelial leak and tissue congestion. The congestive escalator continues to operate after remission of the primary disease once there is a high degree of venous obstruction. The congestive escalator promotes the formation of an enlarging septal network. This functions as a *pressurized common channel* that distributes pressure gradients, potentially throughout the entire organ, facilitating edema, ascites, hepatic vein exudative fibrosis, retrograde portal vein flow, and portal hypertension.

Hepatic Fibrosis—A New Classification

Hepatic fibrosis has often been described as a response to necrosis and inflammation driven by TGF-beta and other substances acting on hepatic stellate cells [9, 10]. The importance of tissue collapse as a cause of increased collagen concentration indicates a need for a more comprehensive classification of hepatic fibrosis (see Table 15.1 and Fig. 15.6) [1]. This classification divides fibrosis into architectural matrix, acquired matrix, and pseudo-fibrosis.

Architectural matrix refers to the original matrix derived from portal stroma, vessel walls, reticulin fibers, interstitial fibers, and hepatic capsule. If this matrix has recognizable features, it is classified as collapsed architectural matrix; if degradation has occurred but is likely derived from pre-existing structures, it is classified as a splayed or degraded architectural matrix. Replacement architectural matrix refers to newly synthesized regenerative tissue with architectural purpose, especially new arteries and sinusoidal reticulin fibers.

Comments—architectural matrix. In normal livers, the vast majority of liver collagen occurs in architectural structures, especially portal tract stroma and vessel walls. In cirrhosis, the septa contain most of the collagen, largely within approximated architectural structures as well as some collagen of less certain origin that may be degraded architecture or acquired matrix.

The importance of "collapse fibrosis" has been noted for over a century [11–17] but most attempts to quantify this type are recent [2, 18]. In normal liver, hepatocytes occupy about 95% of tissue area, while collagen occupies 2%–3%. If all hepatocytes are removed and the tissue collapses, the collagen percent area (CPA) reaches 50%, the remainder being spaces occupied by blood or edema. Cabibi et al. noted that CPA increased six-fold in acute hepatitis, when acquired collagen synthesis would not be expected [18]. In cirrhosis, portal tract number and CPA were also increased by about six-fold (Fig. 15.6) [2]. This increase is similar in magnitude to

Table 15.1 Classification of matrix in chronic liver disease

Matrix types	Notes
Architectural matrix	
Collapsed architectural matrix	i.e., *collapse fibrosis, condensation fibrosis.* Previously normal architecture concentrated in regions of hepatocellular loss or atrophy. e.g., increased numbers of portal tracts, hepatic veins, or reticulin fibers per unit area
Splayed architectural matrix	i.e., degraded, or disorganized remnants of original architecture easily mistaken for a new matrix. ("splay" is a shortened form of "display," often used to denote *unfolding cloth* or *spreading fibers of the yarn)*
Replacement architectural matrix	e.g., post-regenerative reticulin and arteries (the stromal contents of most bud-derived cirrhotic nodules)
Acquired matrix, i.e., *"true fibrosis"*	
Exudative fibrosis	i.e., *congestive fibrosis.* This is usually initiated by transmural pressure gradients. It may be subdivided by location, e.g., capsular, peribiliary, interstitial, or vascular (intimal, intraluminal)
Thrombosis-associated fibrosis	This may be initiated by a thrombus (with or without inflammatory stimulus) but is usually seen after collagenization and resorption of fibrin
Inflammatory fibrosis	Fibrosis secondary to inflammation is usually associated with stellate cell activation. This may overlap mechanistically with exudative or thrombosis-associated fibrosis
Pseudo-fibrosis	
Sampling issues	i.e., *selection bias or sampling error.* e.g., small sample size or normal portal tracts and hepatic veins misinterpreted as fibrosis (usually by computer algorithm)

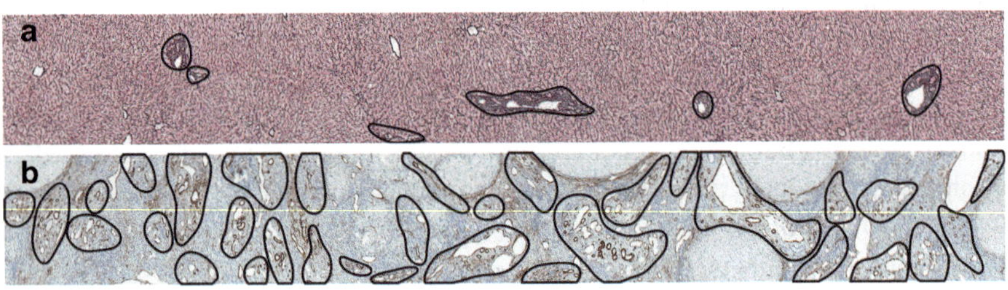

Fig. 15.6 Tissue collapse in cirrhosis. Collapse can be quantitated by the density of original histologic elements. The photos are taken at the same magnification. (**a**) Normal liver with six portal tracts outlined (reticulin stain). (**b**) Severe cirrhosis (stage 4C) with 33 visible portal tracts (actin stain)

collagen measurements in cirrhosis reported without correcting for collapse, suggesting that these studies were reporting increased collagen concentration rather than content. Initial data suggest that the concentrating effect of tissue collapse can account for a large percentage of the measurable collagen in human cirrhosis [2].

Splayed and replacement architectural matrices tend to be small and delicate fibers that are quantitatively less significant than collapse fibrosis (Fig. 15.7). Replacement matrix is easily recognized as normal-appearing fibers found in bud-derived nodules.

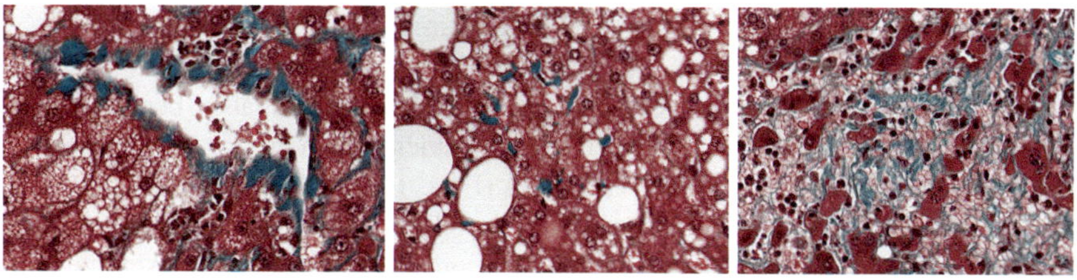

Fig. 15.7 Architectural matrix may be misinterpreted as an acquired matrix. Here are three hepatic veins of similar size (65–71 μm diameter, elastic-trichrome stain) in a liver with early-stage NASH. (Left) A nearly normal vein shows large discrete collagen bundles. (Center) This post-inflammatory vein has suffered intimal resorption with intra-luminal migration of hepatocytes. This demonstrates that the original thicker wall fibers were more resistant than intimal fibers to resorptive processes. The circular arrangement of fibers confirms this is a vein wall and not an acquired matrix (fibrosis). (Right) This disorganized mass of splayed fibers can be recognized as a hepatic vein by the linear arrangement of larger bundles. The delicate fibers at the lower right corner are probably splayed fibers from a small branch of this vein [35]

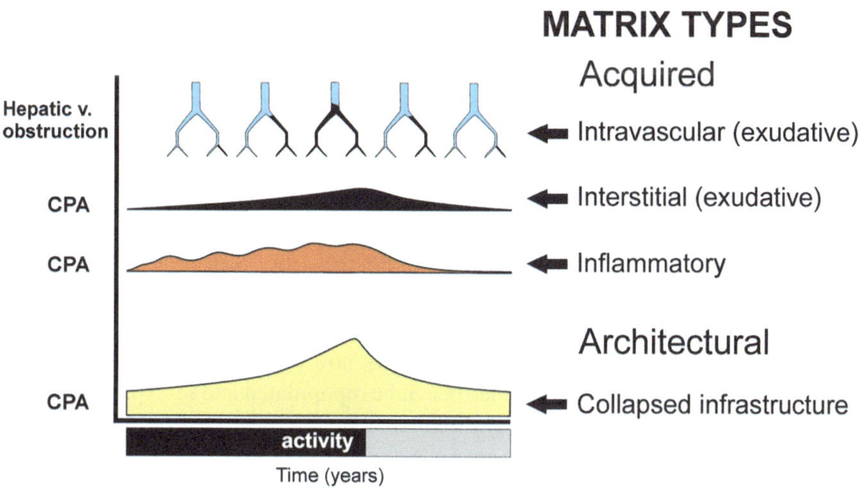

Fig. 15.8 Natural history of matrix components with time. This diagram lists four subtypes of the matrix to demonstrate the change in collagen concentration during the progression and regression of cirrhosis. Additional matrix subtypes are listed in Table 15.1. Acquired matrix is a true increase in content, largely related to congestion or inflammation followed by exudation and/or collagen synthesis; Architectural matrix demonstrates an increased concentration when there is collapse after removal of hepatocytes. During regression, the concentrations of both acquired matrix and collapsed structural matrix are reduced by dilution as the hepatocyte population is restored. Both types of matrices may also be chemically resorbed. *CPA*, Collagen percent area

Acquired matrix (i.e., "true fibrosis") is an increase in the quantity of newly synthesized matrix with exudative, thrombotic, and inflammatory subtypes (Fig. 15.8).

Exudative (or congestive) fibrosis is an acquired matrix likely to be caused by congestion and plasma exudation, generally at sites of transmural pressure gradients. It is the most easily recognized form of the acquired matrix because of its layered histologic appearance as well as being located where arterial sources are approximated to low-pressure drainage sites (Figs. 15.2 and 15.3).

Thrombosis-associated fibrosis is easily recognized in the acute phase but after resorption of fibrin and secondary collagenization, it may be indistinguishable without special techniques. If initiated by inflammation, accurate classification may be impossible, though the possibility of a thrombotic contribution is of therapeutic significance [19, 20].

Inflammatory fibrosis is considered to be that driven by toxin- or virus-induced necrosis and inflammation. This is the most studied form of fibrosis, largely based on observations from rodent studies. Many rodent models have shown that inflammatory fibrosis is resorbed within days or weeks of discontinuing the toxin [13, 21]. Similarly, in human viral cirrhosis, fibrosis concentration decreases in a large percentage of patients treated with anti-viral agents for 1–6 years [22, 23]; this can be explained, at least in part, by re-expansion of collapsed architecture [24].

Pseudo-fibrosis represents tissue that is erroneously assigned because of selection bias, sampling, or segmentation errors during computerized digital image analysis. With careful selection of tissue and curation prior to automated digital analysis, these forms of error can be minimized [25].

Regression

During periods of inactivity, cirrhosis often regresses with the disappearance of septa. This sequence illustrates the fate of PELs, now aggregated and recognized as septa. The collapsed architectural matrix can be repopulated and re-expanded by the generation of new hepatocytes [26]. Smaller fibers are degraded (seen as splayed matrix) or resorbed while broad fibers are more resistant and may persist to be seen as remnants or incorporated into functional replacement architecture (Fig. 15.7) [1, 22]. Acquired matrix may also be resorbed [10].

Hepatocellular regeneration is largely accomplished by activation of the bud maturation sequence (Fig. 15.9) [22, 26, 27]. Buds of liver cells, derived from progenitor cells, enlarge to form bud-derived nodules that become functioning parenchyma within the septa. Repopulation by budding is not always successful, probably because of persistent hepatic vein obstruction (Figs. 15.2 and 15.3) [28].

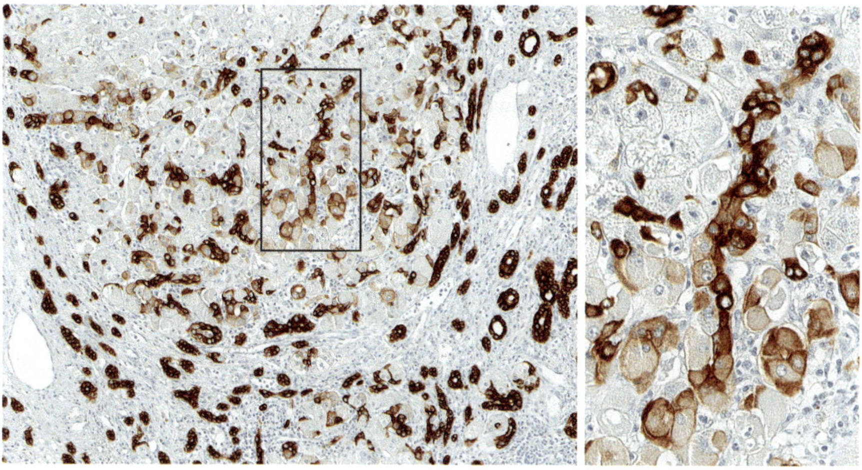

Fig. 15.9 The bud maturation sequence demonstrated in a bud-derived nodule (HCV cirrhosis). (Left) This hepatocellular nodule is surrounded by septa that contain numerous small ducts and other portal tract elements. The buds are in the early stage of development, comprised of new hepatocytes that are single or in small clusters attached to their ducts of origin. (Right) At higher magnification, a duct remnant is seen as a string of cholangiocytes with minimal residual tubular structure at the top. New hepatocytes are weakly K7 positive until they mature in clusters of three or more cells. A few residual cholangiocytes are also seen as short strings (K7 stain)

Discussion

The vascular theory of cirrhosis is based on numerous pieces of evidence that form a coherent framework for understanding the progression and regression of disease. The main driving force is an imbalance of blood flow and outflow capacity. Balance is required to keep the tissue milieu within physiological limits. Imbalance creates a positive feedback loop of congestive injury that drives parenchymal extinction with collapse and microvascular destruction.

These concepts facilitate an understanding of the hemodynamic forces that cause cirrhosis as well as portal hypertension. The importance of vascular damage in driving the development of cirrhosis suggests how current therapeutic efforts directed to controlling inflammation and thrombosis, stabilizing endothelium, and reducing pressure gradients have been fruitful [29–34].

Tissue collapse creates an increased collagen concentration that may be mistaken for acquired collagen content. The new classification of hepatic fibrosis may clarify the differential diagnosis. Another effect is to raise awareness of how the stage of chronic liver disease should be measured and interpreted in a clinical setting. Existing staging systems (e.g., Laennec, METAVIR, or CPA) may be satisfactory to estimate the severity of disease as they should measure visible collagen whether caused by collapse or new collagen synthesis. Collagen of either origin may be a useful marker of the clinical stage, but interpretation requires some adjustment. In clinical trials, the interpretation of an improvement in stage or CPA can be explained by resorbed fibrosis or regeneration and repopulation (or by sampling error). Misinterpreted structures become the target for misguided therapeutic efforts. For example, advanced histologic stage may be considered a call for a treatment designed to dissolve collagen. However, a more appropriate treatment might be one that stimulates regeneration.

References

1. Wanless IR. The role of vascular injury and congestion in the pathogenesis of cirrhosis: the congestive escalator and the parenchymal extinction sequence. Curr Hepatol Rep. 2020;19(1):40–53.
2. Wanless IR, Ravi PY. Importance of tissue collapse and unmasking of reticulin on the assessment of fibrosis in acute and chronic liver disease. Hepatogy. 2016;63:214A.
3. Goodman ZD, Ishak KG. Occlusive venous lesions in alcoholic liver disease: a study of 200 cases. Gastroenterology. 1982;83:786–96.
4. Nakanuma Y, Ohta G, Doishita K. Quantitation and serial section observations of focal veno-occlusive lesions of hepatic veins in liver cirrhosis. Virchows Arch A Pathol Anat Histopathol. 1985;405:429–38.
5. Burt AD, MacSween RNM. Hepatic vein lesions in alcoholic liver disease: retrospective biopsy and necropsy study. J Clin Pathol. 1986;39(1):63–7.
6. Wanless IR, Shiota K. The pathogenesis of nonalcoholic steatohepatitis and other fatty liver diseases: a four-step model including the role of lipid release and hepatic venular obstruction in the progression to cirrhosis. Semin Liver Dis. 2004;24(1):99–106.
7. Xu S, Wang Y, Tai DCS, Wang S, Cheng CL, Peng Q, et al. qFibrosis: a fully-quantitative innovative method incorporating histological features to facilitate accurate fibrosis scoring in animal model and chronic hepatitis B patients. J Hepatol. 2014;61(2):260–9.
8. Edwards L, Wanless IR. Mechanisms of liver involvement in systemic disease. Best Pract Res Clin Gastroenterol. 2013;27(4):471–83.
9. Trautwein C, Friedman SL, Schuppan D, Pinzani M. Hepatic fibrosis: concept to treatment. J Hepatol. 2015;62(1 Suppl):S15–24.
10. Campana L, Iredale JP. Regression of liver fibrosis. Semin Liver Dis. 2017;37(1):1–10.
11. Adami JG, Nicholls AG. The principles of pathology. Philadelphia: Lea and Febiger; 1909. p. 473.
12. Mallory FB. Cirrhosis of the liver. Five different types of lesion from which it may arise. Bull Johns Hopkins Hosp. 1911;22:69–75.
13. Cameron GR, Karunaratne WAE. Carbon tetrachloride cirrhosis in relation to liver regeneration. J Pathol Bacteriol. 1936;42:1–21.
14. Hartroft WS, Ridout JH. Pathogenesis of the cirrhosis produced by choline deficiency; escape of lipid from fatty hepatic cysts into the biliary and vascular systems. Am J Pathol. 1951;27(6):951–89.
15. Meister HP, Szanto PB, Schoolman H. Quantitative-morphologic evaluation of postnecrotic cirrhosis. Virchows Arch Pathol Anat Physiol Klin Med. 1963;336:447–64.
16. Popper H. Pathologic aspects of cirrhosis. Am J Pathol. 1977;87:228–64.
17. Scheuer PJ, Maggi G. Hepatic fibrosis and collapse: histological distinction by orecin staining. Histopathology. 1980;4(5):487–90.
18. Cabibi D, Calvaruso V, Giuffrida L, Ingrao S, Balsamo L, Giannone AG, et al. Comparison of histochemical staining methods and correlation with transient elastography in acute hepatitis. Pathobiology. 2015;82:48–52.
19. Levy GA, MacPhee PJ, Fung LS, Fisher MM, Rappaport AM. The effect of mouse hepatitis virus infection on the microcirculation of the liver. Hepatology. 1983;3(6):964–73.

20. Bryce C, Grimes Z, Pujadas E, Ahuja S, Beasley MB, Albrecht R, et al. Pathophysiology of SARS-CoV-2: the Mount Sinai COVID-19 autopsy experience. Mod Pathol. 2021;34(8):1456–67.
21. Troeger JS, Mederacke I, Gwak GY, Dapito DH, Mu X, Hsu CC, et al. Deactivation of hepatic stellate cells during liver fibrosis resolution in mice. Gastroenterology. 2012;143(4):1073–83. e22
22. Wanless IR, Nakashima E, Sherman M. Regression of human cirrhosis: morphologic features and the genesis of incomplete septal cirrhosis. Arch Pathol Lab Med. 2000;124:1599–607.
23. Schiff ER, Lee SS, Chao YC, Kew Yoon S, Bessone F, Wu SS, et al. Long-term treatment with entecavir induces reversal of advanced fibrosis or cirrhosis in patients with chronic hepatitis B. Clin Gastroenterol Hepatol. 2011;9(3):274–6.
24. Sun Y, Wang B, You H, Jia J, Chen S, Wu X, et al. Artery density in human liver: a valuable measure for staging chronic liver disease. J Hepatol. 2019;70:e141.
25. Standish RA, Cholongitas E, Dhillon A, Burroughs AK, Dhillon AP. An appraisal of the histopathological assessment of liver fibrosis. Gut. 2006;55(4):569–78.
26. Stueck AE, Wanless IR. Hepatocyte buds derived from progenitor cells repopulate regions of parenchymal extinction in human cirrhosis. Hepatology. 2015;61(5):1696–707.
27. Falkowski O, An HJ, Ianus IA, Chiriboga L, Yee H, West AB, et al. Regeneration of hepatocyte 'buds' in cirrhosis from intrabiliary stem cells. J Hepatol. 2003;39(3):357–64.
28. Wanless IR. The mechanism of irreversibility of late stage cirrhosis. J Hepatol. 2018;68(Suppl 1):S692.
29. Turco L, Schepis F, Villa E. The role of anticoagulation in treating portal hypertension. Curr Hepatol Rep. 2018;17(3):200–8.
30. Moctezuma-Velazquez C, Abraldes JG, Montano-Loza AJ. The use of statins in patients with chronic liver disease and cirrhosis. Curr Treat Options Gastroenterol. 2018;16(2):226–40.
31. Turco L, Villanueva C, La Mura V, Garcia-Pagan JC, Reiberger T, Genesca J, et al. Lowering portal pressure improves outcomes of patients with cirrhosis, with or without ascites: a meta-analysis. Clin Gastroenterol Hepatol. 2020;18(2):313–327.e6.
32. Hide D, Warren A, Fernandez-Iglesias A, Maeso-Diaz R, Peralta C, Le Couteur DG, et al. Ischemia/reperfusion injury in the aged liver: the importance of the sinusoidal endothelium in developing therapeutic strategies for the elderly. J Gerontol A Biol Sci Med Sci. 2020;75(2):268–77.
33. Boyer-Diaz Z, Aristu-Zabalza P, Andres-Rozas M, Robert C, Ortega-Ribera M, Fernandez-Iglesias A, et al. Pan-PPAR agonist lanifibranor improves portal hypertension and hepatic fibrosis in experimental advanced chronic liver disease. J Hepatol. 2021;74(5):1188–99.
34. Gines P, Krag A, Abraldes JG, Sola E, Fabrellas N, Kamath PS. Liver cirrhosis. Lancet. 2021;398(10308):1359–76.
35. Wanless IR. Quantitative SHG-microscopy: unraveling the nano-architecture of the cirrhotic liver. Clin Res Hepatol Gastroenterol. 2020;44(1):1–3.

chronic liver diseases: a systematic review and meta-analysis.

16 Liver Fibrosis and Its Regression in the Context of Portal Hypertension

Massimo Pinzani

Introduction

Portal hypertension (PH) is a major complication of liver disease, which results from a variety of pathological conditions that increase the resistance to the portal blood flow into the liver. The primary cause of portal hypertension in cirrhosis is an increase in intrahepatic vascular resistance due to massive structural changes associated with fibrosis and increased vascular tone in the hepatic microcirculation. It is reported that intrahepatic vasoconstriction accounts for at least 25% of increased intrahepatic vascular resistance. Phenotypic changes in hepatic cells, such as hepatic stellate cells (HSC) and liver sinusoidal endothelial cells (LSECs), are known to play critical roles in increased intrahepatic vascular resistance and have been extensively studied over the past 30 years. In response to liver injury, HSC undergoes a phenotypical modulation toward myofibroblast-like cells driving fibrogenesis and becoming highly contractile [1–4]. Therefore, activated HSCs play a crucial role in the development of portal hypertension. LSECs are the first line of defense protecting the liver from injury and exert diverse effects on liver functions including blood clearance, vascular tone, immunity, hepatocyte homeostasis, and angiogenesis/sinusoidal remodeling. Therefore, LSEC dysfunction occurring in cirrhosis leads to impaired vasomotor control favoring vasoconstriction, inflammation, fibrosis, and impaired liver regeneration, all of which facilitate the development of liver cirrhosis and PH.

The progression of chronic liver diseases (CLDs) toward cirrhosis is not exclusively due to fibrogenesis. The formation of new vessels (neoangiogenesis) and the establishment of an abnormal angioarchitecture of the liver is a process strictly related to intrahepatic vascular remodeling with capillarization of sinusoids, and the

M. Pinzani (✉)
University College London, Institute for Liver and Digestive Health, Division of Medicine, Royal Free Hospital, London, UK
e-mail: m.pinzani@ucl.ac.uk

R. de Franchis (ed.), *Portal Hypertension VII*,
https://doi.org/10.1007/978-3-031-08552-9_16

development of intrahepatic shunts, which would lead to increased hepatic resistance and decreased effective hepatocyte perfusion thus reiterating the hypoxic stimulus leading to angiogenesis and fibrogenesis [5].

Although the evolution of CLD toward cirrhosis has common mechanisms and common consequences in CLD due to different causes (viral, toxic, metabolic, autoimmune, etc.), there are important etiology-dependent features. These differences are crucial to understand how PH develops in different CLDs and to assess the limits of fibrosis reversibility in the cirrhotic liver that may have an impact on PH.

Etiology-Driven Fibrosis and Cirrhosis: A Key for Understanding Cirrhotic Portal Hypertension

In recent years, the increasing awareness that different CLDs have distinct patterns of fibrotic development has led to the concept of "etiology-driven fibrosis and cirrhosis" [6, 7]. These different patterns of fibrogenic evolution are related to different factors and particularly: (1) the topographic localization of tissue damage, (2) the relative concentration of pro-fibrogenic factors, and (3) the prevalent pro-fibrogenic mechanism(s), as well as the prevalent cellular effectors. Accordingly, time to disease progression, distribution of fibrosis, and onset/progression of PH are correlated and depend on the etiological agent. For example, the chronic viral hepatitis pattern of fibrosis (also defined as "post-necrotic") is the result of portal-central (vein) bridging necrosis, thus originating portal-central septa. In addition, this form of fibrogenic evolution is characterized by the presence of "interface" hepatitis and the development of portal-to-portal septa and septa ending blind in the parenchyma, and by the rapid derangement of the vascular connections with the portal system (early portal hypertension). On the other hand, in alcoholic and metabolic liver diseases, fibrogenesis is concentrated initially around centrilobular (zone 3) sinusoids (capillarization) and around groups of hepatocytes (chicken-wire pattern) with a progressive pan-lobular extension and final development of a cirrhotic morphology. Finally, biliary fibrosis, due to the co-proliferation of reactive bile ductules and periductular myofibroblast-like cells at the portal-parenchymal interface, tends to follow a portal-to-portal direction. This leads to the formation of portal–portal septa surrounding liver lobules, where the central vein and its connections with the portal tract are preserved until the late stages.

The differences in the fibrotic evolution in two paradigmatic CLDs, i.e., chronic viral hepatitis and alcoholic/non-alcoholic steatohepatitis, are illustrated in Fig. 16.1. As indicated, the direction of fibrotic development follows opposite directions in the two conditions. Importantly, in post-necrotic fibrosis, there is a rapid portal-to-central evolution (fibrosis and neoangiogenesis) with early involvement of the centrilobular vein (CLV), while capillarization of sinusoids remains limited to the periportal area. On the contrary, in the case of pericentral fibrosis, capillarization of sinusoids proceeds central-to-portal and progressively become pan-lobular. Consequently, at least until cirrhotic changes are fully established, sinusoidal hemodynamics are differently affected in the two types of fibrotic evolution (Fig. 16.2).

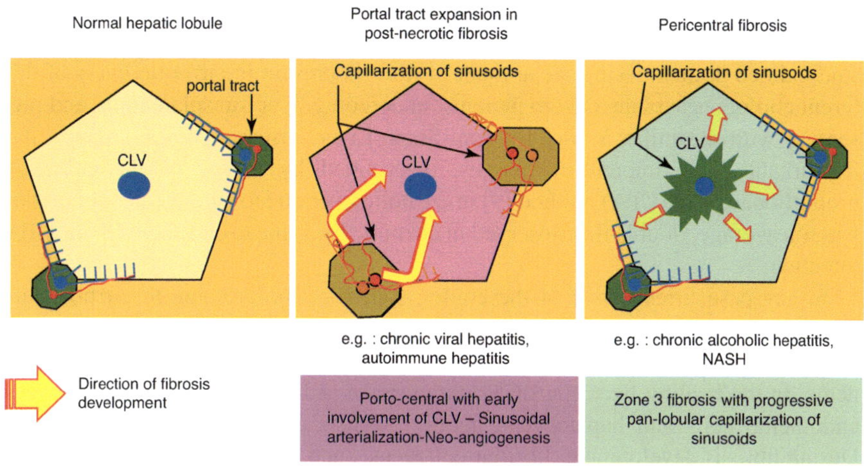

Fig. 16.1 Impact of the pattern of fibrosis progression on the origin of portal hypertension

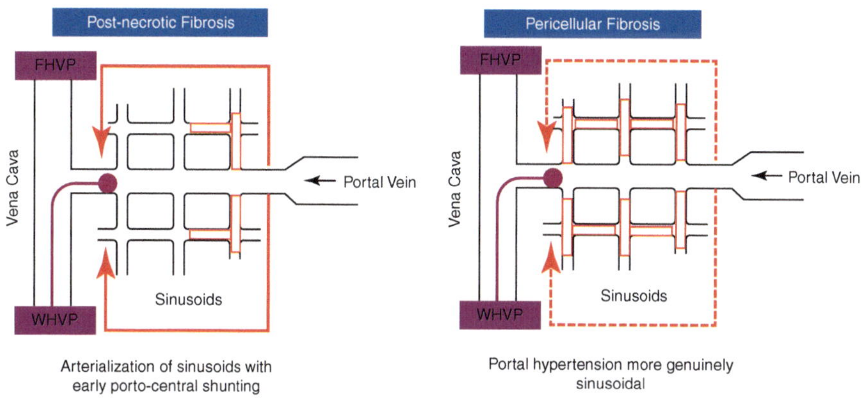

Fig. 16.2 Relationship between HVPG (= WHVP-FHVP) and the pattern of fibrosis development

Importantly, neoangiogenesis, which is more typical of the post-necrotic form, is characterized by largely ineffective attempts to connect the portal with the hepatic vein circulation with dead-end vascular branches often characterized by microthrombosis. In addition, even when successfully established, these portal-central anastomoses follow irregular patterns and are embedded in a developing scar tissue characterized by the presence of contractile cells (i.e., activated HSC and myofibroblasts). Therefore, even when cirrhosis is fully established, the increase in intrahepatic resistance in the post-necrotic form also reflects the impact of ineffective neoangiogenesis and the chaotic derangement of intralobular hemodynamics.

The knowledge of these aspects of the pathophysiology of CLD provides important insights on the correlation between the progression of the disease, the etiological agents, the dynamics of the necro-inflammatory infiltrate, the distribution of fibrosis, and the onset and progression of PH, depending on the etiological agent leading to cirrhosis. A proof of concept of these considerations was provided by a study aimed at quantifying the amount of fibrosis present in cirrhotic livers of different etiologies explanted from patients, undergoing liver transplantation, and presenting with a similar Model for End-Stage Liver Disease (MELD) scores [8]. Remarkably, the amount of fibrosis, determined by means of the collagen-proportionate area (CPA) method [9] in cirrhotic liver due to chronic alcohol intake, was, on average, double that observed in cirrhotic liver due to chronic HCV or HBV infection.

In any case, irrespective of the etiology, all CLD progressing to cirrhosis are characterized not only by an increase in the quantity of the extracellular matrix (ECM) but also by changes in the quality and distribution of different ECM components. In the healthy liver, the ECM in the space of Disse, i.e., the space between endothelial cells and hepatocytes, mainly consists of collagen IV and laminin. During fibrosis development, fibrillar collagens, such as collagen I and III, become progressively prevalent and distributed also in areas normally occupied by basal membrane-like structures. The possibility of decellularizing normal and fibrotic human liver tissue has recently provided insights into healthy and disease-specific hepatic "matrisome," i.e., the biochemical and biomechanical properties of the ECM, and has improved the understanding of how cells interact with and respond to healthy and pathologic tissue microenvironment [10]. In particular, the unique disease-specific ECM environment affects both cell differentiation and function thus recapitulating those processes that are important in the progression to cirrhosis and in the development of hepatocellular carcinoma. Indeed, the analysis of the matrisome in cirrhosis has revealed significant qualitative differences across etiologies and has highlighted the presence of etiology-specific protein signatures [11]. Accordingly, it may become possible to identify specific ECM shedding fragments in plasma or urine to be employed as disease-specific biomarkers for the stratification of different types of cirrhosis.

Based on the divergent patterns of fibrotic evolution across etiologies, it is possible that the development of portal hypertension and its clinical features are also etiology-dependent. Since the hepatic vein pressure gradient [HVPG, i.e., the wedged hepatic venous pressure (WHVP) minus the free hepatic vein pressure (FHVP)] is the only direct standard measurement on which clinical decisions and noninvasive parameters are founded, it may be necessary to "reset" HVPG thresholds according to the etiology of cirrhosis. In particular, as illustrated in Fig. 16.2, the type of fibrosis development may differently affect the value of the WHVP because of different alterations of sinusoidal hemodynamics, at least in the early phases of cirrhosis. In this context, pericellular fibrosis with extensive capillarization of sinusoids (highlighted in red in Fig. 16.2) may more genuinely reflect sinusoidal pressure than post-necrotic fibrosis. Along these lines, recent data indicate that the classic HVPG thresholds, developed mostly in HCV cirrhosis, do not reflect the risk of clinical manifestation in NASH cirrhosis in which severe complications may develop when HVPG is still below 10 mmHg [12].

Is Fibrosis Regression Different in Different CLDs?

The regression of fibrosis—defined as a reduction in the content of fibrillary ECM—is most critical in patients with cirrhosis and PH, with the expectation that clinical outcomes may improve, while for patients without cirrhosis, a reasonable endpoint is simply attenuation of further progression, such that cirrhosis never fully develops. A crucial problem when discussing the issue of fibrosis regression in cirrhosis or even cirrhosis reversal is the lack of a clear understanding of the relationship between the biology of fibrosis in a cirrhotic liver and the clinical manifestations with particular reference to PH. Indeed, this understanding would represent the ideal basis for a more realistic stratification of cirrhosis.

In terms of reversibility, while it is doubtful that an accurately defined cirrhotic liver is able to reverse to normal, there is sound evidence concerning the capacity of the healing liver to reabsorb scar tissue following an effective causative treatment (i.e., sustained viral response to treatment, abstinence from alcohol, etc.). However, scar tissue in the liver of patients with CLD lasting 30 or more years is likely characterized by different stages of biochemical and biological evolution. Indeed, fibrotic deposition related to recent disease and characterized by the presence of thin reticulin fibers, often in the presence of a diffuse inflammatory infiltrate, is likely fully reversible, whereas long-standing fibrosis, branded by extensive collagen cross-linking by tissue transglutaminase, presence of elastin, dense acellular ECM, and decreased expression and/or activity of specific metalloproteinases, is not fully reversible [13, 14]. In addition, there is no convincing evidence that the abnormalities of the intrahepatic vasculature can regress in the human cirrhotic liver. Actually, the available evidence suggests that direct communications between the portal and hepatic vein systems (the so-called veno-portal adhesions) persist even in cases of extensive fibrosis regression, and evident "arterialized" sinusoids appear in the context of intrahepatic arteriovenous shunts [15]. In other words, within the same liver are present different types of scar tissues *with* different potential and dynamics of reversibility once the etiological agent is removed and/or antifibrogenic strategy is established. In addition, substantial experimental evidence suggests that long-term fibrogenesis occurring in human CLD is characterized by a progressive resistance to apoptosis of hepatic stellate cells/myofibroblasts with the consequent immovability of a critical mass of pro-fibrogenic cells [16]. These considerations lead to the concept that fibrogenesis occurring in a cirrhotic liver may still progress in spite of SVR obtained with anti-HCV DDAs according to a sort of kinetic energy and, consequently, the clinical manifestations of PH can persist for a significant amount of time after SVR [17–19]. In any case, in this clinical context, it would be key to make a stratification of cirrhosis according to the biological features of fibrogenesis. Indeed, in case fibrogenesis is still characterized by extensive necro-inflammation, obtaining a SVR will lead to a sharp reduction of this component with a consequent decrease intrahepatic resistance and thus portal pressure [20–21]. In particular, improvement of liver inflammation, aminotransferases, and liver function early during treatment can explain the rapid decreases in HVPG and liver stiffness reported in some studies. Similar considerations can be made for the significant decrease in HVPG observed after a few weeks of abstinence in patients with alcoholic cirrhosis [22]. Overall, these observations call for a more critical calibration of HVPG values and thresholds according to the general features of the chronic inflammatory/fibrogenic process, which may reflect etiological differences. The combined use of HVPG with noninvasive techniques, liver elastography, in particular, may help in this direction. Indeed, as shown in a recent report [23], for the same standard HVPG cutoffs of 10 and 12 mmHg, the corresponding values obtained by transient elastography are significantly more elevated (almost double) in alcoholic cirrhosis compared to HCV cirrhosis. In addition, the use of spleen elastography may further refine this combination considering that, when it is accurately feasible, spleen stiffness shows a better correlation with HVPG compared with liver stiffness for values of HVPG ranging from 5 to 20 mmHg and beyond [24].

Coming to the question raised in this paragraph: "Is fibrosis regression different in different CLD?" The answer is, in my opinion, yes, although this is not based on precise clinical data but rather on contingent considerations related to the features of fibrosis evolution. These considerations are summarized in Table 16.1 and are based on the concepts developed in this chapter. The translation of these concepts would require more accurate follow-up studies comparing CLD due to different etiologies.

In conclusion, after more than 30 years of intense research activity and a huge number of acquisitions, the relationship between PH and liver fibrosis still bears more questions than answers. Somehow, we are dealing with the clinical management of PH in cirrhosis still using the tip of the iceberg of the current knowledge on the biology of liver fibrosis in CLD of different etiologies. The way ahead is based on the translation of this knowledge in clinical tools which allow a more realistic and accurate stratification of cirrhosis.

Table 16.1 Reversibility of liver fibrosis according to the pattern

Fibrosis pattern	Early portal to central septa	Neo-angiogenesis	Panlobular capillarization of sinusoids	Potential reversibility
POST-NECROTIC (HBV, HCV, Autoimmune)	++++	++++	++	++
PERICELLULAR-PERISINUSOIDAL (NASH, ASH, Hemochromatosis)	–	+	++++	+++
BILIARY (PBC. PSC)	+	++	–	+++
CENTRILOBULAR (Chronic heart failure, upstream vascular disorders)	–	±	–	++++

References

1. Bataller R, Brenner DA. Liver fibrosis. J Clin Invest. 2005;115:209–18.
2. Friedman SL. Mechanisms of hepatic fibrogenesis. Gastroenterology. 2008;134:1655–69.
3. Pinzani M, Macias-Barragan J. Update on the pathophysiology of liver fibrosis. Expert Rev Gastroenterol Hepatol. 2010;4:459–72.
4. Wynn TA. Common and unique mechanisms regulate fibrosis in various fibroproliferative diseases. J Clin Invest. 2007;117:524–9.
5. Fernández M, Semela D, Bruix J, Colle I, Pinzani M, Bosch J. Angiogenesis in liver disease. J Hepatol. 2009;50:604–20.
6. Pinzani M, Rombouts K. Liver fibrosis: from the bench to clinical targets. Dig Liver Dis. 2004;36:231–42.
7. Tsochatzis EA, Bosch J, Burroughs AK. New therapeutic paradigm for patients with cirrhosis. Hepatology. 2012;56:1983–92.
8. Hall A, Germani G, Isgrò G, et al. Fibrosis distribution in explanted cirrhotic livers. Histopathology. 2012;60:270–7.
9. Calvaruso V, Burroughs AK, Standish R, et al. Computer-assisted image analysis of liver collagen: relationship to Ishak scoring and hepatic venous pressure gradient. Hepatology. 2009;49:1236–44.
10. Mazza G, Al-Akkad W, Telese A, Longato L, Urbani L, Robinson B, Hall A, et al. Rapid production of human liver scaffolds for functional tissue engineering by high shear stress oscillation-decellularization. Sci Rep. 2017;7:5534.
11. Mazza G, Telese A, Al-Akkad W, Frenguelli L, Levi A, Marrali M, Longato L, et al. Cirrhotic human liver extracellular matrix 3D scaffolds promote Smad-dependent TGF-beta1 epithelial Mesenchymal transition. Cell. 2019;9
12. Ferrusquia-Acosta J, Bassegoda O, Turco L, Reverter E, Pellone M, Bianchini M, Perez-Campuzano V, et al. Agreement between wedged hepatic venous pressure and portal pressure in non-alcoholic steatohepatitis-related cirrhosis. J Hepatol. 2021;74:811–8.
13. Hayasaka A, Ilda S, Suzuki N, Kondo F, Miyazaki M, Yonemitsu H. Pyridinoline collagen cross-links in patients with chronic viral hepatitis and cirrhosis. J Hepatol. 1996;24:692–8.
14. Issa R, Zhou X, Constandinou CM, Fallowfield J, Millward-Sadler H, Gaca MD, Sands E, Suliman I, Trim N, Knorr A, Arthur MJ, Benyon RC, Iredale JP. Spontaneous recovery from micronodular cirrhosis: evidence for incomplete resolution associated with matrix cross-linking. Gastroenterology. 2004;126:1795–808.
15. Shiratori Y, Imazeki F, Moriyama M, Yano M, Arakawa Y, Yokosuka O, et al. Histological improvement of fibrosis in patients with hepatitis C who have sustained response to interferon therapy. Ann Intern Med. 2000;132:517–24.
16. Novo E, Marra F, Zamara E, Valfrè di Bonzo L, Monitillo L, Cannito S, Petrai I, Mazzocca A, Bonacchi A, De Franco RS, Colombatto S, Autelli R, Pinzani M, Parola M. Overexpression of Bcl-2 by activated human hepatic stellate cells: resistance to apoptosis as a mechanism of progressive hepatic fibrogenesis in humans. Gut. 2006;55:1174–82.
17. Di Marco V, Calvaruso V, Ferraro D, Bavetta MG, Cabibbo G, Conte E, Cammà C, Grimaudo S, Pipitone RM, Simone F, Peralta S, Arini A, Craxì A. Effects of eradicating hepatitis C virus infection in patients with cirrhosis differ with stage of portal hypertension. Gastroenterology. 2016;151:130–9.
18. Díez C, Berenguer J, Ibañez-Samaniego L, Llop E, Pérez-Latorre L, Catalina MV, Hontañón V, Jiménez-Sousa MA, Aldámiz-Echevarría T, Martínez J, Calleja JL, Albillos A, Bellón JM, Resino S, González-García J, Bañares R. Persistence of clinically significant portal hypertension after eradication of hepatitis C virus in patients with advanced cirrhosis. Clin Infect Dis. 2020;71:2726–9.
19. Puigvehí M, Londoño MC, Torras X, Lorente S, Vergara M, Morillas RM, Masnou H, Serrano T, Miquel M, Gallego A, Lens S, Carrión JA. Impact of sustained virological response

with DAAs on gastroesophageal varices and Baveno criteria in HCV-cirrhotic patients. J Gastroenterol. 2020;55:205–16.
20. Deterding K, Höner Zu Siederdissen C, Port K, Solbach P, Sollik L, Kirschner J, Mix C, Cornberg J, Worzala D, Mix H, et al. Improvement of liver function parameters in advanced HCV-associated liver cirrhosis by IFN-free antiviral therapies. Aliment Pharmacol Ther. 2015;42:889–901.
21. Aqel BA, Pungpapong S, Leise M, Werner KT, Chervenak AE, Watt KD, Murphy JL, Ryland K, Keaveny AP, McLemore R, et al. Multicenter experience using simeprevir and sofosbuvir with or without ribavirin to treat hepatitis C genotype 1 in patients with cirrhosis. Hepatology. 2015;62:1004–12.
22. Spahr L, Goossens N, Furrer F, Dupuis M, Vijgen S, Elkrief L, Giostra E, Rubbia-Brandt L, Frossard JL. A return to harmful alcohol consumption impacts on portal hemodynamic changes following alcoholic hepatitis. Eur J Gastroenterol Hepatol. 2018;30:967–74.
23. Ryu SR, Yoo JJ, Kang SH, Jeong SW, Kim MY, Cho YK, Chang Y, Kim SG, Jang JY, Kim YS, Baik SK, Kim YJ, Park SY, Baymbajav B. The cut-off value of transient elastography to the value of hepatic venous pressure gradient in alcoholic cirrhosis. Clin Mol Hepatol. 2021;27:197–206.
24. Colecchia A, Montrone L, Scaioli E, Bacchi-Reggiani ML, Colli A, Casazza G, Schiumerini R, Turco L, Di Biase AR, Mazzella G, Marzi L, Arena U, Pinzani M, Festi D. Measurement of spleen stiffness to evaluate portal hypertension and the presence of esophageal varices in patients with HCV-related cirrhosis. Gastroenterology. 2012;143:646–54.

Angiogenesis and Progression of ACLD

17

Seth M. Buryska, Kyle E. Robinson, and Vijay Shah

Sinusoidal Architecture and Functional Components

Liver Vascular Structure

The liver is a highly vascularized organ, receiving 75% of its blood from the portal vein and 25% from the hepatic artery. The portal vein divides into septal branches as the hepatic artery concomitantly divides into axillary branches that run adjacent to the septal branches. They then converge at the sinusoidal interface before draining into the hepatic veins [1]. Sinusoidal endothelial cells comprise the walls of the vasculature in the sinusoid, forming a fenestrated barrier that lacks a basement membrane [2, 3]. This fenestrated barrier acts as a mechanical sieve, separating the contents of the blood from the Space of Disse and other parenchymal liver cells [4, 5].

The Sinusoid Microenvironment

Sinusoidal endothelial cells and hepatic stellate cells (HSCs) will be the primary focus of this review as research over the last two to three decades has revealed the inherent relationship between these components concerning liver health, homeostasis, regeneration, angiogenesis, angiocrine signaling, and the pathophysiology of advanced chronic liver disease (ACLD). These mechanisms are multifaceted, incompletely understood, and extend beyond the aforementioned constituents.

S. M. Buryska · K. E. Robinson · V. Shah (✉)
Division of Gastroenterology and Hepatology, Mayo Clinic, Rochester, MN, USA
e-mail: buryska.seth@mayo.edu; robinson.kyle@mayo.edu; shah.vijay@mayo.edu

R. de Franchis (ed.), *Portal Hypertension VII*,
https://doi.org/10.1007/978-3-031-08552-9_17

Sinusoidal Endothelial Cell Functions

Outside of acting as a selective physical barrier, SECs play a pivotal role in maintaining liver health and act as gatekeepers in regulating homeostasis and regeneration within the sinusoid microenvironment. SECs are regulators of intrahepatic pressure through their production of nitric oxide (NO) via the endogenous nitric oxide synthase (eNOS) pathway. NO synthesis increases in response to a variety of environmental stimuli, including mechanical forces such as shear stress [6]. Paracrine factors like vascular endothelial growth factor (VEGF) produced by surrounding HSCs and hepatocytes also promote NO production [7–9]. A more prominent presence of vasodilators (NO, PGI2) in comparison to vasoconstrictors (ET-1, Thromboxane A) establishes the low-resistant state in healthy livers [10].

NO from SECs also plays a role in maintaining the sinusoid microenvironment. HSCs remain in a quiescent, fat-storing state during periods of sufficient NO production [11, 12]. Angiocrine signaling from SECs promotes hepatocyte homeostasis and self-renewal through the production of hepatocyte growth factor (HGF), Wnt2, and Wnt9b [13, 14]. The signaling required for the maintenance of the sinusoid microenvironment is interdependent and reciprocal, as VEGF from HSCs (as well as hepatocytes) promotes the development and maintenance of fenestrae, thereby maintaining the SEC phenotype [8, 9, 15]. Later sections will review how dysfunction of SECs changes angiocrine pathways, thus perturbing homeostatic mechanisms [1, 13, 16, 17].

Hepatic Stellate Cell Functions

Hepatic stellate cells are pericyte-like cells located in the space of Disse adjacent to SECs. They possess an intermediate phenotype; Vitamin A/fat-storing cells in their quiescent state vs. a proliferative, contractile, fibrogenic myofibroblast phenotype in their activated state [18]. HSCs possess a variety of functions ancillary to SECs, mediated through paracrine cross-talk with the neighboring cells [18]. Their other functions include retinol transport and storage [19]; collagen production and extracellular matrix (ECM) regulation via TGF-β [20, 21]; mitogenic and motogenic responses to platelet-derived growth factor (PDGF) [22, 23]. Later sections will reveal how activation of HSCs plays a pivotal role in fibrogenesis. Furthermore, link between HSC activation and SEC dysfunction will be elaborated upon [1, 24, 25]. Appropriate cross-talk between SECs and HSCs is necessary to maintain hepatic homeostasis. Disruption of this cooperative relationship initiates multiple pathogenic modalities that contribute to the development of ACLD [1, 2, 11, 18, 24, 25].

The role of the sinusoid microenvironment during angiogenesis provides invaluable insight into the phenomenon of how the liver shifts from a regenerative response to a pathogenic modality. Angiogenesis in both liver regeneration and liver disease will be the focus of the following sections.

Relationship of Angiogenesis and Angiocrine Signaling During Liver Health and Regeneration

Overview of Angiogenesis: Friend or Foe?

Angiogenesis, the formation of new blood vessels from preexisting vasculature, is a nearly ubiquitous process in the body during tissue repair and regeneration from injury [26, 27]. Angiogenesis is necessary for liver regeneration to ensue. SECs and HSCs crosstalk facilitates this process [1, 18, 28, 29]. Regeneration is effectively a cycle. In response to injury, the liver's goal is to reconstitute its mass and restore homeostasis. In order to achieve this goal, angiogenesis is necessary. For angiogenesis to occur, angiocrine signals from SECs are necessary to regulate and facilitate the regenerative cycle. Furthermore, there must be a cooperative effort between angiocrine signaling and angiogenesis.

Modulators of Angiogenesis and Regeneration: Angiocrine Signals from Sinusoidal Endothelial Cells

The role of SECs during liver regeneration parallels the roles of endothelial cells (EC) during embryonic development. ECs initially release angiocrine signals which stimulate organogenesis. They subsequently respond to paracrine signaling which stimulates EC proliferation, thus, stimulating angiogenesis [30, 31]. There are two stages of regeneration that are spatiotemporal in nature [32–36].

Hepatic regeneration occurs in two phases: (I) hepatocyte proliferation (inductive angiogenesis) and (II) SEC proliferation (proliferative angiogenesis) [35–37]. In phase I, hepatocyte proliferation is induced by the upregulated production of hepatocyte growth factor (HGF) [37, 38]. The exact mechanism of HGF release currently is unclear. There is evidence that SECs upregulate HGF production in response to VEGF [37]. There is also evidence that SECs coordinate the recruitment of bone marrow-derived progenitor cells (BM-PCs), substances rich in HGF [38–40]. In any event, deletion of the SEC transcription factors associated with HGF expression consequently attenuates liver regeneration, elucidating the regulatory role of angiocrine signaling to promote the first phase of regeneration; hepatocyte proliferation (Fig. 17.1) [36, 37].

Angiopoietin-2 (Ang2) is a potential angiocrine signal which then links the first and second stages of liver regeneration [41]. Downregulation of Ang2 in early phases permits organogenesis, and its upregulation in the second phase stimulates angiogenesis [41]. The role of angiocrine regulation in the regenerative cycle is an area of ongoing research.

In summary, the angiocrine pathways disseminated from SECs interweave into the process of hepatic regeneration. Appropriate SEC function is necessary to facilitate these response pathways that ultimately restore appropriate liver mass, function, and homeostasis.

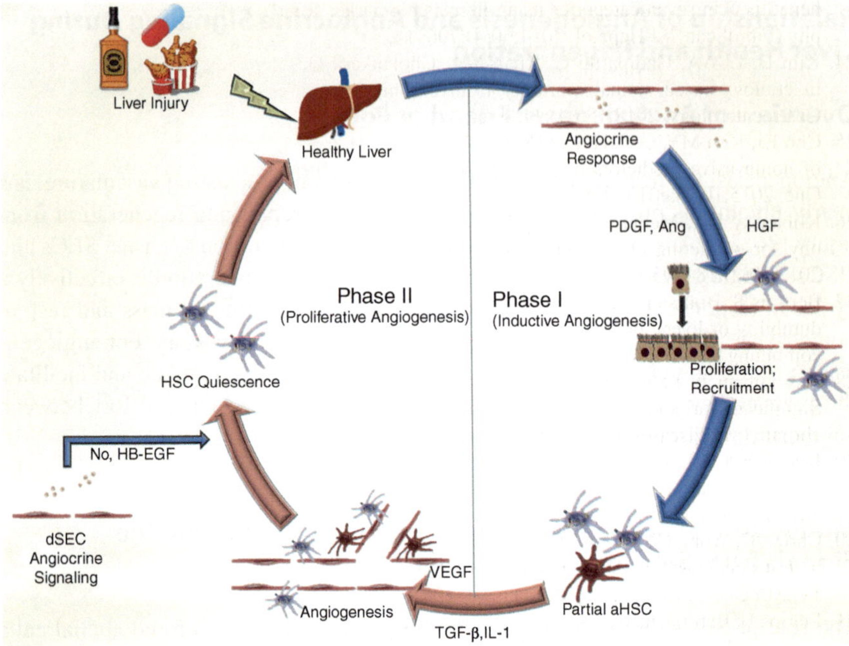

Fig. 17.1 Interplay of angiocrine signaling and angiogenesis during liver regeneration. In response to liver injury, SECs initiate the first phase of liver regeneration by the release of angiocrine signals that mediate HSC recruitment (PDGF, Ang) and hepatocyte proliferation (HGF). A portion of HSCs subsequently activates, releasing pro-angiogenic signals (VEGF) and also control the hepatocyte proliferation response (TGF-β, IL-1). These processes facilitate the second phase of regeneration: angiogenesis. Once the liver has reconstituted its mass and is subsequently vascularized it returns to its homeostatic state by reverting aHSCs to their quiescent state by the release of angiocrine signals from differentiated SECs (dSEC). *HGF* hepatocyte growth factor, *PDGF* platelet-derived growth factor, *Ang* Angiopoietins, *VEGF* vascular endothelial growth factor, *TGF* transforming growth factor, *IL* interleukin, *NO* nitric oxide, *HB-EGF* heparin-binding epidermal growth factor-like growth factor, *dSEC* differentiated sinusoidal endothelial cell, *aHSC* activated hepatic stellate cell

Relationship of Sinusoidal Endothelial Cells and Hepatic Stellate Cells during Angiogenesis and Regeneration

Crosstalk between SECs and HSCs is necessary to facilitate angiogenesis and vascular remodeling. Given the pericyte-like nature of HSCs, our understanding of their role in angiogenesis derives from pericytes located in other areas of the body, exerting a paracrine effect that induces EC proliferation, migration, differentiation, vascular branching, and stabilization. These pericyte-like cells simultaneously respond to angiocrine signals from SECs that promote HSC migration to sites of developing vasculature to aid in angiogenesis (Fig. 17.1) [18, 42–46]. Angiocrine factors that mediate pericyte recruitment are platelet-derived growth factor (PDGF) and angiopoietins (Ang) [44, 45]. Mouse models deficient in PDGF signaling demonstrated reduced pericyte recruitment, resulting in abnormalities to vessel diameter, endothelial hyperplasia, and vascular leak [46, 47]. Pericyte recruitment, vessel stabilization, and maturation are associated with angiopoietin pathways [45, 48].

In addition to ensuring optimal revascularization, HSCs are a source of two molecules necessary to regulate hepatocyte proliferation: HGF and TGF-β. HGF and TGF-β paradoxically exert proliferative and antiproliferative effects respectively [38, 49, 50]. During the initial stage, HSC TGF-β expression attenuates, permitting hepatocyte proliferation [41]. In response, the activation of HSCs is necessary to regulate hepatocyte proliferation because HSCs upregulate paracrine factors like IL-1 and TGF-β, subsequently inducing antiproliferative effects [51]. A certain degree of HSC activation is necessary to facilitate the regenerative response as variations to the degrees of activation result in suboptimal sinusoidal revascularization [18, 51, 52].

Once the liver has completed its regenerative response and returns to a homeostatic state, HSCs must return to their quiescent state normally observed in the homeostatic environment [1]. This return to baseline is achieved by regulatory angiocrine molecules from SECs, highlighting the relationship between angiocrine signaling and angiogenesis during regeneration (Fig. 17.1).

The failure of activated HSCs to revert to quiescence consequently contributes to fibrogenesis (via TGF-β), pathological angiogenesis, and drives the development of ACLD. This pathophysiological mechanism represents how changes in the sinusoid microenvironment subsequently induce a decompensating cascade that results in the development of ACLD and PHT.

The shift from a regenerative to degenerative response is a result of changes to the complex, interdependent relationship between neighboring cells: SECs and HSCs. Chronic liver injury perturbs the symbiotic relationship of angiocrine signaling (from SECs). The following sections will demonstrate how the disrupted relationship between angiogenesis and angiocrine signaling engenders pathological angiogenesis.

Endothelial Cell Dysfunction: Implications for ACLD and Portal Hypertension

The shift to a dedifferentiated SEC phenotype, termed capillarization, is a known precursor to the development of liver fibrosis. Capillarization is described as the loss of fenestration and deposition of a basement membrane in the Space of Disse [53, 54]. In healthy livers, differentiated SECs (dSECs) exert angiocrine signals that control vascular tone, immune response, hepatocyte growth, and HSC quiescence. In contrast, when these barriers shift to their capillarized, dedifferentiated forms, this invokes multiple interrelated pathophysiological mechanisms. The following sections will demonstrate how endothelial cell dysfunction permits lasting hepatic stellate cell activation, thereby propagating fibrotic, hypoxic, and inflammatory changes (Fig. 17.2) [6, 10, 15, 30, 55].

As previously stated, VEGF stimulates the release of NO via activation of eNOS, and NO subsequently regulates a pathway that phosphorylates downstream protein targets to maintain SEC phenotype [8, 56]. Changes in this pathway result in decreased vasoregulation and are associated with a pathogenic shift in the sinusoid microenvironment.

Decrease Vasodilators

During capillarization NO production is attenuated by the binding of inhibitors (e.g., caveolin) to eNOS, inhibiting a synthetic response to external stimuli [57, 58]. Even with upregulated VEGF, the capillarized phenotype does not secrete a sufficient amount of NO in response [59–61]. Oxidative stress from bacterial endotoxins, viruses, drugs, and ethanol is another possible contributor to the decrease in the availability of vasodilators [62, 63]. The increased superoxide free radicals present in cirrhotic livers undergo a reaction with NO to form peroxynitrite ($ONOO^-$), further attenuating the availability of NO [64].

Increased Vasoconstrictors

In juxtaposition to the decreasing amounts of vasodilators during SEC dysfunction, there concomitantly is an increased prevalence of vasoconstrictors like ET-1 and thromboxane A in diseased livers [10, 65]. The shift from a sinusoid microenvironment with a higher proportion of vasodilators to vasoconstrictors inverts during the capillarization process, compromising vasoregulation, and contributes to an increase in intrahepatic resistance, thereby contributing to the development of ACLD (Fig. 17.2).

Pathogenic Shift in the Sinusoid Microenvironment

The capillarization of SECs leads to not only an imbalance of vasoregulatory molecules but also to the loss of SEC's ability to maintain HSC quiescence. In correspondence with a more constricted vascular network, the shift to an activated, contractile HSC myofibroblast phenotype further constricts blood flow, raising intrahepatic resistance [11, 66, 67]. Once activated, HSCs are less responsive to vasodilators and more susceptible to vasoconstrictors, enhancing their contractility [66, 68].

SECs in their differentiated form are able to revert activated HSCs into their quiescent state through activation of the NO-dependent pathway (via an sGC activator) [56]. In capillarization, the NO-dependent pathway is downregulated, and cSECs are unable to return activated HSCs to their quiescent state. The release of this regulatory linchpin due to capillarization is a critical component that allows for persisting HSC activation and the subsequent release of pro-angiogenic factors that drives pathological angiogenesis (Fig. 17.2).

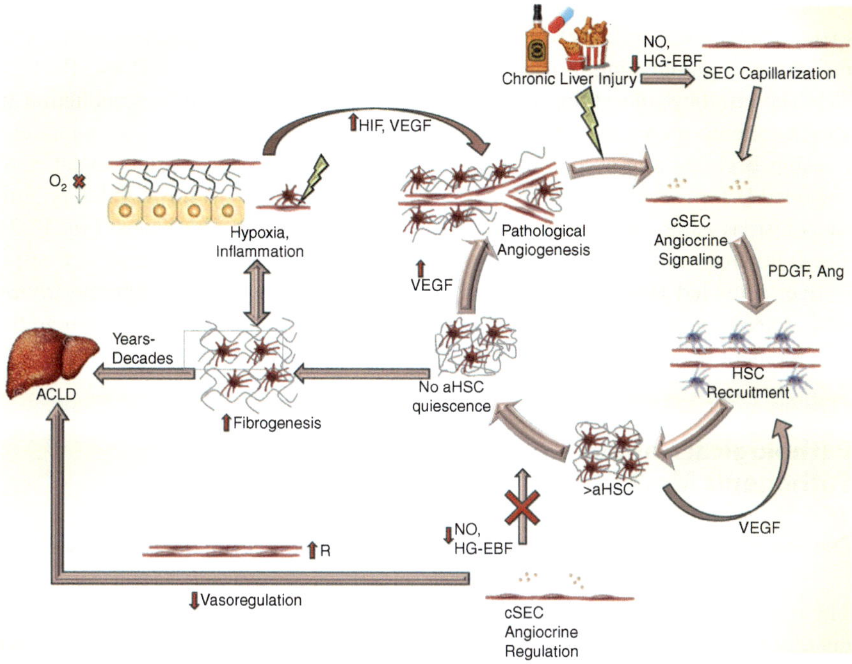

Fig. 17.2 Relationship between pathological angiogenesis and ancillary pathophysiological mechanisms. Chronic liver injury results in the capillarization of SECs. cSEC angiocrine signaling is still able to recruit HSCs. However, the regulatory capacity of angiocrine signaling is lost, resulting in overactivation of HSCs and a failure of aHSCs to return to quiescence. In turn, this stimulates the release of pro-angiogenic factors causing a positive feedback loop. Pathological angiogenesis drives the progression of liver disease by decreasing vasoregulation and permitting the activation of fibrogenic pathways, concomitantly activating ancillary pathophysiological mechanisms. These ancillary mechanisms reciprocally stimulate fibrogenesis and further perpetuate the pathologic angiogenic cycle. Consequently, this gives rise to fibrogenesis and disease progression. Over time, the culmination of these pathogenic modalities results in ACLD. *PDGF* platelet-derived growth factor, *Ang* angiopoietins, *VEGF* vascular endothelial growth factor, *NO* nitric oxide, *HB-EGF* heparin-binding epidermal growth factor-like growth factor, *cSEC* capillarized sinusoidal endothelial cell, *aHSC* activated hepatic stellate cell, *HIF* hypoxic inducible factors

Recent efforts have investigated potential etiologies responsible for capillarization. A study by Duan et al. demonstrated how the activation of the Notch pathway invokes a genetic shift that results in capillarization due to the inhibition of the aforementioned NO-dependent pathway. Activation of the Notch pathway also attenuates the production of hepatocyte mitogens like HGF, thereby retarding regenerative responses [69]. Interestingly, an sGC activator was again able to reactivate this pathway and restore a differentiated phenotype [69].

Another study may have identified a putative mediator responsible for SEC maintenance of HSC quiescence: heparin-binding epidermal growth factor-like growth factor (HB-EGF). The study found that cSECs were actually bone marrow-derived endothelial progenitor cells that failed to mature. In turn, these immature cSECs secreted HB-EGF at a lower rate, permitting HSC activation [70].

In summary, differentiated SECs prevent HSC activation and promote reversion to quiescence, but capillarized SECs do not [71]. The result is increased intrahepatic resistance, a quintessential mechanism in the development of ACLD and PHT [1, 57]. The etiology underlying this phenomenon is a decrease in the production of specific angiocrine signals like NO and HB-EGF, thus disrupting the symbiotic relationship between angiocrine signaling and angiogenesis. Angiocrine signaling is responsible for regulating angiogenesis in healthy livers. When this regulatory ability is lost, as is the case with cSECs, pathological angiogenesis ensues (Fig. 17.2). The following sections will demonstrate how pathological angiogenesis is a result of over-activated HSCs invoking increased rates of angiogenesis, thereby driving HSC activation, and ultimately stimulating ancillary mechanisms that promote the development of ACLD and PHT.

Pathological Angiogenesis and Ancillary Pathogenic Modalities

Driver of Pathological Angiogenesis: Hepatic Stellate Cell Activation

The activation of HSCs is a salient component of pathological angiogenesis, subsequently invoking multiple ancillary mechanisms (Fig. 17.2). As previously stated, differentiated SECs help to keep HSCs in their quiescent state, while capillarized SECs are unable to do so [56]. Other surrounding cells such as hepatocytes, Kupffer cells, and lymphocytes release soluble factors that contribute to the development of activated HSCs (aHSCs) [24]. ECM stiffness is also a stimulant for HSC activation [72].

Once activated, HSCs upregulate the secretion of pro-angiogenic factors like VEGF and angiopoietin 1 (Ang-1), stimulating the proliferation of SECs [8, 73, 74]. The upregulation of VEGF stimulates the activation, proliferation, and chemoattraction of HSCs to sites of newly forming vasculature [75, 76]. SECs then secrete certain angiocrine signals like PDGF that further stimulate HSC migration and recruitment to newly developing vessels, creating a positive feedback loop (Fig. 17.2) [18]. This positive feedback loop cultivates pathologic angiogenesis.

In pathologic angiogenesis, there is an over-recruitment of HSCs and a loss of angiocrine regulation, permitting the development of a proliferative, myofibroblast-like phenotype. This activated HSC phenotype results in the activation of pro-fibrogenic pathways [24, 77]. During liver injury, TGF-β is the most pro-fibrogenic factor released by aHSCs, stimulating the production of non-fibrillar and fibrillar matrix components [24, 78]. In summary, a lack of angiocrine regulation affords HSC activation, stimulating pathological angiogenesis and collaterally stimulating fibrogenesis (Fig. 17.2).

In conjunction with the concept of pathological angiogenesis comes the term sinusoidal remodeling. Due to the mechanisms mentioned above, there is a high density of aHSCs around sites of developing sinusoids. Due to the contractile phenotype of aHSCs, as well as their lack of response to vasodilators, these remodeled sinusoids perpetuate rises in intrahepatic pressure and decrease blood flow [18, 66,

68]. Furthermore, the fibrotic changes in the remodeled sinusoid further inhibit oxygen perfusion, consequently engendering another pathophysiologic modality: hypoxia [25, 79].

Driver of Pathological Angiogenesis: Hypoxia

While aHSCs can stimulate pathological angiogenesis via the proliferation of SECs in a direct manner, their activation and subsequent effects on the sinusoidal microenvironment can secondarily stimulate pathological angiogenesis. Indeed, there is an emerging link between fibrogenesis, pathological angiogenesis, and hypoxia [80].

Typically, a hypoxic environment stimulates the group of transcription factors called hypoxia-inducible factors (HIFs). These, in turn, upregulate genes associated with pro-angiogenic factors such as VEGF that facilitate the reperfusion of hypoxic cells [81–84]. VEGF upregulation in response to hypoxia stimulates angiogenesis. However, in the setting of chronic liver injury, this means that hypoxia collaterally drives fibrogenesis by stimulating pathological angiogenesis (Fig. 17.2). The HIF isomer HIF-1α is intrinsically involved in the development of fibrogenesis [85, 86] as it regulates numerous genes involved in fibrosis development [87–89]. In addition, HIF-1α knockout mice who underwent a bile duct ligation (an animal model of liver fibrosis), exhibited decreased regulation of profibrotic mediators and had substantially less fibrotic changes compared to the control group [86].

To summarize, cSECs and aHSCs cultivate an increasingly hypoxic environment. Consequently, HIFs stimulate many pro-fibrogenic and pro-angiogenic pathways, perpetuating the pathological effects of one another. Indeed, this decompensating cascade carries collateral effects in other ancillary pathophysiological modalities, including inflammation (Fig. 17.3).

Driver of Pathological Angiogenesis: Inflammation

Macrophages act as a double-edged sword when determining whether a regenerative or pathogenic response ensues. The liver possesses the most macrophages of any solid organ, contextualizing their importance in liver processes. Macrophages usually contribute to liver homeostasis and regeneration, stimulate angiogenesis, and facilitate tissue remodeling by secreting soluble molecules [55]. However, their role changes during pathological angiogenesis.

In essence, pathological angiogenesis drives the increased infiltration of macrophages through mechanisms like the production of adhesion molecules from activated HSCs, as well as hepatocyte degeneration [24, 90, 91]. In turn, the recruited macrophages secrete a variety of chemokines and cytokines that further augment the inflammatory response. Furthermore, macrophages stimulate angiogenesis by releasing pro-angiogenic molecules like VEGF, consequently perpetuating pathological angiogenesis [92]. Macrophages possess pro-fibrogenic properties, secreting TGF-β, PDGF, and activate HSCs which enhances the aforementioned

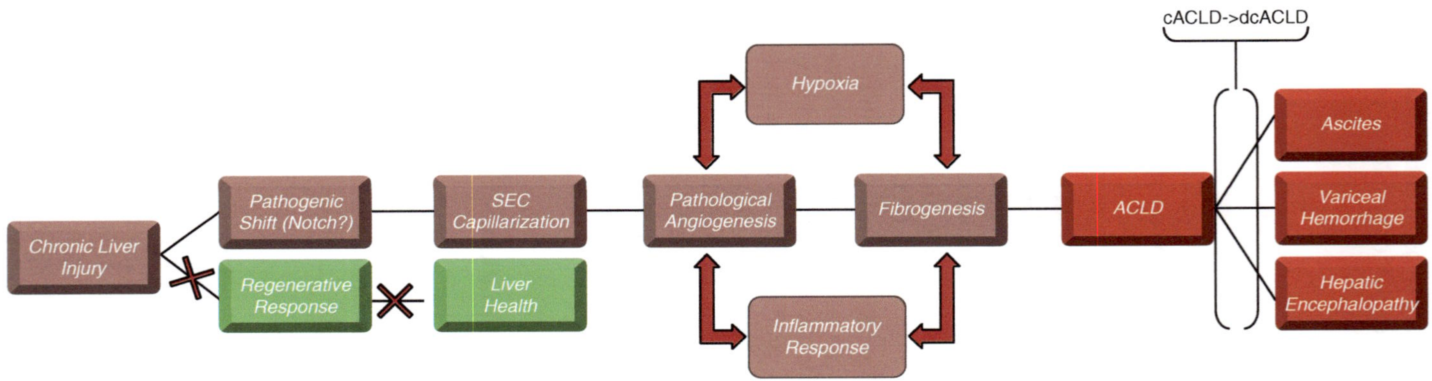

Fig. 17.3 Pathological angiogenesis; decompensating cascade in development of ACLD. In the setting of chronic liver injury, there is a diversion from a regenerative response to a pathogenic one. Epigenetic analyses unveiled that shifts in gene expression, such as in the Notch pathway, are associated with this pathogenic shift. Consequently, the capillarization of SECs (cSECs) occurs. From this comes pathological angiogenesis, as defined in Fig. 17.2. Pathological angiogenesis, fibrogenesis, hypoxia, and inflammatory responses perpetuate the effects of one another, cultivating a decompensating cascade that drives the development of ACLD. Over time, this decompensating cascade surmounts the liver's ability to compensate for chronic injury and decompensating events ensue. *SEC* sinusoidal endothelial cell, *ACLD* advance chronic liver disease, *cACLD* compensated advance chronic liver disease, *dcACLD* decompensated advance chronic liver disease

pathogenic effects. This stimulates inflammatory response pathways through another positive-feedback loop [93, 94].

These processes are emblematic of how disturbances to angiocrine signaling in the sinusoid microenvironment can trigger a myriad of pathophysiological mechanisms whose effects build off one another (Fig. 17.3). The culmination of these effects continually increases intrahepatic resistance and promotes fibrotic changes, eventually resulting in the development of ACLD. Over time, ACLD will continue to worsen and result in decompensating events like ascites, variceal hemorrhaging, and hepatic encephalopathy (Fig. 17.3).

Clinical Manifestations and Measures of ACLD

Portal Hypertension

The pathological angiogenesis mechanisms mentioned above eventually result in abnormally developed nodules separated by fibrotic tissue, known as cirrhosis [95]. Cirrhotic livers can then be divided into two main categories: compensated (cACLD) and decompensated (dcACLD). cACLD is then further divided into four stages of fibrosis: F_0-F_4 [96]. Monitoring portal pressure has diagnostic and prognostic purposes, as a hepatic venous pressure gradient (HVPG) >10 mmHg is indicative of clinically significant portal hypertension where the development of esophageal varices may be observed. An HVPG value >12 mmHg is about the threshold value that divides cACLD from dcACLD (Fig. 17.4). Those with portal

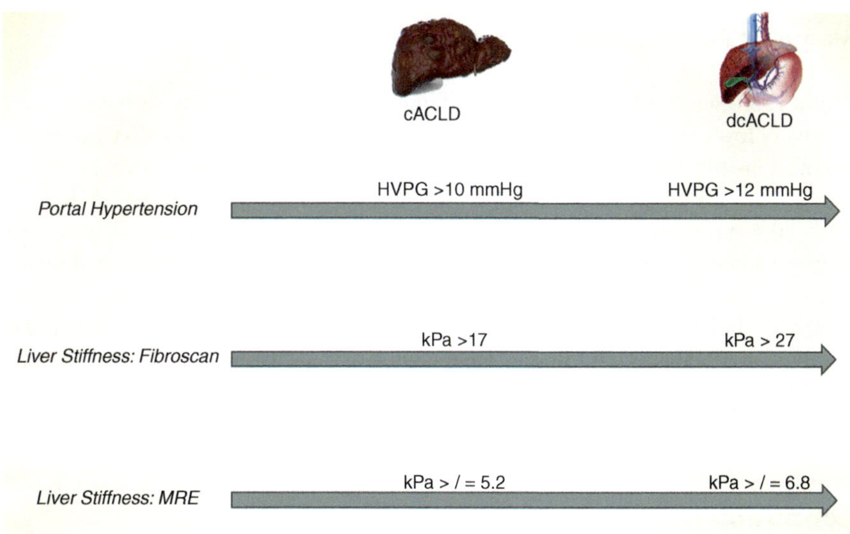

Fig. 17.4 Clinical measures for compensated and decompensated advanced chronic liver disease. Clinical measures to differentiate cACLD from dcACLD

pressures higher than this value are at a significantly increased risk of developing acute decompensating events like ascites, variceal hemorrhage, and hepatic encephalopathy [96].

Transient Elastography

The progression of liver disease and fibrosis results in increased liver stiffness. An affordable, non-invasive way to diagnose stiffness and diagnose disease severity is transient elastography. Commonly known as a FibroScan, this test measures the mechanical property of stiffness, a surrogate measure to assess fibrosis. A FibroScan measures the speed at which mechanical or sound waves propagate through the liver. Stiff, fibrotic livers will propagate the waves at a faster rate. Multiple studies have demonstrated that FibroScans can accurately differentiate the four stages of fibrosis, with values ranging from ~6 kPa to ~18 kPa in a stepwise fashion [97]. FibroScans have also been used to distinguish cohorts of patients with cACLD (kPa > ~17–18) from those with dcACLD (kPa > 27) [98]. FibroScans have additional prognostic value with respect to predicting the development of PHT, esophageal varices, variceal hemorrhage, ascites, and HCC [99].

This prognostic value may be due to stiffness being a biomechanical property associated with disease progression. Increasing extracellular matrix stiffness perpetuates HSC activation, augmenting fibrogenic response pathways, thereby increasing ECM deposition and stiffness. This creates another pathogenic positive feedback loop [72].

Magnetic Resonance Elastography

While FibroScans retain the value of being a reliable, non-invasive measure at a relatively low cost to the patient, magnetic resonance elastography (MRE) scans are a more sensitive and specific measure. In a recent meta-analysis involving approximately 1500 MRE patients and approximately 3600 FibroScan patients, MRE was 10%–20% more specific and 10%–20% more sensitive in the diagnosis of significant and advanced fibrosis as well as cirrhosis [100]. MRE can also stratify diseased livers into the four stages of fibrosis, with values ranging from 3.0 kPa to >5 kPa in a stepwise fashion. MREs can also differentiate cirrhotic livers based on whether or not they are compensated. Studies have demonstrated that the average stiffness for cACLD (~5.2 kPa) is significantly lower compared to stiffness values for dcACLD (~6.8 kPa) (Fig. 17.4) [101].

Conclusion

The effects of SEC capillarization, HSC activation, sinusoidal remodeling, hypoxic, and inflammatory changes all contribute to pathological angiogenesis. Pathological angiogenesis affords the activation of pro-fibrogenic pathways and results in the loss of vasoregulation; classical mechanisms responsible for PHT in cirrhotic livers. The pathogenic shift to a capillarized SECs phenotype results in changes to respective angiocrine regulatory pathways. This affords for the unregulated activation and proliferation of HSCs, subsequently invoking hypoxic and inflammatory pathways. These pathways stimulate the proliferation of capillarized SECs (pathological angiogenesis), thereby augmenting the previously mentioned pathogenic modalities through various positive-feedback loops. Pathological angiogenesis and its ancillary pathogenic modalities contribute to the development PHT: a major risk factor for acute decompensation in advanced chronic liver diseases. Angiocrine, paracrine, and autocrine regulatory pathways control the regulation of pathological angiogenesis, fibrosis, and disease progression. Therapeutic targets aimed at remedying abnormalities in these pathways may prove to be a quintessential tool in inhibiting fibrotic changes, attenuating the development of PHT and decreasing the mortality of advanced chronic liver disease.

References

1. Kostallari E, Shah VH. Angiocrine signaling in the hepatic sinusoids in health and disease. Am J Physiol Gastrointest Liver Physiol. 2016;311(2):246–51.
2. Arii S, Imamura M. Physiological role of sinusoidal endothelial cells and Kupffer cells and their implication in the pathogenesis of liver injury. J Hepato-Biliary-Pancreat Surg. 2000;7(1):40–8.
3. Braet F, De Zanger R, Baekeland M, Crabbé E, Van Der Smissen P, Wisse E. Structure and dynamics of the fenestrae-associated cytoskeleton of rat liver sinusoidal endothelial cells. Hepatology. 1995;21(1):180–9.
4. Burt AD, Le Bail B, Balabaud C. Morphological investigation of sinusoidal cells. Semin Liver Dis. 1993;13:21–38.
5. Ishibashi H, Nakamura M, Komori A, Migita K, Shimoda S. Liver architecture, cell function, and disease. Semin Immunopathol. 2009;31(3):399–409.
6. Shah V, Haddad FG, Garcia-Cardena G, Frangos JA, Mennone A, Groszmann RJ, et al. Liver sinusoidal endothelial cells are responsible for nitric oxide modulation of resistance in the hepatic sinusoids. J Clin Invest. 1997;100(11):2923–30.
7. He H, Virginia VJ, Xiaolin G, Richard VC, Mario MB, Ruth CB. Vascular endothelial growth factor signals endothelial cell production of nitric oxide and prostacyclin through Flk-1/KDR activation of c-Src. J Biol Chem. 1999;274:25130–5.
8. DeLeve LD, Wang X, Hu L, McCuskey MK, McCuskey RS. Rat liver sinusoidal endothelial cell phenotype is maintained by paracrine and autocrine regulation. Am J Physiol Gastrointest Liver Physiol. 2004;287(4):G757–63.
9. Yamane A, Seetharam L, Yamaguchi S, Gotoh N, Takahashi T, Neufeld G, Shibuya M. A new communication system between hepatocytes and sinusoidal endothelial cells in liver through vascular endothelial growth factor and Flt tyrosine kinase receptor family (Flt-1 and KDR/Flk-1). Oncogene. 1994;9:2683–90.
10. Kamoun WS, Karaa A, Kresge N, Merkel SM, Korneszczuk K, Clemens MG. LPS inhibits endothelin-1–induced endothelial NOS activation in hepatic sinusoidal cells through a negative feedback involving caveolin-1. Hepatology. 2006;43(1):182–90.
11. Deleve LD, Wang X, Guo Y. Sinusoidal endothelial cells prevent rat stellate cell activation and promote reversion to quiescence. Hepatology. 2009;48:920–30.
12. Geerts A. History, heterogeneity, developmental biology, and functions of quiescent hepatic stellate cells. Semin Liver Dis. 2001;21(3):311–36.
13. Bocca C, Novo E, Miglietta A, Parola M. Angiogenesis and fibrogenesis in chronic liver diseases. Cell Mol Gastroenterol Hepatol. 2015;1(5):477–88.

14. Wang B, Zhao L, Fish M, Logan CY, Nusse R. Self-renewing diploid Axin2+ cells fuel homeostatic renewal of the liver. Nature. 2015;524(7564):180–5.
15. Yokomori H, Oda M, Yoshimura K, Nagai T, Ogi M, Nomura M, et al. Vascular endothelial growth factor increases fenestral permeability in hepatic sinusoidal endothelial cells. Liver Int. 2003;23(6):467–75.
16. Laurie LD. Liver sinusoidal endothelial cells in hepatic fibrosis. Hepatology. 2015;61(5):1740–6.
17. Ni Y, et al. Pathological process of liver sinusoidal endothelial cells in liver disease. World J Gastroenterol. 2017;23(43):7666–77.
18. Lee JS, Semela D, Iredale J, Shah VH. Sinusoidal remodeling and angiogenesis: a new function for the liver-specific pericyte? Hepatology. 2007;45(3):817–25.
19. Blomhoff R. Hepatic retinol metabolism: role of the various cell types. Nutr Rev. 1987;45(9):257–63.
20. Davis BH. Transforming growth factor beta responsiveness is modulated by the extracellular collagen matrix during hepatic ito cell culture. J Cell Physiol. 1988;136(3):547–53.
21. Czaja MJ, Weiner FR, Flanders KC, Giambrone MA, Wind R, Biempica L, et al. In vitro and in vivo association of transforming growth factor-beta 1 with hepatic fibrosis. J Cell Biol. 1989;108(6):2477–82.
22. Pinzani M, Gesualdo L, Sabbah GM, Abboud HE. Effects of platelet-derived growth factor and other polypeptide mitogens on DNA synthesis and growth of cultured rat liver fat-storing cells. J Clin Invest. 1989;84(6):1786–93.
23. Bachem MG, Riess U, Gressner AM. Liver fat storing cell proliferation is stimulated by epidermal growth factor/transforming growth factor alpha and inhibited by transforming growth factor beta. Biochem Biophys Res Commun. 1989;162(2):708–14.
24. Greuter T, Shah VH. Hepatic sinusoids in liver injury, inflammation, and fibrosis: new pathophysiological insights. J Gastroenterol. 2016;51:511–9.
25. Lee UE, Friedman SL. Mechanisms of hepatic fibrogenesis. Best Pract Res Clin Gastroenterol. 2011;25(2):195–206.
26. Carmeliet P. Angiogenesis in health and disease. Nat Med. 2003;9:653–60.
27. Yancopoulos GD, Davis S, Gale N, Rudge J, Wiegand S, Holash J. Vascular specific growth factors and blood vessel formation. Nature. 2000;407:242–8.
28. Taura K, et al. Hepatic stellate cells secrete angiopoietin 1 that induces angiogenesis in liver fibrosis. Gastroenterology. 2008;135(5):1729–38.
29. Carmeliet P. Manipulating angiogenesis in medicine. J Intern Med. 2004;255:538–61.
30. Matsumoto K, Yoshitomi H, Rossant J, Zaret KS. Liver organogenesis promoted by endothelial cells prior to vascular function. Science. 2001;294(5542):559–63.
31. Makita T, Sucov HM, Gariepy CE, Yanagisawa M, Ginty DD. Endothelins are vascular-derived axonal guidance cues for developing sympathetic neurons. Nature. 2008;452(7188):759–63.
32. Cleaver O, Melton DA. Endothelial signalling during development. Nat Med. 2003;9(6):661–8.
33. Coultas L, Chawengsaksophak K, Rossant J. Endothelial cells and VEGF in vascular development. Nature. 2005;438(7070):937–45.
34. Crivellato E, Nico B, Ribatti D. Contribution of endothelial cells to organogenesis: a modern reappraisal of an old Aristotelian concept. J Anat. 2007;211(4):415–27.
35. Ramasamy SK, Kusumbe AP, Adams RH. Regulation of tissue morphogenesis by endothelial cell-derived signals. Trends Cell Biol. 2015;25(3):148–57.
36. Zhang XJ, et al. Angiocrine hepatocyte growth factor signaling controls physiological organ and body size and dynamic hepatocyte proliferation to prevent liver damage during regeneration. Am J Pathol. 2019;190(2):358–71.
37. Ding BS, et al. Inductive angiocrine signals from sinusoidal endothelium are required for liver regeneration. Nature. 2010;468(7321):310–5.
38. DeLeve LD. Liver sinusoidal endothelial cells and liver regeneration. J Clin Invest. 2013;123(5):1861–6.
39. DeLeve LD, Wang X, Wang L. VEGF-sdf1 recruitment of CXCR7+ bone marrow progenitors of liver sinusoidal endothelial cells promotes rat liver regeneration. Am J Physiol Gastrointest Liver Physiol. 2016;310(9):G739–46.

40. Mendt M, Cardier JE. Role of SDF-1 (CXCL12) in regulating hematopoietic stem and progenitor cells traffic into the liver during extramedullary hematopoiesis induced by G-CSF, AMD3100 and PHZ. Cytokine. 2015;76:214–21.
41. Hu J, et al. Endothelial cell-derived angiopoietin-2 controls liver regeneration as a spatiotemporal rheostat. Science. 2014;343(6169):416–9.
42. Ribatti D, Nico B, Crivellato E. The role of pericytes in angiogenesis. Int J Dev Biol. 2011;55(3):261–8.
43. Bergers G, Song S. The role of pericytes in blood-vessel formation and maintenance. Neuro-Oncology. 2005;7(4):452–64.
44. Hellstrom M, Kalen M, Lindahl P, Abramsson A, Betsholtz C. Role of PDGF-B and PDGFR-beta in recruitment of vascular smooth muscle cells and pericytes during embryonic blood vessel formation in the mouse. Development. 1999;126(4):3047–55.
45. Stoeltzing O, Ahmad SA, Liu W, McCarty MF, Wey JS, Parikh AA, et al. Angiopoietin-1 inhibits vascular permeability, angiogenesis, and growth of hepatic colon cancer tumors. Cancer Res. 2003;63(12):3370–7.
46. Hellstrom M, Gerhardt H, Kalen M, Li X, Eriksson U, Wolburg H, et al. Lack of pericytes leads to endothelial hyperplasia and abnormal vascular morphogenesis. J Cell Biol. 2001;153(3):543–53.
47. Lindahl P, Johansson BR, Leveen P, Betsholtz C. Pericyte loss and microaneurysm formation in PDGF-B-deficient mice. Science. 1997;277(5323):242–5.
48. Suri C, Jones PF, Patan S, Bartunkova S, Maisonpierre PC, Davis S, et al. Requisite role of angiopoietin-1, a ligand for the TIE2 receptor, during embryonic angiogenesis. Cell. 1996;87(7):1171–80.
49. Chen L, et al. HSCs play a distinct role in different phases of oval cell-mediated liver regeneration. Cell Biochem Funct. 2012;30(7):588–96.
50. Yin C, Evason KJ, Asahina K, Stainier DY. Hepatic stellate cells in liver development, regeneration, and cancer. J Clin Invest. 2013;123:1902–10.
51. Mabuchi A, Mullaney I, Sheard PW, Hessian PA, Mallard BL, Tawadrous MN, et al. Role of hepatic stellate cell/hepatocyte interaction and activation of hepatic stellate cells in the early phase of liver regeneration in the rat. J Hepatol. 2004;40(6):910–6.
52. Balabaud C, Bioulac-Sage P, Desmouliere A. The role of hepatic stellate cells in liver regeneration. J Hepatol. 2004;40(6):1023–6.
53. Schaffner F, Poper H. Capillarization of hepatic sinusoids in man. Gastroenterology. 1963;44:239–42.
54. Horn T, Christoffersen P, Henriksen JH. Alcoholic liver injury: defenestration in noncirrhotic livers–a scanning electron microscopic study. Hepatology. 1987;7(01):77–82.
55. Krenkel O, Tacke F. Liver macrophages in tissue homeostasis and disease. Nat Rev Immunol. 2017;17(5):306–21.
56. Xie G, Wang X, Wang L, Atkinson RD, Kanel GC, Gaarde WA, DeLeve LD. Role of liver sinusoidal endothelial cell differentiation in progression and regression of rat hepatic fibrosis. Gastroenterology. 2012;142(4):918–27.
57. Rockey DC, Chung JJ. Reduced nitric oxide production by endothelial cells in cirrhotic rat liver: endothelial dysfunction in portal hypertension. Gastroenterology. 1998;114(2):344–51.
58. Shah V, Toruner M, Haddad F, Cadelina G, Papapetropoulos A, Choo K, Sessa WC, et al. Impaired endothelial nitric oxide synthase activity associated with enhanced caveolin binding in experimental cirrhosis in the rat. Gastroenterology. 1999;117(5):1222–8.
59. Corpechot C, Barbu V, Wendum D, Kinnman N, Rey C, Poupon R, Housset C, et al. Hypoxia-induced VEGF and collagen I expressions are associated with angiogenesis and fibrogenesis in experimental cirrhosis. Hepatology. 2002;35(5):1010–21.
60. Kwon SH, Jeong SW, Jang JY, Lee JE, Lee SH, Kim SG, Kim YS, et al. Cyclooxygenase-2 and vascular endothelial growth factor in chronic hepatitis, cirrhosis and hepatocellular carcinoma. Clin Mol Hepatol. 2012;18:287–94.
61. Abraldes JG, Rodriguez-Vilarrupla A, Graupera M, Zafra C, Garcia-Caldero H, Garcia-Pagan JC, Bosch J. Simvastatin treatment improves liver sinusoidal endothelial dysfunction in CCl4 cirrhotic rats. J Hepatol. 2007;46(6):1040–6.

62. Gracia-Sancho J, Lavina B, Rodriguez-Vilarrupla A, et al. Increased oxidative stress in cirrhotic rat livers: a potential mechanism contributing to reduced nitric oxide bioavailability. Hepatology. 2008;47:1248–56.
63. Hernandez-Guerra M, Garcia-Pagan JC, Turnes J, et al. Ascorbic acid improves the intrahepatic endothelial dysfunction of patients with cirrhosis and portal hypertension. Hepatology. 2006;43(3):485–91.
64. Trebicka J, Hennenberg M, Laleman W, Shelest N, Biecker E, Schepke M, Nevens F, et al. Atorvastatin lowers portal pressure in cirrhotic rats by inhibition of RhoA/rho-kinase and activation of endothelial nitric oxide synthase. Hepatology. 2007;46(1):242–53.
65. Gracia-Sancho J, Lavina B, Rodriguez-Vilarrupla A, et al. Enhanced vasoconstrictor prostanoid production by sinusoidal endothelial cells increases portal perfusion pressure in cirrhotic rat livers. J Hepatol. 2007;47(2):220–7.
66. Kawada N, Tran-Thi TA, Klein H, et al. The contraction of hepatic stellate (ito) cells stimulated with vasoactive substances. Possible involvement of endothelin 1 and nitric oxide in the regulation of the sinusoidal tonus. Eur J Biochem. 1993;213(2):815–23.
67. Iwakiri Y. Endothelial dysfunction in the regulation of cirrhosis and portal hypertension. Liver Int. 2012;32(2):199–213.
68. Perri RE, Langer DA, Chatterjee S, et al. Defects in cGMP-PKG pathway contribute to impaired NO-dependent responses in hepatic stellate cells upon activation. Am J Physiol Gastrointest Liver Physiol. 2006;290(3):G535–42.
69. Duan JL, Ruan B, Yan XC, et al. Endothelial notch activation reshapes the angiocrine of sinusoidal endothelia to aggravate liver fibrosis and blunt regeneration in mice. Hepatology. 2018;68(2):677–90.
70. Maretti-Mira AC, Wang X, Wang L, DeLeve LD. Incomplete differentiation of engrafted bone marrow endothelial progenitor cells initiates hepatic fibrosis in the rat. Hepatology. 2019;69(3):1259–72.
71. DeLeve LD. Liver sinusoidal endothelial cells in hepatic fibrosis. Hepatology. 2015;61(5):1740–6.
72. Wells RG. The role of matrix stiffness in hepatic stellate cell activation and liver fibrosis. J Clin Gastroenterol. 2005;39(4 Suppl 2):S158–61.
73. Jin X, Aimaiti Y, Chen Z, Wang W, Li D. Hepatic stellate cells promote angiogenesis via the TGF-β1-Jagged1/VEGFA axis. Exp Cell Res. 2018;373(1–2):34–43.
74. Taura K, De Minicis S, Seki E, et al. Hepatic stellate cells secrete angiopoietin 1 that induces angiogenesis in liver fibrosis. Gastroenterology. 2008;135(5):1729–38.
75. Novo E, Povero D, Busletta C, et al. The biphasic nature of hypoxia-induced directional migration of activated human hepatic stellate cells. J Pathol. 2012;226:588–97.
76. Novo E, Cannito S, Zamara E, et al. Proangiogenic cytokines as hypoxia-dependent factors stimulating migration of human hepatic stellate cells. Am J Pathol. 2007;170(6):1942–53.
77. Cao Y, Szabolcs A, Dutta SK, Yaqoob U, Jagavelu K, Wang L, et al. Neuropilin-1 mediates divergent R-Smad signaling and the myofibroblast phenotype. J Biol Chem. 2010;285(41):31840–8.
78. Friedman SL. Cytokines and fibrogenesis. Semin Liver Dis. 1999;19(2):129–40.
79. Chow AM, Gao DS, Fan SJ, et al. Liver fibrosis: an intravoxel incoherent motion (IVIM) study. J Magn Reson Imaging. 2012;36(1):159–67.
80. Rosmorduc O, Housset C. Hypoxia: a link between fibrogenesis, angiogenesis, and carcinogenesis in liver disease. Semin Liver Dis. 2010;30(3):258–70.
81. Wang GL, Semenza GL. Characterization of hypoxia-inducible factor 1 and regulation of DNA binding activity by hypoxia. J Biol Chem. 1993;268:21513–8.
82. Semenza GL, Wang GL. A nuclear factor induced by hypoxia via de novo protein synthesis binds to the human erythropoietin gene enhancer at a site required for transcriptional activation. Mol Cell Biol. 1992;12:5447–54.
83. Liu Y, Cox SR, Morita T, Kourembanas S. Hypoxia regulates vascular endothelial growth factor gene expression in endothelial cells: identification of a 5′ enhancer. Circ Res. 1995;77:638–43.

84. Roth KJ, Copple BL. Role of hypoxia-inducible factors in the development of liver fibrosis. Cell Mol Gastroenterol Hepatol. 2015;1(6):589–97.
85. Yoshiji H, Kuriyama S, Yoshii J, et al. Vascular endothelial growth factor and receptor interaction is a prerequisite for murine hepatic fibrogenesis. Gut. 2003;52(9):1347–54.
86. Moon JO, Welch TP, Gonzalez FJ, Copple BL. Reduced liver fibrosis in hypoxia-inducible factor-1α-deficient mice. Am J Physiol Gastrointest Liver Physiol. 2009;296(3):G582–92.
87. Yoshida D, Kim K, Noha M, et al. Hypoxia inducible factor 1-alpha regulates of platelet derived growth factor-B in human glioblastoma cells. J Neuro-Oncol. 2006;76:13–21.
88. Calvani M, Rapisarda A, Uranchimeg B, Shoemaker RH, Melillo G. Hypoxic induction of an HIF-1α–dependent bFGF autocrine loop drives angiogenesis in human endothelial cells. Blood. 2006;107(7):2705–12.
89. Forsythe JA, Jiang BH, Iyer NV, Agani F, Leung SW, Koos RD, Semenza GL. Activation of vascular endothelial growth factor gene transcription by hypoxia-inducible factor 1. Mol Cell Biol. 1996;16(9):4604–13.
90. Ramirez-Pedraza M, Fernández M. Interplay between macrophages and angiogenesis: a double-edged sword in liver disease. Front Immunol. 2019;10:2882.
91. Hellerbrand C, Wang SC, Tsukamoto H, Brenner DA, Rippe RA. Expression of intracellular adhesion molecule 1 by activated hepatic stellate cells. Hepatology. 1996;24(3):670–6.
92. Ehling J, Gremse F, Möckel D, Kiessling F, Lammers T, Bartneck M, et al. CCl2-dependent infiltrating macrophages promote angiogenesis in progressive liver fibrosis. Gut. 2014;63:1960–71.
93. Wynn TA, Vannella KM. Macrophages in tissue repair, regeneration, and fibrosis. Immunity. 2016;44:450–62.
94. Vannella KM, Wynn TA. Mechanisms of organ injury and repair by macrophages. Annu Rev Physiol. 2017;79:593–617.
95. Anthony PP, Ishak KG, Nayak NC, Poulsen HE, Scheuer PJ, Sobin LH. The morphology of cirrhosis. Recommendations on definition, nomenclature, and classification by a working group sponsored by the World Health Organization. J Clin Pathol. 1978;31:395–414.
96. Garcia-Tsao G, Friedman S, Iredale J, Pinzani M. Now there are many (stages) where before there was one: in search of a pathophysiological classification of cirrhosis. Hepatology. 2010;51(4):1445–9.
97. Guo L, Zheng L, Hu L, Zhou H, Yu L, Liang W. Transient elastography (FibroScan) performs better than non-invasive markers in assessing liver fibrosis and cirrhosis in autoimmune hepatitis patients. Med Sci Monit. 2017;23:5106.
98. Foucher J, Chanteloup E, Vergniol J, Castera L, Le Bail B, Adhoute X, Bertet J, Couzigou P, de Ledinghen V. Diagnosis of cirrhosis by transient elastography (FibroScan): a prospective study. Gut. 2006;55(3):403–8.
99. Gherlan GS. Liver ultrasound elastography: more than staging the disease. World J Hepatol. 2015;7(12):1595.
100. Xiao H, Shi M, Xie Y, Chi X. Comparison of diagnostic accuracy of magnetic resonance elastography and Fibroscan for detecting liver fibrosis in chronic hepatitis B patients: a systematic review and meta-analysis. PLoS One. 2017;12(11):e0186660.
101. Asrani SK, Talwalkar JA, Kamath PS, Shah VH, Saracino G, Jennings L, Gross JB, Venkatesh S, Ehman RL. Role of magnetic resonance elastography in compensated and decompensated liver disease. J Hepatol. 2014;60(5):934–9.

Drugs to Modify Liver Fibrosis Progression and Regression

18

Marina Vilaseca and Jordi Gracia-Sancho

Introduction

Liver fibrosis is a histological feature that appears in different types of chronic liver disease. It is mainly characterized by excessive accumulation of extracellular matrix fibers formed by matrix proteins including collagens [1]. Many different etiologies end up with liver fibrosis such as drug-induced liver injury (DILI), alcoholic liver disease (ALD), non-alcoholic fatty liver disease (NAFLD), and autoimmune hepatitis (AH), among others.

Despite the etiology, the main cells involved in the production of these matrix fibers are the same, Hepatic Stellate Cells (HSCs) that acquire an activated myofibroblast-like profile during the development of the disease, increasing their proliferation rate, collagen synthesis, and contractility [1]. Nevertheless, many other cell types in the liver become deregulated such as liver endothelial sinusoidal cells (LSECs), Kupffer cells (KCs) or hepatocytes, becoming the target of different therapeutic strategies.

Importantly, some of the liver cell changes that occur during the progression of the disease fade away once the causal agent is removed, starting the process of liver fibrosis regression, where the accumulation of extracellular matrix fibers decreases, at least in part, and liver cells phenotype improves.

In this regard, with the achievement of the hepatitis C virus (HCV) cure, the scientific community has realized that only eliminating the causal agent is not

M. Vilaseca
Laboratory for Cellular Epigenomics, Riken, Yokohama, Kanagawa, Japan

J. Gracia-Sancho (✉)
Liver Vascular Biology, IDIBAPS Research Institute, Hospital Clínic Barcelona, CIBEREHD, Barcelona, Spain

Hepatology, Department for Biomedical Research, University of Berne and Inselspital, Bern, Switzerland
e-mail: jgracia@recerca.clinic.cat

R. de Franchis (ed.), *Portal Hypertension VII*,
https://doi.org/10.1007/978-3-031-08552-9_18

enough and there is an actual need to foster the process of liver fibrosis regression. In fact, it has been shown that liver fibrosis does not regress in some patients with effective antiviral therapy, and it may even progress into a worse scenario [2].

With this focus in mind, many research teams are trying to develop new drugs to treat liver fibrosis, both during progression and regression. In fact, there are more than 250 clinical trials that are active or recruiting. For this reason, we present a review with different strategies that are being used to prevent or improve liver fibrosis.

Basics of Liver Fibrosis Progression and Regression

To evaluate different strategies that target liver fibrosis, we first need to define the basics of liver fibrosis progression and regression.

We know that there are many different causal agents that can induce liver fibrosis, such as toxins, viruses, oxidative stress, and free fatty acids, among others [3]. These stimuli will induce liver inflammation that, when maintained in time or when repetitive, can provoke chronic liver injury increasing extracellular matrix (ECM) protein synthesis. During the progression of liver fibrosis, inflammatory cells like KCs will start secreting inflammatory cytokines and growth factors that will recruit monocytes and macrophages from the bloodstream, increasing the inflammatory state and inducing HSCs activation. HSCs are located in the perisinusoidal space of Disse and, when activated, they change their phenotype increasing their proliferation state, their contractility and ECM protein synthesis. Importantly, established ECM can also induce HSCs activation [4], creating a positive loop for HSCs activation.

But not only HSCs and KCs will undergo specific changes during the progression of the disease. LSECs, which are specialized endothelial cells with fenestrae that maintain liver homeostasis, will start changing their phenotype and start losing their fenestrae in the process called capillarization of the liver sinusoids [4–6].

On the other hand, it is known that the liver itself is able to improve liver fibrosis once the insult is removed. Hepatocytes and surrounding non-parenchymal cells switch their phenotype into a more restorative and anti-inflammatory one, favoring spontaneous liver fibrosis regression. This has been shown in patients with Non-Alcoholic Steatohepatitis (NASH) or with HCV, once they changed their lifestyle or after having long-term virus suppression [7]. However, only removing the causal agent is not effective or fast enough to totally cure patients, and there is still the need to treat those patients to speed up the process of liver fibrosis regression.

Although there is still no current treatment specific for liver fibrosis, there are many therapies under study that are focusing on finding a cure specific for liver fibrosis, using different strategies such as targeting HSCs pathways or synthesis and degradation of ECM (Fig. 18.1).

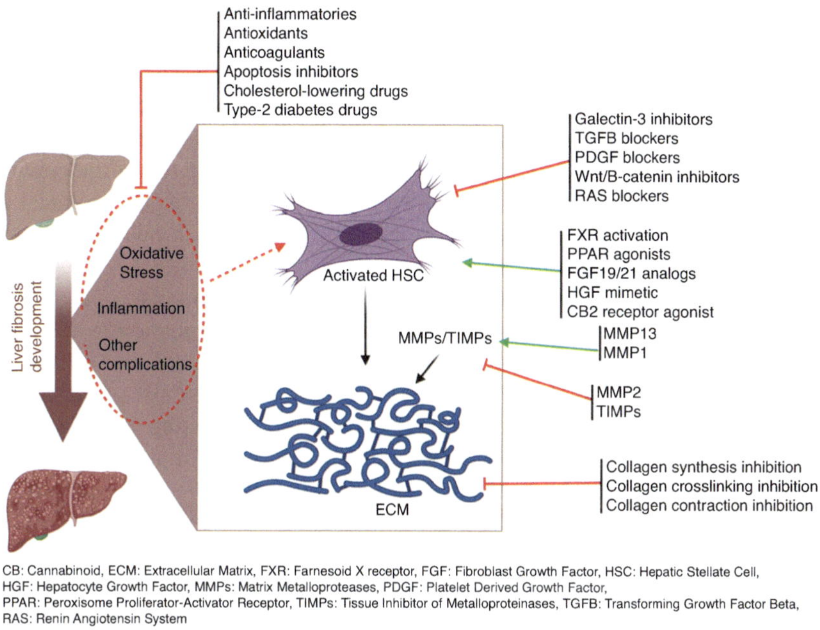

Fig. 18.1 Therapeutic strategies to target liver fibrosis

Antifibrotic Drugs Targeting HSCs

HSCs are located in the space of Disse, between hepatocytes and LSECs. In a healthy condition, they store retinol within lipid droplets, but in response to liver injury HSCs transdifferentiate into cells producing collagen which will lead to liver fibrosis, therefore becoming one of the main targets for treating liver fibrosis. Many pathways are involved in regulating HSCs phenotype and their activity, and some of them have been already the target of different therapeutic strategies. From those different therapeutic strategies, we can mainly categorize them considering the specific pathways they target. Here we summarize the main approaches that have been tested in preclinical studies and clinical trials.

Activation of HSC Pathways

A therapeutic strategy focused on modulating HSCs is targeting nuclear receptors. Nuclear receptors are transcriptional regulators and contribute to regulate HSCs under healthy and pathological conditions. Two of them, the Farnesoid X receptor (FXR) and the Peroxisome proliferator-activator receptor (PPAR), showed promising results for improving HSCs phenotype and fibrosis.

FXR activation in HSCs is associated with less collagen production [8] and some drugs have already been developed or are evaluated in current clinical trials. FXR agonists like Tropifexor (NCT02855164) and Cilofexor [9] have been mainly studied for the treatment of NASH and NAFLD with already good results on their tolerability and decreasing hepatic steatosis. Moreover, another FXR agonist, Obeticholic acid (OCA), showed also good results in patients with PBC on top of the previous ones (NCT02548351). Based on the published results and the ongoing clinical trials, it seems that these drugs could be an interesting approach to treat liver fibrosis although more studies are still required.

Since several years ago PPARs agonists have gained importance in treating NASH and NAFLD, mostly due to PPAR deregulation contribution to increased inflammation and fibrogenesis [10]. Although there are three different PPAR isoforms (PPARα, PPARß/d, and PPARγ), PPARγ has been the most promising for targeting liver fibrosis. There have been many studies analyzing PPAR agonists such as Rosiglitazone, Saroglitazar, Elafibranor, and Pioglitazone, some of them showing promising results both in preclinical and ongoing clinical studies (NCT02704403, NCT04584242, NCT00885313). Furthermore, the pan-PPAR agonist Lanifibranor showed positive results in preclinical models decreasing fibrosis and portal hypertension [11] and in a phase 2b clinical trial [12], and there are now two ongoing clinical trials on NASH and NAFLD patients (NCT04849728 and NCT03459079).

Analogs of human Fibroblast Growth Factor 19 and 21 (FGF19/21) have been described to regulate glucose and lipid metabolism and have been under clinical trials for NASH patients with an improvement in fibrosis resolution [13]. FGF19, or its murine analogue FGF15, is released via FXR activation and regulates the synthesis of bile acids in the liver, while FGF21 is released during fasting and by activation of PPARα receptor. They are anti-steatosic, regulate oxidative stress and autophagy, and have anti-inflammatory and anti-fibrotic effects [13]. Aldafermin (NGM282) is a FGF19 analogue that was evaluated in clinical trials in NASH patients with a significant improvement in NASH score without worsening fibrosis [14, 15]. There are several FGF21 analogues, being Pegbelfermin (BMS-986036) and Efruxifermin the ones with more data on NASH patients. Pegbelfermin showed beneficial effects in NASH patients, decreasing their hepatic fat content and it is currently evaluated in clinical trials (NCT02413372, NCT03486899, NCT03486912). The same was observed in Efruxifermin-treated patients, where liver steatosis was decreased up to 70% in all patients (NCT03976401). There are currently two active clinical trials recruiting patients with NASH, with or without cirrhosis (NCT04767529, NCT05039450).

Hepatocyte Growth Factor (HGF) is a molecule with potential therapeutic action due to its interaction with the profibrogenic molecule transforming growth factor ß (TGFß). There are some preclinical studies showing the potential of HGF overexpression in promoting liver cirrhosis improvement in rat models [16], but further studies are required to assess their effect in patients. Importantly, Refanalin—a small HGF mimetic—has been developed recently and unpublished preliminary data showed interesting results on attenuating profibrotic genes and proteins in preclinical models of liver fibrosis.

Another important pathway involved in HSCs activation is the hepatic endocannabinoid system with two different receptors with different effects. CB1 receptor activation promotes hepatic inflammation but CB2 receptor activation has anti-inflammatory effects [17]. CB2 receptor agonist effectively decreased inflammation and fibrosis and improved liver regression in preclinical models [18] and even showed antioxidant effects [19] but it has not been tested in patients yet.

Inhibition of HSC Pathways

A variety of profibrotic pathways can induce and modulate HSC's activation during the progression of the disease. Therefore, many therapies are focused on inhibiting some of those.

Galectin-3 (mainly produced in KCs) has been described in many different fibrotic diseases to play a role in inducing fibrotic pathways [20]. Some galectin-3 inhibitors have been already tested in preclinical settings such as GR-MD-02, Belapectin and Davanat (GM-CT-01) [21] showing decreased liver fibrosis in rodent models. Clinical trials at different stages are evaluating the safety of GR-MD-02 on NASH (NCT02421094, NCT01899859, NCT04365868) and portal hypertension patients (NCT024663967). GB1211 is another galectin-3 inhibitor that has been under clinical trials testing its anti-fibrotic effects in grade 2–4 NASH patients (NCT03809052).

Different profibrotic cytokines that induce HSCs activation are expressed by inflammatory cells during liver fibrosis development, such TGFß, platelet-derived growth factor (PDGF), or connective tissue growth factor (CTGF). Suppressing their expression has been a challenge due to their broad systemic effects. In fact, blocking TGFß systemically can induce inflammation and increased risk of neoplasia, but its neutralization showed decreased liver fibrosis in a CCl_4-induced hepatic fibrosis model [22]. Hydronidone is a TGFß blocker that has been tested in chronic viral hepatitis B patients in China and showed improvement in liver fibrosis and liver inflammation after 52 weeks of treatment (NCT02499562).

PDGF, which stimulates HSCs proliferation and migration, has some advantages compared to TGFß inhibitor treatments. TGFß receptors are found in HSCs, but also in hepatocytes and macrophages [23]; on the other hand, PDGF receptors are found only in HSCs and myofibroblasts making it a more direct target for drugs, with less side effects [24].

The Wnt/ß-catenin is another pathway associated with increased fibrosis and could be a potential target for liver fibrosis treatment [25]. PRI-724, a ß-catenin inhibitor, prevented HSC activation and collagen synthesis in mice [26]. There is an ongoing clinical trial with Hepatitis B or C patients (NCT03620474).

Nitazoxanide (NTZ) is an already approved drug for Cryptosporidium infection [27] but it has also shown strong anti-fibrotic activity due to its interaction with HSCs in preclinical studies using fibrosis and NASH models [28]. Furthermore, it has been assessed together with Elafibranor in a NASH model and showed potential synergic beneficial effects [29]. A phase 2 clinical trial on NASH patients evaluating its safety and efficacy has been recently completed (NCT03656068).

The Renin-Angiotensin System (RAS) regulates blood pressure and vascular resistance in the liver. RAS upregulation during liver fibrosis has been described and it is known that Angiotensin II plays a role in HSCs activation [30, 31]. Therefore, different angiotensin receptor blockers such as Candesartan, Ramipril, and Losartan showed promise in reducing liver fibrosis or NASH [32–34] and they are now under clinical trials showing good results in HCV patients (NCT00298714). Indeed, the effects of different drugs under RAS like angiotensin-converting enzyme inhibitors and angiotensin receptor blockers were analyzed on patients with liver fibrosis and showed reduced content of fibrosis and other fibrotic markers [35].

Integrins are proteins that mediate cell–cell interaction and cell–ECM contact. Some integrins can even activate matrix metalloproteases and other fibrogenic mediators such as TGFß1. Therefore, inhibition of different integrins has been an interesting strategy with many different drugs being developed and tested nowadays. Due to the extended information that is available on this topic, we recommend this recent review that describes in detail all different approaches regarding integrins [36].

Antifibrotic Dugs Targeting the ECM

Collagen Synthesis Inhibition

During liver fibrosis progression, there is a disturbance in the balance between collagen synthesis and degradation that leads to increased fibrogenesis, chronic inflammation, and endothelial de-differentiation, ending up with fibrotic dysfunctional liver tissue.

Since one of the main proteins forming ECM in liver fibrosis is type I collagen [37], some therapies have been focusing on targeting Col1a1 gene [38] or its collagen-specific chaperone [39]. Preclinical studies showed how their inhibition leads to decreased hepatic collagen deposition and diminished inflammation without any important side effects [40, 41]. Importantly, this effect was also observed when the drug was administered using vitamin A-coupled liposomes, that mainly target HSCs [39], confirming the importance of targeting HSCs when targeting collagen synthesis inhibition. All these positive results led to clinical trials such as phase 1b/2 clinical trials with subjects with moderate to extensive hepatic fibrosis. They tested different doses of nanoparticles containing siRNA against collagen-specific chaperone HSP47 (ND-L02-s0201) showing promising results (NCT02227459) [42].

There are other approaches to decrease collagen synthesis, for example inhibiting LARP6 that binds the 5′ stem-loop structure (5’SL) and promotes collagen synthesis. Tested in preclinical studies, HSCs were treated with C9, a compound able to dissociate LARP6 from 5’SL and showed reduced secretion of type I collagen [43].

Targeting MMPs and TIMPs

Matrix metalloproteases (MMPs) are enzymatic proteins that degrade ECM proteins and play a critical role in regulating liver fibrosis homeostasis. Therefore, their regulation during the progression and regression of liver fibrosis is critical.

MMPs are usually secreted into the extracellular matrix and tightly regulated transcriptionally, post-transcriptionally and at the protein level, being tissue inhibitors of metalloproteinases or TIMPs the ones controlling their activity [44].

There are different MMPs, mainly produced by HSCs and KCs, being MMP13 the major collagenase with dynamic expression during fibrosis development [45]. Its anti-fibrotic effects have been analyzed and showed that overexpression of MMP13 prevented liver fibrosis in a rat liver fibrosis model [46]. Other MMPs, like MMP1, have been under investigation, where several studies analyzed different ways of inducing MMP1 in preclinical settings showing promising approaches for the treatment of liver fibrosis [47–49], or MMP2 siRNA delivered with vitamin A-coupled liposomes that showed in vitro reduction in HSCs activation and decreased collagen deposition [50].

Another approach based on MMPs is to inhibit TIMPs. Some attempts in preclinical studies had shown their potential in reducing liver fibrosis by keeping higher levels of MMPs or by deactivating HSCs [51, 52]. Despite good results on the bench side, they have not been studied in the clinics as a therapeutic target but MMPs have been explored as potential biomarkers for liver diseases [53].

Inhibition of Collagen Cross-Linking

Lysyl oxidases (LOXs) and tissue transglutaminases (TGs) are the main proteins involved in collagen cross-linking. LOXs secrete amine oxidases that deamidate the free amino group of lysine or hydroxy-lysine residues found in collagens and elastins generating irreversible connections or cross-linking. These cross-linked proteins impede the normal degradation of the collagen fibers by MMPs, therefore increasing insoluble collagen deposition in the liver [54].

It is known that LOXs activity increases during the progression of liver fibrosis and that their inhibition could be a strategy to decrease ECM stiffness and therefore, decrease HSCs activation and ECM content. Importantly, LOXs have other roles apart from ECM regulation, including gene regulator, for example, regulating TGFß function [55].

Previous studies showed that blocking Lysyl oxidase Like 2 (LOXL-2) inhibited liver fibrosis in CCl_4-induced rats [56]; however, its translation to the clinical scenario using Simtuzumab/GS-6624 has not been as good as expected [57].

TGs can covalently link two proteins and induce cross-linking [58] and some specific inhibitors for TG2 have been developed and tested in preclinical settings [59]; however, TGs inhibitors have not reached clinical trials.

Inhibiting Collagen Contraction

Discoid domain receptors (DDR) are collagen receptors that are increased during inflammation and chronic injuries. They are mainly expressed in hepatocytes and cholangiocytes and they are activated by different types of collagen [60]. Their main function is to mediate collagen contraction and induce fibrogenic responses. Although its potential role as a drug in fibrotic processes is proposed [61], there is still the need to test DDR-targeting drugs in liver fibrosis settings.

Antifibrotic Drugs Targeting Inflammation and Oxidative Stress

Apart from focusing on HSCs or ECM, there are many other strategies to target liver fibrosis. Anti-inflammatory drugs can directly inhibit inflammation by neutralizing different inflammatory cytokines that will ultimately prevent liver cell damage. Moreover, reactive oxygen species (ROS) are generated during the progression of the disease and can promote necrosis and apoptosis of liver cells and even amplify the ongoing inflammatory process [62]. Therefore, strategies targeting these pathological processes can indirectly reduce fibrogenesis by mainly decreasing liver cell damage.

Anti-Inflammatory Strategies

Cenicriviroc is a chemokine receptor 2/5 antagonist that inhibits hepatic inflammation by blocking binding sites for CCL2 and CCL5 inflammatory cytokines. A clinical trial on NASH patients showed good results in patients receiving Cenicriviroc by reducing their fibrosis degree without worsening steatohepatitis [63]. Nevertheless, such positive results were not confirmed in a recently finished phase 3 trial (NCT03028740).

One of the main cytokines involved in liver inflammation is tumor necrosis factor alpha (TNFα). Pentoxifylline is a phosphodiesterase inhibitor that blocks TNFα inflammatory response and oxidative stress [64]. Furthermore, Pirfenidone was initially used for idiopathic pulmonary fibrosis treatment and, although its mechanism of action is not known yet, it has a clear effect on reducing TNFα among other cytokines [65]. A phase 2 study (NCT04099407) evaluated its antifibrotic effects in patients with chronic liver disease and showed reduced liver fibrosis [66]. Some other strategies based on blocking pro-inflammatory cytokines showed promising results in preclinical studies, such as IL-4ra antisense oligonucleotides [67] or CCR5 inhibitors [68].

Focused also on targeting inflammation, but not as an anti-inflammatory itself, Relaxin is a vasodilator that has been reported recently as a potential therapeutic agent for liver fibrosis in preclinical studies. These studies showed that Relaxin is able to deactivate HSCs in different models of fibrosis and NASH and triggers a phenotypic switch on profibrogenic immune cells [69, 70].

Antioxidants

Chronic liver diseases are characterized by increased content in ROS due to increased production and decreased elimination leading to increased liver damage and liver fibrosis [71]. Antioxidants have been under investigation for being potentially beneficial in this setting with different drugs and different approaches tested in preclinical and clinical settings.

NADPH oxidase (NOX) is one of the main producers of ROS and it is known that NOX1, NOX2, and NOX4 play an important role in HSC's activation [72]. GKT137831, a NOX1/4 inhibitor has shown to improve liver fibrosis by reducing ROS production in preclinical studies [73] and it has been tested in clinical trials for PBC patients (NCT03226067).

Another important producer of ROS inside the cell is the mitochondrial respiratory chain [74]. Therefore, mitochondrial-targeted drugs such as Mitoquinone could be an interesting approach to directly reduce mitochondrial ROS production. Preclinical studies showed an important effect on HSCs deactivation and decreased liver fibrosis [75, 76], and also decreased liver damage in HCV patients [77].

Resveratrol is a polyphenol found naturally in plants and fruits, and it has many beneficial effects on oxidative stress, inflammation, endothelial dysfunction, and liver fibrosis [78, 79]. Due to its wide effects, it has been a central focus in many preclinical studies, showing beneficial results in liver transplantation, liver ischemia, and protection against different liver fibrosis etiologies [80, 81]. However, it has been tested only once in the clinics, where the effect of resveratrol was assessed in NASH patients (NCT02030977) [82].

Anticoagulants

Although liver fibrosis has always been associated with a higher risk of bleeding, there is growing evidence that bleeding events are mainly gastrointestinal [83] and, conversely, a prothrombotic state exists within the liver [84]. Anticoagulants could have beneficial effects as they are thrombin and factor Xa (FXa) inhibitors, known activators of HSCs, and prevent liver fibrogenesis [85]. To elucidate the beneficial effects that anticoagulation could bring to chronic liver diseases, some studies have been done in different settings. Initially, traditional anticoagulation such as low molecular weight heparin (LMWH) like Enoxaparin was evaluated in preclinical models of liver fibrosis with contradictory results on liver fibrosis [86, 87]. Different clinical trials in patients with liver fibrosis and with [88] or without portal vein thrombosis [89, 90] demonstrated its efficacy and safety without bleeding complications during the treatment, although the fibrosis score was not assessed.

Nowadays, direct oral anticoagulants (DOAC) are the new anticoagulant therapy under study in cirrhotic patients, as they do not require monitoring, can be taken orally, and display similar efficacy and safety profiles compared to LMWH [91]. Therefore, they have been already tested and have shown safety and beneficial effects in chronic liver disease in several preclinical and clinical trials, using different approaches like Rivaroxaban [92, 93] (NCT04874428, NCT03201367) Apixaban (NCT04874428), Dabigatran [94], and Edoxaban [95].

Apoptosis Signal-Regulating Kinase 1 (ASK1) Inhibitor

ASK1 is activated by oxidative stress during hepatocyte apoptosis and necrosis which leads to inflammation and liver fibrosis development [96]. The selective ASK1 inhibitor, Selonsertib, showed improvement in NASH patients in phase 2 clinical trial [97] but no effects were observed in phase 3 trials (NCT03053050, NCT03053063) [98].

Pan-Caspase Inhibitor

Emricasan (IDN-6556) is a pan-caspase inhibitor that prevents hepatocyte apoptosis and inflammatory processes that lead to liver fibrosis [99]. Importantly, its beneficial effects were also observed in patients with liver fibrosis [100] but it did not improve HVPG in NASH patients [101].

Cholesterol-Lowering Drugs

Statins such as Atorvastatin and Simvastatin were designed to treat cardiovascular diseases due to their effect in inhibiting the activity of HMG-CoA reductase [102]. However, their potential for treating other diseases like liver fibrosis has only started. It is already known that statins are safe and effective for treating compensated liver diseases [103] but they have also shown beneficial effects on oxidative stress, inflammation, and vasoprotection in animal models of chronic liver diseases [104] and preventing progression of acute-on-chronic liver failure (ACLF) [105]. Some clinical trials are testing the efficacy of statins in different clinical settings like ALD (NCT04971577), to prevent ACLF in patients with decompensated cirrhosis (NCT03780673) or prevent decompensation and death in compensated patients with Simvastatin (NCT03654053) and Atorvastatin (NCT04072601).

New developed drugs that improve cholesterol content and, therefore, are more focused on metabolic diseases are Stearoyl-coenzyme A desaturase 1 (SCD1) inhibitor: Aramchol, and thyroid hormone receptor beta (THR-ß) agonists: VK2809 and Resmetirom. They all are potent activators of lipid metabolism which leads to improvements mainly in patients with NASH and NAFLD [106, 107]. Some clinical trials are ongoing with SCD1 inhibitor Aramchol in NAFLD (NCT01094158) [108] and NASH patients (NCT04104321). In the case of THR-ß agonists, they have been shown to reduce liver fat content and they are being assessed for efficacy and safety in NASH patients (NCT04173065, NCT03900429).

Type-2 Diabetes Drugs

Liraglutide and Semaglutide are glucagon-like peptide (GLP)-1 receptor agonists that stimulate insulin production and its secretion [109]. Initially, these drugs were designed to treat diabetic patients but also showed good results on renal conditions in those patients, giving hope to also treat NASH and NAFLD patients. In fact, a preclinical study demonstrated the anti-fibrotic effects of Liraglutide in CLD [110] and GLP1-R agonists have been tested already in several clinical trials (NCT01237119, NCT02970942) and have shown improvement in NASH [111]. Metformin is also an insulin sensitizer in type 2 diabetes that improves endothelial dysfunction and vascular protection by reducing oxidative stress and enhancing nitric oxide bioavailability. It has been tested in preclinical studies showing reduced liver fibrosis [112].

New Strategies: Antifibrotic Cell Therapy

Cell therapy is a new strategy to treat chronic liver diseases [113]. In fact, different cell types are being studied in different settings, being liver fibrosis one of the most challenging due to the presence of cellular necrosis and apoptosis [113]. However, it has been demonstrated that infusion of bone marrow cells such as mesenchymal stem cells (MSCs), hematopoietic progenitor cells, and macrophages are critical to improve liver microenvironment repair. It has been demonstrated that infusion of MSCs decreased liver fibrosis in preclinical studies [114, 115] and in cirrhotic patients [116]. Other cell therapies, such as endothelial progenitor cells [117] (NCT03109236) and induced pluripotent stem cells [118], have shown promising results both in animal models and in clinical trials.

Although in the last years the development of new cell therapies has been in constant evolution, there is still the need to study those therapies in clinical trials including higher numbers of participants and to clarify some concerns such as long-term effectiveness and their tumorigenic risk.

References

1. Tsuchida T, Friedman SL. Mechanisms of hepatic stellate cell activation. Nat Rev Gastroenterol Hepatol. 2017;14(7):397–411. https://doi.org/10.1038/nrgastro.2017.38.
2. Lens S, Baiges A, Alvarado-Tapias E, LLop E, Martinez J, Fortea JI, et al. Clinical outcome and hemodynamic changes following HCV eradication with oral antiviral therapy in patients with clinically significant portal hypertension. J Hepatol. 2020;73(6):1415–24. https://doi.org/10.1016/j.jhep.2020.05.050.
3. Bataller R, Brenner DA. Liver fibrosis. J Clin Invest. 2005;115(2):209–18.
4. Guixé-Muntet S, Ortega-Ribera M, Wang C, Selicean S, Andreu I, Kechagia JZ, et al. Nuclear deformation mediates liver cell mechanosensing in cirrhosis. JHEP Rep. 2020;2(5):100145. https://pubmed.ncbi.nlm.nih.gov/32939447
5. Manicardi N, Fernández-Iglesias A, Abad-Jordà L, Royo F, Azkargorta M, Ortega-Ribera M, et al. Transcriptomic profiling of the liver sinusoidal endothelium during cirrhosis reveals

stage-specific secretory signature. Cancer. 2021;13(11):2688. https://pubmed.ncbi.nlm.nih.gov/34072510
6. Gracia-Sancho J, Caparrós E, Fernández-Iglesias A, Francés R. Role of liver sinusoidal endothelial cells in liver diseases. Nat Rev Gastroenterol Hepatol. 2021;18(6):411–31. https://doi.org/10.1038/s41575-020-00411-3.
7. Rockey DC, Friedman SL. Fibrosis regression after eradication of hepatitis C virus: from bench to bedside. Gastroenterology. 2021;160(5):1502–1520.e1. https://doi.org/10.1053/j.gastro.2020.09.065.
8. Verbeke L, Mannaerts I, Schierwagen R, Govaere O, Klein S, Vander Elst I, et al. FXR agonist obeticholic acid reduces hepatic inflammation and fibrosis in a rat model of toxic cirrhosis. Sci Rep. 2016;6:33453.
9. Patel K, Harrison SA, Elkhashab M, Trotter JF, Herring R, Rojter SE, et al. Cilofexor, a nonsteroidal FXR agonist, in patients with noncirrhotic NASH: a phase 2 randomized controlled trial. Hepatology. 2020;72(1):58–71. https://doi.org/10.1002/hep.31205.
10. Francque S, Szabo G, Abdelmalek MF, Byrne CD, Cusi K, Dufour J-F, et al. Nonalcoholic steatohepatitis: the role of peroxisome proliferator-activated receptors. Nat Rev Gastroenterol Hepatol. 2021;18(1):24–39. https://doi.org/10.1038/s41575-020-00366-5.
11. Boyer-Diaz Z, Aristu-Zabalza P, Andrés-Rozas M, Robert C, Ortega-Ribera M, Fernández-Iglesias A, et al. Pan-PPAR agonist lanifibranor improves portal hypertension and hepatic fibrosis in experimental advanced chronic liver disease. J Hepatol. 2021;74(5):1188–99.
12. Francque SM, Bedossa P, Parziu V, Anstee QM, Bugianesi E, Sanyal AJ, et al. A randomized, controlled trial of the Pan-PPAR agonist lanifibranor in NASH. N Engl J Med. 2021;385(17):1547–58.
13. Henriksson E, Andersen B. FGF19 and FGF21 for the treatment of NASH-two sides of the same coin? differential and overlapping effects of FGF19 and FGF21 from mice to human. Front Endocrinol. 2020;11:601349. https://pubmed.ncbi.nlm.nih.gov/33414764
14. Harrison SA, Rossi SJ, Paredes AH, Trotter JF, Bashir MR, Guy CD, et al. NGM282 improves liver fibrosis and histology in 12 weeks in patients with nonalcoholic steatohepatitis. Hepatology. 2020;71(4):1198–212. https://pubmed.ncbi.nlm.nih.gov/30805949
15. Harrison SA, Neff G, Guy CD, Bashir MR, Paredes AH, Frias JP, et al. Efficacy and safety of Aldafermin, an engineered FGF19 analog, in a randomized, double-blind, placebo-controlled trial of patients with nonalcoholic steatohepatitis. Gastroenterology. 2021;160(1):219–231.e1.
16. Cao X, Jin S, Sun L, Zhan Y, Lin F, Li Y, et al. Therapeutic effects of hepatocyte growth factor-overexpressing dental pulp stem cells on liver cirrhosis in a rat model. Sci Rep. 2017;7(1):15812. https://doi.org/10.1038/s41598-017-14995-5.
17. Melgar-Lesmes P, Perramon M, Jiménez W. Roles of the hepatic endocannabinoid and apelin systems in the pathogenesis of liver fibrosis. Cell. 2019;8(11):1311. https://pubmed.ncbi.nlm.nih.gov/31653030
18. Teixeira-Clerc F, Belot M-P, Manin S, Deveaux V, Cadoudal T, Chobert M-N, et al. Beneficial paracrine effects of cannabinoid receptor 2 on liver injury and regeneration. Hepatology. 2010;52(3):1046–59. https://pubmed.ncbi.nlm.nih.gov/20597071
19. Calleja MA, Vieites JM, Montero-Meterdez T, Torres MI, Faus MJ, Gil A, et al. The antioxidant effect of β-caryophyllene protects rat liver from carbon tetrachloride-induced fibrosis by inhibiting hepatic stellate cell activation. Br J Nutr. 2013;109(3):394–401. https://www.cambridge.org/core/article/antioxidant-effect-of-caryophyllene-protects-rat-liver-from-carbon-tetrachlorideinduced-fibrosis-by-inhibiting-hepatic-stellate-cell-activation/8EA3298D7E51DAA8940D41C5C28E5B57
20. Slack RJ, Mills R, Mackinnon AC. The therapeutic potential of galectin-3 inhibition in fibrotic disease. Int J Biochem Cell Biol. 2021;130:105881. https://www.sciencedirect.com/science/article/pii/S1357272520301989
21. Traber PG, Chou H, Zomer E, Hong F, Klyosov A, Fiel M-I, et al. Regression of fibrosis and reversal of cirrhosis in rats by galectin inhibitors in thioacetamide-induced liver disease. PLoS One. 2013;8(10):e75361. https://pubmed.ncbi.nlm.nih.gov/24130706

22. Fan X, Zhang Q, Li S, Lv Y, Su H, Jiang H, et al. Attenuation of CCl4-induced hepatic fibrosis in mice by vaccinating against TGF-β1. PLoS One. 2013;8(12):e82190. https://pubmed.ncbi.nlm.nih.gov/24349218
23. Bedossa P, Peltier E, Terris B, Franco D, Poynard T. Transforming growth factor—beta 1 (TGF-β1) and TGF-β1 receptors in normal, cirrhotic, and neoplastic human livers. Hepatology. 1995;21(3):760–6. https://doi.org/10.1002/hep.1840210325.
24. Dijk F, Olinga P, Poelstra K, Beljaars L. Targeted therapies in liver fibrosis: combining the best parts of platelet derived growth factor BB and interferon gamma. Front Med. 2015;2:72. https://www.frontiersin.org/article/10.3389/fmed.2015.00072
25. Nishikawa K, Osawa Y, Kimura K. Wnt/β-catenin signaling as a potential target for the treatment of liver cirrhosis using antifibrotic drugs. Int J Mol Sci. 2018;19(10):3103. https://pubmed.ncbi.nlm.nih.gov/30308992
26. Tokunaga Y, Osawa Y, Ohtsuki T, Hayashi Y, Yamaji K, Yamane D, et al. Selective inhibitor of Wnt/β-catenin/CBP signaling ameliorates hepatitis C virus-induced liver fibrosis in mouse model. Sci Rep. 2017;7(1):325. https://pubmed.ncbi.nlm.nih.gov/28336942
27. Rossignol J-F. Nitazoxanide: a first-in-class broad-spectrum antiviral agent. Antivir Res. 2014;110:94–103. https://pubmed.ncbi.nlm.nih.gov/25108173
28. Belanger C, Foucart C, Negro E, Dubernet M, Hum DW, Staels B, et al. Drug repurposing screen identifes novel small molecule compounds with potent antifibrotic properties. J Hepatol. 2017;66:S605.
29. Walczak R, Carole B, Benoit N, Descamps E, Nathalie D, Megnien S, et al. Elafibranor and nitazoxanide synergize to reduce fibrosis in a NASH model. J Hepatol. 2018;68:S352–3.
30. Simões E, Silva AC, Miranda AS, Rocha NP, Teixeira AL. Renin angiotensin system in liver diseases: friend or foe? World J Gastroenterol. 2017;23(19):3396–406. https://pubmed.ncbi.nlm.nih.gov/28596676
31. Bataller R, Gines P, Nicolas JM, Gorbig MN, Garcia-Ramallo E, Gasull X, et al. Angiotensin II induces contraction and proliferation of human hepatic stellate cells. Gastroenterology. 2000;118:1149–56.
32. Salama ZA, Sadek A, Abdelhady AM, Darweesh SK, Morsy SA, Esmat G. Losartan may inhibit the progression of liver fibrosis in chronic HCV patients. Hepatobiliary Surg Nutr. 2016;5(3):249–55. https://pubmed.ncbi.nlm.nih.gov/27275467
33. Murad HA, Gazzaz ZJ, Ali SS, Ibraheem MS. Candesartan, rather than losartan, improves motor dysfunction in thioacetamide-induced chronic liver failure in rats. Braz J Med Biol Res. 2017;50(11):e6665. https://pubmed.ncbi.nlm.nih.gov/28953991
34. Sturzeneker MCS, de Noronha L, Olandoski M, Wendling LU, Precoma DB. Ramipril significantly attenuates the development of non-alcoholic steatohepatitis in hyperlipidaemic rabbits. Am J Cardiovasc Dis. 2019;9(2):8–17. https://pubmed.ncbi.nlm.nih.gov/31131153
35. Kim G, Kim J, Lim YL, Kim MY, Baik SK. Renin–angiotensin system inhibitors and fibrosis in chronic liver disease: a systematic review. Hepatol Int. 2016;10(5):819–28. https://doi.org/10.1007/s12072-016-9705-x.
36. Schuppan D, Ashfaq-Khan M, Yang AT, Kim YO. Liver fibrosis: direct antifibrotic agents and targeted therapies. Matrix Biol. 2018;68–69:435–51. https://www.sciencedirect.com/science/article/pii/S0945053X18301604
37. Karsdal MA, Nielsen SH, Leeming DJ, Langholm LL, Nielsen MJ, Manon-Jensen T, et al. The good and the bad collagens of fibrosis—their role in signaling and organ function. Adv Drug Deliv Rev. 2017;121:43–56. https://www.sciencedirect.com/science/article/pii/S0169409X17301217
38. Leber N, Kaps L, Aslam M, Schupp J, Brose A, Schäffel D, et al. SiRNA-mediated in vivo gene knockdown by acid-degradable cationic nanohydrogel particles. J Control Release. 2017;248:10–23. https://www.sciencedirect.com/science/article/pii/S0168365916309221
39. Sato Y, Murase K, Kato J, Kobune M, Sato T, Kawano Y, et al. Resolution of liver cirrhosis using vitamin A–coupled liposomes to deliver siRNA against a collagen-specific chaperone. Nat Biotechnol. 2008;26(4):431–42. https://doi.org/10.1038/nbt1396.

40. Jiménez Calvente C, Sehgal A, Popov Y, Kim YO, Zevallos V, Sahin U, et al. Specific hepatic delivery of procollagen α1(I) small interfering RNA in lipid-like nanoparticles resolves liver fibrosis. Hepatology. 2015;62(4):1285–97. https://pubmed.ncbi.nlm.nih.gov/26096209
41. Molokanova O, Schönig K, Weng S-Y, Wang X, Bros M, Diken M, et al. Inducible knockdown of procollagen I protects mice from liver fibrosis and leads to dysregulated matrix genes and attenuated inflammation. Matrix Biol. 2018;66:34–49. https://www.sciencedirect.com/science/article/pii/S0945053X17302548
42. Sakamoto N, Ogawa K, Suda G, Morikawa K, Sho T, Nakai M, et al. Clinical phase 1b study results for safety, pharmacokinetics and efficacy of ND-L02-s0201, a novel targeted lipid nanoparticle delivering HSP47 SIRNA for the treatment of Japanese patients with advanced liver fibrosis. J Hepatol. 2018;18:S242.
43. Stefanovic B, Manojlovic Z, Vied C, Badger C-D, Stefanovic L. Discovery and evaluation of inhibitor of LARP6 as specific antifibrotic compound. Sci Rep. 2019;9(1):326. https://pubmed.ncbi.nlm.nih.gov/30674965
44. Sternlicht MD, Werb Z. How matrix metalloproteinases regulate cell behavior. Annu Rev Cell Dev Biol. 2001;17:463–516. https://pubmed.ncbi.nlm.nih.gov/11687497
45. Watanabe T, Niioka M, Hozawa S, Kameyama K, Hayashi T, Arai M, et al. Gene expression of interstitial collagenase in both progressive and recovery phase of rat liver fibrosis induced by carbon tetrachloride. J Hepatol. 2000;33(2):224–35. https://doi.org/10.1016/S0168-8278(00)80363-3.
46. Abe H, Kamimura K, Kobayashi Y, Ohtsuka M, Miura H, Ohashi R, et al. Effective prevention of liver fibrosis by liver-targeted hydrodynamic gene delivery of matrix metalloproteinase-13 in a rat liver fibrosis model. Mol Ther Nucleic Acids. 2016;5:e276. https://doi.org/10.1038/mtna.2015.49.
47. Iimuro Y, Brenner DA. Matrix metalloproteinase gene delivery for liver fibrosis. Pharm Res. 2008;25(2):249–58. https://pubmed.ncbi.nlm.nih.gov/17577645
48. Liu T, Wang P, Cong M, Zhang D, Liu L, Li H, et al. Matrix metalloproteinase-1 induction by diethyldithiocarbamate is regulated via Akt and ERK/miR222/ETS-1 pathways in hepatic stellate cells. Biosci Rep. 2016;36(4):e00371. https://pubmed.ncbi.nlm.nih.gov/27412967
49. Du C, Jiang M, Wei X, Qin J, Xu H, Wang Y, et al. Transplantation of human matrix metalloproteinase-1 gene-modified bone marrow-derived mesenchymal stem cell attenuates CCL4-induced liver fibrosis in rats. Int J Mol Med. 2018;41(6):3175–84. https://pubmed.ncbi.nlm.nih.gov/29512750
50. Li Y, Liu F, Ding F, Chen P, Tang M. Inhibition of liver fibrosis using vitamin A-coupled liposomes to deliver matrix metalloproteinase-2 siRNA in vitro. Mol Med Rep. 2015;12(3):3453–61. https://pubmed.ncbi.nlm.nih.gov/26017616
51. Cong M, Liu T, Wang P, Fan X, Yang A, Bai Y, et al. Antifibrotic effects of a recombinant adeno-associated virus carrying small interfering RNA targeting TIMP-1 in rat liver fibrosis. Am J Pathol. 2013;182(5):1607–16. https://doi.org/10.1016/j.ajpath.2013.01.036.
52. Parsons CJ, Bradford BU, Pan CQ, Cheung E, Schauer M, Knorr A, et al. Antifibrotic effects of a tissue inhibitor of metalloproteinase-1 antibody on established liver fibrosis in rats. Hepatology. 2004;40(5):1106–15. https://doi.org/10.1002/hep.20425.
53. Geervliet E, Bansal R. Matrix metalloproteinases as potential biomarkers and therapeutic targets in liver diseases. Cell. 2020;9(5):1212. https://pubmed.ncbi.nlm.nih.gov/32414178
54. Kothapalli CR, Ramamurthi A. Lysyl oxidase enhances elastin synthesis and matrix formation by vascular smooth muscle cells. J Tissue Eng Regen Med. 2009;3(8):655–61. https://pubmed.ncbi.nlm.nih.gov/19813219
55. Liu SB, Ikenaga N, Peng Z-W, Sverdlov DY, Greenstein A, Smith V, et al. Lysyl oxidase activity contributes to collagen stabilization during liver fibrosis progression and limits spontaneous fibrosis reversal in mice. FASEB J. 2016;30(4):1599–609. https://doi.org/10.1096/fj.14-268425.
56. Ikenaga N, Peng Z-W, Vaid KA, Liu SB, Yoshida S, Sverdlov DY, et al. Selective targeting of lysyl oxidase-like 2 (LOXL2) suppresses hepatic fibrosis progression and accelerates its reversal. Gut. 2017;66(9):1697–708. https://pubmed.ncbi.nlm.nih.gov/28073888

57. Harrison SA, Abdelmalek MF, Caldwell S, Shiffman ML, Diehl AM, Ghalib R, et al. Simtuzumab is ineffective for patients with bridging fibrosis or compensated cirrhosis caused by nonalcoholic steatohepatitis. Gastroenterology. 2018;155(4):1140–53. https://doi.org/10.1053/j.gastro.2018.07.006.
58. Klöck C, Khosla C. Regulation of the activities of the mammalian transglutaminase family of enzymes. Protein Sci. 2012;21(12):1781–91. https://pubmed.ncbi.nlm.nih.gov/23011841
59. Daneshpour N, Griffin M, Collighan R, Perrie Y. Targeted delivery of a novel group of site-directed transglutaminase inhibitors to the liver using liposomes: a new approach for the potential treatment of liver fibrosis. J Drug Target. 2011;19(8):624–31. https://doi.org/10.3109/1061186X.2010.531731.
60. Song S, Shackel NA, Wang XM, Ajami K, McCaughan GW, Gorrell MD. Discoidin domain receptor 1: isoform expression and potential functions in cirrhotic human liver. Am J Pathol. 2011;178(3):1134–44. https://pubmed.ncbi.nlm.nih.gov/21356365
61. Moll S, Desmoulière A, Moeller MJ, Pache J-C, Badi L, Arcadu F, et al. DDR1 role in fibrosis and its pharmacological targeting. Biochim Biophys Acta Mol Cell Res. 2019;1866(11):118474. https://www.sciencedirect.com/science/article/pii/S0167488919300758
62. Arauz J, Ramos-Tovar E, Muriel P. Redox state and methods to evaluate oxidative stress in liver damage: from bench to bedside. Ann Hepatol. 2016;15(2):160–73. https://www.sciencedirect.com/science/article/pii/S1665268119306702
63. Friedman SL, Ratziu V, Harrison SA, Abdelmalek MF, Aithal GP, Caballeria J, et al. A randomized, placebo-controlled trial of cenicriviroc for treatment of nonalcoholic steatohepatitis with fibrosis. Hepatology. 2018;67(5):1754–67. https://pubmed.ncbi.nlm.nih.gov/28833331
64. Wen WX, Lee SY, Siang R, Koh RY. Repurposing pentoxifylline for the treatment of fibrosis: an overview. Adv Ther. 2017;34(6):1245–69. https://doi.org/10.1007/s12325-017-0547-2.
65. Komiya C, Tanaka M, Tsuchiya K, Shimazu N, Mori K, Furuke S, et al. Antifibrotic effect of pirfenidone in a mouse model of human nonalcoholic steatohepatitis. Sci Rep. 2017;7(1):44754. https://doi.org/10.1038/srep44754.
66. Poo JL, Torre A, Aguilar-Ramírez JR, Cruz M, Mejía-Cuán L, Cerda E, et al. Benefits of prolonged-release pirfenidone plus standard of care treatment in patients with advanced liver fibrosis: PROMETEO study. Hepatol Int. 2020;14(5):817–27. https://doi.org/10.1007/s12072-020-10069-3.
67. Weng S-Y, Wang X, Vijayan S, Tang Y, Kim YO, Padberg K, et al. IL-4 receptor alpha signaling through macrophages differentially regulates liver fibrosis progression and reversal. EBioMedicine. 2018;29:92–103. https://www.sciencedirect.com/science/article/pii/S235239641830032X
68. Coppola N, Perna A, Lucariello A, Martini S, Macera M, Carleo MA, et al. Effects of treatment with Maraviroc a CCR5 inhibitor on a human hepatic stellate cell line. J Cell Physiol. 2018;233(8):6224–31. https://doi.org/10.1002/jcp.26485.
69. Fallowfield JA, Ramachandran P. A relaxin-based nanotherapy for liver fibrosis. Nat Nanotechnol. 2021;16(4):365–6. https://doi.org/10.1038/s41565-020-00832-w.
70. Bennett RG, Heimann DG, Singh S, Simpson RL, Tuma DJ. Relaxin decreases the severity of established hepatic fibrosis in mice. Liver Int. 2014;34(3):416–26. https://doi.org/10.1111/liv.12247.
71. Gracia-Sancho J, Laviña B, Rodríguez-Vilarrupla A, García-Calderó H, Fernández M, Bosch J, et al. Increased oxidative stress in cirrhotic rat livers: a potential mechanism contributing to reduced nitric oxide bioavailability. Hepatology. 2008;47(4):1248–56. https://doi.org/10.1002/hep.22166.
72. Paik Y-H, Iwaisako K, Seki E, Inokuchi S, Schnabl B, Osterreicher CH, et al. The nicotinamide adenine dinucleotide phosphate oxidase (NOX) homologues NOX1 and NOX2/gp91(phox) mediate hepatic fibrosis in mice. Hepatology. 2011;53(5):1730–41. https://pubmed.ncbi.nlm.nih.gov/21384410
73. Aoyama T, Paik Y-H, Watanabe S, Laleu B, Gaggini F, Fioraso-Cartier L, et al. Nicotinamide adenine dinucleotide phosphate oxidase in experimental liver fibrosis: GKT137831 as a novel

potential therapeutic agent. Hepatology. 2012;56(6):2316–27. https://pubmed.ncbi.nlm.nih.gov/22806357
74. Murphy MP, Smith RAJ. Drug delivery to mitochondria: the key to mitochondrial medicine. Adv Drug Deliv Rev. 2000;41(2):235–50. https://www.sciencedirect.com/science/article/pii/S0169409X99000691
75. Vilaseca M, Garcia-Caldero H, Lafoz E, Ruart M, Lopez-Sanjurjo CI, Murphy MP, et al. Mitochondria-targeted antioxidant mitoquinone deactivates human and rat hepatic stellate cells and reduces portal hypertension in cirrhotic rats. Liver Int. 2017;37(7):1002–12.
76. Turkseven S, Bolognesi M, Brocca A, Pesce P, Angeli P, di Pascoli M. Mitochondria-targeted antioxidant mitoquinone attenuates liver inflammation and fibrosis in cirrhotic rats. Am J Physiol Gastrointest Liver Physiol. 2019;318(2):G298–304. https://doi.org/10.1152/ajpgi.00135.2019.
77. Gane EJ, Weilert F, Orr DW, Keogh GF, Gibson M, Lockhart MM, et al. The mitochondria-targeted anti-oxidant mitoquinone decreases liver damage in a phase II study of hepatitis C patients. Liver Int. 2010;30(7):1019–26.
78. Bishayee A, Darvesh AS, Politis T, McGory R. Resveratrol and liver disease: from bench to bedside and community. Liver Int. 2010;30(8):1103–14. https://doi.org/10.1111/j.1478-3231.2010.02295.x.
79. Gracia-Sancho J, Villarreal G Jr, Zhang Y, García-Cardeña G. Activation of SIRT1 by resveratrol induces KLF2 expression conferring an endothelial vasoprotective phenotype. Cardiovasc Res. 2010;85(3):514–9. https://pubmed.ncbi.nlm.nih.gov/19815564
80. di Pascoli M, Diví M, Rodríguez-Vilarrupla A, Rosado E, Gracia-Sancho J, Vilaseca M, et al. Resveratrol improves intrahepatic endothelial dysfunction and reduces hepatic fibrosis and portal pressure in cirrhotic rats. J Hepatol. 2013;58(5):904–10. http://www.ncbi.nlm.nih.gov/pubmed/23262250
81. Heebøll S, Thomsen KL, Pedersen SB, Vilstrup H, George J, Grønbæk H. Effects of resveratrol in experimental and clinical non-alcoholic fatty liver disease. World J Hepatol. 2014;6(4):188–98. https://pubmed.ncbi.nlm.nih.gov/24799987
82. Theodotou M, Fokianos K, Moniatis D, Kadlenic R, Chrysikou A, Aristotelous A, et al. Effect of resveratrol on non-alcoholic fatty liver disease. Exp Ther Med. 2019;18(1):559–65. https://pubmed.ncbi.nlm.nih.gov/31316594
83. Basili S, Raparelli V, Napoleone L, Talerico G, Corazza GR, Perticone F, et al. Platelet count does not predict bleeding in cirrhotic patients: results from the PRO-LIVER study. Am J Gastroenterol. 2018;113(3):368–75. https://journals.lww.com/ajg/Fulltext/2018/03000/Platelet_Count_Does_Not_Predict_Bleeding_in.13.aspx
84. Northup PG. Hypercoagulation in liver disease. Clin Liver Dis. 2009;13(1):109–16. https://www.sciencedirect.com/science/article/pii/S1089326108000950
85. Dhar A, Sadiq F, Anstee QM, Levene AP, Goldin RD, Thursz MR. Thrombin and factor Xa link the coagulation system with liver fibrosis. BMC Gastroenterol. 2018;18(1):60. https://pubmed.ncbi.nlm.nih.gov/29739329
86. Cerini F, Vilaseca M, Lafoz E, García-Irigoyen O, García-Calderó H, Tripathi DM, et al. Enoxaparin reduces hepatic vascular resistance and portal pressure in cirrhotic rats. J Hepatol. 2016;64(4):834–42. https://doi.org/10.1016/j.jhep.2015.12.003.
87. Fortea JI, Zipprich A, Fernandez-Mena C, Puerto M, Bosoi CR, Almagro J, et al. Enoxaparin does not ameliorate liver fibrosis or portal hypertension in rats with advanced cirrhosis. Liver Int. 2018;38(1):102–12. https://doi.org/10.1111/liv.13510.
88. Cui S, Shu R, Yan S, Wu H, Chen Y, Wang L, et al. Efficacy and safety of anticoagulation therapy with different doses of enoxaparin for portal vein thrombosis in cirrhotic patients with hepatitis B. Eur J Gastroenterol Hepatol. 2015;27(8):914–9. https://journals.lww.com/eurojgh/Fulltext/2015/08000/Efficacy_and_safety_of_anticoagulation_therapy.7.aspx
89. Amitrano L, Guardascione MA, Menchise A, Martino R, Scaglione M, Giovine S, et al. Safety and efficacy of anticoagulation therapy with low molecular weight heparin for portal vein thrombosis in patients with liver cirrhosis. J Clin Gastroenterol. 2010;44(6):448–51. https://

journals.lww.com/jcge/Fulltext/2010/07000/Safety_and_Efficacy_of_Anticoagulation_Therapy.17.aspx

90. Villa E, Cammà C, Marietta M, Luongo M, Critelli R, Colopi S, et al. Enoxaparin prevents portal vein thrombosis and liver decompensation in patients with advanced cirrhosis. Gastroenterology. 2012;143(5):1253–1260.e4. https://doi.org/10.1053/j.gastro.2012.07.018.
91. Violi F, Vestri A, Menichelli D, di Rocco A, Pastori D, Pignatelli P. Direct oral anticoagulants in patients with atrial fibrillation and advanced liver disease: an exploratory meta-analysis. Hepatol Commun. 2020;4(7):1034–40. https://doi.org/10.1002/hep4.1513.
92. Vilaseca M, Garcia-Caldero H, Lafoz E, Garcia-Irigoyen O, Avila M, Reverter JC, et al. The anticoagulant rivaroxaban lowers portal hypertension in cirrhotic rats mainly by deactivating hepatic stellate cells. Hepatology. 2017;65(6):2031–44.
93. Pannach S, Babatz J, Beyer-Westendorf J. Successful treatment of acute portal vein thrombosis with rivaroxaban. Thromb Haemost. 2013;110(4):626–7.
94. Lee K-C, Hsu W-F, Hsieh Y-C, Chan C-C, Yang Y-Y, Huang Y-H, et al. Dabigatran reduces liver fibrosis in thioacetamide-injured rats. Dig Dis Sci. 2019;64(1):102–12. https://doi.org/10.1007/s10620-018-5311-1.
95. Nagaoki Y, Aikata H, Daijyo K, Teraoka Y, Shinohara F, Nakamura Y, et al. Efficacy and safety of edoxaban for treatment of portal vein thrombosis following danaparoid sodium in patients with liver cirrhosis. Hepatol Res. 2018;48(1):51–8. https://doi.org/10.1111/hepr.12895.
96. Sumida Y, Yoneda M. Current and future pharmacological therapies for NAFLD/NASH. J Gastroenterol. 2018;53(3):362–76. https://pubmed.ncbi.nlm.nih.gov/29247356
97. Loomba R, Lawitz E, Mantry PS, Jayakumar S, Caldwell SH, Arnold H, et al. The ASK1 inhibitor selonsertib in patients with nonalcoholic steatohepatitis: a randomized, phase 2 trial. Hepatology. 2018;67(2):549–59. https://pubmed.ncbi.nlm.nih.gov/28892558
98. Harrison SA, Wong VW-S, Okanoue T, Bzowej N, Vuppalanchi R, Younes Z, et al. Selonsertib for patients with bridging fibrosis or compensated cirrhosis due to NASH: results from randomized phase III STELLAR trials. J Hepatol. 2020;73(1):26–39. https://doi.org/10.1016/j.jhep.2020.02.027.
99. Gracia-Sancho J, Manicardi N, Ortega-Ribera M, Maeso-Díaz R, Guixé-Muntet S, Fernández-Iglesias A, et al. Emricasan ameliorates portal hypertension and liver fibrosis in cirrhotic rats through a hepatocyte-mediated paracrine mechanism. Hepatol Commun. 2019;3(7):987–1000. https://pubmed.ncbi.nlm.nih.gov/31304452
100. Frenette CT, Morelli G, Shiffman ML, Frederick RT, Rubin RA, Fallon MB, et al. Emricasan improves liver function in patients with cirrhosis and high model for end-stage liver disease scores compared with placebo. Clin Gastroenterol Hepatol. 2019;17(4):774–783.e4. https://doi.org/10.1016/j.cgh.2018.06.012.
101. Garcia-Tsao G, Bosch J, Kayali Z, Harrison SA, Abdelmalek MF, Lawitz E, et al. Randomized placebo-controlled trial of emricasan for non-alcoholic steatohepatitis-related cirrhosis with severe portal hypertension. J Hepatol. 2020;72(5):885–95. https://doi.org/10.1016/j.jhep.2019.12.010.
102. Zhou Q, Liao JK. Statins and cardiovascular diseases: from cholesterol lowering to pleiotropy. Curr Pharm Des. 2009;15(5):467–78. https://pubmed.ncbi.nlm.nih.gov/19199975
103. Bosch J, Gracia-Sancho J, Abraldes JG. Cirrhosis as new indication for statins. Gut. 2020;69(5):953. http://gut.bmj.com/content/69/5/953.abstract
104. Trebicka J, Hennenberg M, Odenthal M, Shir K, Klein S, Granzow M, et al. Atorvastatin attenuates hepatic fibrosis in rats after bile duct ligation via decreased turnover of hepatic stellate cells. J Hepatol. 2010;53(4):702–12. https://doi.org/10.1016/j.jhep.2010.04.025.
105. Tripathi DM, Vilaseca M, Lafoz E, Garcia-Calderó H, Viegas Haute G, Fernández-Iglesias A, et al. Simvastatin prevents progression of acute on chronic liver failure in rats with cirrhosis and portal hypertension. Gastroenterology. 2018;155(5):1564–77. https://doi.org/10.1053/j.gastro.2018.07.022.

106. Guaraldi G, Maurice JB, Marzolini C, Monteith K, Milic J, Tsochatzis E, et al. New drugs for NASH and HIV infection: great expectations for a great need. Hepatology. 2020;71(5):1831–44. https://doi.org/10.1002/hep.31177.
107. Sinha RA, Bruinstroop E, Singh BK, Yen PM. Nonalcoholic fatty liver disease and hypercholesterolemia: roles of thyroid hormones, metabolites, and agonists. Thyroid. 2019;29(9):1173–91. https://pubmed.ncbi.nlm.nih.gov/31389309
108. Safadi R, Konikoff FM, Mahamid M, Zelber-Sagi S, Halpern M, Gilat T, et al. The fatty acid–bile acid conjugate aramchol reduces liver fat content in patients with nonalcoholic fatty liver disease. Clin Gastroenterol Hepatol. 2014;12(12):2085–2091.e1. https://www.sciencedirect.com/science/article/pii/S1542356514006739
109. Manandhar B, Ahn J-M. Glucagon-like peptide-1 (GLP-1) analogs: recent advances, new possibilities, and therapeutic implications. J Med Chem. 2015;58(3):1020–37. https://doi.org/10.1021/jm500810s.
110. de Mesquita FC, Guixé-Muntet S, Fernández-Iglesias A, Maeso-Díaz R, Vila S, Hide D, et al. Liraglutide improves liver microvascular dysfunction in cirrhosis: evidence from translational studies. Sci Rep. 2017;7(1):3255. https://pubmed.ncbi.nlm.nih.gov/28607430
111. Newsome PN, Buchholtz K, Cusi K, Linder M, Okanoue T, Ratziu V, et al. A placebo-controlled trial of subcutaneous semaglutide in nonalcoholic steatohepatitis. N Engl J Med. 2020;384(12):1113–24. https://doi.org/10.1056/NEJMoa2028395.
112. Tripathi DM, Erice E, Lafoz E, García-Calderó H, Sarin SK, Bosch J, et al. Metformin reduces hepatic resistance and portal pressure in cirrhotic rats. Am J Physiol Gastrointest Liver Physiol. 2015;309(5):G301–9. https://doi.org/10.1152/ajpgi.00010.2015.
113. Dwyer BJ, Macmillan MT, Brennan PN, Forbes SJ. Cell therapy for advanced liver diseases: repair or rebuild. J Hepatol. 2021;74(1):185–99. https://doi.org/10.1016/j.jhep.2020.09.014.
114. Pietrosi G, Fernández-Iglesias A, Pampalone M, Ortega-Ribera M, Lozano JJ, García-Calderó H, et al. Human amniotic stem cells improve hepatic microvascular dysfunction and portal hypertension in cirrhotic rats. Liver Int. 2020;40(10):2500–14. https://doi.org/10.1111/liv.14610.
115. Itaba N, Kono Y, Watanabe K, Yokobata T, Oka H, Osaki M, et al. Reversal of established liver fibrosis by IC-2-engineered mesenchymal stem cell sheets. Sci Rep. 2019;9(1):6841. https://doi.org/10.1038/s41598-019-43298-0.
116. Lyra AC, Soares MBP, da Silva LFM, Braga EL, Oliveira SA, Fortes MF, et al. Infusion of autologous bone marrow mononuclear cells through hepatic artery results in a short-term improvement of liver function in patients with chronic liver disease: a pilot randomized controlled study. Eur J Gastroenterol Hepatol. 2010;22(1):33–42. https://journals.lww.com/eurojgh/Fulltext/2010/01000/Infusion_of_autologous_bone_marrow_mononuclear.5.aspx
117. Lan L, Liu R, Qin L-Y, Cheng P, Liu B-W, Zhang B-Y, et al. Transplantation of bone marrow-derived endothelial progenitor cells and hepatocyte stem cells from liver fibrosis rats ameliorates liver fibrosis. World J Gastroenterol. 2018;24(2):237–47. https://pubmed.ncbi.nlm.nih.gov/29375209
118. Povero D, Pinatel EM, Leszczynska A, Goyal NP, Nishio T, Kim J, et al. Human induced pluripotent stem cell-derived extracellular vesicles reduce hepatic stellate cell activation and liver fibrosis. JCI Insight. 2019;5(14):e125652. https://pubmed.ncbi.nlm.nih.gov/31184999

Part V

New Scenarios 2: Management of ACLD after Removal of the Primary Etiological Factor

19 Therapies for Alcohol-Related Liver Disease and for Non-Alcoholic Fatty Liver Disease

Hitoshi Yoshiji, Tadashi Namisaki, Kosuke Kaji, and Sven Francque

Abbreviations

ACLD	Advanced Chronic Liver Disease
aHCV	Advanced HCV
ALD	Alcohol-Related Liver Disease
aNAFLD	Advanced NAFLD
AUD	Alcohol Use Disorder
CI	Confidence Interval
CSPH	Clinically Significant Portal Hypertension
EMA	European Medicines Agency
FDA	Food and Drug Administration
HCV	Hepatitis C Virus
HVPG	Hepatic Venous Pressure Gradient

H. Yoshiji (✉)
Department of Gastroenterology, Nara Medical University, Nara, Japan

Third Department of Internal Medicine (Department of Gastroenterology, Endocrinology and Metabolism), Nara Medical University, Nara, Japan
e-mail: yoshijih@naramed-u.ac.jp

T. Namisaki · K. Kaji
Department of Gastroenterology, Nara Medical University, Nara, Japan
e-mail: tadashin@naramed-u.ac.jp; kajik@naramed-u.ac.jp

S. Francque (✉)
Department of Gastroenterology and Hepatology, Antwerp University Hospital, Edegem, Belgium

Laboratory of Experimental Medicine and Paediatrics, Faculty of Medicine and Health Science, University of Antwerp, Wilrijk, Belgium

Translational Science in Inflammation and Immunology, Faculty of Medicine and Health Sciences, University of Antwerp, Antwerp, Belgium
e-mail: Sven.Francque@uza.be

R. de Franchis (ed.), *Portal Hypertension VII*,
https://doi.org/10.1007/978-3-031-08552-9_19

IQR	Interquartile Range
MELD	Model of End-Stage Liver Disease
NAFL	Non-Alcoholic Fatty Liver
NAFLD	Non-Alcoholic Fatty Liver Disease
NASH	Non-Alcoholic Steatohepatitis
NIT	Non-invasive Test
OCA	Obeticholic Acid
PHT	Portal Hypertension
RCT	Randomised Controlled Trial
WHO	World Health Organisation

Alcohol-Related Liver Disease

Natural History

The management of viral liver diseases has improved dramatically over recent decades with the development of new antiviral agents. Contrarily, the proportion of alcohol-related liver cirrhosis increased over the years. The World Health Organisation (WHO) reports that approximately half of all cirrhosis-related deaths worldwide were attributable to chronic alcohol consumption in 2018 [1]. Alcoholic liver disease (ALD) encompasses a wide spectrum of conditions comprised of characteristic histological features: steatosis, pure alcoholic steatosis, steatohepatitis, progressive liver fibrosis, cirrhosis and hepatocellular carcinoma [2]. ALD is the most common cause of cirrhosis in the western world [3]. A positive correlation between cumulative alcohol intake and the severity of liver fibrosis has been reported [4, 5]. The cornerstone of treatment for ALD still remains achieving total alcohol abstinence and preventing relapse [6]. When drinking continues, pharmacologic options for ALD are limited. Harm reduction is an alternative approach to the treatment of the alcohol-use disorder (AUD). In this chapter, the impact of alcohol abstinence on the natural history of ALD is initially reviewed. We discuss the management of patients with ALD who fail to maintain abstinence from alcohol and a potential novel therapeutic agent for ALD.

Impact of Alcohol Abstinence on the Natural History of ALD

Recently, a systematic review has been published assessing the natural history of histologically proven alcohol-related liver disease [2]. Nine studies of 918 ALD patients demonstrate a histological progression of ALD with paired liver biopsies at a mean interval of 7 years [7–15]. The overall annual progression from pre-cirrhotic fibrosis to cirrhosis was 4% (95% confidence interval (CI) 2%–11%). Annual progression rates to cirrhosis were 1% (95% CI 0%–8%) for patients with histologically normal liver, 3% (95% CI 2%–4%) for steatosis, 10% (95% CI 6%–17%) for steatohepatitis and 8% (95% CI 3%–19%) for any grade of pre-cirrhotic fibrosis (Fig. 2 in [2]). Only two reports are available on the effect of alcohol abstinence on liver fibrosis progression [7, 16]. A study examining the relationship between alcohol consumption and disease progression showed that the progression rates from steatohepatitis to cirrhosis were 18% and 23% in abstinent and non-abstinent patients over 1.7 years, respectively [7]. In an analysis of 11 abstinent alcoholic patients with alcoholic hepatitis who underwent serial liver biopsies, 3 of the 11 patients returned to normal in 4 to 7 months, but 2 of the 11 patients developed liver cirrhosis in 12–14 months [16].

Three subtypes of mortality were analysed: total mortality, non-liver-related mortality and liver-specific mortality. Twenty-three studies showed mortality outcomes [8, 11, 13, 17–34]. In the eight studies (1091 patients) reporting mortality in alcohol-related steatosis [8, 17, 20, 22, 28, 30, 31, 34], annual total was 6% (95% CI 4%–7%), annual non-liver mortality was 4% (95% CI 3%–6%) and annual liver-specific mortality was 1.0% (95% CI 1%–2%) (Fig. 3 in [2]). In the seven studies (732 patients) reporting mortality in alcohol-related steatohepatitis [12, 16, 19, 21, 26, 28, 29], annual total mortality was 11% (95% CI 6%–19%), annual non-liver mortality was 4% (95% CI 2%–9%) and annual liver specific mortality was 7% (95% CI 3%–14%) (Fig. 4 in [2]). In the seven studies comprising 930 patients and reporting mortality in alcohol-related cirrhosis [16, 19–22, 25, 32], annual total mortality was 8% (95% CI 5%–13%), annual non-liver mortality was 2% (95% CI 1%–4%) and annual liver-specific mortality was 6% (95% CI 3%–10%) (Fig. 5 in [2]). These findings indicate that steatohepatitis shows the highest rates of progression to cirrhosis and highest mortality among the spectrum of ALD. Liver-related factors are the leading cause of death in steatohepatitis and cirrhosis.

There are only three studies including patients with alcohol-related cirrhosis available with regard to the mortality between abstinent and non-abstinent patients. These studies included a total of 519 patients with information on alcohol consumption comprised of 187 who abstained in the follow-up period and 332 who continued to drink. The average annual mortality was 4.7% (interquartile range (IQR) 4%–7%) and 8.0% (IQR 6.2%–11.2%) in abstinent and non-abstinent patients, respectively. However, the difference was not statistically significant.

Management of Patients with ALD who Fail to Maintain Abstinence from Alcohol

Complete abstinence still represents a therapeutic goal for patients with ALD. Systematic review and meta-analysis demonstrate that at least 1.5 years of alcohol abstinence is required for significant improvement of long-term survival of patients with alcoholic cirrhosis [35]. Acamprosate has been approved for the treatment of ALD in United States, Europe and Japan (Table 19.1) [36, 37]. Patients with alcohol dependence are often unable to successfully quit drinking alcohol. Harm reduction international, a nongovernmental organisation and global leader in harm reduction and drug policy reform, demonstrates that harm reduction aims to

Table 19.1 Food and Drug Administration or European Medicines Agency-approved pharmacological agents to treat alcohol dependence

Drugs	Mechanism of action	Approval	Europe	USA	Japan
Nalmefene	κ-opioid receptor partial agonist and δ- and μ-opioid receptor antagonist	Reduction in heavy drinking	○	—	○
Disulfiram	Aldehyde dehydrogenase Inhibitor	Alcohol abstinence	○	○	○
Naltrexone (oral)	A competitive, non-selective, specific opioid antagonist with high affinity for μ-receptor	Alcohol abstinence	○	○	—
Naltrexone (intramuscular injection)	A competitive, non-selective, specific opioid antagonist with high affinity for μ-receptor	Alcohol abstinence	—	○	—
Acamprosate	A γ-amino butyric acid receptor agonist and glutamate system modulator	Alcohol abstinence	○	○	○

overcome this issue by reducing the risk of negative effects associated with ongoing alcohol and drug use [38]. The relative risk of cirrhosis increases in subjects who consume more than 120 g of alcohol per week and rises more steeply [39]. Clinical practice guidance from the American Association for the Study of Liver Diseases suggests that harm reduction may be useful in some contexts for all patients with alcoholic liver disease [40]. The European Association for the Study of the Liver Clinical Practice Guideline shows that disulfiram, naltrexone, acamprosate and nalmefene are approved to treat alcohol dependence [41]. All these drugs except for nalmefene are approved for abstinence from alcohol. Nalmefene is approved for the reduction of heavy drinking, but not tested in patients with cirrhosis [42]. A RCT conducted in Japan demonstrates that nalmefene showed a significant reduction in the number of heavy drinking days and total alcohol consumption compared to placebo after 24 weeks (Fig. 19.1) [43]. A recent study shows that patients treated with nalmefene for 12 weeks showed trends to improvement in liver stiffness and controlled attenuation parameter for quantification of hepatic steatosis [44]. Further research is required to elucidate the impact of nalmefene on patients with alcoholic cirrhosis.

Potential Impact of Novel Pharmacological Therapies for ALD Currently in Clinical Development

A cornerstone of treatment for ALD is the achievement and maintenance of alcohol abstinence, since the efficacy of medical treatments for ALD is limited in those who continue to drink. Excessive alcohol use exerts direct detrimental effects on the intestinal barrier integrity by regulating the tight junction protein expression and mucus layer [45]. In ALD, impaired intestinal barrier function and the alterations in intestinal microbiota composition lead to increased gut permeability and subsequent translocation of endotoxin, resulting in the activation of hepatic stellate cells

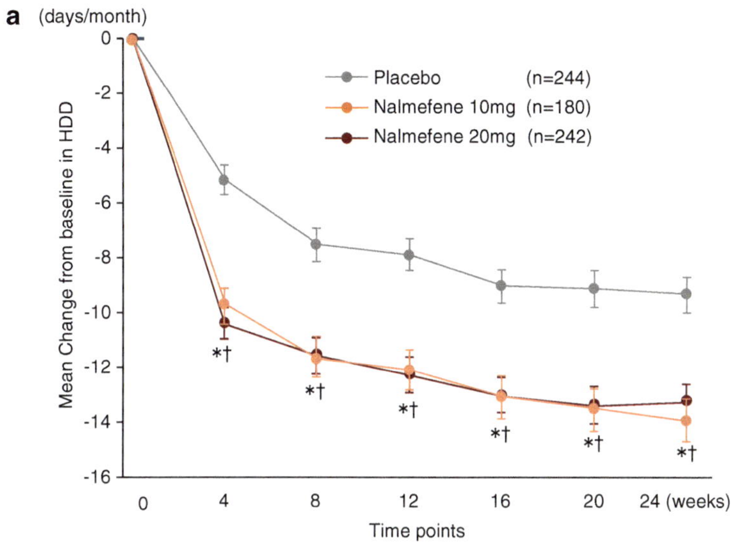

Placebo	244	244	238	234	230	224	222
Nalmefene 20 mg	242	242	212	206	199	193	189
Nalmefene 10 mg	180	180	162	154	148	144	141

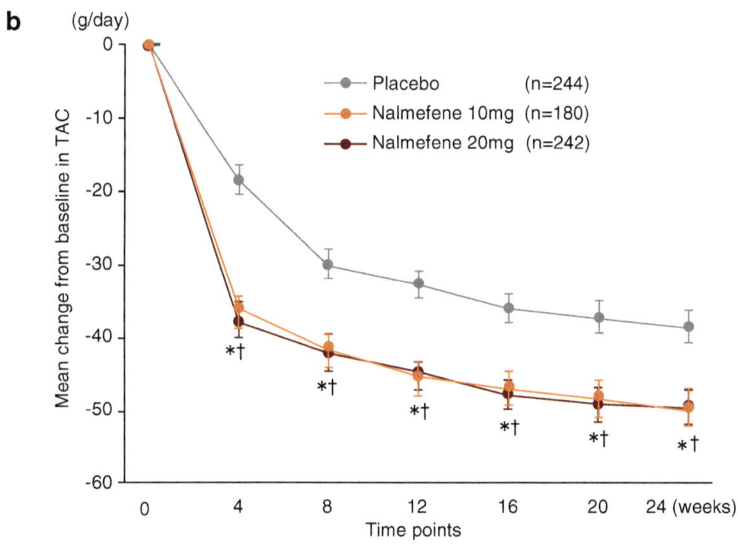

Placebo	244	244	238	234	230	224	222
Nalmefene 20 mg	242	242	212	206	199	193	189
Nalmefene 10 mg	180	180	162	154	148	144	141

Fig. 19.1 Nalmefene in alcohol-dependent patients with a high and very high drinking risk: Randomised controlled trial. (**a**) Change from baseline in the number of heavy drinking days (HDD) in study participants receiving nalmefene or placebo. (**b**) Change from baseline in total alcohol consumption (TAC) in study participants receiving nalmefene or placebo (Modified from [43])

and the development of liver fibrosis [46]. As this microbial translocation triggers the release of inflammatory mediators into the bloodstream, gut-derived endotoxin plays a pivotal role in the pathogenesis of cirrhosis and its complications such as hepatic encephalopathy [47], spontaneous bacterial peritonitis [48], portal hypertension (PHT) [49] and sarcopenia. Excessive alcohol consumption causes an increase in microbial translocation markers including lipopolysaccharide-binding protein and soluble CD14 in patients with ALD [50] and the levels of both markers decrease with alcohol withdrawal [50]. Rifaximin, a non-absorbable antibiotic, improves hepatic encephalopathy with decreased endotoxin activity in patients with cirrhosis [51]. Rifaximin has been shown to reduce serum levels of soluble CD163 and soluble mannose receptor, markers of paracellular permeability with partial modification of gut microbiota in patients with cirrhosis [52]. These decreases show highly positive correlations with improvements in endotoxin activity. Rifaximin decreases hepatic venous pressure gradient (HVPG) values through a reduction of plasma endotoxin levels in patients with decompensated alcoholic cirrhosis [53]. Long-term rifaximin administration reduces the risk of development of PHT and improves survival of patients with decompensated alcoholic cirrhosis [54]. Furthermore, a new study is investigating anti-fibrotic and molecular aspects of rifaximin in patients with ALD [55]. In addition to its direct bactericidal effect, rifaximin may therefore have beneficial effects in patients with ALD.

Conclusions and Future Perspectives

Complete alcohol abstinence remains the only therapeutic intervention to inhibit disease progression in patients with ALD, particularly alcoholic cirrhosis. Rifaximin may have beneficial effects in patients with ALD. Potential pharmacological approaches for the treatment of alcoholic cirrhosis need to be developed. Harm reduction approach represents an effective tool in the treatment of individuals suffering from alcohol withdrawal. Further studies need to be carried out to investigate the effects of harm reduction on patients with alcoholic cirrhosis who fail to maintain abstinence from alcohol.

Non-Alcoholic Fatty Liver Disease

Natural History

Non-alcoholic Fatty Liver Disease (NAFLD) has become the most prevalent chronic liver disease [56, 57]. A distinction is usually made between Non-alcoholic Fatty Liver (NAFL) and Non-alcoholic Steatohepatitis (NASH), the latter requiring the combined presence of steatosis, lobular inflammation and ballooning [58]. NASH is considered the more severe form, with a risk of progressive fibrosis and hence development of cirrhosis and its complications [59–61]. The process of steatohepatitis is highly dynamic and fluctuating over time, implying a potentially fluctuating and variable course of fibrosis progression [59].

Current estimates point towards a NAFLD prevalence of 25%–30% in populations with the so-called Western lifestyle [57]. Important regional differences exist, in part driven by ethnic factors [56]. NASH probably affects 10%–20% of NAFLD patients, 45%–50% of whom will develop progressive fibrosis, with about 10%–25% evolving to cirrhosis over 10–20 years [62].

Factors associated with more rapid fibrosis progression are active steatohepatitis, the presence of diabetes, older age, ethnicity (Hispanics being at a higher risk) and several genetic variants [63]. However, the impact of these factors on disease progression and regression in patients who have already developed advanced chronic liver disease (ACLD) is less well established. Of note, several of the risk factors for NAFLD and for more rapid progression have also been associated with more rapid progression of liver fibrosis of other aetiologies as discussed elsewhere.

Hepatic Decompensation in Patients with Compensated NAFLD-ACLD

A recent meta-analysis including 13 studies encompassing 770 patients with cirrhosis showed a relative risk of 12.78 of a liver-related event and of 3.42 for all-cause mortality compared to F0 patients [64]. Very few prospective cohort studies to assess outcome in NAFLD patients have been performed so far. The largest study to date comprises 1773 adults followed up for a median of 4 years [65]. Prognosis was mainly determined by baseline fibrosis stage, with an incidence rate of liver-related events of 2.69 per 100 patient years in patients with cirrhosis. Hepatic encephalopathy was the most frequent event. Other studies were retrospective in nature.

Clinical trials in NASH cirrhosis also give some insights. In the trials with Simtuzumab, a humanised lysyl oxidase-like 2 antibody that failed to reduce fibrosis, 258 compensated NASH cirrhotic patients were followed for a median period of 30.9 months, 19% of whom developed a hepatic decompensation event. Of note, body mass index, but not diabetes, was predictive of clinically significant PHT (CSPH) and both did not predict clinical events [66]. In the patients with F3 fibrosis at baseline, 22% evolved to cirrhosis during a median follow-up of 29 months, with a decompensation event in three patients. In the trials with Selonsertib, an apoptosis signal-regulating kinase inhibitor, 27/877 F4 patients developed a liver-related event during a mean follow-up of 16 months, with ascites being the most frequent one [67].

Prognosis of Decompensated NAFLD-ACLD

The placebo-treated NASH cirrhosis patients with model of end-stage liver disease (MELD) ≥15 in the 3-month trial with the pan-caspase inhibitor Emricasan did not show significant evolution in liver function parameters in this short time interval [68]. Rinaldi et al. report 102 cryptogenic cirrhosis (presumed NASH) patients of mixed stage at entry, with Child-Pugh class at entry being the only independent predictor of decompensation, and with no differences with a matched hepatitis C (HCV) cirrhosis cohort [69]. A study by Sanyal et al. also showed no difference in mortality between NASH or HCV related cirrhosis in Child-Pugh B and C patients

[70]. The study by Saunders et al. already indicated a high mortality rate of patients presenting with decompensation regardless of aetiology, cryptogenic cirrhosis representing 35% of the cohort [71]. In the PREDICT study, NAFLD accounted for 7.6% of the cases, which were equally distributed over the three identified patterns of prognosis [72]. Globally these data point towards the fact that, although the metabolic co-morbidities seem to accelerate disease progression in other liver diseases, once the stage of decompensation is reached, there are no significant differences between NASH and other aetiologies.

Specifics of PHT in NAFLD

In the trials with Simtuzumab, although the prognostic importance of CSPH was confirmed, of the 50 patients with a clinical event, 7 (14%) had an HVPG <10 mmHg [66]. Rodrigues et al also reported in a retrospective analysis of 89 patients with CSPH, that 16% did not have cirrhosis, with NASH as the most frequent underlying aetiology [73]. Another prospective study of 109 compensated NAFLD-ACLD also confirmed those with CSPH to be at the highest risk of decompensation but observed hepatic decompensation in 3.1% of patients without CSPH after a mean follow-up of 5 years [74].

Bassegoda et al. compared 548 patients with advanced NAFLD (aNAFLD) to 444 patients with advanced HCV matched for disease severity, age and gender. Significantly more patients with aNAFLD compared to aHCV were decompensated at baseline, a difference mainly driven by the presence of ascites [75]. For the same MELD or Child-Pugh score, the aNAFLD patients had a lower HVPG. In aNAFLD with HVPG <10 mmHg, 9% had a decompensation, compared to 0 in the aHCV with HVPG <10 mmHg and also 6% had large varices. Analysing the HVPG in relation with the presence of decompensation, decompensation was present at lower HVPG values in the aNAFLD group compared to the aHCV group, and for every level of HVPG, the rate of decompensation was higher. Obesity did not have a role in the prevalence of decompensation.

In patients undergoing transjugular intrahepatic portosystemic shunt it was also demonstrated that HVPG, compared to the direct measurement of the portal pressure gradient, tended to underestimate the real portal pressure gradient in NASH [76].

All these observations point towards an additive role for (pre-sinusoidal) vascular mechanisms in advanced disease stages of NASH and suggest that HVPG measurement may underestimate the severity of the PHT in aNAFLD, but this obviously requires further study.

Potential Impact of Therapies for NAFLD

Weight Loss in NAFLD

The effects of weight loss in NAFLD have mainly been studied in the context of non-cirrhotic NAFLD. Weight loss has been shown to improve histological features of NAFLD, with a requirement of 10% of weight loss to induce regression of fibrosis [77]. In addition, other studies clearly indicate that weight loss improves liver histology or non-invasive indicators of liver damage upon lifestyle intervention [78–80] or weight-lowering treatment [81, 82].

Bariatric surgery also has been shown to substantially improve liver histology, including fibrosis [83, 84]. There is, however, paucity of data on the effect of weight loss once the stage of F4 has been reached. In the largest cohort to date, only three patients with cirrhosis were included, 1 of whom regressed [83]. A question of particular interest is how to position bariatric surgery once a diagnosis of NASH cirrhosis is established, especially in advanced cases or decompensated patients. Given the paucity of data, the correct timing of bariatric surgery in obese patients to date largely depends on local expertise [85].

Histology and Clinical Events as Endpoints in Clinical Trials

In patients with non-cirrhotic NASH, clinical liver-related events are not likely to occur in a time frame of a few years. Therefore, histological endpoints have been proposed as reasonably likely (but not validated) surrogate endpoints predictive of clinically meaningful benefit [86, 87]. NASH resolution without worsening of fibrosis and $a \geq 1$-stage decrease in fibrosis score according to NASH Clinical Research Network without worsening of NASH are the two key regulatory endpoints in non-cirrhotic NAFLD patients for Phase 3 trials [88]. The European Medicines Agency (EMA) tends to require $a \geq 2$-point reduction in fibrosis for drugs with a purely anti-fibrotic profile [89]. Clinical events, including progression to cirrhosis, need to confirm the efficacy seen on histology in the long run.

In NAFLD patients with cirrhosis, time to a composite endpoint of death from any cause and liver-related events is considered the Food and Drug Administration (FDA) regulatory endpoint [88]. EMA also considers reversal of cirrhosis as an acceptable endpoint, as far as this endpoint is backed by secondary endpoints based on biomarkers and clinical events data [90].

Several trials have meanwhile failed in Phase 2 or 3. Positive results on NASH resolution but not fibrosis regression have been reported for Semaglutide (a glucagon-like peptide 1 receptor agonist) [91], Aramchol (an acetyl-coA carboxylase inhibitor, albeit only in per protocol analysis) [92], vitamin E (borderline significant) [93] and resmetirom (a thyroid hormone receptor β agonist, significant when patients with weight loss ≥9.5% were excluded) [94]. Fibrosis regression has been reported with the farnesoid X receptor agonist obeticholic acid (OCA) in Phase 3 [95]. Recently, the pan-peroxisome proliferator-activated receptor agonist lanifibranor showed positive results on both these endpoints and on the composite endpoint of NASH resolution and fibrosis improvement after 24 weeks of treatment [96]. All these trials concern non-cirrhotic NASH patients.

Based on the results of a Phase 2 trial (that did not include F4 patients) [97], pegbelfermin, a pegylated fibroblast growth factor 21 analogue, is evaluated in patients with cirrhosis with $a \geq 1$ stage reduction in fibrosis (hence reversal of cirrhosis) as primary endpoint [98] (NCT03486912), as well as NGM282 (aldafermin) [99], a fibroblast growth factor 19 analogue (NCT04210245).

HVPG

Changes in HVPG have also served as endpoint, with few positive results to date. The galectin-3 inhibitor GR-MD-02 was tested in 162 NASH cirrhosis patients with PHT and no or mild varices. It did not reduce HVPG in the overall population but had a significant effect in non-CSPH patients [100]. Also, in the subgroup of patients without varices at baseline, it improved HVPG after 52 weeks of treatment. Changes in HVPG is one of the secondary endpoints in the 2-year Phase 3 trial (NCT04365868).

Emricasan was tested in 263 patients with NASH cirrhosis. Overall, there was no effect on HVPG, but in the subgroup with high HVPG (defined as ≥16 mmHg), a significant reduction of HVPG could be noted [101].

Varices

The development of varices in patients who do not have varices at baseline is one of the events that is captured in some of the trials in patients with NASH cirrhosis as part of composite clinical end point [102].

GR-MD-02 reduced the occurrence of new varices in patients without varices at baseline in the aforementioned Phase 2 trial [100]. Development of varices in patients with CSPH but without varices at baseline is the primary endpoint in the Phase 3 trial (NCT04365868).

Hepatic Decompensation in Patients With Compensated NASH-ACLD

The development of hepatic decompensation events is, mostly in a composite endpoint with death of all causes, an endpoint approved by the regulatory authorities in Phase 3 trials, as outlined before. OCA has been shown to reduce PHT in preclinical models [103]. A Phase 3 of OCA in patients with compensated cirrhosis with time to a decompensation event or death as primary endpoint is still ongoing and is currently the only Phase 3 trial in cirrhotic patients (all other Phase 3 trials are in non-cirrhotic patients and add evolution towards cirrhosis into their event-driven endpoint) (NCT03439254). Lanifibranor was also shown to significantly improve intrahepatic vascular resistance and portal pressure in animal models of PHT (related to improvements in endothelial cell phenotype markers [104]). This kind of data provides additional rationale to test these drugs in NASH cirrhosis for their impact on PHT.

Outcomes in Trials for Decompensated NAFLD-ACLD

To date, few studies have included patients with decompensated NASH cirrhosis from an efficacy perspective. The FDA guidance currently states that patients with MELD >12 cannot be included in Phase 3 NASH cirrhosis trials. EMA does not exclude these patients from clinical trials, but requires mechanistic data, as well as clinical efficacy and safety data. Besides Phase 1 trials, no compounds are currently being tested in decompensated patients.

Emricasan has been tested in decompensated NASH cirrhosis patients, defined as MELD ≥12 and ≤20, history of variceal bleeding requiring transfusing or moderate ascites treated with diuretics [68]. There was, however, no benefit of the drug in terms of event rates.

Management of NAFLD Patients With ACLD Who Achieve Weight Loss/on Pharmacological Therapies

Monitoring the Evolution of Liver Disease in Patients Who Achieved Weight Loss/on Pharmacological Therapies

Invasive Methods

HVPG has been confirmed to be of prognostic value [66]. The threshold of HVPG decrease that translates into a meaningful benefit, is not well established in NASH cirrhosis [105]. The only available study [66] analysed the impact of HVPG reductions of ≥20% and ≥20% or to <10 mmHg and linked them to a more than fivefold increased risk of liver-related events, which differs from the 10%-decrease suggested by Baveno VI to denote a meaningful benefit [106].

Non-Invasive Methods

Non-invasive tests (NITS), including laboratory tests on several types of specimens, a combination of these parameters (with or without clinical parameters) in several scores, and imaging modalities are currently extensively studies in the context of monitoring disease evolution [107–109]. Paired biopsy studies, whether observational or interventional, in which changes of NITs can be linked to changes in histology, will help identifying which parameters reliably reflect disease evolution and how they can be used. Long-term follow-up studies, including the large Phase 3 studies, also have the potential of providing the community with the evidence needed to develop recommendations for non-invasive follow-up and monitoring of treatment response.

Screening for Varices in Patients Who Achieved Weight Loss/on Pharmacological Therapies

Although weight loss has been shown to improve fibrosis and some pharmacological therapies show promise, the persistence of fibrosis in many patients [83] and the current absence of reliable NITs to monitor disease progression or regression and hence therapeutic response, imply that currently patients need to be followed up according to their initial diagnosis stage, regardless of indicators of improvement. To date, no data are available to justify NASH-specific recommendations that differ from other aetiologies of advanced liver disease.

General Considerations

NAFLD is undoubtedly a frequent cause of advanced liver disease and a non-negligible co-factor in many patients with liver disease of another aetiology. Its pathophysiology and natural history are, however, still poorly understood. Adequate assessment of disease activity and stage rely on a liver biopsy, limiting the acquisition of large volume data to generate evidence-based recommendations. Pharmacological therapy is still in its infancy, especially in patients with NASH-induced cirrhosis. The metabolic drivers of the disease are also recognised to be factors that influence disease progression in other liver diseases and therefore need to be assessed and appropriately treated. The ongoing studies, including the efforts of the biomarker consortia, the longitudinal collaborative cohorts and the large Phase 3 trials will most probably generate data that will allow establishing NASH-specific guidelines.

References

1. Yeverino-Gutiérrez ML, González-González MDR, González-Santiago O. Mortality from alcohol-related liver cirrhosis in Mexico (2000–2017). Front Public Health. 2020;8:524356. https://doi.org/10.3389/fpubh.2020.524356.
2. Parker R, Aithal GP, Becker U, Gleeson D, Masson S, Wyatt JI, et al. Natural history of histologically proven alcohol-related liver disease: a systematic review. J Hepatol. 2019;71:586–93. https://doi.org/10.1016/j.jhep.2019.05.020.
3. World Health Organization. Age-standardized death rates of liver cirrhosis. Global Health Observatory. n.d.. http://www.who.int/gho/alcohol/harms_consequences/deaths_liver_cirrhosis/en/index.html. Accessed October 29, 2021.
4. Farooq MO, Bataller R. Pathogenesis and management of alcoholic liver disease. Dig Dis. 2016;34:347–55. https://doi.org/10.1159/000444545.
5. Bataller R, Gao B. Liver fibrosis in alcoholic liver disease. Semin Liver Dis. 2015;35:146–56. https://doi.org/10.1055/S-0035-1550054.
6. Addolorato G, Mirijello A, Barrio P, Gual A. Treatment of alcohol use disorders in patients with alcoholic liver disease. J Hepatol. 2016;65:618–30. https://doi.org/10.1016/j.jhep.2016.04.029.
7. Parés A, Caballería J, Bruguera M, Torres M, Rodés J. Histological course of alcoholic hepatitis. Influence of abstinence, sex and extent of hepatic damage. J Hepatol. 1986;2:33–42. https://doi.org/10.1016/s0168-8278(86)80006-x.
8. Teli MR, Day CP, Burt AD, Bennett MK, James OF. Determinants of progression to cirrhosis or fibrosis in pure alcoholic fatty liver. Lancet. 1995;346:987–90. https://doi.org/10.1016/s0140-6736(95)91685-7.
9. Sørensen TI, Orholm M, Bentsen KD, Høybye G, Eghøje K, Christoffersen P. Prospective evaluation of alcohol abuse and alcoholic liver injury in men as predictors of development of cirrhosis. Lancet. 1984;2:241–4. https://doi.org/10.1016/s0140-6736(84)90295-2.
10. Worner TM, Lieber CS. Perivenular fibrosis as precursor lesion of cirrhosis. JAMA. 1985;254:627–30.
11. Marbet UA, Bianchi L, Meury U, Stalder GA. Long-term histological evaluation of the natural history and prognostic factors of alcoholic liver disease. J Hepatol. 1987;4:364–72. https://doi.org/10.1016/s0168-8278(87)80547-0.
12. Mathurin P, Beuzin F, Louvet A, Carrié-Ganne N, Balian A, Trinchet JC, et al. Fibrosis progression occurs in a subgroup of heavy drinkers with typical histological features. Aliment Pharmacol Ther. 2007;25:1047–54. https://doi.org/10.1111/j.1365-2036.2007.03302.x.
13. Motoo Y, Wakatsuki T, Nakanuma Y. Long-term histologic follow-up study of alcoholic liver disease. Intern Med. 1992;31:33–8. https://doi.org/10.2169/internalmedicine.31.33.
14. Oancea R, Serbanescu M, Lazar V, Milcu V, Antonescu O. The criteria of histological activity and the prognosis in precirrhotic alcoholic hepatopathies. Med Interna. 1991;43:103–11.
15. Nakano M, Worner TM, Lieber CS. Perivenular fibrosis in alcoholic liver injury: ultrastructure and histologic progression. Gastroenterology. 1982;83:777–85.
16. Galambos JT. Natural history of alcoholic hepatitis. 3. Histological changes. Gastroenterology. 1972;63:1026–35.
17. Bouchier IA, Hislop WS, Prescott RJ. A prospective study of alcoholic liver disease and mortality. J Hepatol. 1992;16:290–7. https://doi.org/10.1016/s0168-8278(05)80659-2.

18. Alves PS, Correia JP, Borda d'Agua C, Portugal L, Capaz V, Rodrigues ML, et al. Alcoholic liver diseases in Portugal. Clinical and laboratory picture, mortality, and survival. Alcohol Clin Exp Res. 1982;6:216–24. https://doi.org/10.1111/j.1530-0277.1982.tb04965.x.
19. Brunt PW, Kew MC, Scheuer PJ, Sherlock S. Studies in alcoholic liver disease in Britain. I. Clinical and pathological patterns related to natural history. Gut. 1974;15:52–8. https://doi.org/10.1136/gut.15.1.52.
20. Chedid A, Mendenhall CL, Gartside P, French SW, Chen T, Rabin L. Prognostic factors in alcoholic liver disease. Am J Gastroenterol. 1991;86:210–6.
21. Morgan MY, Sherlock S. Sex-related differences among 100 patients with alcoholic liver disease. Br Med J. 1977;1:939–41. https://doi.org/10.1136/bmj.1.6066.939.
22. Orrego H, Blake JE, Blendis LM, Medline A. Prognosis of alcoholic cirrhosis in the presence and absence of alcoholic hepatitis. Gastroenterology. 1987;92:208–14. https://doi.org/10.1016/0016-5085(87)90861-4.
23. Powell WJ, Klatskin G. Duration of survival in patients with Laennec's cirrhosis. Influence of alcohol withdrawal, and possible effects of recent changes in general management of the disease. Am J Med. 1968;44:406–20. https://doi.org/10.1016/0002-9343(68)90111-3.
24. Orrego H, Israel Y, Blake JE, Medline A. Assessment of prognostic factors in alcoholic liver disease: toward a global quantitative expression of severity. Hepatology. 1983;3:896–905. https://doi.org/10.1002/hep.1840030602.
25. Orholm M, Sørensen TI, Bentsen K, Høybye G, Eghøje K, Christoffersen P. Mortality of alcohol abusing men prospectively assessed in relation to history of abuse and degree of liver injury. Liver. 1985;5:253–60. https://doi.org/10.1111/j.1600-0676.1985.tb00246.x.
26. Ginés P, Quintero E, Arroyo V, Terés J, Bruguera M, Rimola A, et al. Compensated cirrhosis: natural history and prognostic factors. Hepatology. 1987;7:122–8. https://doi.org/10.1002/hep.1840070124.
27. Nissenbaum M, Chedid A, Mendenhall C, Gartside P. Prognostic significance of cholestatic alcoholic hepatitis. VA Cooperative Study Group #119. Dig Dis Sci. 1990;35:891–6. https://doi.org/10.1007/BF01536804.
28. Dam-Larsen S, Becker U, Franzmann M-B, Larsen K, Christoffersen P, Bendtsen F. Final results of a long-term, clinical follow-up in fatty liver patients. Scand J Gastroenterol. 2009;44:1236–43. https://doi.org/10.1080/00365520903171284.
29. Sagedal S, Hartmann A, Nordal KP, Osnes K, Leivestad T, Foss A, et al. Impact of early cytomegalovirus infection and disease on long-term recipient and kidney graft survival. Kidney Int. 2004;66:329–37.
30. Deleuran T, Grønbaek H, Vilstrup H, Jepsen P. Cirrhosis and mortality risks of biopsy-verified alcoholic pure steatosis and steatohepatitis: a nationwide registry-based study. Aliment Pharmacol Ther. 2012;35:1336–42. https://doi.org/10.1111/j.1365-2036.2012.05091.x.
31. Haflidadottir S, Jonasson JG, Norland H, Einarsdottir SO, Kleiner DE, Lund SH, et al. Long-term follow-up and liver-related death rate in patients with non-alcoholic and alcoholic related fatty liver disease. BMC Gastroenterol. 2014;14:166. https://doi.org/10.1186/1471-230X-14-166.
32. Lackner C, Spindelboeck W, Haybaeck J, Douschan P, Rainer F, Terracciano L, et al. Histological parameters and alcohol abstinence determine long-term prognosis in patients with alcoholic liver disease. J Hepatol. 2017;66:610–8. https://doi.org/10.1016/J.JHEP.2016.11.011.
33. Masson S, Emmerson I, Henderson E, Fletcher EH, Burt AD, Day CP, et al. Clinical but not histological factors predict long-term prognosis in patients with histologically advanced non-decompensated alcoholic liver disease. Liver Int. 2014;34:235–42. https://doi.org/10.1111/liv.12242.
34. Semb S, Neermark S, Dam-Larsen S, Franzmann MB, Albrectsen J, Kallemose T, et al. Presence of alcoholic steatohepatitis, but no selective histological feature, indicates an increased risk of cirrhosis and premature death. Scand J Gastroenterol. 2016;51:1367–74. https://doi.org/10.1080/00365521.2016.1203016.

35. Xie Y-D, Feng B, Gao Y, Wei L. Effect of abstinence from alcohol on survival of patients with alcoholic cirrhosis: a systematic review and meta-analysis. Hepatol Res. 2014;44:436–49. https://doi.org/10.1111/hepr.12131.
36. Antonelli M, Ferrulli A, Sestito L, Vassallo GA, Tarli C, Mosoni C, et al. Alcohol addiction - the safety of available approved treatment options. Expert Opin Drug Saf. 2018;17:169–77. https://doi.org/10.1080/14740338.2018.1404025.
37. Higuchi S, Japanese Acamprosate Study Group. Efficacy of acamprosate for the treatment of alcohol dependence long after recovery from withdrawal syndrome: a randomized, double-blind, placebo-controlled study conducted in Japan (Sunrise Study). J Clin Psychiatry. 2015;76:181–8. https://doi.org/10.4088/JCP.13m08940.
38. Harm Reduction International. What is harm reduction? n.d.. https://www.hri.global/what-is-harm-reduction. Accessed October 29, 2021.
39. Simpson RF, Hermon C, Liu B, Green J, Reeves GK, Beral V, et al. Alcohol drinking patterns and liver cirrhosis risk: analysis of the prospective UK Million Women Study. Lancet Public Health. 2019;4:e41–8. https://doi.org/10.1016/S2468-2667(18)30230-5.
40. Crabb DW, Im GY, Szabo G, Mellinger JL, Lucey MR. Diagnosis and treatment of alcohol-associated liver diseases: 2019 practice guidance from the American Association for the Study of Liver Diseases. Hepatology. 2020;71:306–33. https://doi.org/10.1002/hep.30866.
41. Carvalho AF, Heilig M, Perez A, Probst C, Rehm J. Alcohol use disorders. Lancet. 2019;394:781–92. https://doi.org/10.1016/S0140-6736(19)31775-1.
42. Lackner C, Tiniakos D. Fibrosis and alcohol-related liver disease. J Hepatol. 2019;70:294–304. https://doi.org/10.1016/J.JHEP.2018.12.003.
43. Miyata H, Takahashi M, Murai Y, Tsuneyoshi K, Hayashi T, Meulien D, et al. Nalmefene in alcohol-dependent patients with a high drinking risk: randomized controlled trial. Psychiatry Clin Neurosci. 2019;73:697–706. https://doi.org/10.1111/pcn.12914.
44. Mueller S, Luderer M, Zhang D, Meulien D, Brach BS, Schou MB. Open-label study with nalmefene as needed use in alcohol-dependent patients with evidence of elevated liver stiffness and/or hepatic steatosis. Alcohol Alcohol. 2020;55:63–70. https://doi.org/10.1093/ALCALC/AGZ078.
45. Meroni M, Longo M, Rametta R, Dongiovanni P. Genetic and epigenetic modifiers of alcoholic liver disease. Int J Mol Sci. 2018;19:3857. https://doi.org/10.3390/ijms19123857.
46. Cruz-Jentoft AJ, Baeyens JP, Bauer JM, Boirie Y, Cederholm T, Landi F, et al. Sarcopenia: European consensus on definition and diagnosis: report of the European Working Group on Sarcopenia in Older People. Age Ageing. 2010;39:412–23. https://doi.org/10.1093/ageing/afq034.
47. Kaji K, Takaya H, Saikawa S, Furukawa M, Sato S, Kawaratani H, et al. Rifaximin ameliorates hepatic encephalopathy and endotoxemia without affecting the gut microbiome diversity. World J Gastroenterol. 2017;23:8355–66. https://doi.org/10.3748/wjg.v23.i47.8355.
48. Juanola O, Ferrusquía-Acosta J, García-Villalba R, Zapater P, Magaz M, Marín A, et al. Circulating levels of butyrate are inversely related to portal hypertension, endotoxemia, and systemic inflammation in patients with cirrhosis. FASEB J. 2019;33:11595–605. https://doi.org/10.1096/fj.201901327R.
49. Lumsden AB, Henderson JM, Kutner MH. Endotoxin levels measured by a chromogenic assay in portal, hepatic and peripheral venous blood in patients with cirrhosis. Hepatology. 1988;8:232–6. https://doi.org/10.1002/hep.1840080207.
50. Donnadieu-Rigole H, Pansu N, Mura T, Pelletier S, Alarcon R, Gamon L, et al. Beneficial effect of alcohol withdrawal on gut permeability and microbial translocation in patients with alcohol use disorder. Alcohol Clin Exp Res. 2018;42:32–40. https://doi.org/10.1111/acer.13527.
51. Bass NM, Mullen KD, Sanyal A, Poordad F, Neff G, Leevy CB, et al. Rifaximin treatment in hepatic encephalopathy. N Engl J Med. 2010;362:1071–81. https://doi.org/10.1056/NEJMOA0907893.
52. Kaji K, Saikawa S, Takaya H, Fujinaga Y, Furukawa M, Kitagawa K, et al. Rifaximin alleviates endotoxemia with decreased serum levels of soluble CD163 and mannose receptor and

partial modification of gut microbiota in cirrhotic patients. Antibiotics (Basel). 2020;9(4):145. https://doi.org/10.3390/antibiotics9040145.
53. Vlachogiannakos J, Saveriadis AS, Viazis N, Theodoropoulos I, Foudoulis K, Manolakopoulos S, et al. Intestinal decontamination improves liver haemodynamics in patients with alcohol-related decompensated cirrhosis. Aliment Pharmacol Ther. 2009;29:992–9. https://doi.org/10.1111/j.1365-2036.2009.03958.x.
54. Vlachogiannakos J, Viazis N, Vasianopoulou P, Vafiadis I, Karamanolis DG, Ladas SD. Long-term administration of rifaximin improves the prognosis of patients with decompensated alcoholic cirrhosis. J Gastroenterol Hepatol. 2013;28:450–5. https://doi.org/10.1111/jgh.12070.
55. Madsen BS, Trebicka J, Thiele M, Israelsen M, Arumugan M, Havelund T, et al. Antifibrotic and molecular aspects of rifaximin in alcoholic liver disease: study protocol for a randomized controlled trial. Trials. 2018;19:143. https://doi.org/10.1186/s13063-018-2523-9.
56. Estes C, Anstee QM, Arias-Loste MT, Bantel H, Bellentani S, Caballeria J, et al. Modeling NAFLD disease burden in China, France, Germany, Italy, Japan, Spain, United Kingdom, and United States for the period 2016–2030. J Hepatol. 2018;69:896–904. https://doi.org/10.1016/j.jhep.2018.05.036.
57. Younossi ZM, Blissett D, Blissett R, Henry L, Stepanova M, Younossi Y, et al. The economic and clinical burden of nonalcoholic fatty liver disease in the United States and Europe. Hepatology. 2016;64:1577–86. https://doi.org/10.1002/hep.28785.
58. European Association for the Study of the Liver (EASL); European Association for the Study of Diabetes (EASD); European Association for the Study of Obesity (EASO). EASL-EASD-EASO Clinical Practice Guidelines for the management of non-alcoholic fatty liver disease. J Hepatol. 2016;64:1388–402. https://doi.org/10.1016/J.JHEP.2015.11.004.
59. Schuppan D, Surabattula R, Wang XY. Determinants of fibrosis progression and regression in NASH. J Hepatol. 2018;68:238–50. https://doi.org/10.1016/J.JHEP.2017.11.012.
60. Kleiner DE, Brunt EM, Wilson LA, Behling C, Guy C, Contos M, et al. Association of histologic disease activity with progression of nonalcoholic fatty liver disease. JAMA Netw Open. 2019;2:e1912565. https://doi.org/10.1001/jamanetworkopen.2019.12565.
61. Singh S, Allen AM, Wang Z, Prokop LJ, Murad MH, Loomba R. Fibrosis progression in nonalcoholic fatty liver vs nonalcoholic steatohepatitis: a systematic review and meta-analysis of paired-biopsy studies. Clin Gastroenterol Hepatol. 2015;13:643–654.e9. https://doi.org/10.1016/j.cgh.2014.04.014.
62. Rinella M, Charlton M. The globalization of nonalcoholic fatty liver disease: prevalence and impact on world health. Hepatology. 2016;64:19–22. https://doi.org/10.1002/hep.28524.
63. Eslam M, Valenti L, Romeo S. Genetics and epigenetics of NAFLD and NASH: clinical impact. J Hepatol. 2018;68:268–79. https://doi.org/10.1016/j.jhep.2017.09.003.
64. Taylor RS, Taylor RJ, Bayliss S, Hagström H, Nasr P, Schattenberg JM, et al. Association between fibrosis stage and outcomes of patients with nonalcoholic fatty liver disease: a systematic review and meta-analysis. Gastroenterology. 2020;158(6):1611–1625.e12. https://doi.org/10.1053/j.gastro.2020.01.043.
65. Sanyal AJ, van Natta ML, Clark J, Neuschwander-Tetri BA, Diehl A, Dasarathy S, et al. Prospective study of outcomes in adults with nonalcoholic fatty liver disease. N Engl J Med. 2021;385(17):1559–69. https://doi.org/10.1056/NEJMoa2029349.
66. Sanyal AJ, Harrison SA, Ratziu V, Abdelmalek MF, Diehl AM, Caldwell S, et al. The natural history of advanced fibrosis due to nonalcoholic steatohepatitis: data from the simtuzumab trials. Hepatology. 2019;70:1913–27. https://doi.org/10.1002/hep.30664.
67. Harrison SA, Wong VW-S, Okanoue T, Bzowej N, Vuppalanchi R, Younes Z, et al. Selonsertib for patients with bridging fibrosis or compensated cirrhosis due to NASH: results from randomized phase III STELLAR trials. J Hepatol. 2020;73(1):26–39. https://doi.org/10.1016/j.jhep.2020.02.027.
68. Frenette C, Kayali Z, Mena E, Mantry PS, Lucas KJ, Neff G, et al. Emricasan to prevent new decompensation in patients with NASH-related decompensated cirrhosis. J Hepatol. 2021;74(2):274–82. https://doi.org/10.1016/j.jhep.2020.09.029.

69. Rinaldi L, Nascimbeni F, Giordano M, Masetti C, Guerrera B, Amelia A, et al. Clinical features and natural history of cryptogenic cirrhosis compared to hepatitis C virus-related cirrhosis. World J Gastroenterol. 2017;23(8):1458–68. https://doi.org/10.3748/wjg.v23.i8.1458.
70. Sanyal AJ, Banas C, Sargeant C, Luketic VA, Sterling RK, Stravitz RT, et al. Similarities and differences in outcomes of cirrhosis due to nonalcoholic steatohepatitis and hepatitis C. Hepatology. 2006;43(4):682–9. https://doi.org/10.1002/hep.21103.
71. Saunders JB, Walters JR, Davies AP, Paton A. A 20-year prospective study of cirrhosis. Br Med J (Clin Res Ed). 1981;282(6260):263–6. https://doi.org/10.1136/bmj.282.6260.263.
72. Trebicka J, Fernandez J, Papp M, Caraceni P, Laleman W, Gambino C, et al. The PREDICT study uncovers three clinical courses of acutely decompensated cirrhosis that have distinct pathophysiology. J Hepatol. 2020;73(4):842–54. https://doi.org/10.1016/j.jhep.2020.06.013.
73. Rodrigues SG, Montani M, Guixé-Muntet S, de Gottardi A, Berzigotti A, Bosch J. Patients with signs of advanced liver disease and clinically significant portal hypertension do not necessarily have cirrhosis. Clin Gastroenterol Hepatol. 2019;17(10):2101–2109.e1. https://doi.org/10.1016/j.cgh.2018.12.038.
74. Paternostro R. Distinct features of the portalhypertensive syndrome in patients with advanced chronic liver disease due to non-alcoholic steatohepatitis. J Hepatol. 2020;73:S754–5.
75. Bassegoda O, Olivas P, Turco L, Mandorfer M, Serra-Burriel M, Tellez L, et al. Decompensation in advanced non-alcoholic fatty liver disease may occur at lower hepatic venous pressure gradient levels that in patients with viral disease. Clin Gastroenterol Hepatol. 2021; https://doi.org/10.1016/j.cgh.2021.10.023.
76. Ferrusquía-Acosta J, Bassegoda O, Turco L, Reverter E, Pellone M, Bianchini M, et al. Agreement between wedged hepatic venous pressure and portal pressure in non-alcoholic steatohepatitis-related cirrhosis. J Hepatol. 2021;74:811–8. https://doi.org/10.1016/j.jhep.2020.10.003.
77. Vilar-Gomez E, Martinez-Perez Y, Calzadilla-Bertot L, Torres-Gonzalez A, Gra-Oramas B, Gonzalez-Fabian L, et al. Weight loss through lifestyle modification significantly reduces features of nonalcoholic steatohepatitis. Gastroenterology. 2015;149:367–378.e5. https://doi.org/10.1053/j.gastro.2015.04.005.
78. Dixon JB, Bhathal PS, Hughes NR, O'Brien PE. Nonalcoholic fatty liver disease: improvement in liver histological analysis with weight loss. Hepatology. 2004;39(6):1647–54. https://doi.org/10.1002/hep.20251.
79. Promrat K, Kleiner DE, Niemeier HM, Jackvony E, Kearns M, Wands JR, et al. Randomized controlled trial testing the effects of weight loss on nonalcoholic steatohepatitis. Hepatology. 2010;51:121–9. https://doi.org/10.1002/hep.23276.
80. Wong VWS, Chan RSM, Wong GLH, Cheung BHK, Chu WCW, Yeung DKW, et al. Community-based lifestyle modification programme for non-alcoholic fatty liver disease: a randomized controlled trial. J Hepatol. 2013;59:536–42. https://doi.org/10.1016/j.jhep.2013.04.013.
81. Patel AA, Torres DM, Harrison SA. Effect of weight loss on nonalcoholic fatty liver disease. J Clin Gastroenterol. 2009;43(10):970–4. https://doi.org/10.1097/MCG.0b013e3181b57475.
82. Newsome PN, Sejling A-S, Sanyal AJ. Semaglutide or placebo for nonalcoholic steatohepatitis. Reply. N Engl J Med. 2021;385:e6. https://doi.org/10.1056/NEJMc2106921.
83. Lassailly G, Caiazzo R, Ntandja-Wandji LC, Gnemmi V, Baud G, Verkindt H, et al. Bariatric surgery provides long-term resolution of nonalcoholic steatohepatitis and regression of fibrosis. Gastroenterology. 2020;159:1290–1301.e5. https://doi.org/10.1053/j.gastro.2020.06.006.
84. Cazzo E, Pareja JC, Chaim EA. Nonalcoholic fatty liver disease and bariatric surgery: a comprehensive review. Sao Paulo Med J. 2017;135(3):277–95. https://doi.org/10.1590/1516-3180.2016.0306311216.
85. Tsochatzis E, Coilly A, Nadalin S, Levistky J, Tokat Y, Ghobrial M, et al. International Liver Transplantation Consensus statement on end-stage liver disease due to nonalcoholic steatohepatitis and liver transplantation. Transplantation. 2019;103:45–56. https://doi.org/10.1097/TP.0000000000002433.

86. Rinella ME, Tacke F, Sanyal AJ, Anstee QM, participants of the AASLD/EASL Workshop. Report on the AASLD/EASL joint workshop on clinical trial endpoints in NAFLD. J Hepatol. 2019;71(4):823–33. https://doi.org/10.1016/j.jhep.2019.04.019.
87. Francque S, Vonghia L. Pharmacological treatment for non-alcoholic fatty liver disease. Adv Ther. 2019;36:1052–74. https://doi.org/10.1007/s12325-019-00898-6.
88. U.S. Department of Health and Human Services Food and Drug Administration Center for Drug Evaluation and Research (CDER). Noncirrhotic nonalcoholic steatohepatitis with liver fibrosis: developing drugs for treatment. Guidance for industry. Fed Regist. 2018;83:9.
89. Konerman MA, Jones JC, Harrison SA. Pharmacotherapy for NASH: current and emerging. J Hepatol. 2018;68:362–75. https://doi.org/10.1016/j.jhep.2017.10.015.
90. European Medicines Agency. Reflection paper on regulatory requirements for the 5 development of medicinal products for chronic non6 infectious liver diseases (PBC, PSC, NASH). 2018. https://www.ema.europa.eu/en/draft-reflection-paper-regulatory-requirements-development-medicinal-products-chronic-non-infectious
91. Newsome PN, Buchholtz K, Cusi K, Linder M, Okanoue T, Ratziu V, et al. A placebo-controlled trial of subcutaneous semaglutide in nonalcoholic steatohepatitis. N Engl J Med. 2021;384(12):1113–24. https://doi.org/10.1056/nejmoa2028395.
92. Ratziu V, de Guevara L, Safadi R, Poordad F, Fuster F, Flores-Figueroa J, et al. Aramchol in patients with nonalcoholic steatohepatitis: a randomized, double-blind, placebo-controlled phase 2b trial. Nat Med. 2021;27(10):1825–35. https://doi.org/10.1038/s41591-021-01495-3.
93. Sanyal AJ, Chalasani N, Kowdley KV, McCullough A, Diehl AM, Bass NM, et al. Pioglitazone, vitamin E, or placebo for nonalcoholic steatohepatitis. N Engl J Med. 2010;362(18):1675–85. https://doi.org/10.1056/NEJMoa0907929.
94. Harrison SA, Bashir MR, Guy CD, Zhou R, Moylan CA, Frias JP, et al. Resmetirom (MGL-3196) for the treatment of non-alcoholic steatohepatitis: a multicentre, randomised, double-blind, placebo-controlled, phase 2 trial. Lancet. 2019;394(10213):2012–24. https://doi.org/10.1016/S0140-6736(19)32517-6.
95. Younossi ZM, Ratziu V, Loomba R, Rinella M, Anstee QM, Goodman Z, et al. Obeticholic acid for the treatment of non-alcoholic steatohepatitis: interim analysis from a multicentre, randomised, placebo-controlled phase 3 trial. Lancet. 2019;394:2184–96. https://doi.org/10.1016/S0140-6736(19)33041-7.
96. Francque SM, Bedossa P, Ratziu V, Anstee QM, Bugianesi E, Sanyal AJ, et al. A randomized, controlled trial of the Pan-PPAR agonist lanifibranor in NASH. N Engl J Med. 2021;385:1547–58. https://doi.org/10.1056/NEJMoa2036205.
97. Sanyal A, Charles ED, Neuschwander-Tetri BA, Loomba R, Harrison SA, Abdelmalek MF, et al. Pegbelfermin (BMS-986036), a PEGylated fibroblast growth factor 21 analogue, in patients with non-alcoholic steatohepatitis: a randomised, double-blind, placebo-controlled, phase 2A trial. Lancet. 2019;392(10165):2705–17. https://doi.org/10.1016/S0140-6736(18)31785-9.
98. Abdelmalek MF, Charles ED, Sanyal AJ, Harrison SA, Neuschwander-Tetri BA, Goodman Z, et al. The FALCON program: two phase 2b randomized, double-blind, placebo-controlled studies to assess the efficacy and safety of pegbelfermin in the treatment of patients with nonalcoholic steatohepatitis and bridging fibrosis or compensated cirrhosis. Contemp Clin Trials. 2021;104:106335. https://doi.org/10.1016/j.cct.2021.106335.
99. Harrison SA, Neff G, Guy CD, Bashir MR, Paredes AH, Frias JP, et al. Efficacy and safety of aldafermin, an engineered FGF19 analog, in a randomized, double-blind, placebo-controlled trial of patients with nonalcoholic steatohepatitis. Gastroenterology. 2021;160(1):219–231.e1. https://doi.org/10.1053/j.gastro.2020.08.004.
100. Chalasani N, Abdelmalek MF, Garcia-Tsao G, Vuppalanchi R, Alkhouri N, Rinella M, et al. Effects of belapectin, an inhibitor of galectin-3, in patients with nonalcoholic steatohepatitis with cirrhosis and portal hypertension. Gastroenterology. 2020;158(5):1334–1345.e5. https://doi.org/10.1053/j.gastro.2019.11.296.
101. Garcia-Tsao G, Bosch J, Kayali Z, Harrison SA, Abdelmalek MF, Lawitz E, et al. Randomized placebo-controlled trial of emricasan for non-alcoholic steatohepatitis-related cirrhosis

with severe portal hypertension. J Hepatol. 2020;72(5):885–95. https://doi.org/10.1016/j.jhep.2019.12.010.

102. D'Amico G, Pasta L, Morabito A, D'Amico M, Caltagirone M, Malizia G, et al. Competing risks and prognostic stages of cirrhosis: a 25-year inception cohort study of 494 patients. Aliment Pharmacol Ther. 2014;39(10):1180–93. https://doi.org/10.1111/apt.12721.
103. Verbeke L, Farre R, Trebicka J, Komuta M, Roskams T, Klein S, et al. Obeticholic acid, a farnesoid X receptor agonist, improves portal hypertension by two distinct pathways in cirrhotic rats. Hepatology. 2014;59:2286–98. https://doi.org/10.1002/hep.26939.
104. Boyer-Diaz Z, Aristu-Zabalza P, Andrés-Rozas M, Robert C, Ortega-Ribera M, Fernández-Iglesias A, et al. Pan-PPAR agonist lanifibranor improves portal hypertension and hepatic fibrosis in experimental advanced chronic liver disease. J Hepatol. 2021;74:1188–99. https://doi.org/10.1016/j.jhep.2020.11.045.
105. Mandorfer S, Hernandéz-Gea V, Reiberger T, Garciá-Pagán JC. Hepatic venous pressure gradient response in non-selective beta-blocker treatment—is it worth measuring? Curr Hepatol Rep. 2019;18:174–86.
106. de Franchis R, Baveno VI Faculty. Expanding consensus in portal hypertension: report of the Baveno VI Consensus Workshop: stratifying risk and individualizing care for portal hypertension. J Hepatol. 2015;63(3):743–52. https://doi.org/10.1016/j.jhep.2015.05.022.
107. Selvaraj EA, Mózes FE, Jayaswal ANA, Zafarmand MH, Vali Y, Lee JA, et al. Diagnostic accuracy of elastography and magnetic resonance imaging in patients with NAFLD: a systematic review and meta-analysis. J Hepatol. 2021;75(4):770–85. https://doi.org/10.1016/j.jhep.2021.04.044.
108. Mózes FE, Lee JA, Selvaraj EA, Jayaswal ANA, Trauner M, Boursier J, et al. Diagnostic accuracy of non-invasive tests for advanced fibrosis in patients with NAFLD: an individual patient data meta-analysis. Gut. 2022;71(5):1006–19. https://doi.org/10.1136/gutjnl-2021-324243.
109. Francque S, Vonghia L. The future of diagnosing NASH—could a simple blood test be the key? Expert Rev Gastroenterol Hepatol. 2017;11(11):995–7. https://doi.org/10.1080/17474124.2017.1374851.

Management of ACLD After HBV-Suppression and HCV-Cure

20

Jidong Jia, Sabela Lens, Hitoshi Yoshiji, Sven Francque, Emmanouil A. Tsochatzis, and Mattias Mandorfer

J. Jia
Beijing Key Laboratory of Translational Medicine on Liver Cirrhosis, Beijing, China

National Clinical Research Center of Digestive Diseases, Beijing, China

Liver Research Center, Beijing Friendship Hospital, Capital Medical University, Beijing, China
e-mail: jia_jd@ccmu.edu.cn

S. Lens
Liver Unit, Hospital Clínic de Barcelona, IDIBAPS, Universidad de Barcelona, Barcelona, Spain

Centro de Investigación Biomédica Red de enfermedades hepáticas y digestivas (CIBERehd), Madrid, Spain
e-mail: slens@clinic.cat

H. Yoshiji
Third Department of Internal Medicine (Department of Gastroenterology, Endocrinology and Metabolism), Nara Medical University, Kashihara, Japan
e-mail: yoshijih@naramed-u.ac.jp

S. Francque
Department of Gastroenterology and Hepatology, Antwerp University Hospital, Antwerp, Belgium

Translational Science in Inflammation and Immunology, Faculty of Medicine and Health Sciences, University of Antwerp, Antwerp, Belgium
e-mail: Sven.Francque@uza.be

E. A. Tsochatzis
Sheila Sherlock Liver Unit, UCL Institute for Liver and Digestive Health, Royal Free Hospital and University College London, London, UK
e-mail: e.tsochatzis@ucl.ac.uk

M. Mandorfer (✉)
Vienna Hepatic Hemodynamic Lab, Division of Gastroenterology and Hepatology, Department of Internal Medicine III, Medical University of Vienna, Vienna, Austria
e-mail: mattias.mandorfer@meduniwien.ac.at

R. de Franchis (ed.), *Portal Hypertension VII*,
https://doi.org/10.1007/978-3-031-08552-9_20

Hepatitis B

Impact of HBV-Suppression on the Course of ACLD

A systematic review and meta-analysis on the impact of HBV-suppression on the course of ACLD was conducted under the supervision of J.J. (for further contributors, see the Acknowledgements section) and adhering to standard methodology. In brief, clinical studies on antiviral therapies that were published after the first use of lamivudine (3TC) were considered. Importantly, studies which (1) used interferon (IFN)-based regimens, (2) exclusively included non-ACLD patients/did not report information on the group of ACLD patients, or (3) did not report information on at least one of the outcomes of interest (liver fibrosis, PH, as well as related noninvasive surrogate parameters, varices, hepatic decompensation, or death) were excluded. Finally, 51 studies were considered. The next paragraph summarizes preliminary findings of the meta-analysis as well as important data from uncontrolled studies; the resulting manuscript is expected to be published after Baveno VII.

With a relative risk (RR) of 1.60 (95%CI: 0.90–2.85), the meta-analysis did not observe an impact of nucleos(t)ide analogue (NA)-treatment (vs. untreated) on histological improvement. However, this finding has to be interpreted in the context of large sequential biopsy studies that strongly suggest a high rate of liver fibrosis regression including the reversal of advanced fibrosis and cirrhosis during long-term NA-treatment [1]. In the study by Marcellin et al. [1] overlooking approximately 1 year of adefovir dipivoxil (ADV)/tenofovir disoproxil fumarate (TDF) followed by 4 years of TDF therapy, 87% (304/348) of those who had a follow-up (FU) liver biopsy after approximately 5 years of NA-treatment (348/641) showed an improvement of liver fibrosis. Among patients with pre-treatment cirrhosis (only patients with compensated cirrhosis were included), 74% (71/96) had resolved cirrhosis. Importantly, body mass index (BMI)/obesity (i.e., BMI $\geq$30 kg/m^2) were the only factors associated with persistent cirrhosis, underlining the importance of metabolic co-factors in HBV-induced ACLD treated with NA.

Moreover, an uncontrolled study investigated changes in HVPG during NA therapy [2]. In a small series of 19 patients with biopsy-proven cirrhosis and pre-treatment clinically significant PH (CSPH), 1 year of 3TC treatment decreased HVPG in all but one patient who had viral breakthrough. Mean HVPG of 14.4 pre-treatment decreased to 12.4 mmHg on-treatment, (i.e., by 13.9%). In the subset of patients with an HVPG $\geq$12 mmHg, a HVPG-decrease $\geq$20% or to $<$12 mmHg (i.e., the definition of HVPG-response that was used in this and other studies [3]) was observed in 76% (10/13) of evaluable patients.

The latter observation is in line with the findings of an investigation on the dynamics of varices in 107 patients treated sequentially with 3TC/ADV/TDF [4]. In patients without pre-treatment endoscopic evidence of CSPH (i.e., no varices), de-novo development of varices was rare (7.5% (6/80)); among 27 patients with small varices (i.e., CSPH) pre-treatment, regression (67% (18/27)) was considerably more common than progression (4% (1/27)). Importantly, in six out of seven patients with growth/progression of EV, this worsening occurred in the context of viral breakthrough or hepatocellular carcinoma (HCC) development.

Finally, the systematic review and meta-analysis conducted for Baveno VII confirmed the observation of several individual studies that NA therapy prevents hepatic decompensation (RR: 0.47; 95%CI: 0.32–0.67) including variceal bleeding (RR: 0.35; 95%CI: 0.17–0.71) and reduces the risk of liver transplantation or death (RR: 0.34; 95%CI: 0.23–0.52).

Accordingly, there is a broad body of evidence derived from uncontrolled studies (liver histology, HVPG, and varices) and controlled studies (direct endpoints) that HBV-suppression leads to potentially meaningful decreases in liver disease/PH severity in the majority of patients and reduces the risk of hepatic decompensation and death by more than 50%.

Monitoring the Evolution of Liver Disease in Patients With HBV-Suppression

Invasive methods for monitoring disease evolution in HBV-infected patients include liver biopsy, measurement of HVPG, and endoscopy. However, in clinical practice, endoscopy is usually the only invasive modality used to monitor disease regression/progression in patients with a sustained (off-treatment) or maintained (on-treatment) virological response.

Noninvasive approaches including both serum and imaging markers for diagnosing of liver fibrosis/cirrhosis as well as PH have been extensively investigated [5, 6]. Among them, liver stiffness measurement (LSM) has been found to be well-correlated with HVPG at values below the threshold for CSPH and may be used in conjunction with platelet count (PLT) to assess the probability of CSPH in compensated ACLD (cACLD) patients [7]. However, the concept of cACLD is limited to progressive disease, as acknowledged by Baveno VII. The diagnostic ability of LSM for monitoring the regression of pre-treatment CSPH has been questioned following the observations in HCV-infected patients [8]. While these initial concerns are addressed by the individual patient data-based meta-analysis summarized below, no such data in HBV-infected patients are available. However, serial measurements of LSM by vibration controlled transient elastography (VCTE) in patients on long-term NA-therapy have been shown to provide some information on the regression of histological liver fibrosis in the majority of published studies [9–14]; of note, they usually did not provide ACLD-specific information. A recent, large, prospective study [14] indicated that an LSM-decrease >30% was accompanied by 51% increased likelihood of liver fibrosis improvement. In contrast, LSM-increases decreased the likelihood of liver fibrosis improvement after accounting for regression to the mean. However, the association between changes in LSM and histological fibrosis is strongly "confounded" by hepatic inflammation, and it seems fairly unlikely that changes in LSM are sufficiently accurate to impact clinical decision-making, in particular in patients with ACLD, for whom specific data are limited. In contrast, re-staging of PH (see below) and re-stratification of risk are more promising applications for on-treatment LSM in patients with ACLD, as it has been shown to predict liver-related events (including HCC) in patients with compensated cirrhosis [15].

Blood biomarkers have some utility for diagnosing ACLD in HBV-infected patients, but their diagnostic ability for CSPH and varices is less clear. In a small study by Wu and colleagues [16], which, as a limitation, also included a relevant proportion of decompensated patients (i.e., patients who by definition have CSPH, and thus, do not belong to the target population of a NIT in this context [6]), von Willebrand factor (VWF) showed an area under the receiver operating characteristic curve (AUROC) of around 0.885 for CSPH, while the AUC for varices was <0.8. Of note, treatment/HBV-suppression status has not been reported in this study. However, VWF (combined with PLT to the VITRO score) has shown encouraging results for ruling in/out CSPH and for stratifying hepatic decompensation risk in cACLD patients achieving HCV-cure [17].

Screening for Varices and Risk Stratification in Patients With HBV-Suppression

With Baveno VI [18], criteria for ruling out varices needing treatment have been proposed in order to avoid unnecessary endoscopies (VNT; this concept was based on the assumption that therapeutic measures primarily aim at preventing variceal bleeding rather than hepatic decompensation). Since the paradigm has shifted from prevention of variceal bleeding to prevention of hepatic decompensation with Baveno VII, we are referring to high-risk varices in this article. Based on the Baveno VI criteria, endoscopy can be avoided in patients with a LSM by VCTE <20 kPa and a PLT >150G/L with less than 5% of VNT being missed. Several modifications of these criteria (e.g., "Expanded Baveno VI criteria") have been proposed, aiming at increasing the number of spared endoscopies, while maintaining a proportion of missed VNT <5%. Two recent studies validated the "Baveno VI" criteria in the context of HBV-suppression [19]. While only a minority of patients included in the study by Thabut et al. [19] had HBV-induced cirrhosis, the second study by Wang and co-workers [20] exclusively focused on this etiology. In the latter study, previous decompensation was not a formal exclusion criterion; however, 94.7% of patients were Child-Turcotte-Pugh (CTP) A at the time of inclusion. Importantly, the Baveno VI criteria did not miss any of the 70 patients with high-risk varices. However, by applying these strategies, CSPH and/or low-risk varices are often not detected, which may be problematic for centers using carvedilol/non-selective beta blockers (NSBB) for preventing variceal growth [21], and most importantly, hepatic decompensation [22]. This was also evident in the Wang et al. study, as 40.5% of Baveno VI favorable patients had low-risk varices [20]. However, it is important to note that these preventive strategies have not been investigated after the removal/suppression of the primary etiological factor—a setting in which (as indicated by the above-mentioned data and the meta-analysis) the risk of events is rare. Even if preventive strategies were similarly effective as in active disease (i.e., comparable relative risk reduction (RRR)—which is uncertain), the absolute risk reduction (ARR) would be considerably lower, resulting in a substantially higher number needed to treat (NNT). Considering the uncertainty regarding the RRR and the evident increase in the NNT in the low-risk setting of HBV-suppression, a less aggressive treatment approach (risk stratification/variceal screening based on Baveno VI criteria) seems reasonable, until further data become available.

Hepatitis C

Impact of HCV-Eradication on the Course of ACLD

A majority of long-term follow-up studies in patients with cACLD still used IFN-based regimens, since highly effective IFN-free combination therapies for chronic hepatitis C (CHC) only became available approximately 7 years before Baveno VII (i.e., approval of simeprevir/daclatasvir as combination partners for sofosbuvir). Since IFN was basically contraindicated in decompensated patients, no long-term data are available. Of note, studies including patients from the IFN era have to be interpreted with caution, as liver disease severity—in particular the presence of CSPH [23]—used to be a strong risk factor for treatment failure and also determines the prognosis of patients with HCV-induced cACLD [24, 25]. Accordingly, it can be expected that the pre-treatment severity of PH/decompensation risk was inherently different between those achieving and not achieving SVR. Although a recent study questioned the relevance of this potential source of bias [26], it still seems likely to interfere with the interpretation of the impact of SVR on the risk of liver-related events—in particular hepatic decompensation—in cACLD.

Prominent examples of long-term studies dating back to the IFN era are an Italian multicenter retrospective analysis by Bruno et al. [27], which comprised 920 patients with biopsy-proven cirrhosis of whom 124 (13.5%) achieved SVR. Among those with SVR, no patient developed decompensation during a mean follow-up of 8.6 years. In the non-SVR group, 107 patients had a decompensating event, which resulted in an incidence rate of 1.88/100 person-years. It is important to note that patients were censored at the time of HCC development, which seems to be instrumental for studies aiming at investigating the risk for PH-related events. Other remarkable retrospective cohort studies confirmed the beneficial effects of SVR on hepatic decompensation in patients with cACLD, however, also reported hepatic decompensation events despite SVR [28, 29]. The prospective HALT-C trial [30] which included cACLD patients, found a profoundly decreased risk of decompensation at year 7.5 after enrolment in patients who achieved SVR (0.9% vs. breakthrough/relapse: 4.7% vs. non-response: 11.7%), even after adjusting for differences in baseline characteristics (PLT and serum albumin; SVR vs. non-response: aHR: 0.13). Despite the attempt to adjust for the severity of PH by PLT, it is unclear whether this resulted in a fair comparison. Possibly the most informative study arising from the IFN era was conducted by Di Marco and colleagues [31] and prospectively investigated long-term outcomes in patients with biopsy-proven cirrhosis.

Analyses were further stratified by the pre-treatment presence or absence of small varices, i.e., evidence of CSPH. Importantly, SVR was protective of hepatic decompensation in both strata, however, while patients without varices (lower prevalence of CSPH) seemed to be at negligible risk (0 out of 67 patients), decompensating events occurred at a rate of 1.7/100 person-years in patients with small varices (i.e., patients with CSPH) who achieved SVR. In conclusion, findings of studies using IFN-based regimens suggested that achieving SVR reduces the risk of decompensation in patients with or without pre-treatment varices and that the risk of decompensation is negligible in patients who are successfully treated before CSPH becomes evident. The latter conclusion is also supported by studies performing paired HVPG-measurements that indicated that the resolution of subclinical PH is common and that progression to CSPH did not occur [32], an observation which was also confirmed recently by a study using IFN-free regimens [25, 33].

The combination of several direct-acting antivirals (DAA) to IFN-free regimens uncoupled the severity of underlying liver disease/PH and the probability of SVR [34] in patients with cACLD, allowing for a less biased evaluation of the impact of SVR on outcomes. Even in decompensated patients, these regimens have proven highly effective, and thus, for the first time also provided robust data on the outcomes of these patients after HCV-cure. Nevertheless, several potential sources of bias that are hard to account for (e.g., linkage to care, compliance, and alcohol consumption) may still limit the significance of comparisons between those undergoing antiviral therapy/achieving HCV-cure, and those who do not. Initial evidence for the impact of SVR to IFN-free therapies on hepatic decompensation in cACLD was mostly derived from registry studies.

A Scottish registry confirmed that SVR to IFN-free regimens was accompanied by a substantial reduction in the risk of hospital admissions due to decompensating events (0.188/100 patient-years vs. 1.215/100 patient-years) among patients with initially compensated cirrhosis, which was diagnosed by a variety of methods [35]. This contrasts the findings of the ANRS CO 22 HEPATHER study in the subgroup of patients with pre-treatment compensated cirrhosis [36]. In the latter study, cirrhosis was diagnosed by a combination of decreased PLT and prothrombin time index in nearly half of patients, while in the majority of the remaining patients, the diagnosis was established by various NIT—of note, the accuracy of this approach is unclear. In another registry-based study utilizing Veteran Affairs health care data, the impact of SVR on the incidence of acute variceal bleeding was reduced in the overall subgroup of patients with cirrhosis (aHR: 0.73); however, the reduction did not attain statistical significance in the subgroup of patients with prior varices but without a history of bleeding (aHR: 0.77) [37]. Importantly, all of these registry-based analyses have to be interpreted with caution, as the accuracy of the characterization of patients and definition/ascertainment of outcomes may be limited and some studies had a rather short duration of follow-up. In more extensively characterized cohorts specifically focusing on cACLD, the rates of decompensation were very low, e.g., around 0.3 [38, 39] per 100 patient-years. Moreover, Tosetti et al. [40] provided data on follow-up events in 148 cACLD patients with pre-treatment evidence of CSPH (i.e., LSM $\geq$20 kPa and/or low-risk varices; those with high-risk varices were excluded) and observed only one decompensation event (ascites development), the latter occurring in the context of HCC. Most recently, D'Ambrosio et al. reported a 5-year cumulative incidence of liver-related events of 10.2% in patients with CTP A cirrhosis ($n = 480$); importantly, HCC accounted for 78% of events, while hepatic decompensation (ascites $n = 6$ and variceal bleeding $n = 3$) was comparatively rare and ascites development was precipitated by portal vein thrombosis (which is not prevented by SVR [41]) and alcohol consumption in two cases. Accordingly, cohort studies conclusively indicate that hepatic decompensation in cACLD patients who achieved HCV-cure is a rare event.

Despite the observed reduction in risk, it is evident that a proportion of cACLD patients will develop decompensation despite SVR, underlining the need for simple, noninvasive risk stratification approaches to facilitate individualized surveillance and therapy (see discussion on RRR/ARR in the section on HBV-suppression).

Of note, only limited information on occurrence/growth/regression of varices after HCV-cure with IFN-based therapies is available [19, 42, 43], which interferes with the determination of the ideal timing of re-endoscopy after SVR. Moreover, long-term data on the risk of variceal bleeding (as a function of variceal status) and its evolution over time are needed.

Finally, the impact of HCV-cure in decompensated patients seems to be less profound/well-established [44]. Since a universally agreed definition of re-compensation only became available with Baveno VII, we abstained from evaluating the impact of SVR on this outcome due to a high heterogeneity between studies. Accordingly, incidence and implications of re-compensation (as defined by Baveno VII criteria) after HCV-cure require further study.

Co-Factors Modifying the Course of ACLD After HCV-Eradication and Adjunctive Therapies

Alcohol consumption is a major determinant of hepatic decompensation in cACLD patients with SVR to IFN-free therapy [17, 44]. Due to its high prevalence, it may even compromise the beneficial effect of the introduction of DAA on a population level [45]. In contrast, the evidence supporting the role of obesity and diabetes in this specific context is less clear [8, 17, 36, 44]. However, the absence of evidence should not be mistaken for evidence of absence as the importance of these co-factors for the progression of CLD is well established (see Chap. 4) and their impact on hepatic decompensation after HCV-cure is more difficult to establish due to the considerably lower number of decompensating events, and thus, statistical power. Finally, diabetes has been shown to adversely impact all-cause mortality in cACLD patients achieving SVR [46]. Accordingly, from a more holistic medical perspective, it is obvious that these co-factors need to be addressed before and after removal of the primary etiological factor.

The role for adjunctive therapies to promote the regression of advanced liver fibrosis in patients who achieved HCV-cure is currently under investigation [47].

Pooled Analysis on the Evolution of PH and NIT After HCV-Eradication

A systematic literature search on studies reporting post-treatment HVPG in patients with chronic HCV-infection was performed; we attempted to contact all first/corresponding/last authors. Individual patient data were provided by all except for three studies ($n = 2$ using pegylated interferon and ribavirin [48, 49] and $n = 1$ using sofosbuvir/ribavirin [50]) that comprised a total of 117 patients. Accordingly, individual patient data was available for 675 patients: Specifically, 80 patients from Reiberger et al. [32], 100 patients from Lens et al. [51], 90 patients from Mandorfer and Kozbial [33]/Schwabl and Mandorfer [52]/Mandorfer et al. [25], 226 patients from Lens [53]/Lens et al. [8], 8 patients from Puente et al. [54], 112 patients from Mauro et al. [55], 33 patients from Abadia et al. [56], and 26 patients from Diez et al. [57] were considered. After applying selection criteria (paired HVPG-measurement pre- and post-treatment, pre-treatment HVPG ≥6 mmHg, and SVR), 418 patients were included in the cohort for investigating the evolution of PH. The diagnostic performance of LSM/PLT for diagnosing CSPH was evaluated in a subgroup of 324 patients with paired information on NIT. The following text passage provides preliminary results/findings of the pooled analysis. Median time from end-of-treatment to post-treatment evaluation was 28.4 (interquartile range (IQR): 24–44) and 28.8 (IQR: 25–45) weeks in the HVPG and HVPG/NIT cohorts, respectively. Findings on the evolution of PH were in line with the previous publications, and thus, are not presented in detail. In the HVPG/NIT cohort, pre-treatment prevalence of CSPH was 85%, with 241 patients having cACLD (CSPH prevalence in the latter group of patients: 80%). Importantly, in cACLD patients, the strength of correlation between LSM and HVPG increased statistically significantly ($P = 0.012$) from pre- (Spearman's ρ: 0.53) to post-treatment (ρ: 0.68), while that of PLT remained unchanged (ρ: −0.51 vs. −0.54; $P = 0.613$). The increase in strength of correlation may be attributed to the treatment-induced decrease in PH severity (HVPG: 13.1 ± 4.4 vs. 10.4 ± 4.7 mmHg)/CSPH prevalence (80% vs. 54%), as LSM loses its correlation with HVPG at high values [6]. Accordingly, our data refute the paradigm, that LSM/PLT are less accurate in patients with SVR. However, when evaluating the relationship of pre-/post-treatment LSM/PLT with HVPG using a prediction model derived in viremic patients would result in an overestimation of HVPG by post-treatment LSM/PLT. Accordingly, models and decision rules have to be adopted for the use in patients who achieved HCV-cure. A non-linear model based on post-treatment LSM/PLT developed for this purpose yielded an AUROC of 0.89 for post-treatment CSPH and nomograms derived from this model allow the estimation of the exact probability of CSPH in a given patient. Post-treatment LSM <12 kPa and PLT >150 G/L showed a sensitivity of 99.2% for CSPH, and thus, can be used to exclude CSPH. The Baveno VI criteria also yielded a high sensitivity for post-treatment CSPH (94.7%), with a CSPH prevalence of only 14% in patients meeting these criteria. Accordingly, these patients are reasonably unlikely to benefit from NSBB therapy (see discussion on RRR/ARR in the section on HBV-suppression). Finally, LSM ≥25 kPa was highly specific (93.1%), with a post-treatment CSPH prevalence of 87.7%, i.e., CSPH can be ruled-in in these patients, arguing for maintaining carvedilol/NSBB therapy.

Since the LSM <12 kPa and PLT >150 G/L criterion may be applied to identify patients without post-treatment CSPH who are candidates for being discharged from PH surveillance, this decision rule has also been evaluated in the context of direct clinical outcomes (hepatic decompensation) in a large multicenter cohort of cACLD patients (i.e., merging the derivation and validation cohorts from Semmler et al. [17]): No hepatic decompensation occurred in patients meeting these criteria, if HCC development was considered as a competing event. Accordingly, there is no room for preventive strategies (no risk—no risk reduction achievable) in terms of hepatic decompensation in cACLD patients meeting these criteria, which also argues for their discharge from surveillance (no NIT or endoscopy), if improvements in NIT are consistent (i.e., persist throughout repeated assessments) and no co-factors are present. Importantly, in an unpublished cohort of 1972 unselected cACLD patients with SVR, 37.9% had post-treatment LSM <12 kPa and PLT >150G/L, indicating that a high proportion of patients are candidates for the proposed de-escalation in care, resulting in a considerable reduction in resource utilization.

Screening for Varices and Risk Stratification After HCV-Eradication

In 891 patients with biopsy-proven HCV- (81%) and/or HBV-induced (16.6%) compensated cirrhosis, favorable "Baveno VI" status after SVR/HBV-suppression was associated with the absence of PH progression (as defined by growth of varices or PH-related bleeding) during FU (0% vs. 8.1% at 5 years); in particular, no patient (0/80) with favorable "Baveno VI" status after virological response had large varices and even small varices were uncommon (7.5% (6/80)) in these patients. In contrast, lack of a virological response and the presence of small EV (which indicate CSPH) pre-treatment were independently associated with progression of PH. However, at the time of PH progression, all patients had unfavorable "Baveno VI" status. Finally, "Baveno VI" status was also predictive of survival. This prospective study convincingly demonstrated that LSM/PLT-based "Baveno VI" criteria are useful for ruling out VNT and risk stratification in patients with viral eradication/suppression.

Regarding the risk stratification for clinical events after HCV-cure, it is important to note that HCC may develop independently of (the dynamics of) PH, indicating that hepatic decompensation and HCC should be evaluated separately, rather than using a composite endpoint (e.g., liver-related events). The consideration of this concept is key to add more granularity to investigations on risk stratification approaches, as the clinical implications of low PH-related/HCC risk are drastically different: While patients at negligibly low risk for PH-related events may not benefit from further surveillance by NIT/endoscopy or preventive measures such as carvedilol/NSBB therapy, HCC surveillance (i.e., liver ultrasound) may not be cost-effective in those at very low risk for HCC.

Several studies support the use of other NIT for stratifying hepatic decompensation risk after SVR; however, in the context of predictive modelling/risk stratification, validation in a separate cohort is essential. In this context, the study by Semmler et al. [17] is very valuable, as it comprised a validation cohort. An algorithm combining post-treatment LSM and von Willebrand factor (VWF)/PLT ratio (VITRO) was capable of ruling-in or -out post-treatment CSPH and discriminating between populations at negligible vs. high risk for hepatic decompensation. Importantly, a majority (57.3%) of patients were assigned to the low-risk group (LSM <12.4 kPa and/or VITRO <0.95), and thus, do not require PH surveillance. Moreover, NIT should be simple and broadly available. Although VWF/VITRO fulfills the above-mentioned criteria (simple/broadly available) and provides accurate information regarding PH severity/risk [58], most hepatologists may not be familiar with its use, which was the main reason for relying on LSM/PLT and the results of the pooled analysis (see above) for the Baveno VII recommendations.

Implications for Patient Management After HCV-Eradication

The implications of the above-mentioned evidence for patient management after HCV-cure (i.e., the resulting Baveno VII recommendations) are summarized in Fig. 20.1. In conclusion, NIT provide the opportunity to re-stratify risk in cACLD patients who achieved SVR, and thus, facilitate personalized surveillance and treatment. HCV-cure—and to some extent, HBV-suppression (Baveno VI criteria only)—may serve as models for future investigations regarding the application of NIT/patient management after the removal of other primary etiological factors.

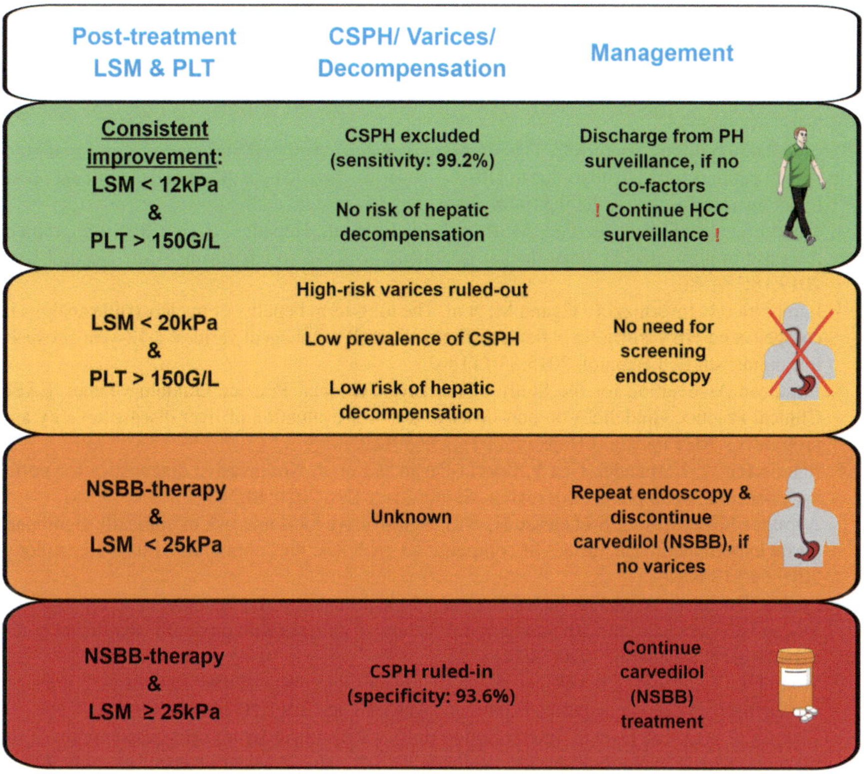

Fig. 20.1 Baveno VII recommendations for PH management after HCV-cure. LSM liver stiffness measurement, CSPH clinically significant portal hypertension, HCC hepatocellular carcinoma, NSBB non-selective betablocker, PH portal hypertension, PLT platelet count

Acknowledgements We would like to acknowledge the contribution of the Clinical Epidemiology and EBM Unit, Beijing Friendship Hospital, Capital Medical University; National Clinical Research Center of Digestive Diseases, Beijing, China (KONG Yuanyuan, Ph.D.; LI Min, Ph.D.; WU Shanshan, Ph.D.; WEI Wei, M.P.H.; ZHANG Qian, Ph.D.; MENG Tongtong, B.Sc.) to the meta-analysis on HBV virus suppression on the course of ACLD. In this context, we would also like to thank the staff of the Liver Research Center, Beijing Friendship Hospital, Capital Medical University; Beijing Key Laboratory of Translational Medicine on Liver Cirrhosis; National Clinical Research Center of Digestive Diseases, Beijing, China (LV Tingting, M.D.; ZHAO Lianghui, M.D.; PEI Yufeng, M.D.; CHEN Sha, M.D.; YOU Hong, M.D. Ph.D.).

Moreover, we would like to express our gratitude to the authors of Reiberger et al. [32], Lens et al. [51], Mandorfer and Kozbial [33]/Schwabl and Mandorfer [52]/Mandorfer et al. [25], Lens [53]/Lens et al. [8], Puente et al. [54], Mauro et al. [55], Abadia et al. [56], and Diez et al. [57] for contributing their individual patient data on the evolution of PH and NIT after HCV-cure and to SEMMLER Georg, M.D. for data analysis and designing Fig. 20.1.

References

1. Marcellin P, Gane E, Buti M, et al. Regression of cirrhosis during treatment with tenofovir disoproxil fumarate for chronic hepatitis B: a 5-year open-label follow-up study. Lancet. 2013;381:468–75.
2. Manolakopoulos S, Triantos C, Theodoropoulos J, et al. Antiviral therapy reduces portal pressure in patients with cirrhosis due to HBeAg-negative chronic hepatitis B and significant portal hypertension. J Hepatol. 2009;51:468–74.
3. Mandorfer M, Hernández-Gea V, Reiberger T, et al. Hepatic venous pressure gradient response in non-selective beta-blocker treatment—is it worth measuring? Curr Hepatol Rep. 2019;18:174–86.
4. Lampertico P, Invernizzi F, Vigano M, et al. The long-term benefits of nucleos(t)ide analogs in compensated HBV cirrhotic patients with no or small esophageal varices: a 12-year prospective cohort study. J Hepatol. 2015;63:1118–25.
5. European Association for the Study of the Liver; Clinical Practice Guideline Panel. EASL Clinical Practice Guidelines on non-invasive tests for evaluation of liver disease severity and prognosis - 2021 update. J Hepatol. 2021;75:659–89.
6. Mandorfer M, Hernández-Gea V, García-Pagán JC, et al. Noninvasive diagnostics for portal hypertension: a comprehensive review. Semin Liver Dis. 2020;40:240–55.
7. Abraldes JG, Bureau C, Stefanescu H, et al. Noninvasive tools and risk of clinically significant portal hypertension and varices in compensated cirrhosis: the "anticipate" study. Hepatology. 2016;64:2173–84.
8. Lens S, Baiges A, Alvarado-Tapias E, et al. Clinical outcome and hemodynamic changes following HCV eradication with oral antiviral therapy in patients with clinically significant portal hypertension. J Hepatol. 2020;73:1415–24.
9. Wong GL, Wong VW, Choi PC, et al. On-treatment monitoring of liver fibrosis with transient elastography in chronic hepatitis B patients. Antivir Ther. 2011;16:165–72.
10. Liang X, Xie Q, Tan D, et al. Interpretation of liver stiffness measurement-based approach for the monitoring of hepatitis B patients with antiviral therapy: a 2-year prospective study. J Viral Hepat. 2018;25:296–305.
11. Wu SD, Liu LL, Cheng JL, et al. Longitudinal monitoring of liver fibrosis status by transient elastography in chronic hepatitis B patients during long-term entecavir treatment. Clin Exp Med. 2018;18:433–43.
12. Dong XQ, Wu Z, Li J, et al. Declining in liver stiffness cannot indicate fibrosis regression in patients with chronic hepatitis B: a 78-week prospective study. J Gastroenterol Hepatol. 2019;34:755–63.
13. Kong Y, Sun Y, Zhou J, et al. Early steep decline of liver stiffness predicts histological reversal of fibrosis in chronic hepatitis B patients treated with entecavir. J Viral Hepat. 2019;26:576–85.
14. Ji D, Chen Y, Shang Q, et al. Unreliable estimation of fibrosis regression during treatment by liver stiffness measurement in patients with chronic hepatitis B. Am J Gastroenterol. 2021;116:1676–85.
15. Wu S, Kong Y, Piao H, et al. On-treatment changes of liver stiffness at week 26 could predict 2-year clinical outcomes in HBV-related compensated cirrhosis. Liver Int. 2018;38:1045–54.
16. Wu H, Yan S, Wang G, et al. von Willebrand factor as a novel noninvasive predictor of portal hypertension and esophageal varices in hepatitis B patients with cirrhosis. Scand J Gastroenterol. 2015;50:1160–9.
17. Semmler G, Binter T, Kozbial K, et al. Noninvasive risk stratification after HCV eradication in patients with advanced chronic liver disease. Hepatology. 2021;73(4):1275–89.
18. de Franchis R, Baveno VI Faculty. Expanding consensus in portal hypertension: report of the Baveno VI consensus workshop: stratifying risk and individualizing care for portal hypertension. J Hepatol. 2015;63:743–52.

19. Thabut D, Bureau C, Layese R, et al. Validation of Baveno VI criteria for screening and surveillance of esophageal varices in patients with compensated cirrhosis and a sustained response to antiviral therapy. Gastroenterology. 2019;156:997–1009. e5
20. Wang H, Wen B, Chang X, et al. Baveno VI criteria and spleen stiffness measurement rule out high-risk varices in virally suppressed HBV-related cirrhosis. J Hepatol. 2021;74:584–92.
21. Mandorfer M, Peck-Radosavljevic M, Reiberger T. Prevention of progression from small to large varices: are we there yet? An updated meta-analysis. Gut. 2017;66:1347–9.
22. Villanueva C, Albillos A, Genescà J, et al. β blockers to prevent decompensation of cirrhosis in patients with clinically significant portal hypertension (PREDESCI): a randomised, double-blind, placebo-controlled, multicentre trial. Lancet. 2019;393:1597–608.
23. Reiberger T, Rutter K, Ferlitsch A, et al. Portal pressure predicts outcome and safety of antiviral therapy in cirrhotic patients with hepatitis C virus infection. Clin Gastroenterol Hepatol. 2011;9:602–8.
24. Lens S, Rincon D, Garcia-Retortillo M, et al. Association between severe portal hypertension and risk of liver decompensation in patients with hepatitis C, regardless of response to antiviral therapy. Clin Gastroenterol Hepatol. 2015;13(10):1846–53.
25. Mandorfer M, Kozbial K, Schwabl P, et al. Changes in hepatic venous pressure gradient predict hepatic decompensation in patients who achieved sustained virologic response to interferon-free therapy. Hepatology. 2020;71:1023–36.
26. Krassenburg LAP, Zanjir WR, Georgie F, et al. Evaluation of sustained virologic response as a relevant surrogate endpoint for long-term outcomes of hepatitis C virus infection. Clin Infect Dis. 2021;72:780–6.
27. Bruno S, Stroffolini T, Colombo M, et al. Sustained virological response to interferon-alpha is associated with improved outcome in HCV-related cirrhosis: a retrospective study. Hepatology. 2007;45:579–87.
28. Veldt BJ, Heathcote EJ, Wedemeyer H, et al. Sustained virologic response and clinical outcomes in patients with chronic hepatitis C and advanced fibrosis. Ann Intern Med. 2007;147:677–84.
29. Cardoso AC, Moucari R, Figueiredo-Mendes C, et al. Impact of peginterferon and ribavirin therapy on hepatocellular carcinoma: incidence and survival in hepatitis C patients with advanced fibrosis. J Hepatol. 2010;52:652–7.
30. Morgan TR, Ghany MG, Kim HY, et al. Outcome of sustained virological responders with histologically advanced chronic hepatitis C. Hepatology. 2010;52:833–44.
31. Di Marco V, Calvaruso V, Ferraro D, et al. Effects of eradicating hepatitis C virus infection in patients with cirrhosis differ with stage of portal hypertension. Gastroenterology. 2016;151:130–9. e2
32. Reiberger T, Payer BA, Ferlitsch A, et al. A prospective evaluation of pulmonary, systemic and hepatic haemodynamics in HIV-HCV-coinfected patients before and after antiviral therapy with pegylated interferon and ribavirin. Antivir Ther. 2012;17:1327–34.
33. Mandorfer M, Kozbial K, Schwabl P, et al. Sustained virologic response to interferon-free therapies ameliorates HCV-induced portal hypertension. J Hepatol. 2016;65:692–9.
34. Mandorfer M, Kozbial K, Freissmuth C, et al. Interferon-free regimens for chronic hepatitis C overcome the effects of portal hypertension on virological responses. Aliment Pharmacol Ther. 2015;42:707–18.
35. McDonald SA, Pollock KG, Barclay ST, et al. Real-world impact following initiation of interferon-free hepatitis C regimens on liver-related outcomes and all-cause mortality among patients with compensated cirrhosis. J Viral Hepat. 2020;27:270–80.
36. Carrat F, Fontaine H, Dorival C, et al. Clinical outcomes in patients with chronic hepatitis C after direct-acting antiviral treatment: a prospective cohort study. Lancet. 2019;393:1453–64.
37. Moon AM, Green PK, Rockey DC, et al. Hepatitis C eradication with direct-acting anti-virals reduces the risk of variceal bleeding. Aliment Pharmacol Ther. 2020;51:364–73.
38. Pons M, Rodriguez-Tajes S, Esteban JI, et al. Non-invasive prediction of liver-related events in patients with HCV-associated compensated advanced chronic liver disease after oral antivirals. J Hepatol. 2020;72:472–80.

39. Morisco F, Federico A, Marignani M, et al. Risk factors for liver decompensation and HCC in HCV-cirrhotic patients after DAAs: a multicenter prospective study. Cancers (Basel). 2021;13:3810.
40. Tosetti G, Degasperi E, Farina E, et al. Decompensation in direct-acting antiviral cured hepatitis C virus compensated patients with clinically significant portal hypertension: too rare to warrant universal Beta-blocker therapy. Am J Gastroenterol. 2021;116:1342–4.
41. Mandorfer M, Turon F, Lens S, et al. Risk of non-tumoural portal vein thrombosis in patients with HCV-induced cirrhosis after sustained virological response. Liver Int. 2021;41(12):2954–64.
42. Puigvehi M, Londono MC, Torras X, et al. Impact of sustained virological response with DAAs on gastroesophageal varices and Baveno criteria in HCV-cirrhotic patients. J Gastroenterol. 2020;55:205–16.
43. Giannini EG, De Maria C, Crespi M, et al. Course of oesophageal varices and performance of noninvasive predictors following hepatitis C virus clearance in compensated advanced chronic liver disease. Eur J Clin Investig. 2020;50:e13231.
44. Krassenburg LAP, Maan R, Ramji A, et al. Clinical outcomes following DAA therapy in patients with HCV-related cirrhosis depend on disease severity. J Hepatol. 2021;74:1053–63.
45. Alavi M, Law MG, Valerio H, et al. Declining hepatitis C virus-related liver disease burden in the direct-acting antiviral therapy era in New South Wales, Australia. J Hepatol. 2019;71:281–8.
46. Backus LI, Belperio PS, Shahoumian TA, et al. Impact of sustained virologic response with direct-acting antiviral treatment on mortality in patients with advanced liver disease. Hepatology. 2019;69:487–97.
47. Lawitz EJ, Shevell DE, Tirucherai GS, et al. BMS-986263 in patients with advanced hepatic fibrosis: 36-week results from a randomized, placebo-controlled phase 2 trial. Hepatology. 2022;75(4):912–23.
48. Rincon D, Ripoll C, Lo Iacono O, et al. Antiviral therapy decreases hepatic venous pressure gradient in patients with chronic hepatitis C and advanced fibrosis. Am J Gastroenterol. 2006;101:2269–74.
49. Roberts S, Gordon A, McLean C, et al. Effect of sustained viral response on hepatic venous pressure gradient in hepatitis C-related cirrhosis. Clin Gastroenterol Hepatol. 2007;5:932–7.
50. Afdhal N, Everson GT, Calleja JL, et al. Effect of viral suppression on hepatic venous pressure gradient in hepatitis C with cirrhosis and portal hypertension. J Viral Hepat. 2017;24:823–31.
51. Lens S, Rincón D, García-Retortillo M, et al. Association between severe portal hypertension and risk of liver decompensation in patients with hepatitis C, regardless of response to antiviral therapy. Clin Gastroenterol Hepatol. 2015;13:1846–53.
52. Schwabl P, Mandorfer M, Steiner S, et al. Interferon-free regimens improve portal hypertension and histological necroinflammation in HIV/HCV patients with advanced liver disease. Aliment Pharmacol Ther. 2017;45:139–49.
53. Lens S, Alvarado-Tapias E, Marino Z, et al. Effects of all-oral anti-viral therapy on HVPG and systemic hemodynamics in patients with hepatitis C virus-associated cirrhosis. Gastroenterology. 2017;153:1273–83.
54. Puente Á, Cabezas J, López Arias MJ, et al. Influence of sustained viral response on the regression of fibrosis and portal hypertension in cirrhotic HCV patients treated with antiviral triple therapy. Rev Esp Enferm Dig. 2017;109:17–25.
55. Mauro E, Crespo G, Montironi C, et al. Portal pressure and liver stiffness measurements in the prediction of fibrosis regression after sustained virological response in recurrent hepatitis C. Hepatology. 2018;67:1683–94.
56. Abadía M, Montes ML, Ponce D, et al. Management of betablocked patients after sustained virological response in hepatitis C cirrhosis. World J Gastroenterol. 2019;25:2665–74.
57. Díez C, Berenguer J, Ibañez-Samaniego L, et al. Persistence of clinically significant portal hypertension after eradication of hepatitis C virus in patients with advanced cirrhosis. Clin Infect Dis. 2020;71:2726–9.
58. Mandorfer M, Simbrunner B. Prevention of first decompensation in advanced chronic liver disease. Clin Liver Dis. 2021;25:291–310.

Management of ACLD After Removal/ Suppression of the Etiological Factor: Consensus Statements of Panel 3

21

Mattias Mandorfer, Emmanouil A. Tsochatzis, Sven Francque, Jidong Jia, Sabela Lens, and Hitoshi Yoshiji

M. Mandorfer (✉)
Vienna Hepatic Hemodynamic Lab, Division of Gastroenterology and Hepatology, Department of Internal Medicine III, Medical University of Vienna, Vienna, Austria
e-mail: mattias.mandorfer@meduniwien.ac.at

E. A. Tsochatzis
UCL Institute for Liver and Digestive Health and Sheila Sherlock Liver Unit, Royal Free Hospital and UCL, London, UK
e-mail: e.tsochatzis@ucl.ac.uk

S. Francque
Department of Gastroenterology and Hepatology, Antwerp University Hospital, Antwerp, Belgium

Translational Science in Inflammation and Immunology, Faculty of Medicine and Health Sciences, University of Antwerp, Antwerp, Belgium
e-mail: Sven.Francque@uza.be

J. Jia
Beijing Key Laboratory of Translational Medicine on Liver Cirrhosis, National Clinical Research Center of Digestive Diseases, Liver Research Center, Beijing Friendship Hospital, Capital Medical University, Beijing, China
e-mail: jia_jd@ccmu.edu.cn

S. Lens
Liver Unit, Hospital Clínic de Barcelona, Universidad de Barcelona, IDIBAPS, Barcelona, Spain

CIBERehd, Madrid, Spain
e-mail: slens@clinic.cat

H. Yoshiji
Third Department of Internal Medicine (Department of Gastroenterology, Endocrinology and Metabolism), Nara Medical University, Nara, Japan
e-mail: yoshijih@naramed-u.ac.jp

R. de Franchis (ed.), *Portal Hypertension VII*,
https://doi.org/10.1007/978-3-031-08552-9_21

3.1 Removal/suppression of the primary etiological factor includes sustained virological response (SVR) in patients with hepatitis C virus (HCV) infection, hepatitis B virus (HBV) suppression in the absence of hepatitis D virus (HDV) coinfection in patients with chronic HBV infection, and long-term abstinence from alcohol in patients with alcohol-related liver disease (ALD). (A1) (New)
3.2 The definition and impact of the removal/suppression of the primary etiological factor in other ACLDs are less well established. (A1) (New)
3.3 Overweight/obesity, diabetes, and alcohol consumption are the important contributors to liver disease progression even after removal/suppression of the primary etiological factor and should be addressed. (A1) (Changed)
3.4 Removal/suppression of the primary etiological factor leads to potentially meaningful decreases in HVPG in the majority of patients and substantially reduces the risk of hepatic decompensation. (A1) (Changed)
3.5 Absence/resolution of CSPH following removal/suppression of the primary etiological factor prevents hepatic decompensation. (B1) (Changed)
3.6 The optimal percent/absolute decrease in HVPG associated with a reduction in hepatic decompensation following the removal/suppression of the primary etiological factor in cACLD patients with CSPH has yet to be established. (B1) (New)
3.7 In the absence of cofactors, patients with HCV-induced cACLD who achieve SVR and show consistent post-treatment improvements with LSM values of <12 kPa and PLT >150 × 10^9/L can be discharged from portal hypertension surveillance (LSM and endoscopy), as they do not have CSPH and are at negligible risk of hepatic decompensation. In these patients, hepatocellular carcinoma surveillance should continue until further data are available. (B1) (New)
3.8 The Baveno VI criteria (i.e., LSM <20 kPa and PLT >150 × 10^9/L) can be used to rule out high-risk varices in patients with HCV- and HBV-induced cACLD who achieved SVR and viral suppression, respectively. (B1) (New)
3.9 cACLD patients on NSBB therapy with no evident CSPH (LSM <25 kPa) after removal/suppression of the primary etiological factor should be considered for repeat endoscopy, preferably after 1–2 years. In the absence of varices, NSBB therapy can be discontinued. (C2) (New)

Research Agenda

- Establish the definition and impact of the removal/suppression of the primary etiological factor in cACLD other than HCV and HBV infection and ALD, particularly in non-alcoholic fatty liver disease (NAFLD).
- Identification of factors responsible for liver disease progression despite removal/suppression of the primary etiological factor.
- Establish the optimal percent/absolute decrease in HVPG associated with a reduction in hepatic decompensation following the removal/suppression of the primary etiological factor in cACLD patients with CSPH.

- Diagnostic ability of NIT in monitoring disease regression and determining the presence of CSPH after removal/suppression of a non-viral primary etiological factor.
- Evaluation and validation of other noninvasive risk stratification algorithms (e.g., LSM/VITRO and SSM) in patients in whom the primary etiological factor has been removed/suppressed.
- Estimates of the regression of varices after removal/suppression of the primary etiological factor and long-term data on the risk of hepatic decompensation (and more specifically, variceal bleeding) and its evolution over time in patients with cACLD.

Part VI

New Scenarios 3: Impact of Non-etiological Novel Therapies in the Course of Cirrhosis

Results of the Baveno VII Questionnaire on the "Impact of Non-etiological Therapies in the Course of Cirrhosis"

22

Agustín Albillos and Jonel Trebicka

The questionnaire of our panel explored the use by the experts in clinical practice of different non-etiological therapies of cirrhosis (i.e., statins, anticoagulants) as well as the experts' opinions regarding the value of interventions on the gut-liver axis. A questionnaire was sent to all Baveno expert faculty ($n = 64$) and was completely filled in by 53 (83%) and its main results are presented in the following paragraphs.

Use of Statins

Do you routinely use statins in patients with advanced chronic liver disease (aCLD)?

The majority of the faculty (83%) use statins in patients with cirrhosis only if there is another indication within the label. 95% use statins routinely to halt or decrease liver fibrosis (11%), and reduce the risk of hepatocellular carcinoma (4%) or NAFLD cirrhosis (5%). Only 4% never use statins in patients with cirrhosis.

Do you use statins in patients with liver cirrhosis FOR the primary variceal bleeding prophylaxis setting?

A. Albillos
Department of Gastroenterology and Hepatology, Hospital Universitario Ramón y Cajal, Instituto Ramón y Cajal de Investigación Sanitaria (IRYCIS), Centro de Investigación Biomédica Red de Enfermedades Hepáticas y Digestivas (CIBEREHD), Universidad de Alcalá, Madrid, Spain
e-mail: agustin.albillos@uah.es

J. Trebicka (✉)
Department of Internal Medicine B, University of Münster, Münster, Germany

European Foundation for Study of Chronic Liver Failure, Barcelona, Spain
e-mail: jonel.trebicka@ukmuenster.de

R. de Franchis (ed.), *Portal Hypertension VII*,
https://doi.org/10.1007/978-3-031-08552-9_22

85% of the faculty never use statins for the primary prophylaxis of variceal bleeding. 11% of the faculty use statins for this indication, except in Child C patients, due to the broad beneficial effects of these agents in cirrhosis.

Do you use statins in patients with liver cirrhosis FOR the secondary variceal bleeding prophylaxis setting?

Most (72%) of the faculty never use statins to prevent rebleeding. 28% use statins in this setting, except in Child C patients.

If you use or would use statins in patients with aCLD or for bleeding prophylaxis, which statins and dose would you use?

Simvastatin was the statin used by the majority (88%) of the faculty. Among the responders who use simvastatin, half use it at a dose of 20 mg/d in all patients (44%) whereas the other half (44%) do it at a dose of 40 mg/d in Child A and of 20 mg/d in Child B/C (44%). Ten percent use atorvastatin instead of simvastatin at doses of 20 to 40 mg/d.

Which adverse events have you experienced during statins treatment?

Myopathy was the most frequent adverse event reported (88%), followed by DILI (33%).

Taken together these answers, most of the faculty experts only use statins in patients with cirrhosis when there is an approved indication for their use. However, experts agree that the use of these agents could be of benefit to halt fibrosis and cirrhosis progression. Simvastatin is the statin mostly used at a dose of 20 mg/d.

Use of Anticoagulants

Have you ever used anticoagulation in patients with cirrhosis to prevent decompensation?

92% of the faculty answered negatively to the question.

When do you use anticoagulants for prophylaxis of PVT in cirrhosis?

The answers of the faculty were almost unanimously against the prophylactic use of anticoagulants (96%), but 6% used them in patients with reduced portal vein velocity.

In patients with cirrhosis and newly diagnosed occlusive thrombosis of the main trunk of the portal vein, when do you use anticoagulants?

72% of the faculty answered that they always use anticoagulants in this situation, 26% only if the patient is on the liver transplantation waiting list, and 17% if the patient has symptoms or SMV extension. None answered that they ever use anticoagulation.

In patients with cirrhosis and newly diagnosed non-occlusive (<50%) thrombosis of the main trunk of the portal vein, when do you use anticoagulants?

The use of anticoagulation for non-occlusive PVT was not as extensive since diagnosis as in the case of occlusive PVT: 39% of the faculty limit anticoagulation to patients on the liver transplantation waiting list, 17% if persistency or progression of the thrombosis at a 3-month Doppler-US follow-up, and 11% in cases with symptoms or SMV extension. 33% would always use anticoagulation in this situation.

In patients with cirrhosis and newly diagnosed thrombosis of one of the branches of the portal vein, when do you use anticoagulants:

The proportions of responders using anticoagulation in this situation were rather similar to those in the case of partial PVT: 33% always, 39% in patients in the liver transplantation waiting list and 20% after assessing persistency or progression at 3-month follow-up. Of note, 11% never use anticoagulation in this situation.

Which would be your choice in case of incomplete clot resolution after 6 months of therapeutic anticoagulation for PVT?

Most (72%) of the faculty would extend the duration of anticoagulation for at least 3 months, but 37% would place TIPS in case of persistence. 13% would have just withdrawn anticoagulation and 6%switch to a different class of anticoagulants.

Which are the anticoagulants you use to treat PVT in cirrhosis?

Most of the faculty use LMWH followed by VKA (57%) or DOACs (39%). 15%, 11%, and 9% of the responder only use LMWH, warfarin, or DOACs, respectively.

Which is your standard of care of primary or secondary prophylaxis of variceal bleeding in patients with PVT on therapeutic anticoagulation?

A majority of the faculty (88%) initiate or continues variceal bleeding prophylaxis according to the current guidelines in patients undergoing anticoagulation, being NSBB the most frequently used prophylaxis. In the case of secondary prophylaxis, 20% of the responders initiate anticoagulation only after variceal eradication.

Summarizing the above-mentioned results, most of the faculty limit the use of anticoagulants in cirrhosis to patients with portal vein thrombosis, with less than 10% using them to modify cirrhosis progression or to prevent PVT. Most of the experts consider anticoagulation in all cases with occlusive (>50%) portal vein thrombosis, whereas they restrict its use in cases of partial portal vein thrombosis to patients in the LT waiting list or persistency or progression at 3-month follow-up. LMWH for 3–6 months followed by VKA or DOACs is the most frequent schedule of anticoagulation used.

Addressing the Gut-Liver Axis

Do you use RIFAXIMIN in your clinical practice in hepatology?

Rifaximin was widely used by the faculty for the prevention of HE recurrence (100%), treatment of acute (41%), or minimal (43%) HE. Nine percent also use it for SBP prevention.

Do you use PROBIOTICS in your clinical practice in hepatology?

Probiotics (VSL-3, Lactobacillus Casei) were used by 6.5% of the faculty.

Do you think microbiota composition could be useful in the near future for compensated cirrhosis screening?

The answers provided are summarized below. Percentages indicate the % of respondents who selected the answer.

No, it is difficult to incorporate into practice	22%
No, microbiota can be affected by aspects unrelated to the disease process	15%
No, there are other accurate methods to do that	22%
Yes, it has potential for this purpose, but further research is needed	41%

Do you think microbiota composition could be useful in the near future for surveillance of patients with cirrhosis at risk of decompensation?

No, it is difficult to incorporate into practice	22%
No, microbiota can be affected by aspects unrelated to the disease process	15%
No, there are other accurate methods to do that	12%
Yes, it has potential for this purpose, but further research is needed	51%

What will be needed to incorporate microbiota analysis into clinical practice of advanced liver disease?

Standardization of collections and analyses	85%
Validity across populations from several countries	80%
Roles of etiologies	72%
Independent addition to the clinical biomarkers	63%
Others: Research tool, absence of evidence	13%

Which do you think will be the role of fecal microbiota transplantation in clinical practice in hepatology?

None	26%
Management of recurrent hepatic encephalopathy	70%
Management of alcoholic hepatitis	28%
Management of NAFLD/NASH	20%
Management of PSC	9%
Management of ACLF	24%

Taken together, these responses indicate that half of the faculty agree that microbiota composition analysis has the potential in the long term to identify patients with cirrhosis or at risk of decompensation, but much more research is still needed. Interestingly, almost the other half of the faculty was skeptical about the use of this tool in clinical practice in the future. Most of the responders were more optimistic regarding the therapeutical use of *fecal microbiota transplantation*, especially in patients with recurrent hepatic encephalopathy.

Statins in Compensated and Decompensated Cirrhosis: Approaching the Bedside

23

Jonel Trebicka

Abbreviations

BDL	Bile duct ligation
CRP	C-reactive protein
CVD	Cardiovascular disease
ECM	Extracellular matrix
eNOS	Endothelial nitric oxide synthase
FPP	Farnesyl-pyrophosphate
GGPP	Geranylgeranyl-pyrophosphate
HCC	Hepatocarcinoma
HCV	Viral hepatitis C
HMG-CoA	3-Hydroxy-3-methyl-glutaryl-coenzyme A
HSC	Hepatic stellate cell
KLF2	Kruppel-like factor 2
MS	Metabolic syndrome
NAFLD	Non-alcoholic fatty liver disease
NASH	Non-alcoholic steatohepatitis
NOX2	Nicotinamide adenine dinucleotide phosphate oxidase isoform 2
PNPLA3	Patatin-like phospholipase domain-containing protein 3
PPARα	Activated peroxisome proliferator-activated receptor alpha
SREBP	Sterol regulatory element binding proteins
UDCA	Ursodeoxycholic acid

J. Trebicka (✉)
Department of Internal Medicine B, University of Münster, Münster, Germany

European Foundation for the Study of Chronic Liver Failure—EF Clif, Barcelona, Spain
e-mail: jonel.trebicka@ukmuenster.de

R. de Franchis (ed.), *Portal Hypertension VII*,
https://doi.org/10.1007/978-3-031-08552-9_23

General Underlying Mechanisms of Statins

Statins are a class of competitive inhibitors of 3-hydroxy-3-methyl-glutaryl--coenzyme A (HMG-CoA) reductase, the rate-limiting enzyme for endogenous cholesterol, and isoprenoid synthesis in the mevalonate pathway. Its inhibition leads to decreased production of precursors and subsequently reduction of cholesterol biosynthesis [1] (Fig. 23.1).

The so-called pleiotropic effects are mediated by the reduction of isoprenoids, such as Farnesyl-pyrophosphate (FPP) and Geranylgeranyl-pyrophosphate (GGPP), necessary for the activation of small GTPases like Rho- and Ras-proteins (Fig. 23.1). Statins diversely inhibit RhoA and Rac1 prenylation, modulating endothelial Nitric Oxide Synthase (eNOS), NO availability and enhancing stability of eNOS mRNA [2] (Fig. 23.1). This statin-dependent eNOS restoration can be also mediated by increased Krüppel-like factor 2 (KLF2) expression [3]. Similar effects were observed in hepatic injury followed by fibrosis and cirrhosis. Moreover, statins decrease oxidative stress in the liver and enhance eNOS expression and activity by inhibiting RhoA membrane association [2, 4]. Taken together, inhibition of GTPase prenylation, restoration of eNOS, and NO availability are the key roles of statins in the improvement of vascular and endothelial function (Fig. 23.1).

But also, the decreased cholesterol synthesis itself is beneficial several-fold. Decreased cholesterol level leads to an increase of Sterol Regulatory Element-Binding Proteins (SREBP), which act as transcription factors for the LDL receptor

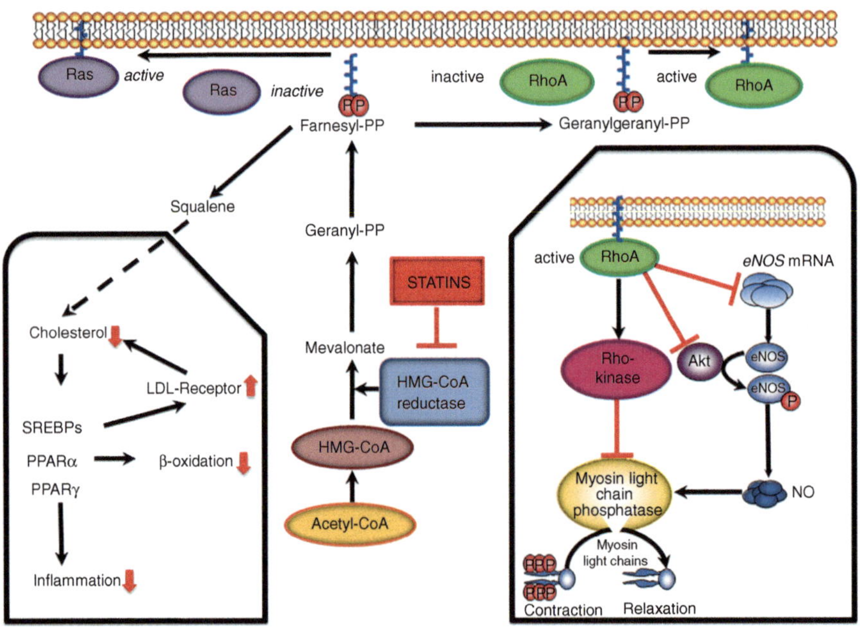

Fig. 23.1 Pleiotropic effects of statins

that induces higher plasma LDL clearance due to an increased LDL receptor-mediated uptake and after lysosomal degradation of LDL [5]. Also, the decreased hepatic triglyceride synthesis is possibly associated with activated peroxisome proliferator-activated receptor α (PPARα) and increased β-oxidation activity [5] (Fig. 23.1). Moreover, statins seem to beneficially modify PPARγ activity, which attenuates the production of tumor necrosis factor α (TNFα), interleukins 1-beta (IL1β) and 6 (IL6) and C-reactive protein (CRP) [6, 7].

In summary, due to various mechanisms, dependent or not on the cholesterol levels, statins modify pathological conditions as outlined in the following section.

Statins in Cardiovascular Diseases and Interaction with Liver Disease

Statins also decrease Low-Density Lipoprotein (LDL) cholesterol levels, which are pro-atherogenic. For this reason, statins have become the standard of care in the treatment of diabetes and Cardiovascular Diseases (CVD) to decrease or even reverse atherosclerosis. However, liver diseases, they are often underused even in high-risk patients [8]. Especially in CVD, several studies have shown a clear effect of statins on inflammation. Besides the Pravastatin Inflammation/CRP Evaluation (PRINCE) study, also another study demonstrated that statins improve inflammation and decrease interleukin-6 (IL-6) levels, the main regulator of CRP [6]. Furthermore, statins directly inhibit the expression of major histocompatibility complex class II molecules by interferon g (IFN-g) in CD4+ helper T cells (TH1 cells), leading to a shift toward anti-inflammatory TH2 cell actions and beneficially modifying atherosclerosis [9]. This anti-inflammatory effect is extremely important for the liver disease since especially in the last years' systemic inflammation has been identified as a marker of disease progression [10] and is persistent besides portal hypertension [11], with strong effects on other organs leading to dysfunction [12, 13]. This has also been recently demonstrated by the PREDICT study, in which portal hypertension and systemic inflammation are the two main mechanisms leading to acute decompensation of liver cirrhosis [14].

Moreover, as outlined above statins in addition improve eNOS and NO in the endothelium, which decreases leukocytes' chemotaxis, adhesion, and inflammation, key mechanisms aggravating atherosclerosis [9]. The decreased infiltration of plaques by macrophages and downregulation of proteolytic enzymes are associated also with decreasing NADPH oxidase isoform 2 (NOX2) [15]. While statins might inhibit the immune cells activity, they increase the number of circulating endothelial progenitor cells, which are important in the neovascularization of ischemic tissue, thereby contributing to the restoration of endothelial function [16]. Endothelial function is extremely important and differently regulated in portal hypertension and cirrhosis not only in the liver but also outside [17]. Endothelial function in cirrhosis with portal hypertension is impaired, promoting vasoconstriction in the liver and sustained vasodilation in the splanchnic region [17].

Furthermore, statins elicit anti-thrombotic effects by down-regulating platelet CD40L, by inhibiting tissue factor activity and thrombin generation [5, 9]. A growing number of studies suggest the importance of Intrahepatic microvascular thrombosis for fibrosis progression and portal hypertension. This observation connecting thrombosis, liver cirrhosis, and portal hypertension was first described as "parenchymal extinction" in pathological specimens of human liver cirrhosis [18]. While the development and progression of splanchnic venous thrombotic events implicate disease progression in cirrhosis, their pathogenesis remains unclear. According to Virchow's triad, coagulation/platelets and vascular wall are the drivers of thrombosis. Even in patients with TIPS, when the flow is restored and, to a large extent, also the shear stress due to portal hypertension, the prothrombotic milieu is increased, probably due to platelet activation [19].

Recent data demonstrate that statins have an important role in the microbiota and their use may be associated with less gut dysbiosis [20]. It is known that microbiota influences progression of the different diseases. In addition, it may also be a driver of the development of portal hypertension and decompensation of liver cirrhosis [21, 22]. The translocated bacteria or bacterial components drive systemic inflammation and potentially also thrombosis and thereby may aggravate the progression of liver disease and development of complications [19, 23].

Adverse Effects and Hepatotoxicity

Although similar mechanisms as in CVD are involved in the development and progression of liver disease, caution is required due to hepatotoxicity. Hepatic cells make a considerable contribution to cholesterol production and therefore are a major target of statins. The pharmacological activity and the hepatic metabolism of statins depend on their molecular structure and physical properties such as lipophilicity, solubility, and absorption. Simvastatin, lovastatin, fluvastatin, and atorvastatin are metabolized by cytochrome P450, while pravastatin, rosuvastatin, and pitavastatin remain almost unaffected by any hepatic metabolic processes.

The effect of statins on aminotransferase levels in the treatment of cardiovascular diseases was investigated in several studies with contradictory results. This inconsistency may be explained by pharmacogenetics and differences in the statins used. Although statins are generally well-tolerated, reports about statin-induced liver injury can be found mainly for atorvastatin and simvastatin. However, this might be coincidental since these two statins are also the two most commonly prescribed ones [24].

The question of whether statins have a hepatotoxic effect is considerably more relevant in patients with acute or chronic liver dysfunction. In a retrospective cohort study, lovastatin showed no increased risk of adverse hepatic effects in a total of 93,106 patients with liver disease. Another prospective randomized, double-blind, placebo-controlled multicenter trial, investigated the safety of high-dose pravastatin in chronic liver disease. After 36 weeks of treatment, alanine aminotransferase (ALT) levels were even lower in the pravastatin-treated group [25]. Furthermore,

HMG-CoA reductase inhibitors were also found to be safe in patients after liver transplantation [26].

Few studies revealed that statin hepatotoxicity is a rare condition and might mimic an autoimmune phenotype of liver injury [27, 28]. However, in patients with chronic kidney diseases, the incidence of severe adverse events seems to be higher [29]. However, in patients with decompensated liver cirrhosis simvastatin seems to elicit rhabdomyolysis and hepatotoxic effects at the dose of 40 mg daily [30]. Myopathy, and less common rhabdomyolysis, are known adverse effects of statin and are rare in normal circumstances—about 2–3 cases/year per 100,000 patients treated [31]. Again, in another cirrhosis trial these adverse events were relatively frequent [27]. The reasons might be related to the dose of statins, genetic predisposition (e.g., SCLO1B1 polymorphism), but also alcoholic etiology of liver disease, being the most prevalent in this study [27]. This was again confirmed in a small uncontrolled Phase IIa study [32].

In a recent meta-analysis on Pharmacokinetics (PK), cardiovascular outcomes, and safety profiles of statins in cirrhosis, the authors conclude that rosuvastatin and pitavastatin showed minimal PK changes, while atorvastatin caused more pronounced PK changes in Child-Pugh A cirrhosis, while no data was available for the most used simvastatin [33]. Yet, simvastatin 40 mg had a pooled frequency for rhabdomyolysis of 2%, and incidence 40-fold higher than that reported in non-cirrhotic patients, while there was no rhabdomyolysis observed in patients on simvastatin 20 mg, atorvastatin 20 mg, or pravastatin 40 mg. In the experience so far published, no overt liver failure was reported. Still, in most cases, the benefits of statins outweigh any potential hepatotoxic risks [34, 35]. Another option unexplored in clinics is the use of novel statin drugs, as suggested by using a compound containing atorvastatin and a NO-donor [36], which significantly reduced myopathy in an animal model.

Distinct and Common Mechanisms of Statins in Liver Diseases

CVD in many patients is associated with Metabolic Syndrome (MS), which is the common ground for the development of NAFLD and NASH. Statins have been considered for treatment in NASH, and in recent years they have been generally evaluated as safe—even at high doses—leading to a wider use in patients [25, 37–39]. Nevertheless, studies assessing the beneficial effects of statins are scarce, mostly investigating only a small number of patients with different endpoints. These studies showed that statin therapy either attenuates inflammation and steatosis [40] or shows a trend toward decreased fibrosis while other studies found no change in fibrosis [40, 41]. The different study outcomes may be due to the different statins used in the respective trials. While atorvastatin at 10 mg/day for 24 months elicited positive effects in NASH, simvastatin 20 mg/day over 12 months had no effect in a similar cohort of patients [41]. Moreover, genetic predispositions were disregarded by most of these studies and may provide additional explanations for the different outcomes regarding fibrosis. For example, a large multicenter study revealed that statin use in high-risk patients is beneficial, except in patients carrying the PNPLA3 I148M risk alleles [42]. Thus, genetic screening may be advisable for NAFLD patients to ascertain the optimal therapy for each patient.

Recent studies revealed improvements in NASH and MS after statin treatment [42, 43]. Statins decrease LDL cholesterol levels in serum, and as a result, oxidized LDL levels play an important role in NASH. As highlighted in Fig. 23.1, statin therapy leads to decreased hepatic steatosis by decreasing LDL and activating SREBPs, PPARα, and β-oxidation [5]. However, the anti-inflammatory effect of statins in NAFLD and NASH is partly attributed to the activation of PPARγ and subsequent downregulation of pro-inflammatory mediators [6, 7]. Additionally, the inhibition of small GTPase prenylation and diminished downstream signaling contribute to the anti-inflammatory features of statins [7]. Recent experimental NASH studies further suggest a beneficial impact of statins in fibrosis. They inhibit the paracrine signaling of hepatocytes on Hepatic Stellate Cells (HSC), thereby inhibiting Hepatic StellateCells (HSC) activation and fibrogenesis during experimental NASH.

Additionally to the hepatic and general metabolic improvement which are known to be tightly linked to chronic viral hepatitis C (HCV) infection, statins might exert a direct anti-replicative effect on HCV [44] Previous small-scale studies investigated the effect of statin monotherapy and revealed only mild antiviral effects [45], while cohort studies could confirm the benefit [46]. Especially in combination with direct-acting antiviral agents statins may have an added value to the antiviral efficacy and mitigate the progression of HCV-related diseases, such as cirrhosis or HCC [47, 48]. Additionally, HCV patients with compensated cirrhosis under statin treatment seem to have a lower risk for decompensation as well as lower mortality [49]. This is a rather decreasing indication and will probably not play a significant role in the future as shown recently [50]; other studies report a highly significant decrease in HCC risk resulting from statin use [51, 52]. The data from the Electronically Retrieved Cohort of HCV Infected Veterans (ERCHIVES) database shows a dose-dependent reduction of fibrosis progression, going along with the decreasing HCC incidence. Remarkably, reduction of HCC incidence was about 47% in all treated patients. Also, this study clearly showed differences in statin efficacy, whereby atorvastatin and fluvastatin had the strongest effects on fibrosis progression, as well as HCC incidence [51]. Nonetheless, it remains uncertain whether these effects are due to fibrosis reduction, direct effects on HCC progression or a combination of both.

Statins decrease not only the risk of development of liver cirrhosis but also may alter the hepatic resistance. Cirrhosis, the common end-stage of chronic liver injury, is characterized by profound liver remodeling and portal hypertension. Portal hypertension in cirrhosis arises by a mechanically increased intrahepatic resistance by narrowing of hepatic outflow. An additional dynamic component of increased intrahepatic resistance is dominated by an imbalance of vascular tone-regulating pathways, showing a shift towards vasocontraction [17]. Furthermore, RhoA and Rho-kinase signaling are responsible for the increased tone of the hepatic vasculature contributing to the activation of HSC, the major contributor to ECM synthesis upon chronic liver injury. Statins modulate the mechanic and the dynamic intrahepatic pathways [17]. Both pathways represent targets of statin therapy in liver cirrhosis with portal hypertension (Fig. 23.1). Atorvastatin inhibits the translocation of RhoA and thus the activity of Rho-kinase. This effect decreases collagen production and hepatic stellate cell activation in early fibrosis as well as proliferation, cytokine production, and contraction of activated hepatic stellate cells in cirrhosis. Importantly, statins might induce senescence in activated HSC leading to a decreased turnover of these highly active cells. Simultaneously, statins improve endothelial dysfunction by the upregulated activity of eNOS and NO availability in cirrhotic livers and further decrease portal pressure [2, 53, 54].

In several studies, acute and chronic effects of statins on portal pressure, complications, or overall outcome of patients with cirrhosis were investigated (Table 23.1). Statins seem to significantly decrease hepatic vascular resistance in cirrhosis with portal hypertension, in addition to the extrahepatic effect of beta-blockers [4, 55]. Another study, so far published only as an abstract confirmed the beneficial effects of statins, even in patients identified as non-responder to non-selective beta-blockers [56]. However, this was not observed in a randomized placebo-controlled trial in the primary prophylaxis setting [57]. Simvastatin may decrease portal pressure by around 10% after only 1 month [4]. The same group intended to show a decrease in rebleeding during the secondary prophylaxis but were unable to demonstrate a lower number of variceal bleeds [27]. However, and most interestingly statins improved overall survival in this study [27]. A meta-analysis summarizing the effect of statins on lowering portal pressure and the related clinical effects defined as the risk of variceal hemorrhage demonstrated a clear overall portal pressure lowering effect, while showing only a tendency for a decreased risk of variceal hemorrhage [58].

Besides decreased portal pressure, cirrhotic patients under statins may also benefit from improved liver function. Interestingly, statin effects are enhanced with increased severity of portal hypertension [59]. In addition, simvastatin improved survival in patients after variceal hemorrhage suggesting multiple beneficial effects of statins. As demonstrated in an animal model [60], the severity of hemorrhage might be also lower in patients receiving statins. Even acute decompensation induced artificially in animal models using LPS could be prevented by the use of statins [61]. This was also retrospectively observed in a large set of patients in the US, presented as abstract so far [62].

Table 23.1 Retrospective Cohort Studies and Randomized Clinical Trials of Statins in patients with cirrhosis and clinical endpoints

Retrospective cohort studies

Study	Patients source	Patients description	Study design	Number of patients	Type of statin	Follow up period	Endpoints	Results	Comments
F. Chang, Hepatology 2017	Taiwan National Health Insurance	Hepatitis B, hepatitis C, and alcohol-related cirrhosis	Retrospective cohort study	1174 statin users vs. 6453 non statin users	NA	Approx. Median of follow up of 3 years	Decompensation	Prevented decompensation aHR 0.39 (0.30–0.50)	Lower risk of ascites, variceal bleeding and hepatic encephalopathy
							Death	Decreased mortality aHR 0.46 (0.34–0.63)	Analysis by etiology in HBV, HCV, and OH cirrhosis. Dose-response relationship
							HCC development	Decreased HCC aHR 0.52 (0.35–0.76)	
Bang, Aliment Pharmacol Ther 2017	Danish National Patient Registry	Alcohol related cirrhosis	Retrospective cohort study	794 statin users vs. 4623 non-statin users	Simvastatin 79%	Approx. Median of follow up of 4 years	Decompensation	Prevented decompensation HR 0.29 (0.24–0.34)	Adjusted by adhesion to treatment but not for liver function scores. HE not evaluated
					Atorvastatin 8%		Death	Decreased mortality HR 0.57 (0.45–0.71)	
					Rosuvastatin 6%				

Retrospective cohort studies

Study	Patients source	Patients description	Study design	Number of patients	Type of statin	Follow up period	Endpoints	Results	Comments
Mohanty, Gastroenterology 2016	US Veterans Health Admin	Hepatitis C-related compensated cirrhosis	Retrospective cohort study	1323 statins users vs. 12,522 non-statin users	Simvastatin 85%	Median of 2.5 years for statin users, 1.5 years for non-users	Decompensation	Prevented decompensation aHR 0.55 (0.39–0.77)	Adjusted for liver tests and scores
					Lovastatin 10%		Death	Decreased mortality aHR 0.56 (0.46–0.69)	Lower risk of ascites and variceal hemorrage
					Pravastatin 3%				
					Rosuvastatin 1%				
					Fluvastatin 1%				
Kumar, Dig Dis Sci 2014	Partners Research Patient Data Registry	NASH, OH, hepatitis C, and hepatitis B-related cirrhosis	Retrospective cohort study	81 statin users vs. 162 non-statin users	Simvastatin 49%	3 years for statin users, 2.5 years for non-statin users	Decompensation	Prevented decompensation HR 0.58 (0.34–0.98)	Low number of patients included, risk of selection, and reporting biass.
					Atorvastatin 30%		Death	Decreased mortality HR 0.66 (0.33–0.86)	Biopsy proven cirrhosis
C. M-Feagans, Aliment Pharmacol Ther 2013	US Veterans Health Admin	Hepatitis C and alcohol-related cirrhosis	Retrospective cohort study	2468 statin users vs. 16,408 non-statin users	Simvastatin 90%	3.3 years	Infections	Prevented infections aHR 0.67 (0.47–0.95)	Adjusted for age and comorbidities. No data of liver function
					Lovastatin 9%				

(continued)

Table 23.1 (continued)

Retrospective cohort studies									
Study	Patients source	Patients description	Study design	Number of patients	Type of statin	Follow up period	Endpoints	Results	Comments
Mahmud, Hepatology 2021 (abstract)	Veteran's Affairs cirrhosis cohort (VOCAL)	NASH, alcohol, hepatitis C, and hepatitis B related cirrhosis	Retrospective cohort study	22,876 statin users vs. 46,515 non-statin users vs. 15,572 new statin initiators	NA	Follow-up of 5 years	Acute on chronic liver failure (ACLF)	Prevented ACLF HR 0.64 (0.59–0.71)	Adjusted for possible time-dependent confounders, other lipid-lowering drugs
									Misclassification of exposures and outcomes
									Primarily male, enriched in psychosocial comorbidities

Study	Patients source	Patients description	Study design	Number of patients	Type of statin	Follow up period	Endpoints	Results	Comments
Randomized clinical trials									
Abraldes, Gastroenterology, 2009	University Hospitals	Cirrhosis and portal hypertension (HVPG>12 mmHg)	Multicenter randomized clinical trial (3 centers)	27 patients on statin treatment vs. 28 patients on placebo	Simvastatin	1 month	Change in HVPG	Decreased HVPG from 18.5 to 17.1 ($p = 0.003$), not decrease in placebo group	Simvastatin administration improved quantitative tests of liver funtion (indocyanine green clearance)
									Non-severe adverse events related to medication
P. Pollo-Flores, Digestive, and Liver Disease, 2015	University Hospital	Cirrhosis and portal hypertension (HVPG>5 mmHg)	Single-center randomized clinical trial	14 patients under statins treatment vs. 20 patients on placebo	Simvastatin	3 months	Change in HVPG	Reduced HVPG in patients under statin treatment compared to placebo: −2 vs. 0 mmHg, $p = 0.02$	Previous variceal bleeding independent variable associated with response to simvastatin
									Non-severe adverse events related to medication

(continued)

Table 23.1 (continued)

Randomized clinical trials

Study	Patients source	Patients description	Study design	Number of patients	Type of statin	Follow up period	Endpoints	Results	Comments
Abraldes, Gastroenterology, 2016	University Hospitals	Cirrhosis and variceal bleeding 5-10 days before inclusion	Multicenter randomized clinical trial (14 centers)	69 patients under statin treatment vs. 78 patients on placebo	Simvastatin	2 years	Composite endpoint (rebleeding or death)	Not significant decrease in risk of rebleeding or death	Decrease in liver-related death
							Death	Decreased mortality HR 0.39 (0.15-0.98)	Not significant decrease in the primary endopoint, or in specific complications of cirrhosis
Bishnu, Eur J Gastroenterol Hepatol, 2018	University Hospital	Cirrhosis and portal hypertension	Single-center randomized clinical trial	11 patients atorvastatin + propranolol vs. 12 placebo + propranolol	Atorvastatin	1 month	Change in HVPG	Decreased HVPG 4.81 ± 2.82 vs. 2.58 ± 1.88 mmHg	No significant differences in clinical outcomes after 1-year follow-up

Study	Patients source	Patients description	Study design	Number of patients	Type of statin	Follow up period	Endpoints	Results	Comments
Randomized clinical trials									
Vijayaraghavan, Am J Gastro, 2020	University Hospital	Cirrhosis and portal hypertension	Single-center randomized clinical trial	110 patients simvastatin + carvedilol vs. 110 placebo + carvedilol	Simvastatin	3 months	HVPG reduction of ≥20% or < 12 mm hg) at 3 months	Endpoint: HVPG-endpoint 36/59 [61%] vs. 36/62 [58.1%]; mean HVPG reduction 17.8% vs. 17.3%, odds ratio OR: 0.88; (0.43–1.83)	No significant differences in clinical outcomes after 1-year follow-up

NA not available, *HCC* hepatocellular carcinoma, *aHR* adjusted hazard ratio, *HBV* hepatitis B virus, *HCV* hepatitis C virus, *OH* alcohol, *HR* hazard ratio, *HE* hepatic encephalopathy, *NASH* nonalcoholic steatohepatitis, *HVPG* hepatic venous pressure gradient

Summary/Conclusion

Statins exhibit pleiotropic effects in many liver diseases. One of these effects is the inhibition of isoprenoid synthesis as a consequence of decreased HMG-CoA reductase activity since the resulting modulated GTPase activity plays a major role in the treatment of most chronic liver diseases (Fig. 23.1).

Statins are cost-effective and generally well-tolerated by patients and the benefits of statin treatment in most patients outweigh their potential hepatotoxic risk. Especially in patients with severe chronic liver injury and high risk of CVD, statin treatment is very promising since it not only prevents the development of CVD but also could help to prevent the progression of liver fibrosis to cirrhosis and the development of HCC, decrease portal pressure, lower inflammation and the related acute decompensation and even ACLF. Therefore, reasons for statin use in chronic liver diseases are more convincing than reasons against, rendering statin treatment a definite advantage.

References

1. Schierwagen R, Uschner FE, Magdaleno F, Klein S, Trebicka J. Rationale for the use of statins in liver disease. Am J Physiol Gastrointest Liver Physiol. 2017;312:G407–12.
2. Trebicka J, Hennenberg M, Laleman W, Shelest N, Biecker E, Schepke M, Nevens F, et al. Atorvastatin lowers portal pressure in cirrhotic rats by inhibition of RhoA/rho-kinase and activation of endothelial nitric oxide synthase. Hepatology. 2007;46:242–53.
3. Marrone G, Shah VH, Gracia-Sancho J. Sinusoidal communication in liver fibrosis and regeneration. J Hepatol. 2016;65:608–17.
4. Abraldes JG, Rodriguez-Vilarrupla A, Graupera M, Zafra C, Garcia-Caldero H, Garcia-Pagan JC, Bosch J. Simvastatin treatment improves liver sinusoidal endothelial dysfunction in CCl4 cirrhotic rats. J Hepatol. 2007;46:1040–6.
5. Goldstein JL, Brown MS. A century of cholesterol and coronaries: from plaques to genes to statins. Cell. 2015;161:161–72.
6. Gruzdeva O, Uchasova E, Dyleva Y, Akbasheva O, Karetnikova V, Barbarash O. Early effects of treatment low-dose atorvastatin on markers of insulin resistance and inflammation in patients with myocardial infarction. Front Pharmacol. 2016;7:324.
7. Schierwagen R, Maybuchen L, Hittatiya K, Klein S, Uschner FE, Braga TT, Franklin BS, et al. Statins improve NASH via inhibition of RhoA and Ras. Am J Physiol Gastrointest Liver Physiol. 2016;311:G724–33.
8. Blais P, Lin M, Kramer JR, El-Serag HB, Kanwal F. Statins are underutilized in patients with nonalcoholic fatty liver disease and dyslipidemia. Dig Dis Sci. 2016;61:1714–20.
9. Kwak B, Mulhaupt F, Myit S, Mach F. Statins as a newly recognized type of immunomodulator. Nat Med. 2000;6:1399–402.
10. Trebicka J, Amoros A, Pitarch C, Titos E, Alcaraz-Quiles J, Schierwagen R, Deulofeu C, et al. Addressing profiles of systemic inflammation across the different clinical phenotypes of acutely decompensated cirrhosis. Front Immunol. 2019;10:476.
11. Trebicka J, Reiberger T, Laleman W. Gut-liver Axis links portal hypertension to acute-on-chronic liver failure. Visc Med. 2018;34:270–5.
12. Praktiknjo M, Monteiro S, Grandt J, Kimer N, Madsen JL, Werge MP, William P, et al. Cardiodynamic state is associated with systemic inflammation and fatal acute-on-chronic liver failure. Liver Int. 2020;40:1457–66.

13. Monteiro S, Grandt J, Uschner FE, Kimer N, Madsen JL, Schierwagen R, Klein S, et al. Differential inflammasome activation predisposes to acute-on-chronic liver failure in human and experimental cirrhosis with and without previous decompensation. Gut. 2021;70(2):379–87.
14. Trebicka J, Fernandez J, Papp M, Caraceni P, Laleman W, Gambino C, Giovo I, et al. The PREDICT study uncovers three clinical courses of acutely decompensated cirrhosis that have distinct pathophysiology. J Hepatol. 2020;73:842–54.
15. Pignatelli P, Carnevale R, Pastori D, Cangemi R, Napoleone L, Bartimoccia S, Nocella C, et al. Immediate antioxidant and antiplatelet effect of atorvastatin via inhibition of Nox2. Circulation. 2012;126:92–103.
16. Oikonomou E, Siasos G, Zaromitidou M, Hatzis G, Mourouzis K, Chrysohoou C, Zisimos K, et al. Atorvastatin treatment improves endothelial function through endothelial progenitor cells mobilization in ischemic heart failure patients. Atherosclerosis. 2015;238:159–64.
17. Iwakiri Y, Trebicka J. Portal hypertension in cirrhosis: pathophysiological mechanisms and therapy. JHEP Rep. 2021;3:100316.
18. Wanless IR, Wong F, Blendis LM, Greig P, Heathcote EJ, Levy G. Hepatic and portal vein thrombosis in cirrhosis: possible role in development of parenchymal extinction and portal hypertension. Hepatology. 1995;21:1238–47.
19. Queck A, Carnevale R, Uschner FE, Schierwagen R, Klein S, Jansen C, Meyer C, et al. Role of portal venous platelet activation in patients with decompensated cirrhosis and TIPS. Gut. 2020;69:1535–6.
20. Vieira-Silva S, Falony G, Belda E, Nielsen T, Aron-Wisnewsky J, Chakaroun R, Forslund SK, et al. Statin therapy is associated with lower prevalence of gut microbiota dysbiosis. Nature. 2020;581:310–5.
21. Trebicka J, Bork P, Krag A, Arumugam M. Utilizing the gut microbiome in decompensated cirrhosis and acute-on-chronic liver failure. Nat Rev Gastroenterol Hepatol. 2021;18:167–80.
22. Trebicka J, Macnaughtan J, Schnabl B, Shawcross DL, Bajaj JS. The microbiota in cirrhosis and its role in hepatic decompensation. J Hepatol. 2021;75(Suppl 1):S67–81.
23. Schierwagen R, Alvarez-Silva C, Madsen MSA, Kolbe CC, Meyer C, Thomas D, Uschner FE, et al. Circulating microbiome in blood of different circulatory compartments. Gut. 2019;68:578–80.
24. Bjornsson ES. Hepatotoxicity of statins and other lipid-lowering agents. Liver Int. 2017;37:173–8.
25. Lewis JH, Mortensen ME, Zweig S, Fusco MJ, Medoff JR, Belder R, Pravastatin in Chronic Liver Disease Study Investigators. Efficacy and safety of high-dose pravastatin in hypercholesterolemic patients with well-compensated chronic liver disease: results of a prospective, randomized, double-blind, placebo-controlled, multicenter trial. Hepatology. 2007;46:1453–63.
26. Martin JE, Cavanaugh TM, Trumbull L, Bass M, Weber F Jr, Aranda-Michel J, Hanaway M, et al. Incidence of adverse events with HMG-CoA reductase inhibitors in liver transplant patients. Clin Transpl. 2008;22:113–9.
27. Abraldes JG, Villanueva C, Aracil C, Turnes J, Hernandez-Guerra M, Genesca J, Rodriguez M, et al. Addition of simvastatin to standard therapy for the prevention of Variceal Rebleeding does not reduce Rebleeding but increases survival in patients with cirrhosis. Gastroenterology. 2016;150:1160–70. e1163
28. Russo MW, Hoofnagle JH, Gu J, Fontana RJ, Barnhart H, Kleiner DE, Chalasani N, et al. Spectrum of statin hepatotoxicity: experience of the drug-induced liver injury network. Hepatology. 2014;60:679–86.
29. Palmer SC, Craig JC, Navaneethan SD, Tonelli M, Pellegrini F, Strippoli GF. Benefits and harms of statin therapy for persons with chronic kidney disease: a systematic review and meta-analysis. Ann Intern Med. 2012;157:263–75.
30. Pose E, Napoleone L, Amin A, Campion D, Jimenez C, Piano S, Roux O, et al. Safety of two different doses of simvastatin plus rifaximin in decompensated cirrhosis (LIVERHOPE-SAFETY): a randomised, double-blind, placebo-controlled, phase 2 trial. Lancet Gastroenterol Hepatol. 2020;5:31–41.

31. Collins R, Reith C, Emberson J, Armitage J, Baigent C, Blackwell L, Blumenthal R, et al. Interpretation of the evidence for the efficacy and safety of statin therapy. Lancet. 2016;388:2532–61.
32. Munoz AE, Pollarsky F, Marino M, Cartier M, Miguez C, Vazquez H, Alvarez D, et al. Safety of chronic simvastatin treatment in patients with decompensated cirrhosis: many adverse events but no liver injury. Dig Dis Sci. 2021;66:3199–208.
33. Sung S, Al-Karaghouli M, Kalainy S, Cabrera Garcia L, Abraldes JG. A systematic review on pharmacokinetics, cardiovascular outcomes and safety profiles of statins in cirrhosis. BMC Gastroenterol. 2021;21:120.
34. Bjornsson E, Jacobsen EI, Kalaitzakis E. Hepatotoxicity associated with statins: reports of idiosyncratic liver injury post-marketing. J Hepatol. 2012;56:374–80.
35. Pose E, Trebicka J, Mookerjee RP, Angeli P, Gines P. Statins: old drugs as new therapy for liver diseases? J Hepatol. 2019;70:194–202.
36. Rodriguez S, Raurell I, Torres-Arauz M, Garcia-Lezana T, Genesca J, Martell M. A nitric oxide-donating statin decreases portal pressure with a better toxicity profile than conventional statins in cirrhotic rats. Sci Rep. 2017;7:40461.
37. Chalasani N, Aljadhey H, Kesterson J, Murray MD, Hall SD. Patients with elevated liver enzymes are not at higher risk for statin hepatotoxicity. Gastroenterology. 2004;126:1287–92.
38. Chalasani N. Statins and hepatotoxicity: focus on patients with fatty liver. Hepatology. 2005;41:690–5.
39. Pastori D, Polimeni L, Baratta F, Pani A, Del Ben M, Angelico F. The efficacy and safety of statins for the treatment of non-alcoholic fatty liver disease. Dig Liver Dis. 2015;47:4–11.
40. Ekstedt M, Franzen LE, Mathiesen UL, Holmqvist M, Bodemar G, Kechagias S. Statins in non-alcoholic fatty liver disease and chronically elevated liver enzymes: a histopathological follow-up study. J Hepatol. 2007;47:135–41.
41. Nelson A, Torres DM, Morgan AE, Fincke C, Harrison SA. A pilot study using simvastatin in the treatment of nonalcoholic steatohepatitis: a randomized placebo-controlled trial. J Clin Gastroenterol. 2009;43:990–4.
42. Dongiovanni P, Petta S, Mannisto V, Mancina RM, Pipitone R, Karja V, Maggioni M, et al. Statin use and non-alcoholic steatohepatitis in at risk individuals. J Hepatol. 2015;63:705–12.
43. Eslami L, Merat S, Malekzadeh R, Nasseri-Moghaddam S, Aramin H. Statins for non-alcoholic fatty liver disease and non-alcoholic steatohepatitis. Cochrane Database Syst Rev. 2013:CD008623.
44. Kapadia SB, Chisari FV. Hepatitis C virus RNA replication is regulated by host geranylgeranylation and fatty acids. Proc Natl Acad Sci U S A. 2005;102:2561–6.
45. Bader T, Fazili J, Madhoun M, Aston C, Hughes D, Rizvi S, Seres K, et al. Fluvastatin inhibits hepatitis C replication in humans. Am J Gastroenterol. 2008;103:1383–9.
46. Butt AA, Yan P, Bonilla H, Abou-Samra AB, Shaikh OS, Simon TG, Chung RT, et al. Effect of addition of statins to antiviral therapy in hepatitis C virus-infected persons: results from ERCHIVES. Hepatology. 2015;62:365–74.
47. Simon TG, King LY, Zheng H, Chung RT. Statin use is associated with a reduced risk of fibrosis progression in chronic hepatitis C. J Hepatol. 2015;62:18–23.
48. Yang YH, Chen WC, Tsan YT, Chen MJ, Shih WT, Tsai YH, Chen PC. Statin use and the risk of cirrhosis development in patients with hepatitis C virus infection. J Hepatol. 2015;63:1111–7.
49. Mohanty A, Tate JP, Garcia-Tsao G. Statins are associated with a decreased risk of Decompensation and death in veterans with hepatitis C-related compensated cirrhosis. Gastroenterology. 2016;150:430–440.e431.
50. Gu W, Hortlik H, Erasmus H-P, Schaaf L, Zeleke Y, Uschner FE, Ferstl P, et al. Diagnosis of cirrhosis is associated with premature death in hospital admissions. Lancet Reg Health Eur. 2021;12:100240.
51. Simon TG, Bonilla H, Yan P, Chung RT, Butt AA. Atorvastatin and fluvastatin are associated with dose-dependent reductions in cirrhosis and hepatocellular carcinoma, among patients with hepatitis C virus: results from ERCHIVES. Hepatology. 2016;64:47–57.

52. Singh S, Singh AG, Singh PP, Murad MH, Iyer PG. Statins are associated with reduced risk of esophageal cancer, particularly in patients with Barrett's esophagus: a systematic review and meta-analysis. Clin Gastroenterol Hepatol. 2013;11:620–9.
53. Trebicka J, Hennenberg M, Odenthal M, Shir K, Klein S, Granzow M, Vogt A, et al. Atorvastatin attenuates hepatic fibrosis in rats after bile duct ligation via decreased turnover of hepatic stellate cells. J Hepatol. 2010;53:702–12.
54. Klein S, Klosel J, Schierwagen R, Korner C, Granzow M, Huss S, Mazar IG, et al. Atorvastatin inhibits proliferation and apoptosis, but induces senescence in hepatic myofibroblasts and thereby attenuates hepatic fibrosis in rats. Lab Investig. 2012;92:1440–50.
55. Zafra C, Abraldes JG, Turnes J, Berzigotti A, Fernandez M, Garca-Pagan JC, Rodes J, et al. Simvastatin enhances hepatic nitric oxide production and decreases the hepatic vascular tone in patients with cirrhosis. Gastroenterology. 2004;126:749–55.
56. Alvarado-Tapias E, Ardévol A, Pavel O, Montanes R, Murzi M, Oblitas Susanibar E, Poca M, et al. Hemodynamic effects of Carvedilol plus simvastatin in cirrhosis with portal hypertension and no-response to βBlockers: a double-blind randomized trial. Hepatology. 2016;64:74A.
57. Vijayaraghavan R, Jindal A, Arora V, Choudhary A, Kumar G, Sarin SK. Hemodynamic effects of adding simvastatin to Carvedilol for primary prophylaxis of Variceal bleeding: a randomized controlled trial. Am J Gastroenterol. 2020;115:729–37.
58. Wan S, Huang C, Zhu X. Systematic review with a meta-analysis: clinical effects of statins on the reduction of portal hypertension and variceal haemorrhage in cirrhotic patients. BMJ Open. 2019;9:e030038.
59. Pollo-Flores P, Soldan M, Santos UC, Kunz DG, Mattos DE, da Silva AC, Marchiori RC, et al. Three months of simvastatin therapy vs. placebo for severe portal hypertension in cirrhosis: a randomized controlled trial. Dig Liver Dis. 2015;47:957–63.
60. Meireles CZ, Pasarin M, Lozano JJ, Garcia-Caldero H, Gracia-Sancho J, Garcia-Pagan JC, Bosch J, et al. Simvastatin attenuates liver injury in rodents with biliary cirrhosis submitted to hemorrhage/resuscitation. Shock. 2017;47:370–7.
61. Tripathi DM, Vilaseca M, Lafoz E, Garcia-Caldero H, Viegas Haute G, Fernandez-Iglesias A, Rodrigues de Oliveira J, et al. Simvastatin prevents progression of acute on chronic liver failure in rats with cirrhosis and portal hypertension. Gastroenterology. 2018;155:1564–77.
62. Mahmud N, Chapin S, Goldberg DS, Reddy KR, Kaplan DE. Statins are associated with reduced development of acute-on-chronic liver failure in a large national cohort of patients with cirrhosis. J Hepatol. 2022;76(5):1100–8.

Anticoagulation for Portal Vein Thrombosis in Cirrhosis: An Evidence-Based Approach to When and How

24

Antonio Guerrero, Luis Téllez, and Agustín Albillos

Portal Vein Thrombosis (PVT) is a well-established complication of cirrhosis. Most frequently occurring at the decompensated stage, PVT is increasingly being recognized as associated with poor outcomes. In patients with cirrhosis, the risk of thrombosis in the splanchnic vascular bed is determined by blood flow stasis due to the following factors: portal hypertension, hypercoagulability as a consequence of thrombotic shifting of a rebalanced coagulation system, and endothelial injury. Recognition of the thrombophilic nature of cirrhosis and the consequences of PVT, especially in patients awaiting Liver Transplantation (LT), has prompted the widespread use of anticoagulation in the past decade, having overcome initial fears of the frequent presence of thrombocytopenia and increased risk of gastrointestinal bleeding. In contrast, outside the LT setting, anticoagulation to recanalize a thrombosed portal vein in cirrhosis remains controversial due to the transient nature of many cases of PVT, contradictory information regarding the impact of PVT on outcomes and survival, and anticoagulation concerns.

In this chapter, we review current evidence regarding the efficacy and safety of anticoagulation in patients with cirrhosis and PVT. Such evidence is, nevertheless, limited by a lack of randomized controlled trials and is based mainly on the findings of observational retrospective studies. These studies also have the limitation of a

A. Guerrero · L. Téllez · A. Albillos (✉)
Department of Gastroenterology and Hepatology, Hospital Universitario Ramón y Cajal, Instituto Ramón y Cajal de Investigación Sanitaria (IRYCIS), Centro de Investigación Biomédica Red de Enfermedades Hepáticas y Digestivas (CIBEREHD), Universidad de Alcalá, Madrid, Spain
e-mail: agustin.albillos@uah.es

R. de Franchis (ed.), *Portal Hypertension VII*,
https://doi.org/10.1007/978-3-031-08552-9_24

small study population and that they vary considerably in terms of the target population (compensated vs. decompensated cirrhosis), thrombosis severity (i.e., partial vs. complete), and extension (i.e., involvement of the portal vein trunk, superior mesenteric vein, intrahepatic branches of the portal vein), and length of follow-up (i.e., short- vs. long-term). This evidence has been analyzed in several aggregated data meta-analyses [1–4] and an individual patient data meta-analysis [5].

Prevalence and Natural History

The reported prevalence of PVT in cirrhosis varies widely among studies and largely depends on the stage of cirrhosis and the presence or absence of hepatocellular carcinoma [6]. Prevalence increases in parallel with the severity of cirrhosis: 10% in compensated cirrhosis, 17% in Child-Pugh B/C class cirrhosis, and up to 26% in patients awaiting LT [7]. The annual incidence of PVT in patients with advanced cirrhosis is approximately 10%–17% [8].

In the absence of any treatment, PVT in cirrhosis features a dynamic natural history. Hence, PVT can progress, remain stable, or improve, and this behavior needs to be considered in clinical decision-making. According to a systematic review of 14 studies biased by notable heterogeneity [9], spontaneous recanalization of PVT occurs in 5% to 70% of cases averaging at around 40%, Knowledge of the factors predicting spontaneous recanalization is rather limited, but data from cohorts of untreated patients with cirrhosis and PVT suggest a higher rate of recanalization in patients with compensated cirrhosis and only partial PVT (Table 24.1) [10–22]. The spontaneous recanalization rate was 22% in a study in which most included patients had Child-Pugh B/C cirrhosis [14] and 70% in a large longitudinal study in patients who mostly had Child-Pugh A cirrhosis with partial PVT [12]. These data warn that a conservative approach with close follow-up is needed before initiating anticoagulation in patients with compensated cirrhosis and partial PVT.

Table 24.1 Natural history of portal vein thrombosis in patients with cirrhosis not receiving anticoagulation

Study author, year (reference)	Type of saudí	Degree of PVT	Severity of liver disease	Number of patients	PVT recanalization (total or partial)	Thrombus progression
Hidaka et al., 2018 [10]	Multicenter, randomized, double-blind, placebo-controlled, full-text	50% MPV: 56% ($n = 20$) PV branches: 30% ($n = 11$) SMV: 14% ($n = 5$)	Child A: 17% Child B/C: 83%	36	7 (19.4%)	7 (19.4%)
Chen et al., 2016 [11]	Single-center, retrospective, controlled, full-text	>50% MPV: 72% ($n = 26$) >50% PV branches (right $n = 25$, left $n = 24$) >50% splenic vein: 14% ($n = 5$) >50% SMV: 52% ($n = 19$)	Child A: 23% Child B/C: 77%mean MELD: 8.9	36	4 (25%)	6 (37.5%)
Nery et al., 2015 [12]	Multi-center, prospective, single-arm, full-text	Non-oclusive: 100%	Child A/B: 100%	101	70 (70%)	4 (13.8%)
Chung et al., 2014 [13]	Single-center, retrospective, matched, controlled, full-text	NA	Child A: 50% Child B/C: 50%	14	5 (36%)	3 (21.4%)
Girleanu et al., 2014 [14]	Single-center, prospective, single-arm, full-text	Partial: 100% MPV ($n = 18$) MPV + SMV ($n = 11$)	Child A: 32% Child B/C: 68%mean MELD: 12.7	22	5 (22.73%)	6 (27.27%)
Risso et al., 2014 [15]	Single-center, retrospective, controlled, abstract	NA	NA	20	8 (40%)	NA
John et al., 2013 [16]	Single-center, prospective, single-arm, full-text	Partial: 55% ($n = 38$) Occlusive: 45% ($n = 32$)	Mean MELD: 14.4	70	22 (31.4%)	3 (4.3%)
Maruyama et al., 2013 [17]	Single-center, prospective, single-arm, full-text	Partial: 36% ($n = 17$) Occlusive: 64% ($n = 31$)	Child A: 27%% Child B/C: 73%median MELD: 12.6	48	5 (12%)	3 (2,4%)

(continued)

Table 24.1 (continued)

Study author, year (reference)	Type of saudí	Degree of PVT	Severity of liver disease	Number of patients	PVT recanalization (total or partial)	Thrombus progression
Caracciolo et al., 2013 [18]	Single-center, retrospective, matched, controlled, abstract	Partial: 100%	Child A/B: 100%	14	8 (57.6%)	NA
Luca et al., 2012 [19]	Single-center, retrospective, single-arm, full-text	Partial: 100% MPV ($n = 15$) MPV + SMV ($n = 18$)	Mean child: 8 Mean MELD: 12	42	19 (45%)	20 (48%)
Senzolo et al., 2012 [20]	Single-center, prospective, controlled, full-text	Partial: 67% ($n = 14$) Complete: 33% ($n = 7$)	Child A: 24% Child B/C: 76%mean MELD: 13.7	21	1 (5%)	15 (71.4%)
Garcovich et al., 2011 [21]	Single-center, retrospective, matched, controlled, abstract	Partial: 80% ($n = 12$) Occlusion >75%: 20% ($n = 3$)	Child A/B: 100%	15	5 (33%)	NA
Francoz et al., 2005 [22]	Single-center, retrospective, controlled, full-text	Partial	Child A: 20% Child B/C: 80%mean MELD: 11.8	10	0 (0%)	6 (60%)

PVT portal vein thrombosis, *SMV* superior mesenteric vein, *MPV* main portal vein, *PV* portal vein

Impact of PVT in Cirrhosis

The development of portal hypertension and collateral circulation reduces the risk of intestinal ischemia secondary to thrombotic occlusion of the portal vein. This means that PVT in cirrhotic patients is usually asymptomatic, and most cases are detected incidentally in a routine ultrasound examination during hepatocellular carcinoma surveillance [23]. Symptoms of intestinal ischemia are usually associated with occlusive thrombosis of the superior mesenteric vein [24].

The contribution of PVT to cirrhosis progression and the development of acute complications is a matter of dispute. The largest prospective and retrospective studies found no effects of partial or complete PVT on hepatic decompensation or survival [12, 17, 19, 25]. The exception is acute variceal bleeding, as PVT increases the risk of its occurrence and negatively impacts its course, having been identified as a predictor of 5-day treatment failure and 6-week mortality [26–28].

The potential contribution of PVT to the natural course of cirrhosis is further supported by the impact of portal vein recanalization through anticoagulation on patient outcomes and survival. Recanalization has been associated with a reduction in liver-related outcomes and increased survival in decompensated (Child-Pugh B/C) cirrhosis in cohorts of patients with cirrhosis on anticoagulants for PVT [29–31]. Additionally, a seminal study has shown that prevention of PVT with prophylactic low molecular weight heparin (LMWH) leads to reduced hepatic decompensation and increased survival [32]. Further, it is likely that the benefits of anticoagulation treatment may go beyond PV recanalization. Indeed, a recent individual patient data meta-analysis of 5 nonrandomized studies that compared anticoagulation with no treatment including 500 patients with cirrhosis and PVT-205 on anticoagulants–has shown that anticoagulation improves the survival of patients with partial or complete PVT, an effect that seems independent of PV recanalization [5]. These findings suggest that the presence of PVT could help identify a subset of patients with cirrhosis likely to benefit from long-term anticoagulation.

In the LT setting, there is general agreement that extensive complete thrombosis poses technical difficulties for transplant surgery, increases graft ischemia times, and compromises survival, especially in cases that require nonphysiological reconstruction of the portal vein [33]. In a meta-analysis of 44 studies involving 98,558 LT recipients, 30-day and 1-year mortality were higher in patients with complete PVT compared to those with partial PVT or a patent PV (OR 5.65, 95% CI 2–15.96, $p < 0.0001$; and OR 2.48, 95% CI 0.99–6.17, $p = 0.38$, respectively) [34] Thus, the threshold for anticoagulation is lower in LT candidates with PVT in whom the goal of therapy is to promote PV recanalization and avoid thrombus progression to ensure normal physiological blood flow to the graft.

Anticoagulation for PVT in Cirrhosis

Goals and Efficacy

The goal of anticoagulation is to recanalize the PV and secondarily to avoid thrombosis worsening and progression. This is especially important in candidates or potential candidates for LT in whom progression of thrombosis may compromise or give rise to technical difficulties during surgery.

Table 24.2 summarizes the meta-analysis of data from retrospective cohorts of patients with cirrhosis and PVT treated with anticoagulants, including the results of the most recent meta-analysis of 33 studies with 1696 participants in total. Based on these data, anticoagulation served to increase the rate of portal vein recanalization by about 2.5 fold, from 18.8% recorded in the no-anticoagulation group to 59.0%. The rate of complete recanalization rose from 18.3% to 48.8% in response to anticoagulation. Thrombosis progressed in 10.4% of the patients on anticoagulants compared to 37.0% of those not receiving them [3].

Two recent aggregated data meta-analyses have shown that anticoagulation improves survival in patients with cirrhosis and PVT [3, 35]. This finding is reinforced by the results of an individual patient data meta-analysis that showed that this benefit was independent of portal vein recanalization and observed both in patients with partial and complete PVT [5].

Table 24.2 Summary of the meta-analysis of studies of cohorts of patients with cirrhosis and portal vein thrombosis

Study author, year (reference)	Patients cohorts	Patients and study number	Main result
Qi et al., 2015 [1]	Anticoagulation AC vs. no treatment	Studies: 16 (3 full texts non comparative 4 full texts comparative 9 abstract non comparative)patients: No data Disease stage: No data	PVT recanalization (AC vs. no treatment): OR = 4.16; 95%CI 1.88–9.20, $P = 0.0004$ Bleeding: AC-related bleeding 0%–18%, pooled rate 3.3% (95% CI, 1.1–6.7)
Lofredo et al., 2017 [2]	AC vs. no treatment	Studies: 8 (full texts comparative) Patients: 353 Disease stage: MELD (overall) 10.9	PVT recanalization (AC vs. no treatment): 71% vs. 42% (OR = 4.8; 95% CI, 2.7–8.7; $P = 0.0001$) Bleeding: 11% for both groups
Le Wang et al., 2021 [3]	AC vs. no treatment	Studies: 33 15 full texts non comparative 7 full texts comparative 8 abstract non comparative 3 abstract comparative) Patients: 1696 Disease stage: No data	PVT recanalization (AC vs. no treatment): 59% vs. 21% (OR = 2.61; 95%CI 1.9–3.4; $P = 0.00001$) Bleeding: RR = 0.78; 95%CI 0.47–1.30; $P = 0.34$ Overall survival: RR = 1.11; 95%CI 1.03–1.21; $P = 0.01$ Predictive factors: Early initiation of AC increased recanalization (RR = 1.58; 95% CI 1.21–2.07; $P = 0.0007$) lower recanalization in child B/C (RR = 0.77; 95%CI 0.62–0.95; $P = 0.02$) and higher MELD (MD = −1.48; 95% CI −2.20–0.76; $P = 0.0001$)
Ghazaleh et al., 2021 [4]	AC vs. no treatment	Studies: 9 5 full texts non comparative 4 abstract comparative) Patients: 474 Disease stage: MELD (overall) 12.9	PVT recanalization (AC vs. no treatment): 65% vs. 25% (RR 2.31, 95%CI 1.8–2.9; $P < 0.00001$) Bleeding: 10.3% vs. 0.22.7% (RR 0.43, 95%CI 0.09–1.99; $P = 0.28$)

AC anticoagulation, *PVT* portal vein thrombosis, *RR* relative risk, *OR* odds ratio

Despite the caveats inherent to the retrospective nature of the studies reviewed here, it has been possible to identify several factors influencing the efficacy of anticoagulation: (1) Thrombus severity and location. The rate of recanalization is lower incomplete PVT and in thrombosis affecting the superior mesenteric vein. (2)

Cirrhosis stage. Compensated (Child-Pugh A) cirrhosis is linked to a higher recanalization rate than decompensated (Child-Pugh B/C) cirrhosis [3]. (3) Anticoagulation timing. Early initiation of anticoagulation gives rise to an increased recanalization rate, and this recanalization is more likely when anticoagulation is started within 6 months of diagnosis [3, 20, 30, 36].

Anticoagulation Duration

Anticoagulation should be kept up for a minimum of 6 months, as recanalization occurs after a mean of 5 months of therapy [31, 37]. Almost two-thirds of patients show recanalization 3 to 6 months after the start of anticoagulation, and one-third do so after 6 to 12 months [37].

The decision to continue with anticoagulation beyond recanalization depends on the severity of the potential consequences of re-thrombosis and a favorable balance of benefits and risks in the individual patient. Thrombosis recurrence is frequent after anticoagulation discontinuation, the pooled rate observed in a recent meta-analysis being as high as 46.7% (95%CI 37.7–69.3) [3]. Re-thrombosis is an early event with most cases occurring from 2 to 5 months after stopping anticoagulation [30, 36]. The consequences of re-thrombosis can be dramatic in patients on the LT waiting list, as it can jeopardize surgery. Other than in LT candidates, continuing anticoagulation should be also considered in patients with symptoms of intestinal ischemia at the time of PVT diagnosis. It is important to remark that cirrhosis is an independent and permanent risk factor for PVT, which explains the high likelihood of recurrence after stopping anticoagulation. For this reason, close observation of portal vein patency at 2–3 months is mandatory after anticoagulation withdrawal.

Type of Anticoagulation

Most of the studies mentioned above used the anticoagulant LMWH followed by a vitamin K Antagonist (VKA). This kind of sequential therapy might be supported by the results of one meta-analysis that identified LMWH as effective for portal vein recanalization, whereas both LMWH and VKA were effective in reducing PVT progression [2].

With regard to selecting the LMWH dose, most studies initiate enoxaparin at 1.5 mg/kg daily which is later switched to 1 mg/kg daily. A single study compared a twice-daily dose of 1.0 mg/kg with 1.5 mg/kg daily. The former gave rise to fewer bleeding events and a similar efficacy [38], but this is not a pragmatic approach for most patients. Their oral route of administration makes VKAs a preferable option for long-term treatment, although for patients with cirrhosis the target therapeutic range has not yet been defined and is difficult to maintain. Moreover, the baseline therapeutic International Normalized Ratio (INR) is subject to frequent modification. VKAs are usually dosed to target INR values in the range 2.0–3.0. In a short series of patients with cirrhosis receiving VKA treatment, a platelet count below

50,000/μL was associated with an increased bleeding risk [30]. In effect, as a general principle, anticoagulation should be discontinued or not used in patients with platelet counts below 30,000/μL.

In clinical practice, Direct Oral Anticoagulants (DOACs) are being increasingly used due to a safer profile and easier administration than VKAs for conventional long-term anticoagulation. Evidence of the safe and effective use of DOACs in patients with cirrhosis comes from retrospective studies in which anticoagulation was indicated for atrial fibrillation and to a lesser extent for PVT [39–41]. Experience is limited to Child-Pugh A and B cirrhosis, as these agents are not recommended for Child-Pugh C patients. In a meta-analysis of four studies involving 3483 patients with advanced liver disease and a multicenter retrospective report on 2694 patients with cirrhosis, all anticoagulated for atrial fibrillation, DOACs were found associated with a lower risk of total and major bleeding episodes and similarly effective to prevent an ischemic stroke than VKAs [39, 40]. This points to a similar safety profile and efficacy of DOACs in cirrhotic and non-cirrhotic patients. Experience with DOACs in patients with cirrhosis and splanchnic thrombosis is rather limited, but suggests a similar efficacy compared to VKAs for recanalization and lowering the bleeding risk (Table 24.3) [41–46].

Safety of Anticoagulation

A conclusion drawn from extensive experience with anticoagulation for more than a decade is that it is fairly safe in patients with cirrhosis and PVT. In the most recent meta-analysis, pooled rates of overall and major bleeding events were 10.3% (95% CI 6.4–15.0) and 2.8% (95% CI 1.4–4.6), respectively, and there were no differences detected in the studies comparing patients receiving and not receiving anticoagulation therapy [3]. The pooled risk of variceal bleeding is 2.0% (95%CI 1.0–3.3), and lower in the anticoagulated cohorts [2, 3]. Most published reports come from units with expertise in the management of cirrhosis and are aware of the need to ensure variceal bleeding prophylaxis before initiating anticoagulation. This could be the reason for the greater rate of variceal bleeding observed in patients with cirrhosis receiving anticoagulants for indications other than PVT [47, 48]. In consequence, it is mandatory to check for the presence of gastroesophageal varices and initiate (or revise) bleeding prophylaxis before starting anticoagulation for PVT in patients with cirrhosis.

A summary of general recommendations for anticoagulation in patients with cirrhosis and portal vein thrombosis is shown in Table 24.4.

Table 24.3 Anticoagulation with DOACs in patients with cirrhosis and portal vein thrombosis

Study author, year (reference)	Type of study	Population included	Indication of anticoagulation	Type of anticoagulation	Number of patients	Efficacy	Safety
Intagliata et al., 2016 [42]	Retrospective (DOACs vs. traditional AC)	Child A (45%) Child B (55%) MELD 12	Splanchnic thrombosis, non-splanchnic venous thromboembolism, and atrial fibrillation (PVT = 12)	Apixaban (55%) Rivaroxaban (45%)	DOAC, $n = 20$ vs. traditional AC, $n = 19$	NA	Total bleeding DOAC 20% vs. traditional 16% $P = 0.91$ Major bleeding DOAC 5% vs. traditional 11%
Hum et al., 2017 [43]	Retrospective (DOACs vs. traditional AC)	Child A (41%) Child B (44%) Child C (15%) MELD 8.9	Splanchnic thrombosis, non-splanchnic venous thromboembolism, and atrial fibrillation (PVT = 4)	Apixaban (47%) Rivaroxaban (63%)	DOAC, $n = 27$ vs. traditional AC, $n = 18$	NA	Total bleeding: DOAC 30% vs. traditional 55% $P = 0.12$ Major bleeding: DOAC 4% vs. traditional 28%, $P = 0.03$

(continued)

Table 24.3 (continued)

Study author, year (reference)	Type of study	Population included	Indication of anticoagulation	Type of anticoagulation	Number of patients	Efficacy	Safety
De Gottardi et al., 2017 [41]	Retrospective (DOACs only)	Child A/B (100%) MELD 10	Splanchnic thrombosis, non-splanchnic venous thromboembolism, and atrial fibrillation (PVT = 22)	Apixaban (11%) Rivaroxaban (83%) Davigatran (5%)	DOAC, $n = 36$	NA	Total bleeding: $n = 5$, 14% Major bleeding: $n = 1$, 3%
Hanafy et al., 2018 [44]	Randomized controlled trial (DOACs vs. traditional AC)	Child A/B (100%) MELD 10 Patients who had undergone splenectomy due to hypersplenism	Acute PVT	Ribaroxaban (100%)	DOAC, $n = 40$ vs. traditional AC, $n = 40$	PVT recanalization DOAC: 100% Traditional: 45%	Major bleeding DOAC 0% vs. traditional 42% $P = 0.001$ Mortality: DOAC 0% vs. traditional 36% $P = 0.001$
Ai et al., 2020 [45]	Prospective (DOACs vs. no treatment)	Child A (100%)	Chronic PVT	Rivaroxaban (65%) Davigatran (35%)	DOAC, $n = 40$ vs. no treatment, $n = 40$	PVT recanalization: DOAC: 28% No treatment: 2.6%	Any bleeding: DOAC 7.5% vs.no treatment 2.5% $P = 0,61$

Yong et al., 2021 [46]	Prospective (AC vs. TIPS vs. no treatment) *AC included DOACs	Child A (35%) Child B (54%) Child C (11%) MELD 11	Chronic PVT	Rivaroxaban (100%)	AC only, $n = 63$ (traditional, $n = 59$; DOAC $n = 4$) TIPS only, $n = 88$ TIPS + AC, $n = 197$ (DOAC + TIPS, $n = 18$)	PVT recanalization: AC: 40% No treatment: 12% TIPS: 100% TIPS + AC: 95%	Major bleeding: AC 22% No treatment: 19% TIPS: 12% TIPS + AC: 15%

AC anticoagulation, *DOACs* direct oral anticoagulants, *PVT* portal vein thrombosis

Table 24.4 Summary of general recommendations for anticoagulation in patients with cirrhosis and portal vein thrombosis

Candidates for anticoagulation
• anticoagulation is recommended in patients with cirrhosis and (1) recent (<6 months) completely or partially occlusive (>50%) thrombosis of the portal vein trunk with or without extension to the superior mesenteric vein, or (2) symptomatic portal vein thrombosis, independently of the extension.
• anticoagulation is recommended in potential liver transplant candidates with portal vein thrombosis, independently of the degree of occlusion and extension.
• anticoagulation should be considered in patients with cirrhosis and minimally occlusive (<50%) thrombosis of the portal vein trunk that (1) progresses on short-term follow-up (1–3 months) or (2) compromises the superior mesenteric vein.
Duration of anticoagulation
• anticoagulation should be (1) maintained until portal vein recanalization or for a minimum of 6 months, (2) continued after recanalization in patients awaiting liver transplantation, and (3) considered to be continued after recanalization in all others, while balancing benefits in preventing recurrence and increasing survival and the risk of bleeding.
• close follow-up with imaging at 3 months to identify recurrence is indicated in patients in whom anticoagulation is stopped.
Type of anticoagulant
• anticoagulation is preferably initiated with LMWH and maintained with either LMWH, VKA, or DOAC.
• advantages of LMWH are that its use is based on the solid data. VKA carry challenges regarding INR monitoring in patients with cirrhosis.
• advantages of DOACs are that they are easier to use but less data are available. DOACs likely have different safety-efficacy profiles in patients with cirrhosis, although no recommendation can be made in favor of a specific DOAC in this setting.
Risks of anticoagulation
• currently evidence suggests that there are no major safety concerns for DOACs in patients with Child-Pugh class A cirrhosis. Due to the possibility of accumulation, DOACs should be used with caution in Child-Pugh class B patients, as well as inpatient with creatinine clearance below 30 mL/min. The use of DOACs in Child-Pugh class C patients is not recommended outside study protocols.
• patients with low platelet count (i.e., <50,000/μL) are at higher risk of complications on anticoagulation and require more caution.
• variceal bleeding prophylaxis, preferably with non-selective beta-blockers, unless contraindicated, is mandatory in patients with cirrhosis and esophagogastric varices and portal vein thrombosis requiring anticoagulation.

Modified from reference [49]

References

1. Qi X, De Stefano V, Li H, Dai J, Guo X, Fan D. Anticoagulation for the treatment of portal vein thrombosis in liver cirrhosis: a systematic review and meta-analysis of observational studies. Eur J Intern Med. 2015;26:23–9.
2. Loffredo L, Pastori D, Farcomeni A, Violi F. Effects of anticoagulants in patients with cirrhosis and portal vein thrombosis: a systematic review and meta-analysis. Gastroenterology. 2017;153:480–487.e481.
3. Wang L, Guo X, Xu X, De Stefano V, Plessier A, Noronha Ferreira C, et al. Anticoagulation favors thrombus recanalization and survival in patients with liver cirrhosis and portal vein thrombosis: results of a meta-analysis. Adv Ther. 2021;38:495–520.

4. Ghazaleh S, Beran A, Aburayyan K, Nehme C, Patel D, Khader Y, et al. Efficacy and safety of anticoagulation in non-malignant portal vein thrombosis in patients with liver cirrhosis: a systematic review and meta-analysis. Ann Gastroenterol. 2021;34(1):104–10.
5. Guerrero A, Del Campo L, Piscaglia F, et al. Anticoagulation improves overall survival through portal vein recanalization in patients with cirrhosis and portal vein thrombosis: individual patient data meta-analysis (IMPORTAL study). J Hepatol. 2021;75(Suppl 1):S216.
6. Tsochatzis EA, Senzolo M, Germani G, Gatt A, Burroughs AK. Systematic review: portal vein thrombosis in cirrhosis. Aliment Pharmacol Ther. 2010;31:366–74.
7. Ponziani FR, Zocco MA, Senzolo M, Pompili M, Gasbarrini A, Avolio AW. Portal vein thrombosis and liver transplantation: implications for waiting list period, surgical approach, early and late follow-up. Transplant Rev (Orlando). 2014;28:92–101.
8. Violi F, Corazza GR, Caldwell SH, Talerico G, Romiti GF, Napoleone L, et al. Incidence and recurrence of portal vein thrombosis in cirrhotic patients. Thromb Haemost. 2019;119:496–9.
9. Qi X, Guo X, Yoshida EM, Méndez-Sánchez N, De Stefano V, Tacke F, et al. Transient portal vein thrombosis in liver cirrhosis. BMC Med. 2018;16(1):83.
10. Hidaka H, Kokubu S, Sato T, Katsushima S, Izumi N, Igura T, et al. Antithrombin III for portal vein thrombosis in patients with liver disease: a randomized, double-blind, controlled trial. Hepatol Res. 2018;48:E107–16.
11. Chen H, Liu L, Qi X, He C, Wu F, Fan D, et al. Efficacy and safety of anticoagulation in more advanced portal vein thrombosis in patients with liver cirrhosis. Eur J Gastroenterol Hepatol. 2016;28:82–9.
12. Nery F, Chevret S, Condat B, de Raucourt E, Boudaoud L, Rautou PE, et al. Causes and consequences of portal vein thrombosis in 1,243 patients with cirrhosis: results of a longitudinal study. Hepatology. 2015;61:660–7.
13. Chung JW, Kim GH, Lee JH, Ok KS, Jang ES, Jeong SH, et al. Safety, efficacy, and response predictors of anticoagulation for the treatment of nonmalignant portal vein thrombosis in patients with cirrhosis: a propensity score matching analysis. Clin Mol Hepatol. 2014;20:384–91.
14. Girleanu I, Stanciu C, Cojocariu C, Boiculese L, Singeap AM, Trifan A. Natural course of nonmalignant partial portal vein thrombosis in cirrhotic patients. Saudi J Gastroenterol. 2014;20(5):288–92.
15. Risso A, Stradella D, Martini S, Rizzetto M, Salizzoni M. Liver transplantation in cirrhotic patients with portal vein thrombosis: a single Centre experience. Dig Liver Dis. 2014;46:e40.
16. John BV, Konjeti R, Aggarwal A, Lopez R, Atreja A, Miller C, et al. Impact of untreated portal vein thrombosis on pre and post liver transplant outcomes in cirrhosis. Ann Hepatol. 2013;12:952–8.
17. Maruyama H, Okugawa H, Takahashi M, Yokosuka O. De novo portal vein thrombosis in virus-related cirrhosis: predictive factors and long-term outcomes. Am J Gastroenterol. 2013;108:568–74.
18. Caracciolo G, Garcovich M, Zocco MA, et al. Clinical outcome of partial portal vein thrombosis in cirrhotic patients: to observe or to treat? Dig Liver Dis. 2013;45:S171.
19. Luca A, Caruso S, Milazzo M, Marrone G, Mamone G, Crino F, et al. Natural course of extrahepatic nonmalignant partial portal vein thrombosis in patients with cirrhosis. Radiology. 2012;265:124–32.
20. Senzolo M, Sartori TM, Rossetto V, Burra P, Cillo U, Boccagni P, et al. Prospective evaluation of anticoagulation and transjugular intrahepatic portosystemic shunt for the management of portal vein thrombosis in cirrhosis. Liver Int. 2012;32:919–27.
21. Garcovich M, Zocco MA, Ainora ME, et al. Clinical outcome of portal vein thrombosis (PVT) in cirrhotic patients: observe or treat? Hepatology. 2011;54:1261A–2A.
22. Francoz C, Belghiti J, Vilgrain V, Sommacale D, Paradis V, Condat B, et al. Splanchnic vein thrombosis in candidates for liver transplantation: usefulness of screening and anticoagulation. Gut. 2005;54(5):691–7.
23. Tessler FN, Gehring BJ, Gomes AS, Perrella RR, Ragavendra N, Busuttil RW, et al. Diagnosis of portal vein thrombosis: value of color Doppler imaging. AJR Am J Roentgenol. 1991;157:293–6.

24. Amitrano L, Guardascione MA, Brancaccio V, Margaglione M, Manguso F, Iannaccone L, et al. Risk factors and clinical presentation of portal vein thrombosis in patients with liver cirrhosis. J Hepatol. 2004;40(5):736–41.
25. Noronha Ferreira C, Marinho RT, Cortez-Pinto H, Ferreira P, Dias MS, Vasconcelos M, et al. Incidence, predictive factors and clinical significance of development of portal vein thrombosis in cirrhosis: a prospective study. Liver Int. 2019;39:1459–67.
26. D'Amico G, De Franchis R, Cooperative Study Group. Upper digestive bleeding in cirrhosis. Post-therapeutic outcome and prognostic indicators. Hepatology. 2003;38(3):599–612.
27. Amitrano L, Guardascione MA, Scaglione M, Menchise A, Martino R, Manguso F, et al. Splanchnic vein thrombosis and variceal rebleeding in patients with cirrhosis. Eur J Gastroenterol Hepatol. 2012;24(12):1381–5.
28. Huang XQ, Ni LY, Jiang SY, Xia RQ, Ma LL, Wang J, et al. Impact of portal vein thrombosis on the efficacy of endoscopic treatment in preventing re-bleeding from ruptured gastroesophageal varices in hepatitis B-related liver cirrhosis. Zhonghua Gan Zang Bing Za Zhi. 2020;28(9):747–52.
29. La Mura V, Braham S, Tosetti G, Branchi F, Bitto N, Moia M, et al. Harmful and beneficial effects of anticoagulants in patients with cirrhosis and portal vein thrombosis. Clin Gastroenterol Hepatol. 2018;16:1146–1152.e1144.
30. Delgado MG, Seijo S, Yepes I, Achécar L, Catalina MV, García-Criado A, et al. Efficacy and safety of anticoagulation on patients with cirrhosis and portal vein thrombosis. Clin Gastroenterol Hepatol. 2012;10:776–83.
31. Senzolo M, Riva N, Dentali F, Rodriguez-Castro K, Sartori MT, Bang SM, et al. Long-term outcome of splanchnic vein thrombosis in cirrhosis. Clin Transl Gastroenterol. 2018;9:176.
32. Villa E, Cammà C, Marietta M, Luongo M, Critelli R, Colopi S, et al. Enoxaparin prevents portal vein thrombosis and liver decompensation in patients with advanced cirrhosis. Gastroenterology. 2012;143:1253–1260.e1254.
33. Englesbe MJ, Schaubel DE, Cai S, Guidinger MK, Merion RM. Portal vein thrombosis and liver transplant survival benefit. Liver Transpl. 2010;16(8):999–1005.
34. Zanetto A, Rodriguez-Kastro KI, Germani G, Ferrarese A, Cillo U, Burra P, et al. Mortality in liver transplant recipients with portal vein thrombosis - an updated meta-analysis. Transpl Int. 2018;31:1318–29.
35. Valeriani E, Di Nisio M, Riva N, Cohen O, Porreca E, Senzolo M, et al. Anticoagulant treatment for splanchnic vein thrombosis in liver cirrhosis: a systematic review and meta-analysis. Thromb Haemost. 2021;121(7):867–76.
36. Rodriguez-Castro KI, Vitale A, Fadin M, Shalaby S, Zerbinati P, Sartori MT, et al. A prediction model for successful anticoagulation in cirrhotic portal vein thrombosis. Eur J Gastroenterol Hepatol. 2019;31:34–42.
37. Pettinari I, Vukotic R, Stefanescu H, Pecorelli A, Morelli M, Grigoras C, et al. Clinical impact and safety of anticoagulants for portal vein thrombosis in cirrhosis. Am J Gastroenterol. 2019;114:258–66.
38. Cui SB, Shu RH, Yan SP, Wu H, Chen Y, Wang L, et al. Efficacy and safety of anticoagulation therapy with different doses of enoxaparin for portal vein thrombosis in cirrhotic patients with hepatitis B. Eur J Gastroenterol Hepatol. 2015;27:914–9.
39. Serper M, Weinberg EM, Cohen JB, Reese PP, Taddei TH, Kaplan DE. Mortality and hepatic decompensation in patients with cirrhosis and atrial fibrillation treated with anticoagulation. Hepatology. 2020;73(1):219–32.
40. Violi F, Vestri A, Menichelli D, Di Rocco A, Pastori D, Pignatelli P. Direct Oral anticoagulants in patients with atrial fibrillation and advanced liver disease: an exploratory meta-analysis. Hepatol Commun. 2020;4(7):1034–40.
41. De Gottardi A, Trebicka J, Klinger C, Plessier A, Seijo S, Terziroli B, et al. Antithrombotic treatment with direct-acting oral anticoagulants in patients with splanchnic vein thrombosis and cirrhosis. Liver Int. 2017;37:694–9.

42. Intagliata NM, Henry ZH, Maitland H, Shah NL, Argo CK, Northup PG, et al. Direct oral anticoagulants in cirrhosis patients pose similar risks of bleeding when compared to traditional anticoagulation. Dig Dis Sci. 2016;61:1721–7.
43. Hum J, Shatzel JJ, Jou JH, Deloughery TG. The efficacy and safety of direct oral anticoagulants vs traditional anticoagulants in cirrhosis. Eur J Haematol. 2017;98(4):393–7.
44. Hanafy AS, Abd-Elsalam S, Dawoud MM. Randomized controlled trial of rivaroxaban versus warfarin in the management of acute non-neoplastic portal vein thrombosis. Vasc Pharmacol. 2019;113:86–91.
45. Ai MH, Dong WG, Tan XP, Xu L, Xu C, Zhang Q, et al. Efficacy and safety study of direct-acting oral anticoagulants for the treatment of chronic portal vein thrombosis in patients with liver cirrhosis. Eur J Gastroenterol Hepatol. 2020;32(10):1395–400.
46. Lv Y, Bai W, Li K, Wang Z, Guo W, Luo B, et al. Anticoagulation and transjugular intrahepatic portosystemic shunt for the management of portal vein thrombosis in cirrhosis: a prospective observational study. Am J Gastroenterol. 2021;116(7):1447–64.
47. Lee SR, Lee HJ, Choi EK, Han KD, Jung JH, Cha MJ, et al. Direct oral anticoagulants in patients with atrial fibrillation and liver disease. J Am Coll Cardiol. 2019;73:3295–308.
48. Sasso R, Rockey DC. Anticoagulation therapy in patients with liver cirrhosis is associated with an increased risk of variceal hemorrhage. Am J Med. 2019;132:758–66.
49. De Franchis R, Bosch J, Garcia-Tsao G, Reiberger T, Ripoll C, On Behalf of the Baveno VII Faculty. Renewing consensus in portal hypertension. J Hepatol. 2022;76(4):959–74.

Novel Approaches and Disease Modifiers to Alter the Course of Cirrhotic Portal Hypertension

25

Emma Vanderschueren, Schalk Van der Merwe, and Wim Laleman

Summary In this chapter, we will give an aggregated overview of potential novel approaches and non-etiological disease modifiers as well as not yet discussed potential future novel diagnostic platforms to assess portal hypertension.

Introduction

Cirrhosis in the presence of an ongoing injuring factor is an evolving process paralleled by progressive portal hypertension which along the way is further facilitated by important accomplices like progressive gut-derived Bacterial Translocation (BT), hepatic and systemic inflammation and cirrhosis-associated Immune Dysfunction (CAID). As pressure becomes more elevated, an increasing number of events occur at subsequent pressure thresholds which not only help us to define different disease-stage but also allow us to implement disease-stage-related intervention, such as prevention of Clinically Significant Portal Gypertension (CSPH), prevention of first decompensation and prevention of further decompensation [1–7].

"To measure is to know" was already dogmatically posed by Lord Kelvin in the nineteenth century as it forms the only objective manner to monitor and modify a given physical condition.

In portal hypertension, measurement of the Hepatic Venous Pressure Gradient (HVPG) by hepatic vein catheterization has proven fundamental in stratifying risk and targeting therapy. HVPG-guided therapy has been shown to achieve a greater reduction in PHT, which may contribute to reduce the risk of (re)bleeding and of

E. Vanderschueren · S. Van der Merwe · W. Laleman (✉)
Department of Gastroenterology and Hepatology, Section of Liver and Biliopancreatic Disorders, University Hospitals Leuven, Leuven, Belgium
e-mail: emma.vanderschueren@uzleuven.be; schalk.vandermerwe@uzleuven.be; wim.laleman@uzleuven.be

R. de Franchis (ed.), *Portal Hypertension VII*,
https://doi.org/10.1007/978-3-031-08552-9_25

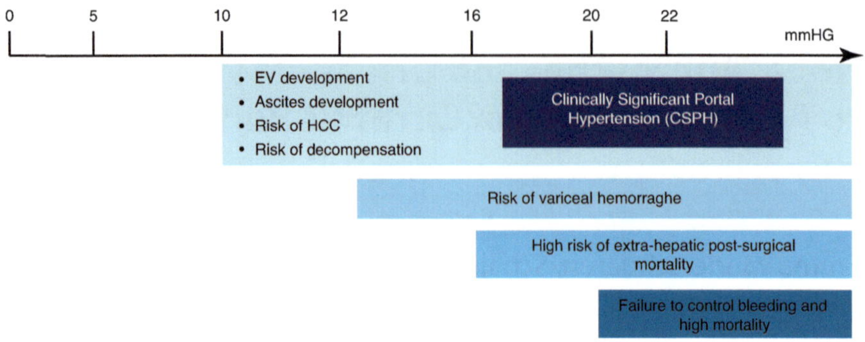

Fig. 25.1 Prognostic relevance of different portal pressure (hepatic venous pressure gradients, HVPG) thresholds. EV esophageal varices, HCC hepatocellular carcinoma

further decompensation of cirrhosis, thus contributing to better survival and bringing us closer to "precision medicine" [7, 8, 9] (Fig. 25.1).

Considering all these elements, how can we potentially intervene in addition to etiological treatment (see earlier) and expand/optimize the armamentarium for quantitative measurement of portal hypertension beyond HVPG?

Future Developments: Non-etiological-Disease Modifiers

Targeting Regression and Progression of Cirrhosis: Basic Science and Preclinical Insights in Regard to Liver Fibrosis and Angiogenesis

Progressive hepatic fibrosis in combination with nodular regeneration are classical mechanisms dominating the increased intrahepatic vascular resistance to portal flow and thus PHT. Recent data also point out a major role in these latter for intrahepatic angiogenesis and sinusoidal remodeling. While angiogenesis is defined as the formation of new vessels deriving from existing ones, sinusoidal remodelling is characterized by increased mural coverage of vessels by contractile HSC. Over the last decade, advances in the pathogenesis of liver fibrogenesis and its resolution, the development non-invasive tools to assess hepatic fibrosis and the growing public awareness of the health impact of global hepatological killers (such as (non-)alcoholic fatty liver disease, hepatitis B & C, ….) have continuously fuelled the quest for anti-fibrotic strategies. In the absence of successful etiological treatment, these approaches should either intend to halt the progression of the underlying chronic liver disease to cirrhosis (and thus prevent clinically significant portal hypertension (CSPH) to arise) or reverse CSPH in conditions where there is no etiological treatment and/or CSPH persists despite these latter efforts [10–14]. The quest for anti-fibrotic drugs/approaches has been vindicated by the illustration of reversibility of hepatic fibrosis (and cirrhosis?)

following successful antiviral therapy for hepatitis B and C with improved clinical outcomes, reduced portal pressure, and decreased all-cause mortality as a clinical trade-off [15, 16]. Advances in basic science and preclinical insights are covered by eminent experts in the field (see Chaps. 16–19).

Targeting the Active Dynamic Component of PHT: Basic Science and Preclinical Insights

At present, non-selective β-blockers, vasopressin analogues and somatostatin analogues are the mainstays of direct intervention for PHT, but these strategies are far from satisfactory and only target splanchnic hyperemia. In contrast, safe and reliable strategies to reduce the increased intrahepatic resistance in cirrhotic patients still represent a pending issue. In recent years, several preclinical and clinical trials have focused on this latter component and other therapeutic avenues. The reader is referred to an in-depth and exhaustive review [17] where novel preclinical data are highlighted. Potential interesting targets involve adrenoreceptor drugs (alfa-2-antagonists, beta-3 agonists, neuropeptide-Y), nitric oxide (PDE5-inhibitors, Sc-activators), endothelin-A-antagonist, urotensin-antagonists, Rho-kinase-inhibitor-modified albumin carrier, TNFα/NF-κB pathway-inhibitors, thromboxane-receptor antagonists, FXR-agonists.

Multi-Targeted Intervention Translated to the Patient Level: Clinical Insights–Ready for Prime Time?

As mentioned earlier, BT, hepatic and systemic inflammation together with PHT have been identified as the fundamental elements to drive the course of cirrhosis towards decompensation. By targeting one or several of these essentials one could imagine altering the evolution of a given cirrhotic patient. Currently ongoing efforts to do have been bundled in different Horizon2020 projects such MICROB-PREDICT [18], DECISION [19], LIVER-HOPE [20] and GALAXY [21].

The following high-potential therapeutic avenues – albeit in different manners and intensity - have reached adolescence nd might prove to evolve to maturity in clinical practice: statins, albumin, aspirin, anticoagulation, rifaximin and Faecal Microbiota Transplantation (FMT).

Statins

Statins are classically known and prescribed as lipid-lowering drugs, but they additionally have shown to carry hepatoprotective potential by decreasing hepatic inflammation and fibrosis as well as lowering PHT. The molecular mechanisms, preclinical data and clinical evidence in support of statins in the prevention of decompensation will be discussed by Prof. Jonel Trebicka (Chap. 24) [22–25].

Albumin

Albumin is to be considered as pleiotropic molecule, a hepatological "jack of all trades", as it carries different important biological functions such as preservation of colloid osmotic pressure, transport, antioxidative capacity, endothelial stabilization and last but not least important immunomodulatory functions such as reduction of cytokine-induced tissue injury, inhibition of proinflammatory cytokines by bacterial DNA-stimulated immune cells, pro-resolution activities and anti-inflammatory action without impairing immune defence [26, 27].

Short-term albumin administration is well accepted and established in the treatment/prevention of circulatory dysfunction in case of AKI-HRS, spontaneous bacterial peritonitis, as well as large-volume paracentesis in case of refractory ascites [28]. A potential new paradigm is long-term albumin administration. Evidence has been recently provided via the ANSWER trial [29] that long-term albumin administration to patients with cirrhosis and uncomplicated ascites improves survivaland prevents liver-related complications (such as HRS, hepatic encephalopathy, SBP, ….), eases the management of ascites and reduces hospitalizations.

However, inconsistencies between different trials [29–31] call for further investigations aiming at confirming the beneficial effects of albumin, clarifying optimal dosing and administration schedule, and identifying ideal patient profiles. The different trials in this matter and their dissimilarities are summarized in Table 25.1.

Aspirin

Recent evidence has strengthened the importance of platelets beyond mere haemostatic properties as they are protagonists in hepatic injury and fibrosis. Moreover, they have been shown to drive a pro-inflammatory partnership with liver sinusoidal endothelial cells and Kupffer cells as well as a pro-hepatocarcinogenic environment. Consequently, the emerging notion of the role of platelets has shifted from passive bystanders to alert dynamic sentinels interacting with immune and non-immune cells [32, 33]. From this perspective, there could be a rational for antiplatelet interventions.

Simon et al. reviewed a nationwide Swedish registry of patients with chronic hepatitis B or hepatitis C diagnosed from 2005 through 2015 and who did not have a history of aspirin use (50,275 patients). Fifteen percent of the patients had cirrhosis. Patients who were starting to take low-dose aspirin (14,205 patients) were identified and subjected to propensity matching. Using Cox proportional-hazards regression modelling, the authors showed that the estimated risk of developing HCC and liver-related mortality was 31% and 27% lower for aspirin users vs. non-users. Moreover, there was an inverse relationship between the duration of aspirin use and the risk of HCC and liver-related mortality. No major differences were observed in terms of gastrointestinal bleeding (7.8% vs. 6.9%, aspirin vs. non-user) [33].

Recently, Queck et al. [34] illustrated that platelets in decompensated cirrhosis develop a higher activated state in the portal venous system compared to in the hepatic vein, driven by BT and oxidative stress, which may explain the increased risk of portal venous thrombosis in these patients.

Table 25.1 Comparison of conceptual differences between different studies evaluating "long-term" albumin administration

	ANSWER	MACHT	ATTIRE
Study population	**Stable cirrhosis and uncomplicated ascites grade 2–3—outpatient**	**Cirrhosis with ascites waiting for liver transplantation (LT)**	**Hospitalized patients for acute complications of cirrhosis**
Trial design	**Multi-centre open-label RCT (n = 431)**	**Randomized placebo-controlled (n = 173)**	**Multi-Centre open-label RCT (n = 777)**
Intervention	**HA 40 g twice a week for 2 weeks, then 40 g every week**	**HA 40 g every 15 days plus midodrine**	**HA targeted to achieve and maintain serum albumin level of > 3.0 g/dl from day 3**
Primary endpoint	18-month mortality	Composite incidence of any complication (renal failure, hyponatraemia, infections, HE or GI bleeding)	Composite of incidence of all-cause infections, renal dysfunction and death between day 3 and 15
Duration of interventional R/	**17.6 months**	**63 days**	**14 days**
Effect on albumin concentration	Significant increase (0.7–0.8 g/dl) in the albumin arm	No differences between groups	Significant increase in the albumin arm (above 3.0 g/dL)
Baseline MELD	**12–13**	**17–18**	**19**
Patients waiting for LT	34 (8%)	173 (100%)	–
Impact on complications of cirrhosis and survival	**Positive**	**Negative**	**Negative**

Finally, a multi-centre retrospective propensity scores matching analysis of 587 patients undergoing TIPS studied the impact of post-TIPS initiated aspirin (163 ASA + vs. 424 ASA –) [35]. The authors found that patients initiated on ASA after TIPS had a better transplant-free survival than non-ASA-users. There was no effect on post-shunt hepatic encephalopathy, relapse of the indication for TIPS or impact on the need for revision of TIPS.

Anticoagulation

Traditionally, cirrhosis was considered a hypocoagulable state with increased bleeding risk. This dogma has critically been revised over the last decade since coagulation in cirrhosis appears to be a complex and fragile balance between endogenous procoagulant (hypercoagulable) and anticoagulant (hypocoagulable) factors. This is corroborated by an increased risk of venous thrombosis cirrhosis on the one hand and by the fact that variceal bleeding is unrelated to deranged haemostasis on the other hand. For an in-depth review on the coagulation cascade in cirrhosis, the reader is referred elsewhere [36]. Accumulating evidence indicates that the hypercoagulant, prothrombotic state in cirrhosis promotes accelerated fibrogenesis and conversely that a hypocoagulable state slows down fibrosis [37–39].

At present, two mechanisms are considered mutually enforcing to explain that intrahepatic thrombus formation is not just a consequence but rather an active player in the progression of the disease. *First*, there is the 'parenchymal extinction'-theory by Wanless [40]. This hypothesis sets off from a persisting inflammatory injury-causing venous thrombosis. The consequent hepatocyte ischaemia and death leads to a Parenchymal Extinction Lesion (PEL) which in turn induces tissue to collapse so that adjacent portal tracts and hepatic veins are approximated and replaced by fibrous septa. When PELs accumulate and become confluent, cirrhosis evolves. *Secondly*, there is direct thrombin-mediated hepatic Stellate Cell (HSC) activation via Protease-Activated Receptors (PARs). More specifically, thrombin, in addition to its haemostatic role, drives a wide range of biological activities, as a serine protease, which enables signalling to a variety of cell types, including the HSC, through G-protein coupled PARs [41] as well as platelets [42] as discussed in the previous paragraph.

The only and most solid clinical proof-of-concept at present that coagulation is involved in cirrhosis progression, beyond Virchow's triad and thrombosis, is the study by Villa et al. [43]. In this single-centre non-blinded randomized controlled trial, 70 cirrhotic outpatients (Child B7-C10) without portal vein thrombosis were randomized to daily enoxaparin 4000 IU/d vs. no anticoagulation for 48 weeks. Enoxaparin treatment showed to be safe and appeared to delay the occurrence of hepatic decompensation and improve survival. Beneficial effects were considered related to improved intestinal microcirculation. The use, safety, and clinical impact of warfarin and newer drugs such as direct oral anticoagulation (DOACs) are discussed in Chap. 24.

Rifaximin

Rifaximin (RFX), an oral non-systemic antibiotic with minimal GI absorption and broad-spectrum antibacterial coverage, is established as an effective add-on to lactulose for the prevention of overt hepatic encephalopathy recurrence after the second episode [44]. However, experimental data suggest additional effects beyond toxin-lowering capacities. By modulating the gut microbiome, RFX might interfere with hepatic and systemic inflammation, bacterial translocation, and portal hypertension (see Fig. 25.2), which form the main drivers of evolving cirrhosis and the pathophysiological cascade leading to decompensation, acute-on-chronic liver failure and end-organ dysfunction [45, 46]. This premise is the subject of the ongoing Horizon-2020 project LIVER-HOPE [46]. RFX will be discussed in Chap. 26.

Faecal Microbiota Transplantation (FMT) and Gut Liver Axis

Modalities of the gut liver axis, potential biomarkers, and associated biomarkers as well as FMT will be discussed exhaustively by Prof. Jasmohan Bajaj and Prof. Aleksander Krag in Chap. 26.

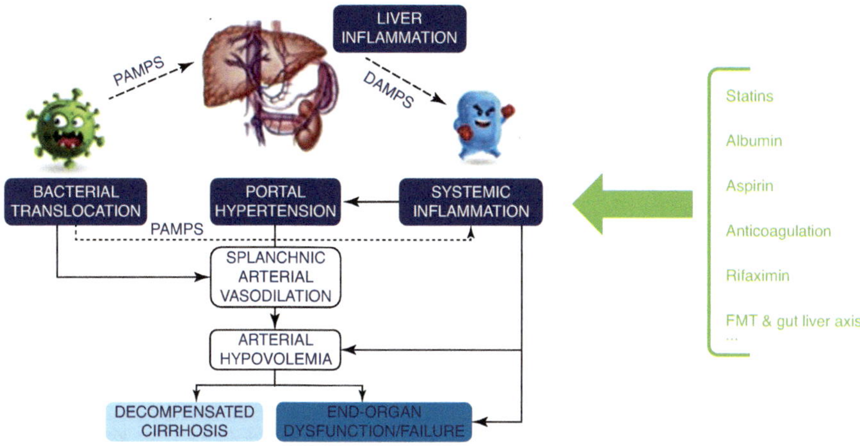

Fig. 25.2 Potential non-etiological disease modifiers

Expanding the Horizon in Portal Pressure Measurement: EUS-Guided Portal Pressure Gradient (EUS-PPG) Measurement

CSPH drives the development of gastroesophageal varices and other liver-related decompensations [7]. The current "*pragmatic*" approach in terms of primary prophylaxis of gastroesophageal varices (GOV) bleeding is based on clinical, biochemical, endoscopic and elastography findings (Fig. 25.3).

However, clinical–haemodynamic correlations clearly subscribe to an *imperative* need for quantifying or measuring PHT given its impact on risk stratification and individualized care [8, 9]. HVPG has laid the foundations for 'precision medicine' or an "à la carte" approach in PHT and remains therefore the golden standard to pursue this concept (Fig. 25.4).

Yet, practical implementation and broad dissemination of HVPG in non-academic clinical practice have proven difficult. Therefore, any additional tool that could expand the horizon in PHT assessment should be explored and objectively tested.

"***Endo-hepatology***" refers to a novel concept integrating different endoscopic ultrasound (EUS)-applications with regard to the assessment of liver diseases and portal hypertension [47]. These involve EUS-guided transgastric liver biopsy (EUS-LB), EUS-guided portal pressure gradient measurement (EUS-PPG), and evaluation of gastroesophageal varices. Endo-hepatology could be performed as a potential "one-stop-clinic" where a more comprehensive diagnostic testing can be performed in a single outpatient visit.

EUS-PPG, in contrast to HVPG, represents a tool assessing hepatic venous and portal pressure directly by puncturing these vessels transgastrically under EUS-guidance with a 25G FNA needle coupled to a digital pressure transducer at the level of the needle. By subtracting the hepatic venous pressure from the portal

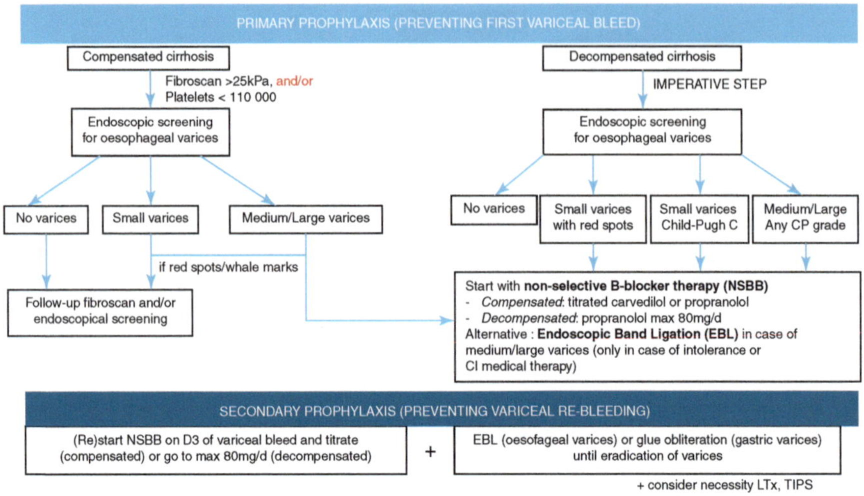

Fig. 25.3 Current pragmatic approach in the prevention of GOV-bleeding before BAVENO VI

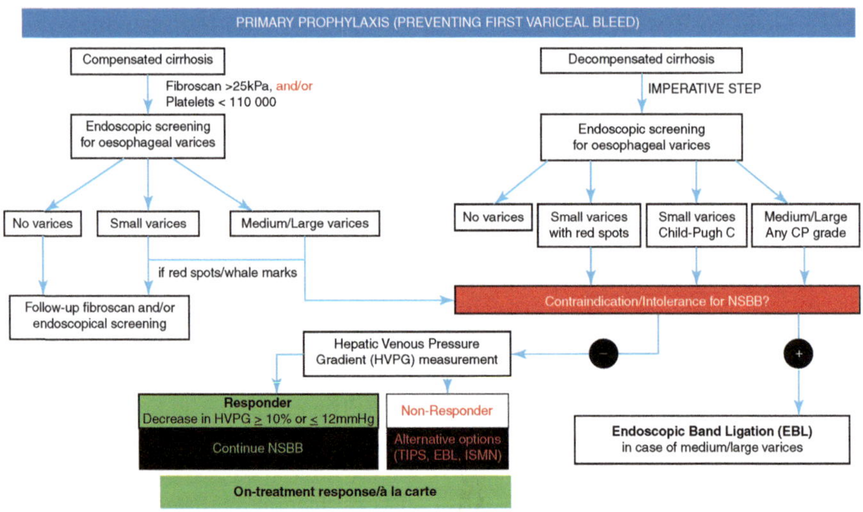

Fig. 25.4 The considered ideal pathway in current daily practice in the prevention of GOV-bleeding before BAVENO VI

pressure, the EUS-portal PPG is determined. A schematical representation is given in Fig. 25.5.

Potential conceived differences and similarities between HVPG and EUS-PPG are summarized in Table 25.2.

The first preclinical experience was published in 2016. In a healthy porcine model, Huang et al. [48] established the clinical feasibility of EUS-PPG and its correlation with HVPG.

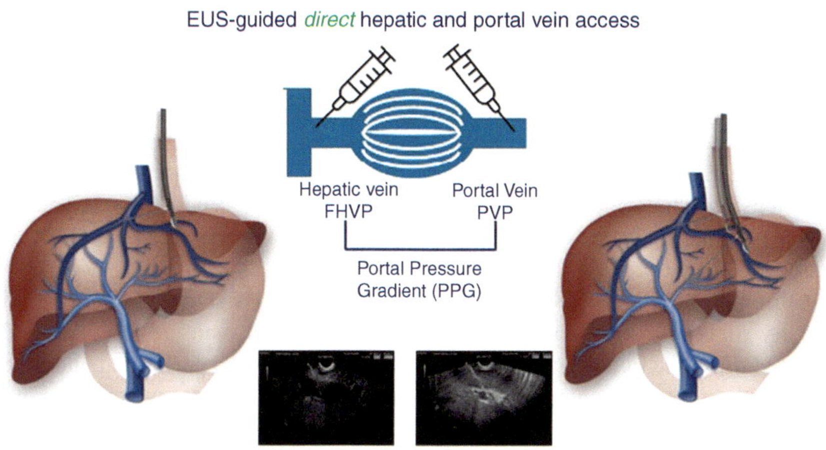

Fig. 25.5 Schematic representation of the EUS-PPG procedure

Table 25.2 Comparison of different aspects of HVPG and EUS-PPG

	HVPG	EUS-PPG
Current position	Golden standard	Novel platform
Portal pressure measurement	Indirectly	Directly
Basic type of procedure	Angiography (hepatic vein catheterization)	Endoscopy (endoscopic UltraSound)
Equipment	Dedicated X-ray machine	Conventional EUS-platform
Performing physician	Interventional radiologists, dedicated hepatologists	Gastroenterologists Dedicated hepatologists
Setting	More typical for tertiary units	Secondary and tertiary units
Training	Yes	Yes
Types of portal hypertension assessed	Sinusoidal	Sinusoidal Presinusoidal
Contra-indications	• Allergy to iodinated contrast • Platelets $<20 \times 10^9$/L or PT < 30%	• Interposed ascites in the puncture path • Anatomic anomalies preventing vessel access • Platelets $<50 \times 10^9$/L or PT < 50% • Contra-indications for upper GI endoscopy
Additive procedures in the same intervention	Transjugular liver biopsy	Transgastric liver biopsy and screening of GOV
Patient sedation	Local anaesthesia or mild sedation	Mild sedation
Procedure time	Comparable	
Feasible as an outpatient procedure	Yes	
Grade of evidence	Validated	Remains to be validated against HVPG

One year later, the same authors [49] published the human pilot study ($n = 28$) assessing feasibility, safety, and correlation with clinical parameters, but not HVPG. Following this study, the currently only available platform, EchoTip Insight (Cook Medical), was FDA-approved. Ever since real-life cohort studies [50, 51] have confirmed safety and feasibility. However, until now there is only one study correlating EUS-PPG with HVPG, albeit in 12 patients with non-cirrhotic portal hypertension (sinusoidal obstruction syndrome ($n = 10$), Budd Chiari syndrome ($n = 2$)). In this study, the device proved safe, feasible and accurate in terms of correlation [52].

Currently, the ENCOUNTER study (NCT04987034) is actively recruiting. The primary study objective is to evaluate the correlation between the calculated PPG obtained using the EchoTip® Insight™ and the HVPG method performed under general anesthesia (to guarantee safety). Patients will serve as their own controls, with both measurements obtained during the same treatment episode making it possible to directly compare the results. In a subgroup undergoing TIPS, direct portal pressure measurement will be compared to EUS-guided portal pressure. Results are awaited in 2022. Provided that EUS-PPG can show adequate correlation to the considered golden standard HVPG, further follow-up studies will need to be performed under minimal sedation.

If these prerequisites are to be realized, EUS-PPG may hold the potential to become a valuable tool given the wide availability of endosonography in GI-practice. A potential future diagnostic algorithm is depicted in Fig. 25.6.

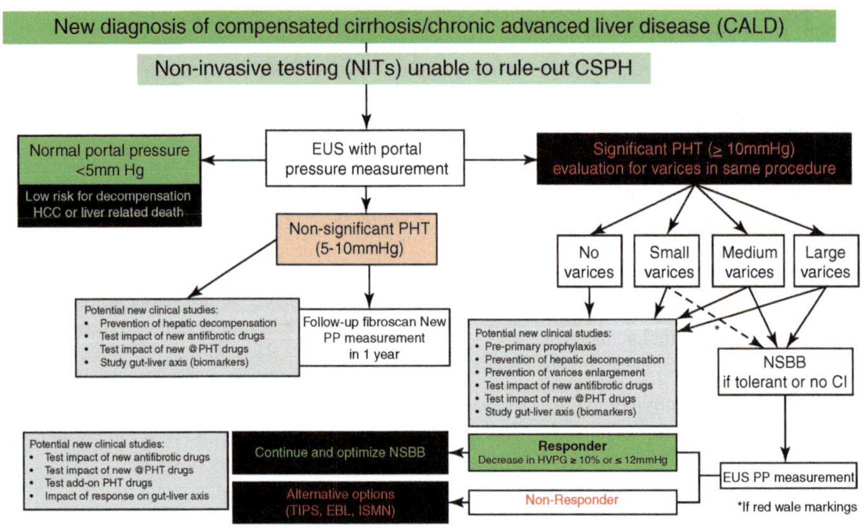

Fig. 25.6 A potential future diagnostic algorithm for personalized medicine in portal hypertension.

References

1. Trebicka J, Fernandez J, Papp M, et al. The PREDICT study uncovers three clinical courses of acute decompensated cirrhosis that have distinct pathophysiology. J Hepatol. 2020;73:842–54.
2. Trebicka J, Reiberger T, Laleman W. Gut-liver axis links portal hypertension to acute-on-chronic liver failure. Visc Med. 2018;34:270–5.
3. Laleman W, Claria J, Van der Merwe S, Moreau R, Trebicka J. Systemic inflammation and acute-on-chronic liver failure: too much, not enough. Can J Gastroenterol Hepatol. 2018;2018:027152. https://doi.org/10.1155/2018/1027152.
4. Audebert CA, Peeters G, Segers P, et al. Closed-loop lumped parameters modeling of hemodynamics during cirrhogenesis in rats. IEEE Biomed Eng. 2018;65:2311–22.
5. Laleman W, Landeghem L, Wilmer A, et al. Portal hypertension: from pathophysiology to clinical practice. Liver Int. 2005;25:1079–90.
6. Albillos A, Lario M, Alvarez-Mon M. Cirrhosis-associated immune dysfunction: distinctive features and clinical relevance. J Hepatol. 2014;61:1385–96.
7. Garcia-Tsao G, Abraldes JG, Berzigotti A, Bosch J. Portal hypertensive bleeding in cirrhosis: risk stratification, diagnosis and management: 2016 practice guidance by the American Association for the Study of Liver Diseases. Hepatology. 2017;65:310–35.
8. Bureau C, Péron JM, Alric L, et al. "A la carte" treatment of portal hypertension: adapting medical therapy to hemodynamic response for the prevention of bleeding. Hepatology. 2002;36:1361–6.
9. Villanueva C, Graupera I, Aracil C, et al. A randomized trial to assess whether portal pressure guided therapy to prevent variceal rebleeding improves survival in cirrhosis. Hepatology. 2017;65:1693–707.
10. Torok N, Dranoff JA, Schuppan D, Friedman SL. Strategies and endpoints of antifibrotic drug trials: summary and recommendations from the AASLD Emerging Trends Conference, Chicago, June 2014. Hepatology. 2015;62:627–34.
11. Trautwein C, Friedman SL, Schuppan D, Pinzani M. Hepatic fibrosis: concept to treatment. J Hepatol. 2015;62:S15–24.
12. Thabut D, Shah V. Intrahepatic angiogenesis and sinusoidal remodeling in chronic liver disease: new targets for the treatment of portal hypertension. J Hepatol. 2010;53:976–80.
13. Hytiroglou P, Theise ND. Regression of human cirrhosis: an update, 18 years after the pioneering article by Wanless et al. Virchows Arch. 2018;473:15–22.
14. Desmet VJ, Roskams T. Cirrhosis reversal: a duel between dogma and myth. J Hepatol. 2004;40:860–7.
15. Marcellin P, Gane E, Buti M, et al. Regression of cirrhosis during treatment with tenofovir disoproxil fumarate for chronic hepatitis B: a 5-year open-label follow-up study. Lancet. 2013;381:468–75.
16. D'Ambrosio R, Aghemo A, Rumi MG, et al. A morphometric and immunohistochemical study to assess the benefit of a sustained virological response in hepatitis C virus patients with cirrhosis. Hepatology. 2012;56:532–43.
17. Schwabl P, Laleman W. Novel treatment options for portal hypertension. Gastroenterol Rep (Oxf). 2017;5:90–103.
18. https://microb-predict.eu/
19. https://decision-for-liver.eu/
20. https://www.liverhope-h2020.eu/index_nl
21. https://www.sdu.dk/en/forskning/galaxy
22. Trebicka J, Hennenberg M, Laleman W, et al. Atorvastatin lowers portal pressure in cirrhotic rats by inhibition of RhoA/Rho-kinase and activation of endothelial nitric oxide synthase. Hepatology. 2007;46:24253.
23. Trebicka J, Hennenberg M, Odenthal M, et al. Atorvastatin attentuates hepatic fibrosis in rats after bile duct ligation via decreased turnover of hepatic stellate cells. J Hepatol. 2010;53:702–12.

24. Abradles JG, Villanueva C, Aracil C, et al. Addition of simvastatin to standard therapy for the prevention of variceal rebleeding does not reduce rebleeding but increases survival in patients with cirrhosis. Gastroenterology. 2016;150:1160–70.
25. Kim RG, Loomba R, Prokop LJ, et al. Statin use and risk of cirrhosis and related complications in patients with chronic liver diseases: a systematic review and meta-analysis. Clin Gastroenterol Hepatol. 2017;15:1521–30.
26. Bernardi M, Angeli P, Claria J, et al. Albumin in decompensated cirrhosis: new concepts and perspectives. Gut. 2020;69:1127–38.
27. Garcia-Martinez R, Caraceni P, Bernardi M, et al. Albumin: pathophysiological basis of its role in the treatment of cirrhosis and its complications. Hepatology. 2013;58:1836–46.
28. Angeli P, Bernardi M, Villanueva C, et al. EASL Clinical Practice Guidelines for the management of patients with decompensated cirrhosis. J Hepatol. 2018;69:406–60.
29. Caraceni P, Riggio O, Angeli P, et al. Long-term albumin administration in decompensated cirrhosis (ANSWER): an open-label randomized trial. Lancet. 2018;391(10138):2417–29.
30. China L, Freemantle N, Forrest E, et al. A randomized trial of albumin infusions in hospitalized patients with cirrhosis. N Engl J Med. 2021;384:808–17.
31. Sola E, Sole C, Simon-Talero M, et al. Midodrine and albumin for prevention of complications in patients with cirrhosis awaiting liver transplantation. A randomized placebo-controlled trial. J Hepatol. 2018;69:1250–9.
32. Ramadori P, Klag T, Malek NP, et al. Platelets in chronic liver disease, from bench to bedside. JHEP Rep. 2019;1:448–59.
33. Simon TG, Duberg AS, Aleman S, et al. Association of aspirin with hepatocellular carcinoma and liver-related mortality. N Engl J Med. 2020;382:1018–28.
34. Queck A, Carnevale R, Uschner FE, et al. Role of portal venous platelet activation in patients with decompensated cirrhosis and TIPS. Gut. 2020;69:1535–6.
35. Seifert LL, Schindler P, Sturm L, et al. Aspirin improves transplant-free survival after transjugular intrahepatic portosystemic shunt implantation: a multicenter analysis. J Hepatol. 2022; https://doi.org/10.1007/s12072-022-10330-x.
36. Lisman T, Caldwell SH, Burroughs AK, et al. Hemostasis and thrombosis in patients with liver disease: the ups and downs. J Hepatol. 2010;53:362–71.
37. Wright M, Goldin R, Hellier S, et al. Leiden polymorphism and the rate of fibrosis development in chronic hepatitis C virus infection. Gut. 2003;52:1206–10.
38. Papatheodoridis GV, Papakonstantinou E, Andrioti E, et al. Thrombotic risk factors and extent of liver fibrosis in chronic viral hepatitis. Gut. 2003;52:404–9.
39. Yee TT, Griffioen A, Sabin CA, et al. The natural history of HCV in a cohort of haemophilic patients infected between 1961 and 1985. Gut. 2000;47:845–51.
40. Wanless IR, Wong F, Blendis LM, et al. Hepatic and portal vein thrombosis in cirrhosis: possible role in development of parenchymal extinction and portal hypertension. J Hepatol. 2010;21:1238–47.
41. Fiorucci S, Antonelli E, Distrutti E, et al. PAR1 antagonism protects against experimental liver fibrosis. Role of proteinase receptors in stellate cell activation. Hepatology. 2004;39:365–75.
42. Marra F, DeFranco R, Grappone C, et al. Expression of the thrombin receptor in human liver: up-regulation during acute and chronic injury. Hepatology. 1998;27:462–71.
43. Villa E, Camma C, Marietta M, et al. Enoxaparin prevents portal vein thrombosis and liver decompensation in patients with advanced cirrhosis. Gastroenterology. 2012;143:1253–60.
44. Bass NM, Mullen K, Sanyal A, et al. Rifaximin treatment in hepatic encephalopathy. NEJM. 2010;362:1071–81.
45. Trebicka J, Macnaughtan J, Schnabl B, et al. The microbiota in cirrhosis and its role in hepatic decompensation. J Hepatol. 2021;75(Suppl 1):S67–81.
46. Caraceni P, Vargas V, Sola E, et al. The use of rifaximin in patients with cirrhosis. Hepatology. 2021;74:1660–73.
47. Rudnick SR, Conway JD, Russo MW. Current state of endohepatology: diagnosis and treatment of portal hypertension and its complications with endoscopic ultrasound. World J Hepatol. 2021;13:887–95.

48. Huang JY, Samarasena JB, Tsujino T, et al. EUS-guided portale pressure gradient measurement with a novel 25-gauge needle device versus transjgular approach: a comparison animal study. Gastrointest Endosc. 2016;84:358–62.
49. Huang JY, Samarasena JB, Tsujino T, et al. EUS-guided portal pressure gradient measurement with a simple novel device: a human pilot study. Gastrointest Endosc. 2017;85:996–1001.
50. Kolb JM, Chang KJ, Wallace MB, et al. EUS-guided portal pressure gradient predicts clinical parameters of liver disease: the first multi-center experience of endohepatology. Gastrointest Endosc. 2021;93:ID3522387. (abstract)
51. Choi AY, Shah S, Kolb JM, et al. EUS-guided portal pressure gradient measurements predict fibrosis on liver histology. Gastrointest Endosc. 2021;93:ID3522384. (abstract)
52. Zhang W, Peng C, Zhang S, et al. EUS-guided portal pressure gradient measurement in patients with acute or subacute portal hypertension. Gastrointest Endosc. 2021;93:565–72.

Targeting the Gut Microbiome in Cirrhosis

26

Aleksander Krag and Jasmohan S. Bajaj

Introduction

The last decade has driven huge scientific attention to studies of the human gut microbiome. In 2020, more than 21,000 new publications emerged on PubMed, an increase from 1100 in 2010. Also, the literature on the gut microbiome in human cirrhosis is expanding [1, 2]. Technological breakthroughs in sequencing with decreasing costs and high throughput platforms have propelled the development. The components of the human gut microbiome are bacteria, viruses, fungi, and archaea. Most of the literature is based on the studies of bacteria and there is only limited data on fungi and viruses in patients with cirrhosis. The significance of gut bacteria in health and disease is generally accepted [3], however, our ability to study gut viruses is limited by the low level of annotations, high signals from bacteriophages, and limited data on functional aspects. Similarly, it is still early days in the study of gut fungi, the methodology is not robust and varies across platforms, further it is still debated if the fungi colonize the gut or just pass

A. Krag (✉)
Centre for Liver Research, Department of Gastroenterology and Hepatology, Odense University Hospital, Odense, Denmark

Institute of Clinical Research, University of Southern Denmark, Odense, Denmark
e-mail: Aleksander.krag@rsyd.dk

J. S. Bajaj
Division of Gastroenterology, Hepatology and Nutrition, Virginia Commonwealth University and McGuire VA Medical Center, Richmond, VA, USA
e-mail: jasmohan.bajaj@vcuhealth.org

R. de Franchis (ed.), *Portal Hypertension VII*,
https://doi.org/10.1007/978-3-031-08552-9_26

through and mainly represent fungi in the food [4]. Consequently, this chapter will focus on gut bacteria in chronic liver diseases. The gut-liver axis is considered a key driver of disease progression and plays a central role in the development of complications in cirrhosis and consequently holds an untapped potential for therapeutic interventions [2, 5].

Methods to Study the Microbiome

Table 26.1 outlines the available methods to measure the presence of gut bacteria in stool samples. qPCR is fast and very cheap and allows resolution at the species or strain level. However, the use of qPCR is not useful for studies of the integrated community due to the low throughput. Most available data on the gut bacteria in cirrhosis is based on 16s rRNA sequencing, which has become cheap and widely available. The limitation is the relatively low throughput and resolution. There is a trend toward metagenomic studies; they are more expensive but allow sequencing of genomic DNA, thus holding the potential to capture the entire community at species level [4]. While such genomic approaches answer the question "who is there?" they do not allow inference regarding "what's going on." Thus, the next frontier is to study functional aspects such as meta transcriptomics but also proteins and metabolites in the gut [6, 7].

A key challenge is the lack of consensus on standardization in sampling and processing. This hampers the comparison across studies. Several initiatives are underway to accommodate this, however, there is still a long way to go, i.e., we still do not know how to define a healthy microbiome [4]. Further to this, it has become clear that analysis and interpretation of the complex sequencing data are very challenging. Beyond the bioinformatics, which most MDs do not comprehend, there are several potential confounding factors that are often not accounted for [8, 9]. Main confounders include drugs [9, 10], food, alcohol, stool frequency, BMI, sex, age, and world region [4, 11].

Table 26.1 Methods to measure presence of gut bacteria in stools samples

Method	Target	Throughput	Taxonomic resolution/spices	Gene function
qPCR	Specific region	1	Species/strain	–
16S rRNA	Amplifies 16s rRNA	100–500	Genus	–
Metagenomics	Genomic DNA	0.5–1 million genes	Species/strain	Presence of genes
Meta transcriptomics	Expressed genes	0.5–1 million genes	Species/strain	Expression of genes

Approaches to Target the Microbiome

The gut microbiome can be manipulated in several ways as outlined in Table 26.2 [3, 11, 12]. In general, approaches can be nontargeted (i.e., FMT or antibiotics) or targeted (i.e., phage therapy). The nontargeted therapy has proven very effective in gut infections such as Cl. Difficile, but several other indications are being assessed [13]. However, with a deeper understanding of disease mechanisms and effectors related to specific bacteria targeted therapies such ad phage therapy will likely emerge as more effective, safe, and with less side effects [14].

The clinical interest in the microbiome also includes the potential for microbiome-based biomarkers to inform clinical decision-making [1]. Several studies observed a strong association between changes in microbiome and severity of liver disease and complications such as hepatic encephalopathy and acute on chronic liver failure [15]. However, the regulatory barriers for new biomarkers to enter clinical medicine are high and must meet specific criteria [16], i.e., clinical benefit must be scientifically proven under the new "medical device regulation" in the EU [17]. But biomarkers of the gut microbiome should be explored further as clinical tools to inform stage of disease (diagnostic), risk of progression (prognostic), likelihood to benefit from intervention (predictive), and efficacy of intervention [1].

Table 26.2 Approaches to target the gut microbiome

Intervention	Example	Concept
Prebiotics [41]	Dietary fibers	Feed the bug
Probiotics	VSL3 [42]	Compete with the bug
Postbiotics [43]	SCFA	Skip the bug
Synbiotics [5]	Pre + pro combo	Feed and compete
Diet [44]	Low FODMAP	Control substrate
Antibiotics	Rifaximin [45]	Kill the bug
FMT [46]	FMT [47]	Substitute the bugs
Phage therapy	Target cytolytic E. faecalis [14]	Bug the bug
Bio-engineered bacteria [48]	Genetically modified E. Nissle 1917	Cheat the bug/"sense-kill"
Drugs [10]	Metformin [49]	Drug the bug

Gut Microbiome in Cirrhosis

Cirrhosis is a microcosm of several influences including etiology of disease, disease stage, medication, and hospital exposure that leads to characteristic changes in the gut-liver axis [18]. Patients with cirrhosis suffer from several complications that are directly or indirectly related to the gut microbiota, such as Hepatic Encephalopathy (HE), Spontaneous Bacterial Peritonitis (SBP), and Multidrug-Resistant (MDR) infections [2, 19]. Microbial alterations have been found in the stool, saliva, intestinal mucosa, skin, and blood in patients with cirrhosis compared to healthy controls in studies from across the world [2].

Moreover, there is increasing evidence that in addition to structural changes in the microbiome, their products, and specific characteristics may be related to the development and progression of cirrhosis [20–22]. These include metabolites such as bile acids, trimethylamines, tryptophan metabolites, and ammonia-related biochemicals and specific genes that predispose toward antibiotic resistance and virulence [23–26].

Microbial composition and function are also modulated by interventions and medications (Fig. 26.1) such as alcohol intake, changes in diet or cultural factors, medications such as proton pump inhibitors and antibiotics, as well as liver transplant [27]. Microbes can be used to detect the presence of cirrhosis, and define the risk of clinically relevant outcomes, such as hospitalization, rehospitalization, death, and ACLF [28–31]. Moreover, modulating microbes directly or indirectly in cirrhosis could have beneficial consequences that could be a common thread linking decompensation and further decompensation [2].

Fig. 26.1 Factors that influence the structure and function of the gut microbiome

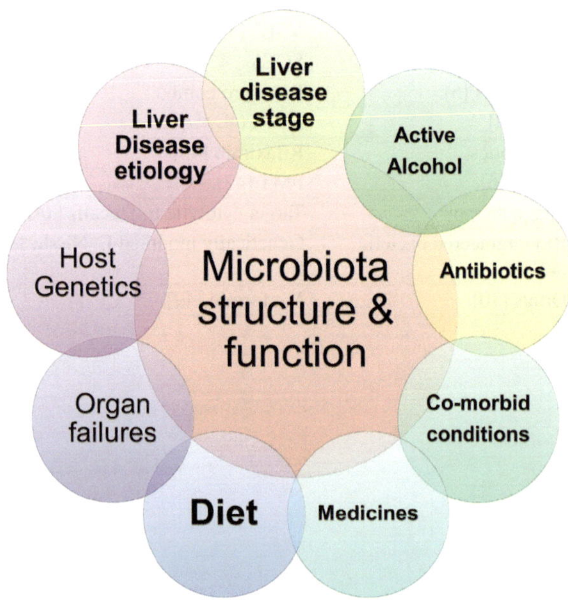

Rifaximin is an unusual compound that has activity beyond that of a simple antibiotic [32]. It has shown clear efficacy and safety in the prevention of recurrence of HE and one RCT has also shown to improve survival in inpatients admitted with HE. The mechanism of action is evolving with greater emphasis on ammonia reduction even in the germ-free setting, changing of bacterial-phage linkages focused on urease and ammonia production, and impacts on salivary and immune markers [33–35]. The evidence regarding complications of cirrhosis other than HE is not as convincing but retrospective and population-based studies or smaller RCTs have shown protection from infections and further decompensation [32, 36, 37].

FMT has been studied in several aspects of cirrhosis (details in Table 26.3), including alcohol use disorder, hepatic encephalopathy, and alcohol-related hepatitis [27, 38]. Most experiences are either small RCTs or case series with comparisons with historical controls [38]. Regardless of the study design, FMT from donors that have been well-characterized under the regulatory agency guidelines is safe. In cases where the donors are inadequately screened, there has been an instance of MDRO transmission to the patient with cirrhosis.

Regarding the outcomes of these trials, there is a trend toward cognitive improvement, changes in microbial structure and function, reduction in antibiotic-resistance genes, and reduction in alcohol craving in cirrhosis [39]. In alcohol-related hepatitis, there is also an improvement in survival based on the historical controls [38, 40]. However, further studies are needed and are underway.

Table 26.3 Studies on fecal microbiota transplantation in human cirrhosis

Study and design	Samples/groups compared	Route and duration of FMT	Findings and significance
Alcohol-related disorders			
Bajaj et al. Hepatology 2020 Double-blind RCT [50]	Men with AUD and cirrhosis who were not successful on abstinence using current therapies	– one-time enema vs. placebo – reduced short-term alcohol craving and consumption with higher SCFA in FMT – lower AUD-related hospitalizations long-term in FMT vs. placebo	Reduction of addictive behavior resulting in long-term reduction in AUD-related hospitalizations over 6 months
Phillips et al. CGH 2017 [38] Case series with historical controls	Men with steroid resistant alcohol-related hepatitis	– 1-week of daily NJT FMT from many donors – 1-year open-label study with historical controls	Higher survival versus controls
Phillips et al. Indian J Gastro 2018 [40] Open-label trial	Men with alcohol-related hepatitis	– 1 week of daily NJT FMT from many donors versus standard therapy – 3-month follow-up	3-month survival was higher in FMT group while 1-month survival was similar

(continued)

Table 26.3 (continued)

Study and design	Samples/groups compared	Route and duration of FMT	Findings and significance
Cirrhosis			
Kao et al. Hepatol 2016 [51] Case report	One patient with HE	– 1 FMT via colonoscopy followed by 3-weekly enemas – safe and well-tolerated with improvement in cognitive function	Case report of FMT in cirrhosis with brain function improvement
Bajaj et al. Hepatol 2017 and 2018 Randomized clinical trial of SOC vs. FMT+ antibiotics [52, 53]	20 HE patients on lactulose and rifaximin	– one 90 mL of enema after 5 days of broad-spectrum antibiotics – safe and well-tolerated, improvement in hospitalizations, dysbiosis, and SCFAs post antibiotics after FMT	First randomized trial to study this in cirrhosis and HE and under investigational new drug under FDA
Mehta et al. 2018 Indian J Gastro [54] Case series	10 HE patients open label	– one FMT via colonoscopy – sustained clinical response at week 20 in 6 patients	Further evidence about safety and potential efficacy
Bajaj et al. Randomized trial of oral capsules vs. placebo Hepatology and JCI Insight 2019 [47, 55]	20 HE patients on lactulose and rifaximin	– 15 capsules of FMT vs. placebo once – brain function improved and outcomes got better in those with secondary BA formation	Oral capsular FMT is also safe in HE and success can be linked to secondary BA formation

References

1. Trebicka J, Bork P, Krag A, Arumugam M. Utilizing the gut microbiome in decompensated cirrhosis and acute-on-chronic liver failure. Nat Rev Gastroenterol Hepatol. 2021;18(3):167–80.
2. Trebicka J, Macnaughtan J, Schnabl B, Shawcross DL, Bajaj JS. The microbiota in cirrhosis and its role in hepatic decompensation. J Hepatol. 2021;75(Suppl 1):S67–81.
3. Albillos A, de Gottardi A, Rescigno M. The gut-liver axis in liver disease: pathophysiological basis for therapy. J Hepatol. 2020;72(3):558–77.
4. Shanahan F, Ghosh TS, O'Toole PW. The healthy microbiome-what is the definition of a healthy gut microbiome? Gastroenterology. 2021;160(2):483–94.
5. Swanson KS, Gibson GR, Hutkins R, et al. The International Scientific Association for Probiotics and Prebiotics (ISAPP) consensus statement on the definition and scope of synbiotics. Nat Rev Gastroenterol Hepatol. 2020;17(11):687–701.
6. Integrative HMPRNC. The integrative human microbiome project. Nature. 2019;569(7758):641–8.
7. Trost K, Ahonen L, Suvitaival T, et al. Describing the fecal metabolome in cryogenically collected samples from healthy participants. Sci Rep. 2020;10(1):885.

8. Rothschild D, Weissbrod O, Barkan E, et al. Environment dominates over host genetics in shaping human gut microbiota. Nature. 2018;555(7695):210–5.
9. Vieira-Silva S, Falony G, Belda E, et al. Statin therapy is associated with lower prevalence of gut microbiota dysbiosis. Nature. 2020;581(7808):310–5.
10. Maier L, Pruteanu M, Kuhn M, et al. Extensive impact of non-antibiotic drugs on human gut bacteria. Nature. 2018;555(7698):623–8.
11. Schmidt TSB, Raes J, Bork P. The human gut microbiome: from association to modulation. Cell. 2018;172(6):1198–215.
12. Delzenne NM, Bindels LB. Food for thought about manipulating gut bacteria. Nature. 2020;577(7788):32–4.
13. Allegretti JR, Mullish BH, Kelly C, Fischer M. The evolution of the use of faecal microbiota transplantation and emerging therapeutic indications. Lancet. 2019;394(10196):420–31.
14. Duan Y, Llorente C, Lang S, et al. Bacteriophage targeting of gut bacterium attenuates alcoholic liver disease. Nature. 2019;575(7783):505–11.
15. Acharya C, Bajaj JS. Altered microbiome in patients with cirrhosis and complications. Clin Gastroenterol Hepatol. 2019;17(2):307–21.
16. FDA-NIH Biomarker Working Group. BEST (Biomarkers, EndpointS, and other Tools) Resource. Silver Spring, MD: Food and Drug Administration; 2016.
17. Wilkinson B, van Boxtel R. The medical device regulation of the European Union intensifies focus on clinical benefits of devices. Ther Innov Regul Sci. 2020;54(3):613–7.
18. Gines P, Krag A, Abraldes JG, Sola E, Fabrellas N, Kamath PS. Liver cirrhosis. Lancet. 2021;398:1359–76.
19. Bajaj JS, Kamath PS, Reddy KR. The evolving challenge of infections in cirrhosis. N Engl J Med. 2021;384(24):2317–30.
20. Hosseinkhani F, Heinken A, Thiele I, Lindenburg PW, Harms AC, Hankemeier T. The contribution of gut bacterial metabolites in the human immune signaling pathway of non-communicable diseases. Gut Microbes. 2021;13(1):1–22.
21. Krautkramer KA, Fan J, Backhed F. Gut microbial metabolites as multi-kingdom intermediates. Nat Rev Microbiol. 2021;19(2):77–94.
22. Kakiyama G, Pandak WM, Gillevet PM, et al. Modulation of the fecal bile acid profile by gut microbiota in cirrhosis. J Hepatol. 2013;58(5):949–55.
23. Bajaj JS, Kakiyama G, Cox IJ, et al. Alterations in gut microbial function following liver transplant. Liver Transpl. 2018;24(6):752–61.
24. Jimenez B, Montoliu C, MacIntyre DA, et al. Serum metabolic signature of minimal hepatic encephalopathy by (1)H-nuclear magnetic resonance. J Proteome Res. 2010;9(10):5180–7.
25. Roager HM, Licht TR. Microbial tryptophan catabolites in health and disease. Nat Commun. 2018;9(1):3294.
26. Shamsaddini A, Gillevet PM, Acharya C, et al. Impact of antibiotic resistance genes in gut microbiome of patients with cirrhosis. Gastroenterology. 2021;161(2):508–521.e7.
27. Bajaj JS, Khoruts A. Microbiota changes and intestinal microbiota transplantation in liver diseases and cirrhosis. J Hepatol. 2020;72(5):1003–27.
28. Caussy C, Tripathi A, Humphrey G, et al. A gut microbiome signature for cirrhosis due to nonalcoholic fatty liver disease. Nat Commun. 2019;10(1):1406.
29. Bajaj JS. Gut microbiota as biosensors in patients with cirrhosis. Cell Mol Gastroenterol Hepatol. 2019;8(2):231–3.
30. Bajaj JS, Vargas HE, Reddy KR, et al. Association between intestinal microbiota collected at hospital admission and outcomes of patients with cirrhosis. Clin Gastroenterol Hepatol. 2019;17(4):756–65.e3.
31. Sole C, Guilly S, Da Silva K, et al. Alterations in gut microbiome in cirrhosis as assessed by quantitative metagenomics: relationship with acute-on-chronic liver failure and prognosis. Gastroenterology. 2021;160(1):206–18. e13
32. Caraceni P, Vargas V, Sola E, et al. The use of rifaximin in patients with cirrhosis. Hepatology. 2021;74(3):1660–73.

33. Kang DJ, Kakiyama G, Betrapally NS, et al. Rifaximin exerts beneficial effects independent of its ability to alter microbiota composition. Clin Transl Gastroenterol. 2016;7(8):e187.
34. Bajaj JS, Sikaroodi M, Shamsaddini A, et al. Interaction of bacterial metagenome and virome in patients with cirrhosis and hepatic encephalopathy. Gut. 2021;70(6):1162–73.
35. Lv XY, Ding HG, Zheng JF, Fan CL, Li L. Rifaximin improves survival in cirrhotic patients with refractory ascites: a real-world study. World J Gastroenterol. 2020;26(2):199–218.
36. Salehi S, Tranah TH, Lim S, et al. Rifaximin reduces the incidence of spontaneous bacterial peritonitis, variceal bleeding and all-cause admissions in patients on the liver transplant waiting list. Aliment Pharmacol Ther. 2019;50(4):435–41.
37. Vlachogiannakos J, Viazis N, Vasianopoulou P, Vafiadis I, Karamanolis DG, Ladas SD. Long-term administration of rifaximin improves the prognosis of patients with decompensated alcoholic cirrhosis. J Gastroenterol Hepatol. 2013;28(3):450–5.
38. Philips CA, Pande A, Shasthry SM, et al. Healthy donor fecal microbiota transplantation in steroid-ineligible severe alcoholic hepatitis: a pilot study. Clin Gastroenterol Hepatol. 2017;15(4):600–2.
39. Bajaj JS, Shamsaddini A, Fagan A, et al. Fecal microbiota transplant in cirrhosis reduces gut microbial antibiotic resistance genes: analysis of two trials. Hepatol Commun. 2021;5(2):258–71.
40. Philips CA, Phadke N, Ganesan K, Ranade S, Augustine P. Corticosteroids, nutrition, pentoxifylline, or fecal microbiota transplantation for severe alcoholic hepatitis. Indian J Gastroenterol. 2018;37(3):215–25.
41. Gibson GR, Hutkins R, Sanders ME, et al. Expert consensus document: the International Scientific Association for Probiotics and Prebiotics (ISAPP) consensus statement on the definition and scope of prebiotics. Nat Rev Gastroenterol Hepatol. 2017;14(8):491–502.
42. Dhiman RK, Rana B, Agrawal S, et al. Probiotic VSL#3 reduces liver disease severity and hospitalization in patients with cirrhosis: a randomized, controlled trial. Gastroenterology. 2014;147(6):1327–37. e3
43. Salminen S, Collado MC, Endo A, et al. The International Scientific Association of Probiotics and Prebiotics (ISAPP) consensus statement on the definition and scope of postbiotics. Nat Rev Gastroenterol Hepatol. 2021;18(9):649–67.
44. Marco ML, Sanders ME, Ganzle M, et al. The International Scientific Association for Probiotics and Prebiotics (ISAPP) consensus statement on fermented foods. Nat Rev Gastroenterol Hepatol. 2021;18(3):196–208.
45. Bureau C, Thabut D, Jezequel C, et al. The use of rifaximin in the prevention of overt hepatic encephalopathy after transjugular intrahepatic portosystemic shunt: a randomized controlled trial. Ann Intern Med. 2021;174(5):633–40.
46. Ng SC, Kamm MA, Yeoh YK, et al. Scientific frontiers in faecal microbiota transplantation: joint document of Asia-Pacific Association of Gastroenterology (APAGE) and Asia-Pacific Society for Digestive Endoscopy (APSDE). Gut. 2020;69(1):83–91.
47. Bajaj JS, Salzman NH, Acharya C, et al. Fecal microbial transplant capsules are safe in hepatic encephalopathy: a phase 1, randomized, placebo-controlled trial. Hepatology. 2019;70(5):1690–703.
48. Ochoa-Sanchez R, Oliveira MM, Tremblay M, et al. Genetically engineered E. coli Nissle attenuates hyperammonemia and prevents memory impairment in bile-duct ligated rats. Liver Int. 2021;41(5):1020–32.
49. Forslund K, Hildebrand F, Nielsen T, et al. Disentangling type 2 diabetes and metformin treatment signatures in the human gut microbiota. Nature. 2015;528(7581):262–6.
50. Bajaj JS, Gavis E, Fagan A, et al. A randomized clinical trial of fecal microbiota transplant for alcohol use disorder. Hepatology. 2020;73(5):1688–700.
51. Kao D, Roach B, Park H, et al. Fecal microbiota transplantation in the management of hepatic encephalopathy. Hepatology. 2016;63(1):339–40.
52. Bajaj JS, Kakiyama G, Savidge T, et al. Antibiotic-associated disruption of microbiota composition and function in cirrhosis is restored by fecal transplant. Hepatology. 2018;68(4):1549–58.

53. Bajaj JS, Kassam Z, Fagan A, et al. Fecal microbiota transplant from a rational stool donor improves hepatic encephalopathy: a randomized clinical trial. Hepatology. 2017;66(6):1727–38.
54. Mehta R, Kabrawala M, Nandwani S, et al. Preliminary experience with single fecal microbiota transplant for treatment of recurrent overt hepatic encephalopathy—a case series. Indian J Gastroenterol. 2018;37(6):559–62.
55. Bajaj JS, Salzman N, Acharya C, et al. Microbial functional change is linked with clinical outcomes after capsular fecal transplant in cirrhosis. JCI Insight. 2019;4(24):e133410.

Impact of Non-etiological Novel Therapies in the Course of Cirrhosis: Consensus Statements of Panel 4

27

Agustín Albillos, Jonel Trebicka, Jasmohan S. Bajaj, Aleksander Krag, and Wim Laleman

4.1 The use of statins should be encouraged in patients with cirrhosis and an approved indication for statins since these agents may decrease portal pressure (A1) and improve overall survival. (B,1) (Changed)

4.2 In patients with Child B and C cirrhosis, statins should be used at a lower dose (simvastatin at max. 20 mg/d) and patients should be followed closely for muscle and liver toxicity. (A,1) In Child C cirrhosis the benefit of statins has not been proven yet and their use should be more restrictive (D,1). (Changed)

4.3 The use of aspirin should not be discouraged in patients with cirrhosis and an approved indication for aspirin, since it may reduce the risk of hepatocellular carcinoma, liver-related complications, and death. (B,2) (New)

A. Albillos (✉)
Department of Gastroenterology and Hepatology, Hospital Ramón y Cajal, Universidad de Alcalá, Madrid, Spain
e-mail: agustin.albillosm@uah.es

J. Trebicka
Translational Hepatology, Department of Internal Medicine I, University Clinic Frankfurt, Frankfurt, Germany
e-mail: Jonel.trebicka@kgu.de

J. S. Bajaj
Division of Gastroenterology, Hepatology and Nutrition, Virginia Commonwealth University and McGuire VA Medical Center, Richmond, VA, USA
e-mail: jasmohan.bajaj@vcuhealth.org

A. Krag
Department of Gastroenterology and Hepatology, University of Southern Denmark, Odense, Denmark
e-mail: aleksander.krag@rsyd.dk

W. Laleman
Department of Gastroenterology and Hepatology, University Hospital Gasthuisberg, KU Leuven, Leuven, Belgium
e-mail: wim.laleman@uzleuven.be

R. de Franchis (ed.), *Portal Hypertension VII*,
https://doi.org/10.1007/978-3-031-08552-9_27

4.4 Long-term albumin administration may reduce complications of cirrhosis and improve transplant-free survival in patients with uncomplicated ascites, but a formal recommendation cannot be given until further data become available. (B,2) (New)

4.5 Short-term albumin administration is indicated for SBP (A,1), AKI > stage 1A (C,1), large-volume paracentesis (A,1), and HRS-AKI combined with terlipressin (B,1). (New)

4.6 Primary antibiotic prophylaxis is recommended in selected patients (i.e., GI hemorrhage, Child-C cirrhosis with low protein ascites) at high risk for SBP (B,1) (New)

4.7 Secondary antibiotic prophylaxis is indicated in patients with previous SBP. (A,1) (New)

4.8 Rifaximin is indicated for the secondary prophylaxis of hepatic encephalopathy. (A,1) (New)

4.9 Rifaximin should be considered for prophylaxis of overt hepatic encephalopathy in patients with previous overt hepatic encephalopathy undergoing elective TIPS. (B,2) (New)

4.10 Rifaximin is not indicated beyond these indications, including primary or secondary prophylaxis of spontaneous bacterial peritonitis. (C,1) (New)

4.11 Anticoagulation should not be discouraged in patients with cirrhosis and an approved indication for anticoagulation, since anticoagulation may reduce liver-related outcomes in patients with and without portal vein thrombosis and may improve overall survival. (B,1) (Changed)

4.12 The safety and efficacy of DOACs to prevent cardiovascular events in patients with Child A and B cirrhosis are equivalent to those in patients without cirrhosis (B,2). DOACs are not recommended in patients with Child C cirrhosis outside study protocols. (B,2) (New)

Research Agenda

- The gut microbiome can be targeted by several means including pre-, pro-, syn- and post-biotics, diet, FMT, phage therapy, drugs, bioengineered bacteria, and antibiotics. There is a need for interventional trials with outcome assessments that includes functional aspects and clinical outcomes.
- The composition of the gut microbiome (e.g., high relative abundance of Enterobacteriaceae) in various body fluids (stool, saliva, blood, bile, intestinal mucosa, skin) is associated with the severity of cirrhosis, complications, and presence of organ failures and ACLF. Components of the gut microbiome should be explored for biomarkers to inform the stage of disease (diagnostic), to predict the risk of progression (prognostic), likelihood to benefit from intervention (predictive), and efficacy of the intervention.

- Fecal microbiota transplant (by enema or by the oral route) seems to be safe in patients with cirrhosis and hepatic encephalopathy but efficacy studies are pending.
- Anti-fibrotic strategies including the FXR-pathway, the renin–angiotensin system, and angiogenesis should be further explored in cirrhosis and portal hypertension.

Part VII

Clinical Settings 1: Preventing (First) Decompensation

Prevention of First Decompensation: Questionnaire

28

Vincenza Calvaruso, Cristina Ripoll, and Jaime Bosch

Introduction

This session approaches open issues regarding the first decompensation of cirrhosis and its prevention. Previous Baveno Consensus Conferences have focused on the prevention/treatment of variceal bleeding, however, since Baveno VI a more holistic approach to the patient with chronic advanced chronic liver disease/cirrhosis was introduced. Indeed, the most common complication of portal hypertension in compensated cirrhotic patients is not bleeding, but ascites, followed by hepatic encephalopathy. Altogether, these complications denote what has been considered "decompensated cirrhosis," which leads to a marked reduction in survival [1, 2]. Therefore, Baveno VI emphasized the need to use the development of decompensation as the clinically relevant endpoint in studies centered on patients with compensated advanced chronic liver disease [1]. An important consequence of this change

V. Calvaruso (✉)
Gastroenterology and Hepatology Unit, Department of Health Promotion Sciences Maternal and Infantile Care, Internal Medicine and Medical Specialities, PROMISE, University of Palermo, Palermo, Italy
e-mail: vincenza.calvaruso@unipa.it

C. Ripoll
First Department of Internal Medicine, Martin Luther University Halle-Wittenberg, Halle (Saale), Germany

Internal Medicine IV, Jena University Hospital Friedrich-Schiller University, Jena, Germany
e-mail: Cristina.Ripoll@med.uni-jena.de

J. Bosch
Department of Hepatology, University Clinic for Visceral Surgery and Medicine, Inselspital Bern University Hospital, Bern, Switzerland

Department of Biomedical Research, University of Bern, Bern, Switzerland
e-mail: jaume.bosch@idibaps.org

R. de Franchis (ed.), *Portal Hypertension VII*,
https://doi.org/10.1007/978-3-031-08552-9_28

in the main endpoint of therapy is that a treatment that may prevent both variceal bleeding and ascites will score automatically better than one effective only in bleeding or only in ascites.

Accordingly, this Session is devoted first, to revise (and if needed, modify) the definition of decompensation, and second, to consider the best approach to prevent first clinical decompensation.

In order to approach these aims, we initially conducted a survey to sense how international experts in the field, represented by the Baveno faculty members, perceive what is a manifestation of decompensation and what is not, how these should be defined, and which approach they feel is most appropriate for its medical prevention. This questionnaire was developed specifically for that purpose. The questions and the answers received are presented in the following sections. A second questionnaire was sent given the delay between the first questionnaire caused by the postponement of the meeting due to the pandemic, in order to approach novel issues that had arisen in the past 18 months. The information that can be withdrawn from this survey are expert opinions and have therefore the lowest level of evidence. However, it is useful to identify areas where there is enough consensus and others where there is uncertainty, in which further research is necessary.

Defining Consensus

An arbitrary definition of what represents Consensus was applied in accordance with previous Baveno Meetings [1]:

1. Agreement of at least 70% of responders: Moderate Consensus
2. Agreement of at least 80% of responders: Consensus
3. Agreement of at least 90% of responders: Strong Consensus
4. Less than 70% agreement: Lack of Consensus

The Questionnaire

The Questionnaire was sent to the members of the Baveno VII faculty. Of these, 81% ($n = 50$) response rate was obtained. There were two blocks of questions. The first block of questions deals with the definition of compensation and clinical decompensation in ACLD, while the second block deals with the prevention of the first decompensation.

Analysis of Responses

Definition of Clinical Decompensation

The main finding of the survey is that there is no consensus among the Baveno Faculty regarding what events should be included in the definition of clinical decompensation [3–5]. About 40% of responses suggested adding jaundice to the usual definition based on the development of ascites, variceal bleeding, and Hepatic Encephalopathy (HE), alone or in combination, and a few suggested including infections. Even less agreement was present as to what serum bilirubin level defined jaundice. Of note, over 70% of responders think that bacterial infections should not define decompensation (Table 28.1). So, for accepting any of these as a manifestation of decompensation there should be clear evidence, preferentially from high-quality clinical trials and/or meta-analysis.

There was moderate consensus regarding what should and should not be considered ascites (Table 28.2, Q4–8). Of note, there was consensus in considering that the presence of small amounts of perihepatic fluid detected on ultrasound was not enough to define the presence of ascites. Also, the sole presence of lower limb edema without ascites was not considered sufficient to define decompensation. In contrast, a strong agreement was obtained regarding the presence of pleural effusion

Table 6.1 Definition of Clinical Decompensation: Questionnaire Part 1

Questions	Answer choices	Responses (%)
(1) What defines clinical decompensation of cirrhosis:	(a) Development of any of the following, alone or in combination: Ascites, variceal bleeding, hepatic encephalopathy (HE)	36
	(b) Same as in (a) plus jaundice	44
	(c) Same as in (a) plus bacterial infection	16
	(d) Other	4
(2) If you suggested that jaundice should be included as a manifestation of decompensation, how should it be defined?	(a) Total serum bilirubin over 2.5 mg/dL	35
	(b) Total serum bilirubin over 5 mg/dL	22
	(c) Other:	4
	(d) I don't know	2
	(e) Jaundice should not define decompensation	37
(3) If you suggested bacterial infections should be included as a manifestation of decompensation, how should it be defined?	(a) Any bacterial infection	6
	(b) Any bacterial infection plus fever or increased PCR	8
	(c) Any bacterial infection except a positive urine culture without fever or increased PCR	13
	(d) Bacterial infection should not define decompensation	73

Table 28.2 Definition of Clinical Decompensation: Questionnaire Part 2

(4) How would you define the presence of ascites:	(a) Presence of fluid in the peritoneal cavity, confirmed by ultrasound or paracentesis	78
	(b) Presence of a small amount of perihepatic fluid detected on ultrasound but which is not clinically evident	22
	(c) Other	0
(5) Do you consider that ascites that develop in the context of an acute event (i.e., postoperative, GIB, infection) that resolves within 3 months define clinical decompensation?	(a) Yes	78
	(b) No	16
	(c) I don't know	6
(6) Can the presence of edema without ascites on clinical examination or ultrasound be considered an equivalent to finding ascites?	(a) Yes	6
	(b) No	84
	(c) I don't know	10
(7) Can the presence of pleural effusion (with a serum-pleural effusion albumin gradient over 11 g/L) without ascites on clinical examination or ultrasound be considered an equivalent to finding ascites in patients with cirrhosis?	(a) Yes	86
	(b) No	10
	(c) I don't know	4
(8) Do you consider that patients on diuretics because of previous ascites, who no longer have ascites on ultrasound should still be defined as decompensated?	(a) Yes	77
	(b) No	22
	(c) I don't know	6
(9) Do you think that an episode of bleeding (hematemesis or melena) due to an esophageal ulcer in a patient receiving prophylactic EBL defines clinical decompensation?	(a) Yes	20
	(b) No	70
	(c) I don't know	10
(10) Do you consider a patient without other complications of cirrhosis but with chronic blood loss due to portal hypertensive gastropathy to be clinically decompensated?	(a) Yes	42
	(b) No	46
	(c) I don't know	12
(11) Do you consider a patient without other complications of cirrhosis but with chronic blood loss due to GAVE to be clinically decompensated?	(a) Yes	18
	(b) No	74
	(c) I don't know	8
(12) How would you define hepatic encephalopathy for stating that a patient has decompensated cirrhosis?	(a) Presence of minimal HE	10
	(b) Presence of HE west haven grade I	32
	(c) Presence of overt HE, west haven grade II or over not related to use of sedatives	56
	(d) I don't know	2

with increased serum–pleural effusion albumin gradient as **equivalent of ascites in patients with cirrhosis** [6]. With regards to variceal bleeding (Q9–11), there is no consensus in considering portal hypertensive gastropathy with chronic blood loss as an event defining decompensation [7]. Expert-opinion agreed that bleeding from an esophageal ulcer in a patient under prophylactic endoscopic band ligation does not define clinical decompensation. There are difficulties also in defining HE. There is a strong consensus on not including "minimal HE" as a manifestation of decompensation, but there is no consensus on the degree of HE needed: most experts require overt HE (West Haven classification grade II or greater) [8], which is probably due to the diagnostic conundrums of hepatic encephalopathy grade I.

Another group of questions addresses the issue of "re-compensation" in patients with a previous decompensating event but who resolved these complications (Table 28.3). There was no consensus regarding whether a patient with a previous episode of variceal bleeding but that had no additional bleeding, ascites, or HE in the following 24 months could be considered as compensated. Furthermore, no consensus was achieved when exploring if patients that have not had further manifestations of decompensation after more than 2 years of effectively suppressing the etiologic agent responsible for the cirrhosis could be considered compensated even if in this case at least 60% of the experts consider these patients as compensated [9]. This further underlines the uncertainty in this area and need for further information from prospective studies with a longer follow-up.

In the second questionnaire, the issue of "acute decompensation" was approached. There was no consensus regarding whether "decompensation of cirrhosis" and "acute decompensation" should be distinguished, as most experts retained that both are included in the concept of decompensation although many cases of acute

Table 28.3 Definition of Clinical Decompensation: Questionnaire Part 3

(13) A patient with a previous episode of variceal bleeding that had had no additional bleeding, ascites, or HE in the past 24 months can be considered compensated?	(a) Yes	40
	(b) No	46
	(c) I don't know	14
(14) If the same patient has been for over 24 months without additional bleeding, ascites, or HE after receiving treatment to effectively suppress the etiologic agent of his liver disease, can be considered compensated?	(a) Yes	62
	(b) No	22
	(c) I don't know	16
(16) Should one differentiate between "Decompensation of cirrhosis" and "acute decompensation"?	(1) Yes	35
	(2) No, since both are included in the concept of decompensation, although many cases of "acute decompensation" may indicate ACLF	58
	(3) I don't know	7
(17) Should one differentiate between first Decompensation and subsequent episodes of decompensation?	(1) Yes, always	70
	(2) Only when the new decompensation incorporates an other manifestation of decompensation	30
	(3) I don't know	0
(15) do you think the definition of decompensation based only on clinical judgment is sufficient or should it be an evidence-based definition?	(a) Clinically based is OK	26
	(b) Should be evidence-based, according to findings of prospective studies	74

decompensation may indicate ACLF. By contrast, the experts agreed that differentiation between first decompensation and subsequent episodes (Q16 and Q17) should be done.

Finally, experts agreed that the definition of decompensation should be evidence-based rather than eminence-based (relying on the opinion of experts) (Q15).

Prevention of the First Decompensation of Cirrhosis

After the PREDESCI study [10] showed that it is possible to prevent the first clinical decompensation in patients at risk by means of continued treatment with beta-blockers, it is important to define how to select patients for therapy. In the original trial, this was assessed by demonstrating the presence of an HVPG ≥10 mmHg in a patient with compensated ACLD without varices requiring therapy (large varices or small with red whale signs). When asking the experts on how to select treatment candidates there was no consensus on starting beta-blockers only on the basis of noninvasive data strongly suggesting the presence of CSPH (for instance, transient elastography above 21 kPa) [11] (Table 28.4, Q 16). There was a strong consensus, however, in preferring the use of carvedilol rather than propranolol for preventing decompensation in patients with ACLD [12, 13].

Use of HVPG Monitoring

There was an extremely strong consensus on that HVPG monitoring is useful in evaluating if a new drug may have the potential for the treatment of portal hypertension (proof of concept phase II studies), and in randomized controlled trials of new drugs for portal hypertension [14, 15]. However, there was no consensus that HVPG monitoring could be useful or should be used to assess treatment response in clinical

Table 28.4 Prevention of the First Decompensation of Cirrhosis: Questionnaire Part 4

16) The recent PREDESCI study demonstrates that it is feasible to prevent the first clinical decompensation by means of beta-blockers in compensated cirrhotic patients with clinically significant portal hypertension (CSPH) but without high-risk varices (varices needing treatment). Since CSPH can be reasonably assumed to be present, for instance when the Fibroscan is above 21 KPa, would you: (Please, answer the following questions)		
1. Would you treat with beta-blockers your compensated patients with noninvasive data strongly suggesting the presence of CSPH?	(a) Yes	62
	(b) No	38
2. Which beta-blocker will you use:	(a) Propranolol	20
	(b) Carvedilol	80
3. Would you use HVPG monitoring to see if the patient's HVPG has decreased by more than 10% of baseline or below 10 mmHg?.	(a) Yes	33
	(b) No	63
	(c) I do not know	4

Table 28.5 Use of HVPG Monitoring: Questionnaire Part 5

(17) In compensated cirrhosis, HVPG monitoring is useful in the following conditions? (several answers possible):		
1. In evaluating if a new drug may have potential for portal hypertension;	(a) Yes	96
	(b) No	4
	(c) I do not know	0
2. In phase II-III RCTs of new drugs:	(a) Yes	92
	(b) No	6
	(c) I do not know	2
3. In guiding response to therapy:	(a) Yes	58
	(b) No	36
	(c) I do not know	6
IV. Not useful in compensated patients:	(a) Yes	23
	(b) No	62
	(c) I do not know	15

practice, emphasizing once more the need for reliable noninvasive biomarkers of treatment response or agents which are so effective that make monitoring futile. Still, there was a strong consensus considering that HVPG measurements do have a role in clinical studies evaluating new drugs for portal hypertension in patients with compensated ACLD [16] Table 28.5.

Summary

Experts from the Baveno VII faculty were asked to respond to a questionnaire on the definition of compensation, clinical decompensation, and prevention of decompensation.

Experts could not agree on what should be included in the definition of decompensation besides the development of ascites, variceal bleeding, or hepatic encephalopathy. There was agreement that ankle edema without ascites or a small amount of perihepatic fluid detectable on ultrasound (not detectable clinically) is not a manifestation of clinical decompensation. Similarly, chronic blood loss from portal hypertensive gastropathy or GAVE, as well as acute bleeding from an esophageal ulcer related to a prophylactic banding ligation were not considered equivalents to variceal bleeding for defining decompensation. Minimal hepatic encephalopathy was not considered to indicate decompensation, however, no consensus regarding the minimal West Haven grade necessary for this definition was obtained.

There was no consensus on defining whether lack of further manifestation of decompensation over a 2-year period can be considered as re-compensated, even in patients that have received effective therapy for the cause of their liver disease. Regarding prevention of decompensation, there was strong agreement considering carvedilol as the best treatment option, but there was no consensus on whether the decision to start treatment can be done on a clinical basis or using noninvasive tests. Finally, there was an extremely strong consensus in considering HVPG monitoring useful in the development and assessment of new drugs for portal hypertension (in the context of pilot or phase II and III clinical trials), but not in assessing treatment response in clinical practice.

References

1. de Franchis R, Baveno VI Faculty. Expanding consensus in portal hypertension: report of the Baveno VI consensus workshop: stratifying risk and individualizing care for portal hypertension. J Hepatol. 2015;63:743–52.
2. D'Amico G, Garcia-Tsao G, Pagliaro L. Natural history and prognostic indicators of survival in cirrhosis: a systematic review of 118 studies. J Hepatol. 2006;44:217–31.
3. Ginés P, Quintero E, Arroyo V, Terés J, Bruguera M, Rimola A, Caballería J, Rodés J, Rozman C. Compensated cirrhosis: natural history and prognostic factors. Hepatology. 1987;7(1):122–8.
4. D'Amico G, Abraldes JG, Rebora P, Valsecchi MG, Garcia-Tsao G. Ordinal outcomes are superior to binary outcomes for designing and evaluating clinical trials in compensated cirrhosis. Hepatology. 2020;72(3):1029–42.
5. Arvaniti V, D'Amico G, Fede G, Manousou P, Tsochatzis E, Pleguezuelo M, Burroughs AK. Infections in patients with cirrhosis increase mortality four-fold and should be used in determining prognosis. Gastroenterology. 2010;139(4):1246–56, 1256.e1–5
6. Machicao VI, Balakrishnan M, Fallon MB. Pulmonary complications in chronic liver disease. Hepatology. 2014;59(4):1627–37.
7. Ripoll C, Garcia-Tsao G. Management of gastropathy and gastric vascular ectasia in portal hypertension. Clin Liver Dis. 2010;14(2):281–95.
8. American Association for the Study of Liver Diseases; European Association for the Study of the Liver. Hepatic encephalopathy in chronic liver disease: 2014 practice guideline by the European Association for the Study of the Liver and the American Association for the Study of Liver Diseases. J Hepatol. 2014;61(3):642–59.
9. El-Sherif O, Jiang ZG, Tapper EB, Huang KC, Zhong A, Osinusi A, et al. Baseline factors associated with improvements in decompensated cirrhosis after direct-acting antiviral therapy for hepatitis C virus infection. Gastroenterology. 2018;154(8):2111–2121.e8.
10. Villanueva C, Albillos A, Genescà J, Garcia-Pagan JC, Calleja JL, Aracil C, et al. β blockers to prevent decompensation of cirrhosis in patients with clinically significant portal hypertension (PREDESCI): a randomised, double-blind, placebo-controlled, multicentre trial. Lancet. 2019;393(10181):1597–608.
11. Robic MA, Procopet B, Metivier S, Peron JM, Selves J, Vinel JP, et al. Liver stiffness accurately predicts portal hypertension related complications in patients with chronic liver disease: a prospective study. J Hepatol. 2011;55:1017–24.
12. Reiberger T, Ulbrich G, Ferlitsch A, Payer BA, Schwabl P, Pinter M, et al. Carvedilol for primary prophylaxis of variceal bleeding in cirrhotic patients with haemodynamic non-response to propranolol. Gut. 2013;62(11):1634–41.
13. Sinagra E, Perricone G, D'Amico M, Tinè F, D'Amico G. Systematic review with meta-analysis: the haemodynamic effects of carvedilol compared with propranolol for portal hypertension in cirrhosis. Aliment Pharmacol Ther. 2014;39(6):557–68.
14. Abraldes JG, Tarantino I, Turnes J, Garcia-Pagan JC, Rodes J, Bosch J. Hemodynamic response to pharmacological treatment of portal hypertension and longterm prognosis of cirrhosis. Hepatology. 2003;37:902–8.

15. Abraldes JG, Trebicka J, Chalasani N, D'Amico G, Rockey DC, Shah VH, Bosch J, Garcia-Tsao G. Prioritization of therapeutic targets and trial design in cirrhotic portal hypertension. Hepatology. 2019;69(3):1287–99.
16. Ripoll C, Groszmann R, Garcia-Tsao G, Grace N, Burroughs A, Planas R, Escorsell A, et al. Hepatic venous pressure gradient predicts clinical decompensation in patients with compensated cirrhosis. Gastroenterology. 2007;133(2):481–8.

29 Definition of First Decompensation in Cirrhosis

Susana G. Rodrigues, Rafael Bañares, Alessandra Dell'Era, Jaime Bosch, and Cristina Ripoll

Introduction

Establishing prognosis in patients with chronic diseases is necessary for several reasons. First, many patients are eager to know what can happen in the course of their disease, and providing this information, allows them to make informed decisions regarding their lifestyle, work, and health care, as well as for better organizing their life. Second, the clinical management of the patient should be adjusted to the individual risk of developing relevant events. Furthermore, stratification of patients in homogeneous risk groups is relevant for clinical trials in

S. G. Rodrigues · J. Bosch
Department of Hepatology, University Clinic for Visceral Surgery and Medicine, Inselspital, Bern University Hospital, University of Bern, Bern, Switzerland

Department for BioMedical Research, University of Bern, Bern, Switzerland
e-mail: susana.GomesRodrigues@insel.ch; jaime.bosch@dbmr.unibe.ch

R. Bañares
Digestive Disease Department, Hospital General Universitario Gregorio Marañón, Madrid, Spain

Universidad Complutense, CIBERehd, Madrid, Spain
e-mail: rbanares@ucm.es

A. Dell'Era
Department of Biomedical and Clinical Sciences, "Luigi Sacco", University of Milan, Milan, Italy

Gastroenterology and Endoscopy Unit, ASST Fatebenefratelli Sacco, Milan, Italy
e-mail: Alessandra.dellera@unimi.it

C. Ripoll (✉)
Internal Medicine IV, Jena University Hospital Friedrich-Schiller University, Jena, Germany
e-mail: cristina.ripoll@med.uni-jena.de

R. de Franchis (ed.), *Portal Hypertension VII*,
https://doi.org/10.1007/978-3-031-08552-9_29

order to avoid heterogeneity in the study sample and choose appropriate end-points for the particular risk group.

Several considerations should be taken into account concerning the prognosis of individual patients. First, the information that is used for the estimation of prognosis is population-based, which can significantly differ from patient to patient. Second, most information obtained from prognostic studies is based on the evaluation at a given moment. In clinical practice, however, patients are seen repeatedly so that the physician can observe subtle dynamic changes, which can refine their prognostic estimation.

Risk stratification for research purposes fulfills other aims than prognostication for the individual patient. In this case, the aim is to identify a homogenous subgroup of patients who bear a similar risk for the development of clinically relevant end-points. In research, a balance between obtaining a homogeneous population, which increases the internal validity, and external validity (applicability of the results to the general patient population) has to be attained.

Although the concept of compensated and decompensated cirrhosis has been around for about 70 years [1, 2], D'Amico et al. published a landmark study in which the importance of differentiating between compensated and decompensated cirrhosis was underlined [3]. Indeed, the transition from the compensated to the decompensated stage marks an inflection point in the natural history of the disease, beyond which the survival of the patient is markedly reduced, namely from a median survival of over 12 years to only 2 years [3]. In this study, the development of decompensation was defined by the development of ascites, variceal bleeding, hepatic encephalopathy, and jaundice. This data was derived from a large cohort of patients with untreated viral and alcohol-associated cirrhosis who were followed for over 20 years. Furthermore, this study identified predictors of decompensation and death and emphasized the need to evaluate appropriate end-points according to the stage of the disease, namely decompensation for compensated patients and death for decompensated patients. (Fig. 29.1)

Even among compensated patients, several risk groups can be distinguished according to the presence of clinically significant portal hypertension [4] or the presence or absence of varices [3, 5]. This has been used to distinguish two sub-stages in compensated cirrhosis. Although it is clear that in the compensated patient with clinically significant portal hypertension, the main aim is to prevent decom-pensation, the relevant end-point or the appropriate surrogate marker in patients without CSPH remains to be established.

The concept of decompensation implies that the progression of underlying pathophysiological mechanisms, namely portal hypertension, liver insufficiency, and systemic inflammation reaches a threshold in which compensatory mechanisms are insufficient to maintain corporal homeostasis and clinical events occur as a consequence of these pathophysiological mechanisms. Indeed, the underlying mechanisms do not follow an on/off mechanism but rather undergo a gradual progression. Consequently, the impact on prognosis is most likely also not a black-and-white phenomenon. The velocity at which the progression occurs can be influenced by external factors so that it can be accelerated or slowed.

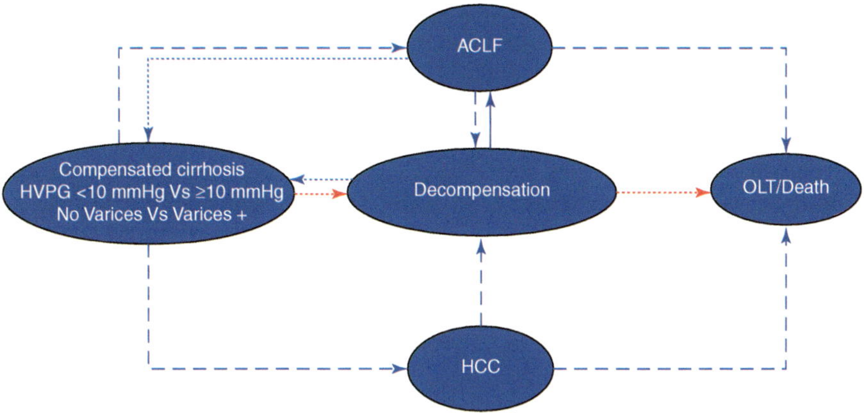

Fig. 29.1 Summary of the natural history of cirrhosis. In the natural history of cirrhosis, compensated patients will decompensate before eventually dying. Competing events in the context of compensated cirrhosis are the development of ACLF and HCC. These events may in turn lead to the transition to the decompensated phase or directly to death or liver transplantation (OLT). Recompensation from the decompensated phase to the compensated phase may also occur

Most of the information regarding the natural history of cirrhosis has been derived from studies in an era in which, contrary to the current situation, viral hepatitis was a major etiology of cirrhosis and was difficult to treat. Also, many of these studies come from the pretransplant times, and the complications of cirrhosis were managed by less efficient treatments than at present, as shown by the marked improvement in the prognosis of the complications of portal hypertension. This, together with the increased prevalence of other entities, such as NAFLD/NASH, casts reasonable doubts on the validity of previous studies, calling for large-scale, multicentric prospective studies to update our knowledge of natural history and to develop new, accurate, and preferably simple noninvasive tests and biomarkers reflecting both progression and regression of cirrhosis.

Definition of First Decompensation in Compensated Cirrhosis

After the publication of the D'Amico study, numerous studies centered on compensated cirrhosis have been published. However, the definition of decompensation is not homogeneous across the studies, so the concept of compensated and decompensated cirrhosis remains unclear. A systematic review of the studies published since 2005 in the context of compensated cirrhosis has shown that most studies (>90%) included at least the presence of ascites, hepatic encephalopathy, and variceal bleeding (or bleeding due to portal hypertension) in the definition of decompensation. Although most studies included at least these events, many studies additionally considered other events, resulting in a great heterogeneity in the definition of decompensation. The reasons for this are further evaluated here.

Inclusion of Subclinical (or Not Clinically Detectable) Forms of the Major Complications of Cirrhosis

First, the inclusion or not of the paucisymptomatic manifestations of classical complications in the definition of clinical decompensation such as subclinical ascites, covert hepatic encephalopathy, and chronic anemia due to portal hypertensive gastropathy may lead to gray areas in the definition of decompensation. This will be further elaborated in Chap. 32.

Jaundice

Whether jaundice alone defines pure decompensation of cirrhosis remains debatable. Jaundice in the general population can frequently arise due to many causes (hemolytic anemia, bile duct obstruction) without being associated with liver disease. In patients with cirrhosis, on top of these relatively common entities, it can also be associated with the progression of the disease. It was described as the first decompensating event in 3% of compensated patients in a large cohort of patients with cirrhosis mainly due to HCV [5]. Jaundice is, however, very frequently present in superimposed liver injury or Acute Chronic Liver Failure (ACLF) [6], which was not considered in this study. Clinical experience suggests that isolated jaundice without other complications is an infrequent first decompensating event. In fact, when jaundice appears in patients with cirrhosis outside the setting of a superimposed injury, it generally reflects liver insufficiency in an already decompensated patient. Thus, it should be considered a further decompensation of cirrhosis. Furthermore, although outright jaundice is clear to all observers, many factors influence the recognition of low-level jaundice, which is therefore subjective. A cut-off of bilirubin to define jaundice has never been evaluated in clinical studies. This will be further elaborated on in Chap. 31.

Complications That Require Classical Decompensation (Ascites) by Definition

Another source of uncertainty is the inclusion of entities, which by definition require the presence of classical complications, namely ascites. These complications (such as spontaneous bacterial peritonitis, hepatorenal syndrome, dilutional hyponatremia, etc.) usually refer to a further stage in the natural history of the disease called further decompensation. Occasionally, these present at the first decompensation, however, they all require the presence of ascites for their development so the inclusion of these complications in the definition of the first decompensation is redundant and can only lead to confusion. The significance of the development of these entities on top of ascites in the context of a first decompensation or ACLF remains to be determined.

Complications That Can Follow Other Pathophysiological Pathways

The traditional complications of cirrhosis hepatic encephalopathy, variceal bleeding, and ascites develop as a consequence of the characteristic pathophysiological changes that take place along the natural history of liver disease namely portal hypertension, liver insufficiency, and systemic inflammation [7, 8]. Although these complications are typical of decompensated cirrhosis, they are not specific to cirrhosis, despite being rare outside of the setting of cirrhosis. Ascites or hepatic hydrothorax due to portal hypertension (i.e., with a SAAG above the 1.1 g/dL) is usually due to cirrhosis and less frequently due to right heart failure. About 10% of variceal bleeding episodes are due to non-cirrhotic portal hypertension (see Chap. 58), and a few variceal bleeds are due to "down-hill" varices in mediastinal neoplasms, and hepatic encephalopathy can occur in the absence of cirrhosis, for instance in urea-cycle disorders and congenital portosystemic shunts.

A more common problem is that a number of complications that can take place in the context of cirrhosis can also arise in patients without cirrhosis following other pathophysiological mechanisms such as hepatocellular carcinoma, portal vein thrombosis, sarcopenia, or infections. These complications can have an impact on prognosis by favoring decompensation, but also by themselves (independent of the presence of cirrhosis). The role of infections and sarcopenia will be further elaborated in Chaps. 30 and 33.

New Entities in the Natural History of Cirrhosis

In recent years, increasing recognition is being given to new entities whose role in the natural history of liver cirrhosis remains to be characterized namely ACLF and recompensation (Fig. 29.1).

ACLF denotes the patient with cirrhosis that acutely deteriorates and develops organ failure. Although from a conceptual point of view one would consider that identification of superimposed liver damage is essential, its identification is not necessary for its diagnosis. Indeed, in up to 30% of cases, the cause is unknown. In Europe, the accepted definition of acute chronic liver failure is based on the CANONIC study [6]. This study included patients who were hospitalized due to "acute decompensation" and the patients were followed up to identify risk factors for death. The definition of "acute" decompensation is independent of the classical compensated/decompensated definition but rather refers to patients who develop an acute event including gastrointestinal bleeding, hepatic encephalopathy, infection, or ascites, without considering previous acute or non-acute decompensation [9]. Although not explicitly stated in the definition, hospitalization due to these events was intrinsically one of the criteria, which selected patients for the study. As expected, patients developing an increasing number of organ failures during follow-up had worse survival. ACLF can occur in both compensated and decompensated patients. The CANONIC study observed that compensated patients with ACLF had worse survival than decompensated patients. It was hypothesized, that this would be due to a more severe insult being necessary to lead to multiorgan failure. Patients who survive ACLF have worse survival than those who just have "acute" decompensation (according to the definition of the authors). The studies do not analyze the results according to previous compensated or decompensated status,

Recompensation is an increasingly recognized entity, which is less clearly defined than ACLF. The idea of recompensation is that after effectively suppressing or controlling the cause of cirrhosis, a patient can revert from the decompensated phase (with worse prognosis) to the compensated phase (with a considerably better prognosis) no longer needing treatment for the manifestations of decompensation which were previously present (ascites, variceal bleeding, hepatic encephalopathy). This can only be accomplished by an improvement of the structural changes that take place in cirrhosis accompanied by a regain of function. The idea that recompensation may occur arises from the observation that histological cirrhosis disappears in about two-thirds of the patients that have long-term (5-years or more) follow-up liver biopsies, after successful treatment of viral cirrhosis (sustained virological response in the case of HCV or sustained virological suppression in the case of HBV [10–13]). These findings challenged the classical dogma that cirrhosis was not a reversible condition. Careful pathological studies have allowed the identification of histological markers of impending cirrhosis regression (see Chap. 16). Although regression of cirrhosis should always lead to recompensation, these two concepts are not synonymous. First, decompensated patients are less likely to achieve regression of cirrhosis, and, second, despite control of the underlying liver disease, not all patients experience regression, with some never recompensating and others only transiently recompensating. The long-term prognosis of recompensated patients in comparison to the prognosis of the classical compensated patient and the implications regarding screening measures remains largely unknown [14, 15] .

Concluding Remarks

From a research point of view, it is important that all studies which evaluate the natural history of compensated cirrhosis or its change after interventions, use the same event of interest, in this case, decompensation. The definition of decompensation needs to be homogeneous among the different studies. Indeed, one of the first accomplishments of the Baveno Consensus Workshop in the setting of variceal bleeding was to homogenize definitions of events and standardize end-points, which had a clear impact on improving the quality and relevance of research in this field.

The choice of events to define decompensation should be based on the identification of subgroup(s) in which its incidence is frequent enough so that studies can be feasible. Furthermore, the events that define decompensation should have a similar prognostic impact that is a similar increase in the risk of death. Finally, it is important that these studies have sufficient follow-up periods for the events to take place.

Presently, the most accepted definition of decompensation is the development of clinically evident ascites, variceal bleeding, or overt hepatic encephalopathy (alone or in combination). Indeed, there was consensus among the Baveno VII faculty that this is the preferred definition of first decompensation (Chap. 28). Up to date, there have been no studies that compare the impact of including any additional complication in the definition of decompensation. Whether other complications should also be included in this definition, remains a matter for further research (see Chap. 36).

References

1. Manguel M. Decompensated hepatic cirrhosis. Dia Med. 1949;21(74):2957–61.
2. Epstein FH, Lesser GT, Berger EY. Renal function in decompensated cirrhosis of the liver. Proc Soc Exp Biol Med. 1950;75(3):822–4.
3. D'Amico G, Garcia-Tsao G, Pagliaro L. Natural history and prognostic indicators of survival in cirrhosis: a systematic review of 118 studies. J Hepatol. 2006;44(1):217–31.
4. Ripoll C. Hepatic venous pressure gradient and outcomes in cirrhosis. J Clin Gastroenterol. 2007;41(Suppl 3):S330–5.
5. D'Amico G. The clinical course of cirrhosis. Population based studies and the need of personalized medicine. J Hepatol. 2014;60(2):241–2.
6. Moreau R, Jalan R, Gines P, et al. Acute-on-chronic liver failure is a distinct syndrome that develops in patients with acute decompensation of cirrhosis. Gastroenterology. 2013;144(7):1426–37, 37.e1–9
7. Schrier RW, Arroyo V, Bernardi M, Epstein M, Henriksen JH, Rodes J. Peripheral arterial vasodilation hypothesis: a proposal for the initiation of renal sodium and water retention in cirrhosis. Hepatology. 1988;8(5):1151–7.
8. Bernardi M, Moreau R, Angeli P, Schnabl B, Arroyo V. Mechanisms of decompensation and organ failure in cirrhosis: from peripheral arterial vasodilation to systemic inflammation hypothesis. J Hepatol. 2015;63(5):1272–84.
9. D'Amico G, Bernardi M, Angeli P. Towards a new definition of decompensated cirrhosis. J Hepatol. 2022;76(1):202–7.
10. Poynard T, McHutchison J, Manns M, et al. Impact of pegylated interferon alfa-2b and ribavirin on liver fibrosis in patients with chronic hepatitis C. Gastroenterology. 2002;122(5):1303–13.
11. George SL, Bacon BR, Brunt EM, Mihindukulasuriya KL, Hoffmann J, Di Bisceglie AM. Clinical, virologic, histologic, and biochemical outcomes after successful HCV therapy: a 5-year follow-up of 150 patients. Hepatology. 2009;49(3):729–38.
12. Marcellin P, Gane E, Buti M, et al. Regression of cirrhosis during treatment with tenofovir disoproxil fumarate for chronic hepatitis B: a 5-year open-label follow-up study. Lancet. 2013;381(9865):468–75.
13. Chang TT, Liaw YF, Wu SS, et al. Long-term entecavir therapy results in the reversal of fibrosis/cirrhosis and continued histological improvement in patients with chronic hepatitis B. Hepatology. 2010;52(3):886–93.
14. Villanueva C, Albillos A, Genesca J, et al. Beta blockers to prevent decompensation of cirrhosis in patients with clinically significant portal hypertension (PREDESCI): a randomised, double-blind, placebo-controlled, multicentre trial. Lancet. 2019;393(10181):1597–608.
15. Groszmann RJ, Garcia-Tsao G, Bosch J, et al. Beta-blockers to prevent gastroesophageal varices in patients with cirrhosis. N Engl J Med. 2005;353(21):2254–61.

30 Evaluation of the Impact of the Sole Presence of Infection (Without Accompanying Decompensation) in the Natural History of Compensated Cirrhosis

Yuly P. Mendoza, Cristina Ripoll, and Susana G. Rodrigues

Introduction

Bacterial Infection (BI) is an important cause of hospitalization and death in patients with cirrhosis [1, 2]. In decompensated patients, infections lead to a fourfold increase in mortality [1]. BI occurs in 25%–35% of inpatients with cirrhosis [3, 4]. Unsurprisingly, observational studies that have shown BIs are frequent and associated with increased mortality, were performed mainly in cirrhosis patients hospitalized for an episode of sepsis or decompensation without describing the previous stage of the liver disease [1, 4–6]. Therefore, in compensated patients, there is less evidence regarding BI incidence, the interaction of BI with other critical events, and the prognostic implications of BI.

Y. P. Mendoza
Hepatology, University Clinic for Visceral Surgery and Medicine, Inselspital Bern University Hospital, University of Bern, Bern, Switzerland

Hepatology, Department for BioMedical Research, University Clinic for Visceral Surgery and Medicine, University of Bern, Bern, Switzerland

Graduate School for Health Sciences (GHS), University of Bern, Bern, Switzerland
e-mail: yuly.mendozajaimes@insel.ch

C. Ripoll
Internal Medicine IV, Universitätsklinikum, Friedrich-Schiller University, Jena, Germany
e-mail: Cristina_ripoll@med.uni-jena.de

S. G. Rodrigues (✉)
Hepatology, Department for BioMedical Research, University Clinic for Visceral Surgery and Medicine, University of Bern, Bern, Switzerland
e-mail: susana.GomesRodrigues@insel.ch

R. de Franchis (ed.), *Portal Hypertension VII*,
https://doi.org/10.1007/978-3-031-08552-9_30

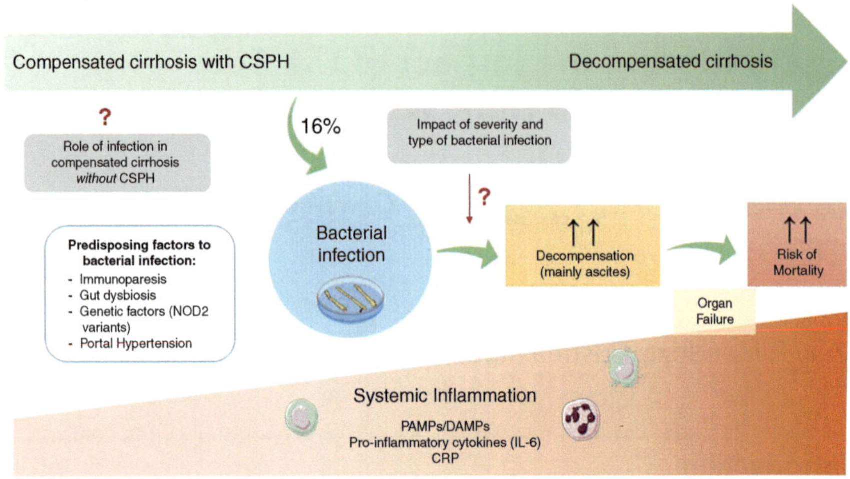

Fig. 30.1 Bacterial infections and systemic inflammation predispose to decompensation and death in cirrhosis. Some predisposing factors increase the risk of BI; once BI occurs, it triggers or favors the development of decompensation, consequently, a higher risk of death and organ failure. CSPH clinically significant portal hypertension, DAMPs damage-associated molecular patterns, PAMPs pathogen-associated molecular patterns, CRP C-reactive protein, IL-6 interleukin 6

Systemic inflammation is known to be present, particularly in more advanced stages of cirrhosis, and findings suggest that systemic inflammation is already present in compensated patients [7–9]. However, the precise role that systemic inflammation and bacterial infection play in the natural history of compensated patients is not well established. Herein, we aim to delineate the latest evidence regarding prognostic implications of systemic inflammation and BIs in compensated cirrhosis (Fig. 30.1).

The Role of Systemic Inflammation in Compensated Cirrhosis

Pathophysiological Background: Systemic Inflammation

Portal hypertension (PH) is a crucial driver of decompensation and mortality in patients with cirrhosis [10]. Systemic inflammation due to translocation of enteric organisms or bacterial products from the intestinal lumen and the ensuing release of pro-inflammatory cytokines is believed to be present in decompensated patients and lead to further decompensation [11–14]. However, the exact interplay between systemic inflammation, bacterial translocation, and PH in compensated cirrhosis remains to be established. Findings suggest that systemic inflammation is already present in compensated patients and may mediate the pathogenesis of decompensation and, in combination with PH, may be key drivers of cirrhosis progression [7–9] and perhaps BI.

Systemic Inflammation in Compensated Cirrhosis

Levels of circulating pro-and anti-inflammatory cytokines in patients with cirrhosis, mainly in compensated stage ($N = 88$), were significantly higher than those of non-cirrhotic patients with liver disease regardless of underlying etiology [7]. Recently, the impact of systemic inflammation in the substages of compensated cirrhosis has been explored [8, 9]. These studies included patients with mild PH (HVPG of 6–9 mmHg) and clinically significant PH (CSPH; HVPG ≥10 mmHg without gastroesophageal varices (GEV)). Patients with CSPH can be further divided into those with or without GEV [15].

Turco et al. [8] investigated systemic inflammation across distinct prognostic substages of cirrhosis. Of note, C-reactive Protein (CRP) levels were incrementally higher across the distinct substages of compensated cirrhosis ($n = 157$). Furthermore, elevated CRP was an independent predictor of decompensation [8]. In contrast, in a single-center cohort, Costa et al. [9] failed to demonstrate significant elevation of CRP across the substages of compensated patients ($n = 78$), showing that CRP and IL-6 were progressively increased only in the decompensated substages ($n = 90$). Interestingly, IL-6 was the *only* independent predictor of the first decompensation at 12-months follow-up in compensated cirrhosis [9].

Despite study limitations, these results suggest that inflammation may play a role in the pathogenesis of cirrhosis progression. Larger longitudinal prospective studies with adequate follow-up periods are needed to confirm the prognostic value of these markers of inflammation in the early phase of cirrhosis.

The Impact of Bacterial Infections (BI) on Compensated Cirrhosis

Pathophysiological Background of BI in Compensated Cirrhosis

A long-recognized feature of patients with decompensated cirrhosis is that they are at risk of developing BI [1, 2, 4]. In decompensated patients, cirrhosis-associated immune dysfunction plays a key role in disease progression and BI [16, 17]. However, there is no data regarding this aspect in compensated disease. Other factors that may contribute to BI are gut microbiome changes and disruption of the intestinal barriers [18]. A study with patients at different stages of cirrhosis, including strictly compensated patients, showed progressive changes in the gut microbiome that increased with worsening disease [19]. There is data from preclinical models and patients without cirrhosis in the setting with alcoholic and metabolic liver disease demonstrating dysbiosis and increased intestinal permeability [19–22]. Interestingly, propranolol was noted to improve these abnormalities in cirrhotic rats [23] which is probably one of the reasons for some of the beneficial effects of beta-blockers in cirrhosis [24]. Moreover, genetic immune alterations can possibly increase the risk of BIs in cirrhosis. The presence of a *NOD2* risk variant was a major risk factor for BI in compensated cirrhosis [25].

To summarize, a combination of factors, such as systemic inflammation, intestinal dysbiosis, PH, and genetic factors may promote BI in compensated cirrhosis.

Epidemiology of BI in Compensated Cirrhosis

In a large prospective cohort of 1672 patients with compensated viral cirrhosis without previous history of decompensation, the 5-year cumulative incidence of BI was 13% [26]. This is consistent with other more recent studies involving patients with compensated cirrhosis of different etiologies, which reported 14% and 16% of patients developing BI episodes during a median follow-up of 20 and 36 months, respectively [27, 28]. Moreover, a high rate was reported in a population-based cohort [29], in which 32% of patients with compensated disease had at least one episode of BI during the follow-up. Thus, BIs are a frequent event among patients with compensated cirrhosis, in fact, in compensated cirrhosis and clinically significant portal hypertension (CSPH), Villanueva et al. [28] described that BIs were as frequent as ascites, the most common decompensation event in cirrhosis. Nevertheless, this epidemiological data may underestimate the true incidence of BI in compensated cirrhosis because it refers to patients from tertiary centers and does not take into account those managed in primary care facilities. Overall, although BI can be frequent in compensated cirrhosis, it is unclear which factors, liver-related (CSPH) or infection-related (severity, location, microorganisms) contribute to disease progression.

Definitions of Severity of BI

Severity of BI in patients with cirrhosis has been characterized by Systemic Inflammatory Response Syndrome (SIRS), sepsis, severe sepsis and septic shock, the development of organ failure, and the use of Child-Pugh and MELD scores [30]. However, SIRS criteria have some limitations, particularly in cirrhosis where leukopenia, tachypnea, and bradycardia can be present without infection [4]. More recently, the terms sepsis and septic shock were refined. According to the new consensus, sepsis (Sepsis-3) was defined as a "life-threatening organ dysfunction caused by a dysregulated host response to infection" [31]. In patients with *mostly* decompensated cirrhosis, sepsis-3 criteria were more accurate than SIRS criteria in predicting the clinical outcome of infections [30]. Although it is plausible that more severe infections may lead more frequently to the development of decompensation, to date no studies have explored this issue.

Location of BI and Type of Microorganism in Compensated Cirrhosis

Studies in compensated cirrhosis reported the most frequent sites of BI were the urinary tract and lung, followed by bacteremia and skin infections [26–28]. These contrasts with decompensated patients where Spontaneous Bacterial Peritonitis (SBP) (per definition only possible in decompensated patients) followed by pneumonia, urinary tract infections, bacteremia, skin, and soft tissue infections were more frequent [3]. This prevalence may change according to geographic regions [32].

In a study including mostly decompensated patients, survival was worse in those with bacteremia, SBP, and pneumonia compared to other infections, with a survival of 14.2 vs. 20.8 months, respectively [3]. Interestingly, Nahon et al. [26] found that patients who remained compensated had mainly urinary tract infections, whereas patients who decompensated had BI in other sites. Whether or not the site of infection contributes to decompensation is a topic for further research.

Besides the location of the infection, the type of microorganism isolated may eventually also influence the outcome of the infection in compensated patients. Microbiological data in compensated patients are in accordance with cohorts of decompensated patients reporting a worse short-term prognosis in polymicrobial or monomicrobial with Staphylococcaceae, and nosocomial infections involving multidrug-resistant (MDR) bacteria [26].

Factors Associated With BI in Compensated Cirrhosis

Some patient-associated risk factors for BI in compensated cirrhosis patients have been described. A large multicenter cohort of viral compensated cirrhosis [26] demonstrated that Proton Pump Inhibitors (PPIs) were associated with a higher risk of BI occurrence (HR of 1.71, $P = 0.003$), although not adjusted for confounding factors. This may be an argument for re-evaluating the indications for long-term PPI therapies commonly adopted in patients with cirrhosis. Nonselective Beta-blockers (NSBBs) were associated with improvement of BT and, consequently, reduced the incidence of BIs, including SBP [33] in decompensated cirrhosis [24]. However, two prospective studies involving compensated patients did not find a significant decrease in BIs treated with NSBBs [26, 28]. Further research in this field is warranted.

Sarcopenia led to higher mortality rates in "stable" cirrhosis patients with BIs than those without sarcopenia (50% vs. 16%; $P = 0.01$) [34]. Furthermore, Villanueva et al. [28] reported that lower BMI was linked to a higher risk of BI in compensated patients.

Impact of BI on the Natural History of Compensated Cirrhosis

Bacterial infections are frequent in patients with compensated cirrhosis. Their impact on the natural history of compensated cirrhosis has been analyzed in six cohort studies over the last 5 years. These are summarized in Table 30.1.

The first of three observational studies evaluating specifically the prognostic role of BI on compensated patients without previous decompensation showed a higher risk of long-term decompensation and death in those patients with HBV or HCV

Table 30.1 Studies evaluating the prognostic impact of bacterial infections in compensated cirrhosis

Author	Design	Number of cirrhosis patients	Etiology of cirrhosis	Definition of infection	Follow-up time	Incidence rate of BI as first event in cirrhosis		Impact of first episode of BI on decompensation	Impact of BI on mortality
						Non-SBP	With SBP		
Villanueva C. 2021	Prospective	201[a]	Alcohol, HCV, NASH, others	Non-SBP BI	36 months (median)	33 (16%)		Higher risk of DE HR 2.93, 95%CI 1.02–8.42; $p = 0.047$	Higher risk of mortality HR 6.93; 95%CI 2.64–18.18; $p < 0.001$[b]
Reichert M. 2021	Secondary analysis of prospective cohort	425	Alcohol, NASH, viral, others	BI treated with antibiotics	20 months (median)	12/257 (4.7%) at baseline	26/180 (14%)[c]	Lack of DE during follow-up among those with isolated BI at baseline	– at baseline: Isolated BI have no impact on survival – at baseline: BI with first DE (HR 1.91; 95%CI 0.59–6.21)[d] – during FU: Development of BI (HR 8.06; 95%CI 4.02–16.14)[e]
Nahon P. 2017	Prospective	1672	HCV and HBV	SBP and other infections	43 months (median)		140 (8.4%)	In HCV, first episode of BI increased 5-year probability of DE (45.2% vs. 14.7%, $p < 0.001$)	First BI episode impaired 5-year probability of survival (in HCV: 60.2% vs. 90.4%; in HBV: 69.2% vs. 97.6%, both $p < 0.001$)

Hassan EA 2019	Retrospective	620[f]	HCV	Cellulitis	Hospitalization stays 2–25 days (range)	–	–	–	In-hospital mortality in patients with cellulitis was 0% in compensated vs. 27.3% in decompensated
Dionigi E 2017	Retrospective	92	Alcohol, viral, other	SBP and other infections lead to hospitalization	12.7 months (mean)	–	–	–	No difference in Survival between patients (child-Pugh A) with and without infections was seen ($P = 0.165$).
Sargenti K 2015	Retrospective	332	Alcohol, viral, NASH, others	SBP and other infections lead to hospitalization	36 months (median)	–	66 (20%)	–	BI was not an independent predictor of mortality/OLT (combined outcome)

[a]Compensated patients with CSPH (HVPG ≥10 mmHg)
[b]To assesses the impact of BI on survival, all the episodes that occurred before or after decompensation were considered (accounting for death and OLT as competing events)
[c]For the incidence of BI, we consider patients compensated at baseline who developed the first episode of BI during follow-up: 26 patients (14%) among 180 compensated patients without BI at baseline and without decompensation during follow-up
[d]BI was independent predictor of survival with liver transplantation as a competing risk
[e]Development of BI during follow-up was only considered among those who did not have these complications at baseline
[f]The number included compensated and decompensated cirrhosis patients (no data about each stage)
BI bacterial infection, *DE* decompensation, *HR* Hazard ratio, *CI* confidence interval, *NASH* nonalcoholic steatohepatitis, *HCV* hepatitis C virus, *HBV* hepatitis B virus, *SBP* Spontaneous bacterial peritonitis, *FU* follow-up, *OLT* Orthotopic liver transplantation

cirrhosis who developed a BI during follow-up [26]. In HCV cirrhosis patients, the first episode of BI increased the probability of decompensation (45.2% vs. 14.7%, $p < 0.001$) and reduced survival (60.2% vs. 90.4%, $p < 0.001$) at 5 years. BI was the direct cause of death in 16.5%, and the remaining mortality (more than 80%) was attributed to disease progression (decompensation) and hepatocellular carcinoma.

Reichert et al. [27] recently performed a secondary analysis of a prospectively registered cohort including 425 compensated cirrhosis patients of several etiologies. They showed that isolated BI (without concurrent decompensation or SBP) at baseline had no impact on survival in patients who remained compensated. However, transplant-free survival was significantly reduced in patients with BI and *simultaneous* first decompensation than in patients with decompensation, but without BI (log-rank P-value< 0.001). However, this result may be limited by a short follow-up with a median of 20 months and the small sample size of patients with the BI at baseline ($n = 12$) in the group of patients who remain compensated. This data suggests that the increase in mortality following BI extends beyond the BI episode per se and is related to the presence of concomitant decompensation.

Recently, Villanueva et al. [28] performed a secondary analysis from a randomized controlled trial (PREDESCI study) with 201 patients with compensated cirrhosis and CSPH (HVPG ≥10 mmHg). Development of bacterial infections was a pre-specified end-point of the study, specifically documented. A time-to-event assessment of the development of BI during follow-up showed a higher risk of decompensation (HR 2.93; 95%CI 1.02–8.42; $p = 0.047$) in patients who developed BI. The authors further reported that BIs increased mortality (HR 6.93; 95%CI 2.64–18.18; $p < 0.001$), with 71% of deaths occurring after decompensation. This finding, keeping with the other abovementioned studies, suggests that higher mortality may occur due to the deterioration of the liver function *after* the decompensation triggered by BI.

None of the studies analyzed whether the incidence of BI in compensated patients varied according to the presence or not of CSPH. Only one study included exclusive patients with CSPH [28], who may inherently have a higher risk of disease progression. Thus, it remains uncertain to what extent BIs impact compensated patients without CPSH. Further investigation is needed.

The remaining three studies were retrospective and not designed to clarify the impact of BI on survival in compensated patients. They showed no negative survival impact of a BI episode in compensated patients. Although Dionigi et al. [3] reported that in their overall series BIs were associated with a decrease in survival independently of markers of liver function (MELD >15), in the subgroup of 88 compensated patients, there was no survival difference between patients with and without infection ($P = 0.165$), although a lack of significance may just reflect the type-II error. Sargenti et al. [29] also failed to demonstrate that BI was an independent predictor of survival in compensated cirrhosis. Furthermore, a recent study reported in-hospital mortality of 27.3% in decompensated patients, but no deaths in compensated patients [35].

Criticism to the Available Literature and Further Directions

The studies in this field warrant several comments. These included different designs and outcomes; short duration of follow-up (median ranging from 13 to 43 months); heterogeneous definitions of compensation (according to Child-Pugh score A or absence of the previous decompensation); mixed different substages of compensated cirrhosis (according to the presence of CSPH); and diverse definitions of BI (some of them include SBP, but per definition, SBP is only possible in decompensated patients with ascites). Therefore, the real impact of infection as a driver toward decompensation cannot be proven in all the stages of compensated cirrhosis, and more data are required before firm conclusions can be drawn. Nevertheless, recent robust evidence demonstrates that BIs are prevalent and influence the survival of compensated patients with CSPH due to a higher risk of developing decompensation and, consequently, a higher risk of death on follow-up. Thus, BI affects the natural history of compensated cirrhosis with CSPH as a risk factor favoring decompensation and favoring death once decompensation has occurred.

Large prospective natural history studies with adequate follow-up periods, including all the substages of compensated cirrhosis and evaluating the prognostic impact of the different types of infections in this population, are needed. Furthermore, a future investigation should assess the impact of changes in innate immunity, intestinal microbiome, mucosal barrier, and systemic inflammation in the development of decompensation in patients with compensated cirrhosis.

References

1. Arvaniti V, D'Amico G, Fede G, Manousou P, Tsochatzis E, Pleguezuelo M, et al. Infections in patients with cirrhosis increase mortality four-fold and should be used in determining prognosis. Gastroenterology. 2010;139(4):1246–56, 56.e1–5
2. Fernández J, Gustot T. Management of bacterial infections in cirrhosis. J Hepatol. 2012;56(Suppl 1):S1–12.
3. Dionigi E, Garcovich M, Borzio M, Leandro G, Majumdar A, Tsami A, et al. Bacterial infections change natural history of cirrhosis irrespective of liver disease severity. Am J Gastroenterol. 2017;112(4):588–96.
4. Jalan R, Fernandez J, Wiest R, Schnabl B, Moreau R, Angeli P, et al. Bacterial infections in cirrhosis: a position statement based on the EASL special conference 2013. J Hepatol. 2014;60(6):1310–24.
5. Singal AK, Salameh H, Kamath PS. Prevalence and in-hospital mortality trends of infections among patients with cirrhosis: a nationwide study of hospitalised patients in the United States. Aliment Pharmacol Ther. 2014;40(1):105–12.
6. Wong F, Piano S, Singh V, Bartoletti M, Maiwall R, Alessandria C, et al. Clinical features and evolution of bacterial infection-related acute-on-chronic liver failure. J Hepatol. 2021;74(2):330–9.
7. Tilg H, Wilmer A, Vogel W, Herold M, Nölchen B, Judmaier G, et al. Serum levels of cytokines in chronic liver diseases. Gastroenterology. 1992;103(1):264–74.
8. Turco L, Garcia-Tsao G, Magnani I, Bianchini M, Costetti M, Caporali C, et al. Cardiopulmonary hemodynamics and C-reactive protein as prognostic indicators in compensated and decompensated cirrhosis. J Hepatol. 2018;68(5):949–58.

9. Costa D, Simbrunner B, Jachs M, Hartl L, Bauer D, Paternostro R, et al. Systemic inflammation increases across distinct stages of advanced chronic liver disease and correlates with decompensation and mortality. J Hepatol. 2021;74(4):819–28.
10. Ripoll C, Groszmann R, Garcia-Tsao G, Grace N, Burroughs A, Planas R, et al. Hepatic venous pressure gradient predicts clinical decompensation in patients with compensated cirrhosis. Gastroenterology. 2007;133(2):481–8.
11. Bernardi M, Moreau R, Angeli P, Schnabl B, Arroyo V. Mechanisms of decompensation and organ failure in cirrhosis: from peripheral arterial vasodilation to systemic inflammation hypothesis. J Hepatol. 2015;63(5):1272–84.
12. Albillos A, de la Hera A, González M, Moya JL, Calleja JL, Monserrat J, et al. Increased lipopolysaccharide binding protein in cirrhotic patients with marked immune and hemodynamic derangement. Hepatology. 2003;37(1):208–17.
13. Engelmann C, Clària J, Szabo G, Bosch J, Bernardi M. Pathophysiology of decompensated cirrhosis: portal hypertension, circulatory dysfunction, inflammation, metabolism and mitochondrial dysfunction. J Hepatol. 2021;75(Suppl 1):S49–s66.
14. Trebicka J, Fernandez J, Papp M, Caraceni P, Laleman W, Gambino C, et al. The PREDICT study uncovers three clinical courses of acutely decompensated cirrhosis that have distinct pathophysiology. J Hepatol. 2020;73(4):842–54.
15. Garcia-Tsao G, Abraldes JG, Berzigotti A, Bosch J. Portal hypertensive bleeding in cirrhosis: risk stratification, diagnosis, and management: 2016 practice guidance by the American association for the study of liver diseases. Hepatology. 2017;65(1):310–35.
16. Bernsmeier C, van der Merwe S, Périanin A. Innate immune cells in cirrhosis. J Hepatol. 2020;73(1):186–201.
17. Albillos A, Lario M, Álvarez-Mon M. Cirrhosis-associated immune dysfunction: distinctive features and clinical relevance. J Hepatol. 2014;61(6):1385–96.
18. Van der Merwe S, Chokshi S, Bernsmeier C, Albillos A. The multifactorial mechanisms of bacterial infection in decompensated cirrhosis. J Hepatol. 2021;75(Suppl 1):S82–s100.
19. Bajaj JS, Heuman DM, Hylemon PB, Sanyal AJ, White MB, Monteith P, et al. Altered profile of human gut microbiome is associated with cirrhosis and its complications. J Hepatol. 2014;60(5):940–7.
20. Keshavarzian A, Farhadi A, Forsyth CB, Rangan J, Jakate S, Shaikh M, et al. Evidence that chronic alcohol exposure promotes intestinal oxidative stress, intestinal hyperpermeability and endotoxemia prior to development of alcoholic steatohepatitis in rats. J Hepatol. 2009;50(3):538–47.
21. Bjarnason I, Peters TJ, Wise RJ. The leaky gut of alcoholism: possible route of entry for toxic compounds. Lancet. 1984;1(8370):179–82.
22. Le Roy T, Llopis M, Lepage P, Bruneau A, Rabot S, Bevilacqua C, et al. Intestinal microbiota determines development of non-alcoholic fatty liver disease in mice. Gut. 2013;62(12):1787–94.
23. Pérez-Paramo M, Muñoz J, Albillos A, Freile I, Portero F, Santos M, et al. Effect of propranolol on the factors promoting bacterial translocation in cirrhotic rats with ascites. Hepatology. 2000;31(1):43–8.
24. Reiberger T, Ferlitsch A, Payer BA, Mandorfer M, Heinisch BB, Hayden H, et al. Non-selective betablocker therapy decreases intestinal permeability and serum levels of LBP and IL-6 in patients with cirrhosis. J Hepatol. 2013;58(5):911–21.
25. Reichert MC, Ripoll C, Casper M, Greinert R, Vandieken E, Grünhage F, et al. Common NOD2 risk variants as major susceptibility factors for bacterial infections in compensated cirrhosis. Clin Transl Gastroenterol. 2019;10(1):e00002.
26. Nahon P, Lescat M, Layese R, Bourcier V, Talmat N, Allam S, et al. Bacterial infection in compensated viral cirrhosis impairs 5-year survival (ANRS CO12 CirVir prospective cohort). Gut. 2017;66(2):330–41.
27. Reichert MC, Schneider C, Greinert R, Casper M, Grünhage F, Wienke A, et al. Isolated bacterial infection without decompensation has no impact on survival of compensated patients with cirrhosis. Liver Int. 2021;41(6):1370–8.

28. Villanueva C, Albillos A, Genescà J, Garcia-Pagan JC, Brujats A, Calleja JL, et al. Bacterial infections adversely influence the risk of decompensation and survival in compensated cirrhosis. J Hepatol. 2021;75(3):589–99.
29. Sargenti K, Prytz H, Nilsson E, Kalaitzakis E. Predictors of mortality among patients with compensated and decompensated liver cirrhosis: the role of bacterial infections and infection-related acute-on-chronic liver failure. Scand J Gastroenterol. 2015;50(7):875–83.
30. Piano S, Bartoletti M, Tonon M, Baldassarre M, Chies G, Romano A, et al. Assessment of Sepsis-3 criteria and quick SOFA in patients with cirrhosis and bacterial infections. Gut. 2018;67(10):1892–9.
31. Singer M, Deutschman CS, Seymour CW, Shankar-Hari M, Annane D, Bauer M, et al. The third international consensus definitions for sepsis and septic shock (Sepsis-3). JAMA. 2016;315(8):801–10.
32. Piano S, Singh V, Caraceni P, Maiwall R, Alessandria C, Fernandez J, et al. Epidemiology and effects of bacterial infections in patients with cirrhosis worldwide. Gastroenterology. 2019;156(5):1368–80.e10.
33. Senzolo M, Cholongitas E, Burra P, Leandro G, Thalheimer U, Patch D, et al. Beta-blockers protect against spontaneous bacterial peritonitis in cirrhotic patients: a meta-analysis. Liver Int. 2009;29(8):1189–93.
34. Lucidi C, Lattanzi B, Di Gregorio V, Incicco S, D'Ambrosio D, Venditti M, et al. A low muscle mass increases mortality in compensated cirrhotic patients with sepsis. Liver Int. 2018;38(5):851–7.
35. Hassan EA, Rehim A, Abdel-Malek MO, Ahmed AO, Abbas NM. Are there differences in risk factors, microbial aspects, and prognosis of cellulitis between compensated and decompensated hepatitis C virus-related cirrhosis? Clin Mol Hepatol. 2019;25(3):317–25.

31 Evaluation of the Role of Jaundice in the Definition of Decompensation in the Compensated Patients

Vincenza Calvaruso, Cristina Ripoll, Jaime Bosch, and Alessandra Dell'Era

Jaundice is defined as the yellowish staining of the skin, sclera, and mucous membranes caused by bilirubin that becomes clinically evident when the serum bilirubin levels rise above 3 mg/dL. The word derives from Old French "jaunice," from "jaune" (yellow).

Bilirubin is a major marker of liver dysfunction and has been included in several scores and models to predict mortality in patients with Advanced Chronic Liver Disease (ACLD), such as the Child-Pugh score [1] and the MELD-score [2] (with its modifications). Development of jaundice has been traditionally considered a

V. Calvaruso
Gastroenterology and Hepatology Unit, Department of Health Promotion Sciences Maternal and Infantile Care, Internal Medicine and Medical Specialities (PROMISE), University of Palermo, Palermo, Italy
e-mail: Vincenza.calvaruso@unipa.it

C. Ripoll
Internal Medicine IV, Jena University Hospital Friedrich-Schiller University, Jena, Germany
e-mail: Cristina_ripoll@med.uni-jena.de

J. Bosch
Department of Visceral Surgery and Medicine, Inselspital, University of Bern, Bern, Switzerland

Department of Biomedical Research, Visceral Surgery and Medicine, Inselspital, University of Bern, Bern, Switzerland

IDIBAPS, CIBERehd, University of Barcelona, Barcelona, Spain
e-mail: jaime.bosch@dbmr.unibe.ch

A. Dell'Era (✉)
Department of Biomedical and Clinical Sciences, University of Milan, Milan, Italy

Gastroenterology and Endoscopy Unit, ASST Fatebenefratelli Sacco, Milan, Italy
e-mail: Alessandra.dellera@unimi.it

R. de Franchis (ed.), *Portal Hypertension VII*,
https://doi.org/10.1007/978-3-031-08552-9_31

typical finding in decompensated cirrhosis. Jaundice, however, is not specific and is frequently seen in patients with acute liver injury of many different causes.

The distinction between compensated and decompensated cirrhosis is of utmost importance as it has prognostic relevance [3] with significantly different mortality risk. Patients who are compensated will first develop decompensation before eventually dying. Longitudinal trials in the context of compensated cirrhosis should include decompensation among the relevant endpoints, in addition to the development of hepatocellular carcinoma and superimposed liver injury/ACLF (acute-on-chronic liver failure). A homogenous definition of what is understood as decompensation is paramount to be able to compare results from different studies.

In compensated cirrhosis, the development of jaundice is an event of unclear significance, which probably reflects the heterogeneity of its potential causes. Some consider that jaundice in a compensated patient is per se a sign of decompensation. Indeed, when jaundice is associated with other manifestations of decompensated disease, and the patient has other markers indicating a worsening of liver function, such as reduced albumin or prolonged prothrombin time, it is undoubtedly a sign of further decompensation.

However, on the other hand, compensated patients may develop jaundice without any other manifestation of decompensation. In this situation, the patient should undergo a comprehensive evaluation to assess all the possible concomitant causes. The first step to determine the cause of jaundice is abdominal imaging, usually an abdominal ultrasound to rule out obstructive jaundice. In patients with cirrhosis, however, this may be difficult to assess due to the fact that the increased liver stiffness may prevent dilatation of the intrahepatic bile ducts. Other aspects have to be taken into consideration after ruling out biliary obstruction, starting with the etiology of the liver disease since jaundice may be found in patients with cholestatic diseases. Second, the compensated patient may suffer from superimposed liver injury as the cause of jaundice. Indeed, the clinical experience taught us that in a patient with ACLD presenting with jaundice, without associated complications from portal hypertension, the most frequent causes of the jaundice are bacterial infections and/or alcoholic hepatitis, followed by Drug-induced Liver Injury (DILI) or viral infections. Some mild forms of superimposed liver injury may present with only transient jaundice but more frequently, superimposed liver injury occurs in the context of ACLF that is characterized by the development of several organ failures, including liver, kidney, brain, circulation, coagulation, and lung, triggered by events causing systemic inflammation (CANONIC) [4]. In these patients, jaundice is frequent and accompanies other manifestations of decompensation of the liver disease. ACLF can develop in patients with compensated cirrhosis, although less frequently than in decompensated patients [4], and it appears that its course has even a worse prognosis than in patients with decompensated cirrhosis. ACLF has been proposed to have pathophysiology and natural history different from that of decompensation, being mostly driven by systemic inflammation rather than by portal hypertension and liver failure. As ACLF has only been characterized in recent years, most studies evaluating the natural history of ACLD have not taken jaundice into consideration.

A recent review by D'Amico et al. [5] which evaluated 92 studies, assessed the incidence of decompensation according to the different combinations of decompensating events, including ascites, bleeding, jaundice, encephalopathy, hepatocellular carcinoma, development of varices, worsening of prothrombin time prolongation, aggravation of Child-Pugh score, and need of diuretics). The authors concluded that in most studies decompensation was defined by the development of ascites, variceal bleeding, and encephalopathy. When jaundice was also considered among the events defining decompensation, the cut-off of bilirubin was not uniform across the studies. As already commented, jaundice rarely has been reported as a first decompensating event. The fact that the development of jaundice in cirrhosis is associated with a 5-year survival of about 20% [3], underlines its relevance as an end-stage disease marker.

In the context of Baveno VII, a literature search of papers published between 2005 and 2021 dealing with the first decompensation retrieved 116 articles, of which only 32 included jaundices as a potential decompensating event in previously compensated patients. From these 32 articles, only 11 actually defined what was meant by jaundice [3, 6–15] (Table 31.1). The definition among the different studies was heterogeneous, ranging from values of bilirubin above 2 mg/dL to above 5 mg/dL or greater than three times the upper normal limit. In one paper [14] the definition also included the presence of normal bile ducts at the ultrasound. In seven papers jaundice was not defined [16–21]. No study discussed whether transient jaundice was considered as a decompensating event. The incidence of jaundice as a decompensating event was described only in 18 studies [3, 6–22] (Table 31.1).

Table 31.1 Jaundice as a decompensating event

Author	No. of compensated patients	Jaundice definition: bilirubin levels	Jaundice as first event	Jaundice and other events	Decompensation definition
Chon YE et al. [6]	1126	>2 mg/dL		2.9%	Jaundice, VB, ascites, HE
D'Amico G et al. [3]	377	≥3 mg/dL	3%	15%	Ascites, VB, HCC, HE, jaundice
Das K et al. [16]	253	n.d.	1%		Ascites, VB, HE, jaundice
Karagozian R et al. [9]	153	Clinical exam or ≥ 2.5 mg/dL		73.9%	Jaundice, ascites, HE, VB
Macías J et al. [22]	297	n.d.	0.7%		Ascites, PHGB, HE, jaundice, SBP
Procopet B et al. [12]	280	>3 mg/dL		8.2%	Ascites, jaundice, VB, HCC, PVT, HE, hydrothorax, infection

(continued)

Table 31.1 (continued)

Author	No. of compensated patients	Jaundice definition: bilirubin levels	Jaundice as first event	Jaundice and other events	Decompensation definition
Ampuero J et al. [13]	135	≥3 mg/dL	0.7%		HE, VB, ascites, jaundice
Sangiovanni A et al. [17]	214	n.d.	9%	17%	HCC, ascites, jaundice, UGIB
Gomez EV et al. [18]	402	n.d.	8.9%		Ascites, VB, HE, SBP
Pineda JA et al. [19]	154	n.d.	1.9%		Ascites, HE, PHGB, HCC, jaundice
Gheorghe L et al. [15]	166	>4 mg/dL		30.1%	Ascites, jaundice, PHGB, HE, HCC, PVT, SBP, HRS
Bruno R et al. [20]	69	n.d.		39.1%	Ascites, jaundice, HE
Macías J et al. [21]	892	n.d.	2.5% (biopsy)/1.9% (LSM)		Ascites, PHGB, HCC, SBP, HE, jaundice
Giron-Gonzales J et al. [14]	50	≥5 mg/dL and US normal bile ducts	0%		Ascites, SBP, PHGB, HE, jaundice, HCC
Fartoux L et al. [8]	102	>51 μmol/L		6.9%	HCC, VB, ascites, HE, jaundice
Kondo T et al. [10]	110	>3.0 mg/dL		12.7%	VB, ascites, HE, SBP, jaundice
Radu C et al. [11]	29	>3.0 mg/dL		0%	VB, ascites, HE, jaundice, infection, SBP, HRS, HCC, death or LT
Reichert M et al. [7]	257	>3.0 mg/dL		16.7%	VB, HE, ascites, jaundice

n.d. not defined, *VB* variceal bleeding, *HE* hepatic encephalopathy, *HCC* hepatocellular carcinoma, *PHGB* portal hypertensive gastrointestinal bleeding, *SBP* spontaneous bacterial peritonitis, *PVT* portal vein thrombosis, *UGIB* upper gastrointestinal bleeding, *HRS* hepatorenal syndrome, *LSM* liver stiffness measurement, *US* ultrasound, *LT* liver transplantation

In papers where jaundice was stated as a first single decompensation event, its prevalence ranged from 0.7% to 3% [3, 13, 16–19, 21, 22]. In two papers [11, 14], no cases of jaundice were reported. The remaining articles reporting jaundice as a first decompensation event did not specify if it occurred alone or in association with other traditional manifestations of decompensation [6, 9, 10, 12, 15, 20] although it might be so as suggested by the much greater range of prevalence of jaundice, from 2.9% to 30.1%. In two articles [6, 9], jaundice was the most frequent decompensating event, while in all others this was ascites [3, 12, 15, 16, 18–22], and in one paper [13] hepatic encephalopathy, variceal bleeding, ascites, and jaundice were reported with the same frequency.

The majority of the studies reviewed included patients with cirrhosis related to HCV [8, 9, 17, 18], HBV [6, 16], and HBV/HDV [15] alone or with HIV-coinfection [14, 19–22]. None of the studies describing the incidence of jaundice as the first decompensation ruled out specifically superimposed liver injury.

In conclusion, the prevalence and significance of jaundice in patients with compensated cirrhosis remain unclear. In the survey sent to members of the Baveno VII faculty (see Chap. 28), there was no agreement on whether or not jaundice should be included in the events defining the first decompensation of a previously compensated patient. The review of the available literature shows that jaundice "per se" is most often not included among the decompensating events, and when it is, then its prevalence is frequently not reported in the results, or it is unclear whether it is reported as an isolated event or associated with other manifestations of decompensation. Furthermore, the lack of a uniform definition of jaundice hampers the comparison of the results of the different studies which do report its incidence. Prospective studies are needed to evaluate its incidence as the first clinical manifestation of advanced chronic liver disease and if its impact on prognosis justifies considering it a decompensating event.

References

1. Child CG, Turcotte JG. Surgery and portal hypertension. Major Probl Clin Surg. 1964;1:1–85.
2. Wiesner RH, McDiarmid SV, Kamath PS, Edwards EB, Malinchoc M, Kremers WK, et al. MELD and PELD: application of survival models to liver allocation. Liver Transpl. 2001;7(7):567–80.
3. D'Amico G, Pasta L, Morabito A, D'Amico M, Caltagirone M, Malizia G, et al. Competing risks and prognostic stages of cirrhosis: a 25-year inception cohort study of 494 patients. Aliment Pharmacol Ther. 2014;39(10):1180–93.
4. Moreau R, Jalan R, Gines P, Pavesi M, Angeli P, Cordoba J, et al. Acute-on-chronic liver failure is a distinct syndrome that develops in patients with acute decompensation of cirrhosis. Gastroenterology. 2013;144(7):1426–37, 37.e1–9
5. D'Amico G, Perricone G. Prediction of decompensation in patients with compensated cirrhosis: does etiology matter? Curr Hepatol Rep. 2019;18(2):144–56. https://doi.org/10.1007/s11901-019-00473-1.
6. Chon YE, Jung ES, Park JY, Kim DY, Ahn SH, Han K-H, et al. The accuracy of noninvasive methods in predicting the development of hepatocellular carcinoma and hepatic decompensation in patients with chronic hepatitis B. J Clin Gastroenterol. 2012;46(6):518–25.
7. Reichert MC, Schneider C, Greinert R, Casper M, Grünhage F, Wienke A, et al. Isolated bacterial infection without decompensation has no impact on survival of compensated patients with cirrhosis. Liver Int. 2021;41(6):1370–8.
8. Fartoux L, Degos F, Trépo C, Goria O, Calès P, Tran A, et al. Effect of prolonged interferon therapy on the outcome of hepatitis C virus-related cirrhosis: a randomized trial. Clin Gastroenterol Hepatol. 2007;5(4):502–7.
9. Karagozian R, Grace ND, Qamar AA. Hematologic indices improve with eradication of HCV in patients with cirrhosis and predict decompensation. Acta Gastroenterol Belg. 2014;77(4):425–32.

10. Kondo T, Maruyama H, Sekimoto T, Shimada T, Takahashi M, Okugawa H, et al. Impact of portal hemodynamics on Doppler ultrasonography for predicting decompensation and long-term outcomes in patients with cirrhosis. Scand J Gastroenterol. 2016;51(2):236–44.
11. Radu C, Stefanescu H, Procopet B, Lupsor Platon M, Tantau M, Grigorescu M. Is spleen stiffness a predictor of clinical decompensation in cirrhotic patients? J Gastrointestin Liver Dis. 2014;23:223–4.
12. Procopet B, Farcau O, Balagel M, Crisan D, Stefanescu H, Pop A, et al. The metabolic syndrome is not correlated with the short-term risk of decompensation in patients with cirrhosis. J Gastrointestin Liver Dis. 2014;23(4):397–403.
13. Ampuero J, Montoliú C, Simón-Talero M, Aguilera V, Millán R, Márquez C, et al. Minimal hepatic encephalopathy identifies patients at risk of faster cirrhosis progression. J Gastroenterol Hepatol. 2018;33(3):718–25.
14. Girón-González JA, Brun F, Terrón A, Vergara A, Arizcorreta A. Natural history of compensated and decompensated HCV-related cirrhosis in HIV-infected patients: a prospective multicentre study. Antivir Ther. 2007;12(6):899–907.
15. Gheorghe L, Iacob S, Simionov I, Vadan R, Gheorghe C, Iacob R, et al. Natural history of compensated viral B and D cirrhosis. Rom J Gastroenterol. 2005;14(4):329–35.
16. Das K, Das K, Datta S, Pal S, Hembram JR, Dhali GK, et al. Course of disease and survival after onset of decompensation in hepatitis B virus-related cirrhosis. Liver Int. 2010;30(7):1033–42.
17. Sangiovanni A, Prati GM, Fasani P, Ronchi G, Romeo R, Manini M, et al. The natural history of compensated cirrhosis due to hepatitis C virus: a 17-year cohort study of 214 patients. Hepatology. 2006;43(6):1303–10.
18. Gomez EV, Rodriguez YS, Bertot LC, Gonzalez AT, Perez YM, Soler EA, et al. The natural history of compensated HCV-related cirrhosis: a prospective long-term study. J Hepatol. 2013;58(3):434–44.
19. Pineda JA, Aguilar-Guisado M, Rivero A, Girón-González JA, Ruiz-Morales J, Merino D, et al. Natural history of compensated hepatitis C virus-related cirrhosis in HIV-infected patients. Clin Infect Dis. 2009;49(8):1274–82.
20. Bruno R, Sacchi P, Puoti M, Maiocchi L, Patruno S, Carosi G, et al. Natural history of compensated viral cirrhosis in a cohort of patients with HIV infection. J Acquir Immune Defic Syndr. 2007;46(3):297–303.
21. Macías J, Márquez M, Téllez F, Merino D, Jiménez-Aguilar P, López-Cortés LF, et al. Risk of liver decompensation among HIV/hepatitis C virus-coinfected individuals with advanced fibrosis: implications for the timing of therapy. Clin Infect Dis. 2013;57(10):1401–8.
22. Macías J, Camacho A, Von Wichmann MA, López-Cortés LF, Ortega E, Tural C, et al. Liver stiffness measurement versus liver biopsy to predict survival and decompensations of cirrhosis among HIV/hepatitis C virus-coinfected patients. AIDS. 2013;27(16):2541–9.

Evaluation of the Role of Minimal Perihepatic Ascites, Minimal Hepatic Encephalopathy, and Bleeding Due to Portal Hypertensive Gastroenteropathy in the Definition of Decompensation

32

Luis Ibáñez-Samaniego and Rafael Bañares

Introduction

Cirrhosis is a complex and heterogeneous disease in which several substages integrating different histologic, hemodynamic, and clinical features, have been defined [1]. From the clinical point of view, the most important landmark in the natural history of the disease is the transition from compensated to a decompensated stage which is clearly associated with changes in the pathophysiological mechanisms of the disease and a worsening in prognosis [2].

However, which clinical events define decompensation is not fully elucidated. While classic forms of decompensation (clinical ascites, overt hepatic encephalopathy, and variceal bleeding) are widely accepted as markers of decompensation, there is no agreement regarding the significance of other manifestations of cirrhosis. This is especially relevant for (1) small amount of ascites only detected in imaging procedures/minimal perihepatic ascites, (2) minimal hepatic encephalopathy (MHE) and (3) bleeding from Portal Hypertensive Gastroenteropathy (PHG).

Several papers have partially addressed this problem, but important methodological problems preclude the obtention of robust information. First, very few studies have been designed to answer this specific question. Additionally, these studies have included a heterogeneous population of patients with cirrhosis (in terms of age, gender, etiology of liver disease, liver function) and have not differentiated between compensated or decompensated patients. Importantly, the definition of the three entities (minimal perihepatic ascites, MHE, and bleeding from PHG) is not

L. Ibáñez-Samaniego · R. Bañares (✉)
Digestive Diseases Department, Hospital General Universitario Gregorio Marañón, Facultad de Medicina, Universidad Complutense, Madrid, Spain

Centro de Investigación Biomédica en Red (CIBERehd), ISCiii, Madrid, Spain
e-mail: rbanares@ucm.es

R. de Franchis (ed.), *Portal Hypertension VII*,
https://doi.org/10.1007/978-3-031-08552-9_32

homogeneous among the studies. Furthermore, most of the studies are retrospective, and therefore with a greater risk of bias. Finally, the evaluation and definitions of endpoints (i.e., decompensation, liver-related death, etc.) are also heterogeneous or not fully explored in the studies. It should be emphasized that very few studies have included appropriate competing risk analyses.

This chapter aims to critically evaluate the available information regarding the impact of the presence of minimal perihepatic ascites, minimal hepatic encephalopathy, and chronic bleeding from portal hypertensive gastroenteropathy in the natural history of compensated cirrhosis and, specifically, in the definition of the first decompensation.

Minimal Hepatic Encephalopathy

MHE has been associated with a decreased quality of life and to an increased risk of falls, traffic accidents, and risk of developing Overt Hepatic Encephalopathy (OHE) [3, 4]. Moreover, the prevalence of MHE increases with the severity of the liver disease being less frequent in Child-Pugh A patients as compared with Child-Pugh B or C. Regarding the influence of MHE in relevant outcomes (i.e., classic decompensation or death), the analyzed studies show different (and somehow conflicting) results (Table 32.1). The discrepancies found in the different studies may be explained by certain biases.

1. First, MHE diagnosis does not always rely on the same diagnostic criteria. Although previous recommendations suggested the combination of at least one psychometric and one neuropsychological test for diagnosis, a recent consensus position paper accepted that MHE diagnosis could be based on the patient's performance on a single and valid (culturally and nationally) neurophysiological test, according to local expertise [5]. In addition, several studies analyzed together patients with covert hepatic encephalopathy (comprising MHE and Grade I HE), precluding the possibility to identify the specific impact of MHE.
2. Next, the population of patients included in the studies is very heterogenous regarding age, etiology of liver disease, the severity of liver dysfunction, previous history of hepatic decompensation, comorbid diseases, and use of concomitant medication that may potentially affect cognitive performance. This fact may influence the diagnosis of MHE and, most importantly, survival.
3. Third, although most of the analyzed studies were prospective, the length of follow-up was too short to demonstrate an impact of MHE on relevant outcomes (i.e., decompensation and death) in compensated patients.
4. Finally, the studies included only a small number of compensated patients with well-characterized MHE.

Despite these limitations, the presence of properly identified MHE (diagnosed by different approaches: PHEs, CFF, computer-aided neuropsychological tests, EEG, etc.) could be associated with an increased risk of cirrhosis stage progression in compensated patients (58.3% vs. 20.5%) [6] or to greater mortality or need for liver transplantation [7]. However, further prospective studies specifically designed

Table 32.1 Studies assessing the influence of minimal hepatic encephalopathy on relevant outcomes

Author	Year	Study design	Study Population	MHE diagnosis	End-points	Statistical analyses	Characteristics of patients
Ampuero et al. [6]	2018	Multicenter cohort with prospective follow up	Patients with cirrhosis (exclusion of active alcohol drinkers, TIPS, recent GI bleeding or HCC and drugs which potentially could interfere with the evaluation of MHE). Patients with previous decompensation were included and classified in different stages (compensated without varices, compensated with varices, one previous decompensation and, subsequent decompensations)	PHES **AND** CFF, as recommended in the AASLD and EASL guidelines.	**Decompensation** (ascites, OHE, Variceal bleeding, jaundice), **development of varices** (stage 1 to 2), **LT, liver related death**	Competing risk models (nonliver death and LT due to HCC)	320 patients. 52% alcohol related, 32% HCV related cirrhosis.
Labenz et al. [7]	2020	Multicenter cohort with prospective follow up	Patients with cirrhosis (exclusion of active alcohol drinkers, TIPS, OHE in the previous 6 weeks, and psychotropics or opioids, NYHA III-IV, COPD, renal failure, neurological comorbidities, active malignancy, electrolyte disorders). Patients with previous OHE were included if were on treatment with rifaximin or lactulose	PHES, CFF, S-ANT1 (animal naming test), CCHE test (a test previously described by the authors).	**A**) Ocurrence of an episode of **OHE** that needed hospitalization or development of OHE during hospitalization for other reason. **B**) **Death or liver transplantation** (for terminal liver failure)	Competing risk models (OHE vs. death/LT)	224 patients, 33% alcohol, 20% viral hepatitis, 33% mixed, 14% NAFLD)

(continued)

Table 32.1 (continued)

Author	Year	Study design	Study Population	MHE diagnosis	End-points	Statistical analyses	Characteristics of patients
Thomsen KL et al. [19]	2016	Single-center cohort with prospective follow-up	Patients with cirrhosis 18-75 years (exclusion of active alcohol drinkers, severe decompensation of cirrhosis in the last 30 days, use of probiotics, norfloxacin/ antibiotic and immunomodulatory agents	,mHE was considered if at least one of the following test was abnormal (PHES, CFF or EEG) HE grade 1 according to west-haven criteria	Development of a **complication requiring hospitalization** or **death**	Logistic regression to identify predictors of mortality.	106 patients. 63% men. MELD 15 (6), child-Pugh A: 5%, B: 67%, C: 28%. MELD was similar between patients with HE grade 1, MHE or unimpaired. 98% of patients with HE grade 1 had moderate to severe ascites
Patidar KR et al. [20]	2014	Single-center cohort with prospective follow-up	Patients with cirrhosis 18–65 years (exclusion of previous or current OHE, infection, or gastrointestinal hemorrhage within the past 6 weeks, with hepatocellular carcinoma, who were on psychoactive medications, and with recent illicit drug and alcohol use within 6 months)	CHE was diagnosed if patients scored abnormally on ≥2 psychometric tests (the number connection test-A (>35 s), the number connection test-B (>99 s), the digit symbol test (<68 raw score), and block design (<28 raw score))	**Episodes of OHE** (which included outpatient and inpatient diagnoses), **hospitalizations, transplant, and death**	Cox models for hospitalization and transplant/ death	170 patients. 58% male, 55 (8) years. MELD 9.2 (3.4). Etiologies: HCV 60% NASH 20% Alcohol 4%)

Ampuero et al. [4]	2015	Single-center estimation cohort with prospective follow up. Multicenter validation cohort	Patients with cirrhosis exclusion of HCC; ongoing treatment for viral cirrhosis; recent (<3 mo) alcohol abuse; infection; recent (<6 wk) antibiotic use or gastrointestinal bleeding; and a history of recent (<6 wk) use of benzodiazepines, anti-epileptic, or psychotropic drugs	PHES AND CFF	Survival. Decompensation	Kaplan Meier. Logistic regression	Estimation cohort: 117 patients. 57.8 years. 74% males. Child A: 77% Child B: 36% Child C: 4% MELD 9.8 (4). Alcohol 53%. Validation cohort 114 patients. Higher MELD and greater proportion of child C patients in the validation cohort
Wang et al. [21]	2017	Single-center cohort with prospective follow-up	Consecutive patients with cirrhosis 18–65 years (exclusion: Episode of OHE in the last 6 months, malignancy including liver cancer; TIPS, systemic disease, significant head injury, neurological or psychiatric diseases; alcohol or psychoactive medication intake within 6 months; antibiotics, lactulose, probiotics or L-ornithine-L-aspartate intake or albumin infusion within 6 weeks	Serial testing with PHES every 3 months.	**Complications of cirrhosis**: OHE, infections, GI bleeding, HRS, HCC, severe ascites requiring albumin infusion. **Death/liver trasplantation**	Sample-size calculation powered to detect differences for the first episode of OHE	366 patients. Age 47.2 (8.6) years. Male 73%. Child A: 43.2%, Child B: 50.5%

(continued)

Table 32.1 (continued)

Author	Year	Study design	Study Population	MHE diagnosis	End-points	Statistical analyses	Characteristics of patients
Dhiman RK et al. [22]	2010	Single-center cohort with prospective follow-up	Consecutive outpatient patients with cirrhosis exclusion criteria: OHE or history of overt HE, recent alcohol intake, infection, recent(6 weeks) antibiotic use or gastrointestinal bleeding, history of recent (6 weeks) use of drugs affecting psychometric performances, TIPS, electrolyte imbalance, creatinine [>1.5 mg/dL), presence of HCC, severe medical problems such as congestive heart failure, pulmonary disease, neurological or psychiatric disorder, etc., that could influence performance of neuropsychiatric tests. **83 healthy controls age and sex matched**	PHES and CFF but using Z-scores	Survival	Conventional survival analysis and COX regression models	104 patients. 83% male. 48 years. Child A: 21%, Child B: 61%, Child C: 18%. Alcohol 59%, HBV 11%, HCV 9%

Barone et al. [23]	2018	Single-center cohort with prospective follow-up	Consecutive outpatient patients with cirrhosis (exclusion of **current or previous presence of overt HE,** current or past alcohol consumption, gastrointestinal bleeding and spontaneous bacterial peritonitis in the previous 6 weeks, significant comorbidities such as cardiac, respiratory or renal failure; previous transjugular intrahepatic portosystemic shunt, electrolyte imbalance as hyponatremia (Na <125 mg/dL), neurological diseases, not liver-related metabolic encephalopathies, BMI <18.5 kg/m^2, more than 5% loss of body weight in the last 3 months, illiteracy, color blindness or severe visual disturbances (cataracts, diabetic retinopathy), hepatocellular carcinoma or other malignancies. **150 healthy controls**	**CFF only** (<39 Hz)	Development of **OHE and survival**	Cox regression models	134 cirrhotic patients. Age 62.3 (9.9) years, Male 72% Child A: 68% Child B: 29% Child C: 3%. 150 controls, age, sex distribution and scholarity were similar to those observed in the whole cirrhotic population

(continued)

Table 32.1 (continued)

Author	Year	Study design	Study Population	MHE diagnosis	End-points	Statistical analyses	Characteristics of patients
Miwa T et al. [24]	2021	Single-center retrospective analysis of a prospectively collected database	Patients with cirrhosis >20 years old. Exclusion: **Presence or history of OHE**; dementia or other neurological or psychiatric disorders; active infection or other systemic inflammation in the past 6 weeks; gastrointestinal bleeding in the past 6 weeks, presence of HCC or other malignancies; presence of acute liver failure; andpresence of severe comorbidities	**Computer-aided simple neuropsychiatric test** composed of four subtests, including the number connection tests A and B, digit symbol test, and block design test. Patients were diagnosed as having MHE when the results of two or more subtests were abnormal	Evaluation of the relationship between zinc levels and the **occurrence of an episode of OHE** in patients with MHE. **Mortality** in patients with MHE	Competing risk (mortality/liver transplantation)	100 patients Age 73 years, 61% male Alcohol 26% HCV 25% HBV 16% Child A 59% Child B 26% Child C 15%

Hanai et al. [25]	2019	Retrospective single-center	Patients with cirrhosis exclusion criteria included presence of OHE or history of OHE in the past 6 weeks; infection or spontaneous bacterial peritonitis in the past 6 weeks; presence of uncontrolled hepatocellular carcinoma (HCC) or other malignancies; previous TIPS; gastrointestinal bleeding in the past 6 weeks; presence of neurological diseases such as Alzheimer's disease; presence of psychiatric disorders; use of benzodiazepines, anti-epileptic, or psychotropic drugs; presence of severe comorbidities such as heart, respiratory, and/or renal failure	Computer-aided **neuropsychiatric test** test: Number connection test-A, number connection test-B, digit symbol test, and block design test MHE was diagnosed if the results of more than two subtests were abnormal	All-cause mortality.	Propensity score (age, sex, etiology, presence of HCC, child, MELD, albumin, bilirubin, INR, ammonia)	269 patients. Age 71. 64% male. HBV 14%, HCV 39% Alcohol 15%. HCC 68%. Child 5 (IQR 5–7), child A 67% Child B 22% Child C 11%. MELD 8 (7–10)

(continued)

Table 32.1 (continued)

Author	Prevalence of MHE in study population	Follow-up	Association between MHE and decompensation development	Liver trasplant	Survival	Potential bias	Comments
Ampuero et al. [6]	MHE present in 57/314 (18.2%). MHE could not be explored in all patients (illiterate, visual defects)	Visits every 6 months. Patients were censored at the time of cirrhosis progression or after **5 years** of follow up Mean F-U 3.5 (1.8) years	Overall, cirrhosis progression was observed in 38.1% (122/320). MHE was linked to cirrhosis progression [(64.9% (37/57) vs. 31.9% (82/257); logRank 29,952, $P = 0.0001$]. MHE was more frequently associated with cirrhosis progression in stage 1 (58.3%; 7/12 vs. 20.5%; 16/78; $P = 0.005$), stage 2 (50%; 2/4 vs. 44.4%; 17/38; $P = 0.365$), stage 3 (80%; 16/20 vs. 38.2%; 26/68; $P = 0.002$), and stage 4 (57.1%; 12/21 vs. 31.5%; 23/73; $P = 0.005$). In competing risk regression, MHE was independently associated with disease progression in cirrhotic patients after adjusting by other already known factors (age, child, MELD, platelets)	Overall: 10.9% (35/320)	Overall: 19.1% (61/320). The influence of MHE on survival is not explored.	Authors do not explore the role of betablockers or diuretic on the diagnosis of MHE. MHE is considered as a static process at the start of follow up but patients may receive treatments or have events like resume of alcohol consumption during FU that could influence the progression/stabilization of the disease. 50% of patients included in the study have a previous history of alcohol consumption (which may influence cognitive performance) but this characteristic has not been fully explored in the multivariate analysis. 19 patients with HCV cirrhosis received treatment during FU (9 achieved SVR)	The presence of MHE identifies a higher risk of cirrhosis progression in decompensated patients at short (1 year), mid (3 years) and long (5 years) term follow up. In compensated patients, MHE only predicts mid and long-term risk of progression

Labenz et al. [7]	MHE present in 10.1% of patients (but data is analysed using the pathological cutoffs of each test, not by the presence of MHE).	Visits every 6 months. Patients were censored at the time of OHE, LT/death or after **2 years** of follow up. Median F-U 364 days (IQR 202–508)	17% developed an episode of OHE. 23.6% developed any decompensation	Death and LT for final liver failure are considered as the same outcome (20.1% of patients). Pathological PHES, CFF and CCHE are associated with lower survival at 2 years (PHES: 48% vs. 90%; $p < 0.05$)(CFF: 58% vs. 72%; $p < 0.05$) (CCHE: 45% vs. 88%; $p < 0.05$). In multivariate cox regression, PHES, S-ANT and CCHE were independent predictors of mortality	Both compensated and decompensated stages, are analyzed together so there is not information about the influence of MHE at each stage. 13% of patients were lost to follow up. The authors do not explore the reversibility of MHE or the effect of treatments started during FU that could influence on MHE/survival status	A normal result in PHES **or** CFF test can exclude the occurrence of an episode of OHE in the next 180 days the presence of a pathological PHES or CFF increased the risk of developing an episode of OHE during FU. A pathological result in both PHES and CFF tests is associated with a lower 2-years survival. Authors propose that PHES and CFF are useful tools to identify patients who would benefit from LT despite lower MELD

(continued)

Table 32.1 (continued)

Author	Prevalence of MHE in study population	Follow-up	Association between MHE and decompensation development	Liver trasplant	Survival	Potential bias	Comments
Thomsen KL et al. [19]	37% MHE 42% HE grade 1. 21% unimpaired.	Up to **1 year** Mean F-U 234 (99) days	Increased omplications requiring hospital admission (unimpaired/ mHE/grade 1 HE: 26/36/64% respectively; $P = 0.02$) No significant differences were found between patients with MHE vs. unimpaired Only infections were higher in the group of HE grade 1 (not differences for HE, VB, or ascites)	Not evaluated	Mortality was significantly different between the three groups (unimpaired/ mHE/grade 1 HE: 4/5/20% respectively; $P = 0.04$) with grade 1 HE patients having greater mortality than mHE patients ($P = 0.04$). Grade 1 HE and bilirubin were the only predictors of mortality	Very low number of patients, short follow up to identify the risk of decompensation/ mortality in patients with MHE. The population included in the study was mainly at the decompensated stage (only 8% and 4% of the patients with MHE and HE grade 1 respectively were child A), 28% of patients with MHE had previous episodes of MHE.	Mainly focused on the important prognostic differences between MHE and HE grade 1. Patients with HE grade 1 seem to be at a different stage of the disease as they have increased risk of hospitalization and death. In conclusion, in the short term the presence of MHE does not seem to imply a higher risk of decompensation or mortality

Patidar KR et al. [20]	56% presented CHE	Mean FU 13.0 (4.6) months median 8.6 months IQR (0.1-20.6)	Development of OHE: 37.9% of CHE vs. 17.3% of non-CHE; $p = 0.001$. There were more liver-related (36 vs. 16) and liver-unrelated hospitalizations (35 vs. 12) in the CHE group as compared with the no-CHE group. The multivariate cox regression analysis identified that CHE was significantly associated with first OHE event, first hospitalization, death, and death/transplant	Higher rate of death in the CHE group (17) than in the no-CHE group (4), with the most common etiology reported as liver related, followed by sepsis and multiorgan failure. The number of transplants was not significantly different between the two groups (8 and 4 in the CHE and no-CHE groups, respectively, $P = 0.435$)	CHE is considered as a single entity without differentiating MHE No data about DAA in hepatitis C patients Very short follow up for most patients. Although included in multivariate analysis, liver dysfunction is significantly greater in patients with CHE that may act as a confounder for the need of hospitalization or death Surprisingly high proportion of patients with CHE in a compensated population of cirrhotics	The main message is that even in the most stable-appearing patients, CHE is a frequent issue that deserves attention. The presence of CHE, despite liver dysfunction, is associated with poor clinical outcomes
Ampuero et al. [4]	**CFF < 39 Hz**: Estimation cohort 36.5% (35/96). Validation 47.4% (54/114). **PHES < −4 points**: Estimation 25.9% (29/112), validation 29.8% (34/114)	From 2003–2007 to December 2013. Estimation cohort: 5 (2.8) years Validation cohort: 4.4 (3.9) years	**OHE**: **Estimation cohort**: CFF < 39 Hz 37.1% vs. CFF > 39 Hz 24.6% (log-rank, 4.896; $P = 0.027$); **validation cohort**: CFF < 39 Hz 38.9% vs. CFF > 39 Hz 18.3% patients (log-rank, 9.576; $P = 0.002$) No association between CFF and the occurrence of other complications of cirrhosis	**Survival:** **Estimation cohort:** CFF <39 Hz 68.6% vs. CFF > 39 Hz 82% (log-rank, 5.073; $P = 0.024$). **Validation cohort:** CFF <39 Hz 57.4% vs. CFF > 39 Hz 70% (log-rank, 4.752; $p = 0.029$) PHES was not associated with mortality	Small sample size (especially for patients with MELD >15)	The most saliient point from this study is that CFF < 39 Hz might help to identify a subgroup of patients with intermediate MELD score (10–15 points) that present high mortality. The influence of CFF was confirmed in two different cohorts

(continued)

Table 32.1 (continued)

Author	Prevalence of MHE in study population	Follow-up	Association between MHE and decompensation development	Liver trasplant	Survival	Potential bias	Comments
Wang et al. [21]	36% of patients presented MHE	11.2 (13) months	Cirrhosis-related complications requiring hospitalization: CHE 28.2% vs. unimpaired 18.3%; $p = 0.008$	None	Mortality: CHE 11.5% vs. unimpaired 4.7%; $p = 0.012$	Short follow up period	CHE is highly prevalent in clinically stable cirrhotic patients and is associated with an increased risk of complications, hospitalizations and mortality. CHE can evolve into OHE, persist or spontaneously resolve (20%) without medication intervention within a short period

Dhiman RK et al. [22]	48%	Up to Feb-2009 (recruitment from jan-2007 to dec-2007)			Overall mortality at the end of FU: MHE 39.1% vs. non-MHE 22.9%. Probability of survival at 1 and 2 years: PHES<-6: 68/26%. PHES > −6: 89/75%	Norms for general population are constructed calculating Z-scores in the healthy control group (n = 83). However, in India there are already norms for PHES. For CFF authors perform a similar approach (do not use the standard cut-off of 39 Hz)	Diagnosis of MHE was not related to mortality in the multivariate analysis. PHES score < −6 points was associated with mortality

(continued)

Table 32.1 (continued)

Author	Prevalence of MHE in study population	Follow-up	Association between MHE and decompensation development	Liver trasplant	Survival	Potential bias	Comments
Barone et al. [23]	31% presented MHE	Median FU 36 months (IQR 12–47)	Development of OHE: MHE 46.3% vs. unimpaired 11.8%?	ND	ND		**CFF is able to predict** the development of the **first episode of overt HE** and represents an independent predictor of **survival in well compensated cirrhotic patients** independently from MELD and child-Pugh scores. Authors suggest that the screening of patients with CFF could be useful to identify a population that deserves a more careful clinical monitoring

Miwa T et al. [24]	100% presented MHE at inclusion	9.9 (IQR 6.3–23)	OHE: 16% incidence rates of OHE: 10.5% at 1-year, 23% at 3 years. The risk of developing OHE was greater in patients with zinc deficiency	0%	29% mortality ¡ median survival time: 40.2 months. The 1-, 3-, and 5-year survival rates were 83.8%, 52.6%, and 43.4%, respectively. Mortality was higher in patients with zinc deficiency	Short follow up period. Retrospective Lack of a group of unimpaired patients. Unclear if patients received antiviral treatment during follow up	This study was focused to describe the impact of zinc level on the development of OHE in patients with MHE. However, it provides data regarding the risk of OHE and mortality in the overall cohort
Hanai et al. [25]	20.8% presented MHE	13.4 months (IQR 4–28)	Not evaluated		Mortality: Overall: MHE 41% vs. non-MHE 20.6% ($p = 0.02$). Propensity-score adjusted: MHE 42% vs. non-MHE 24%. Survival rates at 1, 2, 3 years: MHE 68/55/41% vs. non-MHE 81/78/71%	Retrospective	MHE increases mortality in Japanese population, independently of liver failure and HCC

FU follow-up, *IQR* interquartile range, *CP* Child-Pugh, *NA* not available, *HCV* hepatitis C virus, *HBV* hepatitis B virus, *NASH* nonalcoholic steato hepatitis, *PHES* Psychometric Hepatic Encephalopathy Score, *CFF* critical flicker frequency, *CHE* cover hepatic encephalopathy, *OHE* overt hepatic encephalopathy, *DAA* direct antiviral agent, *HCC* hepatocellular carcinoma, *GI* gastrointestinal, *HRS* hepatorenal syndrome

to evaluate the prognostic influence of well-characterized MHE in compensated patients are necessary to fully address this issue. Importantly, competing risk analysis considering non-liver events is mandatory.

Minimal Perihepatic Ascites

The wide use of imaging techniques in the evaluation of patients with cACLD increases the possibility to identify minimal amounts of intra-abdominal fluid (specially perihepatic), that is not clinically detectable. The significance of this finding in the natural history of cirrhosis and the development of decompensation or death is controversial among published studies (Table 32.2).

There are several data that may explain these conflicting results

1. First, most studies evaluated patients with grade 1 ascites according to the EASL definition (ascites only detected in ultrasonography). However, this concept may include patients who really have a minimal amount of fluid but also patients with a relatively large amount of fluid not detected in clinical evaluation. The current evidence does not allow distinguishing between these two clinical scenarios.
2. Part of the available information relies on retrospective studies, in which the evaluation of diagnosis and outcomes may be biased.
3. There are some dynamic factors intervening during the follow-up of the patients that could modulate the risk of progression to other classical forms of hepatic decompensation. Not surprisingly, there are factors that can upregulate (activation of systemic inflammation) [8] or downregulate (adequate etiological treatment) [9], the risk of disease progression in patients with grade 1 ascites.

Overall, the impact of subclinical ascites on survival is controversial. While two retrospective studies [9, 10] did not demonstrate an increase in mortality in patients with grade 1 ascites, a recent prospective study with long-term FU [8] showed an increased risk of mortality in patients with grade 1 ascites. This finding aligns with the results of previous studies [11, 12] in which the presence of grade 1 ascites was associated with an intermediate risk of death.

Therefore, further studies are needed to determine if the presence of minimal amounts of intra-abdominal fluid may be considered a dynamic intermediate (or preclinical) stage of decompensation.

Table 32.2 Studies assessing the influence of minimal perihepatic ascites on the natural history of cirrhosis

Author	Year	Study Design	Study Population	Definition of grade 1 ascites	End-points	Statistical analyses	Characteristics of patients	Prevalence of subclinical ascites
Theodorokapoulos et al. [10]	2021	Retrospective multicenter cohort study	Consecutive cirrhotic patients with ascites: Study groups: Grade 1 vs. grade 2/3 vs. no ascites Exclusion criteria: HIV+ and severe cardiopulmonary disease or renal failure	ICA-EASL	Clinical decompensation or mortality	Kaplan-Meier. Cox and logistic regression	Age was similar between groups MELD and CP increased accordingly to ascites severity	Grade 1: 100. Grade 2/3: 145 No ascites: 175
Bruno et al. [11]	2013	Prospective multicenter cohort from Italy	Out-patient	ICA-EASL	Mortality after the first decompensation as defined by (1) ascites (overt or ultrasound detected alone (UD), (2) gastroesophageal variceal bleeding (GEVB), (3) hepatic encephalopathy (HE), (4) jaundice, and (5) hepatorenal syndrome (HRS)	Competing events considering HCC as a competing event. For survival analysis: Time to death or LT after decompensation. Cox multivariate models.	455 patients (66% males. Mean age 61 years) CTP A: 19%, B: 61%, C: 20%. HCV 41% (none received antiviral treatment during FU) alcohol 30%	73 patients (17.7%) had grade 1 ascites.

(continued)

Table 32.2 (continued)

Author	Year	Study Design	Study Population	Definition of grade 1 ascites	End-points	Statistical analyses	Characteristics of patients	Prevalence of subclinical ascites
Shah et al. [26]	2018	Prospective single center cohort study	Cirrhotic patients who survived a first episode of liver decompensation defined as (1) ascites (overt or ultrasound detected alone (UD)), (2) Gastroesophageal Variceal bleeding (GEVB), and (3) hepatic encephalopathy (HE)	ICA-EASL	Mortality after the first episode of decompensation	Time to event (death or liver transplantation). Cox regression models, Kaplan Meier	110 patients. (85% males. Mean age, age 50 (11) years). Alcohol 48%, NASH 26% CP stage A 3%. B 56%, C 41%	Grade 1 ascites 19%
Yim et al. [9]	2016	Multicenter retrospective study	HBV-cirrhotic patients who started antiviral treatment. Exclusion of previous history of antiviral treatment for HBV ($n = 80$), variceal bleeding ($n = 37$), HCC ($n = 358$), and other malignancies ($n = 13$). Three groups: No ascites vs. grade 1 ascites vs. grades 2/3 ascites + taking diuretics.	ICA-EASL	Primary: Liver related mortality. Secondary: Development of HCC, virologic response, HBeAg seroclearance, and HBeAgseroconversion.	Patients who underwent liver transplantation were censored at the time of transplantation.Survival curves were constructed using the Kaplan-Meier method, and differences were assessed using the log-rank test. Additional survival curves were constructed censoring HCC or including it as a competing risk. Cox regression analysis to determine predictors of mortality.	501 patients. (68% male. Age 52 (10) years. CPS 7 (2), MELD 12 (5)	No ascites 67% Grade 1 ascites 10% grade 2/3 ascites. 23%

Zipprich et al. [12]	2012	Single-center retrospective study.	Consecutive cirrhotic patients who underwent HVPG measurement Exclusion criteria included primary biliary cirrhosis, previous placement of TIPS, hepatocellular carcinoma (HCC), splenic or portal vein thrombosis, concurrent illnesses expected to decrease life expectancy to less than 1 year	ICA-EASL	Death	Kaplan Meier. Cox regression	443 patients.	No ascites 35% Grade 1 ascites 9%, grade 2/3 56%
Tonon et al. [8]	2020	Post-hoc analysis of a prospective cohort study	All cirrhotic outpatients attended from march 2003 to September 2017 Exclusion criteria: Presence of HCC or extrahepatic malignancies at inclusion; severe extrahepatic disease; previous LT; clinical signs of bacterial infection at evaluation.	Patients with no history of ascites, with ascites only detectable by US who never received diuretics before inclusion were labelled as patients with grade 1 ascites	To assess the prevalence, mortality rate, and complication rate in outpatients with grade 1 ascites To compare the natural history of patients with grade 1 ascites with that of patients without ascites or with grade 2 or 3 ascites	The cumulative probability of death and the probability of developing complications of cirrhosis were estimated by the Kaplan-Meier method. Cumulative incidence function for the development of overt ascites was performed to analyze the competing risk, considering death *and LT* as *competing* events for ascites development.	547 patients	47% no ascites 10% grade 1 ascites 43% grade 2/3

(continued)

Table 32.2 (continued)

Author	Follow up	Association between ascites grade 1 and decompensation	Liver trasplant	Survival	Mortality Grade 1 vs. no ascites	Cumulative incidence of failure (death or LT)	Potential bias	Comments
Theodorokapoulos et al. [10]	Patients included since 1993. Until death or LT mean FU 18.9 (31) months. Range 1–241	38% of patients with grade 1 ascites progressed to grade 2/3 during follow up. 29% resolved the ascites. Compared with patients without ascites, patients with grade 1 ascites developed more frequently: OHE 23% vs. 5.7% HRS 10% vs. 0.6%. Pleural effusion 14% vs. 2.3%. Portal gastropathy bleeding 12% vs. 4.6% and infection 32% vs. 15.5%. There were no differences in HCC, new variceal bleeding or portal vein thrombosis development. Grade 2/3 vs. grade 1 only presented an increased risk of new variceal bleeding. The development of other clinical manifestations was similar between grade 1 and grade 2/3	NA	Overall mortality: 39.5%. In multivariate analysis the presence of grade 1 ascites was not associated with mortality. (grade 1 ascites: 36% of mortality. Grade 2/3: 50%. No ascites: 32.8%. Grade 1 vs. no ascites $p = 0.68$. Grade 1 vs. grade 2/3 $p = 0.03$. No ascites vs. grade 2/3 $p = 0.002$)	Grade 1 ascites: 36% vs. no ascites: 32.8%. $p = 0.68$	Data not provided	Assessment of alcohol intake, etiologic treatment of liver disease during FU were not evaluated. Long period of inclusion in which management of liver disease has improved survival significantly.	Grade 1 ascites seems to increase the risk of developing other decompensations compared with no ascites The number of clinical manifestations is similar between grade 1 and grade 2/3 groups Grade 1 ascites was not a risk factor for progression to grade 2/3 ascites in multivariate analysis No survival differences among the three groups

Bruno et al. [11]	Median FU 33 months (range 1–48). 65 patients were lost to FU (included in the survival analysis) Evaluation every 3-months	Not assessed because patients were already decompensated at inclusion	24 patients.	Overall mortality: 27.5%. Mortality of grade 1 ascites: 21% Cumulative incidence of failure (death or LT): 1 year 10%, 2 years 18%, 3 years 25% (similar to the observed in patients with GEVB as first decompensation). Patients with grade 1 ascites died of liver failure in 85% of cases	NA	1 year 10%, 2 years 18%, 3 years 25%	Unclear use of diuretics in patients with grade 1 ascites	Three 3-year mortality rate of patients with ascites grade 1 was approximately eightfold higher than the one reported by other previous investigations in patients with compensated disease
Shah et al. [26]	12 months	Not assessed because patients were already decompensated at inclusion	2 patients	1 year transplant free survival: 78%. Grade 1 ascites: 20% mortality Cumulative incidence of failure (death or LT) for 1 year: 28% (higher than the observed for HE, GEVB or even overt ascites=	Grade 1 ascites 20% vs. no group to compare	1 year: 28%	Small sample size. Single center study Unclear use of diuretics in patients with grade 1 ascites. Unexpected higher mortality for patients with grade 1 ascites as compared to the observed in patients with overt ascites	After presenting the first decompensation, cirrhotic patients died within the first 2-3 months, with no new events happening after that time. Unexpectedly, survival curves are similar for the different types of decompensation. Grade 1 ascites has a high death rate

(continued)

Table 32.2 (continued)

Author	Follow up	Association between ascites grade 1 and decompensation	Liver trasplant	Survival	Mortality Grade 1 vs. no ascites	Cumulative incidence of failure (death or LT)	Potential bias	Comments
Yim et al. [9]	Mean FU 58 (24) months Patients were followed from initiation of antiviral therapy to end of FU, death or LT	After 1 year of antiviral therapy, liver function tests (INR, albumin, bilirubin and CTP score) improved significantly in the grade 1 ascites group and did not differ significantly from the no ascites group. HCC development: Cumulative incidence rates 3%, 7.2%, 12.2% and 21.2% at 1, 2, 3 and 5 years. No significant differences were observed between the grade 1 ascites group and the non ascites group	14 patients underwent liver transplantation during the follow-up period and were censored (2.8%) No ascites:2.4% Grade 1: 2%, grades 2/3 4.4%	Overall mortality (9%; $n = 45$). Survival rate did not differ significantly between the non ascites and grade 1 ascites groups ($P = 0.444$), whereas it was significantly lower in the grades 2/3 group than in the non ascites ($P < 0.001$) and grade 1 ascites ($P = 0.001$) groups. Five-year survival rates were 95.7%, 93%, and 74.6% for non-ascites, grade 1 ascites and grades 2/3 ascites groups, respectively	Raw data not provided	At 5 years No ascites 4.3% Grade 1 7% Grades 2/3 25.4%.	Retrospective study. Unclear use of diuretics and salt restriction compliance during follow up. Statistical approach did not consider competing events	Authors conclude that the presence of grade 1 ascites is an important turning point in the clinical course of liver disease, as commencement of antiviral treatment before or at this point may improve patient prognosis, whereas commencement after this point does not

Zipprich et al. [12]	Mean FU 18.6 (21.1) months	Grade 1 ascites: During follow-up 14 patients were treated with diuretics (9 had stable disease without progression of ascites, 6 progressive disease) and 3 patients received a TIPS due of progressive ascites. Five patients were not treated either with diuretics or with TIPS, four showed stable disease without progression of ascites during follow-up. Kaplan-Meier curves between these two groups of untreated and treated patients showed no significant differences in survival		Patients with grade 1 ascites had a shorter survival compared to patients without ascites ($p < 0.01$) but longer than patients with grades 2/3 ascites ($p = 0.01$).In patients with grade 1 ascites HVPG ($p = 0.009$; HR 1.11; 95% CI 1.03–1.21) and child-Pugh score ($p = 0.031$; HR 0.75; 95% CI 0.58–0.97) were independent factors of survival in both sets of multivariate analysis	Raw data not provided	Data not provided	Retrospective study, single center. Unclear dietary sodium restriction compliance or other disease modifiers during FU	Higher HVPG in patients with grade 1 ascites compared with non ascites. Grade 1 ascites implies an intermediate risk of death between non ascites and grade 2/3 ascites. Mortality among patients with grade 1 ascites treated with diuretics was like the observed in the untreated grade 1 ascites patients

(continued)

Table 32.2 (continued)

Author	Follow up	Association between ascites grade 1 and decompensation	Liver trasplant	Survival	Mortality Grade 1 vs. no ascites	Cumulative incidence of failure (death or LT)	Potential bias	Comments
Tonon et al. [8]	At least 6 months or until death, LT or the end of the follow up median FU 29 (IQR 13–60) months.	Development of overt ascites: 26 patients without ascites (10.0%) and 7 patients with grade 1 ascites (13.0%) with no differences The presence of grade 1 ascites was not found to be an independent predictor of overt ascites development. When death and LT were considered as competing events regards to the development of overt ascites, 60-month development of overt ascites was confirmed to be similar in patients without ascites and with grade 1 ascites Other complications of cirrhosis: Only the development of ACLF was higher in the grade 1 ascites groups. The development of other complications such HCC, GI bleeding, HE was similar between non ascites and grade 1 ascites	102 patients	5-year mortality: Grade 1 ascites 36% vs. grade 2/3 43% p = NS. Mortality rates are higher for grade 1 ascites than non ascites group In multivariate analysis there is not association between grade 1 ascites and mortality	Grade 1 ascites 27.8% vs. non ascites 9.3%		Small number of events in the grade 1 ascites group	Grade 1 ascites implies a higher risk of mortality than patients without ascites. However, the presence of grade 1 ascites was not related to a higher risk of developing other clinical decompensation (included overt ascites) except for ACLF (that may explain the higher mortality rate observed). Authors also show correlation of C-reactive protein increasing according to the degree of decompensation (non ascites < grade 1 < grade 2/3). The authors suggest that systemic inflammation precedes the development of overt ascites. Grade 1 ascites does not necessarily evolve towards overt ascites, but it is associated with a higher probability to develop ACLF and a lower probability of survival

ICA international club of ascites, *EASL* EASL guidelines, *FU* follow-up, *IQR* interquartile range, *CP* Child-Pugh, *NA* not available, *HCV* hepatitis C virus, *HBV* hepatitis B virus, *NASH* nonalcoholic steatohepatitis

Bleeding from Portal Hypertensive Gastroenteropathy

Portal Hypertensive Gastroenteropathy (PHG) is present in 20% to 80% of patients with cirrhosis and its severity and prevalence increases along with the degree of portal hypertension and the severity of liver dysfunction [13]. Although PHG is a frequent cause of morbidity (i.e., chronic gastrointestinal blood loss and iron deficiency anemia) it is frequently underdiagnosed due to the lack of uniform diagnostic criteria and classification. Furthermore, the results of current therapies are suboptimal. The main clinical finding of PHG is chronic gastrointestinal bleeding (defined as a decrease of 2 g/dL of hemoglobin within 6 months without overt bleeding) which has been reported in 3%–60% of patients [14]. Acute gastrointestinal bleeding occurs in 2%–12% of patients with PHG [15]. Mortality attributed to massive bleeding from PHG is extremely infrequent, especially in compensated cirrhosis [14, 16]. PHG has been identified as an independent risk factor for gastric variceal bleeding in two small clinical trials [17, 18]. Besides these findings, there is no information regarding the influence of PHG on the development of hepatic decompensation, need for liver transplantation, or death. Most of the available data, however, comes from old, cross-sectional, retrospective, single-center studies that included a highly selected population. In addition, none of the published studies has specifically evaluated the natural history of PHG in a prospective, longitudinal, large cohort of patients with compensated cirrhosis. Therefore, further well-designed studies are needed to clarify whether the presence of PHG influences the prognosis of patients with cirrhosis.

Table 32.3 summarizes the quality of available evidence regarding the three entities revised in this chapter.

Table 32.3 Summary of quality of evidence

Entity	Agreement on entity definition	Number of papers	Quality of study design	Main results	Comments
Minimal hepatic encephalopathy	Mild	10	Mild	Properly diagnosed MHE may have prognostic relevance	Prospective studies are needed. Competing risk analyses are mandatory
Minimal perihepatic ascites	Mild	6	Mild	The relevance of perihepatic ascites is controversial	Prospective studies are needed. The influence of etiological therapy or systemic inflammation influences the natural history of perihepatic ascites
					Competing risk analyses are mandatory
Chronic bleeding due to portal hypertensive gastroenteropathy	Poor	2	Poor	No clear influence of chronic PHG bleeding on survival or liver transplantation	Prospective studies are needed

References

1. Garcia-Tsao G, Friedman S, Iredale J, Pinzani M. Now there are many (stages) where before there was one: in search of a pathophysiological classification of cirrhosis. Hepatology. 2010;51(4):1445–9.
2. Engelmann C, Claria J, Szabo G, Bosch J, Bernardi M. Pathophysiology of decompensated cirrhosis: portal hypertension, circulatory dysfunction, inflammation, metabolism and mitochondrial dysfunction. J Hepatol. 2021;75(Suppl 1):S49–66.
3. Bajaj JS, Pinkerton SD, Sanyal AJ, Heuman DM. Diagnosis and treatment of minimal hepatic encephalopathy to prevent motor vehicle accidents: a cost-effectiveness analysis. Hepatology. 2012;55(4):1164–71.
4. Ampuero J, Simon M, Montoliu C, Jover R, Serra MA, Cordoba J, et al. Minimal hepatic encephalopathy and critical flicker frequency are associated with survival of patients with cirrhosis. Gastroenterology. 2015;149(6):1483–9.
5. Bajaj JS, Lauridsen M, Tapper EB, Duarte-Rojo A, Rahimi RS, Tandon P, et al. Important unresolved questions in the Management of Hepatic Encephalopathy: an ISHEN consensus. Am J Gastroenterol. 2020;115(7):989–1002.
6. Ampuero J, Montoliu C, Simon-Talero M, Aguilera V, Millan R, Marquez C, et al. Minimal hepatic encephalopathy identifies patients at risk of faster cirrhosis progression. J Gastroenterol Hepatol. 2018;33(3):718–25.
7. Labenz C, Toenges G, Schattenberg JM, Nagel M, Huber Y, Marquardt JU, et al. Outcome prediction of covert hepatic encephalopathy in liver cirrhosis: comparison of four testing strategies. Clin Transl Gastroenterol. 2020;11(6):e00172.
8. Tonon M, Piano S, Gambino CG, Romano A, Pilutti C, Incicco S, et al. Outcomes and mortality of grade 1 ascites and recurrent ascites in patients with cirrhosis. Clin Gastroenterol Hepatol. 2021;19(2):358–66. e8
9. Yim SY, Lee JH, Ahn H, Kim SU, Kim SG, Kim YS, et al. Subclinical ascites does not affect the long-term prognosis in hepatitis B virus-related cirrhosis patients receiving antivirals. J Clin Gastroenterol. 2016;50(8):676–85.
10. Theodorakopoulos T, Kalafateli M, Kalambokis GN, Samonakis DN, Aggeletopoulou I, Tsolias C, et al. Natural history of grade 1 ascites in patients with liver cirrhosis. Ann Gastroenterol. 2021;34(1):93–103.
11. Bruno S, Saibeni S, Bagnardi V, Vandelli C, De Luca M, Felder M, et al. Mortality risk according to different clinical characteristics of first episode of liver decompensation in cirrhotic patients: a nationwide, prospective, 3-year follow-up study in Italy. Am J Gastroenterol. 2013;108(7):1112–22.
12. Zipprich A, Seufferlein T, Dollinger MM. Subclinical ascites defines an intermediate stage between compensated and decompensated cirrhosis. Z Gastroenterol. 2012;50(9):996–1001.
13. Ripoll C, Garcia-Tsao G. Management of gastropathy and gastric vascular ectasia in portal hypertension. Clin Liver Dis. 2010;14(2):281–95.
14. Primignani M, Carpinelli L, Preatoni P, Battaglia G, Carta A, Prada A, et al. Natural history of portal hypertensive gastropathy in patients with liver cirrhosis. The new Italian endoscopic Club for the study and treatment of esophageal varices (NIEC). Gastroenterology. 2000;119(1):181–7.
15. Rockey DC. An update: portal hypertensive Gastropathy and Colopathy. Clin Liver Dis. 2019;23(4):643–58.
16. Merli M, Nicolini G, Angeloni S, Gentili F, Attili AF, Riggio O. The natural history of portal hypertensive gastropathy in patients with liver cirrhosis and mild portal hypertension. Am J Gastroenterol. 2004;99(10):1959–65.
17. Mishra SR, Chander Sharma B, Kumar A, Sarin SK. Endoscopic cyanoacrylate injection versus beta-blocker for secondary prophylaxis of gastric variceal bleed: a randomised controlled trial. Gut. 2010;59(6):729–35.

18. Mishra SR, Sharma BC, Kumar A, Sarin SK. Primary prophylaxis of gastric variceal bleeding comparing cyanoacrylate injection and beta-blockers: a randomized controlled trial. J Hepatol. 2011;54(6):1161–7.
19. Thomsen KL, Macnaughtan J, Tritto G, Mookerjee RP, Jalan R. Clinical and pathophysiological characteristics of cirrhotic patients with grade 1 and minimal hepatic encephalopathy. PLoS One. 2016;11(1):e0146076.
20. Patidar KR, Thacker LR, Wade JB, Sterling RK, Sanyal AJ, Siddiqui MS, et al. Covert hepatic encephalopathy is independently associated with poor survival and increased risk of hospitalization. Am J Gastroenterol. 2014;109(11):1757–63.
21. Wang AJ, Peng AP, Li BM, Gan N, Pei L, Zheng XL, et al. Natural history of covert hepatic encephalopathy: an observational study of 366 cirrhotic patients. World J Gastroenterol. 2017;23(34):6321–9.
22. Dhiman RK, Kurmi R, Thumburu KK, Venkataramarao SH, Agarwal R, Duseja A, et al. Diagnosis and prognostic significance of minimal hepatic encephalopathy in patients with cirrhosis of liver. Dig Dis Sci. 2010;55(8):2381–90.
23. Barone M, Shahini E, Iannone A, Viggiani MT, Corvace V, Principi M, et al. Critical flicker frequency test predicts overt hepatic encephalopathy and survival in patients with liver cirrhosis. Dig Liver Dis. 2018;50(5):496–500.
24. Miwa T, Hanai T, Toshihide M, Ogiso Y, Imai K, Suetsugu A, et al. Zinc deficiency predicts overt hepatic encephalopathy and mortality in liver cirrhosis patients with minimal hepatic encephalopathy. Hepatol Res. 2021;51(6):662–73.
25. Hanai T, Shiraki M, Watanabe S, Imai K, Suetsugu A, Takai K, et al. Prognostic significance of minimal hepatic encephalopathy in patients with liver cirrhosis in Japan: a propensity score-matching analysis. J Gastroenterol Hepatol. 2019;34(10):1809–16.
26. Shah AS, Amarapurkar DN. Natural history of cirrhosis of liver after first Decompensation: a prospective study in India. J Clin Exp Hepatol. 2018;8(1):50–7.

Evaluation of the Role of Sarcopenia in the Definition of Decompensation of the Compensated Patient

33

Susana G. Rodrigues and Chiara Becchetti

Introduction

Advanced Chronic Liver Disease (ACLD) is a dynamic condition where a patient can transition from a compensated to a decompensated state. In recent years, there has been growing interest in the role body composition and nutritional status play, primarily in the outcome of decompensated patients. Sarcopenia is a term initially defined as an age-related loss of skeletal muscle [1]. Nowadays, it more generally describes low muscle mass and reduced muscle mass quality leading to negative effects on physical performance and clinical outcomes [2]. Sarcopenia is widely recognized as a frequent phenomenon in cirrhosis, and it has a proven negative impact on the natural history of patients with decompensated cirrhosis [3, 4]. This has been explored in depth in patients awaiting liver transplants (LT) [5]. Its real prevalence is difficult to estimate since there is high heterogeneity among the different methods of measurement. Nevertheless, in decompensated cirrhosis, its prevalence varies between 25% and 50% [4, 5].

S. G. Rodrigues (✉) · C. Becchetti
Department of Visceral Surgery and Medicine, Inselspital, Bern University Hospital, University of Bern, Bern, Switzerland

Hepatology, Department for BioMedical Research, Visceral Surgery and Medicine, University of Bern, Bern, Switzerland
e-mail: susana.gomesrodrigues@insel.ch

R. de Franchis (ed.), *Portal Hypertension VII*,
https://doi.org/10.1007/978-3-031-08552-9_33

Another relevant emerging concept is frailty. Frailty is a biological syndrome of decreasing physiologic reserve and increasing vulnerability to health stressors [6]. In the field of hepatology, most attention has focused on physical frailty and its link with sarcopenia as a functional reflection of muscle mass [7]. The prevalence of frailty in compensated patients has been reported between 10% and 25% [8, 9].

However, despite the role of sarcopenia and frailty are established prognostic risk factors for mortality [10, 11], and more precisely LT wait-list mortality [12], hepatic encephalopathy [13], and further decompensation [14], their role in compensated liver disease is less clear. Indeed, recent studies are emphasizing that sarcopenia may be present even in the early asymptomatic stages of ACLD [15]. Nevertheless, to what extent sarcopenia impacts the natural history of compensated cirrhosis remains an open question.

Pathophysiological Background: Potential Role of Sarcopenia as a Driver of Decompensation

Sarcopenia has been recently recognized as a systemic syndrome mainly driven by inflammation, proteolysis, and unbalanced muscle homeostasis. Multiple mechanisms contribute to sarcopenia in cirrhosis (Fig. 33.1). These include physical inactivity, reduced dietary intake, low glycogen deposits, and a rapid transition to fasting metabolism [7]. Other factors known to play a role in sarcopenia are endotoxemia, increased aromatase activity to lower testosterone, and mitochondrial dysfunction [16]. Specifically, in the context of liver disease, hyperammonemia plays an important part. A direct toxic effect of ammonia on skeletal muscles, acting on uptake and conversion to glutamate and glutamine, was demonstrated, not only in animal

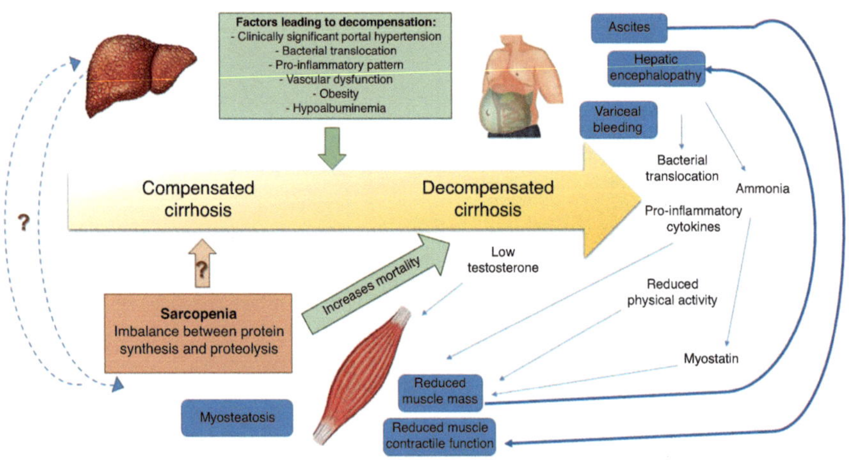

Fig. 33.1 Proposed possible mechanisms of sarcopenia in the early phase of compensated liver disease of cirrhosis

models [17] but also in human liver disease [18]. Additionally, hyperammonemia mediates upregulation of myostatin, which inhibits protein synthesis, activating the ubiquitin-proteasome and autophagy [19]. Recently, the role of the dysbiosis as a co-factor in the development of sarcopenia in chronic liver disease gained more attention, particularly for specific etiologies of liver disease, such as alcoholic liver disease [20] and nonalcoholic fatty liver disease (NAFLD) [21], where dysbiosis frequently leads to endotoxemia [22]. Furthermore, in the setting of NAFLD, fatty muscle infiltration, or myosteatosis, has been investigated in the setting of sarcopenia and, namely, sarcopenic obesity [23].

While in decompensated cirrhosis, sarcopenia is frequently present and qualifies as an independent predictor of increased morbidity, mortality, and reduced quality of life, there are still multiple knowledge gaps on how sarcopenia exactly acts in earlier stages of liver cirrhosis [7]. This is particularly true when we consider that there are very few preclinical and human mechanistic studies on skeletal muscle responses in the setting of cirrhosis. However, emerging clinical observational studies might suggest a role of sarcopenia in developing decompensation and worse prognosis, even in an early phase of compensated cirrhosis [15, 24, 25].

Modalities to Evaluate Muscle Mass and Frailty in Patients with Cirrhosis

One of the most controversial topics in the setting of sarcopenia and liver disease is the ideal method of evaluation. Although in the geriatric population [26] Dual-Energy X-ray Absorptiometry (DEXA) was used to define sarcopenia, in liver disease, different imaging techniques have been adopted. As often required for other clinical settings (e.g., hepatocellular carcinoma diagnosis and staging, portal thrombosis), imaging methods, such as Computer Tomography (CT) scan or Magnetic Resonance (MR), have been widely used in hepatology within the scope of sarcopenia. This has led to heterogeneous definitions, measurements, and study designs in the field, generating confounding factors and limiting comparisons across available studies. Table 33.1A summarizes the most common tools used to assess sarcopenia. Although the gold standard in cirrhosis is still unclear, the majority of the studies have established different cut-offs and techniques. Skeletal Muscle Index (SMI) has emerged as the standard method. SMI is assessed using analysis software on CT scan images, measuring the total cross-sectional area (cm^2) of abdominal skeletal muscle at the third lumbar vertebra. The cut-offs of SMI < 50 cm^2/m^2 for men and < 39 cm^2/m^2 for women better discriminating hard outcomes such as survival, have been used to define sarcopenia in cirrhosis [26]. These measurements derived from a study conducted in North America aimed to define sarcopenia using data generated from different populations of patients with cirrhosis [27]. Nonetheless, the majority of studies used to validate these cut-off values were conducted in decompensated patients or those awaiting LT [28]. In these studies, there were compensated patients included, but many of them had HCC, which per se can influence the presence of sarcopenia. Therefore, it remains uncertain to what extent they may

Table 33.1 Characteristics of the Tools Adopted to Assess Sarcopenia (A) and Frailty (B) in Clinical and Research Settings.

A. Assess sarcopenia

Sarcopenia assessment tools	Advantages	Disadvantages
Mid-arm muscle Circumference (MAMC)	Inexpensive, noninvasive, widely available	Low reproducibility, specialized training, influenced by Subcutaneous Adipose tissue loss,
Bioelectrical impedance Analysis (BIA)	Inexpensive, noninvasive, quick	Fluid overload, highly influenced by the external environment
Dual-energy X-ray Absorptiometry (DEXA)	Inexpensive, noninvasive, widely Available	Influenced by lower limb edema, radiation
Ultrasound (thigh Muscle thickness)	Inexpensive, noninvasive, simple, no Radiation, independent of ascites	Reproducibility, specialized training
Computer tomography (CT)/ magnetic resonance (MR)	Precise, independent of Ascites or edema, evaluation of adipose tissue	Expensive, radiation exposure, reproducibility (different software adopted)

B. Frailty

Tools for assessing frailty	Type of measurements	Advantages	Disadvantages
Grip strength	Using Jamar dynamometer	Inexpensive, noninvasive, quick Classification according to score	Requires instrumental equipment, upper limb test
Gait speed	Measuring the time, the patient takes to walk between two stripes situated on the floor at different distances	Inexpensive, noninvasive, quick	Specialized training
Fried frailty index [6]	Involuntary weight loss of 10 lbs. or more in the last 6 months; Reduced grip strength; Difficulty initiating movements; Reduced walking speed; Fatigue	Inexpensive, noninvasive, Classification according to three classes	Time-consuming, requires instrumental equipment, specialized training
Short performance physical battery (SPPB)	Repeated chair stands; Balance tests (side-by-side, semi-tandem, and tandem balance tests); An eight-foot walk test.	Inexpensive, noninvasive, widely Available, classification according to score	Specialized training

Table 33.1 (continued)

B. Frailty			
Tools for assessing frailty	Type of measurements	Advantages	Disadvantages
Clinical frailty scale (CFS) [47]	Scale of frailty based on 9 stages	Inexpensive, noninvasive, quick, no specialized training needed	No direct testing with the patient, no clear functional measurements
Liver frailty index [29]	Dominant handgrip strength; Time to do 5 chair stands; Seconds holding 3 position balances.	Inexpensive, noninvasive, widely Available, classification according to score and class, widely validated in liver disease	Require instrumental equipment, specialized training

represent a reliable standard for sarcopenia in compensated patients Furthermore, as CT is associated with ionizing radiation, it would not be the ideal method to serially follow up compensated patients treated for sarcopenia. In addition to the evaluation of muscle mass, new semi-automatic software available (i.e., SlicesOmatic®, Analytic Morphomics®, Fujifilm Synapse 3D®, etc.), can reconstruct body composition, evaluating the different tissue types based on tissue-specific CT-based Hounsfield unit thresholds. These techniques are able to discriminate fat and quantify fat in the different tissues, namely subcutaneous adipose tissue, visceral adipose tissue, and intramuscular adipose tissue. Recent studies have focused on these alterations as a prognostic factor in liver disease.

Concerning frailty, in 2017, Lai et al. [29] validated a scoreable to predict mortality for patients on the LT waiting list based on three simple tests (handgrip, sit-up and stand, and balance) evaluating a patient's vulnerability to stress, decreased physiologic reserve, and functional status deficits. The "Liver Frailty Index" (LFI) was widely validated in the setting of liver disease and linked to prognosis [30]. Other measurements adopted for quantity frailty are described in Table 33.1B.

Sarcopenia in Compensated Liver Disease

Sarcopenia in Chronic Liver Disease

Sarcopenia has been recently investigated in the initial phase of chronic liver disease, first in NAFLD and then in viral hepatitis, as an associated factor to fibrosis progression.

Myosteatosis is particularly evident in NAFLD [31]. The overlap in the pathophysiology of NAFLD and sarcopenia makes it challenging to determine whether sarcopenia is a risk factor for NASH or whether it is a complication of NASH. Insulin resistance and increased inflammation play a key role in the development of both conditions. Furthermore, myokines secreted by skeletal muscle (IL-6 and irisin) are recognized to be involved in the regulation of some metabolic variables, such as weight gain control and insulin resistance [32]. It follows that sarcopenia could perhaps play a role early in cirrhosis, particularly when related to NASH [33].

In 2015, a Korean cohort study ($N = 2761$) demonstrated that sarcopenia is associated with significant liver fibrosis (defined as fibrosis stage ≥2 and assessed by NAFLD fibrosis score and FIB-4) in subjects with NAFLD, and the association was independent of obesity (OR 2.12, 95% CI 1.33–3.38, $P < 0.001$) and insulin resistance (OR 2.68, 95% CI 2.06–3.50, $P < 0.001$). Sarcopenia was present in 12.2% of the subjects with NAFLD [34]. Furthermore, SMI correlated negatively with the homeostasis model assessment of insulin resistance (HOMA-IR) ($P < 0.001$). In this cohort of obese patients, risk of having NAFLD in the lowest SMI quartile was five times higher when compared to other quartiles (OR 5.16, CI 1.63–16.33) [35]. Indeed, another study suggested that in patients with type 2 diabetes mellitus (T2DM), among patients with a normal body mass index (BMI), those with sarcopenia had a significantly higher FIB-4 than those without (1.66 vs. 1.38, $p = 0.004$) [36]. One can speculate that there is a mutual relationship between NAFLD and sarcopenia, with the latter acting, in a probably predisposed environment, as a facilitator of fibrosis progression. Concerning prognosis, an epidemiological study from the United States with non-end-stage NAFLD patients suggested NAFLD with sarcopenia vs. without was associated with a higher risk of all-cause mortality (HR = 1.78, 95%CI 1.16–2.73), particularly cardiac and cancer-related death, but not hepatic [37].

With regards to chronic liver disease due to viral hepatitis, a study in chronic hepatitis B patients showed that sarcopenia was associated with significant fibrosis, evaluated using the FIB-4 score, this association remains significant after adjusting for confounding factors. This association was even more evident among subgroups with obesity, insulin resistance, metabolic syndrome, and liver steatosis [38]. Data from the North American NHANES cohort showed that there was a significantly higher prevalence of low muscle mass (calculated as mid-upper arm circumference) in hepatitis C virus (HCV), infected patients, when compared to the uninfected. This association remained valid also in patients without significant fibrosis, suggesting that, in HCV-infected patients, sarcopenia is highly prevalent even in the absence of advanced liver disease [39].

Therefore, based on the available data, the described findings indicate an association between sarcopenia and chronic liver disease, without further elucidating whether sarcopenia accelerates disease progression.

Sarcopenia in Compensated Cirrhosis

In the setting of compensated cirrhosis, data is scarce, specifically addressing the very early phase of the disease (Table 33.2). According to the available literature, sarcopenia is reported between 10% and 30% [40, 41] in the compensated phase, defined either as Child-Pugh (CP) class A or cirrhosis without previous episodes of decompensation. In a study conducted on 452 patients stratified according to CP

Table 33.2 Summary of the main studies available in compensated cirrhosis and sarcopenia

Author (Year)	Country	Type of study	Aim of the study/ outcome	Sarcopenia assessment tool	Compensated patients included, *N*, (%)	Main finding
Hara et al. (2016) [48]	Japan	Observational retrospective	Mortality	BIA	82	Sarcopenia or sarcopenic obesity had a poor prognosis, more pronounced in the subset of patients classified as CP A
Hikoara et. al (2016) [49]	Japan	Observational prospective	Prevalence of sarcopenia	Psoas muscle index CT + hand grip	330	Sarcopenia and pre-sarcopenia are present at every stage of the liver disease (chronic hepatitis, CP A, CP B-C). The incidence rates of sarcopenia and pre-sarcopenia increased with progression of CLD
Benjamin et al. (2017) [50]	India	Observational retrospective	Prevalence of sarcopenia in ALD cirrhosis	L3 SMI CT scan + adipose tissue	47 (31.8)	Compared to healthy control, compensated patients had higher adiposity and comparable muscularity
Lucidi et al. (2018) [51]	Italy	Observational retrospective	Mortality and decompensation	Mid-arm muscle circumference (MAMC) and triceps skin-fold thickness (TSF)	45 (60)	In CP A-B, mortality was higher in patients with low muscle mass compared with those without. The mortality rate and the incidence of complications in malnourished patients classified in CP A-B were similar to those CP C
Kang et al. (2018) [41]	Korea	Observational retrospective	Mortality	L3 SMI CT scan (quartile)	215 (47.6)	Sarcopenia was associated with mortality and specifically with compensated and early decompensated stages of cirrhosis, but not with advanced decompensated stages

(continued)

Table 33.2 (continued)

Author (Year)	Country	Type of study	Aim of the study/ outcome	Sarcopenia assessment tool	Compensated patients included, *N*, (%)	Main finding
Rodrigues et al. (2019) [25]	Switzerland	Observational retrospective	Mortality and decompensation	L3 SMI CT scan	38 [45]	In compensated patients, TATI improves noninvasive prediction of decompensation
Tapper et al. (2019) [15]	USA	Observational prospective	Mortality decompensation and transplant free-survival	T12 SMI CT scan	130 [47]	Decreased normal density muscle mass was associated with mortality, as well as visceral and subcutaneous fat density. Using competing risk analysis subcutaneous fat density was most predictive of decompensation for all stages of CLD
Beer et al (2020) [40]	Austria	Observational retrospective	Decompensation and transplant free-survival	MR/CT transverse psoas muscle thickness	110 (42.9)	Sarcopenia was not predictive of first or further hepatic decompensation. In patients with cACLD and dACLD, sarcopenia was a risk factor for mortality on univariate analysis. At the multivariate sarcopenia remained an independent risk factor for mortality in patients with cACLD
Paternostro et al. (2021) [24]	Austria	Observational retrospective	Decompensation and transplant free-survival	MR transverse psoas muscle thickness	54 (26.6)	Sarcopenia was significantly associated with first/further decompensation both in compensated and decompensated patients and it was a significant predictor of mortality irrespective of HVPG

CT computer tomography, *MR* magnetic resonance, *HVPG* hepatic venous pressure gradient, *PH* portal hypertension, *CLD* chronic liver disease, *cACLD* compensated advanced liver disease, *dACLD* decompensated advanced liver disease, *BIA* bioelectrical impedance analysis, *CP* Child-Pugh, *TATI* total adipose tissue index, *SMI* skeletal muscle index

class, Model for End-Stage Liver Disease (MELD) score, and hepatic venous pressure gradient (HVPG), the authors found that sarcopenia (assessed by CT-scan L3 SMI) was associated with mortality (HR = 2.253, $P < 0.001$), specifically in compensated and early decompensated stages of cirrhosis. Additionally, they showed that in the lowest SMI quartile, namely severe sarcopenia, the classical prognostic indicators (i.e., MELD score, CP classes, and HVPG) failed to predict mortality, suggesting that stratifying these scores according to sarcopenia would better assess the prognosis of cirrhosis, particularly in the early stage [41]. Two other studies, coming from the same group, evaluated sarcopenia through MR scan psoas muscle thickness, and aimed to analyze the impact of sarcopenia on mortality and decompensation rates in patients stratified according to compensated and decompensated cirrhosis. The first study by Beer et al. [40] showed that sarcopenia was not predictive of first or further liver decompensation in both groups. Regarding mortality, sarcopenia was a risk factor in both groups, although it only remained an independent predictor in the compensated group (HR: 2.76, 95% CI: 1.02–7.42). One-year after the same group corroborated this finding demonstrating that the presence of sarcopenia doubled the risk for mortality independently from the HVPG level in compensated cirrhosis [24]. In contrast, another study failed to demonstrate a correlation between HVPG level and the detection of sarcopenia [25]. Nevertheless, the authors demonstrated that CT total adipose tissue area, visceral adipose tissue area, and their ratio measured at L3 level were associated with decompensation in the compensated population. Interestingly all the three aforementioned studies have considered the definition of compensation according to stage 1 and 2 of D'Amico classification [42]. Concerning body composition, Tapper et al. [15] published an extensive analysis on the better prognostication of mortality by adding body composition to MELD, taking into account: muscle density, muscle mass area, visceral fat density, visceral fat area, subcutaneous fat density, subcutaneous fat area, and bone mineral density. The authors sub-analyzed not only CP A patients, but also compensated patients ($N = 111$) and found that similar to the CP A groups the prediction model which included morphomic features significantly outperformed MELD (C-statistic 0.74 [0.62–0.87], $p = 0.001$). Thus, there is some evidence suggesting that sarcopenia seems to play a role even in compensated cirrhosis and this might be independent of the classical drivers of liver disease progression.

Frailty in Compensated Liver Disease

While the role of frailty in predicting mortality [43, 44] and hospital admissions [45] in decompensated liver disease is established, there are extremely few data on the influence of frailty on disease transition from a compensated to the decompensated stage. A multicentre study on 882 patients with cirrhosis demonstrated that the risk of progression of cirrhosis to the next clinical stage (according to D'Amico stages) or death remained significantly higher in patients who were frail as compared to those who were robust (HR 2.47, 95% CI 1.63–3.76, $p < 0.001$) or pre-frail (HR 2.04, 95% CI 1.56–2.65, $p < 0.001$). Interestingly, the association between frailty and increased risk of progression or death was present both in compensated (stages 1 or 2) and decompensated patients (stages 3–5) [8]. Another study conducted on compensated patients with no history of the previous decompensation, either CP A or B, showed that frail patients compared with the robust group have significantly higher cumulative probabilities of developing an episode of decompensation and unplanned hospitalization. This was much more pronounced for subjects belonging to CP B group [9]. It is uncertain to what extent disease progression may be driven by frailty (and/or underlying factors) or if frailty may represent an early manifestation of disease progression.

Criticism to the Available Literature and Further Directions

Several limitations burden the majority of the literature in the field. First, the definition of sarcopenia and the tools adopted to measure it varies extremely between the different studies. Additionally, the stage of liver cirrhosis is not defined in the majority of the studies analyzed. Indeed, only the most recent studies stratify the patients according to compensation/decompensation status. However, few studies specify the definition of compensation according to either D'Amico classification or no previous episodes of decompensations, while the majority assume patients in CP A as compensated without mention of previous episodes of decompensation. Therefore, the real impact of sarcopenia as a driver toward decompensation cannot be proven. Another limitation concerning the available body of evidence is that not all the studies excluded patients with Hepatocellular Carcinoma (HCC), which per se could represent, as several other oncologic diseases, an independent risk factor for sarcopenia [46] and a powerful/significant/robust prognostic risk factor. Overall, larger studies with adequate follow-up periods are needed to properly assess the impact of body composition and functional changes on the natural history of compensated cirrhosis. Future studies should aim to evaluate body composition changes over longer periods and define the prognostic and causal role of sarcopenia and frailty and associated biological processes, in the early stages of liver cirrhosis, mainly its role in disease progression.

References

1. Rosenberg IH. Summary comments. Am J Clin Nutr. 1989;50:1231–3.
2. Bhanji RA, Montano-Loza AJ, Watt KD. Sarcopenia in cirrhosis: looking beyond the skeletal muscle loss to see the systemic disease. Hepatology. 2019;70(6):2193–203.
3. Montano-Loza AJ, Meza-Junco J, Prado CM, Lieffers JR, Baracos VE, Bain VG, et al. Muscle wasting is associated with mortality in patients with cirrhosis. Clin Gastroenterol Hepatol. 2012;10(2):166–73, 73.e1
4. Tandon P, Ney M, Irwin I, Ma MM, Gramlich L, Bain VG, et al. Severe muscle depletion in patients on the liver transplant wait list: its prevalence and independent prognostic value. Liver Transpl. 2012;18(10):1209–16.

5. Montano-Loza AJ. Muscle wasting: a nutritional criterion to prioritize patients for liver transplantation. Curr Opin Clin Nutr Metab Care. 2014;17(3):219–25.
6. Fried LP, Tangen CM, Walston J, Newman AB, Hirsch C, Gottdiener J, et al. Frailty in older adults: evidence for a phenotype. J Gerontol A Biol Sci Med Sci. 2001;56(3):M146–56.
7. Tandon P, Montano-Loza AJ, Lai JC, Dasarathy S, Merli M. Sarcopenia and frailty in decompensated cirrhosis. J Hepatol. 2021;75(Suppl 1):S147–S62.
8. Wang S, Whitlock R, Xu C, Taneja S, Singh S, Abraldes JG, et al. Frailty is associated with increased risk of cirrhosis disease progression and death. Hepatology. 2022;75(3):600–9.
9. Siramolpiwat S, Kiattikunrat K, Soontararatpong R, Pornthisarn B, Vilaichone RK, Chonprasertsuk S, et al. Frailty as tested by the liver frailty index is associated with decompensation and unplanned hospitalization in patients with compensated cirrhosis. Scand J Gastroenterol. 2021;56(10):1210–9.
10. Montano-Loza AJ, Angulo P, Meza-Junco J, Prado CM, Sawyer MB, Beaumont C, et al. Sarcopenic obesity and myosteatosis are associated with higher mortality in patients with cirrhosis. J Cachexia Sarcopenia Muscle. 2016;7(2):126–35.
11. Tapper EB, Finkelstein D, Mittleman MA, Piatkowski G, Lai M. Standard assessments of frailty are validated predictors of mortality in hospitalized patients with cirrhosis. Hepatology. 2015;62(2):584–90.
12. Carey EJ, Lai JC, Sonnenday C, Tapper EB, Tandon P, Duarte-Rojo A, et al. A north American expert opinion statement on sarcopenia in liver transplantation. Hepatology. 2019;70(5):1816–29.
13. Bhanji RA, Moctezuma-Velazquez C, Duarte-Rojo A, Ebadi M, Ghosh S, Rose C, et al. Myosteatosis and sarcopenia are associated with hepatic encephalopathy in patients with cirrhosis. Hepatol Int. 2018;12(4):377–86.
14. Barbero-Becerra VJ, Lopez-Mendez I, Romo-Araiza A, Visag-Castillo V, Chavez-Tapia NC, Uribe M, et al. Sarcopenia in chronic liver diseases: a translational overview. Expert Rev Gastroenterol Hepatol. 2020;14(5):355–66.
15. Tapper EB, Zhang P, Garg R, Nault T, Leary K, Krishnamurthy V, et al. Body composition predicts mortality and decompensation in compensated cirrhosis patients: a prospective cohort study. JHEP Rep. 2020;2(1):100061.
16. Dasarathy S, Merli M. Sarcopenia from mechanism to diagnosis and treatment in liver disease. J Hepatol. 2016;65(6):1232–44.
17. Dasarathy S, Muc S, Hisamuddin K, Edmison JM, Dodig M, McCullough AJ, et al. Altered expression of genes regulating skeletal muscle mass in the portacaval anastomosis rat. Am J Physiol Gastrointest Liver Physiol. 2007;292(4):G1105–13.
18. Kumar A, Davuluri G, Silva RNE, Engelen M, Ten Have GAM, Prayson R, et al. Ammonia lowering reverses sarcopenia of cirrhosis by restoring skeletal muscle proteostasis. Hepatology. 2017;65(6):2045–58.
19. Qiu J, Thapaliya S, Runkana A, Yang Y, Tsien C, Mohan ML, et al. Hyperammonemia in cirrhosis induces transcriptional regulation of myostatin by an NF-kappaB-mediated mechanism. Proc Natl Acad Sci U S A. 2013;110(45):18162–7.
20. Dasarathy S, Brown JM. Alcoholic liver disease on the rise: Interorgan cross talk driving liver injury. Alcohol Clin Exp Res. 2017;41(5):880–2.
21. Mouries J, Brescia P, Silvestri A, Spadoni I, Sorribas M, Wiest R, et al. Microbiota-driven gut vascular barrier disruption is a prerequisite for non-alcoholic steatohepatitis development. J Hepatol. 2019;71(6):1216–28.
22. Nishikawa H, Enomoto H, Nishiguchi S, Iijima H. Liver cirrhosis and sarcopenia from the viewpoint of dysbiosis. Int J Mol Sci. 2020;21(15):5254.
23. Batsis JA, Villareal DT. Sarcopenic obesity in older adults: aetiology, epidemiology and treatment strategies. Nat Rev Endocrinol. 2018;14(9):513–37.
24. Paternostro R, Bardach C, Hofer BS, Scheiner B, Schwabl P, Asenbaum U, et al. Prognostic impact of sarcopenia in cirrhotic patients stratified by different severity of portal hypertension. Liver Int. 2021;41(4):799–809.

25. Rodrigues SG, Brabandt B, Stirnimann G, Maurer MH, Berzigotti A. Adipopenia correlates with higher portal pressure in patients with cirrhosis. Liver Int. 2019;39(9):1672–81.
26. European Association for the Study of the Liver. EASL clinical practice guidelines on nutrition in chronic liver disease. J Hepatol. 2019;70(1):172–93.
27. Carey EJ, Lai JC, Wang CW, Dasarathy S, Lobach I, Montano-Loza AJ, et al. A multicenter study to define sarcopenia in patients with end-stage liver disease. Liver Transpl. 2017;23(5):625–33.
28. Kalafateli M, Mantzoukis K, Choi Yau Y, Mohammad AO, Arora S, Rodrigues S, et al. Malnutrition and sarcopenia predict post-liver transplantation outcomes independently of the model for end-stage liver disease score. J Cachexia Sarcopenia Muscle. 2017;8(1):113–21.
29. Lai JC, Covinsky KE, Dodge JL, Boscardin WJ, Segev DL, Roberts JP, et al. Development of a novel frailty index to predict mortality in patients with end-stage liver disease. Hepatology. 2017;66(2):564–74.
30. Lai JC, Rahimi RS, Verna EC, Kappus MR, Dunn MA, McAdams-DeMarco M, et al. Frailty associated with waitlist mortality independent of ascites and hepatic encephalopathy in a multicenter study. Gastroenterology. 2019;156(6):1675–82.
31. Bhanji RA, Narayanan P, Allen AM, Malhi H, Watt KD. Sarcopenia in hiding: the risk and consequence of underestimating muscle dysfunction in nonalcoholic steatohepatitis. Hepatology. 2017;66(6):2055–65.
32. Dasarathy S. Is the adiponectin-AMPK-mitochondrial axis involved in progression of nonalcoholic fatty liver disease? Hepatology. 2014;60(1):22–5.
33. Li AA, Kim D, Ahmed A. Association of Sarcopenia and NAFLD: an overview. Clin Liver Dis (Hoboken). 2020;16(2):73–6.
34. Lee YH, Kim SU, Song K, Park JY, Kim DY, Ahn SH, et al. Sarcopenia is associated with significant liver fibrosis independently of obesity and insulin resistance in nonalcoholic fatty liver disease: Nationwide surveys (KNHANES 2008-2011). Hepatology. 2016;63(3):776–86.
35. Hong HC, Hwang SY, Choi HY, Yoo HJ, Seo JA, Kim SG, et al. Relationship between sarcopenia and nonalcoholic fatty liver disease: the Korean Sarcopenic obesity study. Hepatology. 2014;59(5):1772–8.
36. Sung MJ, Lim TS, Jeon MY, Lee HW, Kim BK, Kim DY, et al. Sarcopenia is independently associated with the degree of liver fibrosis in patients with type 2 diabetes mellitus. Gut Liver. 2020;14(5):626–35.
37. Golabi P, Gerber L, Paik JM, Deshpande R, de Avila L, Younossi ZM. Contribution of sarcopenia and physical inactivity to mortality in people with non-alcoholic fatty liver disease. JHEP Rep. 2020;2(6):100171.
38. Han E, Lee YH, Kim BK, Park JY, Kim DY, Ahn SH, et al. Sarcopenia is associated with the risk of significant liver fibrosis in metabolically unhealthy subjects with chronic hepatitis B. Aliment Pharmacol Ther. 2018;48(3):300–12.
39. Gowda C, Compher C, Amorosa VK, Lo Re V 3rd. Association between chronic hepatitis C virus infection and low muscle mass in US adults. J Viral Hepat. 2014;21(12):938–43.
40. Beer L, Bastati N, Ba-Ssalamah A, Potter-Lang S, Lampichler K, Bican Y, et al. MRI-defined sarcopenia predicts mortality in patients with chronic liver disease. Liver Int. 2020;40(11):2797–807.
41. Kang SH, Jeong WK, Baik SK, Cha SH, Kim MY. Impact of sarcopenia on prognostic value of cirrhosis: going beyond the hepatic venous pressure gradient and MELD score. J Cachexia Sarcopenia Muscle. 2018;9(5):860–70.
42. D'Amico G, Garcia-Tsao G, Pagliaro L. Natural history and prognostic indicators of survival in cirrhosis: a systematic review of 118 studies. J Hepatol. 2006;44(1):217–31.
43. Lai JC, Ganger DR, Volk ML, Dodge JL, Dunn MA, Duarte-Rojo A, et al. Association of Frailty and sex with Wait List Mortality in liver transplant candidates in the multicenter functional assessment in liver transplantation (FrAILT) study. JAMA Surg. 2021;156(3):256–62.
44. Haugen CE, McAdams-DeMarco M, Holscher CM, Ying H, Gurakar AO, Garonzik-Wang J, et al. Multicenter study of age, frailty, and waitlist mortality among liver transplant candidates. Ann Surg. 2020;271(6):1132–6.

45. Shah S, Goldberg DS, Kaplan DE, Sundaram V, Taddei TH, Mahmud N. Patient frailty is independently associated with the risk of hospitalization for acute-on-chronic liver failure. Liver Transpl. 2021;27(1):16–26.
46. Argiles JM, Busquets S, Stemmler B, Lopez-Soriano FJ. Cachexia and sarcopenia: mechanisms and potential targets for intervention. Curr Opin Pharmacol. 2015;22:100–6.
47. Rockwood K, Song X, MacKnight C, Bergman H, Hogan DB, McDowell I, et al. A global clinical measure of fitness and frailty in elderly people. CMAJ. 2005;173(5):489–95.
48. Hara N, Iwasa M, Sugimoto R, Mifuji-Moroka R, Yoshikawa K, Terasaka E, et al. Sarcopenia and Sarcopenic obesity are prognostic factors for overall survival in patients with cirrhosis. Intern Med. 2016;55(8):863–70.
49. Hiraoka A, Michitaka K, Ueki H, Kaneto M, Aibiki T, Okudaira T, et al. Sarcopenia and two types of presarcopenia in Japanese patients with chronic liver disease. Eur J Gastroenterol Hepatol. 2016;28(8):940–7.
50. Benjamin J, Shasthry V, Kaal CR, Anand L, Bhardwaj A, Pandit V, et al. Characterization of body composition and definition of sarcopenia in patients with alcoholic cirrhosis: a computed tomography based study. Liver Int. 2017;37(11):1668–74.
51. Lucidi C, Lattanzi B, Di Gregorio V, Incicco S, D'Ambrosio D, Venditti M, et al. A low muscle mass increases mortality in compensated cirrhotic patients with sepsis. Liver Int. 2018;38(5):851–7.

β-Blockers to Prevent Decompensation of Cirrhosis in Compensated Patients With Clinically Significant Portal Hypertension

34

Càndid Villanueva, Dhiraj Tripathi, Susana G. Rodrigues, Ferran Torres, Cristina Ripoll, and Jaime Bosch

C. Villanueva (✉)
Hospital de la Santa Creu i Sant Pau, Biomedical Research Institute Sant Pau (IIB Sant Pau), Barcelona, Spain

Universitat Autònoma de Barcelona, Barcelona, Spain

Centro de Investigación Biomédica en Red de Enfermedades Hepáticas y Digestivas (CIBERehd), Barcelona, Spain
e-mail: cvillanueva@santpau.cat

D. Tripathi
Liver Unit, University Hospitals Birmingham NHS Foundation Trust, Birmingham, UK

Institute of Immunology and Immunotherapy, University of Birmingham, Birmingham, UK
e-mail: D.Tripathi@bham.ac.uk

S. G. Rodrigues
Department of Visceral Surgery and Medicine, Inselspital, Bern University Hospital, University of Bern, Bern, Switzerland

Department for BioMedical Research, Visceral Surgery and Medicine, University of Bern, Bern, Switzerland
e-mail: susana.GomesRodrigues@insel.ch

F. Torres
Medical Statistics Core Facility, IDIBAPS, Hospital Clinic, Barcelona, Spain

Biostatistics Unit, Faculty of Medicine, Universitat Autònoma de Barcelona, Barcelona, Spain
e-mail: ferran.torres@idibaps.org

C. Ripoll
Internal Medicine IV, Universitätsklinikum, Friedrich-Schiller University, Jena, Germany
e-mail: Cristina_ripoll@med.uni-jena.de

J. Bosch
Centro de Investigación Biomédica en Red de Enfermedades Hepáticas y Digestivas (CIBERehd), Barcelona, Spain

Department for BioMedical Research, Visceral Surgery and Medicine, University of Bern, Bern, Switzerland

University Clinic for Visceral Surgery and Medicine, Bern, Switzerland
e-mail: jaime.bosch@dbmr.unibe.ch

R. de Franchis (ed.), *Portal Hypertension VII*,
https://doi.org/10.1007/978-3-031-08552-9_34

Abbreviations

CI	Confidence interval
CSPH	Clinically significant portal hypertension
FU	Follow-up
HVPG	Hepatic venous pressure gradient
INR	International normalized ratio
MELD	Model for end stage liver disease
NSBBs	Non-selective β-blockers
OLT	Ortothopic liver transplantation
PH	Portal hypertension
RCT	Randomized controlled trial
SD	Standard deviation
SHR	Subdistribution hazard ratio

The progression of cirrhosis over time from a compensated stage to the development of decompensated determines a markedly declined life expectancy [1, 2]. Portal hypertension (PH) is the main determinant of decompensation [1, 3]. A hepatic venous pressure gradient (HVPG) ≥10 mmHg defines clinically significant PH (CSPH) since decompensation almost never develops until reaching this threshold [4]. Thus, CSPH defines a substage of compensated cirrhosis with a higher risk of decompensation. CSPH is also the threshold for the development of varices, which identifies a substage in compensated cirrhosis with CSPH, since patients that have already developed, varices are known to have an increased risk of decompensation [5, 6].

Progression of Portal Hypertension in Compensated Cirrhosis and Risk of Decompensation

In compensated cirrhosis, hypoalbuminemia, MELD score, and obesity are predictors of decompensation, according to nested studies of the timolol trial [7, 8]. Nevertheless, the severity of PH remains the stronger determinant of the risk of decompensation [5, 6]. This has been confirmed by the timolol and the PREDESCI studies, both large trials dealing with long-term prevention of decompensation in patients with compensated cirrhosis [7, 9]. It is clear from these studies that, once CSPH develops, the risk of decompensation increases with every mmHg increase in HVPG values (see Chap. 35).

An increased hepatic vascular resistance is the primary factor leading to PH in early compensated cirrhosis, and is related to several factors, including sinusoidal remodeling, endothelial dysfunction, accumulation of fibrillar extracellular matrix, vascular occlusion, and nodule formation [10]. In this stage, even mild increases in portal pressure may activate vasodilatory and angiogenic signals, with the development of portosystemic collaterals and progressive splanchnic vasodilatation. The ensuing increase in portal blood flow leads to the development of hyperdynamic circulation [11]. The persistence of etiological factors by facilitating the systemic delivery of pathogen-associated molecular patterns or damage-associated molecular patterns, may favor the release of pro-inflammatory cytokines [12, 13]. This may in turn further increase the intrahepatic vascular resistance and may exacerbate splanchnic vasodilatation and hyperdynamic circulation, entering into a vicious circle exacerbating PH and eventually leading to decompensation [14]. On the other hand, liver fibrosis increases tissue stiffness that through mechanic-sensing pathways further activates hepatic stellate cells and de-differentiates sinusoidal endothelial cells and macrophages, which represents a self-perpetuating mechanism for continued fibrogenesis and disease progression [15].

Patients With Compensated Cirrhosis and CSPH: An Ideal Target Population for NSBBs

Experimental and clinical studies have shown that the development of splanchnic vasodilation and hyperdynamic circulation in cirrhosis is progressive during the course of the disease [11, 14]. In compensated cirrhosis, hyperdynamic circulation is more developed in patients with CSPH than in those with mild-PH (i.e., HVPG between 5 to 10 mmHg) [16]. Among patients with CSPH, hyperdynamic circulation is more accentuated in those that have already developed varices than in those without varices [17, 18]. Furthermore, patients with decompensated cirrhosis have much more evident hyperdynamic circulation than those compensated and also have higher portal pressure [18]. Thus, the progression of cirrhosis from compensation to decompensation is linked to an increased HVPG and to the development of progressive splanchnic vasodilation [17, 18]. Heart rate and cardiac output progressively increase until the end stages of decompensation, where a reduction of cardiac compensatory reserve may develop, which is mainly evident in stress situations such as infections or ACLF, and which may have a negative impact on survival [13, 17, 18].

Non-selective beta-blockers (NSBBs) decrease portal pressure by inducing a β1-adrenergic blockade (reducing heart rate and cardiac output) and a β2-adrenergic blockade causing splanchnic vasoconstriction due to unopposed alpha-adrenergic tone, thus reducing splanchnic blood flow. In keeping with this, NSBBs have a much greater portal-pressure decreasing effect in patients with CSPH and well-established hyperdynamic circulation than in those with HVPG<10 mmHg when it is still poorly developed [16]. In fact, the portal-pressure reducing effects of NSBBs are twice as much in patients with CSPH, than in those with subclinical-PH ($-16 \pm 12\%$ vs. $-8 \pm 9\%$, $P < 0.01$) [16]. Other therapies may be more adequate than NSBBs in patients with mild-PH. Drugs acting upon stellate cell activation, endothelial dysfunction, and/or vasoconstricting pathways might be effective in such patients to arrest disease progression and development of CSPH by means of decreasing liver fibrosis and intra-hepatic vascular resistance [19]. In fact, promising results have been achieved with statins [20–22], while investigation on anti-fibrogenic drugs is evolving [23]. On the other hand, the HVPG decreasing effect induced by NSBBs in compensated patients with CSPH and without high-risk varices, is similar to that achieved in patients treated for primary or secondary prophylaxis of variceal bleeding [24, 25], where the efficacy of NSBBs to prevent bleeding has been well demonstrated [26, 27].

It has been postulated that in advanced cirrhosis with extensive collaterals, the response to NSBBs in terms of reduction in HVPG may be hindered by a concomitant increase of hepatic and portal-collateral resistance due to unopposed alpha-adrenergic vasoconstriction [28]. In fact, under chronic treatment with NSBBs decompensated patients exhibit a smaller lowering effect on HVPG than compensated patients, despite achieving a greater degree of chronic β-blockade (with a greater reduction in heart rate and CO), and a greater decrease in arterial pressure [18]. Such a blunted response to NSBBs may also be partly related to a more severe vascular dysfunction in decompensated patients, with hypo-contractility induced by dysregulation of vasoactive proteins [29]. Whatever the case, current evidence points out that patients with compensated cirrhosis and CSPH can be the population that may benefit the most from treatment with NSBBs.

The Hemodynamic Effects of NSBBs Influence Clinical Outcomes

Observational studies show an improvement in the risk of decompensation when patients exhibit a satisfactory hemodynamic response to NSBBs [24, 30]. In the PREDESCI study, chronic treatment with NSBBs determined a sustained decrease in portal-pressure during long-term follow-up not observed with placebo [9]. The proportion of patients with clinically relevant HVPG decreases, such as a reduction >10% from baseline or to <10 mmHg, was greater with NSBBs [9]. The study showed that such a sustained decrease of HVPG was associated with a significant reduction in the incidence of ascites, underlining the pathogenic relevance of PH in the development of ascites. This was already known from the experience with portal-systemic derivative procedures such as surgical shunts or TIPS, which markedly decrease portal pressure [31–33]. In addition the PREDESCI study showed that such beneficial effect on ascites may also occur with much less pronounced decreases in portal pressure. Concordantly, this study demonstrated that patients remaining compensated had lower HVPG at every yearly control during follow-up than patients developing decompensation [9].

NSBBs to Prevent Decompensation in Patents With Compensated Cirrhosis

A number of RCTs and metanalyses have shown that NSBBs effectively prevent variceal bleeding in patients with cirrhosis and high-risk varices [3, 27, 34]. Recently, the PREDESCI study further demonstrated that NSBBs may also prevent decompensation in compensated cirrhosis with CSPH mainly by preventing the development of ascites [9]. This was the first clinical demonstration that ascites can be effectively prevented by safe drug therapy [9]. Previously, in the timolol-trial NSBBs failed to prevent the development of varices in compensated cirrhosis [7]. However, only patients without varices were included in the timolol trial and a large proportion (almost 40%) had not even developed CSPH. Thus, the timolol trial had a high proportion of low-risk patients with poorly developed hyperdynamic circulation and low potential capacity of response to NSBBs [7]. Nevertheless, the timolol trial provided a relevant advance in knowledge by introducing the concept of CSPH [4]. This was used as entry criteria in the PREDESCI study, which reinforced the value of NSBBs in compensated cirrhosis [4]. The importance of an adequate selection of patients in the PREDESCI study by including only those with an HVPG ≥10 mmHg, is emphasized by the finding of a significant HVPG reduction at each yearly control with NSBBs, which was not observed at any time-point in the timolol trial [7, 9].

A subgroup analysis of the PREDESCI study suggested that treatment with NSBBs was particularly successful in patients with already developed varices (only patients with small varices were included) [9]. Overall, the risk of developing the primary endpoint (survival without decompensation) was greater in patients with varices than in those without (SRH = 1.67, 95%CI = 1.0–3.34). Under placebo the primary endpoint occurred in 34% among those with varices and in only 16% of those without varices. NSBBs decreased the incidence of decompensation both in patients with varices (SHR = 0.39, 95%CI = 0.27–0.88) and in those without, although in these the effect was less marked (SHR = 0.84, 95%CI = 0.29–2.44) [9]. However, special care should be exerted when interpreting subgroup analyses such as this, to avoid wrong or misleading conclusions [35, 36]. The PREDESCI study proved that NSBBs prevent decompensation, mainly by preventing ascites, in patients with compensated cirrhosis and CSPH (both with small-varices or without varices). The fact that the benefit appears greater in those with varices may just reflect their greater baseline risk.

The PREDESCI study also has important implications for compensated patients that have already developed "high-risk varices". Up to now guidelines recommend prophylactic treatment with NSBB or EVL to prevent bleeding in these patients since both therapies have shown similar efficacy in RCTs involving patients with either compensated or decompensated cirrhosis [3, 37, 38]. However, the PREDESCI study by demonstrating that NSBBs can also prevent ascites (a decompensating event occurring more frequently than bleeding) challenges this recommendation. The additional benefit of preventing ascites clearly favors treating with NSBBs all patients who had no contraindication or intolerance, since this advantage is not afforded by endoscopic band ligation.

Further support favoring NSBBs over EVL in compensated patients with high-risk varices has been provided by an individual patient data (IPD) meta-analysis performed for the Baveno-VII consensus workshop. This IPD-meta-analysis precisely investigated whether NSBBs may be more adequate than EVL in patients with high-risk varices and compensated cirrhosis. IPD meta-analyses offer a unique possibility of allowing reanalysis of the individual-level data of patients included in RCTs comparing NSBBs vs. EVL, providing a greater sample size and allowing analyses using a time-to-event and competing-risk approach. IPD-meta-analysis also facilitates stratifying risk according to decompensation, thus properly investigating cirrhosis as a multistate disease and outcomes as time-dependent events [39]. RCTs comparing NSBBs vs. EVL, either in monotherapy or combined, for primary prevention of bleeding were identified and IPD-meta-analysis of the studies providing data was then performed using a competing-risk time-to-event approach. Importantly, the analyses were stratified according to the previous decompensation of cirrhosis. Finally, 11 studies providing IPD of 1400 patients with cirrhosis and high-risk varices were included, 656 with compensated cirrhosis. Preliminary data suggests, in keeping with the PREDESCI study, that NSBBs may be more valuable than EVL in patients with compensated cirrhosis. Overall, the risk of death was similar in patients treated with EVL vs. NSBBs (SHR = 1.04, 95%CI = 0.71–1.52, P = 0.848), with heterogeneity among RCTs (Table 34.1). When considering patients with compensated cirrhosis, mortality risk was reduced almost by half with

Table 34.1 NSBBs vs. EVL to prevent first bleeding in patients with high-risk varices, stratifying according to compensation of cirrhosis (IPD analysis with competing risk)[a]

	EVL[b]	NSBBs[b]	SHR (95%CI)[c]	P-value	Q-statistic	I^2 (95%CI)
Death[d]						
Overall	112/30/60974	115/15/69238	1.04 (0.71–1.52)	0.848	<0.001	77 (60–87)
Compensated	44/11/16141	31/5/20794	1.76 (1.11–2.77)	0.016	0.953	0.0 (0.0–0.0)
First bleeding[e]						
Overall	62/106/56243	80/96/63968	0.85 (0.56–1.29)	0.446	<0.001	75 (55–86)
Compensated	28/36/14410	40/23/19103	0.94 (0.47–1.87)	0.855	0.088	44 (0.0–71)
Ascites[e]						
Overall	142/51/39338	90/51/50760	1.62 (1.09–2.41)	0.016	<0.001	83 (69–91)
Compensated	22/14/6611	13/10/12190	2.66 (1.36–5.19)	0.004	0.626	0.0 (0.0–0.0)

[a]The table shows the pooled values of IPD-metanalysis with competing risk of studies comparing NSBBs vs EVL, showing only pooled values of EVL vs NSBBs. The analyses were stratified according to whether patients had compensated cirrhosis

[b]Descriptive statistics are events/(competing-events)/person-years

[c]Values indicate the subdistribution hazard ratio in patients treated with EVL as compared to NSBBs. CI stands for confidence interval

[d]By competing-risk analysis (OLT as competing event)

[e]By competing-risk analysis (death & OLT as competing events)

NSBBs vs. EVL (SHR = 0.57, 95%CI = 0.36–0.90, $P = 0.016$), without significant heterogeneity (Table 34.1). Addition of EVL to NSBBs did not provide any further benefit. The survival gain favoring NSBBs in compensated patients was mainly due to a much decreased risk of developing ascites (SHR = 0.38, 95%CI = 0.19–0.73; $P = 0.004$) while the risk of first bleeding was similar (SHR = 0.94, 95%CI = 0.47–0.87; $P = 0.855$), without significant heterogeneity in both analyses. Neither the risk of bleeding nor the risk of developing ascites was improved by adding EVL to NSBBs as compared to NSBBs alone. These results strongly support the concept that, in patients with compensated cirrhosis and high-risk varices, NSBBs are preferable over EVL, as, on top of a similar risk of bleeding, NSBBs additionally decrease the risk of developing ascites and significantly improve survival.

Carvedilol, the β-Blocker of Choice Iin Patients With Compensated Cirrhosis

Carvedilol has greater portal pressure decreasing effect than classical-NSBBs, such as propranolol or nadolol, and may achieve hemodynamic response in previous non-responders to classical-NSBBs [40–42]. In the PREDESCI trial, carvedilol was only given to previous non-responders to propranolol in an acute response test [9]. Despite being used only in poor potential candidates, carvedilol achieved a significantly greater long-term reduction of HVPG than that observed with propranolol, which was used in previous responders. Carvedilol achieved greater decreases in HVPG at 12-months (16% ± 3% vs. 10% ± 2%, $p = 0.036$) and 24-months (15% ± 4% vs. 9% ± 3%, $p = 0.048$) [9]. The more pronounced portal-pressure lowering effect of carvedilol is likely due to its anti-α-adrenergic activity together with the enhanced intrahepatic release of NO, which translates into a decrease in intra-hepatic vascular resistance [43, 44], a key factor leading to portal hypertension in compensated cirrhosis [10, 15]. These data suggest that carvedilol may be particularly adequate in compensated cirrhosis. The PREDESCI study showed a trend towards better outcomes and better adherence to therapy with carvedilol as compared with propranolol in compensated patients, suggesting better patient tolerability [9]. In addition, experimental studies suggest that carvedilol may also have pleiotropic actions such as anti-oxidant properties and may improve inflammation and fibrosis [45, 46]. These effects may also be beneficial to avoid decompensation. Furthermore, experience from different RCTs suggests efficacy with carvedilol to prevent enlargement of small varices [47], to prevent first bleeding in patients with high-risk varices [48], and even to improve long-term survival in patients with either compensated or decompensated cirrhosis [49].

Investigating whether carvedilol may be useful to prevent decompensation and to improve survival in compensated cirrhosis, was the aim of another IPD meta-analyses performed for Baveno-VII consensus workshop [50]. This was investigated using a competing-risk time-to-event IPD-meta-analysis of RCTs comparing carvedilol with a control group receiving no active therapy or receiving EVL in patients with high-risk varices. Only compensated patients were included. OLT and death were competing events for prevention of decompensation and OLT for death. Models were adjusted using propensity score for baseline covariates with the IPTW approach. Four RCTs were finally included in the IPD-meta-analysis, leading to 352 patients with compensated cirrhosis included, 181 treated with carvedilol and 171 controls [50]. Baseline characteristics were similar between groups. The study showed that the risk of decompensation of cirrhosis was lower in patients treated with carvedilol vs. controls (SHR = 0.506, 95%CI = 0.289–0.887, $P = 0.017$), without significant heterogeneity (See Fig. 34.1). This was mainly due to a significant reduction in the risk of developing ascites favoring carvedilol (SHR = 0.491, 95%CI = 0.247–-0.974, $P = 0.042$). The risk of death was also significantly decreased with carvedilol vs. controls (SHR = 0.417, 95%CI = 0.194–0.896, $P = 0.025$), again without heterogeneity (Fig. 34.1) [50]. Thus, this IPD-meta-analysis in concordance with the PREDESCI study, showed that long-term therapy with carvedilol can prevent decompensation of cirrhosis with CSPH, significantly improving the survival of compensated patients with CSPH. This suggests that screening patients with compensated cirrhosis for development of CSPH to start therapy with carvedilol can be beneficial.

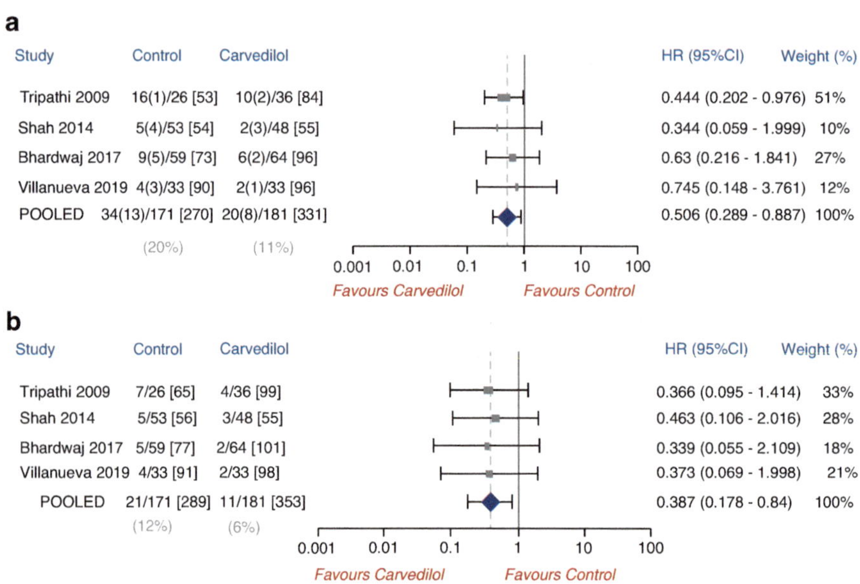

Fig. 34.1 Forest plots for risk of decompensation of cirrhosis (**a**) and risk of death (**b**) in compensated patients treated with carvedilol or with a control therapy (placebo or EVL) by IPD competing risk metanalysis (death and OLT were competing events for decompensation and OLT for death). (**a**) Risk of decompensation of cirrhosis was significantly lower in patients treated with carvedilol than in controls ($P = 0.017$), without significant heterogeneity ($Q = 0.67$, $P = 0.8802$; $I^2 = 0.0\%$, 95%CI = 0.0%–31.5%). (**b**) Risk of death was significantly lower in patients treated with carvedilol than in controls ($P = 0.016$). without significant heterogeneity ($Q = 0.09$, $P = 0.9934$; $I^2 = 0.0\%$, 95%CI = 0.0%–0.0%)

Conclusions

Altogether, current evidence indicates that in patients with compensated cirrhosis and CSPH, NSBBs may prevent the progression to decompensation mainly by improving the risk of developing ascites. This clearly may represent a paradigm shift in the management of patients with compensated cirrhosis, since up to now compensated patients had no indication for therapy until the development of high-risk varices. However, such indication can be expanded based on the findings from the PREDESCI study showing efficacy with NSBBs in compensated patients with CSPH without high-risk varices. Furthermore, a recent IPD-meta-analysis indicates that long-term therapy with carvedilol in addition to prevent decompensation may significantly improve survival in compensated patients with CSPH. In patients with high-risk varices either NSBBs or EVL were previously recommended to prevent first bleeding. However, recent data indicate that also compensated patients with high-risk varices may have greater benefits using NSBBs, which have shown efficacy to prevent ascites in addition to bleeding and have also shown survival benefits over EVL in compensated cirrhosis.

Conflict of Interest The authors have no conflict of interest with regards to this study.

Financial Support This study has been supported in part by grants from the Instituto de Salud Carlos III (EC08/00087, PI10/01552, PI13/02535, PS09/00485, PI14/00876, PI15/00066) and Stiftung fur Leberkrankenheit.

The CIBERehd is funded by the Instituto de Salud Carlos III (ISCiii).

References

1. Tsochatzis EA, Bosch J, Burroughs AK. Liver cirrhosis. Lancet. 2014;383:1749–61.
2. D'Amico G, Garcia-Tsao G, Pagliaro L. Natural history and prognostic indicators of survival in cirrhosis: a systematic review of 118 studies. J Hepatol. 2006;44:217–31.
3. Garcia-Tsao G, Abraldes JG, Berzigotti A, Bosch J. Portal hypertensive bleeding in cirrhosis: risk stratification, diagnosis and management: 2016 practice guidance by the American Association for the Study of Liver Diseases. Hepatology. 2017;65:310–35.
4. Ripoll C, Groszmann R, Garcia-Tsao G, Grace N, Burroughs A, Planas R, et al. Hepatic venous pressure gradient predicts clinical decompensation in patients with compensated cirrhosis. Gastroenterology. 2007;133:481–8.
5. D'Amico G, Pasta L, Morabito A, et al. Competing risks and prognostic stages of cirrhosis: a 25-year inception cohort study of 494 patients. Aliment Pharmacol Ther. 2014;39:1180–93.
6. Garcia-Tsao G, Friedman S, Iredale J, Pinzani M. Now there are many (stages) where before there was one: in search of pathophysiological classification of cirrhosis. Hepatology. 2010;51:1445–9.
7. Groszmann RJ, Garcia-Tsao G, Bosch J, et al. Beta-blockers to prevent gastroesophageal varices in patients with cirrhosis. N Engl J Med. 2005;353:2254–61.
8. Berzigotti A, Garcia-Tsao G, Bosch J, Grace ND, Burroughs AK, Morillas R, et al. Obesity is an independent risk factor for clinical decompensation in patients with cirrhosis. Hepatology. 2011;54:555–61.
9. Villanueva C, Albillos A, Genescà J, Garcia-Pagan J, Calleja JL, Bosch J, et al. β blockers to prevent decompensation of cirrhosis in patients with clinically significant portal hyperten-

sion (PREDESCI): a randomised, double-blind, placebo-controlled multicentre trial. Lancet. 2019;393:1597–608.
10. García-Pagán JC, Gracia-Sancho J, Bosch J. Functional aspects on the pathophysiology of portal hypertension in cirrhosis. J Hepatol. 2012;57:458–61.
11. Sikuler E, Groszmann RJ. Interaction of flow and resistance in maintenance of portal hypertension in a rat model. Am J Physiol. 1986;250:G205–12.
12. Albillos A, de Gottardi A, Rescigno M. The gut-liver axis in liver disease: pathophysiological basis for therapy. J Hepatol. 2020;72:558–77.
13. Bernardi M, Moreau R, Angeli P, Schnabl B, Arroyo V. Mechanisms of decompensation and organ failure in cirrhosis: from peripheral arterial vasodilation to systemic inflammation hypothesis. J Hepatol. 2015;63:1272–84.
14. Iwakiri Y, Groszmann RJ. The hyperdynamic circulation of chronic liver diseases: from the patient to the molecule. Hepatology. 2006;43(Suppl 1):S121–31.
15. Guixé-Muntet S, Ortega-Ribera M, Wang C, Selicean S, Andreu I, Kechagia JZ, Fondevila C, Roca-Cusachs P, Dufour JF, Bosch J, Berzigotti A, Gracia-Sancho J. Nuclear deformation mediates liver cell mechanosensing in cirrhosis. JHEP Rep. 2020;2(5):100145.
16. Villanueva C, Albillos A, Genescà J, et al. Development of hyperdynamic circulation and response to β-blockers in compensated cirrhosis with portal hypertension. Hepatology. 2016;63:197–206.
17. Turco L, Garcia-Tsao G, Magnani I, et al. Cardiopulmonary hemodynamics and C-reactive protein as prognostic indicators in compensated and decompensated cirrhosis. J Hepatol. 2018;68:949–58.
18. Alvarado-Tapias E, Ardevol G-GM, MontañésR PO, Cuyas B, Graupera I, Brujats A, Vilades D, Colomo A, Poca M, Torras X, Guarner C, Concepción M, Aracil C, Torres F, Villanueva C. Short-term hemodynamic effects of b-blockers influence survival of patients with decompensated cirrhosis. J Hepatol. 2020;73:829–41.
19. Bosch J, Abraldes JG, Fernández M, García-Pagán JC. Hepatic endothelial dysfunction and abnormal angiogenesis: new targets in the treatment of portal hypertension. J Hepatol. 2010;53:558–67.
20. Abraldes JG, Villanueva C, Aracil C, Turnes J, Hernandez-Guerra M, Genescan J, Rodriguez M, Castellote J, García-Pagán JC, Torres F, Calleja JL, Albillos A, Boschet J. Addition of simvastatin to standard therapy for the prevention of variceal rebleeding does not reduce rebleeding but increases survival in patients with cirrhosis. Gastroenterology. 2016;150:1160–70.
21. Mohanty A, Tate JP, Garcia-Tsao G. Statins are associated with a decreased risk of decompensation and death in veterans with hepatitis C-related compensated cirrhosis. Gastroenterology. 2016;150:430–40.
22. Kaplan DE, Serper MA, Mehta R, Fox R, John B, Aytaman A, Baytarian M, Hunt K, Albrecht J, Njei B, Taddei TH. Effects of hypercholesterolemia and statin exposure on survival in a large National Cohort of patients with cirrhosis. Gastroenterology. 2019;156:1693–706.
23. Schierwagen R, Klein S, Uschner F, Trebicka J. Novel targets and drug development in portal hypertension. Curr Hepatol Rep. 2019;18:187–96.
24. D'Amico G, Garcia-Pagan JC, Luca A, Bosch J. Hepatic vein pressure gradient reduction and prevention of variceal bleeding in cirrhosis: a systematic review. Gastroenterology. 2006;131:1611–24.
25. Villanueva C, Aracil C, Colomo A, Hernández-Gea V, López-Balaguer JM, Alvarez-Urturi C, et al. Acute hemodynamic response to beta-blockers and prediction of long-term outcome in primary prophylaxis of variceal bleeding. Gastroenterology. 2009;137:119–28.
26. D'Amico G, Pagliaro L, Bosch J. Pharmacological treatment of portal hypertension: an evidence-based approach. Semin Liver Dis. 1999;19:475–505.
27. Rodrigues SG, Mendoza YP, Bosch J. Beta-blockers in cirrhosis: evidence-based indications and limitations. JHEP Rep. 2019;2(1):100063.
28. Kroeger RJ, Groszmann RJ. Increased portal venous resistance hinders portal pressure reduction during the administration of β-adrenergic blocking agents in a portal hypertensive model. Hepatology. 1985;5:97–101.

29. Trebicka J, Von Heydebrand M, Lehmann J, et al. Assessment of response to beta-blockers by expression of βarr2 and RhoA/ROCK2 in antrum mucosa in cirrhotic patients. J Hepatol. 2016;64:1265–73.
30. Hernández-Gea V, Aracil C, Colomo A, et al. Development of ascites in compensated cirrhosis with severe portal hypertension treated with β-blockers. Am J Gastroenterol. 2012;107:418–27.
31. Salerno F, Cammà C, Enea M, Rössle M, Wong F. Transjugular intrahepatic portosystemic shunt for refractory ascites: a meta-analysis of individual patient data. Gastroenterology. 2007;133:825–34.
32. García-Pagán JC, Saffo S, Mandorfer M, Garcia-Tsao G. Where does TIPS fit in the management of patients withcirrhosis? JHEP Rep. 2020;2(4):100122.
33. Bosch J. Small diameter shunts should lead to safe expansion of the use of TIPS. J Hepatol. 2021;74:230–4.
34. Sharma M, Singh S, Desai V, et al. Comparison of therapies for primary prevention of esophageal variceal bleeding: a systematic review and network meta-analysis. Hepatology. 2019;69:1657–75.
35. Peto R. Current misconception 3: that subgroup-specific trial mortality results often provide a good basis for individualising patient care. Br J Cancer. 2011;104(7):1057–8.
36. Ciolino JD, Spino C, Ambrosius WT, Khalatbari S, Cayetano SM, Lapidus JA, Nietert PJ, Oster RA, Perkins SM, Pollock BH, Pomann GM, Price LL, Rice TW, Tosteson TD, Lindsell CJ, Spratt H. Guidance for biostatisticians on their essential contributions to clinical and translational research protocol review. J Clin Transl Sci. 2021;5(1):e161.
37. De Franchis R. Expanding consensus in portal hypertension. Report of the Baveno VI consensus workshop: stratifying risk and individualizing care for portal hypertension. Revising consensus in portal hypertension. J Hepatol. 2015;63:743–52.
38. European Association for the Study of the Liver. EASL Clinical Practice Guidelines for the management of patients with decompensated cirrhosis. J Hepatol. 2018;69:406–60.
39. Riley RD, Lambert PC, Abo-Zaid G. Meta-analysis of individual participant data: rationale, conduct, and reporting. BMJ. 2010;340:c221.
40. Sinagra E, Perricone G, D'Amico M, et al. Systematic review with meta-analysis: the haemodynamic effects of carvedilol compared with propranolol for portal hypertension in cirrhosis. Aliment Pharmacol Ther. 2014;39:557–68.
41. Reiberger T, Ulbrich G, Ferlitsch A, Payer BA, Schwabl P, Pinter M, et al. Carvedilol for primary prophylaxis of variceal bleeding in cirrhotic patients with haemodynamic non-response to propranolol. Gut. 2013;62:1634–41.
42. Zacharias AP, Jeyaraj R, Hobolth L, Bendtsen F, Gluud LL, Morgan MY. Carvedilol versus traditional, non-selective beta-blockers for adults with cirrhosis and gastroesophageal varices. Cochrane Database Syst Rev. 2018;10(10):CD011510. https://doi.org/10.1002/14651858.CD011510.pub2.
43. Frishman WH. Carvedilol. N Engl J Med. 1998;339:1759–65.
44. Bosch J. Carvedilol for portal hypertension in patients with cirrhosis. Hepatology. 2010;51:2214–8.
45. Akbas H, Ozden M, Kanko M, et al. Protective antioxidant effects of carvedilol in a rat model of ischaemia-reperfusion injury. J Int Med Res. 2005;33:528–36.
46. Hamdy N, El Demerdash E. New therapeutic aspect for carvedilol: antifibrotic effects of carvedilol in chronic carbon tetrachloride-induced liver damage. Toxicol Appl Pharmacol. 2012;261:292.
47. Bhardwaj A, Kedarisetty CK, Vashishtha C, et al. Carvedilol delays the progression of small oesophageal varices in patients with cirrhosis: a randomised placebo-controlled trial. Gut. 2017;66:1838–43.
48. Tripathi D, Ferguson JW, Kochar N, Leithead JA, Therapondos G, McAvoy NC, Stanley AJ, Forrest EH, Hislop WS, Mills PR, Hayes PC. Randomized controlled trial of carvedilol versus variceal band ligation for the prevention of the first variceal bleed. Hepatology. 2009;50:825–33.

49. McDowell HR, Chuah CS, Tripathi D, Stanley AJ, Forrest EH, Hayes PC. Carvedilol is associated with improved survival in patients with cirrhosis: a long-term follow-up study. Aliment Pharmacol Ther. 2021;53:531–9.
50. Villanueva C, Torres F, Shah HA, Tripathi D, Sarin SK, Brujats A, Rodrigues SG, Azam Z, Hayes P, Bhardwaj A, Abid S, Jindal A, Alvarado E, Bosch J. Carvedilol improves risk of decompensation and survival in compensated cirrhosis. A competing-risk meta-analysis of individual patient data. J Hepatol. 2021;75:S378–9.

Evaluation of the Effect of CSPH, Reduction of HVPG, and Other Factors Predicting the First Decompensation in Cirrhosis

35

Dhiraj Tripathi, Càndid Villanueva, and Jaime Bosch

Abbreviations

ACLD	Advanced chronic liver disease
HVPG	Hepatic venous pressure gradient
CSPH	Clinically significant portal hypertension
ACLF	Acute on chronic liver failure
NAFLD	Non-alcoholic fatty liver disease
RCT	Randomised controlled trial
RNA	Ribonucleic acid
HCV	Hepatitis C virus

D. Tripathi (✉)
Liver Unit, University Hospitals Birmingham NHS Foundation Trust, Birmingham, UK

Institute of Immunology and Immunotherapy, University of Birmingham, Birmingham, UK
e-mail: D.Tripathi@bham.ac.uk

C. Villanueva
Gastrointestinal Bleeding Unit, Department of Gastroenterology, Hospital de la Santa Creu i Sant Pau, Biomedical Research Institute Sant Pau (IIB Sant Pau), Barcelona, Spain

Universitat Autònoma de Barcelona, Barcelona, Spain

Centro de Investigación Biomédica en Red de Enfermedades Hepáticas y Digestivas (CIBERehd), Barcelona, Spain
e-mail: cvillanueva@santpau.cat

J. Bosch
Hepatology, Department of Visceral Surgery and Medicine, University of Bern, Bern, Switzerland

IDIBAPS-CIBERehd, University of Barcelona, Barcelona, Spain

Department of Biomedical Research, University of Bern, Bern, Switzerland
e-mail: jaime.bosch@dbmr.unibe.ch

R. de Franchis (ed.), *Portal Hypertension VII*,
https://doi.org/10.1007/978-3-031-08552-9_35

BMI	Body mass index
APC	Abdominal porto-systemic collaterals
MELD	Model for end stage liver disease
HCC	Hepatocellular carcinoma
NASH	Non-alcoholic steatohepatitis
NSBB	Non-selective beta blocker
SVR	Sustained virological response
VBL	Variceal band ligation
As	Ascites
HE	Hepatic encephalopathy
PB	Portal hypertensive bleeding
Ja	Jaundice
SBP	Spontaneous bacterial peritonitis
LSM	Liver stiffness measurement
IQR	Interquartile range
SS	Splenic stiffness
LSPS	LSM x spleen diameter/platelet count
TE	Transient elastography
SWE	Shear wave elastography
ARFI	Acoustic radiation force impulse
BI	Bacterial infections
SHR	Subdistribution hazard ratios
tAUC	Time-dependent area under the curve
ABIDE	Aspartate aminotransferase/alanine, aminotransferase ratio, bilirubin, International normalized ratio, type 2 Diabetes, and oesophageal varices

Stages of Cirrhosis and Clinically Significant Portal Hypertension

In ACLD, stages 0–2 (compensated phase) have a median duration of over 10 years, and further progression leads to decompensation (Fig. 35.1) with variceal bleeding, ascites, and hepatic encephalopathy (alone or in combination). Some patients may recompensate to stages 0–2, but a second decompensation invariably leads to the downward spiral to end-stage liver disease, ACLF, or death. Mortality from decompensated cirrhosis is much higher at 40%, 65%, and 80% at 1, 2, and 5 years respectively, compared to 1% in the compensated state [1].

Portal hypertension in cirrhosis results from increased intra–hepatic resistance due to fibrosis and contraction of sinusoidal and peri-sinusoidal cells due to interplay between vascular mediators favoring vasoconstriction with reduced intrahepatic eNOS activity. There is also increased portal inflow due to splanchnic vasodilatation driven by nitric oxide (NO) and sGC-PKG signaling, which perpetuates the initial rise in portal pressure [2, 3]. These hemodynamic changes result in the development of the hyperdynamic circulation. HVPG is an estimation of the true

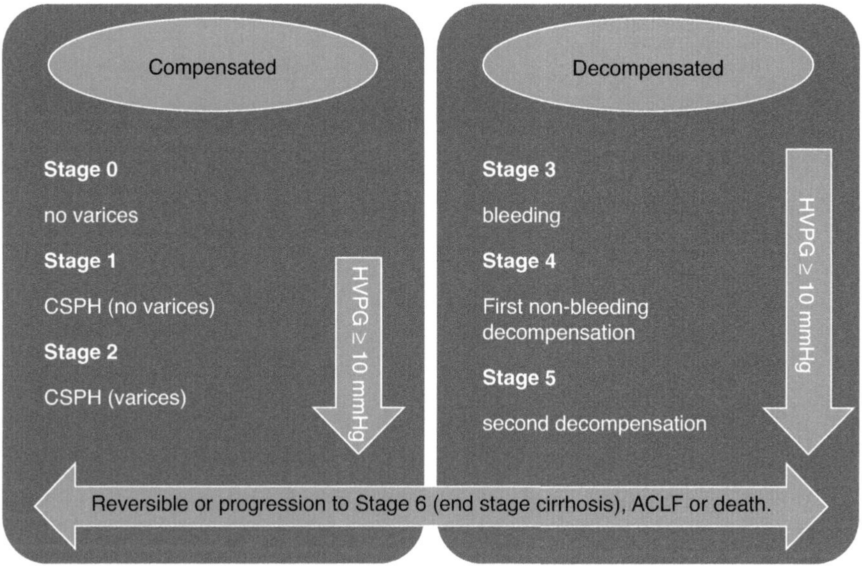

Fig. 35.1 Clinical stages of cirrhosis [1]

portal pressure in sinusoidal portal hypertension and is the Wedge Hepatic Venous Pressure (WHVP) minus Free Hepatic Venous Pressure (FHVP). The method of accurately measuring HVPG is described elsewhere [4].

Normal HVPG is between 1–5 mmHg and is described as CSPH at ≥10 mmHg. The hyperdynamic circulation has not yet fully developed at HVPG <10 mmHg where portal inflow has less of a contribution. Thus, the therapeutic effect of NSBB on portal hypertension is more pronounced at HVPG ≥10 mmHg [5]. Above this threshold varices may develop with progression to ascites or hepatic encephalopathy [6, 7]. Studies consistently show that the more common first decompensating event is ascites [7–9].

Thus, the discovery of surrogate markers predicting decompensation is an important clinical goal. These tools can aid in patient selection for therapies such as beta-blockers (or future therapies), or closely monitor those at low risk.

Hepatic Venous Pressure Gradient in Predicting Decompensation

HVPG as a marker of prognosis in cirrhosis and decompensation has been widely studied. Table 35.1 provides summaries of the important studies investigating the role of HVPG in predicting decompensation [6–16]. Significant heterogeneity exists in baseline characteristics with regards to the presence of CSPH, cirrhosis, varices, use of NSBB as prophylaxis against bleeding, etiology, the definition of decompensation and presence of hepatocellular carcinoma. Decompensation can vary

Table 35.1 Key studies evaluating HVPG for predicting progression of compensated cirrhosis and decompensation

Study	Design	Definition of decompensation	Patient selection and characteristic	Therapies	HVPG threshold predicting clinical events	Comments
Groszmann et al., 2005 [6]	RCT Primary endpoint: Development of varices or variceal hemorrhage Mean follow up: 54.9 months HVPG every 3/12	Not given	Patients without varices and HVPG $\geq$6 mmHg ($n = 213$) Etiology: Alcohol (24%), HCV (59%, non on treatment). Cholestatic diseases excluded	Timolol vs. placebo	HVPG $\geq$10 mmHg at baseline and HVPG >10% increase associated with the primary endpoint HVPG >10% decrease predicted being free of the primary endpoint	No difference in the primary endpoint when all patients were analyzed HVPG response was significantly higher with timolol than with placebo
Ripoll et al., 2007 [7]	Nested cohort study within an RCT Median follow-up: 51.1 months Endpoint: Development of clinical decompensation	As, HE, PB	Patients without varices and HVPG $\geq$6 mmHg ($n = 213$, 154 patients had repeat HVPG)	Timolol vs. placebo	HVPG <10 mmHg is associated with a 90% probability of absence of decompensation (ascites 75%, variceal bleeding 105, hepatic encephalopathy 27%) HVPG decrease <10% associated with decompensation Every HVPG increase in 1 mmHg is associated with a 19% increased risk of clinical decompensation	Decompensation in 29% during follow-up. HVPG, MELD, and albumin predicted decompensation. HVPG has the greatest discriminative ability

Villanueva et al., 2009 [16]	Prospective analysis	Not given	Patients with large varices that had not bled ($n = 105$) Mean follow 25 ± 21 months Etiology: Alcohol (39%), HCV (42%) A mixture of compensated and decompensated patients	Primary prevention: Nadolol ($n = 96$), VBL ($n = 9$)	HVPG response (defined as reduction ≥10% or <12 mmHg) to acute intravenous propranolol. Nadolol commenced. Second HVPG at 1–3 months Lower risk of variceal bleeding in both acute (c statistic, 0.83; 95% CI, 0.75–0.9) and chronic ((c statistic, 0.83; 95% CI, 0.72–0.91) responders The risk of ascites was also lower in acute and chronic responders ($p = 0.001$)	15% bled from varices during follow-up HVPG HVPG reduction ≥10% had the greatest discriminative ability to predict bleeding
Berzigotti et al., 2011 [11]	Retrospective analysis Median follow up: 28 months	As, HE, PB, SBP, Ja	Patients with HVPG ≥10 mmHg ($n = 86$) 73% compensated Etiology: Viral (54%), alcohol (11%)	Primary prevention: NSBB ($n = 33$) VBL ($n = 5$) Secondary prevention ($n = 3$): NSBB + VBL	HVPG ≥16 mmHg and bilirubin predicted the first decompensation with HVPG having the greatest discriminative ability	APC correlated with HVPG ≥16 mmHg, with a trend towards predicting decompensation
Berzigotti et al., 2011 [10]	Post hoc analysis of an RCT Median follow up: 59 months	As, HE, PB	Patients without varices and HVPG ≥6 mmHg and where BMI available ($n = 161$)	Timolol vs. placebo	HVPG (1.14 [95% CI, 1.07–1.20]), albumin (HR 4.54 [2.44–8.33]) and high baseline BMI (hazard Ratio 1.06; 95% CI, 1.01–1.12) independently predicted decompensation	Decompensation in 30% of patients (ascites, 69%; encephalopathy, 31%, variceal bleeding 10%)

(continued)

Table 35.1 (continued)

Study	Design	Definition of decompensation	Patient selection and characteristic	Therapies	HVPG threshold predicting clinical events	Comments
Herná ndez-Gea et al., 2012 [9]	Prospective analysis Median follow up: 53 months	As, HE, PB	Large varices, no previous bleeding or other decompensation ($n = 83$, 78 with HVPG data) Etiology: Alcohol (18%), HCV (62%)	Primary prevention with nadolol[a]	Pre-therapy (with nadolol) acute HVPG response to IV propranolol. HVPG reduction ≥10% defined response. Non-response predicted decompensation (ascites, bleeding) and death Ascites independently predicted nonresponse while refractory ascites, hepatorenal syndrome, and bacterial peritonitis did not Chronic HVPG response at 3 months predicted ascites	In hemodynamic responders, MELD provided additional prognostic information No control group 62% of patients decompensated during follow-up
Ripoll et al., 2012 [15]	Retrospective single center Median follow up: 11 months (compensated); 10 months (decompensated)	As, HE, PB	Compensated ($n = 51$). Decompensated ($n = 66$). HCC in 29% of compensated patients (within Milan criteria) Varices in 48% Aetiology: Viral (62%), alcohol (30%)	NSBB use: Compensated (50%), decompensated (73%)	HVPG × 2 measurements were done at median 13 months (compensated) and 8 months (decompensated) intervals HVPG ≥10 mmHg and MELD ≥10 independent predictors of decompensation in compensated patients. MELD ≥12 independent predictors of death in decompensated patients. MELD had a much narrower variation range than HVPG in compensated cirrhosis	Changes in HVPG and MELD did not influence outcomes after multivariate analysis (only on univariate analysis). This may reflect the short time and loss of patients between HVPG measurements. Baseline single measurement had the greatest discriminate function (c-statistic (95% CI) 0.792 (0.655–0.893)) No association between NSBB and endpoints During follow up 29% developed decompensation

Rincón et al., 2013 [14]	Retrospective single-center study. Median follow-up: 27 months	As, HE, PB	Compensated stage 1 HCV cirrhosis ($n = 145$, 76 with varices, HCC 26%)	37% on antiviral therapy without SVR (those without SVR excluded)	Baseline HVPG was done. HVPG ≥10 mmHg in 72%) HVPG and albumin independently predicted decompensation Each mmHg increase in HVPG led to an 11% increase in the risk of decompensation PI[b] model discriminative for decompensation with good calibration (AUROC: 0.77 (95% CI: 0.64–0.89). PI <2.5 is highly predictive of compensated state even after excluding HCC patients	29% suffered decompensation (ascites most common, and especially varices at baseline). No data on NSBB use
Lens et al., 2015 [12]	Retrospective over four centers. Median follow-up: 5 years	As, HE, PB	Compensated HCV on interferon-based antiviral therapy ($n = 100$). CSPH in 74% (35% achieved SVR)		HVPG at baseline before antiviral therapy and at 12 weeks ($n = 30$) and 23 weeks ($n = 62$) Baseline HVPG but not SVR predicted decompensation and transplant-free survival	31% had varices (4% on NSBB). 19% developed decompensation. Higher if CSPH at baseline

(continued)

Table 35.1 (continued)

Study	Design	Definition of decompensation	Patient selection and characteristic	Therapies	HVPG threshold predicting clinical events	Comments
Reiberger et al., 2012 [13]	Prospective non-randomized study Median follow-up was 19.5 months	As, HE (grade 3, 4), PB, Ja	Oesophageal varices and HVPG ≥12 mmHg. (n = 104) Aetiology: Alcohol (55%), viral (33%) A mixture of compensated and decompensated patients	Primary prevention: Propranolol and carvedilol (for hemodynamic non-responders). VBL in carvedilol non-responders (carvedilol stopped)	Baseline HVPG, and at 4 weeks (hemodynamic response defined as >20% reduction or reduction to <12 mmHg) 56% not respond to propranolol responded to carvedilol. Overall 72% of NSBB had a hemodynamic response	Less bleeding and mortality in hemodynamic responders Less decompensation in: (a) Propranolol non-responders on carvedilol compared to VBL (b) NSBB responders compared to VBL
Villanueva et al., 2019 [8]	RCT (PREDESCI). Median follow-up 37 months	As, HE, PB	Small oesophageal varices (57%) or no varices (43%). HVPG ≥10 mmHg. N = 201 Aetiology: HCV (56%, none treated), alcohol (16%)	Pre-primary prevention: NSBB vs. placebo	NSBB allocation according to acute response (HVPG >10% reduction from baseline). Responders—Propranolol (n = 67). Non-responders—Carvedilol (n = 33) HVPG annually Reduced decompensation and death in NSBB group (HR 0.51 (0.26–0.97), p = 0.0412). HVPG reduction from baseline ≥10% was seen in 51% on NSBB and 29% on placebo. Carvedilol decreased HVPG more than propranolol	Annual endoscopies. Development of high-risk varices treated with variceal band ligation Decompensation mostly ascites The benefit of NSBB is greater in ArLD and if HVPG response (>10% decrease from baseline or to <10 mmHg) Bleeding only in 3%

Jindal et al., 2020 [20]	Retrospective study of prospectively collected data. Median follow up 1.6 ± 0.4 years	As, HE, PB, Ja	$N = 741$ Large varices (24%), small oesophageal varices (66%), or no varices (10%) HVPG ≥6 mmHg: • HVPG 6 to <12 mmHg (group A; $n = 163$) • HVPG 12 to <20 mmHg (group B; $n = 437$) • HVPG ‡ 20 mmHg (group C; $n = 141$)	Pre-primary and primary prevention: Carvedilol in all patients with baseline HVPG ≥12 mmHg	Baseline HVPG ≥12 mmHg and HVPG ≥20 mmHg independent predictors of decompensation (HR 2.73 & 4.48 respectively) Hemodynamic response to carvedilol not associated with decompensation	217 (29%) developed decompensation during follow-up Total leucocyte count (HR 1.07), serum creatinine (HR 1.19) are associated with decompensation-free survival MELD is not associated with decompensation Group C had a higher proportion of NASH cirrhosis than group A (35% vs. 20%) Baseline LSM did not correlate with decompensation

(continued)

Table 35.1 (continued)

Study	Design	Definition of decompensation	Patient selection and characteristic	Therapies	HVPG threshold predicting clinical events	Comments
Mandorfer et al., 2020 [22]	Prospective. Median follow up 35.3 months (IQR 21.8 months)	As, HE, PB	$N = 90$. Varices in 40% (small, 53%; large, 47%). LSM 23.4 kPa. HVPG ≥6 mmHg: • HVPG 6–9 mmHg ($n = 23$) • HVPG 10–15 mmHg ($n = 29$) • HVPG ≥16 mmHg ($n = 38$) $N = 67$ underwent third HVPG Etiology: HCV (100%)—All achieved SVR with antiviral therapy	NSBB (42%)	Baseline HVPG No progression with HVPG <10 mmHg (no decompensation) In those with CSPH at baseline, this persisted in 76%, but a decrease in HVPG ≥10% occurred in 60% Baseline HVPG did not predict decompensation Change in HVPG HVPG change during follow up predicted decompensation • Absolute change (AUROC 0.872) • Relative change (AUROC 0.877) • Where CSPH at baseline, HVPG decrease ≥10% had ess decompensation (2.5% vs. 40.5%) HVPG change after third HVPG ($n = 67$): • HVPG reduction was 24.4% • 46% had CSPH HVPG decrease ≥10% in the second measurement maintained and had no decompensation	Previous decompensation in 14% Three patients underwent a liver transplant Decompensation associated with child-Pugh score, MELD

Turco et al. 2020 [23]	Meta-analysis of primary ($n = 7$) and secondary prophylaxis ($n = 7$) studies or both ($n = 1$). Ten case series and five RCTs. Total number = 1113 unique patients	As, HE, PB	Cirrhotic patients had at least two measurements of HVPG pre-therapy and during NSBB therapy. Most had ArLD Ascites is 40.6%	Propranolol or nadolol was used (carvedilol or propranolol in one study) 332 patients were compensated	HVPG response defined as >20% reduction from baseline or < 12 mmHg in 14 studies In patients without ascites and no previous variceal bleeding, significantly lower decompensation (OR 0.28; 95% CI 0.13–0.58) and death (OR 0.44; 95% CI 0.20–0.98) in responders	A mixture of data from RCTs and observational studies Most studies were performed before effective antiviral therapy

As ascites, *HE* hepatic encephalopathy, *PB* portal hypertensive bleeding, *Ja* jaundice, *SBP* spontaneous bacterial peritonitis, *RCT* randomized controlled trial, *VBL* variceal band ligation, *APC* abdominal portosystemic collaterals, *LSM* liver stiffness measurement, *IQR* interquartile range

[a]Nadolol withdrawn in nine patients due to intolerance and offered VBL

[b]Prognostic Index (PI) = 4 + (0.11 × HVPG − 0.8 × albumin)

considerably from 12.8% to 33.6% for cholestatic diseases and alcohol-related liver diseases respectively [17]. HVPG may underestimate the true portal pressure in cholestatic diseases due to the pre-sinusoidal component. Recent data also suggests patients with NAFLD can decompensate on follow-up even with baseline HVPG slightly <10 mmHg [18] and at a higher frequency for a given HVPG than RNA positive HCV [19], although this requires further validation. There is also variability with regards to antiviral therapy, and one can argue that studies using interferon-based regimens are outdated in the current era of directly acting antiviral therapy. However, a study showed that baseline HVPG influenced decompensation rates rather than interferon-based regimens [12]. One can infer that the results would apply to current antiviral therapies.

Baseline HVPG as a Marker of Risk of Decompensation

A seminal placebo-controlled trial investigating the role of timolol in preventing the development of varices and variceal bleeding in patients without varices and HVPG ≥6 mmHg showed that varices only developed at HVPG ≥10 mmHg [6]. The primary endpoint occurred in 84 out of 213 patients, without any difference between timolol and placebo. A nested cohort study within this RCT showed that HVPG mmHg <10 mmHg was associated with reduced decompensation, in particular ascites [7]. A further post hoc analysis of the timolol study showed that baseline BMI was a predictor of decompensation, although the association was stronger with baseline HVPG and albumin [10].

A retrospective analysis of patients with ACLD (73% compensated) showed that baseline HVPG ≥16 mmHg and bilirubin predicted the first decompensation, with HVPG having the greatest discriminative ability. In this study, the presence of abdominal portosystemic collaterals (APC) was only seen with HVPG ≥10 mmHg, and strongly correlated with HVPG >16 mmHg and suggests that APC on ultrasound scan could be a non-invasive tool to categorize patients with high HVPG [11]. This requires validation in prospective studies.

A single-center retrospective study with a mixture of compensated and decompensated patients showed that baseline HVPG >10 mmHg and MELD >12 predicted decompensation [15]. Furthermore baseline single HVPG had the greatest discriminative ability and patients with HVPG <10 mmHg were unlikely to decompensate, The high decompensation rate of 29% most likely reflects patients with HCC and varices being included.

Another retrospective study identified baseline HVPG and albumin to predict decompensation in compensated patients with HCV infection [14]. Patients with prognostic index <2.5 were very unlikely to decompensate. As in the study by Ripoll et al. [15], the high decompensation rate appears to reflect the inclusion of patients with HCC and varices. Lens and colleagues found that baseline HVPG before HCV treatment but not sustained viral response predicted decompensation and transplant-free survival [12]. These patients had interferon-based antiviral therapies.

A large recent study of 741 consecutive compensated patients with cirrhosis (predominantly NASH, 30.8%) and HVPG ≥6 mmHg, showed that decompensation developed in 29.2% over a mean follow-up of 1.6 ± 0.4 years [20]. Decompensation occurred earlier and more frequently in patients with high HVPG (≥20 mmHg, 35.5% NASH) with higher mortality. Baseline HVPG independently predicted decompensation. Limitations of this study include retrospective uncontrolled design, small numbers of patients in the high HVPG group (n = 18), and short follow-up.

HVPG Response as a Marker of Risk of Decompensation

Studies show the role of HVPG response to drug therapies as a prognostic marker predicting decompensation. In these studies, the protocols involve acute HVPG response and repeat HVPG measurements performed at variable intervals. The latter can make a comparison of studies challenging. There could also be a degree of selection bias since not all patients would have repeat HVPG measurements due to dropout or censoring events such as decompensation, death, or transplantation.

Acute HVPG response to NSBB is consistently reliable in predicting decompensation. A retrospective study found that a 12% reduction in HVPG in response to intravenous propranolol had the greatest discriminative ability for rebleeding and mortality [21]. Acute and chronic (1–3 months) HVPG response to NSBB, defined as reduction ≥10% or to <12 mmHg, was associated with a lower risk of variceal bleeding and ascites in a prospective study of 105 patients [16]. Another prospective series with a similar study design investigated the role of HVPG response to NSBB in a purely compensated cirrhotic population and mirrored these findings [9]. Baseline MELD >9 and chronic hemodynamic nonresponse were associated with ascites development. MELD added additional prognostic data in hemodynamic responders.

In the nested study of the timolol RCT, HVPG reduction of <10% from baseline predicted decompensation [7]. An increase of 1 mmHg in HVPG led to an 11% increased risk of decompensation. Multivariate analysis revealed that a lack of hemodynamic response at 12 months predicted decompensation (HR, 2.6; 95% CI, 1.1–5.6). A retrospective study found that, unlike baseline HVPG, delta HVPG at 1 year did not influence outcomes after multivariate analysis [15]. Furthermore, NSBB therapy did not appear to influence clinical outcomes. Heterogeneity with the inclusion of both compensated and decompensated patients, and those with HCC, along with low sample size and retrospective design are limitations.

A retrospective study of 100 compensated HCV patients on antiviral therapy over 24 weeks found that repeat HVPG decreased significantly and was related to high baseline viral load. After 24 weeks only those with SVR had HVPG reduction [12]. There was a trend toward higher decompensation in patients failing to achieve HVPG <10 mmHg. The small sample size in this study is a limitation. In a prospective study, 90 HCV patients treated with interferon-free therapies underwent hemodynamic studies. All patients had portal hypertension (HVPG ≥6 mmHg) and underwent follow-up HVPG assessment at 8.79 months [22]. Patients with HVPG <10 mmHg at baseline did not progress to CSPH. Follow-up HVPG was associated with decompensation (per mmHg rise, HR 1.18 (95% CI 1.08–1.28; AUROC 0.819)). By contrast, baseline HVPG was not associated with decompensation during follow-up.

A prospective study of patients treated with either propranolol or carvedilol (in propranolol nonresponders) as primary prevention, found less decompensation in the carvedilol group when compared with VBL (p = 0.035) [13]. Haemodynamic responders on propranolol or carvedilol also suffered less decompensation (ascites (p = 0.031) and variceal bleeding (p = 0.012) than those on VBL. There was a history of previous ascites in 10% of patients. In the large hemodynamic study from India mentioned earlier, 20 patients with HVPG ≥12 mmHg were started on carvedilol, and the hemodynamic response was less in the high HVPG group [20]. Moreover, reductions in portal pressure did not influence the risk of decompensation.

The PREDESCI RCT, compared NSBB with placebo in patients with compensated liver disease (Fig. 35.2) [8]. The rigorous protocol comprised a hemodynamic study at baseline to assess for the presence of CSPH and determine acute hemodynamic response to intravenous propranolol. Responders received propranolol and non-responders were given carvedilol. A placebo arm was required for each NSBB. The HVPG measurements were repeated annually. Decompensation was inversely associated with HVPG reduction >10% from baseline or to <10 mmHg at 1 year. Indeed, the primary outcome was significantly reduced in these hemodynamic responders compared to non-responders (HR = 0.32, 95% CI = 0.13 to 0.75; p = 0.008).

A recent meta-analysis of over 1100 patients showed that HVPG response (<12 mmHg or >20% from baseline) to NSBB as part of primary or secondary prevention was associated with significantly fewer clinical events and lower deaths [23].

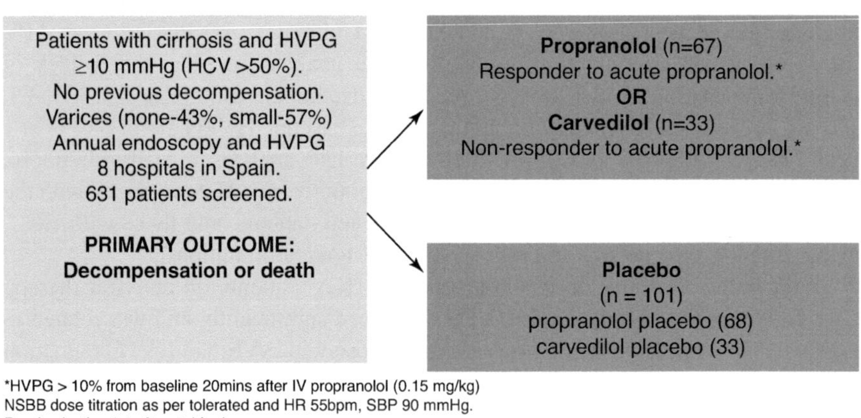

Fig. 35.2 Schema of PREDESCI trial [8]

The Role of Other Factors in Predicting Decompensation

Although HVPG remains the gold standard for predicting decompensation it is invasive, with limited availability in many countries. Therefore, non-invasive markers predicting decompensation have an important role (Table 35.2).

Ripoll found that MELD, albumin, and HVPG predicted decompensation [7], and a nomogram based on this study incorporating platelet count, MELD, albumin, and AST has been proposed [24]. Markers of systemic inflammation predicting decompensation have also been studied. Obesity has also been associated with decompensation [10]. The role of nutrition and etiology is covered elsewhere.

Simple serum markers of fibrosis have been assessed in several studies [25–29]. Prospective studies in patients with compensated cirrhosis have failed to show APRI

Table 35.2 Candidates for noninvasive markers predicting decompensation

Study	Markers	Predictive ability of decompensation	Comments
Ripoll, 2007 [7] $N = 213$	Albumin MELD HVPG	c statistic: Albumin (0.66; 95% CI, 0.58–0.74) MELD (0.64; 95% CI, 0.55–0.72) HVPG (0.71; 95% CI, 0.64–0.78)	Nested study of RCT of timolol vs. placebo (see Table 35.1)
Guha, 2019 [25] Retrospective study $N = 145$	FIB-4 and ALBI which includes the following variables age, AST, albumin, platelets, bilirubin, ALT[a]	Harell's c statistic: 0.805 (95% CI 0.718–0.873) Hazard ratio of high-risk patients was 7.1 (95% CI, 3.07–16.42)	Etiology mainly ArLD (45%) and NASH (30%) Does not take into account the influence of etiology. Lack of calibration of the model. Decompensation in 19.3% over 4.59 years
Colecchia, 2014 Prospective [26] $N = 92$	HVPG, LSM, splenic stiffness, platelet count/spleen diameter ratio, liver stiffness-spleen diameter to platelet ratio score, APRI, liver stiffness x spleen diameter, MELD	AUROC HVPG: 0.83 (95% CI 0.75–0.92) SS: 0.85 (95% CI 0.77–0.93) (independent of the presence of varices) SS < 54 kPa: Sensitivity 97% Specificity 63%, LR-0.05, NPV 97%) for predicting low risk of decompensation MELD and SS predictive model[b]: 0.87 (95% CI 0.80–0.94)	Compensated HCV cirrhosis. No patients on NSBB or antiviral therapy at baseline Varices in 53% (F1) 33% decompensated over 2 year period Calibration done

(continued)

Table 35.2 (continued)

Study	Markers	Predictive ability of decompensation	Comments
Pérez-Latorre, 2014 [37] Retrospective *N* = 60	LSM, HVPG	AUROC (95% CI) for predicting liver decompensation: LSM: 0.85 (.69–1.00). LS < 25 kPa and LS > 40 kPa thresholds for absence and presence of decompensation HVPG: 0.76 (0.59–0.93)	HCV cirrhosis with and without HIV co-infection CSPH in 53% Varices in 38% Decompensation in 13% over 42 months
Sebastiani, 2015 [27] Retrospective cohort study *N* = 146	HVPG, APRI, FIB-4, NAFLD fibrosis score, histology, imaging	Area under curve: • Histologic fibrosis stage, 0.85 (95% CI 0.76–0.93) • HVPG, 0.81 (95% CI 0.70–0.91) • APRI, 0.89 (95% CI 0.82–0.96) • FIB-4, 0.89 (95% CI 0.83–0.95) • NAFLD fibrosis score, 0.79 (95% CI 0.69–0.91)	Only NASH patients (F3/F4 fibrosis in 34%) CSPH in only 18.2% cases Clinical outcomes (decompensation, liver transplantation, HCC, or death) in 16.2% over 5 years Histological steatosis and non-invasive steatosis methods did not predict outcomes
Kitson, 2015 [28] *N* = 95 Prospective	HVPG, LSM, APRI, FIB-4, PSDR	AUROC: • LSM: 0.73 (95% CI 0.61–0.84). LS > 34.5 kPa optimal threshold for the presence of decompensation	Compensated cirrhosis (previous decompensation in 24%) 75% had CSPH Aetiology: Alcohol (41%), HCV (33%) Varices (72%) Cirrhosis in 93% NSBB in 22%
Merchante, 2012 [29] *N* = 239 Prospective	LSM, APRI, FIB4.	AUROC • LSM: 0.72 (95% CI 0.61–0.82) • LS ≥ 40 kPa optimal threshold to predict decompensation • Child-Pugh score: 0.76 (95% CI, 0.65–0.88) • MELD: 0.71 (95% CI 0.61–0.81)	All compensated with no previous decompensation HCV/HIV co-infection. Previous HCV therapy, 39% Varices in 100% (93% small) 13% decompensated over 20 months (most commonly ascites)

Table 35.2 (continued)

Study	Markers	Predictive ability of decompensation	Comments
Merchante, 2015 [36] $N = 275$ Prospective	LSM	AUROC (decompensation and/or HCC): LSM (baseline): 0.609 (0.471–0.748) LSM progression: 0.680 (0.541–0.818) Only LSM progression is associated with the endpoint.	All compensated cirrhosis with no previous decompensation Baseline LS < 40 kPa HCV/HIV co-infection. SVR in 31% No data on varices 6.9% decompensated and/or developed HCC over 32 months follow up
Robic, 2011 [35] Prospective $N = 100$	LSM HVPG	AUROC (95% CI) for portal hypertension related complications: • LSM: 0.830 [0.751–0.910]. No decompensation if LS < 21.1 kPa • HVPG: 0.845 [0.767–0.920]. No patients with HVPG <10 mmHg developed decompensation	65% had cirrhosis (72% had varices) 51% had CSPH 66% had ArLD or viral hepatitis (none on antivirals)
Villaneuva, 2019 [8] RCT $N = 201$ (see Table 35.1) Villanueva et al. 2021 [46]	LSM HVPG Child Pugh score	Cox proportional-hazards regression for decompensation and/or death (hazard ratio, 95% CI)): • Baseline child-Pugh score:4.13 (2.03–8.39) • Baseline HVPG: 4.72 (2.24–9.95) • LSM (AUROC): 0.63 (0.51–0.74). The optimal threshold was 22 KPa, with an NPV of 0.92 but a PPV of only 0.31	All compensated cirrhotic patients with CSPH Decompensation developed in 18% over 37 months No effect of etiology, portosystemic collaterals, presence of varices or not Bacterial infections were associated with decompensation (SHR 2.98 (95% CI, 1.02–8.42) and mortality (SHR 6.93 (95% CI, 2.64–18.18)

(continued)

Table 35.2 (continued)

Study	Markers	Predictive ability of decompensation	Comments
Harrison, 2019 [30] Sanyal, 2019 [18] RCT (Simtuzumab vs. placebo) Bridging fibrosis ($n = 219$) Compensated cirrhosis ($n = 258$)	HVPG (in compensated cirrhosis, 68% had CSPH) ELF FibroSure FibroTest FIB-4 APRI Serum LOXL2 NAFLD fibrosis score Other lab assessments	Variables associated with clinical events in cirrhotic patients (HR with 95%CI): • ELF score: 2.11 (1.53–2.90) • FibroSure/FibroTest, per 0.1 units: 1.21 (1.06–1.38) • NAFLD fibrosis score: 1.78 (1.43–2.21) • FIB-4: 1.24 (1.14–1.35) • APRI: 1.88 (1.45–2.46) sLOXL2, per 10 pg/mL 1.02 (1.01–1.04) • CSPH (HVPG ≥10 mmHg): 2.83 (1.33–6.02) • Failure to achieve ≥20% decrease in HVPG: 5.38 (1.65–17.58) • Failure to achieve HVPG <10 mmHg and/or *a* ≥ 20% decrease: 5.51 (1.69–17.98)	Of cirrhotics 42% had varices, 19% experienced clinical events over a median follow-up of 30.9 months
Eaton et al., 2020 [39] Retrospective ($n = 204$, PSC)	LSM measured by MRE (146 had second MRE) APRI	Variables predicting decompensation (HR with 95% CI): Single LSM > 4.32 kPa (second MRE): 60.41 (17.85–204.47) Change in LSM 0.34 kPa/year: 13.29 (5.23–33.78) Change in APRI/year: 0.76 (0.62–0.93)	LSM progression directly related to baseline LSM (stage 0 fibrosis—0.03 kPa/year vs. stage 4 fibrosis—0.31 kPa/year) Ascites was noted in all patients that developed decompensation ($n = 23$)
Osman et al., 2021 [40] Prospective ($n = 538$, PBC)	LSM measured by transient elastrography ($n = 286$) and MRE ($n = 332$)	Variables predicting decompensation (HR with 95% CI): Transient elastrography: 1.14 (1.05–1.24); optimal threshold 10.2 kPa MRE: 1.68 (1.28–2.19); optimal threshold 4.2 kPa	

Table 35.2 (continued)

Study	Markers	Predictive ability of decompensation	Comments
Gidener et al., 2020 [41] Retrospective (*n* = 194/829 with cirrhosis, NAFLD)	MRE	Variables predicting decompensation or death (HR with 95% CI): MRE: 1.32 (1.13–1.56) per 1 kPA increase after adjusting for age, sex, MELD Na	In non-cirrhotic group: Baseline LSM by MRE predicted risk of cirrhosis (HR 2.93, (95% CI, 1.86–4.62) per 1 kPa increase, AUROC 0.86)
Han et al., 2020 [42] Retrospective (*n* = 39/320 with NAFLD cirrhosis (compensated, 26))	MRE	Variables predicting decompensation or death (OR with 95% CI): MRE: 3.28 (2.04–5.28)	LSM by MRE at a threshold of 3.99 kPa discriminated between cirrhosis and non-cirrhosis (AUROC 0.986) LSM by MRE at a threshold of 6.48 kPa discriminated between compensated and decompensated cirrhosis (AUROC 0.707)
Calzadilla-Bertot et al., 2020 [31] Retrospective (*n* = 299) Biopsy proved NAFLD cirrhosis	ABIDE NAFLD fibrosis score FIB-4 MELD CPS ALBI ALBI-FIB4	5-year prediction of decompensation (tAUC): ABIDE (0.80) ABIDE vs.: NAFLD fibrosis score (0.72) FIB-4 (0.74) MELD (0.69) CPS (0.72) ALBI (0.72) ALBI-FIB4 (0.73)	
Younes et al. 2021 [32] Prospective (*n* = 1173) Biopsy proven NAFLD	NAFLD fibrosis score FIB-4 BARD APRI Hepamet fibrosis score	Variables predicting liver events over a median follow up of 81 months (medial Harrell's c-indices): NAFLD fibrosis score (0.796) FIB-4 (0.783) BARD (0.728) APRI (0.6) Hepamet fibrosis score (0.729)	F3/F4 fibrosis in 24.1%

(continued)

Table 35.2 (continued)

Study	Markers	Predictive ability of decompensation	Comments
Costa et al., 2021 [45] Prospective (n = 168, 78 cACLD)	HVPG MELD CRP Il-6	Variables predicting decompensation (HR with 95% CI) in Baveno stages 0–2: Il-6: 1.06 (1.01–1.1)	In decompensated patients, Il-6 also independently predicted death/ transplantation CRP and IL-6 increased only in decompensated patients
Petta et al., 2021 [38] NAFLD F3–F4 fibrosis and/or LSM > 10 kPa Minimum 6 months follow-up Repeat LSM within 1 year Median FU 35 months	Baseline LSM Delta LSM (improvement if >20% reduction, stable if between −/+ 20% from baseline, impairment if 20% or more increase)	Baseline LSM independently predicted: (a) Decompensation (HR 1.03; 95% CI 1.02–1.04) (b) Liver related death (HR 1.02; 95% CI 1.02–1.03) Delta LSM (n = 533) predicted: (a) Decompensation (HR 1.56; 95% CI 1.05–2.51) (b) HCC (HR 1.72; 95% CI 1.01–3.02) (c) Overall mortality (HR 1.73 (95% CI 1.11–2.69) (d) Liver related mortality (HR 1.96; 95% CI 1.10–3.38)	Threshold baseline LSM 21 kPa (CSPH) independently associated with decompensation (HR 3.71; 95% CI 1.89–6.78) Delta LSM was associated with decompensation in patients without CSPH at baseline but not those with CSPH Age and platelet count also associated with decompensation Retrospective design

LSM liver stiffness measurement, *SHR* subdistribution hazard ratios, *tAUC* time-dependent area under the curve, *ABIDE* aspartate aminotransferase/alanine, aminotransferase ratio, bilirubin, International normalized ratio, type 2 Diabetes, and oesophageal varices

[a]Formula available online: https://jscalc.io/calc/gdEJj89Wz5PirkSL

[b]Formula: exp.(−11:5 + 0.107 * SS + 0.45 * MELD)/[1 + exp.(−11:5 + 0.107 * SS + 0:4 5 * MELD)]

nor FIB-4 to predict decompensation [28, 29]. A recently published retrospective study found FIB-4 and ALBI to predict decompensation [25]. A study on the predictive value of non-invasive markers and HVPG in NASH cirrhosis was published recently [18, 30]. ELF was the serum marker with greater prognostic capacity. Other large retrospective [31] and prospective studies [32] in NAFLD cirrhosis have found noninvasive markers to be highly predictive of decompensation.

Liver stiffness (LS) has been shown to accurately reflect HVPG ≤12 mmHg, but at higher pressures correlation with HVPG is less strict and likely to reflect other factors in the pathogenesis of portal hypertension, in particular increased portal inflow [33]. A large multicentre study found that LSM ≥ 25 kPa correlates with

CSPH in cACLD apart from obese NASH patients. In obese NASH, the ANTICIPATE-NASH model was proposed based on a nomogram [34]. A prospective study showed that LSM < 21.1 kPa predicted freedom from decompensation with similar precision to HVPG [35]. Other studies have confirmed these findings in patients with HCV cirrhosis with or without HIV co-infection and alcohol-related liver disease [28, 29, 36, 37]. The LS thresholds for predicting the presence of decompensation varied between 34.5 and 40 kPa. However, in the PREDESCI trial [8], LS at baseline had low precision in predicting decompensation and/or death. Another study confirmed this finding [20]. However, sequential LSM was found to be accurate for diagnosing CSPH [22]. Baseline (threshold LSM 21 kPa) and changes in repeated LSM were found to predict decompensation, HCC, and mortality in a large retrospective cohort of NAFLD patients with F3–F4 fibrosis [38]. Studies have shown that LSM obtained using MRE was strongly associated with the development of decompensation in different aetiologies [39–42].

Spleen Stiffness (SS) has been suggested to correlate better with portal hypertension at higher portal pressures. In a prospective study [43] of 100 patients with compensated HCV cirrhosis, LS, and SS were more precise than other noninvasive makers (platelet/spleen ratio, LSPS) in predicting CSPH. The "Anticipate" study, revealed that LSM x spleen diameter/platelet count (LSPS) score values >2.65 were associated with an 80% risk of CSPH with AUC of 0.88 [44]. In a study of HCV patients with compensated cirrhosis, SS value of 54 kPa had sensitivity and specificity of 97% and 63% respectively in predicting low risk of decompensation [26]. Using TE and 2D-SWE, there can be greater non-valid or failed reading of SS compared with LS due to small-sized spleens. pSWE such as ARFI can be more reliable since it can compensate for high BMI, ascites, or small spleens. However, the data on variability is somewhat limited.

A prospective study found that IL-6 levels correlated with risk of decompensation (hazard ratio 1.06 (96% CI 1.01–1.10), with CRP and HVPG showing a strong trend, highlighting the importance of markers of systemic inflammation [45].

In a nested study of the PREDESCI trial, Bacterial Infections (BI) were developed in 36 patients that presented with decompensation [46]. BI occurred invariably before decompensation, with the principal sources being community-acquired respiratory and urinary tract and predominantly gram-negative organisms. Decompensation and particularly mortality were associated with BI, with subdistribution hazard ratios of 2.98 (95% CI, 1.02–8.42) and 6.93 (95% CI, 2.64–18.18) respectively. Age, lower albumin, lower BMI, and HCC were noted to be risk factors for BI in compensated cirrhotic patients. NSBB use showed a trend towards reduced risk of developing BI.

Conclusions

The development of CSPH has profound effects on the natural history of ACLD, and much research has been undertaken to understand factors predicting decompensation. Although HVPG, both at baseline and change over time or in response to pharmacotherapy, remains the gold standard, there is an unmet need to identify noninvasive surrogate makers of CSPH and decompensation. Liver and spleen stiffness are promising in this regard, although a lack of large, controlled studies including different etiologies with extended follow-up prevents universal adoption of these tools.

Acknowledgement Jaime Bosch supported in part by Stiftung für Leberkrankheiten, Bern, Switzerland.

References

1. D'Amico G, Garcia-Tsao G, Pagliaro L. Natural history and prognostic indicators of survival in cirrhosis: a systematic review of 118 studies. J Hepatol. 2006;44:217–31.
2. Newby DE, Hayes PC. Hyperdynamic circulation in liver cirrhosis: not peripheral vasodilatation but 'splanchnic steal'. QJM. 2002;95:827–30.
3. Wiest R, Groszmann RJ. The paradox of nitric oxide in cirrhosis and portal hypertension: too much, not enough. Hepatology. 2002;35:478–91.
4. Groszmann RJ, Wongcharatrawee S. The hepatic venous pressure gradient: anything worth doing should be done right. Hepatology. 2004;39:280–2.
5. Villanueva C, Albillos A, Genesca J, et al. Development of hyperdynamic circulation and response to beta-blockers in compensated cirrhosis with portal hypertension. Hepatology. 2016;63:197–206.
6. Groszmann RJ, Garcia-Tsao G, Bosch J, et al. Beta-blockers to prevent gastroesophageal varices in patients with cirrhosis. N Engl J Med. 2005;353:2254–61.
7. Ripoll C, Groszmann R, Garcia-Tsao G, et al. Hepatic venous pressure gradient predicts clinical decompensation in patients with compensated cirrhosis. Gastroenterology. 2007;133:481–8.
8. Villanueva C, Albillos A, Genesca J, et al. Beta-blockers to prevent decompensation of cirrhosis in patients with clinically significant portal hypertension (PREDESCI): a randomized, double-blind, placebo-controlled, multicentre trial. Lancet. 2019;393:1597–608.
9. Hernandez-Gea V, Aracil C, Colomo A, et al. Development of ascites in compensated cirrhosis with severe portal hypertension treated with beta-blockers. Am J Gastroenterol. 2012;107:418–27.
10. Berzigotti A, Garcia-Tsao G, Bosch J, et al. Obesity is an independent risk factor for clinical decompensation in patients with cirrhosis. Hepatology. 2011;54:555–61.
11. Berzigotti A, Rossi V, Tiani C, et al. Prognostic value of a single HVPG measurement and Doppler-ultrasound evaluation in patients with cirrhosis and portal hypertension. J Gastroenterol. 2011;46:687–95.
12. Lens S, Rincon D, Garcia-Retortillo M, et al. Association between severe portal hypertension and risk of liver decompensation in patients with hepatitis C, regardless of response to antiviral therapy. Clin Gastroenterol Hepatol. 2015;13:1846–1853.e1.
13. Reiberger T, Ulbrich G, Ferlitsch A, et al. Carvedilol for primary prophylaxis of variceal bleeding in cirrhotic patients with hemodynamic non-response to propranolol. Gut. 2013;62:1634–41.
14. Rincon D, Lo Iacono O, Tejedor M, et al. Prognostic value of hepatic venous pressure gradient in patients with compensated chronic hepatitis C-related cirrhosis. Scand J Gastroenterol. 2013;48:487–95.
15. Ripoll C, Lastra P, Rincon D, et al. Comparison of MELD, HVPG, and their changes to predict clinically relevant endpoints in cirrhosis. Scand J Gastroenterol. 2012;47:204–11.
16. Villanueva C, Aracil C, Colomo A, et al. Acute hemodynamic response to beta-blockers and prediction of long-term outcome in primary prophylaxis of variceal bleeding. Gastroenterology. 2009;137:119–28.

17. D'Amico G, Perricone G. Prediction of decompensation in patients with compensated cirrhosis: does etiology matter? Curr Hepatol Rep. 2019;18:144–56.
18. Sanyal AJ, Harrison SA, Ratziu V, et al. The natural history of advanced fibrosis due to nonalcoholic steatohepatitis: data from the simtuzumab trials. Hepatology. 2019;70:1913–27.
19. Bassegoda O, Olivas P, Turco L, et al. Decompensation in advanced non-alcoholic fatty liver disease may occur at lower hepatic venous pressure gradient levels than in patients with viral disease. Clin Gastroenterol Hepatol. 2021; https://doi.org/10.1016/j.cgh.2021.10.023.
20. Jindal A, Bhardwaj A, Kumar G, et al. Clinical decompensation and outcomes in patients with compensated cirrhosis and a hepatic venous pressure gradient >/=20 mmHg. Am J Gastroenterol. 2020;115:1624–33.
21. La Mura V, Abraldes JG, Raffa S, et al. Prognostic value of acute hemodynamic response to i.v. propranolol in patients with cirrhosis and portal hypertension. J Hepatol. 2009;51:279–87.
22. Mandorfer M, Kozbial K, Schwabl P, et al. Changes in hepatic venous pressure gradient predict hepatic decompensation in patients who achieved sustained Virologic response to interferon-free therapy. Hepatology. 2020;71:1023–36.
23. Turco L, Villanueva C, La Mura V, et al. Lowering portal pressure improves outcomes of patients with cirrhosis, with or without ascites: a meta-analysis. Clin Gastroenterol Hepatol. 2020;18:313–327.e6.
24. Abraldes JG, Garcia-Tsao G. Simple clinical tools to predict decompensation in patients with compensated cirrhosis: an unmet need. Clin Gastroenterol Hepatol. 2019;17:2179–81.
25. Guha IN, Harris R, Berhane S, et al. Validation of a model for identification of patients with compensated cirrhosis at high risk of decompensation. Clin Gastroenterol Hepatol. 2019;17:2330–2338.e1.
26. Colecchia A, Colli A, Casazza G, et al. Spleen stiffness measurement can predict clinical complications in compensated HCV-related cirrhosis: a prospective study. J Hepatol. 2014;60:1158–64.
27. Sebastiani G, Alshaalan R, Wong P, et al. Prognostic value of non-invasive fibrosis and steatosis tools, hepatic venous pressure gradient (HVPG) and histology in nonalcoholic steatohepatitis. PLoS One. 2015;10:e0128774.
28. Kitson MT, Roberts SK, Colman JC, et al. Liver stiffness and the prediction of clinically significant portal hypertension and portal hypertensive complications. Scand J Gastroenterol. 2015;50:462–9.
29. Merchante N, Rivero-Juarez A, Tellez F, et al. Liver stiffness predicts clinical outcome in human immunodeficiency virus/hepatitis C virus-coinfected patients with compensated liver cirrhosis. Hepatology. 2012;56:228–38.
30. Harrison SA, Abdelmalek MF, Caldwell S, et al. Simtuzumab is ineffective for patients with bridging fibrosis or compensated cirrhosis caused by nonalcoholic steatohepatitis. Gastroenterology. 2018;155:1140–53.
31. Calzadilla-Bertot L, Vilar-Gomez E, Wong VW, et al. ABIDE: an accurate predictive model of liver decompensation in patients with nonalcoholic fatty liver-related cirrhosis. Hepatology. 2021;73(6):2238–50.
32. Younes R, Caviglia GP, Govaere O, et al. Long-term outcomes and predictive ability of non-invasive scoring systems in patients with non-alcoholic fatty liver disease. J Hepatol. 2021;75(4):786–94.
33. Vizzutti F, Arena U, Romanelli RG, et al. Liver stiffness measurement predicts severe portal hypertension in patients with HCV-related cirrhosis. Hepatology. 2007;45:1290–7.
34. Pons M, Augustin S, Scheiner B, et al. Noninvasive diagnosis of portal hypertension in patients with compensated advanced chronic liver disease. Am J Gastroenterol. 2021;116:723–32.
35. Robic MA, Procopet B, Metivier S, et al. Liver stiffness accurately predicts portal hypertension-related complications in patients with chronic liver disease: a prospective study. J Hepatol. 2011;55:1017–24.
36. Merchante N, Téllez F, Rivero-Juárez A, et al. Progression of liver stiffness predicts clinical events in HIV/HCV-coinfected patients with compensated cirrhosis. BMC Infect Dis. 2015;15:557.

37. Perez-Latorre L, Sanchez-Conde M, Rincon D, et al. Prediction of liver complications in patients with hepatitis C virus-related cirrhosis with and without HIV coinfection: comparison of hepatic venous pressure gradient and transient elastography. Clin Infect Dis. 2014;58:713–8.
38. Petta S, Sebastiani G, Viganò M, et al. Monitoring occurrence of liver-related events and survival by transient elastography in patients with nonalcoholic fatty liver disease and compensated advanced chronic liver disease. Clin Gastroenterol Hepatol. 2021;19:806–815.e5.
39. Eaton JE, Sen A, Hoodeshenas S, et al. Changes in liver stiffness, measured by magnetic resonance elastography, associated with hepatic decompensation in patients with primary sclerosing cholangitis. Clin Gastroenterol Hepatol. 2020;18:1576–83. e1
40. Osman KT, Maselli DB, Idilman IS, et al. Liver stiffness measured by either magnetic resonance or transient elastography is associated with liver fibrosis and is an independent predictor of outcomes among patients with primary biliary cholangitis. J Clin Gastroenterol. 2021;55:449–57.
41. Gidener T, Ahmed OT, Larson JJ, et al. Liver stiffness by magnetic resonance elastography predicts future cirrhosis, decompensation, and death in NAFLD. Clin Gastroenterol Hepatol. 2021;19(9):1915–1924.e6.
42. Han MAT, Vipani A, Noureddin N, et al. MR elastography-based liver fibrosis correlates with liver events in nonalcoholic fatty liver patients: a multicenter study. Liver Int. 2020;40:2242–51.
43. Colecchia A, Montrone L, Scaioli E, et al. Measurement of spleen stiffness to evaluate portal hypertension and the presence of esophageal varices in patients with HCV-related cirrhosis. Gastroenterology. 2012;143:646–54.
44. Abraldes JG, Bureau C, Stefanescu H, et al. Noninvasive tools and risk of clinically significant portal hypertension and varices in compensated cirrhosis: the "Anticipate" study. Hepatology. 2016;64:2173–84.
45. Costa D, Simbrunner B, Jachs M, et al. Systemic inflammation increases across distinct stages of advanced chronic liver disease and correlates with decompensation and mortality. J Hepatol. 2021;74:819–28.
46. Villanueva C, Albillos A, Genesca J, et al. Bacterial infections adversely influence the risk of decompensation and survival in compensated cirrhosis. J Hepatol. 2021;75(3):589–99.

Preventing (First) Decompensation: Consensus Statements of Panel 5

36

Jaime Bosch, Cristina Ripoll, Rafael Bañares, Vincenza Calvaruso, Alessandra Dell'Era, Susana Gomes Rodrigues, Diraj Tripathi, and Càndid Vilanueva

J. Bosch (✉)
University Clinic for Visceral Surgery and Medicine, Inselspital-Bern, University Hospital, University of Bern, Bern, Switzerland

Department of Biomedical Research, University of Bern, Bern, Switzerland
e-mail: jaime.bosch@dbmr.unibe.ch

C. Ripoll
Internal Medicine IV, Universitätsklinikum, Friedrich-Schiller University, Jena, Germany
e-mail: Cristina_ripoll@med.uni-jena.de

R. Bañares
Complutense University, Madrid, Spain

Hospital General Universitario Gregorio Marañón, Madrid, Spain

CIBERehd, Madrid, Spain
e-mail: rbanares@ucm.es

V. Calvaruso
Gastroenterology and Hepatology Unit, Department of Health Promotion Sciences, Maternal and Infant Care, Internal Medicine and Medical Specialities (PROMISE), University of Palermo, Palermo, Italy
e-mail: Vincenza.calvaruso@unipa.it

A. Dell'Era
"L. Sacco" Department of Biomedical and Clinical Sciences, University of Milan, Milan, Italy

Gastroenterology and Endoscopy Unit, ASST Fatebenfratelli Sacco, Milan, Italy
e-mail: Alessandra.dellera@unimi.it

S. G. Rodrigues
Department of Biomedical Research, University of Bern, Bern, Switzerland

Hepatology, University Clinic for Visceral Surgery and Medicine, Inselspital Bern University Hospital, University of Bern, Bern, Switzerland
e-mail: susana.GomesRodrigues@insel.ch

R. de Franchis (ed.), *Portal Hypertension VII*,
https://doi.org/10.1007/978-3-031-08552-9_36

D. Tripathi
Liver Unit, University Hospitals, Birmingham NHS Foundation Trust, Birmingham, UK

Institute of Immunology and Immunotherapy, University of Birmingham, Birmingham, UK
e-mail: D.Tripathi@bham.ac.uk

C. Vilanueva
Gastrointestinal Bleeding Unit, Department of Gastroenterology, Hospital de Sant Pau, Autonomous University of Barcelona, Barcelona, Spain

CIBERehd, Barcelona, Spain
e-mail: cvillanueva@santpau.cat

5.1 Compensated cirrhosis is defined by the absence of present or past complications of cirrhosis. The transition from compensated to decompensated cirrhosis leads to increased mortality risk." (A1) (New)
5.2 Compensated cirrhosis can be divided into two stages, based on the absence or presence of clinically significant portal hypertension (CSPH). Patients with CSPH have an increased risk of decompensation. The goal of treatment in compensated cirrhosis is to prevent complications that define decompensation. (A1) (Changed)
5.3 Prevention of decompensation is especially relevant in compensated patients with CSPH and/or esophageal or gastric varices due to their higher risk of developing decompensation. (B1) (New)
5.4 The events that define decompensation in a compensated patient are overt ascites (or pleural effusion with increased SAAG (> 1.1 g/dL), overt hepatic encephalopathy (West Haven grading >= II), and variceal bleeding. (B1) (New)
5.5 Other relevant liver-related events in compensated cirrhosis are the development of superimposed liver injury (see statement 5.12)/ACLF and hepatocellular carcinoma. (B1) (New)
5.6 Insufficient data are available regarding whether a minimal amount of ascites only detected in imaging procedures, minimal hepatic encephalopathy, and occult bleeding from portal hypertensive gastroenteropathy can be considered as decompensation." (D1) (New)
5.7 Limited data suggest that jaundice alone (in non-cholestatic etiologies) may be the first manifestation of cirrhosis in a minority of patients; however, its definition, whether it should be considered true first decompensation or if it reflects superimposed liver injury/ACLF in compensated cirrhosis requires further research. (D1) (New)
5.8 Non-hepatic comorbidities are frequent in patients with compensated cirrhosis, can adversely impact prognosis, and should be specifically dealt with. (A1) (Changed)
5.9 There is insufficient data to draw definitive conclusions on the impact of sarcopenia and frailty on the natural history of compensated cirrhosis." (D1) (New)
5.10 Bacterial infections are frequent in compensated patients with CSPH, can lead to decompensation (ascites, variceal bleeding, hepatic encephalopathy), and, consequently, adversely affect natural history. (B1) (New)

5.11 There is insufficient data as to whether infections are frequent in compensated cirrhosis without CSPH and whether they may per se impact prognosis. (D1) (New)

5.12 Superimposed liver damage, such as (acute) alcoholic hepatitis, acute viral hepatitis (HEV, HAV), HBV flares, or drug-induced liver injury can precipitate decompensation. (A1) (New)

5.13 Other factors such as HCC and major surgery can precipitate decompensation of cirrhosis in patients with CSPH. (B1) (New)

5.14 Treatment with nonselective beta-blockers (NSBBs) (propranolol, nadolol, or carvedilol) should be considered for the prevention of decompensation in patients with CSPH. (B1) (New)

5.15 Carvedilol is the preferred NSBB in compensated cirrhosis, since it is more effective in reducing HVPG (A1), has a tendency toward greater benefit to prevent decompensation and toward better tolerance than traditional NSBBs, and has shown an improvement in survival (B1) compared to no active therapy in compensated patients with CSPH. (Changed)

5.16 The decision to treat with NSBBs should be taken when clinically indicated, independent of the possibility of measuring HVPG. (B2) (Unchanged)

5.17 Patients with compensated cirrhosis who are on NSBB for the prevention of decompensation do not need a screening endoscopy for the detection of varices, since endoscopy will not change management. (B2) (New)

5.18 There is no evidence that endoscopic therapies such as EBL or glue might prevent ascites or hepatic encephalopathy. (D1) (New)

5.19 In compensated patients with high-risk varices who have contraindications or intolerance to NSBB, endoscopic band ligation is recommended to prevent first variceal bleeding. (A1) (Changed)

5.20 There is no indication at present to use NSBB in patients without CSPH. (A1) (Unchanged)

5.21 Although a single study suggested that cyanoacrylate injection is more effective than propranolol in preventing first bleeding in patients with large gastroesophageal varices type 2 or isolated gastric varices type 1, there were no differences in survival. However, NSBB is indicated in these patients to prevent decompensation (B1). Further studies are required in these patients using new therapeutic approaches in addition to NSBBs. (D1) (Changed)

5.22 There is no indication at present for BRTO/BATO/BARTO/TIPS in primary prophylaxis of gastric variceal bleeding in compensated patients. (D1) (New)

Research Agenda

- Competing risks from comorbidities should be taken into account in future studies on compensated cirrhosis. Impact of early detection and treatment of comorbidities.
- Impact of sarcopenia and frailty (and of its treatment) on prognosis and mortality of compensated cirrhosis.

- Prognostic significance of the sole presence of minimal ascites only detected in imaging procedures, minimal hepatic encephalopathy, and chronic bleeding from portal hypertensive gastroenteropathy.
- Prognostic significance of the sole presence of jaundice in compensated cirrhosis, and its definition.
- Role of statins to prevent decompensation.
- Impact of sole bacterial infection in compensated cirrhosis on natural history. Impact of nonbacterial infections in compensated cirrhosis.
- Impact of vaccination (pneumococcal, hemophilus, influenza, coronavirus) on the natural history of compensated cirrhosis.
- Prevention of bacterial infections in patients with CSPH and its impact on the incidence of decompensation.
- Factors predicting which infections will give rise to decompensation and/or worsen prognosis.

Part VIII

Clinical Settings 2: Acute Variceal Bleeding

General Management of Acute Variceal Bleeding

37

Àngels Escorsell

Acute variceal bleeding (AVB) is a severe complication of cirrhosis; its prognosis has improved over the years, due to both, the efficacy of etiological therapies and preventive measures, and the improvement of the general management of the patients in this critical situation. Previous Baveno Consensus Conferences established the basis of this general management including hemodynamic targets, transfusion policies, use of antibiotics and vasoactive drugs, and ways to perform an accurate and safe diagnostic endoscopy. We review here the new recent data which accumulated since the last Baveno in 2015 in patients with cirrhosis presenting AVB.

Management of AVB was clearly established at Baveno Consensus Conferences. All the recommendations introduced at previous Baveno conferences led to a marked decrease in mortality related to AVB. In fact, varices "per se" are not a determinant of a patient's survival as shown by the study of Ardevol demonstrating that the origin of acute bleeding (gastroesophageal varices or peptic ulcer) did not influence the survival of cirrhotic patients, being 81% in variceal bleeders and 83% in bleeders from an acute peptic ulcer at 45 days from acute bleeding [1]. This paper, in agreement with Baveno VI conclusions, stated that the hepatic failure (measured by Child-Pugh or MELD score) and its complications (mainly presence of acute kidney injury, acute-on-chronic liver failure, shock, or presence of bacterial infections) were the real determinants of patient's survival or death. Of course, when looking at AVB, other factors, such as active bleeding, portal pressure, and the presence of hepatocellular carcinoma have an important prognostic value as already stated in previous Baveno meetings.

Unfortunately, although individual guideline adherence to previous Baveno recommendations was correct (endoscopy <12 h in 80% of the cases, antibiotics in 85%, band ligation in 79%, and vasoactive drugs in 91%), only 63% of the patients received care that was adherent to all indicated criteria [2].

À. Escorsell (✉)
Digestive Diseases Unit, FGS Hospital de la Santa Creu i Sant Pau, Barcelona, Spain

CiberEHD, Madrid, Spain
e-mail: mescorsell@santpau.cat

R. de Franchis (ed.), *Portal Hypertension VII*,
https://doi.org/10.1007/978-3-031-08552-9_37

Orotracheal Intubation in AVB

In Baveno VI, experts agreed that in patients with altered consciousness, endoscopy should be performed with the protection of the airway.

Notably, new data further confirm that orotracheal intubation must be indicated with caution, except in cases with severe depression of consciousness such as patients with grade 3–4 hepatic encephalopathy or in those patients vomiting blood. Chaudhuri presented a meta-analysis on upper endoscopy in all patients (including more than 5600 patients) showing that prophylactic intubation significantly increased the risk of death in patients having AVB [3]. Moreover, the meta-analysis performed by Almashhrawi [4] on upper GI bleeding also concluded that pneumonia within 48 h is more likely in upper GI bleeding patients who received prophylactic endotracheal intubation prior to endoscopy.

Last, the recent study by Martinez on bacterial infections in patients with AVB in the era of antibiotic prophylaxis showed that 1/5th of the patients presented bacterial infections despite the use of antibiotic prophylaxis (received by 94% of the patients) [5]. Half of these infections were of respiratory origin and their incidence was directly related not only to the liver function (Child-Pugh score) but also to manipulations of the oro-nose-tracheal area, such as orotracheal intubation to perform upper endoscopy, placement of a nasogastric tube, or balloon tamponade and grade 3–4 HE. Remarkably, the indication for the procedure was severe HE for only 21% of patients. Thus, the study suggests that prophylactic intubation for endoscopy may confer an increased risk of respiratory infection in patients with AVB [5]. Limitations of all those studies include their design and the fact that indications for intubation were not standardized.

Therefore, and according to the limitations of the previous studies, we can conclude that intubation is recommended before endoscopy in patients with altered consciousness (severe HE) and those actively vomiting blood for practical reasons. Once intubated, extubation should be performed as quickly as possible after endoscopy.

Pharmacological Treatment

In suspected variceal bleeding, vasoactive drugs (terlipressin, somatostatin, octreotide) should be started as soon as possible and continued for 2–5 days. This different period of time, 2 to 5 days, results from studies comparing the different duration of vasoactive therapy according to the functional hepatic reserve.

Bacterial Infections

Presently, all patients presenting with AVB need to receive antibiotic prophylaxis as soon as possible.

At Baveno VI consensus conference, it was suggested that prophylactic antibiotics may be avoided in Child-Pugh A patients, which have a very low incidence of bacterial infections and death. Indeed, Tandon and coworkers investigated this issue in a retrospective study analyzing 381 patients with cirrhosis admitted in two tertiary care hospitals in Canada [6]. The proportion of patients who received antibiotics was only 54%. Overall, antibiotic therapy was associated with a lower risk of infection and death. Nevertheless, Child-Pugh A patients displayed a very low incidence of infections and mortality, which was not different in patients under antibiotics or not. In Child-Pugh B patients, the rate of infections was 6% (vs. 14% in patients not receiving antibiotic prophylaxis) with no significant differences in mortality. Finally, the administration of antibiotics in Child-Pugh C patients was associated with a decrease higher than 50% in the incidence of both, infections and death.

The recent study by Martinez et al. analyzed the incidence and characteristics of bacterial infections in more than 1500 patients with AVB receiving antibiotic prophylaxis [5]. Notably, the most frequent bacterial infection was respiratory (almost 50%) occurring early after admission (median: 3 days), and its development was independently associated with Child-Pugh C, grade III-IV HE, orotracheal intubation for endoscopy, nasogastric tube, or esophageal balloon tamponade.

The main drawback of antibiotic prophylaxis is the development of infections due to multidrug-resistant bacteria including Clostridium Difficile infection. In fact, the study by Martinez et al. [5] found that over 50% of the bacteria isolated were resistant to third-generation cephalosporins, used in ¾ of the patients. Hence, antibiotic prophylaxis could be restricted in the future to subpopulations at very high risk for infection; at present, it is mandatory to avoid antibiotic overuse, and early de-escalation policies should be always kept in mind. Moreover, the antibiotic prescription should be always in accordance with local resistance and antimicrobial policies.

Antibiotics are not free of other serious adverse events such as acute kidney injury due to interstitial nephritis associated with quinolones. In this sense, the use of non-absorbed antibiotics such as rifaximin appears to prevent infections while not causing antibiotic resistance due to its low bioavailability in blood after oral administration (<0.4%). Unfortunately, this low availability increases in the case of liver failure [5]. More studies are clearly needed because this strategy could be promising.

Nutrition

Thirty-eight percent of patients with cirrhosis and AVB may be classified as high-risk nutritional patients [7]. A possible side effect of early nutritional support is the postprandial increase in portal pressure. On the other hand, sarcopenia increases mortality, bacterial infections, hyperammonemia, HE, and length of stay after liver transplantation [7].

Sidhu performed a study on AVB treated by endoscopic banding ligation to assess if the very early administration of diet (starting with liquids 1 h after therapy and solid diet after 4 h) has an impact on the outcome of these patients as compared to the slow reintroduction of diet (starting at 4 h with liquids and completing solid diet at 72 h) [8]. It was concluded that early feeding was safe, provided better nutrition, and resulted in a lower incidence of infections in bleeders compared to delayed feeding. More studies are needed to confirm these results.

According to previous studies, nutrition in cirrhotic patients must be normocaloric (35–45 Kcal/g) and ensure energy adequacy and protein intake (1.2–1.5 g/kg daily) and increase of portal pressure should be counterbalanced by the concomitant use of vasoactive drugs [9].

Use of Proton Pump Inhibitors (PPI) in AVB

Acute upper GI bleeding in cirrhotic patients may be due to either peptic lesions or varices. Before endoscopic identification of the bleeding source, high dose omeprazole is indicated in all patients to facilitate endoscopy [10]. PPIs should be discontinued as soon as possible after the diagnosis of AVB. Indeed, PPIs alter the gut flora leading to a gut dysbiosis, and patients with decompensated cirrhosis have increased intestinal permeability and decreased hepatic clearance of PPI, predisposing them to increase the gut dysbiosis.

There are evidence that the prolonged use of PPI (at least more than 28 days) would be associated with an increased incidence of HE [11, 12], either overt or minimal, as well as an increased risk of further decompensation, mainly SBP [13]. More prolonged administrations (>90 days) would cause significant increases in long-term mortality [13, 14]. In addition, the results of previous studies suggest that the risk of hepatic decompensation increases with the dose.

These pieces of evidence lead to the recommendation to regularly review the need for PPI in patients with liver cirrhosis in order to stop when there are no indications for their use. Moreover, if PPI is indicated, the dosage should be reduced to the lowest possible dose.

References

1. Ardevol A, Ibañez-Sanz G, Profitos J, Aracil C, Castellvi JM, Alvarado E, et al. Survival of patients with cirrhosis and acute peptic ulcer bleeding compared with variceal bleeding using current first-line therapies. Hepatology. 2018;67(4):1458–71.
2. Tapper EB, Friderici J, Borman ZA, Alexander J, Bonder A, Nuruzzaman N, et al. A multicenter evaluation of adherence to 4 major elements of the Baveno guidelines and outcomes for patients with acute variceal hemorrhage. J Clin Gastroenterol. 2018;52(2):172–7.
3. Chaudhuri D, Bishay K, Tandon P, Trivedi V, James PD, Kelly EM, et al. Prophylactic endotracheal intubation in critically ill patients with upper gastrointestinal bleed: a systematic review and meta-analysis. JGH Open. 2020;4(1):22–8.
4. Almashhrawi AA, Rahman R, Jersak ST, Asombang AW, Hinds AM, Hammad HT, et al. Prophylactic tracheal intubation for upper GI bleeding: a meta-analysis. World J Metaanal. 2015;3(1):4–10.

5. Martínez J, Hernández-Gea V, Rodríguez-de-Santiago E, Téllez L, Procopet B, Giráldez Á, et al. Bacterial infections in patients with acute variceal bleeding in the era of antibiotic prophylaxis. J Hepatol. 2021;75(2):342–50.
6. Tandon P, Abraldes JG, Keough A, Bastiampillai R, Jayakumar S, Carbonneau M, et al. Risk of bacterial infection in patients with cirrhosis and acute variceal hemorrhage, based on child-pugh class, and effects of antibiotics. Clin Gastroenterol Hepatol. 2015;13(6):1189–1196.e1182.
7. Tsai MH, Huang HC, Peng YS, Chen YC, Tian YC, Yang CW, et al. Nutrition risk assessment using the modified NUTRIC score in cirrhotic patients with acute gastroesophageal variceal bleeding: prevalence of high nutrition risk and its independent prognostic value. Nutrients. 2019;11(9):2152.
8. Sidhu SS, Goyal O, Singh S, Kishore H, Chhina RS, Sidhu SS. Early feeding after esophageal variceal band ligation in cirrhotics is safe: randomized controlled trial. Dig Endosc. 2019;31(6):646–52.
9. Bischoff SC, Bernal W, Dasarathy S, Merli M, Plank LD, Schütz T, et al. ESPEN practical guideline: clinical nutrition in liver disease. Clin Nutr. 2020;39(12):3533–62.
10. Lau JY, Leung WK, Wu JC, Chan FK, Wong VW, Chiu PW, et al. Omeprazole before endoscopy in patients with gastrointestinal bleeding. N Engl J Med. 2007;356(16):1631–40.
11. Nardelli S, Gioia S, Ridola L, Farcomeni A, Merli M, Riggio O. Proton pump inhibitors are associated with minimal and overt hepatic encephalopathy and increased mortality in patients with cirrhosis. Hepatology. 2019;70(2):640–9.
12. Tsai CF, Chen MH, Wang YP, Chu CJ, Huang YH, Lin HC, et al. Proton pump inhibitors increase risk for hepatic encephalopathy in patients with cirrhosis in a population study. Gastroenterology. 2017;152(1):134–41.
13. Hung TH, Lee HF, Tseng CW, Tsai CC, Tsai CC. Effect of proton pump inhibitors in hospitalization on mortality of patients with hepatic encephalopathy and cirrhosis but no active gastrointestinal bleeding. Clin Res Hepatol Gastroenterol. 2018;42(4):353–9.
14. De Roza MA, Kai L, Kam JW, Chan YH, Kwek A, Ang TL, et al. Proton pump inhibitor use increases mortality and hepatic decompensation in liver cirrhosis. World J Gastroenterol. 2019;25(33):4933–44.

Risk Stratification and Prognostic Factors in Variceal Bleeding

38

David Patch

Risk stratification is the process of separating patient populations into different risk categories by combining individual prognostic factors, suitably weighted for their relative importance. This science has developed significantly and in the era of early TIPS and transplantation for Acute on Chronic Liver failure, has a direct clinical impact.

The various Baveno Meetings have reflected this-in Baveno 2, it was agreed that it was important to assess for risk of first bleeding-by endoscopy-but there was no consensus on risk factors for early bleeding and death, or late rebleeding and death [1].

In Baveno 4 (2005), a whole chapter exhaustively explored predictive models in portal hypertension [2]. The relative significance of the decompensation episode (bleeding, jaundice, encephalopathy, and ascites) was described, and a literature review identified >20 prognostic scores, 93 prognostic studies, and 172 candidate variables. The Child-Pugh and MELD scores were the only ones identified as being used in clinical practice [3]. A survey identified variceal haemorrhage as the area where there was most satisfaction with currently available predictive models although this was still low at 26%. The authors rightly pointed out that Child Pugh and Meld scores were derived from patients who had experienced upper GI haemorrhage, and that this may be one of the reasons that they remain valid scoring tools in this area. However, the complexity of the bleeding episode means that these are not the only factors-how is it that a patient who was C-P A pre bleed becomes C-P C when ventilated on ITU post-bleed? Clearly, the severity of the bleeding episode then comes into play-and markers of liver ischaemia (AST rise), the volume of blood loss as measured by transfusion requirement, and anatomical changes such as portal vein thrombosis have also been shown to be prognostically relevant [4].

D. Patch (✉)
Royal Free Hospital, London, UK
e-mail: david.patch@nhs.net

R. de Franchis (ed.), *Portal Hypertension VII*,
https://doi.org/10.1007/978-3-031-08552-9_38

Identification of prognostic markers, and in many cases, prognostic scores, have been an inevitable outcome of just about every clinical paper, and yet they rarely informed decision making directly other than perhaps in areas of futility. Fast forward to Baveno 6 (2015) and the importance of prognostic indicators has hit "prime time"-i.e. clinical decisions are made on the basis of these indicators [5]. Two papers reported that early TIPS placement could improve survival in patients with pre-identified poor prognostic markers -HVPG in the paper by Monescilio [6], and CP score and endoscopic evidence of active bleeding in the paper by Garcia Pagan [7]. These are significant papers not just for their practice-changing outcomes, but also for their demonstration of why identifying prognostic markers is actually important, and not just an academic exercise.

During Baveno 6, there was also a lengthy discussion regarding the time intervals examined in prognostic studies, with most looking at 6-week survival. This highlighted an anomaly- most patients die earlier than 6 weeks-and so outcome indicators are required that are accurate much earlier in the bleed time period. Furthermore, it is not uncommon for patients to die not of bleeding, but of multi-organ failure. Interventions to control bleeding may not be the same as those to manage multi-organ failure, and thus there was also identified a need for more nuanced prognostic indices. This is difficult when the definition of treatment failure has still not been clearly defined. Nonetheless, mortality was proposed to be **the** key major endpoint.

2021

Six years and one pandemic later, the role of pre-emptive TIPS continues to be debated, following the publication of two further trials with conflicting results. What is not in doubt is that the concept of Acute on Chronic Liver Failure (ACLF) has become an established entity-and whilst it is not a new disease, it has focussed the mind as liver transplantation is now proposed as a therapeutic option for those patients who have developed ACLF [8]. Variceal bleeding is not an indication for transplantation in Europe or the USA but a patient who has had a variceal bleed and is on ITU with ACLF is now a candidate for transplantation.

But before one explores these, it is important to remember that most patients with variceal bleeding will present to a district general hospital, where there may not be specialist services, and where the patient will be admitted as a GI haemorrhage. Patients will be usually risk assessed according to non-liver specific scores and whilst these have accuracy, their clinical utility is more questionable as a specific intervention is not defined by any threshold value. The study by Motola-Kuba et al. compared the prognostic accuracy of different non-variceal and liver-specific scoring systems in cirrhotic patients [9]. MELD and AIMS65 best-predicted mortality, whilst Rockall and Glasgow -Blatchford best-predicted re-bleeding.

There also remains the issue of variability in care. In the UK, routine airway protection at endoscopy in a patient with upper GI bleeding—whilst indicated in guidelines- continues to be haphazardly applied. Aspiration has a profound impact on the patient's outcome, yet most studies do not factor this into prognostication—ventilated for airway protection is very different from ventilated for respiratory support. The subtleties of this element of care were exemplified in the study by Hermie et al. in which 32 patients were referred for early TIPS [10]. Haemodynamic instability at admission and a MELD score of ≥19 resulted in 6-week mortality of 78%. The paper does not describe the cause of cardiovascular instability—some patients may have been bleeding or had become septic—but it questioned the role of early TIPS in patients with high MELD scores and shock. One must also ask whether a TIPS in such a patient is a "salvage" as opposed to an "early" TIPS.

Identifying the patient who is unlikely to benefit from intervention is no less important—prolonged ITU stays are avoided, and futile interventions are prevented from occurring. A retrospective study of 144 consecutive salvage TIPS suggested that this procedure was futile in patients with a C-P score > 13 [11]. As CP scores will not change significantly over a 72-h period, this again provides some justification for the early TIPS C-P upper threshold of 13. In a retrospective study of 164 patients treated with salvage TIPS between 2007 and 2017, 6-week survival was <10% when MELD was ≥30 and lactate ≥12 mmol/L [12]. These are important papers, as they provide an evidence base for a "receiving" hospital to decline a transfer request on the basis of futility—recognizing this would be on a case-by-case basis.

The move to early TIPS has such a significant impact on clinical practice as well as care provision that it comes as little surprise that those prognostic markers utilized by Monescillo and Garcia Pagan have received much academic attention. Rudler et al. analyzed 219 prospectively collected patients with variceal bleeding-specifically to look at factors associated with mortality at 6 weeks [13]. There was substantial variability in the diagnosis of active bleeding at endoscopy, and this did not come out as a significant prognostic factor. Re-calibrated MELD score accurately predicted mortality, and the presence of hepatic encephalopathy was also a poor prognostic indicator. Mortality in Pugh's B was low at 7%. An accompanying editorial argues that the stop–start nature of variceal bleeding means that the presence of active variceal bleeding will vary over time, but its presence is still a poor prognostic marker [14]. The authors indicate that there remains a need for good quality, prospective prognostic studies.

The third and fourth randomized controlled trials of early TIPS enrolled Pugh B (≥7 [15] or ≥ 8 [16] and Pugh C < 14) and did not pre-specify the presence of active bleeding at endoscopy for the Pughs B. This followed a study from the Chinese group on identifying the optimal candidates for early TIPS-and whilst patients with CP B and active bleeding did appear to have a survival benefit with early TIPS, as per the Rudler study, the evaluation of active bleeding was associated with significant inter-observer error [17].

Depending on your viewpoint, either fuel or water is added to the debate by the addition of transplantation to the therapeutic options- not for variceal bleeding, but for those who have bled and have developed ACLF3. With respect to transplantation, the early papers proposing transplantation for ACLF have not identified the precipitating cause of the ACLF [8, 18]. The ITU survival paper from Kings however is noteworthy in that they found patients with GI bleeding as the primary cause of acute decompensation had better survival, and whilst three organ failures was a watershed, the GI bleed group with three organ failures still had a hospital survival rate of 50% [19]. I.e. is the variceal bleeding ACLF3 the same as the non-variceal ACLF3?

Looking forward, whilst HVPG remains the gold standard assessment of the severity of portal hypertension, its lack of uptake is a reflection of its "inconvenience". Virtual HVPG [20], and serological markers of portal hypertension (e.g. Von Willebrand factor [21]) are exciting areas of research, and it may be combinations of these techniques provide better identification of the patient who needs more than endoscopic management [22].

Conclusion

MELD and Child-Pugh remains valid tools for predicting outcomes, and for decision-making. The survival outcomes in Child-Pugh B are now so good that any intervention in this group requires prospective studies with large numbers of patients. Identifying, and preventing, the descent into multi-organ failure remains critical to good outcomes.

References

1. Pagliaro L, Franchis RD. Where were we. In: Franchis RD, editor. Portal Hypertension 2. Proceedings of The Second Baveno International Consensus Workshop on Definitions, Methadology and Therapeutic Strategies. Oxford: Blackwell Science Ltd; 1996. p. 2.
2. Garcia-Tsao G, et al. Predictive models in portal hypertension. In: Franchis RD, editor. Portal Hypertension 4. Proceedings of the Fourth Baveno International Consensus Workshop on Methadology of Diagnosis and Treatment. Oxford: Blackwell Publishing; 2006. p. 47–100.
3. D'Amico G, Licata A, Pasta L, Rizzuto G, D'Amico M, Mocciaro F. Prognostic indicators in cirrhosis: the role of Child classification. Overview of 93 published studies and validation in a prospective cohort of patients. Dig Liver Dis. 2005;37:S(29).
4. D'Amico G, De Franchis R. Upper digestive bleeding in cirrhosis. Post-therapeutic outcome and prognostic indicators. Hepatology. 2003;38(3):599–612.
5. Villanueva C, O'Beirne J, Bureau C, Tahbut D. A la carte treatment of acute variceal bleeding. In: Franchis RD, editor. Portal Hypertension 6. Proceedings of the SIxth Baveno Consensus Workshop: Stratifying Risk and Individualising Patient Care. Cham: Springer International Publishing; 2016. p. 261–5.
6. Monescillo A, et al. Influence of portal hypertension and its early decompression by TIPS placement on the outcome of variceal bleeding. Hepatology. 2004;40(4):793–801.
7. García-Pagán JC, et al. Early use of TIPS in patients with cirrhosis and variceal bleeding. N Engl J Med. 2010;362(25):2370–9.
8. Thuluvath PJ, et al. Liver transplantation in patients with multiple organ failures: feasibility and outcomes. J Hepatol. 2018;69(5):1047–56.
9. Motola-Kuba M, et al. Validation of prognostic scores for clinical outcomes in cirrhotic patients with acute variceal bleeding. Ann Hepatol. 2016;15(6):895–901.
10. Hermie L, et al. Model for end-stage liver disease score and hemodynamic instability as a predictor of poor outcome in early transjugular intrahepatic portosystemic shunt treatment for acute variceal hemorrhage. Eur J Gastroenterol Hepatol. 2018;30(12):1441–6.

11. Maimone S, et al. Predictors of re-bleeding and mortality among patients with refractory variceal bleeding undergoing salvage transjugular intrahepatic portosystemic shunt (TIPS). Dig Dis Sci. 2019;64(5):1335–45.
12. Walter A, Rudler M, Olivas P, Moga L, Trépo E, Robic MA, et al. Combination of model for end-stage liver disease and lactate predicts death in patients treated with salvage transjugular intrahepatic portosystemic shunt for refractory variceal bleeding. Hepatology. 2021;74(4):2085–101. https://doi.org/10.1002/hep.31913.
13. Rudler M, et al. Recalibrated MELD and hepatic encephalopathy are prognostic factors in cirrhotic patients with acute variceal bleeding. Liver Int. 2018;38(3):469–76.
14. D'Amico G, D'Amico M, Malizia G. Refining early-TIPS criteria requires good quality prognostic studies. Liver Int. 2018;38(3):412–4.
15. Lv Y, et al. Early TIPS with covered stents versus standard treatment for acute variceal bleeding in patients with advanced cirrhosis: a randomised controlled trial. Lancet Gastroenterol Hepatol. 2019;4(8):587–98.
16. Dunne PDJ, et al. Randomised clinical trial: standard of care versus early-transjugular intrahepatic porto-systemic shunt (TIPSS) in patients with cirrhosis and oesophageal variceal bleeding. Aliment Pharmacol Ther. 2020;52(1):98–106.
17. Lv Y, et al. Identifying optimal candidates for early TIPS among patients with cirrhosis and acute variceal bleeding: a multicentre observational study. Gut. 2019;68(7):1297–310.
18. Sundaram V, et al. Factors associated with survival of patients with severe acute-on-chronic liver failure before and after liver transplantation. Gastroenterology. 2019;156(5):1381–1391.e3.
19. McPhail MJ, et al. Increased survival for patients with cirrhosis and organ failure in liver intensive care and validation of the chronic liver failure-sequential organ failure scoring system. Clin Gastroenterol Hepatol. 2015;13(7):1353–1360.e8.
20. Qi X, et al. Virtual hepatic venous pressure gradient with CT angiography (CHESS 1601): a prospective multicenter study for the noninvasive diagnosis of portal hypertension. Radiology. 2019;290(2):370–7.
21. Zou Z, et al. von Willebrand factor as a biomarker of clinically significant portal hypertension and severe portal hypertension: a systematic review and meta-analysis. BMJ Open. 2019;9(8):e025656.
22. Qi X, et al. Emerging non-invasive approaches for diagnosis and monitoring of portal hypertension. Lancet Gastroenterol Hepatol. 2018;3(10):708–19.

Endoscopic Management: Classic and New Therapies

39

Marvin Ryou and Andres Cardenas

Acute esophageal and gastric Variceal Bleeding (AVB) is a dreaded complication in patients with portal hypertension and the first episode in a patient with cirrhosis constitutes a significant milestone in the progression of the hepatic disease with important implications for prognosis. Endoscopy by means of using Endoscopic Band Ligation (EBL) plays a key role in the management of AVB from esophageal varices. Patients with isolated gastric varices are best managed by injection of cyanoacrylate glue. Currently available methods, combined with vasoactive drugs and antibiotics, allow for the control of bleeding in 90% of cases within the first days of the index bleed.

Endoscopic Therapy for Acute Esophageal Variceal Bleeding

Endoscopy is a key aspect in the management of AVB because it confirms the diagnosis and allows therapy during the same session. Based on the current data, endoscopy for AVB is recommended within the first 12 h of presentation as

M. Ryou
Division of Gastroenterology, Hepatology and Endoscopy, Brigham and Women's Hospital, Harvard Medical School, Boston, MA, USA
e-mail: mryou@BWH.Harvard.edu

A. Cardenas (✉)
GI & Liver Transplant Unit, Institut de Malalties Digestives i Metaboliques, Hospital Clinic, University of Barcelona, Barcelona, Spain

Ciber de Enfermedades Hepáticas y Digestivas (CIBERehd), Barcelona, Spain

Institut d'Investigacions Biomèdiques August Pi-Sunyer (IDIBAPS), Barcelona, Spain
e-mail: acardena@clinic.cat

R. de Franchis (ed.), *Portal Hypertension VII*,
https://doi.org/10.1007/978-3-031-08552-9_39

overall 6-week and 3-month mortality is better in those patients in whom endoscopy and EBL are performed early [1]. In the absence of contraindications (i.e., QT prolongation), pre-endoscopy infusion of erythromycin as a measure to improve gastric emptying and visibility during endoscopy (250 mg IV 30–120 min before endoscopy) should be considered. An important consideration before endoscopy is choosing the adequate type of sedation as patients need to be fully sedated for the procedure to be successful. Intravenous sedation with propofol is safer and better tolerated than benzodiazepines with opiates. The two endoscopic methods available for AVB are endoscopic sclerotherapy (ES) and EBL. ES consists of the injection of a sclerosing agent [sodium morrhuate (5%), ethanolamine oleate (5%), or polidocanol (1%–2%)] into the variceal lumen or adjacent to it. This causes inflammation and thrombosis creating a scar over the site of varix. Although it is easy to perform, there are significant side effects related to the procedure, such as esophageal ulcers, strictures, substernal chest pain, fever, dysphagia, and bacteremia which may arise in up to 30%–55% of cases and can predispose to spontaneous bacterial peritonitis or distal abscesses [2]. Compared to endoscopic band ligation of varices, ES for AVB is associated with higher rebleeding rates and more adverse effects [3]. Clinical trials have shown that EBL is better than ES for all major outcomes including initial control of bleeding, recurrent bleeding, side effects, time to variceal eradication, and even survival [4].

Esophageal Varices: Endoscopic Band Ligation

The concept of using EBL for esophageal varices dates to the 1980s. The idea behind EBL relates to the anatomy and areas of venous return in the gastroesophageal junction, whereby ligation of varices aims to block this drainage. The veins in the palisade zone which extend proximal to the cardias into the lower esophagus are those predisposed to bleeding since this is an area between the portal and systemic circulation with no perforating veins from the submucosa allowing for drainage or decompression to periesophageal veins. Therefore, a method for obliterating these vessels is placing elastic bands on the varices in the distal esophagus. EBL consists of the placement of several elastic bands (range between 4 and 7) on the varices to occlude varix and cause thrombosis which eventually leads to necrosis of the mucosa. The bands fall off within 5–7 days leaving a shallow ulcer that heals and subsequently scars. There are several commercially available multiband devices that have between 6 and 7 preloaded bands. This allows for enough bands to be placed in a single session without having to repeatedly intubate the patient with the gastroscope. Band ligators allow the placement of elastic bands on varix after it is sucked into a clear plastic cylinder attached to the tip of the endoscope (Fig. 39.1). After the index diagnostic endoscopy is performed in a patient with AVB and the suspected varix is identified, the endoscope is withdrawn, and the ligation device is loaded. The device needs to be firmly attached to the shaft of the scope near the knobs

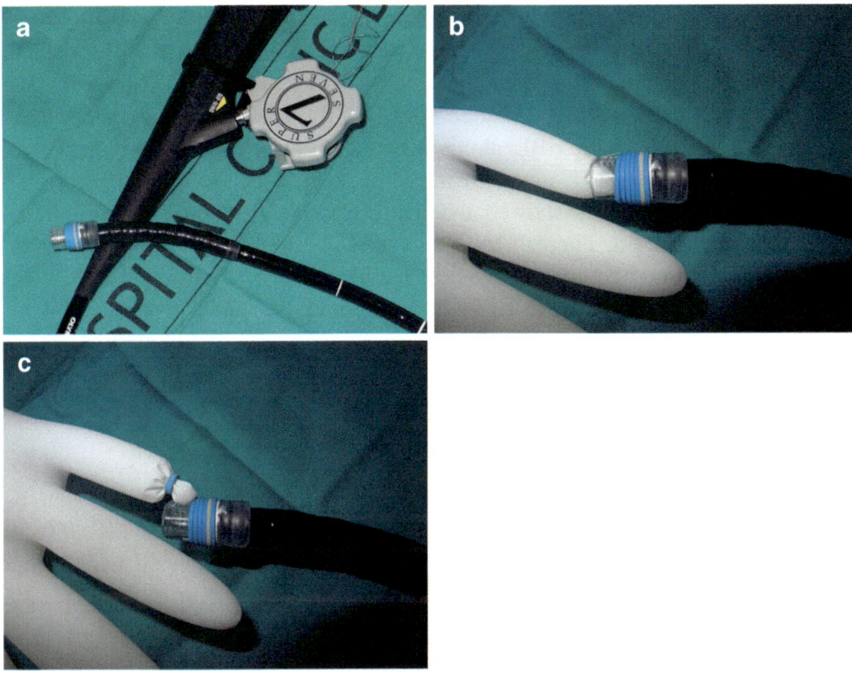

Fig. 39.1 (**a**) Band ligator placed on the shaft of the endoscope, the knob is attached to the working channel. (**b**) The proximal end of the endoscope has a cap with pre-mounted bands that allow for the suction of varix (in this case a glove is used). (**c**) Once the varix is suctioned into the cap the band is fired and the varix is ligated

(Fig. 39.1). After the varix is identified, the tip of the scope is pushed towards it and continuous suction applied so the mucosa of the varix fills the cap and causes a “red out” sign; at this point, the band can be fired and a click must be felt. Afterwards the scope should not be advanced distally to prevent dislodgement of the band. This is the reason why ligation should always commence in the most distal portion of the esophagus near the Gastroesophageal (GE) junction. Bands are applied in a spiral pattern progressing up the esophagus until all major columns of varices of the lower third of the esophagus (no more than 10 cm above the GE junction) are banded. If there is a limited view because of ongoing bleeding an option is to aggressively flush with water, perform suction and start placing bands at the GE junction. This reduces the heavy bleeding and further bands can be fired afterwards.

The procedure is not exempt from adverse events; these include transient dysphagia and chest pain which respond well to liquid analgesics (i.e., acetaminophen) as well as an oral suspension of antacids. Shallow ulcers at the site of bands are frequent and can bleed in up to 4%–5% of cases. If a patient bleeds due to an ulcer after EBL, ES may be performed. Another option is applying a hemostatic powder (Hemospray) to the bleeding site. This powder which is used for non-variceal gastrointestinal bleeding seems to be promising as a hemostatic technique for patients with bleeding portal gastropathy, variceal bleeding, and post EBL bleeding ulcers [5]. EBL is highly effective in the control of AVB with an immediate efficacy in 90% of cases. Several RCTs, that compared EBL and ES in AVB, have clearly shown that treatment with EBL along with vasoconstrictors is associated with higher efficacy, safety, and improved mortality than ES and vasoconstrictors. A meta-analysis of 36 RCTs showed that EBL was associated with a significant improvement in bleeding control (relative risk [RR] 1.08; 95% confidence interval [CI] 1.02–1.15) when compared to sclerotherapy and was associated with fewer adverse events [4]. Therefore, EBL is considered the endoscopic therapy of choice in AVB.

Gastric Varices

The prevalence of Gastric Varices (GV) in patients with cirrhosis is estimated to be 17%–20% [6]. Although less prevalent than esophageal varices (16%–45% at 3 years), hemorrhage from GV is often more severe and associated with higher mortality. Gastric varices are most commonly classified according to their location within the stomach: Gastroesophageal Varices (GOV) and Isolated Gastric Varices (IGV). GOV is further subdivided into GOV-1, which represents an extension of esophageal varices along the lesser curvature of the stomach, and GOV-2, which represent esophageal varices extending into the gastric fundus (Fig. 39.2). IGV is

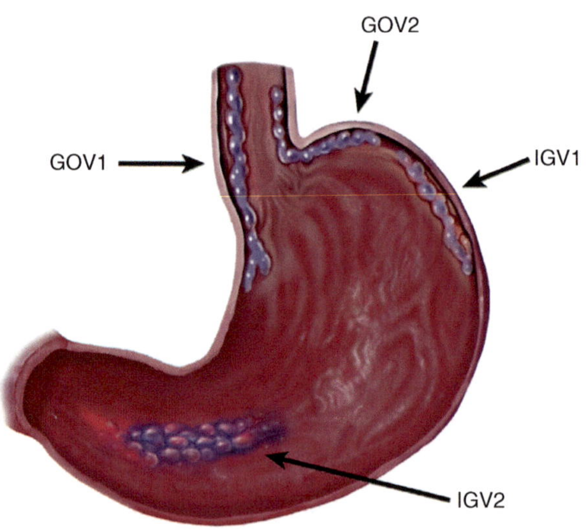

Fig. 39.2 Sarin's Classification of GV [originally from the AGA Institute Gastroslides-Cirrhosis and Portal Hypertension] [6]

further subdivided into IGV-1, which is located in the fundus, and IGV-2, which are ectopic varices located throughout the remainder of the stomach. This classification system was described by Sarin et al. [6], and has important clinical implications for the risk of bleeding and management. GOV1 represents 75%, GOV2 21%, IGV1 less than 2%, and IGV2 4% of all GV. IGV-1 and GOV-2, tend to result in significant hemorrhage (bleeding incidence of 78% and 55%, respectively), and therefore carry significant morbidity and mortality. Endoscopic therapy of GOV-1 tends to be similar to esophageal varices (i.e., banding) since they share the same venous anatomy. Besides the location in the stomach, the size of varices and the presence of stigmata contribute to bleeding risk. Endoscopy is recommended for definitive confirmation of GV bleeding, and endoscopic therapy is recommended for initial hemostasis. In cases of massive bleeding, balloon tamponade (Linton-Nachlas is preferred over Sengstaken–Blakemore tube, although either is acceptable) is recommended to stabilize the patient as a bridge to endovascular therapy (TIPS or BRTO).

Endoscopic Approach to GOV-1: Band or Glue

Due to their similarity to EV, GOV-1 tends to respond well to band ligation. Similar to EV, band ligation is initially performed along the gastric lesser curve (distal-most aspect) and marched proximally. Endoscopic obturation with cyanoacrylate glue (e.g., N-butyl-2-cyanoacrylate) may also be considered, particularly for larger GOV-1 for which band ligation would be difficult. Retrospective studies focusing on acutely bleeding GOV-1 have favored Direct Endoscopic Injection (DEI) of cyanoacrylate over band ligation. In general, however, comparisons of DEI of cyanoacrylate relative to other endoscopic therapies have not been stratified by GV type. Therefore, for GOV-1 a definitive recommendation of one endoscopic therapy over another cannot be made. Most centers will usually treat GOV-1 with band ligation given the availability and relative technical ease.

Endoscopic Approach to GOV-2 and IGV

DEI or freehand technique [7], of cyanoacrylate, has become the definitive endoscopic treatment for GOV2 and IGV-1, particularly for large mass-like varices for which band ligation would be impossible. Specific high-quality data on the use of endoscopic therapy for acute GV bleeding are limited, and in most RCTs half of the patients included in the trials had GOV 2 and IGV-1. Nevertheless, most uncontrolled series report a high rate of hemostasis (>90%) with DEI of cyanoacrylate. Notably, DEI has been shown to be superior to alcohol-based sclerotherapy and band ligation in terms of preventing both early and late rebleeding, as well as complications [8].

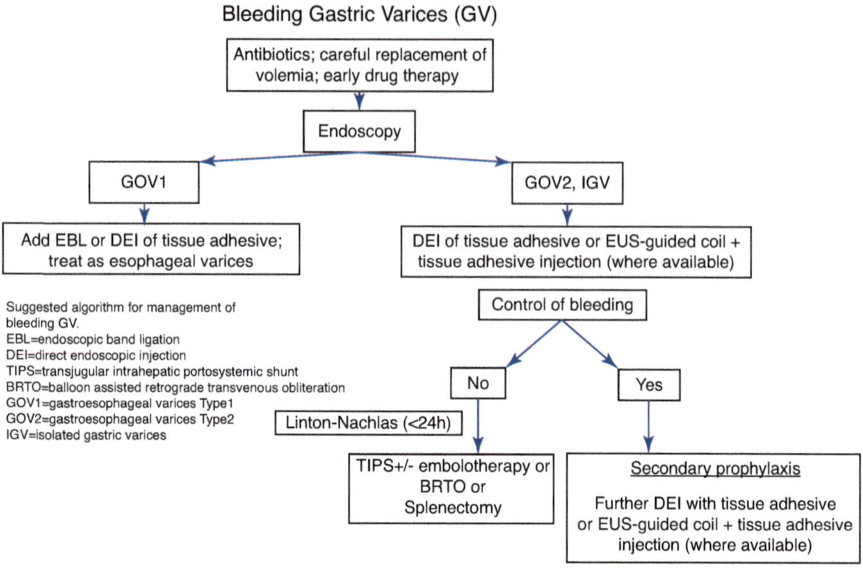

Fig. 39.3 Suggested Algorithm for Management of Bleeding Gastric Varices

There are variations in published methodology for DEI regarding cyanoacrylate formulation and adjunctive agents [9]. Early studies used the N-butyl preparation, which polymerizes faster than the 2-octyl preparation. Lipiodol, a radiopaque plant-based oil that delays cyanoacrylate polymerization, has also been used to minimize premature needle occlusion and/or for radiographic localization, although its use may increase distal embolization risk. When used, the typical ratio of Lipiodol to glue is 1:1. The feared complications of DEI include embolic complications, such as pulmonary embolism and stroke. However, in the largest study of DEI, the rate of pulmonary embolism was 0.7% [10], and overall, the frequency of symptomatic pulmonary embolism requiring anticoagulation or even leading to death is extremely rare. Other reported complications of DEI in this study included rebleeding from (expected) intraluminal extrusion of the glue cast (4.4%), sepsis (1.3%), gastric ulcer formation (0.1%), and peritonitis (0.1%) [10]. DEI is used for initial endoscopic hemostasis, particularly in the acute setting. With multidisciplinary input, subsequent therapeutic options may include repeat cyanoacrylate injection every 2–4 weeks until obliteration [9], EUS-guided therapy (see below) where available, or endovascular therapy (TIPS or BRTO) if bleeding is uncontrolled (Fig. 39.3).

Thrombin converts fibrinogen to a fibrin clot, thus forming a clot inside the GV and occluding blood flow. Studies of endoscopic injection of thrombin for acute gastric variceal bleeding show similar rates of initial hemostasis, 5-day treatment failure, and 6-week when compared to cyanoacrylate glue. A randomized study that compared endoscopic thrombin injection and glue injection showed both had similar efficacy in achieving successful hemostasis, however, a higher incidence of adverse events (i.e., ulcers) was associated with glue injection [11]. Available data indicate that thrombin is safe and effective in the treatment of acute GV bleeding, with hemostasis rates of 70%–100% [9].

Endoscopic Ultrasound (EUS)-Guided Injection Therapies

EUS enhances the precision of injection and expands the options of injectate for GV. EUS provides sonographic guidance for intravascular access with the Fine Needle Aspirate (FNA) needle, as well as Doppler interrogation to provide real-time feedback of hemostasis. Based on the precedent of DEI, the original choice of injectate under EUS guidance was cyanoacrylate, and at least one retrospective comparative study has shown EUS-guided cyanoacrylate injection to be superior to DEI [12]. Borrowing from interventional radiology, hemostatic coils are used for EUS-guided injection. These coils are constructed from soft coiled platinum wires with spaced synthetic fibers, and they pack within the vessel to cause thrombus formation. The most commonly published technique has been to inject 1–3 of these coils within the GV to form a scaffolding onto which an adjunct, such as cyanoacrylate, can be subsequently injected to minimize concerns of embolic phenomenon [13]. Retrospective series and small randomized controlled trials have shown high rates of technical success and control of bleeding and low rates of rebleeding (0%–16%) and adverse events (0%–7%), although one RCT had a 25% rate due to 4/16 patients having asymptomatic pulmonary emboli on per-protocol CT. Two meta-analyses suggested that EUS-guided combination therapy (coil + cyanoacrylate) is superior to EUS-guided monotherapy with cyanoacrylate alone and coils alone and superior to freehand DEI [14, 15]. Table 39.1 summarizes the literature for EUS-based GV therapy.

Figure 39.4 demonstrates the technique for EUS-guided GV therapy. 100–200 cc of water is instilled, which tends to be retained in the gastric fundus if the patient is rolled slightly leftward. The water enhances the delineation of intramural vessels (i.e., gastric varices) from extramural collaterals. A 22 G or 19 G FNA needle is used to access gastric varix under EUS guidance. A 19 G FNA needle can be used when the maximal axial diameter is >5 mm; a 22 G FNA needle can be used for <5 mm. A transesophageal needle approach is preferred, when possible, for optimal ergonomics (i.e., the EUS scope is in a straight configuration), and fluoroscopy can be helpful particularly for operators early in their experience. Multiple coils can be sequentially injected. Doppler interrogation is used to guide the parameters of therapy. When there is a significant reduction of or near-absent Doppler flow, an adjunctive agent such as cyanoacrylate can be injected.

Table 39.1 Published studies reporting the efficacy of EUS-guided injection therapies for the treatment of acute gastric variceal bleeding

	Study type	*N*	Injectatate	# of sessions, mean	Technical success	Clinical success	Rate of adverse events	Rate of rebleeding (%)	All-cause mortality (%)
EUS CYA alone									
Lee (2000)	Prospective	54	CYA, repeated injection	2.2 (1.7)	52/54 (96.3%)	43/54 (79.6%)	22/54 (40.7%)	19/54 (35.2%)	28/54 (51.9%)
Lee (2000)	Prospective	47	CYA, on-demand injection	1.3 (0.5)	45/47 (95.7%)	–	9/47 (19.1%)	33/47 (70.2%)	35/47 (74.5%)
Romero-Castro (2013)	Retrospective	19	CYA	1.5	17/19 (89.5%)	19/19 (100%)	11/19 (57.9%)	–	–
Gubler (2014)	Retrospective	40	CYA	–	40/40 (100%)	36/36 (100%)	2/40 (5%)	6/40 (15%)	6/40 (15%)
Bick (2018)	Retrospective	64	CYA	1.2	–	62/64 (96.9%)	13/64 (20.3%)	5/56 (5.9%)	–
EUS CYA + coil									
Binmoeller (2011)	Retrospective	30	Coil + CYA	1	30/30 (100%)	29/30 (100%)	0/30 (0%)	4/24 (16.6%)	1/30 (3.3%)
Robles-Medranda (2020)	RCT	30	Coil + CYA	1	30/30 (100%)	30/30 (100%)	2/30 (6.6%)	2/30 (6.6%)	–
Fujii Lau (2016)	Retrospective	3	Coil + CYA	1	3/3 (100%)	3/3 (100%)	0/3 (0)	0/3 (0)	–
Lobo (2017)	RCT	29	Coil + CYA	–	16/16 (100%)	–	4/16 (25%)[a]	–	–
Bhat (2016)	Retrospective	152	Coil + CYA	–	151/152 (99.3%)	–	9/125 (7.2%)	20/125 (16%)	–
EUS coil alone or non-CYA									
Romero-Castro (2013)	Retrospective	11	Coil	1.3	10/11 (90.9%)	10/11 (90.9%)	1/11 (9.1%)	–	–

	Study type	*N*	Injectatate	# of sessions, mean	Technical success	Clinical success	Rate of adverse events	Rate of rebleeding (%)	All-cause mortality (%)
Robles-Medranda(2020)	RCT	29	Coil	–	29/29 (100%)	26/29 (89.7%)	1/29 (3.4%)	5/29 (17.2%)	–
Fujii-Lau (2016)	Retrospective	3	Coil	1	3/3 (100%)	3/3 (100%)	0/3 (0)	1/3 (33%)	–
Bazarbashi (2020)	Retrospective	10	Coil + gelfoam	1	10/10 (100%)	10/10 (100%)	0/10 (0)	0/10 (0)	1/10 (10%)

CYA cyanoacrylate, *RCT* randomized controlled trial
[a]All had asymptomatic pulmonary embolism on per-protocol CT

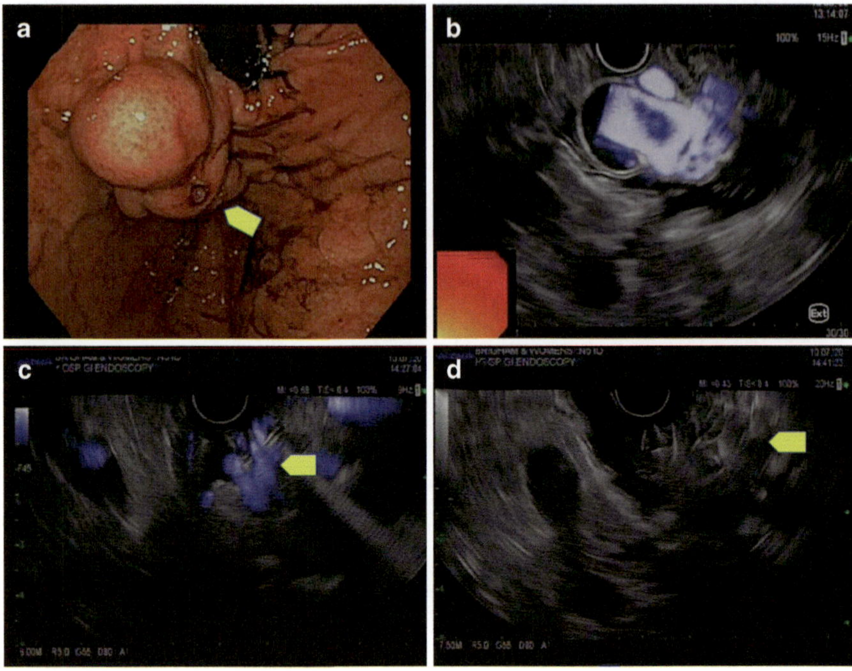

Fig. 39.4 Endoscopic and EUS images of Bleeding IGV-1. (**a**) endoscopic view of IGV-1 with stigmata; (**b**) EUS view of GV with baseline Doppler flow assessment; (**c**) Extrusion of the initial coil with diminution of Doppler flow; (**d**) EUS appearance of intravascular coil + tissue adhesive, no Doppler flow seen

Summary

AVB is a dreaded complication of patients with cirrhosis and portal hypertension. Standard of care mandates for early administration of vasoactive drugs, antibiotics, and endoscopy with EBL within the first 12 h of the index bleed. Patients that fail combined pharmacological and endoscopic therapy may require temporary placement of balloon tamponade or esophageal stents until definitive treatment (preferably TIPS) can be instituted. Endoscopy is recommended for diagnosis and initial treatment of bleeding GV. Endoscopic treatment for bleeding GOV1 varices remains band ligation although DEI of cyanoacrylate can also be considered. DEI of cyanoacrylate or thrombin is recommended for bleeding IGV and GOV2 varices. There is a growing experience that supports EUS-based therapies in terms of acute bleeding control, the durability of hemostasis, and complication rates, however more robust data is needed. EUS-based therapies should be considered for follow-up therapy where available. Current literature supports a combination of coils plus cyanoacrylate (or another adjunct) for EUS-based therapies.FundingAndrés Cárdenas is funded by the Instituto de Salud Carlos III and Plan Estatal de Investigación

Ciéntifica y Técnica y de Innovación–Grant No PI19/00752 and has received funding for this work by "Fundación Marta Balust".

References

1. Chen PH, Chen WC, Hou MC, Liu TT, Chang CJ, Liao WC, et al. Delayed endoscopy increases re-bleeding and mortality in patients with hematemesis and active esophageal variceal bleeding: a cohort study. J Hepatol. 2012;57(6):1207–13.
2. Rolando N, Gimson A, Philpott-Howard J, Sahathevan M, Casewell M, Fagan E, et al. Infectious sequelae after endoscopic sclerotherapy of oesophageal varices: role of antibiotic prophylaxis. J Hepatol. 1993;18(3):290–4.
3. Laine L, el-Newihi HM, Migikovsky B, Sloane R, Garcia F. Endoscopic ligation compared with sclerotherapy for the treatment of bleeding esophageal varices. Ann Intern Med. 1993;119(1):1–7.
4. Onofrio FQ, Pereira-Lima JC, Valença FM, Azeredo-da-Silva ALF, Tetelbom SA. Efficacy of endoscopic treatments for acute esophageal variceal bleeding in cirrhotic patients: systematic review and meta-analysis. Endosc Int Open. 2019;7(11):E1503–14.
5. Ibrahim M, Lemmers A, Devière J. Novel application of Hemospray to achieve hemostasis in post-variceal banding esophageal ulcers that are actively bleeding. Endoscopy. 2014;46(Suppl 1 UCTN):E263.
6. Sarin SK, Lahoti D, Saxena SP, Murthy NS, Makwana UK. Prevalence, classification and natural history of gastric varices: a long-term follow-up study in 568 portal hypertension patients. Hepatology. 1992;16(6):1343–9.
7. Soehendra N, Nam VC, Grimm H, Kempeneers I. Endoscopic obliteration of large esophagogastric varices with bucrylate. Endoscopy. 1986;18(1):25–6.
8. Tan PC, Hou MC, Lin HC, Liu TT, Lee FY, Chang FY, et al. A randomized trial of endoscopic treatment of acute gastric variceal hemorrhage: N-butyl-2-cyanoacrylate injection versus band ligation. Hepatology. 2006;43(4):690–7.
9. Garcia-Pagán JC, Barrufet M, Cardenas A, Escorsell A. Management of gastric varices. Clin Gastroenterol Hepatol. 2014;12(6):919–28.e1. quiz e51–2
10. Cheng LF, Wang ZQ, Li CZ, Lin W, Yeo AE, Jin B. Low incidence of complications from endoscopic gastric variceal obturation with butyl cyanoacrylate. Clin Gastroenterol Hepatol. 2010;8(9):760–6.
11. Lo GH, Lin CW, Tai CM, Perng DS, Chen IL, Yeh JH, et al. A prospective, randomized trial of thrombin versus cyanoacrylate injection in the control of acute gastric variceal hemorrhage. Endoscopy. 2020;52(7):548–55.
12. Bick BL, Al-Haddad M, Liangpunsakul S, Ghabril MS, DeWitt JM. EUS-guided fine needle injection is superior to direct endoscopic injection of 2-octyl cyanoacrylate for the treatment of gastric variceal bleeding. Surg Endosc. 2019;33(6):1837–45.
13. Robles-Medranda C, Oleas R, Valero M, Puga-Tejada M, Baquerizo-Burgos J, Ospina J, et al. Endoscopic ultrasonography-guided deployment of embolization coils and cyanoacrylate injection in gastric varices versus coiling alone: a randomized trial. Endoscopy. 2020;52(4):268–75.
14. McCarty TR, Bazarbashi AN, Hathorn KE, Thompson CC, Ryou M. Combination therapy versus monotherapy for EUS-guided management of gastric varices: a systematic review and meta-analysis. Endosc Ultrasound. 2020;9(1):6–15.
15. Mohan BP, Chandan S, Khan SR, Kassab LL, Trakroo S, Ponnada S, et al. Efficacy and safety of endoscopic ultrasound-guided therapy versus direct endoscopic glue injection therapy for gastric varices: systematic review and meta-analysis. Endoscopy. 2020;52(4):259–67.

Preemptive TIPS (p-TIPS) 40

Pol Olivas and Virginia Hernández-Gea

Despite all the improvements in the management of patients with Acute Variceal Bleeding (AVB) in the last decades, mortality still remains as high as 20% even when the best possible care is fully applied (Band ligation + vasoactive drugs + antibiotic prophylaxis) and hemorrhage is initially controlled [1, 2]. In the last years, early identification of patients with AVB at high risk of bad outcomes has been extensively pursued with the final aim of guiding clinical decision-making. It is the rationale behind the preemptive TIPS (p-TIPS).

Preemptive-TIPS (also known as Early-TIPS) refers to the early and prophylactic insertion of a TIPS in patients at high risk of failure and/or rebleeding after AVB. Natural history studies in the clinical scenario of AVB have shown that the time frame in which both mortality and failure concentrates is the first 72 h, especially the first 24 [3]. Therefore, if we want to modify the natural history of AVB and prevent failure/mortality, the p-TIPS has to be placed as soon as possible and this is why initially it was called early TIPS. However, as this terminology can be easily confused with the salvage TIPS placed early during admission in a patient with refractory bleeding, it was recommended to called its p-TIPS, a term that reflects better the preventive aim behind this strategy.

The key issue to enhancing p-TIPS benefit and limiting complications is the adequate selection of the candidates. Although patients at high risk of a bad outcome can be identified using several prognostic tools (recalibrated MELD score, presence of ascites, ACLF…) [1, 4] only HVPG and Child-Pugh score + endoscopic findings have been validated to guide p-TIPS placement.

P. Olivas · V. Hernández-Gea (✉)
Barcelona Hepatic Hemodynamic Laboratory, Liver Unit, Hospital Clínic-IDIBAPS, Health Care Provider of the European Reference Network on Rare Liver Disorders (ERN-Liver), Barcelona, Spain

Centro de Investigación Biomédica Red de Enfermedades Hepáticas y Digestivas (CIBEREHD), Madrid, Spain
e-mail: olivas@clinic.cat; vihernandez@clinic.cat

R. de Franchis (ed.), *Portal Hypertension VII*,
https://doi.org/10.1007/978-3-031-08552-9_40

HVPG ≥20 mmHg during the AVB episode is associated with a substantially (fivefold) increased risk of poor bleeding control, bleeding mortality, and 1-year mortality [5]. HVPG-guided risk stratification was used in the first study evaluating the efficacy of p-TIPS. In this study Monescillo and colleagues were able to improve survival in patients at high risk of a bad outcome (HVPG ≥20 mmHg) placing a p-TIPS early during admission. Within the first 24 h after admission HVPG was measured and those with HVPG ≥20 mmHg were randomized to receive standard of care (SOC) management or TIPS. This study demonstrated for the first time that placing a TIPS early during admission in a subgroup of patients at high risk of bad outcomes reduces both treatment failure and mortality [6].

As HVPG is not widely available, future studies aimed to identify patients at risk using clinical criteria. In that way, in a 2010 RCT, García-Pagan and colleagues defined high risk using clinical criteria: Child-Pugh score (CP) C (CP-C) <13 points or CP-B + active bleeding (AB) at the initial endoscopy. Mortality in the SOC group was 33% at 6 weeks and 39% at 1-year demonstrating (despite the potential subjectivity of active bleeding identification and the difficulty of evaluating accurately CP during the AVB episode) the usefulness of these criteria in the identification of patients at high risk of mortality. This RCT demonstrated again the survival benefit of the early (<72 h) p-TIPS placement (97% vs. 67% at 6 weeks and 86% vs. 61% at 1-year; $P > 0.001$) [7].

So far, four studies have evaluated the safety and efficacy of preemptive TIPS (One RCT and three observational) using these clinical criteria. In all of them, p-TIPS showed a benefit in preventing failure/rebleeding while the survival benefit was demonstrated in 3 [7–10]. However, whether CP-C and CP-B + AB equally benefit from this strategy has been a long matter of debate in the last years. Indeed, in the European observational study the survival benefit was limited to the CP-C group, probably explained by the small number of CP-B + AB patients included (only 19 treated with TIPS) and the lower mortality rates of this population [9].

More recently, the role of p-TIPS was extensively evaluated in two studies performed in China where p-TIPS was performed regardless of the CP score. These studies demonstrated that CP-A patients do not benefit from p-TIPS and in the CP-B group the benefit concentrates in patients with AB at the initial endoscopy and not in the subgroup without AB [11, 12].

Therefore, an individual patient data meta-analysis including only patients with a clearly high risk of a bad outcome (HVPG ≥ 20 mmHg, CP-C, CP-B + AB) was recently performed. The analysis of 1327 patients, 310 of them treated with p-TIPS (138 CP-B + AB and 172 CP-C) corroborates the survival benefit of early p-TIPS in high-risk patients. This study also showed that among the CP-B + AB patients, the benefit in survival concentrates in those patients with CP-B > 7 points and the authors suggest a redefinition of the high-risk criteria: HVPG ≥ 20 mmHg, CP-C < 14, CP-B > 7 + AB. In addition, the study also confirmed the preemptive-TIPS benefit in the control of bleeding, rebleeding, and ascites without increasing hepatic encephalopathy [13].

After this meta-analysis was performed a new RCT was published [14]. The Authors defined high-risk patients as CP-C < 14, and CP-B (both with and without AB) and did not observe a benefit in survival. Potential reasons for these results may be the suboptimal definition of high-risk patients together with a surprisingly long time to randomization (65 +/− 37 h). Moreover, only 45% of the patients in the TIPS group received p-TIPS within the first 72 h (none of them in the first 24 h) and 21% assigned to the TIPS group did not receive it. The analysis of the subgroup of patients who received the p-TIPS in the first 72 h demonstrated a significant benefit in controlling bleeding/rebleeding (no data on survival were provided) reinforcing the need for early intervention.

Despite the classical contraindications to p-TIPS included as exclusion criteria in all the studies (HCC out of Milan criteria, Child C > 13, significant chronic renal disease, age > 75 years, heart failure, and complete occlusive PVT), identification of futility criteria in the setting of p-TIPS has also been a matter of extensive research. Although the combination of age > 55 years, CP >11 points, and Creatinine ≥1.3 mg/dL is able to identify patients with poor prognosis, they also benefit from pTIPS treatment [13]. Even patients with hyperbilirubinemia, ACLF, and/or presence of HE at admission who fulfill the high-risk criteria (HVPG ≥ 20 mmHg, CP-C, CP-B + AB) benefit from p-TIPS treatment and should not be excluded from receiving it [4, 13].

Another critical issue to guarantee the benefit of the p-TIPS strategy is its placement in centers of expertise. According to an American population study, expertise is highly associated with prognosis. Indeed, survival substantially increased when the volume of TIPS per year overcomes 20 and therefore patients should be referred to experienced centers [15].

In summary, mortality in AVB concentrates in the first 72 h (especially 24 h) and therefore, in order to modify the natural history of AVB, p-TIPS should be placed as early as possible. Patient stratification is key to select a population with the highest benefit, being HVPG ≥ 20 mmHg, CP-C < 14 points, CP-B > 7points + AB the criteria with the best performance to guide therapy.

Based on the described survival benefit, all patients with AVB should be evaluated to receive p-TIPS, and those centers unable to urgently provide this treatment should establish referral protocols to guarantee the best possible treatment.

References

1. Reverter E, Tandon P, Augustin S, Turon F, Casu S, Bastiampillai R, et al. A MELD-based model to determine risk of mortality among patients with acute variceal bleeding. Gastroenterology. 2014;146(2):412–19.e3. http://www.ncbi.nlm.nih.gov/pubmed/24148622
2. Magaz M, Baiges A, Hernández-Gea V. Precision medicine in variceal bleeding: are we there yet? J Hepatol. 2020;72(4):774–84. http://www.ncbi.nlm.nih.gov/pubmed/31981725
3. Graham DY, Smith JL. The course of patients after variceal hemorrhage. Gastroenterology. 1981;80(4):800–9. http://www.ncbi.nlm.nih.gov/pubmed/6970703
4. Trebicka J, Gu W, Ibáñez-Samaniego L, Hernández-Gea V, Pitarch C, Garcia E, et al. Rebleeding and mortality risk are increased by ACLF but reduced by pre-emptive TIPS. J Hepatol. 2020;73(5):1082–91. http://www.ncbi.nlm.nih.gov/pubmed/32339602

5. Moitinho E, Escorsell A, Bandi JC, Salmerón JM, García-Pagán JC, Rodés J, et al. Prognostic value of early measurements of portal pressure in acute variceal bleeding. Gastroenterology. 1999;117(3):626–31. http://www.ncbi.nlm.nih.gov/pubmed/10464138
6. Monescillo A, Martínez-Lagares F, Ruiz-del-Arbol L, Sierra A, Guevara C, Jiménez E, et al. Influence of portal hypertension and its early decompression by TIPS placement on the outcome of variceal bleeding. Hepatology. 2004;40(4):793–801. http://www.ncbi.nlm.nih.gov/pubmed/15382120
7. García-Pagán JC, Caca K, Bureau C, Laleman W, Appenrodt B, Luca A, et al. Early use of TIPS in patients with cirrhosis and variceal bleeding. N Engl J Med. 2010;362(25):2370–9. http://www.nejm.org/doi/abs/10.1056/NEJMoa0910102
8. Garcia-Pagán JC, Di Pascoli M, Caca K, Laleman W, Bureau C, Appenrodt B, et al. Use of early-TIPS for high-risk variceal bleeding: results of a post-RCT surveillance study. J Hepatol. 2013;58(1):45–50. http://www.ncbi.nlm.nih.gov/pubmed/22940408
9. Hernández-Gea V, Procopet B, Giráldez Á, Amitrano L, Villanueva C, Thabut D, et al. Preemptive-TIPS improves outcome in high-risk variceal bleeding: an observational study. Hepatology. 2019;69(1):282–93. http://www.ncbi.nlm.nih.gov/pubmed/30014519
10. Rudler M, Cluzel P, Corvec TL, Benosman H, Rousseau G, Poynard T, et al. Early-TIPSS placement prevents rebleeding in high-risk patients with variceal bleeding, without improving survival. Aliment Pharmacol Ther. 2014;40(9):1074–80. http://www.ncbi.nlm.nih.gov/pubmed/25230051
11. Lv Y, Zuo L, Zhu X, Zhao J, Xue H, Jiang Z, et al. Identifying optimal candidates for early TIPS among patients with cirrhosis and acute variceal bleeding: a multicentre observational study. Gut. 2019;68(7):1297–310. http://www.ncbi.nlm.nih.gov/pubmed/30415233
12. Lv Y, Yang Z, Liu L, Li K, He C, Wang Z, et al. Early TIPS with covered stents versus standard treatment for acute variceal bleeding in patients with advanced cirrhosis: a randomised controlled trial. Lancet Gastroenterol Hepatol. 2019;4(8):587–98. http://www.ncbi.nlm.nih.gov/pubmed/31153882
13. Nicoară-Farcău O, Han G, Rudler M, Angrisani D, Monescillo A, Torres F, et al. Effects of early placement of transjugular portosystemic shunts in patients with high-risk acute variceal bleeding: a meta-analysis of individual patient data. Gastroenterology. 2021;160(1):193–205.e10. http://www.ncbi.nlm.nih.gov/pubmed/32980344
14. Dunne PDJ, Sinha R, Stanley AJ, Lachlan N, Ireland H, Shams A, et al. Randomised clinical trial: standard of care versus early-transjugular intrahepatic porto-systemic shunt (TIPSS) in patients with cirrhosis and oesophageal variceal bleeding. Aliment Pharmacol Ther. 2020;52(1):98–106. http://www.ncbi.nlm.nih.gov/pubmed/32452561
15. Sarwar A, Zhou L, Novack V, Tapper EB, Curry M, Malik R, et al. Hospital volume and mortality after transjugular intrahepatic portosystemic shunt creation in the United States. Hepatology. 2018;67(2):690–9. http://www.ncbi.nlm.nih.gov/pubmed/28681542

Management of Refractory Variceal Bleeding

41

Marika Rudler

Abbreviations

TIPS	Transjugular intrahepatic portosystemic shunt
pTIPS	Pre-emptive TIPS
BT	Balloon tamponade
SEMS	Self-expandable metal stents
MA	Meta-analysis

Introduction

Mortality rate related to variceal bleeding in cirrhosis has improved thanks to the widespread use of antibiotics, vasoactive drugs, and endoscopic treatment. Moreover, a pre-emptive TIPS (pTIPS) placement has shown to improve rebleeding, as well as survival in high-risk patients in whom the condition is stabilized thanks to medical and endoscopic therapy. Nevertheless, 15%–20% of patients may develop refractory bleeding despite pharmacological and endoscopic treatment. Mortality is high in these patients in spite of the use of salvage TIPS, reaching 20% to 50% in different series; the management remains challenging in these situations. This chapter is intended to provide an overview of the most frequent practical situations. Definition of refractory bleeding, Balloon Tamponade (BT) or oesophageal stent indication, and salvage TIPS will be discussed.

M. Rudler (✉)
Unité de Soins Intensifs d'Hépato-Gastro-Entérologie, Groupement Hospitalier APHP-Sorbonne Université, Paris, France

Sorbonne Université, INSERM, Centre de Recherche Saint-Antoine (CRSA), Institute of Cardiometabolism and Nutrition (ICAN), Paris, France
e-mail: marika.rudler@aphp.fr

R. de Franchis (ed.), *Portal Hypertension VII*,
https://doi.org/10.1007/978-3-031-08552-9_41

Definition of Refractory Bleeding (Unchanged from Baveno VI)

Failure to control variceal bleeding includes two different clinical situations: the first one corresponds to massive bleeding with haemodynamic instability despite medical treatment and transfusion; in this case, according to Baveno VI recommendations, TIPS placement seems to be the best therapeutic option [1]. The second one is related to an early recurrence of bleeding, i.e., within the 5 days after index bleeding. If rebleeding is severe, TIPS seems to be the best therapeutic option. Strictly speaking, refractory bleeding is defined by persistent bleeding despite combined pharmacological and endoscopic therapy [1]. The clinical/biological criteria were not modified during the Baveno VI conference. There was a slight modification of this definition between Baveno IV and V [2] (the index named ABRI, based on transfusion requirement, was deleted in Baveno V criteria; haemoglobin drop should be restricted to 24 h; introduction of hypovolemic shock). Both Baveno IV and V criteria for treatment failure were validated in 2 different studies [3, 4], underlining a higher performance than previous Baveno II/III criteria. Therefore, refractory bleeding is defined by at least one of the following: (1) fresh haematemesis 2 h after the start of a specific drug treatment or therapeutic endoscopy; (2) development of hypovolaemic shock; (3) 3 g drop in Hb (9% drop in Ht.) within any period if no transfusion is administered.

In case of refractory bleeding, patients should receive a bridge treatment before a more definitive therapy such as TIPS.

Bridge Treatment: Balloon Tamponade or Self-Expandable Metal Stents (SEMS)

Balloon tamponade (BT) as a bridge to salvage TIPS was the first accepted therapy for refractory variceal bleeding. BT is highly effective in the temporary control of bleeding. Its insertion requires careful patient preparation (sedation and intubation) in order to prevent aspiration. Then, the correct insertion of the balloon has to be checked using air injection through the gastric port before inflation of the gastric balloon. In many cases, oesophageal perforation/necrosis is related to the inflation of the gastric balloon in the oesophagus. BT can only be maintained for 24–48 h to prevent the occurrence of severe adverse events, such as oesophageal rupture or ulcers [5]. More recently, covered oesophageal metal stents have been proposed, as an alternative to oesophageal BT. Sedation and intubation can facilitate oesophageal stent insertion but are not absolutely required. The SX-Ella Danis stent is a Self-expandable Metal Stents (SEMSs) specifically designed for variceal bleeding: it is removable, covered, self-expandable, and does not require endoscopic guidance for its insertion. The insertion is quite easy to perform but also needs some basic training. The main advantages of this stent are related to the possibility of withdrawing the device after 7 days and to maintain enteral feeding.

More than 10 studies have now been published evaluating the performance of stents: many of them are observational and/or retrospective studies. Since Baveno VI, one RCT and 3 meta-analyses (MA) were performed. The only RCT available compared the effectiveness of BT and SEMSs [6] in refractory variceal bleeding; 28 patients were randomized: 15 patients in the BT group and 13 patients in SEMS group were analyzed. Success of therapy (which was a composite endpoint of an absence of bleeding and absence of serious adverse events and survival within the first 15 days) was significantly more frequent in the stent group (66% vs. 20%; $p = 0.025$), and the probability of developing a serious adverse events related to the device was significantly lower in the stent group (8% vs. 40%, $p = 0.049$). These encouraging results did not lead to significant improvement in survival, remaining very high in both groups (60% vs. 46%, $p = 0.46$). The actuarial probability of being free of treatment failure during the first 15 days after inclusion was also similar in both groups ($p = 0.056$).

Three MA were published in 2015 and 2019, which enabled the pooling data from case series [7–9]. In the first MA [7], 155 patients from 12 studies were analyzed. Endoscopic expansion of the SEMS was obtained in 97% of patients, control of bleeding in 96% of patients, and a 30-day survival rate of 68%. Serious adverse events occurred in more than 30% of patients and included stent migration, oesophageal ulceration, or rebleeding. In the same line, the MA by Marot et al. [8] reported that failure to control bleeding occurred in 18% of cases. Fewer than 40% of patients treated with SEMS died after 30 days and only 12% died from recurrent bleeding, suggesting a higher survival rate than previously described with BT. More recently, a systematic review and MA was published [9] comparing BT and SEMS: 23 studies were included for the final analysis, 12 in the analysis of BT and 11 in the analysis of SEMS, including 570 patients who benefited from BT and 188 from SEMS. The main outcomes were failure to control bleeding and mortality in the short-term and medium-term follow-ups. BT studies had a pooled rate of short-term failure to control bleeding of 35.5% and adverse events in over 20%; stenting failed to control bleeding in the short and medium-term in 12.7% and 21.5%, respectively. A stent migration occurred in up to one-quarter of patients. Medium-mortality rates were similar in both therapies. Last year, a retrospective multicentre study was published [10], showing that control of bleeding was achieved in 79% of patients with oesophageal stents alone, i.e., without salvage TIPS as rescue therapy. The most frequent adverse event was stent dislocation (38.2%). Last, the rebleeding rate was more than 79% after removal, suggesting that oesophageal stent placement should only be used as a bridge therapy before a more definite treatment such as TIPS.

Overall, SEMSs seem to be at least as effective as BT and probably safer as a bridge therapy, even if 30- or 42-day mortality between these techniques is similar in all studies. The main advantages of SEMSs are the ease of placement, the ability to maintain a patent oesophageal lumen (which theoretically decreases the probability of aspiration), and the possibility of maintaining haemostasis for a much longer period than BT (7 days). Drawbacks are mainly represented by the relatively high rate of migration. Whether oesophageal stents could replace salvage TIPS is an unsolved question. To date, the largest study evaluating stents as a single therapy for refractory bleeding included 34 patients [10]. Bleeding-related mortality was about 50% and median survival after stent placement was 2.1 months. Given this extremely short survival time, stents should be only considered as a bridge therapy in this setting. An interventional study is ongoing in India, comparing the efficacy of TIPS versus SEMS alone in the management of refractory variceal bleeding in patients with cirrhosis. There is also an important issue regarding the costs (the price of SEMS is more or less 20 times higher than BT in France for instance), reimbursement being not currently implemented in many European countries. Cost-effectiveness data are not available, and such an analysis has to be performed.

Salvage TIPS

According to Baveno VI recommendations, refractory bleeding is best managed by salvage TIPS. Overall, 13 studies evaluated the effectiveness of TIPS in this setting, most of them using uncovered stents. Since Baveno VI, 6 studies reported data on patients treated with either uncovered or covered TIPS. Rebleeding rates (20%–29%) and 6-week survival (70%–90%) were comparable to those previously described.

Which Class of Stents Should We Use: Covered/Uncovered?

Two studies including patients treated with PTFE-covered stents have been published and two are available in abstract form. In the setting of refractory bleeding, few data are available regarding the performance of covered stents. The 4 more recent studies included patients who received covered stents [11–14]. The use of covered stents was associated with a significant improvement in rebleeding in one study including patients treated with covered or non-covered stents [11]. Re-bleeding related to stent dysfunction was associated with uncovered TIPS in the majority of cases. Even so, the use of covered stents was not associated with better survival. In non-urgent situations (refractory ascites, secondary prophylaxis for variceal bleeding), covered stents were found to reduce the risk of shunt obstruction without worsening the occurrence of HE after TIPS. Therefore, covered stents should probably be used in an emergency situation requiring salvage TIPS.

The latter result regarding mortality after salvage TIPS justifies the concept of pTIPS to prevent early rebleeding that could avoid further deterioration and improve survival. Interestingly, in the study of Bouzbib et al., more than half of the patients needing salvage TIPS had a previous episode of bleeding; one-third of these patients had at the time of the first bleeding episode an indication of early TIPS. One could imagine that the bleeding episode requiring salvage TIPS could have been avoided if pTIPS had been performed before.

Is Salvage TIPS a Futile Procedure for Patients in the Most Severe Conditions?

Although data are very limited, mortality is high in patients in whom salvage TIPS failed to control bleeding and in those with multiorgan failure. Factors associated with significant higher mortality are infection, renal failure, the use of catecholamines, balloon tamponade, a high MELD score, and a high Child-Pugh score. Interestingly, until then, very few data were available in patients with end-stage liver disease, i.e., with Child-Pugh C14–C15. These patients are not candidates for an early-TIPS placement. In Maimone study, all patients with Child C14–15 (10 out of 144 patients) died within the 42 days after bleeding. In the study by Bouzbib et al., 16/106 patients were classified as Child-Pugh 14–15; the 1-year transplant-free survival was null in this subgroup of patients. Based on this limited amount of data, one can consider that salvage TIPS may be futile in Child C14–15 patients who will not be candidates for liver transplantation. Nevertheless, liver transplantation can be discussed rapidly after the control of bleeding. A short series of patients suggested a 100% rate of survival in transplant recipients after salvage TIPS in Child-Pugh C14–15 cirrhosis [15]. In line with these results, in the most recent, a largest study published in the setting of refractory bleeding and the use of salvage covered TIPS [14], the authors described a limited 42-day survival (<10%) in patients with either a MELD score > 30 or with lactate at admission>12 mmol/L, also suggesting the futility of TIPS in these patients. As no recommendation can be clearly made, an expert's advice is mandatory for each case.

Embolization During Salvage TIPS Placement

In the more recent studies, embolization was not systematically performed at the time of TIPS placement [11–14]: sometimes, a variceal embolization was systematically performed to occlude the filling of large collaterals feeding gastroesophageal varices and maximize the flow of the liver. In another study, variceal embolization was applied only when flow persisted to the varices during the portography performed after TIPS placement. To date, no study was conducted that aimed to compare the effectiveness of TIPS with or without embolization in terms of rebleeding and thus embolization should be considered in an individualized way.

In summary, variceal embolization concomitant with TIPS placement may be useful in some cases even if no clear recommendation can be made.

Is There Any Therapeutic Alternative in Case of Rebleeding After Salvage TIPS?

Most of the patients who rebleed after salvage TIPS will die. Data on other therapeutic means are lacking in the literature. In most of the cases, rebleeding after TIPS is related to TIPS obstruction: it is, therefore, mandatory to check the patency of TIPS in such situations by haemodynamic catheterization of the stent. A variceal embolization should be considered if possible. In a recent retrospective study [16], acute stent occlusion was treated by balloon angioplasty, an additional stent insertion, or a parallel stent with 100% effectiveness. Further investigation is necessary to evaluate the management strategy of these patients. Finally, in patients with persistent bleeding and high MELD score, liver transplantation is the best option. Usually, patients requiring a salvage TIPS have decompensated cirrhosis with high Child-Pugh and MELD scores, except for very rare instances such as splanchnic vein thrombosis, ectopic varices, or both. In these latter patients, liver transplantation could be prioritized based on a MELD score exception. This would require a thorough assessment by a liver transplant expert mandated for each particular case. If accepted, it would then result in a very short waiting time on the transplant waiting list.

What Place for Balloon-Occluded Retrograde Transvenous Obliteration (BRTO) in Case of Refractory Bleeding?

Balloon-occluded Retrograde Transvenous Obliteration (BRTO) is a treatment that has been included in the American Association for the study of Liver Diseases (AASLD) practice recommendations for the secondary prophylaxis of bleeding related to gastric varices. Typically, BRTO is proposed to patients who have a contra-indication for TIPS placement. Classical contra-indications to BRTO include portal or splenic thrombosis without other portosystemic collaterals to provide adequate mesenteric or splenic venous outflow following BRTO. Technically, BRTO is an endovascular technique that was first developed in Asia as a therapeutic adjunct or alternative to TIPS in the management of gastric varices. A BRTO procedure involves occlusion of outflow veins of the portosystemic shunt, such as a gastrorenal shunt, using an occlusion balloon followed by the injection of a sclerosing agent directly into the varix endovascularly. Several studies compared BRTO and TIPS in the setting of gastric varices; however, to date, no RCT has been published. Moreover, effectiveness has not been clearly evaluated in refractory bleeding. Further studies are warranted.

Conclusion

Although the medical literature provides recent data on refractory bleeding, the 6-week mortality remains very high in patients after salvage TIPS even with the systematic use of covered stents. As new strategies have been developed for the last 10 years, such as the pTIPS policy, significant improvement is expected. Whether this policy reduces the recourse to salvage TIPS has to be investigated.

References

1. de Franchis R, Baveno VI Faculty. Expanding consensus in portal hypertension: report of the Baveno VI consensus workshop: stratifying risk and individualizing care for portal hypertension. J Hepatol. 2015;63(3):743–52.
2. de Franchis R, Baveno V Faculty. Revising consensus in portal hypertension: report of the Baveno V consensus workshop on methodology of diagnosis and therapy in portal hypertension. J Hepatol. 2010;53(4):762–8.

3. Thabut D, Rudler M, Dib N, Carbonell N, Mathurin P, Saliba F, Mallet A, Massard J, Bernard-Chabert B, Oberti F, Cales P, Golmard JL, Bureau C, French Club for the Study of Portal Hypertension (CFEHTP). Multicenter prospective validation of the Baveno IV and Baveno II/III criteria in cirrhosis patients with variceal bleeding. Hepatology. 2015;61(3):1024–32.
4. Ahn SY, Park SY, Tak WY, Lee YR, Kang EJ, Park JG, Lee WK, Lee K, Kweon YO. Prospective validation of Baveno V definitions and criteria for failure to control bleeding in portal hypertension. Hepatology. 2015;61(3):1033–40.
5. Cook D, Laine L. Indications, technique, and complications of balloon tamponade for variceal gastrointestinal bleeding. J Intensive Care Med. 1992;7:212–8.
6. Escorsell À, Pavel O, Cárdenas A, Morillas R, Llop E, Villanueva C, Garcia-Pagán JC, Bosch J, Variceal Bleeding Study Group. Esophageal balloon tamponade versus esophageal stent in controlling acute refractory variceal bleeding: a multicenter randomized, controlled trial. Hepatology. 2016;63(6):1957–67.
7. McCarty TR, Njei B. Self-expanding metal stents for acute refractory esophageal variceal bleeding: a systematic review and meta-analysis. Dig Endosc. 2016;28(5):539–47.
8. Marot A, Trépo E, Doerig C, Moreno C, Moradpour D, Deltenre P. Systematic review with meta-analysis: self-expanding metal stents in patients with cirrhosis and severe or refractory oesophageal variceal bleeding. Aliment Pharmacol Ther. 2015;42(11–12):1250–60.
9. Rodrigues SG, Cárdenas A, Escorsell À, Bosch J. Balloon tamponade and esophageal stenting for esophageal Variceal bleeding in cirrhosis: a systematic review and meta-analysis. Semin Liver Dis. 2019;39(2):178–94.
10. Pfisterer N, Riedl F, Pachofszky T, Gschwantler M, König K, Schuster B, Mandorfer M, Gessl I, Illiasch C, Fuchs EM, Unger L, Dolak W, Maieron A, Kramer L, Madl C, Trauner M, Reiberger T. Outcomes after placement of a SX-ELLA oesophageal stent for refractory variceal bleeding—a national multicentre study. Liver Int. 2019;39(2):290–8.
11. Maimone S, Saffioti F, Filomia R, Alibrandi A, Isgro G, Calvaruso V, et al. Predictors of re-bleeding and mortality among patients with refractory variceal bleeding undergoing salvage transjugular intrahepatic portosystemic shunt (TIPS). Dig Dis Sci. 2019;64(5):1335–46.
12. Bouzbib C, Cluzel P, Sultanik P, Bernard-Chabert B, Massard J, Benosman H, Mallet M, Tripon S, Conti F, Thabut D, Rudler M. Prognosis of patients undergoing salvage TIPS is still poor in the preemptive TIPS era. Clin Res Hepatol Gastroenterol. 2021;45(6):101593.
13. Kumar R, Kerbert AJC, Sheikh MF, Roth N, Calvao JAF, Mesquita MD, Barreira AI, Gurm HS, Ramsahye K, Mookerjee RP, Yu D, Davies NH, Mehta G, Agarwal B, Patch D, Jalan R. Determinants of mortality in patients with cirrhosis and uncontrolled variceal bleeding. J Hepatol. 2021;74(1):66–79.
14. Walter A, Rudler M, Olivas P, Moga L, Trépo E, Robic MA, Ollivier-Hourmand I, Baiges A, Sutter O, Bouzbib C, Peron JM, Le Pennec V, Ganne-Carrié N, Garcia-Pagán JC, Mallet M, Larrue H, Dao T, Thabut D, Hernández-Gea V, Nault JC, Bureau C, Allaire M, Salvage TIPS Group. Combination of MELD and lactate predicts death in patients treated with salvage TIPS for refractory variceal bleeding. Hepatology. 2021;74(4):2085–101.
15. Rudler M, Rousseau G, Thabut D. Salvage transjugular intrahepatic portosystemic shunt followed by early transplantation in patients with Child C14-15 cirrhosis and refractory variceal bleeding: a strategy improving survival. Transpl Int. 2013;26(6):E50–1.
16. Wang X, Luo X, Zhao M, Song J, Li X, Yang L. Prognostic factors for acute transjugular intrahepatic portosystemic shunt occlusion using expanded polytetrafluoroethylene-covered stent. Dig Dis Sci. 2020;65(10):3032–9.

Hepatic Encephalopathy and Acute Variceal Bleeding

42

Dominique Thabut, Charlotte Bouzbib, and Marika Rudler

Gastrointestinal bleeding is a classical precipitating factor of HE, and HE impairs prognosis in patients with variceal bleeding. Moreover, the presence or history of HE is often dissuading physicians to indicate TIPS, despite the fact that this therapeutic option has largely proven to improve outcomes in case of portal hypertension-related bleeding. This chapter will portray the interplay between HE and gastrointestinal bleeding in cirrhotic patients and suggest therapeutic strategies in order to manage those patients at best.

Prevalence and Prognosis of HE in Cirrhotic Patients with Gastrointestinal Bleeding

Gastrointestinal bleeding is a classical precipitating factor of HE; HE is generally multifactorial in nature in this context: liver failure, hyperammonaemia in the context of blood protein digestion, systemic inflammation, and infection. The relationship between gastrointestinal bleeding and an increase in blood ammonia is well established [1], as well as that of inflammation and HE [2]. Data on the prevalence of HE during bleeding are scarce, ranging from 8% to 40% in different series. Some studies do not even provide the data on HE at admission in variceal bleeding [3]. Nevertheless, data on HE is crucial, as the clinical impact of HE remains of major

D. Thabut (✉) C. Bouzbib M. Rudler
Groupement Hospitalier APHP-Sorbonne Université, Service d'Hépato-Gastro-Entérologie, Hôpital Pitié-Salpêtrière, Paris, France

Sorbonne Université, INSERM, Centre de Recherche Saint-Antoine (CRSA), Institute of Cardiometabolism and Nutrition (ICAN), Paris, France

Brain-Liver Pitié-Salpêtrière Study Group (BLIPS), APHP-Sorbonne Université, Paris, France
e-mail: dominique.thabut@aphp.fr; charlotte.bouzbib@aphp.fr; Marika.rudler@aphp.fr

R. de Franchis (ed.), *Portal Hypertension VII*,
https://doi.org/10.1007/978-3-031-08552-9_42

importance: the first episode of HE is associated with survival rates of 35%–45% at 1-year [4]. Moreover, recent data from a prospective series of patients with variceal bleeding [5] suggest that HE at admission is independently associated with mortality.

Treatment and Prevention of HE in Patients with Gastrointestinal Bleeding

Treatment of HE

It has never been proven that HE treatment improves outcomes in patients with gastrointestinal bleeding and concomitant HE. However, it seems reasonable to treat each HE bout with lactulose, as this therapy improves survival in patients with HE [6, 7]. This treatment can be administered either by enema, especially when bleeding is still not controlled, or orally when possible.

Prevention of HE

A recent open-label single-centre randomised study showed that lactulose treatment significantly reduced the incidence of HE in patients with gastrointestinal bleeding (14% vs. 40%, $p < 0.03$), without any effect on survival (8.5% vs. 14%, $p =$ NS) [8]. Another single-centre open-label randomised study also suggested that lactulose significantly reduced HE incidence (3.2% vs. 16.9%, $p < 0.02$); factors independently associated with the occurrence of HE were Child-Pugh score and lactulose treatment [9]. The meta-analysis of those 2 trials confirmed the beneficial effect of lactulose on the prevention of HE during gastrointestinal bleeding (7% vs. 28%, $p < 0.01$) without any survival benefit [10]. Finally, rapid removal of blood from the gastrointestinal tract by mannitol by mouth has also been shown to work in this context, also by comparison with paromomycin plus lactulose [11, 12]. Hence, in patients presenting with gastrointestinal bleeding, it is reasonable to state that rapid removal of blood from the gastrointestinal tract (lactulose oral or enemas) should be used to prevent HE. Broad-spectrum antibiotic prophylaxis also had a beneficial effect on survival, especially in patients with Child-Pugh C cirrhosis. However, the efficacy of antibiotic prophylaxis on HE occurrence has not been studied.

The Question of TIPS in the Setting of Gastrointestinal Bleeding in Cirrhotic Patients with HE or History of HE

The creation of a shunt, sometimes associated with a deterioration in liver function, results in an accumulation of neurotoxic substances and thus promotes the onset of HE. TIPS-related HE occurs in approximately 35% of cases. The majority of studies found in the literature are observational and do not allow for recommendations to be made with a high level of evidence. It is important to note that most randomised controlled trials have compared TIPS to other standard treatments (variceal ligation and ß-blockers or repeated paracentesis), and not all of these studies have shown that TIPS increases the risk of developing HE compared to standard treatment.

Is HE a Contra-Indication to TIPS?

In cirrhotic patients with gastrointestinal bleeding, TIPS can be envisioned in three situations: either as salvage therapy in case of refractory bleeding, or as pre-emptive in high-risk patients; last, TIPS can be a therapeutic strategy in case of failure of secondary prophylaxis.

Salvage TIPS

In cases of refractory gastrointestinal bleeding due to variceal haemorrhage, there is no contraindication to TIPS (known as salvage TIPS) as there is no therapeutic alternative. The various studies on the subject do not always describe the existence of HE before or after TIPS. In addition, none of the studies have investigated the risk factors for HE after TIPS in the context of salvage TIPS.

Pre-Emptive TIPS

Pre-emptive TIPS has been the only strategy which showed to improve survival in patients with AVB for 15 years [13, 14]. Hence, although counter-intuitive, it seems reasonable to envision considering TIPS in patients with AVB and HE.

A previous episode of HE or clinical HE at the time of admission has never been considered as a contra-indication for pTIPS in the published RCTs [3, 13, 14]. In one trial though, patients with recurrent hospital admissions with HE were excluded [3]. In this study coming from the UK, a lower incidence of HE was observed in the endo + drugs group as compared to the pre-emptive TIPS group. The difference was almost significant. However, the incidence of HE in the control group was remarkably low, at 17%, and the overall survival of this group was particularly high [15]. It would be interesting to carefully take a look at the patients who developed HE after TIPS and compare them to those who did not experience any HE, in terms of HE history and liver disease severity. Except for this study, all RCTs [13, 14], as well as all observational studies [16–18], showed a similar incidence of HE after TIPS and after standard treatment. Recent unpublished data suggest that pre-emptive TIPS placement improves prognosis in high-risk patients presenting with HE. Hence, HE or history of HE should not be a contra-indication to pre-emptive TIPS.

TIPS in Case of Failure of Secondary Prophylaxis

In the context of scheduled TIPS for secondary prophylaxis of gastrointestinal bleeding, all studies excluded patients with clinical signs of HE at the time of inclusion. Hence, TIPS should not be performed in this indication in case of overt HE. The analysis of risk factors for HE after TIPS is based on observational cohorts, randomised trials, or meta-analyses comparing TIPS with standard treatment (repeated ascites punctures or ligations and ß-blockers) [19–21]: History of overt HE, age, high

MELD, and Child-Pugh scores, low hepatic venous pressure gradient were significantly associated with a higher prevalence of HE after TIPS. However, there is no evidence for strictly contra-indicating TIPS in patients with a history of HE.

Although the risk factors have been well analysed, there is no method capable of identifying patients who will develop HE after TIPS. Minimal HE, assessed by PHES [22] or CFF [23] tests before TIPS, is associated with a significant increase in the risk of overt HE after TIPS. However, the number of studies remains minimal, and they are only observational with small patient numbers. It is therefore not possible to recommend the systematic use of the PHES or CFF tests before TIPS.

Hence, it is important to identify patients at high risk of developing HE after TIPS by patient interview, (investigating for a history of overt HE), a complete clinical examination and assessment of Child-Pugh and MELD scores, and then decide after assessing the benefits and risks for each patient. Thus, on a case-by-case basis, TIPS, or an alternative treatment (transplant) may be more suitable if available. Therefore, at best, liver transplantation feasibility should be considered for all patients for whom TIPS is indicated [24].

Which TIPS Should Be Used?

Data regarding the type and diameter of TIPS have only been collected in trials including patients with non-urgent TIPS. Hence, findings cannot be extended to the setting of pre-emptive TIPS or salvage TIPS. Stent cover and diameter do not affect the incidence of HE after TIPS. Two meta-analyses of six studies have compared the efficacy of covered and uncovered stents outside the emergency setting [25, 26]. None of them found an increased risk of HE in the short, medium, or long term with the use of covered stents. Three studies evaluated whether 8 mm diameter shunts could decrease the incidence of HE after TIPS compared to 10 mm shunts [27–29]; among the two randomised trials, one had to be stopped due to a high shunt malfunction rate in the 8 mm group, and the other study found a similar incidence of HE in both groups. Finally, several prospective trials have shown that a low hepatic venous pressure gradient after stent placement is a risk factor for HE. However, there is no gradient cut-off above which the risk of HE is zero.

Treatment of HE after TIPS

Data regarding the prevention of post-TIPS HE has only been collected in trials including patients with non-urgent TIPS. Hence, findings cannot be extended to the setting of pre-emptive TIPS or salvage TIPS. In the event of HE after TIPS, then HE should be treated as usual with available treatments such as lactulose and/or rifaximin (see Chap. 41). HE is considered refractory when signs of HE persist despite successful medical treatment. Shunt modifications (recalibration, occlusion) can be proposed accordingly. The studies reporting on this subject are all retrospective and describe only a very minimal number of patients (maximum 16) and the level of evidence is low [30–32]. After reduction of shunt size, improvement or recovery from the signs of HE has been reported in 48%–100% and 67% of cases, respectively. It is suggested to consider reducing shunt size before complete shunt occlusion. Liver transplantation should be promptly considered in the event of refractory HE after TIPS.

Two randomised trials have evaluated the efficacy of prophylactic treatment initiated prior to TIPS. The first compared lactulose, rifaximin, and placebo in 75 patients with or without a history of overt HE [33]. No difference was observed in the cumulative incidence of HE after 1 month. The second study evaluated L-ornithine L-aspartate in 21 patients and did not show superiority over placebo in the occurrence of HE 1 week after TIPS [34]. The patient numbers were small, and the endpoint was assessed very early in this latter study. An observational study evaluating the use of albumin infusion after TIPS did not find a significant decrease in the cumulative incidence of HE at 1 month compared to that described in a former cohort (historical controls) [35]. Finally, results of a recent RCT comparing rifaximin to placebo in the primary prevention of post-TIPS HE in patients with an indication for non-urgent TIPS showed a significant decrease of HE prevalence after TIPS in patients treated with Rifaximin 2 weeks before and 6 months after TIPS [36]. Hence, in non-urgent TIPS, Rifaximin should probably be used as a prevention of post-TIPS HE. The duration of therapy after TIPS remains to be determined.

Conclusion

Gastrointestinal bleeding is a classic precipitating factor of HE. The presence of HE is associated with an impaired prognosis in the case of variceal bleeding, with significantly higher mortality. HE should be prevented and treated if present at the onset of bleeding, and TIPS should not be contra-indicated because of HE in the setting of bleeding.

References

1. Jalan R, Olde Damink SW, Lui HF, Glabus M, Deutz NE, Hayes PC, et al. Oral amino acid load mimicking hemoglobin results in reduced regional cerebral perfusion and deterioration in memory tests in patients with cirrhosis of the liver. Metab Brain Dis. 2003;18(1):37–49.
2. Shawcross DL, Sharifi Y, Canavan JB, Yeoman AD, Abeles RD, Taylor NJ, et al. Infection and systemic inflammation, not ammonia, are associated with grade 3/4 hepatic encephalopathy, but not mortality in cirrhosis. J Hepatol. 2011;54(4):640–9.
3. Dunne PDJ, Sinha R, Stanley AJ, Lachlan N, Ireland H, Shams A, et al. Randomised clinical trial: standard of care versus early-transjugular intrahepatic porto-systemic shunt (TIPSS) in patients with cirrhosis and oesophageal variceal bleeding. Aliment Pharmacol Ther. 2020;52(1):98–106.
4. Jepsen P, Ott P, Andersen PK, Sorensen HT, Vilstrup H. Clinical course of alcoholic liver cirrhosis: a Danish population-based cohort study. Hepatology. 2010;51(5):1675–82.
5. Rudler M, Bureau C, Carbonell N, Mathurin P, Saliba F, Mallat A, et al. Recalibrated MELD and hepatic encephalopathy are prognostic factors in cirrhotic patients with acute variceal bleeding. Liver Int. 2018;38(3):469–76.

6. Gluud LL, Vilstrup H, Morgan MY. Non-absorbable disaccharides versus placebo/no intervention and lactulose versus lactitol for the prevention and treatment of hepatic encephalopathy in people with cirrhosis. Cochrane Database Syst Rev. 2016;4:CD003044.
7. Gluud LL, Vilstrup H, Morgan MY. Nonabsorbable disaccharides for hepatic encephalopathy: a systematic review and meta-analysis. Hepatology. 2016;64(3):908–22.
8. Sharma P, Agrawal A, Sharma BC, Sarin SK. Prophylaxis of hepatic encephalopathy in acute variceal bleed: a randomized controlled trial of lactulose versus no lactulose. J Gastroenterol Hepatol. 2011;26(6):996–1003.
9. Wen J, Liu Q, Song J, Tong M, Peng L, Liang H. Lactulose is highly potential in prophylaxis of hepatic encephalopathy in patients with cirrhosis and upper gastrointestinal bleeding: results of a controlled randomized trial. Digestion. 2013;87(2):132–8.
10. Aires FT, Ramos PT, Bernardo WM. Efficacy of lactulose in the prophylaxis of hepatic encephalopathy in cirrhotic patients presenting gastrointestinal bleeding. Rev Assoc Med Bras (1992). 2016;62(3):243–7.
11. Rolachon A, Zarski JP, Lutz JM, Fournet J, Hostein J. Is the intestinal lavage with a solution of mannitol effective in the prevention of post-hemorrhagic hepatic encephalopathy in patients with liver cirrhosis? Results of a randomized prospective study. Gastroenterol Clin Biol. 1994;18(12):1057–62.
12. Tromm A, Griga T, Greving I, Hilden H, Hüppe D, Schwegler U, et al. Orthograde whole gut irrigation with mannite versus paromomycine + lactulose as prophylaxis of hepatic encephalopathy in patients with cirrhosis and upper gastrointestinal bleeding: results of a controlled randomized trial. Hepato-Gastroenterology. 2000;47(32):473–7.
13. Garcia-Pagan JC, Caca K, Bureau C, Laleman W, Appenrodt B, Luca A, et al. Early use of TIPS in patients with cirrhosis and variceal bleeding. N Engl J Med. 2010;362(25):2370–9.
14. Lv Y, Yang Z, Liu L, Li K, He C, Wang Z, et al. Early TIPS with covered stents versus standard treatment for acute variceal bleeding in patients with advanced cirrhosis: a randomised controlled trial. Lancet Gastroenterol Hepatol. 2019;4(8):587–98.
15. García-Pagán JC, Bosch J, Trebicka J, Abraldes JG, Albillos A, Grønbaek H, et al. Letter: improve survival! Place early pre-emptive TIPSS in high-risk variceal bleeders. Aliment Pharmacol Ther. 2020;52(5):927–8.
16. Hernández-Gea V, Procopet B, Giráldez Á, Amitrano L, Villanueva C, Thabut D, et al. Preemptive-TIPS improves outcome in high-risk variceal bleeding: an observational study. Hepatology. 2019;69(1):282–93.
17. Lv Y, Zuo L, Zhu X, Zhao J, Xue H, Jiang Z, et al. Identifying optimal candidates for early TIPS among patients with cirrhosis and acute variceal bleeding: a multicentre observational study. Gut. 2019;68(7):1297–310.
18. Thabut D, Pauwels A, Carbonell N, Remy AJ, Nahon P, Causse X, et al. Cirrhotic patients with portal hypertension-related bleeding and an indication for early-TIPS: a large multicentre audit with real-life results. J Hepatol. 2017;68(1):73–81.
19. Bai M, Qi X, Yang Z, Yin Z, Nie Y, Yuan S, et al. Predictors of hepatic encephalopathy after transjugular intrahepatic portosystemic shunt in cirrhotic patients: a systematic review. J Gastroenterol Hepatol. 2011;26(6):943–51.
20. Holster IL, Tjwa ET, Moelker A, Wils A, Hansen BE, Vermeijden JR, et al. Covered transjugular intrahepatic portosystemic shunt versus endoscopic therapy + β-blocker for prevention of variceal rebleeding. Hepatology. 2016;63(2):581–9.
21. Sauerbruch T, Mengel M, Dollinger M, Zipprich A, Rössle M, Panther E, et al. Prevention of rebleeding from esophageal varices in patients with cirrhosis receiving small-diameter stents versus hemodynamically controlled medical therapy. Gastroenterology. 2015;149(3):660–668.e661.
22. Nardelli S, Gioia S, Pasquale C, Pentassuglio I, Farcomeni A, Merli M, et al. Cognitive impairment predicts the occurrence of hepatic encephalopathy after transjugular intrahepatic Portosystemic shunt. Am J Gastroenterol. 2016;111(4):523–8.
23. Berlioux P, Robic MA, Poirson H, Métivier S, Otal P, Barret C, et al. Pre-transjugular intrahepatic portosystemic shunts (TIPS) prediction of post-TIPS overt hepatic encepha-

lopathy: the critical flicker frequency is more accurate than psychometric tests. Hepatology. 2014;59(2):622–9.
24. Rudler M, Savier E, Alioua I, Sultanik P, Thabut D. TIPS and liver transplantation should always be discussed together. J Hepatol. 2021;75(4):1000–1.
25. Qi X, Tian Y, Zhang W, Yang Z, Guo X. Covered versus bare stents for transjugular intrahepatic portosystemic shunt: an updated meta-analysis of randomized controlled trials. Ther Adv Gastroenterol. 2017;10(1):32–41.
26. Triantafyllou T, Aggarwal P, Gupta E, Svetanoff WJ, Bhirud DP, Singhal S. Polytetrafluoroethylene-covered stent graft versus bare stent in transjugular intrahepatic portosystemic shunt: systematic review and meta-analysis. J Laparoendosc Adv Surg Tech A. 2018;28(7):867–79.
27. Miraglia R, Maruzzelli L, Tuzzolino F, Petridis I, D'Amico M, Luca A. Transjugular intrahepatic portosystemic shunts in patients with cirrhosis with refractory ascites: comparison of clinical outcomes by using 8- and 10-mm PTFE-covered stents. Radiology. 2017;284(1):281–8.
28. Riggio O, Ridola L, Angeloni S, Cerini F, Pasquale C, Attili AF, et al. Clinical efficacy of transjugular intrahepatic portosystemic shunt created with covered stents with different diameters: results of a randomized controlled trial. J Hepatol. 2010;53(2):267–72.
29. Wang Q, Lv Y, Bai M, Wang Z, Liu H, He C, et al. Eight millimetre covered TIPS does not compromise shunt function but reduces hepatic encephalopathy in preventing variceal rebleeding. J Hepatol. 2017;67(3):508–16.
30. Fanelli F, Salvatori FM, Rabuffi P, Boatta E, Riggio O, Lucatelli P, et al. Management of refractory hepatic encephalopathy after insertion of TIPS: long-term results of shunt reduction with hourglass-shaped balloon-expandable stent-graft. AJR Am J Roentgenol. 2009;193(6):1696–702.
31. Kochar N, Tripathi D, Ireland H, Redhead DN, Hayes PC. Transjugular intrahepatic portosystemic stent shunt (TIPSS) modification in the management of post-TIPSS refractory hepatic encephalopathy. Gut. 2006;55(11):1617–23.
32. Maleux G, Verslype C, Heye S, Wilms G, Marchal G, Nevens F. Endovascular shunt reduction in the management of transjugular portosystemic shunt-induced hepatic encephalopathy: preliminary experience with reduction stents and stent-grafts. AJR Am J Roentgenol. 2007;188(3):659–64.
33. Riggio O, Masini A, Efrati C, Nicolao F, Angeloni S, Salvatori FM, et al. Pharmacological prophylaxis of hepatic encephalopathy after transjugular intrahepatic portosystemic shunt: a randomized controlled study. J Hepatol. 2005;42(5):674–9.
34. Bai M, He C, Yin Z, Niu J, Wang Z, Qi X, et al. Randomised clinical trial: L-ornithine-L-aspartate reduces significantly the increase of venous ammonia concentration after TIPSS. Aliment Pharmacol Ther. 2014;40(1):63–71.
35. Riggio O, Nardelli S, Pasquale C, Pentassuglio I, Gioia S, Onori E, et al. No effect of albumin infusion on the prevention of hepatic encephalopathy after transjugular intrahepatic portosystemic shunt. Metab Brain Dis. 2016;31(6):1275–81.
36. Bureau C, Thabut D, Jezequel C, Archambeaud I, D'Alteroche L, Dharancy S, et al. The use of rifaximin in the prevention of overt hepatic encephalopathy after transjugular intrahepatic portosystemic shunt: a randomized controlled trial. Ann Intern Med. 2021;174(5):633–40.

Management of Coagulation in Acute Variceal Bleeding

43

Bogdan Procopet

Coagulation in Cirrhosis

In the last years, growing evidence broke the dogma of "natural anticoagulation" in patients with advanced liver disease. Depending on the level of liver dysfunction, patients with cirrhosis display abnormalities of the conventional hemostatic test, such as platelet count or Prothrombin Time (PT)/international normalized ratio (INR). All the complex hemostasis mechanism components are altered while the process is fragilely rebalanced (Fig. 43.1). In patients with cirrhosis, in all the three phases of the hemostasis, changes that promote either coagulation or bleeding occur concomitantly.

The primary hemostasis is marked by thrombocytopenia that often triggers prophylactic platelet transfusions before invasive procedures. However, this decision is not considering the counterbalanced increased levels of von Willebrand Factor (vWF), which binds the platelets together with factor VIII to the extracellular matrix forming the platelet plug [1]. The reduction of ADAMTS13 (A Disintegrin And Metalloprotease with ThromboSpondin type 1 motif 13), the cleavage enzyme that limits the effects of multimers of vWF, causes higher levels of vWF [1].

In the secondary hemostasis, all the procoagulant factors (FII, FV, FVII, FIX, FX, FX) and natural anticoagulants (protein C, S, and antithrombin III), synthesized by the liver, are concomitantly decreased. Contrary to this tendency, endothelial derivate factor VIII, a potent prothrombotic factor, generally increases in cirrhosis due to endothelial activation [1].

B. Procopet (✉)
Gastroenterology Department, 3rd Medical Clinic, "Iuliu Hatieganu" University of Medicine and Pharmacy, Cluj-Napoca, Romania

Gastroenterology and Hepatology Department, Regional Institute of Gastroenterology and Hepatology "Octavian Fodor", Cluj-Napoca, Romania
e-mail: bogdan.procopet@umfcluj.ro

R. de Franchis (ed.), *Portal Hypertension VII*,
https://doi.org/10.1007/978-3-031-08552-9_43

	Favor thrombosis	Favor bleeding
Primary hemostasis	↑ vWF ↓ ADAMTS 13	↓ platelet count
Secondary hemostasis	↑ VIII ↓ protein C, S, AT III	↓ II, V, VII, IX, X, XI ↓ fibrinogen /dysfib
Fibrinolysis	↑ PAI-1 ↓ plasminogen	↑ tPA ↓ α2 antiplasmin, XIII, TAFI

Fig. 43.1 The global hemostatic changes that occur in patients with cirrhosis and contribute to a rebalance coagulation (modified after Lisman et al. [1])

In the final fibrinolytic phase of the hemostasis, low levels of plasminogen, α2 antiplasmin, thrombin activatable fibrinolysis inhibitor (TAFI), and factor XIII are encountered. In contrast, tissue Plasminogen Activator (tPA) and plasminogen activator inhibitor type 1 (PAI-1), which blocks fibrinolysis, are elevated [1]. The balance between the factors promoting coagulation and those promoting bleeding can be easily broken under certain precipitating factors, like bacterial infections, resulting in bleeding or thrombotic complication.

Assessment of Hemostasis in Cirrhosis

Classically, the hemostasis is assessed by conventional coagulation tests as prothrombin time (PT)/International Normalized Ratio (INR), activated Partial Thromboplastin Time (aPTT), platelet count, or fibrinogen. The prothrombin time (INR) explores the extrinsic and the final common pathway of thrombin generation. It represents the time until the plasma starts to clot after adding an extrinsic pathway activator (thromboplastin) and calcium. The most significant disadvantage of PT/INR is that it counts only the first 5% of thrombin generation and explores only the procoagulant factors [2].

Under the action of the endothelial derivate thrombomodulin, the protein C becomes activated and, together with protein S, counterbalances the coagulation process by reducing the activation of factors VIII and V [2]. Without thrombomodulin, the PT/INR leaves unexplored the natural anticoagulation pathway. Activated partial thromboplastin time (aPTT) explores the intrinsic and common path and has the same disadvantages as PT/INR.

The Endogenous Thrombin Potential (ETP) is a global coagulation test that explores the amount of thrombin generated after the activation of the coagulation by adding recombinant tissue factor [2]. By adding thrombomodulin, ETP will also consider the natural anticoagulation pathway (through proteins C and S), and, consequently, the amount of generated thrombin will be less. Patients with cirrhosis generate significantly less thrombin than healthy controls when ETP is done without thrombomodulin. However, by adding thrombomodulin, patients with cirrhosis generate a similar amount of thrombin as healthy controls [2]. The ratio of ETP with and without thrombomodulin allows us to investigate how efficiently the anticoagulation pathway may be activated. It appears that in patients with cirrhosis, this ratio is increased progressively with liver dysfunction [3]. In other words, the patients with severe liver impairment develop resistance to thrombomodulin, have lower levels of protein C and higher levels of factor VIII, and, therefore, have a more pronounced prothrombotic profile [3]. Investigated by conventional coagulation tests, these kinds of patients would be wrongly considered as spontaneously anticoagulated. Unfortunately, there is no ETP assay to be used in clinical practice, and, therefore, its applicability is limited to research.

Recently, global Viscoelastic Tests (VETs) of hemostasis are increasingly used as a "point-of-care" assessment of complex hemostatic abnormalities. The advantages of these tests lie in providing real-time, dynamic information about the whole coagulation process, including clot initiation (thrombin generation), clot kinetics, clot strength, and clot stability (lysis) [4]. The assessment of the fibrinolysis seems to be a great advantage of the VETs since accelerated intravascular coagulation and fibrinolysis may occur in patients with decompensated cirrhosis [5]. The backside of the coin is that VETs provide information neither about the natural anticoagulation pathway of the protein C dependent on thrombomodulin or the elevated levels of vWF. Thromboelastography (TEG) and Rotational Thromboelastometry (ROTEM) are the current commercially available VETs techniques, both measuring the viscoelastic properties of clot formation and offering similar information [4].

Several randomized controlled trials concluded that conventional coagulation tests do not predict the risk of bleeding and the amount of transfusion decrease using a VET-guided protocol [4]. Most of these studies evaluated the VET-guided correction strategy of hemostasis abnormalities by blood product transfusions before invasive procedures in patients with cirrhosis. In most studies, the primary endpoint for which the sample size was calculated was reducing transfusion requirements. Therefore, VETs were not able to predict procedure-related bleeding. Similarly, in the only randomized control trial in the setting of acute variceal bleeding, TEG reduced Fresh Frozen Plasma (FFP) and platelet transfusions, without difference in bleeding control in both groups [6]. The only predictive difference was in favor of TEG regarding the 42-days rebleeding rate. By now, there is no evidence that VETs may predict better than conventional coagulation tests the bleeding events. Regarding the prediction of long-term prognosis (bleeding, thrombosis, or survival), by now, the VETs seem to have no role. Calibrating sample-size calculations to detect differences regarding clinical endpoints will, eventually, validate the VETs as practical hemostasis evaluation techniques in patients with cirrhosis in different clinical scenarios.

Correction of Coagulation Parameters in Acute Variceal Bleeding

It should be highlighted from the beginning that variceal bleeding is a complication of increased portal pressure and not a coagulation failure. Therefore, judicious and restrictive transfusion policies were associated with better prognoses in patients with cirrhosis and acute variceal bleeding [7]. Liberal transfusion strategy is associated with a significant increase in portal pressure measured by the hepatic venous pressure gradient [7]. The increase of the portal pressure is not related to the type of blood product but the volume of rapid infusion. Because the increase in portal pressure may lead to rebleeding, it is easy to understand why over-transfusions are associated with a worse prognosis in patients with acute variceal bleeding.

Apart from the risk of increased portal pressure, other adverse effects from blood product transfusions should be considered, like allergies, Transfusion-related Acute Lung Injury (TRALI), infections, or circulatory volume overload.

What is somehow troublesome is that one-third of patients with cirrhosis are transfused with at least one blood product [8]. Among those receiving FFP, 24% received transfusions in the absence of any bleeding or planned invasive procedures, in 31% of the patients, the PT or INR was not even checked before transfusion, while in one-quarter of the patients, the threshold used to trigger transfusion was below any benefit [8]. There is probably a lot of inertia in the clinical recommendation of FFP transfusions without considering the actual benefit and risks of the transfusion.

Fresh Frozen Plasma

FFP is prepared from whole blood by separating the plasma, and it contains all pro and natural anticoagulants in physiologic concentration. It is packed in approximately 250 mL per unit, and the usual dose is 10 mL/kg body weight. There is no randomized controlled trial assessing the role of coagulation correction by FFP transfusion in patients with liver diseases and, it is implausible to be ever carried out. In observational studies, adding FFP to cirrhotic patients with increased INR, does not translate to an increase in ETP with thrombomodulin, which is already normal in the vast majority of the patients before transfusions [9]. Adding FFP to the plasma of patients with cirrhosis is also associated with a significant increase in plasma levels of markers for in vivo activation of coagulation, as Thrombin/antithrombin Complexes (TAT) and prothrombin fragment 1 + 2 (F1 + 2), leading to a prothrombotic effect [10].

From the clinical safety point of view of FFP transfusion in AVB, there is only limited data. In a retrospective study comparing patients with AVB who received FFP with those without transfusion, it seems that patients receiving FFP had a lower 42-days survival, a higher rate of failure to control bleeding, and more extended hospitalization [11]. However, given the retrospective design, the probability of selection bias is relatively high. The patients from the FFP transfusion group had a more advanced liver disease with a higher prevalence of ascites and hepatic encephalopathy. Despite that, the results were also maintained after adjustment for MELD and Child-Pugh scores.

In conclusion, correcting the conventional coagulation test (PT/INR) in the setting of AVB will translate into a worse prognosis due to the increase in portal pressure and being utterly inefficient in the hemostasis process.

Platelet's Transfusion

The platelets are significant players in initiating the hemostatic process and represent scaffolds for pro-coagulation factor complexes that generate thrombin. The platelets adhere at the site of vascular wall damage through vWF and collagen and become activated, amplifying the platelet-to-platelet adhesion through vWF and fibrinogen forming the platelet plug.

There are conflicting findings regarding the association between bleeding complications and thrombocytopenia. There was no consistency regarding platelet count threshold in the studies, suggesting an increased risk of bleeding, depending on the setting and the procedures. Using ETP, Tripodi et al. demonstrated that a minimum of 56,000/μL platelets count is needed to conserve a normal thrombin generation [12]. However, the ETP does not consider the adhesion of platelets, their activation, and neither the changes that occur in cirrhosis and counterbalance the low platelet count, as increased vWF. It seems that platelet transfusion may lead to even a pro-thrombotic effect through the activation of platelets (increasing the CD40 ligand) and coagulation mechanism (an increase of thrombin–antithrombin complexes) [10].

Fibrinogen

Fibrinogen is a significant component of the primary and secondary hemostasis phases. Fibrinogen levels may be corrected by cryoprecipitate or fibrinogen concentrate administration. Although some studies associated low fibrinogen levels with bleeding complications in liver diseases [13], no precise cutoffs that should trigger the transfusion are known [5]. Anyway, clinical efficiency was not yet established.

Recombinant Activated Factor VII

The rationale for using recombinant activated factor VII (rVIIa) would be that in patients with cirrhosis, factor VII levels are usually low and because rVIIa proved to be efficient in the treatment of congenital deficiencies. Two randomized control trials and an individual data metanalysis tested rVIIa in addition to the standard therapy in patients with cirrhosis and upper gastrointestinal bleeding. The primary endpoint was composite, including failure to control bleeding in the first 24 h, prevention of further rebleeding, and survival at day 5. Finally, rVIIa is not more efficient than a placebo in controlling bleeding or improving survival [14]. Despite a potential benefit regarding failure to control bleeding in patients with advanced liver function and active bleeding at endoscopy, there was a trend toward arterial thrombotic events in the treated arm.

Tranexamic Acid

Tranexamic Acid (TA) is an antifibrinolytic drug that inhibits the plasminogen interaction with fibrin. According to the level of liver dysfunction, patients with cirrhosis develop profound changes in the fibrinolytic phase of the hemostasis that favor either hyperfibrinolysis or hypo fibrinolysis (Fig. 43.1). Although markers of hyperfibrinolysis were constantly found in patients with cirrhosis, suggesting a similar but still different entity with disseminated intravascular coagulation ("accelerated intravascular coagulation and fibrinolysis") [5], there are conflicting results regarding the fibrinolytic capacity of the plasma of patients with cirrhosis. Despite the favorable effect of antifibrinolytic drugs mainly in liver transplant transfusion requirements, a recent large RCT comparing TA with placebo in patients with gastrointestinal bleeding failed to show any benefit [15]. More than 12,000 patients were included in this trial, and around 45% were suspected of variceal bleeding. The lack of a clear description of the included population (approximately 40% had major liver comorbidity) and the fact that not all patients underwent diagnostic endoscopy (80%) or therapeutic endoscopy (a little more than 40%) represent major drawbacks of the study. Anyway, the subgroup analysis for variceal bleeding and liver patients gave negative results (risk ratio of 0.99 with 95%CI of 0.7–1.4) [15].

Conclusion

Although the hemostasis is profoundly modified in patients with advanced liver disease, the coagulation process is rebalanced. The conventional coagulation tests do not mirror this rebalanced process accurately. Therefore, correcting them by blood products transfusion did not result in clinical benefit but increased the risk of transfusion-related complications. It is crucial to understand that variceal bleeding is the consequence of increased portal pressure. Therefore, all efforts should be targeted toward decreasing portal pressure rather than correcting the coagulation test abnormalities.

References

1. Lisman T, Hernandez-Gea V, Magnusson M, Roberts L, Stanworth S, Thachil J, et al. The concept of rebalanced hemostasis in patients with liver disease: communication from the ISTH SSC working group on hemostatic management of patients with liver disease. J Thromb Haemost. 2021;19:1116–22.

2. Tripodi A, Salerno F, Chantarangkul V, Clerici M, Cazzaniga M, Primignani M, et al. Evidence of normal thrombin generation in cirrhosis despite abnormal conventional coagulation tests. Hepatology. 2005;41:553–8.
3. Tripodi A, Primignani M, Chantarangkul V, Dell'Era A, Clerici M, de Franchis R, et al. An imbalance of pro- vs anti-coagulation factors in plasma from patients with cirrhosis. Gastroenterology. 2009;137:2105–11.
4. Mallett SV. Clinical utility of viscoelastic tests of coagulation (TEG/ROTEM) in patients with liver disease and during liver transplantation. Semin Thromb Hemost. 2015;41:527–37.
5. Intagliata NM, Argo CK, Stine JG, Lisman T, Caldwell SH, Violi F. Concepts and controversies in haemostasis and thrombosis associated with liver disease: proceedings of the 7th International Coagulation in Liver Disease Conference. Thromb Haemost. 2018;118:1491–506.
6. Rout G, Shalimar, Gunjan D, Mahapatra SJ, Kedia S, Garg PK, et al. Thromboelastography-guided Blood Product Transfusion in Cirrhosis Patients with Variceal Bleeding: A Randomized Controlled Trial. J Clin Gastroenterol. 2020;54:255–62.
7. Villanueva C, Colomo A, Bosch A, Concepción M, Hernandez-Gea V, Aracil C, et al. Transfusion strategies for acute upper gastrointestinal bleeding. N Engl J Med. 2013;368:11–21.
8. Desborough MJR, Hockley B, Sekhar M, Burroughs AK, Stanworth SJ, Jairath V. Patterns of blood component use in cirrhosis: a nationwide study. Liver Int. 2016;36:522–9.
9. Rassi AB, Antonio E, Tripodi A, Ferreira M, Carrilho FJ, Farias AQ. Fresh frozen plasma transfusion in patients with cirrhosis and coagulopathy: effect on conventional coagulation tests and thrombomodulin-modified thrombin generation. J Hepatol. 2020;72:85–94.
10. von Meijenfeldt FA, van den Boom BP, Adelmeijer J, Roberts LN, Lisman T, Bernal W. Prophylactic fresh frozen plasma and platelet transfusion have a prothrombotic effect in patients with liver disease. J Thromb Haemost. 2021;19:664–76.
11. Mohanty A, Kapuria D, Canakis A, Lin H, Amat MJ, Rangel Paniz G, et al. Fresh frozen plasma transfusion in acute variceal haemorrhage: results from a multicentre cohort study. Liver Int. 2021;41:1901–8.
12. Tripodi A, Primignani M, Chantarangkul V, Clerici M, Dell'Era A, Fabris F, et al. Thrombin generation in patients with cirrhosis: the role of platelets. Hepatology. 2006;44:440–5.
13. Giannini EG, Giambruno E, Brunacci M, Torres MCP, Furnari M, Bodini G, et al. Low fibrinogen levels are associated with bleeding after varices ligation in thrombocytopenic cirrhotic patients. Ann Hepatol. 2018;17:830–5.
14. Bendtsen F, D'Amico G, Rusch E, De Franchis R, Andersen PK, Lebrec D, et al. Effect of recombinant factor VIIa on outcome of acute variceal bleeding: an individual patient based meta-analysis of two controlled trials. J Hepatol. 2014;61:252–9.
15. The HALT-IT Collaborators. Effects of a high-dose 24-h infusion of tranexamic acid on death and thromboembolic events in patients with acute gastrointestinal bleeding (HALT-IT): an international randomised, double-blind, placebo-controlled trial. Lancet. 2020;395:1927–36.

44 Gastric Varices and Ectopic Varices

Xuefeng Luo and Li Yang

Introduction

Gastric Varices (GV) are present in approximately 20% of cirrhotic patients [1]. GV may bleed less frequently than Esophageal Varices (EV); however, rupture of GV is associated with more severe hemorrhage, higher mortality, and a greater risk of rebleeding [1]. GV can be categorized into four types based on Sarin's classification described in Chap. 39 [2]. GOV1 shares similar vascular anatomy with EV and follows similar management recommendations. In patients with IGV2, left-sided regional portal hypertension secondary to splenic vein obstruction should be considered [3].

Hemodynamic Features of GV

Imaging evaluation of GV is very important to guide treatment. Generally, GV may drain into the systemic circulation via the esophageal and paraesophageal varices, the left inferior phrenic vein (IPV), or both [4]. The left IPV can terminate inferiorly

X. Luo (✉) · L. Yang
Department of Gastroenterology and Hepatology, Sichuan University-University of Oxford Huaxi Joint Centre for Gastrointestinal Cancer, West China Hospital, Sichuan University, Chengdu, China
e-mail: luo_xuefeng@wchscu.cn; yangli@wchscu.cn

R. de Franchis (ed.), *Portal Hypertension VII*,
https://doi.org/10.1007/978-3-031-08552-9_44

into the left renal vein, transversely into the hepatic vein or inferior vena cava, or directly or ascendingly into the pericardiophrenic vein. In most cases, GOV1 drain into the esophageal and paraesophageal varices, IGV1 drain via the IPV, and GOV2 drain via both ways.

Although the presence of large collaterals may counteract the increased Portal Pressure Gradient (PPG), it cannot avert bleeding from GV. Unlike EV, patients with GV can still bleed when the PPG is below 12 mmHg. A study comprising 292 patients found that patients presenting with GV bleeding had a lower PPG than patients who bled from EV (15.8 vs. 21.4 mmHg) [5]. Therefore, decompression of GV using TIPS or surgical shunt may not be as efficacious as in patients with EV.

Management of Acute Gastric Variceal Bleeding

The medical management of acute gastric variceal bleeding does not differ from EV. In those with GV, once the patient is hemodynamically stable, cross-sectional imaging, preferably contrast-enhanced (CT or MRI) should be considered to evaluate the patency of the portal venous system, screen for liver malignancy, and detect the presence of large portosystemic collaterals. Endoscopic techniques including band ligation, glue injection, and endoscopic ultrasound-guided injection are described in Chap. 39.

Transjugular Intrahepatic Portosystemic Shunt (TIPS) creates an artificial shunt between the hepatic and portal veins in the liver to decompress the portal venous system. It is a well-established effective interventional procedure to control acute variceal bleeding. The use of early or preemptive TIPS (pTIPS) within 72 h (ideally <24 h) in patients with high risk of failure and/or rebleeding has proven to reduce treatment failure and improve survival; however, patients with acute bleeding from GOV2 & IGV1 have not been specifically evaluated [6]. Two ongoing randomized controlled trials evaluating the efficacy of pTIPS in GOV2 & IGV1 (NCT02364297 & NCT03705078) will help answer this question in the near future.

As in EV, failure to control bleeding despite combined pharmacological and endoscopic therapy is best managed by salvage PTFE-covered TIPS. In achieving initial hemostasis for acute GV bleeding, TIPS is equally effective as for EV bleeding. However, GOV1 and IGV1 may rebleed despite adequate decompression following TIPS (post-TIPS PPG $\leq$ 12 mmHg), particularly in cases when the portal flow remains diverted to collaterals. The reduced efficiency of TIPS is partially attributed to the presence of well-developed low-resistance large collaterals, or to the fact that the afferent veins are distant from the intrahepatic shunt. Embolization of GV has been proposed to increase the efficiency of the TIPS procedure (Fig. 44.1). A previous study found that TIPS combined with embolization could lower the risk of rebleeding compared with TIPS alone (13.4% vs. 28% at 2 years) [7] and should be considered in selected cases.

A standard Balloon-occluded Retrograde Transvenous Obliteration (BRTO) procedure involves occlusion of the draining veins of large collaterals, usually

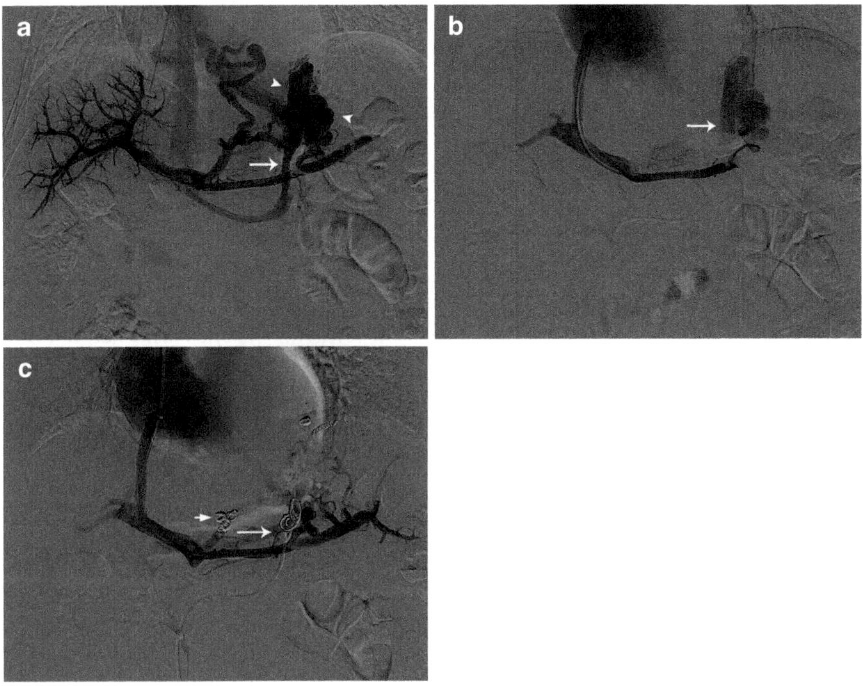

Fig. 44.1 (**a**) Direct portal venogram showing gastric varices (GV) (arrowheads) and the Gastrorenal Shunt (GRS; arrow). (**b**) A second venogram revealed that contrast emptied into the GRS (arrow) despite the TIPS being patent. (**c**) GV was not visualized after balloon-occluded retrograde transvenous obliteration (long arrow) and antegrade embolization with coils (short arrow)

gastrorenal shunt or gastrocaval shunt, followed by the injection of a sclerosing agent directly into the GV (Fig. 44.2). The concept of BRTO was first introduced by Olson et al. in 1984, then further developed in Japan by Kanagawa and his colleagues [8, 9]. BRTO has gained acceptance in Japan and Korea, and more recently in the USA and China. BRTO should be considered as a salvage choice for the management of failure to control or recurrent GV bleeding. Several studies have demonstrated that the rebleeding rate following salvage BRTO is generally less than 5% with an eradication rate of GV as high as 97.9% [10].

Aggravation of EV and ascites following BRTO is one of the major concerns due to the increase in portal hypertension following BRTO. The rates of aggravation of EV ranged from 9.8% to 72.2% with a pooled rate of 33.3% [11]. The high level of variance was probably due to the different degrees of awareness of the possibility of EV aggravation and to the different timeframes of follow-up endoscopy. Improved follow-up strategy, prophylactic ligation alone or in the combination of beta-blockers may reduce the occurrence of the EV and the risk of bleeding and should be considered.

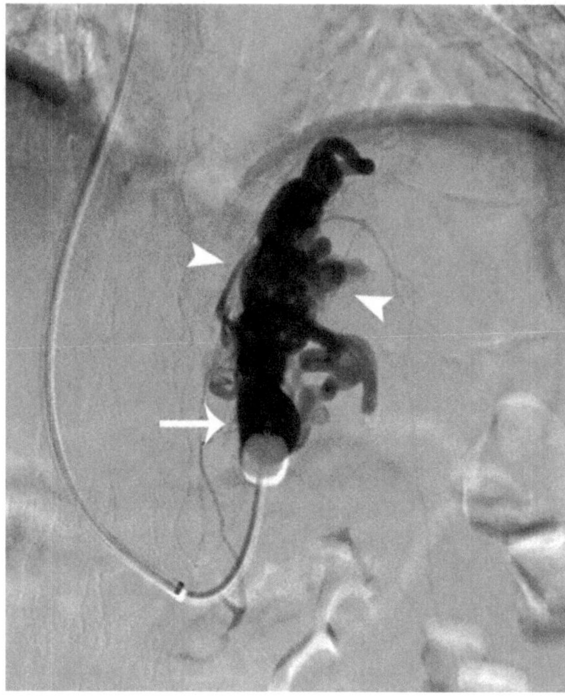

Fig. 44.2 The gastric varices (arrowheads) were filled with polidocanol foam after the balloon was inflated to stop the outflow of the gastrorenal shunt (arrow)

Ectopic Varices

Ectopic varices are dilated porto-portal or portosystemic collateral veins that occur outside the common pathologic variceal sites and constitute 2% to 5% of all variceal bleeding [12]. Ectopic varices can be caused by general or regional portal hypertension with or without splanchnic venous obstruction. A nationwide questionnaire survey in Japan collected 173 cases of ectopic varices and the most frequent sites were rectum (44.5%) and duodenum (32.9%) [13]. The management of ectopic variceal bleeding is challenging as most of the current knowledge comes from case reports and small case series, which include endoscopic treatments (endoscopic band ligation, injection sclerotherapy), embolization using coils or plugs, BRTO, TIPS, and surgical options. Imaging evaluation of ectopic varices and the presence of large collaterals and splanchnic venous thrombosis are very important for treatment planning. Either endovascular or endoscopic treatment should be considered in patients with ectopic varices and treatment should be individualized.

Currently, endoscopic treatment is one of the most common modalities for the management of bleeding from ectopic varices. Endoscopic band ligation, sclerotherapy, or glue injection are all treatment options, and their use depends on the location of varices and local expertise. Despite the rarity of cases, endoscopic modalities for the control of acute ectopic variceal bleeding achieve a high initial hemostasis rate. Rebleeding of ectopic varices may occur and can be treated successfully with additional endoscopic therapy or endovascular treatment.

Percutaneous embolization of ectopic varices is a safe and technically easy treatment option. Coil, plug, and liquid embolization material including glue and sclerosing agents have been reported [12]. Transvenous obliteration via antegrade or retrograde approach may be more advantageous in obliterating complex, multichanneled vascular structures. Still, embolization without decompression of portal hypertension or recanalization of the occluded vein may be less effective to prevent the reoccurrence of ectopic varices or rebleeding.

TIPS is a reliable option, as the underlying cause of bleeding ectopic varices is elevated portal pressure. However, TIPS may be less effective in decompressing ectopic varices with a rebleeding rate ranging from 11% to 37% [14, 15] depending on the localization.

Summary

In conclusion, gastric and ectopic varices in different locations are associated with various hemodynamic features. They should better be managed by a multidisciplinary team with multiple treatment options available after proper radiologic and endoscopic evaluation. Due to the limited number of high-quality studies and the scarcity of cases, strong evidence-based recommendations cannot be made. More randomized controlled trials are needed to better determine the priority of treatments in a particular subset of the population.

References

1. Garcia-Tsao G, Abraldes J, Berzigotti A, Bosch J. Portal hypertensive bleeding in cirrhosis: risk stratification, diagnosis and management—2016 practice guidance by the American Association for the Study of Liver Diseases. Hepatology. 2017;65(1):310–35. https://doi.org/10.1002/hep.28906.
2. Sarin SK, Lahoti D, Saxena SP, Murthy NS, Makwana UK. Prevalence, classification and natural history of gastric varices: a long-term follow-up study in 568 portal hypertension patients. Hepatology. 1992;16(6):1343–9.
3. Luo X, Nie L, Wang Z, Tsauo J, Tang C, Li X. Transjugular endovascular recanalization of splenic vein in patients with regional portal hypertension complicated by gastrointestinal bleeding. Cardiovasc Intervent Radiol. 2014;37(1):108–13.
4. Kiyosue H, Ibukuro K, Maruno M, Tanoue S, Hongo N, Mori H. Multidetector CT anatomy of drainage routes of gastric varices: a pictorial review. Radiographics. 2013;33(1):87–100.
5. Tripathi D, Therapondos G, Jackson E, Redhead DN, Hayes PC. The role of the transjugular intrahepatic portosystemic stent shunt (TIPSS) in the management of bleeding gastric varices: clinical and haemodynamic correlations. Gut. 2002;51(2):270–4.
6. Garcia-Pagan JC, Caca K, Bureau C, Laleman W, Appenrodt B, Luca A, et al. Early use of TIPS in patients with cirrhosis and variceal bleeding. N Engl J Med. 2010;362(25):2370–9.
7. Yu J, Wang X, Jiang M, Ma H, Zhou Z, Yang L, et al. Comparison of transjugular intrahepatic portosystemic shunt (TIPS) alone and combined with embolisation for the management of cardiofundal varices: a retrospective study. Eur Radiol. 2019;29(2):699–706.
8. Olson E, Yune HY, Klatte EC. Transrenal-vein reflux ethanol sclerosis of gastroesophageal varices. AJR Am J Roentgenol. 1984;143(3):627–8.

9. Kanagawa H, Mima S, Kouyama H, Gotoh K, Uchida T, Okuda K. Treatment of gastric fundal varices by balloon-occluded retrograde transvenous obliteration. J Gastroenterol Hepatol. 1996;11(1):51–8.
10. Mukund A, Rangarh P, Shasthry SM, Patidar Y, Sarin SK. Salvage balloon occluded retrograde transvenous obliteration for gastric variceal bleed in cirrhotic patients with endoscopic failure to control bleed/very early rebleed: long-term outcomes. J Clin Exp Hepatol. 2020;10(5):421–8.
11. Park JK, Saab S, Kee ST, Busuttil RW, Kim HJ, Durazo F, et al. Balloon-occluded retrograde transvenous obliteration (BRTO) for treatment of gastric varices: review and meta-analysis. Dig Dis Sci. 2015;60(6):1543–53.
12. Saad WE, Lippert A, Saad NE, Caldwell S. Ectopic varices: anatomical classification, hemodynamic classification, and hemodynamic-based management. Tech Vasc Interv Radiol. 2013;16(2):158–75.
13. Watanabe N, Toyonaga A, Kojima S, Takashimizu S, Oho K, Kokubu S, et al. Current status of ectopic varices in Japan: results of a survey by the Japan Society for Portal Hypertension. Hepatol Res. 2010;40(8):763–76.
14. Haskal ZJ, Scott M, Rubin RA, Cope C. Intestinal varices: treatment with the transjugular intrahepatic portosystemic shunt. Radiology. 1994;191(1):183–7.
15. Vangeli M, Patch D, Terreni N, Tibballs J, Watkinson A, Davies N, et al. Bleeding ectopic varices--treatment with transjugular intrahepatic porto-systemic shunt (TIPS) and embolisation. J Hepatol. 2004;41(4):560–6.

Special Settings: Acute Variceal Bleeding and Portal Vein Thrombosis in Cirrhosis

45

Yong Lv and Guohong Han

Introduction

Nonmalignant portal vein thrombosis (PVT) is a critical but not infrequent occurrence in patients with cirrhosis, with a prevalence ranging from 2% to 23% [1, 2]. The prevalence of PVT increases with the severity of liver disease or portal hypertension [3, 4]. A decrease in portal blood flow velocity is the main risk factor for PVT while acquired or inherited alterations of coagulation do not predict PVT development during follow-up [4–9]. Because of the heterogeneity of severity (partial or complete occlusion), presentation (asymptomatic, mesenteric ischemia, or portal hypertension), and evolution (spontaneous recanalization, stable or extension) of PVT and of the status of liver cirrhosis (compensated or decompensated), the impact of PVT on the natural history of cirrhosis remains poorly defined [3, 10, 11]. However, complete or extensive PVT complicates the liver transplant operation and increases post-transplantation morbidity and mortality [2]. Due to this detrimental

Y. Lv
Department of Liver Diseases and Digestive Interventional Radiology, National Clinical Research Center for Digestive Diseases and Xijing Hospital of Digestive Diseases, Fourth Military Medical University, Xi'an, China

Military Medical Innovation Center, Fourth Military Medical University, Xi'an, China
e-mail: hangh@fmmu.edu.cn

G. Han (✉)
Department of Liver Diseases and Digestive Interventional Radiology, National Clinical Research Center for Digestive Diseases and Xijing Hospital of Digestive Diseases, Fourth Military Medical University, Xi'an, China

Department of Liver Diseases and Interventional Radiology, Digestive Diseases Hospital, Xi'an International Medical Center Hospital, Northwest University, Xi'an, China

Department of Liver Diseases and Digestive Interventional Radiology, Xi'an International Medical Center Hospital of Digestive Diseases, Northwest University, Xi'an, China
e-mail: hangh@fmmu.edu.cn

R. de Franchis (ed.), *Portal Hypertension VII*,
https://doi.org/10.1007/978-3-031-08552-9_45

impact, in candidates for a liver transplant, therapy is recommended with the aim of recanalizing the portal vein and/or preventing recurrence or thrombosis progression and allowing physiological anastomosis during liver transplant surgery.

Acute Variceal Bleeding (AVB) is one of the most serious and feared complications of patients with cirrhosis [12]. In certain circumstances, AVB occurs in cirrhotic patients with PVT. Managing the AVB together with the prevention of thrombotic complications requires a clear evaluation of the risks and benefits of each therapeutic intervention and a fine-tuning of their balance. Because data for the treatment of AVB in cirrhotic patients with PVT are extremely limited, few definitive conclusions can be drawn from the literature. The purpose of this chapter based on an overview of data available is to draw some conclusions regarding the management of AVB in patients with PVT and cirrhosis.

The Impact of PVT on the Prognosis of Patients with Cirrhosis and AVB

Whether PVT impacts the natural history of cirrhotic patient is still a matter of debate. The lack of firm evidence is in part due to the scarcity of well-conducted prospective studies with well-defined populations, staging of PVT, and endpoints. Most of the currently available studies suggest that PVT is a marker of cirrhosis severity and is unlikely to have deleterious effects [3, 13, 14]. A multicenter prospective longitudinal study by Nery [3] showed that the development of PVT did not follow a progression of hepatic decompensation and increased mortality in cirrhosis. Nevertheless, most patients included in this study were compensated and had only partial/nonocclusive PVT. Moreover, the high rate of spontaneous recanalization (70%) might also have influenced the results [3]. Therefore, the real impact of occlusive/progressive PVT on the natural history of cirrhosis needs further evaluation.

In the setting of AVB, few studies evaluated the impact of PVT on the prognosis in such patients. A multicenter, prospective, cohort study performed by D'Amico [15] evaluated the short-term outcomes and prognostic indicators of cirrhotic patients with upper digestive bleeding. In the multivariate logistic regression analysis, the presence of PVT was an independent predictor for 5-day failure (for any sources of bleeding: odds ratio = 3.19, 95% confidence interval: 1.53–6.67, $P = 0.002$; for variceal rebleeding: odds ratio = 3.06, 95% confidence interval: 1.39–6.68, $P = 0.005$) but not 6-week mortality. Amitrano [16] studied 185 cirrhotic patients with AVB treated with current first-line therapy (92% of them receiving endoscopic variceal ligation as emergency endoscopic therapy), 28% of them with hepatocellular carcinoma, and 17% with portal vein thrombosis. By logistic regression analysis, the presence of portal vein thrombosis, together with the Child-Pugh score and white blood cell count, were identified as independent predictors of 5-day treatment failure. Chen [17] studied 101 cirrhotic patients with endoscopy-proven active esophageal variceal bleeding who underwent esophageal variceal ligation. Among them, 25 patients had PVT. They found that PVT along with door-to-endoscopy time and Model for End-Stage Liver Disease (MELD) score were associated with 6-week rebleeding, while hematemesis upon arrival, MELD score, and hepatocellular carcinoma were indicators of 6-week mortality. In a recent retrospective study of 218 cirrhotic patients with AVB, Gao [18] showed that patients with PVT had a higher rate of 14-day and 6-week rebleeding compared with those without PVT (14-day: 8.26% vs. 1.83%, $p = 0.03$; 6-week: 11.92% vs. 1.83%, $p = 0.003$). However, the rates of 5-day failure (3.67% vs. 0.92%, $p = 0.175$), 1-year rebleeding (21.10% vs. 20.18%, $p = 0.867$), and 14-day, 6-week, and 1-year mortality were not significantly different between the two groups (14-day: 3.67% vs. 0.92%, $p = 0.175$; 6-week: 3.67% vs. 0.92%, $p = 0.175$; 1-year: 3.67% vs. 1.83%, $p = 0.408$). Multivariate Cox regression analysis revealed that PVT was associated with high 14-day ($p = 0.05$, HR: 4.622, 95% CI 0.99–21.39) and 6-week rebleeding ($p = 0.012$, HR: 6.732, 95% CI 1.52–29.84) but not 1-year mortality. It should be noted that most of the above-mentioned studies also include patients with hepatocellular carcinoma but whether patients had or not tumoral PVT was not clearly reported, which may have strongly influenced the outcomes. Furthermore, since patients with PVT and cirrhosis generally had worse liver function than those without PVT and a limited number of variables were adjusted in those studies, it is unclear whether the influence of PVT on the prognosis of AVB was real or mediated by its association with more severe liver failure.

Consequently, all patients admitted to the hospital due to AVB should undergo imaging screening for PVT. Current evidence from case series and observational studies has not provided robust data for the impact of PVT on the prognosis of cirrhotic patients with AVB. Based on available evidence, we cannot conclude that PVT has a detrimental impact on the outcome of cirrhotic patients with AVB as survival was not affected. Further large prospective studies remain needed to clarify this topic.

Management of AVB in Patients with Cirrhosis and PVT

There have been no data specifically reported on the management of AVB in patients with PVT. The particular issue that a marked reduction in portal venous blood flow related to vasoconstrictor therapy could cause PVT has not been addressed. Neither has the issue of the early or preemptive Transjugular Intrahepatic Portosystemic Shunt (TIPS) in patients with PVT and active bleeding at endoscopy. Because the data presently available on the management of variceal bleeding in this context are limited, recommendations for clinical practice cannot be based on the solid evidence. The management of AVB in patients with cirrhosis and PVT is mainly based on data from studies in patients with AVB in the absence of PVT, which includes a stepped care approach aimed at resuscitation, restrictive transfusion policy, antibiotic prophylaxis, pharmacologic therapy with vasoconstrictors, and endoscopic therapy [18–21].

The advantage of TIPS is that it reconstructs portal vein flow, in addition to promoting PVT recanalization, and improves portal hypertension globally, therefore reducing the risk of variceal bleeding. Indeed, two Randomized Controlled Trials

(RCTs) have shown that covered TIPS placement in patients with PVT and moderately decompensated cirrhosis was more effective than endoscopic variceal ligation combined with propranolol for the prevention of rebleeding, with a higher probability of PVT resolution without increasing the risk of overt hepatic encephalopathy and adverse effects [35, 36]. One randomized controlled trial showed that post-TIPS anticoagulation is not necessary for patients with cirrhosis and PVT because TIPS placement alone can achieve a high persistent recanalization rate [22]. However, there is no specific randomized controlled clinical trial comparing TIPS and other management strategies in the setting of AVB and PVT. Recent guidelines from the European Association for the Study of the Liver (EASL) [23] and the American Association for the Study of Liver Diseases (AASLD) [24] suggest that TIPS can be considered in patients with acute PVT who do not respond to anticoagulation or in those with chronic PVT with recurrent bleeding not manageable medically or endoscopically. Further studies should clarify whether cirrhotic patients with AVB and PVT could benefit from preemptive TIPS, particularly for those awaiting liver transplantation.

The Anticoagulation Therapy in Patients with AVB and PVT

Anticoagulation has been recommended in patients with PVT in order to prevent the extension of the thrombus to the superior mesenteric vein or to achieve portal vein recanalization when possible [23, 24]. In cirrhosis patients with both variceal bleeding and PVT, clinicians often face a dilemma of either increased risk of rebleeding with anticoagulation treatment or potential worsening of the PVT if the patient is not anticoagulated or the initiation of anticoagulation therapy is delayed until after variceal eradication [25].

On the one hand, the timing of initiation of anticoagulation therapy could affect portal vein recanalization rates. A few observational nonrandomized studies evaluated the safety and efficacy of anticoagulation for PVT in patients with cirrhosis [26–35]. These studies have shown that after anticoagulation therapy, PVT improvement is observed in 60%–100% of patients and that the most important factor predicting PVT recanalization is the delay in initiating anticoagulation. A prospective study of 56 patients indicated that anticoagulation started <6 months from the formation of the thrombus predicted recanalization [34]. Similar results were seen in a retrospective review which concluded that the only factor significantly associated with recanalization was early initiation of therapy, particularly in the first 2-weeks [32].

On the other hand, one of the main concerns for the use of anticoagulants in cirrhosis, especially in patients with AVB, is the risk of increasing the severity of eventual variceal bleedings. In this regard, a small pilot study in patients with variceal bleeding within the context of PVT did not show an increased risk of bleeding in patients receiving acute low-molecular-weight heparin (LMWH) [36]. A multicenter, retrospective, matched study showed that the outcome of upper gastrointestinal bleeding in patients with cirrhosis receiving anticoagulant therapy was related to the degree of multiorgan failure and comorbidities (mainly cardiac disease), rather than to the anticoagulant therapy itself [37]. A meta-analysis by Loffredo [26] showed that the rate of variceal bleeding was significantly lower in anticoagulant-treated patients for PVT compared with untreated patients (odds ratio, 0.23). Furthermore, two recent small retrospective studies showed that LMWH does not increase the risk of bleeding and death in cirrhotic patients undergoing prophylactic endoscopic variceal ligation [38, 39]. Another small study suggested that even in patients with AVB, initiation of anticoagulation immediately after endoscopic hemostasis did not increase the risk of local bleeding [36]. Consequently, the risk of variceal bleeding appears to be low in patients receiving anticoagulation therapy when the recommendations for prophylaxis of variceal bleeding in cirrhosis are followed. Therefore, with regard to variceal bleeding, it appears that anticoagulation does not play a major deleterious role and the risk of variceal bleeding is highly dependent on portal pressure [40]. In our opinion, in patients with cirrhosis and AVB who need anticoagulation therapy, anticoagulation should be initiated as soon as possible after full control of AVB.

The management of AVB occurring in patients taking anticoagulants for PVT raises several difficulties related to the balance between thrombotic risks associated with drug discontinuation and hemorrhagic risks [41–43]. There is a paucity of studies addressing the issue of AVB in anticoagulated patients and a total absence of RCTs comparing different management strategies. Furthermore, the recent introduction of Direct Oral Anticoagulants (DOACs), for whom specific reversal agents are still lacking, further contributes to make the decision-making process even more demanding. However, for patients with PVT and AVB, the immediate risk from AVB may outweigh the risk of thrombosis as a result of stopping anticoagulant therapy [44]. For patients with high-risk conditions on anticoagulation (e.g., atrial fibrillation, mechanical heart valves, or pulmonary embolism), early therapeutic endoscopic intervention may achieve hemostasis with minimal or no cessation of anticoagulant therapy and should be the first aim [42, 43, 45]. There are data to inform the physician's decision to resumption of antithrombotic therapy. In general, once endoscopic hemostasis has been assured, anticoagulants should be restarted as soon as possible [41, 44]. In situations where hemostasis is uncertain, discussion with the patient's cardiologist and/or hematologist is important to ensure an individualized approach for each patient [41, 44].

Conclusion

In conclusion, the impact of PVT on the prognosis of patients with cirrhosis and AVB is still a matter of debate. The management of AVB in cirrhotic patients with PVT has not been the topic for specific randomized controlled clinical trials. Available data do not allow for making solid recommendations for the treatment of AVB in this subgroup of patients. It appears reasonable to apply to these patients the recommendations developed for patients without PVT. After implementing adequate prophylaxis for gastrointestinal bleeding, anticoagulation therapy appears to be tolerable and should be initiated as soon as possible when needed. Future studies are warranted to address the issues of the optimal duration of vasoconstrictor therapy, the role of preemptive TIPS in the treatment of AVB and PVT in such patients, and the optimal timing of endoscopic procedures when AVB occurs in patients who are undergoing anticoagulant therapy, and finally the optimal timing of stopping and restoring anticoagulants.

Competing Interests The authors declare that they have no competing interests.

References

1. Qi X, Han G, Fan D. Management of portal vein thrombosis in liver cirrhosis. Nat Rev Gastroenterol Hepatol. 2014;11(7):435–46.
2. Rodriguez-Castro KI, Porte RJ, Nadal E, Germani G, Burra P, Senzolo M. Management of nonneoplastic portal vein thrombosis in the setting of liver transplantation: a systematic review. Transplantation. 2012;94(11):1145–53.
3. Nery F, Chevret S, Condat B, de Raucourt E, Boudaoud L, Rautou PE, et al. Causes and consequences of portal vein thrombosis in 1,243 patients with cirrhosis: results of a longitudinal study. Hepatology. 2015;61(2):660–7.
4. Turon F, Driever EG, Baiges A, Cerda E, García-Criado Á, Gilabert R, et al. Predicting portal thrombosis in cirrhosis: a prospective study of clinical, ultrasonographic and hemostatic factors. J Hepatol. 2021;75(6):P1367–76. https://doi.org/10.1016/j.jhep.2021.07.020.
5. Zocco MA, Di Stasio E, De Cristofaro R, Novi M, Ainora ME, Ponziani F, et al. Thrombotic risk factors in patients with liver cirrhosis: correlation with MELD scoring system and portal vein thrombosis development. J Hepatol. 2009;51(4):682–9.
6. Amitrano L, Brancaccio V, Guardascione MA, Margaglione M, Iannaccone L, D'Andrea G, et al. Inherited coagulation disorders in cirrhotic patients with portal vein thrombosis. Hepatology. 2000;31(2):345–8.
7. Tripodi A, Primignani M, Chantarangkul V, Dell'Era A, Clerici M, de Franchis R, et al. An imbalance of pro- vs anti-coagulation factors in plasma from patients with cirrhosis. Gastroenterology. 2009;137(6):2105–11.
8. Intagliata NM, Caldwell SH, Tripodi A. Diagnosis, development, and treatment of portal vein thrombosis in patients with and without cirrhosis. Gastroenterology. 2019;156(6):1582–99.
9. Stine JG, Wang J, Shah PM, Argo CK, Intagliata N, Uflacker A, et al. Decreased portal vein velocity is predictive of the development of portal vein thrombosis: a matched case-control study. Liver Int. 2018;38(1):94–101.
10. John BV, Konjeti R, Aggarwal A, Lopez R, Atreja A, Miller C, et al. Impact of untreated portal vein thrombosis on pre and post liver transplant outcomes in cirrhosis. Ann Hepatol. 2013;12(6):952–8.
11. Englesbe MJ, Kubus J, Muhammad W, Sonnenday CJ, Welling T, Punch JD, et al. Portal vein thrombosis and survival in patients with cirrhosis. Liver Transpl. 2010;16(1):83–90.
12. Garcia-Tsao G, Bosch J. Management of varices and variceal hemorrhage in cirrhosis. N Engl J Med. 2010;362(9):823–32.
13. Noronha Ferreira C, Marinho RT, Cortez Pinto H, Ferreira P, Dias MS, Vasconcelos M, et al. Incidence, predictive factors and clinical significance of development of portal vein thrombosis in cirrhosis: a prospective study. Liver Int. 2019;39(8):1459–67.
14. Berry K, Taylor J, Liou IW, Ioannou GN. Portal vein thrombosis is not associated with increased mortality among patients with cirrhosis. Clin Gastroenterol Hepatol. 2015;13(3):585–93.
15. D'Amico G, De Franchis R. Upper digestive bleeding in cirrhosis. Post-therapeutic outcome and prognostic indicators. Hepatology. 2003;38(3):599–612.

16. Amitrano L, Guardascione MA, Manguso F, Bennato R, Bove A, DeNucci C, et al. The effectiveness of current acute variceal bleed treatments in unselected cirrhotic patients: refining short-term prognosis and risk factors. Am J Gastroenterol. 2012;107(12):1872–8.
17. Chen PH, Chen WC, Hou MC, Liu TT, Chang CJ, Liao WC, et al. Delayed endoscopy increases re-bleeding and mortality in patients with hematemesis and active esophageal variceal bleeding: a cohort study. J Hepatol. 2012;57(6):1207–13.
18. Garcia-Tsao G, Abraldes JG, Berzigotti A, Bosch J. Portal hypertensive bleeding in cirrhosis: risk stratification, diagnosis, and management: 2016 practice guidance by the American association for the study of liver diseases. Hepatology. 2017;65(1):310–35.
19. de Franchis R. Expanding consensus in portal hypertension: report of the Baveno VI consensus workshop: stratifying risk and individualizing care for portal hypertension. J Hepatol. 2015;63(3):743–52.
20. European Association for the Study of the Liver. EASL clinical practice guidelines for the management of patients with decompensated cirrhosis. J Hepatol. 2018;69(2):406–60.
21. Tripathi D, Stanley AJ, Hayes PC, Patch D, Millson C, Mehrzad H, et al. U.K. guidelines on the management of variceal haemorrhage in cirrhotic patients. Gut. 2015;64(11):1680–704.
22. Wang Z, Jiang MS, Zhang HL, Weng NN, Luo XF, Li X, et al. Is post-TIPS anticoagulation therapy necessary in patients with cirrhosis and portal vein thrombosis? A Randomized Controlled Trial. Radiology. 2016;279(3):943–51.
23. European Association for the Study of the Liver. EASL clinical practice guidelines: vascular diseases of the liver. J Hepatol. 2016;64(1):179–202.
24. Northup PG, Garcia-Pagan JC, Garcia-Tsao G, Intagliata NM, Superina RA, Roberts LN, et al. Vascular liver disorders, portal vein thrombosis, and procedural bleeding in patients with liver disease: 2020 practice guidance by the American Association for the Study of Liver Diseases. Hepatology. 2021;73(1):366–413.
25. Bull-Henry K. Endoscopy in the coagulopathic patient. Curr Opin Gastroenterol. 2019;35(5):401–7.
26. Loffredo L, Pastori D, Farcomeni A, Violi F. Effects of anticoagulants in patients with cirrhosis and portal vein thrombosis: a systematic review and meta-analysis. Gastroenterology. 2017;153(2):480–7.
27. La Mura V, Braham S, Tosetti G, Branchi F, Bitto N, Moia M, et al. Harmful and beneficial effects of anticoagulants in patients with cirrhosis and portal vein thrombosis. Clin Gastroenterol Hepatol. 2018;16(7):1146–52.
28. Pettinari I, Vukotic R, Stefanescu H, Pecorelli A, Morelli M, Grigoras C, et al. Clinical impact and safety of anticoagulants for portal vein thrombosis in cirrhosis. Am J Gastroenterol. 2019;114(2):258–66.
29. Chen H, Liu L, Qi X, He C, Wu F, Fan D, et al. Efficacy and safety of anticoagulation in more advanced portal vein thrombosis in patients with liver cirrhosis. Eur J Gastroenterol Hepatol. 2016;28(1):82–9.
30. Cui SB, Shu RH, Yan SP, Wu H, Chen Y, Wang L, et al. Efficacy and safety of anticoagulation therapy with different doses of enoxaparin for portal vein thrombosis in cirrhotic patients with hepatitis B. Eur J Gastroenterol Hepatol. 2015;27(8):914–9.
31. De Gottardi A, Trebicka J, Klinger C, Plessier A, Seijo S, Terziroli B, et al. Antithrombotic treatment with direct-acting oral anticoagulants in patients with splanchnic vein thrombosis and cirrhosis. Liver Int. 2017;37(5):694–9.
32. Delgado MG, Seijo S, Yepes I, Achecar L, Catalina MV, Garcia-Criado A, et al. Efficacy and safety of anticoagulation on patients with cirrhosis and portal vein thrombosis. Clin Gastroenterol Hepatol. 2012;10(7):776–83.
33. Lv Y, Bai W, Li K, Wang Z, Guo W, Luo B, et al. Anticoagulation and transjugular intrahepatic portosystemic shunt for the management of portal vein thrombosis in cirrhosis: a prospective observational study. Am J Gastroenterol. 2021;116(7):1447–64.
34. Senzolo M, Sartori TM, Rossetto V, Burra P, Cillo U, Boccagni P, et al. Prospective evaluation of anticoagulation and transjugular intrahepatic portosystemic shunt for the management of portal vein thrombosis in cirrhosis. Liver Int. 2012;32(6):919–27.

35. Senzolo M, Riva N, Dentali F, Rodriguez-Castro K, Sartori MT, Bang SM, et al. Long-term outcome of splanchnic vein thrombosis in cirrhosis. Clin Transl Gastroenterol. 2018;9(8):176.
36. Maruyama H, Takahashi M, Shimada T, Yokosuka O. Emergency anticoagulation treatment for cirrhosis patients with portal vein thrombosis and acute variceal bleeding. Scand J Gastroenterol. 2012;47(6):686–91.
37. Cerini F, Gonzalez JM, Torres F, Puente A, Casas M, Vinaixa C, et al. Impact of anticoagulation on upper-gastrointestinal bleeding in cirrhosis. A retrospective multicenter study. Hepatology. 2015;62(2):575–83.
38. Ponthus S, Spahr L, Casini A, Berney T, Frossard J, Majno P, et al. Safety of variceal band ligation in patients with cirrhosis and portal vein thrombosis treated with anticoagulant therapy: a retrospective study. Eur J Gastroenterol Hepatol. 2020;32(3):395–400.
39. Bianchini M, Cavani G, Bonaccorso A, Turco L, Vizzutti F, Sartini A, et al. Low molecular weight heparin does not increase bleeding and mortality post-endoscopic variceal band ligation in cirrhotic patients. Liver Int. 2018;38(7):1253–62.
40. Nevens F, Bustami R, Scheys I, Lesaffre E, Fevery J. Variceal pressure is a factor predicting the risk of a first variceal bleeding: a prospective cohort study in cirrhotic patients. Hepatology. 1998;27(1):15–9.
41. Radaelli F, Dentali F, Repici A, Amato A, Paggi S, Rondonotti E, et al. Management of anticoagulation in patients with acute gastrointestinal bleeding. Dig Liver Dis. 2015;47(8):621–7.
42. Veitch AM, Vanbiervliet G, Gershlick AH, Boustiere C, Baglin TP, Smith L, et al. Endoscopy in patients on antiplatelet or anticoagulant therapy, including direct oral anticoagulants: British Society of Gastroenterology (BSG) and European Society of Gastrointestinal Endoscopy (ESGE) guidelines. Gut. 2016;65(3):374–89.
43. Chan FKL, Goh K, Reddy N, Fujimoto K, Ho KY, Hokimoto S, et al. Management of patients on antithrombotic agents undergoing emergency and elective endoscopy: joint Asian Pacific Association of Gastroenterology (APAGE) and Asian Pacific Society for Digestive Endoscopy (APSDE) practice guidelines. Gut. 2018;67(3):405–17.
44. Abraham NS. Management of antiplatelet agents and anticoagulants in patients with gastrointestinal bleeding. Gastrointest Endosc Clin N Am. 2015;25(3):449–62.
45. Veitch AM, Baglin TP, Gershlick AH, Harnden SM, Tighe R, Cairns S. Guidelines for the management of anticoagulant and antiplatelet therapy in patients undergoing endoscopic procedures. Gut. 2008;57(9):1322–9.

Clinical Settings 2: Acute Variceal Bleeding—Consensus Statements of Panel 6

46

Virginia Hernández-Gea, Dominique Thabut, Andrés Cardenas, Àngels Escorsell, Guohong Han, Xuefeng Luo, David Patch, Bogdan Procopet, and Marika Rudler

V. Hernández-Gea (✉)
Barcelona Hepatic Hemodynamic Unit, Liver Unit, Hospital Clinic-IDIBAPSm, University of Barcelona, Barcelona, Spain

CIBERehd, Barcelona, Spain
e-mail: vihernandez@clinic.cat

D. Thabut
Hôpital de la Pitié-Salpétrière, APHP, UPMC, Sorbonne University, Paris, France

A. Cardenas
Institut de Malalties Digestives I Metaboliques, Hospital Clínic Barcelona, University of Barcelona, Barcelona, Spain
e-mail: ACARDENA@clinic.cat

À. Escorsell
FGS Hospital de la Santa Creu i Sant Pau, Barcelona, Spain
e-mail: mescorsell@santpau.cat

G. Han
Department of Liver Diseases and Digestive Interventional Radiology, National Clinical Research Center for Digestive Diseases and Xijing Hospital of Digestive Diseases, Fourth Military Medical University, Xi'an, China
e-mail: hangh@fmmu.edu.cn

X. Luo
Department of Gastroenterology and Hepatology, West China Hospital, Sichuan University, Chengdu, China

D. Patch
Royal Free Hospital, London, UK
e-mail: david.patch@nhs.net

B. Procopet
Gastroenterology Department, Regional Institute of Gastroenterology and Hepatology "O Fodor", 3rd Medical Clinic, University of Medicine and Pharmacy "Iuliu Hatieganu", Cluji-Napoca, Romania

M. Rudler
Barcelona Hepatic Hemodynamic Unit, Liver Unit, Hospital Clinic, University of Barcelona, Barcelona, Spain
e-mail: Marika.rudler@aphp.fr

R. de Franchis (ed.), *Portal Hypertension VII*,
https://doi.org/10.1007/978-3-031-08552-9_46

6.1 The goal of resuscitation is to preserve tissue perfusion. Volume restitution should be initiated to restore and maintain hemodynamic stability. (D2) (Unchanged)

6.2 PRBC transfusion should be done conservatively at a target hemoglobin level between 7–8 g/dL, although transfusion policy in individual patients should also consider other factors, such as cardiovascular disorders, age, hemodynamic status, and ongoing bleeding. (A1) (Unchanged)

6.3 Intubation is recommended before endoscopy in patients with altered consciousness and those actively vomiting blood. (D1) (New)

6.4 Extubation should be performed as quickly as safely possible after endoscopy. (D2) (New)

6.5 In suspected variceal bleeding, vasoactive drugs (terlipressin, somatostatin, octreotide) should be started as soon as possible and continued for 2–5 days. (A1) (Changed)

6.6 Hyponatremia has been described in patients on terlipressin, especially in patients with preserved liver function. Therefore, sodium levels should be monitored. (B1) (Unchanged)

6.7 Antibiotic prophylaxis is an integral part of therapy for patients with cirrhosis presenting with upper gastrointestinal bleeding and should be instituted from admission. (A1) (Unchanged)

6.8 The risk of bacterial infection and mortality is very low in patients with Child-Pugh A cirrhosis, but more prospective studies are still needed to assess whether antibiotic prophylaxis can be avoided in this subgroup of patients. (B2) (Unchanged)

6.9 Intravenous ceftriaxone 1 g/24 h should be considered in patients with advanced cirrhosis (A1) in hospital settings with a high prevalence of quinolone-resistant bacterial infections and patients on previous quinolone prophylaxis, and should always be in accordance with local resistance patterns and antimicrobial policies. (D2) (Changed)

6.10 Malnutrition increases the risk of adverse outcomes in patients with cirrhosis and AVB and oral nutrition should be started as soon as possible. (D2) (New)

6.11 Manipulation of the airway, including the use of a nasogastric tube, should be performed with caution because of the risk of pulmonary infection. (D2) (New)

6.12 PPIs, when started before endoscopy, should be stopped immediately after the procedure unless there is a strict indication to continue them. (D2) (New)

6.13 Six-week mortality should be the primary endpoint for studies on the treatment of acute variceal bleeding. (D1) (Unchanged)

6.14 Five-day treatment failure is defined either by the absence of control of bleeding or by rebleeding within the first 5 days. (D1) (Changed)

6.15 Child-Pugh class C, the updated MELD score, and failure to achieve primary haemostasis are the variables most consistently found to predict 6-week mortality. (B2) (Unchanged)

6.16 Child-Pugh and MELD scores are currently the most utilized severity scoring systems. (D2) (Unchanged)

6.17 Following hemodynamic resuscitation, patients with suspected AVB should undergo upper endoscopy within 12 h of presentation (B1). If the patient is unstable, endoscopy should be performed as soon as safely possible. (D1) (Changed)

6.18 The availability of an on-call GI endoscopist proficient in endoscopic hemostasis and on-call support staff with technical expertise in the usage of endoscopic devices, enabling the performance of endoscopy on a 24/7 basis, is recommended. Trainees performing the procedure must always be closely supervised by the GI endoscopist. (D1) (Changed)

6.19 In the absence of contraindications (QT prolongation), pre-endoscopy infusion of erythromycin (250 mg IV 30–120 min before endoscopy) should be considered. (B1) (Unchanged)

6.20 Patients with acute variceal bleeding should be managed in intensive or intermediate care units. (D1) (Unchanged)

6.21 Ligation is the recommended form of endoscopic therapy for acute esophageal variceal bleeding. (A1) (Unchanged)

6.22 Endoscopic therapy with tissue adhesives (e.g., N-butyl-cyanoacrylate/thrombin) is recommended for acute bleeding from isolated gastric varices (IGV) (A1), gastroesophageal varices type 2 (GOV2) that extend beyond the cardia. (D2) (Unchanged)

6.23 EVL or tissue adhesive can be used in bleeding from gastroesophageal varices type 1 (GOV1). (D1) (Unchanged)

6.24 Based on current evidence, the hemostatic powder cannot be recommended as first-line endoscopic therapy for acute variceal bleeding. (D1) (New)

6.25 Endoscopic therapy (argon plasma coagulation, radiofrequency ablation, or band ligation for PHG-GAVE) may be used for local treatment of PHG bleeding. (C2) (New)

6.26 All patients with AVB should undergo abdominal imaging, preferably contrast-enhanced cross-sectional imaging (CT or MRI) to exclude splanchnic vein thrombosis, and hepatocellular carcinoma and to map portosystemic collaterals in order to guide treatment. (D1) (New)

6.27 Preemptive TIPS with PTFE-covered stents within 72 h (ideally <24 h) is indicated in patients bleeding from EV, GOV1, and GOV2 who meet any of the following criteria: Child-Pugh class C < 14 points or Child class B > 7 with active bleeding at initial endoscopy or HVPG >20 mmHg at the time of hemorrhage. (A1) (Changed)

6.28 In patients fulfilling preemptive TIPS criteria, ACLF, HE at admission, and hyperbilirubinemia at admission should not be considered as contraindications to pTIPS. (B1) (New)

6.29 In refractory variceal bleeding, balloon tamponade (BT) or self-expandable metal stents (SEMS) should be used as a bridge therapy to a more definite treatment such as PTFE-covered TIPS. SEMS is as efficacious as BT and a safer option. (B1) (Changed)

6.30 Failure to control variceal bleeding despite combined pharmacological and endoscopic therapy is best managed by salvage PTFE-covered TIPS. (B1) (Changed)

6.31 TIPS may be futile in patients with Child-Pugh ≥14 cirrhosis, or with a MELD score > 30 and lactate >12 mmol/L, unless liver transplantation is envisioned in the short-term (B1). The decision to perform TIPS in such patients should be taken on a case-by-case basis. (D1) (New)

6.32 In patients with AVB and HE, HE bouts should be treated with lactulose (oral or enemas). (D1) (New)

6.33 In patients presenting with AVB, rapid removal of blood from the gastrointestinal tract (lactulose oral or enemas) should be used to prevent HE. (B1) (New)

6.34 Variceal bleeding is due to PHT, and the aim of the treatment should be focused on lowering portal pressure rather than correcting coagulation abnormalities. (B1) (New)

6.35 Conventional coagulation tests, namely, prothrombin time (PT/INR) and activated partial thromboplastin time (aPTT), do not accurately reflect the hemostatic status of patients with advanced liver diseases. (B1) (Changed)

6.36 In the acute variceal bleeding episode, transfusion of fresh frozen plasma is not recommended as it will not correct coagulopathy and may lead to volume overload and worsening of portal hypertension. (B1) (New)

6.37 In the setting of acute variceal bleeding, there is no evidence that platelet count and fibrinogen levels are correlated with the risk of failure to control bleeding or rebleeding. However, in case of failure to control bleeding, the decision to correct the hemostatic abnormalities should be considered on a case-by-case basis. (D2) (New)

6.38 Recombinant factor VIIa and tranexamic acid are not recommended in acute variceal bleeding. (A1) (New)

6.39 In patients with AVB who are on anticoagulants, these should be temporarily discontinued until the hemorrhage is under control. Length of discontinuation should be individualized based on the strength of the indication for anticoagulation. (D2) (New)

6.40 In patients with GOV2, IGV1, and ectopic varices, BRTO could be considered as an alternative to endoscopic treatment or TIPS, provided it is feasible (type and diameter of the shunt) and local expertise is available, as it has demonstrated to be safe and effective. (D2) (New)

6.41 Either endovascular or endoscopic treatment should be considered in patients with ectopic varices. (D1) (New)

6.42 TIPS may be combined with embolization to control bleeding or to reduce the risk of recurrent variceal bleeding from gastric or ectopic varices, particularly in cases when despite a decrease in portosystemic pressure gradient, the portal flow remains diverted to collaterals. (D2) (New)

6.43 In patients with cirrhosis and PVT, management of AVB should be performed according to the guidelines for patients without PVT, when possible. (D1) (New)

Research Agenda

- Role of vasoactive drugs and antibiotics in Child-Pugh A patients.
- Identification of the optimal shorter time frame limit for vasoactive drug therapy
- Definition of active bleeding at endoscopy as high-risk criteria: assessment of its subjectivity, and prognostic value.
- Role of hemostatic powder in acute and refractory variceal bleeding.
- Role of thrombin in gastric variceal bleeding.
- Role of pre-emptive TIPS in patients with gastric varices.
- Management of high risk in patients not fulfilling the high-risk criteria used for pre-emptive TIPS.
- Cost effectiveness analysis on the use of SEMSs.
- Alternatives other than Sengstaken-Blakemore/Linton tubes should be developed as they are in shortage.
- The role of global hemostasis tests, such as viscoelastic tests and thrombin generation assays in the assessement and correction of, hemostasis abnormalities in decompensated cirrhosis and acute variceal bleeding (using clinical endpoints).
- The potential role of prothrombin complex concentrates, fibrinogen, or cryoprecipitate in portal hypertension related bleeding.
- Is there any relation between low platelet count (up to which level) or fibrinogen and the risk of variceal bleeding, failure to control bleeding, or bleeding after endoscopic band ligation?
- Identification of patients that will benefit from variceal embolization during TIPS.
- Role of EUS-guided therapy with tissue adhesive with or without coils in the management of bleeding from gastric varices.
- Definition of the impact of PVT on the prognosis of cirrhotic patients with AVB.
- Identification of the optimal duration of vasoactive therapy in cirrhotic patients with PVT and AVB.
- Role of pre-emptive TIPS in cirrhotic patients with PVT presenting with AVB.
- Management of AVB in patients with cirrhosis and PVT, including management of anticoagulation and timing of endoscopic/invasive procedures.

Part IX

Clinical Settings 3: Preventing Further Decompensation

Concept of Further Decompensation and Recompensation

47

Gennaro D'Amico and Guadalupe Garcia-Tsao

Further Decompensation

Pathophysiology of Decompensation and Further Decompensation

Cirrhosis is associated with two main syndromes: portal hypertension (the earliest to occur) and liver insufficiency (a latter complication). Of the decompensating events associated with cirrhosis, variceal hemorrhage and ascites are those due almost solely to portal hypertension and the consequent hyperdynamic circulation, while hepatic encephalopathy results from a combination of portal hypertension (portosystemic shunting) and/or liver insufficiency (deficient urea cycle metabolic pathway) and jaundice is due solely to liver insufficiency and/or to an acute-on-chronic liver injury.

In cirrhosis, the initial pathogenic mechanism leading to portal hypertension is an increased intrahepatic resistance due to a combination of architectural disruption, specifically fibrosis and capillarization of the hepatic sinusoids with narrowing of their lumen, angiogenesis, endothelial dysfunction (vasoconstriction) and the development of intrahepatic microthrombi [1].

G. D'Amico (✉)
Department of Gastroenterology, Ospedale Cervello, Palermo, Italy

G. Garcia-Tsao
Digestive Diseases Section, School of Medicine, Yale University, New Haven, CT, USA

Digestive Diseases Section, VA-CT Healthcare System, West Haven, CT, USA
e-mail: Guadalupe.garcia-tsao@yale.edu

R. de Franchis (ed.), *Portal Hypertension VII*,
https://doi.org/10.1007/978-3-031-08552-9_47

The increased intrahepatic vascular resistance leads to increased pressure in the portal venous system that induces shear stress in the splanchnic vessels and the release of vasodilators such as nitric oxide. Subsequent splanchnic arterial vasodilation and increased flow into the portal venous system lead to the progression and worsening of portal hypertension leading to the development of Clinically Significant Portal hHpertension (CSPH), the main predictor of the development of ascites, variceal hemorrhage, and encephalopathy [2]. Splanchnic vasodilatation also impacts the systemic circulation leading to a decrease in mean arterial pressure and a decrease in effective arterial blood volume that in turn leads to activation of neurohumoral systems (the renin-angiotensin aldosterone system and sympathetic nervous system) with consequent sodium and water retention and an increase in cardiac output. This hyperdynamic circulatory state typical of the patient with cirrhosis further increases portal pressure [3, 4]. In patients with gastroesophageal varices this increase in pressure and flow will lead to the growth of varices and their eventual rupture (variceal hemorrhage). An increased sinusoidal pressure together with the hypervolemic state leads to the formation of ascites and, with the hyperdynamic state, more portosystemic shunting will occur which, together with liver dysfunction, will lead to hepatic encephalopathy.

These hemodynamic abnormalities are more pronounced in the patient with decompensated cirrhosis and, at this stage, a major additional pathogenic mechanism involved in the translocation of bacteria or bacterial products from the intestinal lumen into the systemic circulation [5]. Bacterial translocation causes the activation of the immune system (macrophages, dendritic cells) and the release of inflammatory cytokines, which cause oxidative stress, and the release of vasodilators with a further worsening of splanchnic vasodilation [6]. Moreover, systemic inflammation can also impair cardiac contractility and decrease cardiac output which will further alter systemic and splanchnic hemodynamics.

Bacterial translocation is not the sole cause of inflammation in patients with cirrhosis, because liver damage per se can cause the release of Danger-Associated Molecular Patterns (DAMPs) from hepatocytes, which can trigger an inflammatory response. Consequently, systemic inflammation is present across all stages of cirrhosis, however, it seems to be most pronounced in decompensated cirrhosis where it drives the development of *further* decompensation [7]. In fact, markers of inflammation have been shown to predict decompensation in patients with compensated cirrhosis and death in patients with decompensated cirrhosis [4, 7].

Worsening of hemodynamics, with more vasodilatation, will result in the development of events such as refractory ascites and hepatorenal syndrome (HRS-AKI) [4, 8], the latter being the complication associated with the highest mortality in patients with cirrhosis [9, 10]. Spontaneous bacterial peritonitis (SBP), an infection of ascitic fluid, and other spontaneous infections such as bacteriuria or spontaneous bacteremia, unique to patients with decompensated cirrhosis, represent the most severe clinically overt manifestation of bacterial translocation and are associated with an acute inflammatory state with maximal alterations in hemodynamics and high risk of HRS-AKI and death.

Definition of Decompensation and Further Decompensation

The natural history of cirrhosis is characterized by two distinct clinical stages: a long asymptomatic stage (compensated cirrhosis) with a median survival that exceeds 12 years and a symptomatic stage (decompensated cirrhosis) with a median survival of 1-2 years [11]. In Baveno VII, and although controversial (Chap. 29), the proposed definition of cirrhosis decompensation consisted of the development of any of the portal hypertension-driven events, that is, variceal hemorrhage, overt ascites (or hepatic hydrothorax) or overt hepatic encephalopathy. This makes clinical sense as therapies that reduce portal pressure will only have an effect in preventing the decompensating events related to portal hypertension. It should be noted that this proposed definition is based on expert consensus but not supported by scientific evidence, leaving the role of jaundice in the definition of decompensation under debate. As a matter of fact, in the only inception cohort study investigating the incidence of decompensation by competing risk analysis, while ascites had the highest incidence, followed by bleeding, jaundice and encephalopathy had the same cumulative evidence, i.e., the same weight, as first decompensating events [12].

It is recognized but not well-defined, that beyond decompensation, there is a stage where mortality is higher than that of patients developing a single decompensating event, and this would represent a stage of "further" decompensation. As described above, these events result from further deterioration of the already altered hemodynamics of the patient with decompensated cirrhosis.

In order to better characterize the stage of further decompensation in patients who present with ascites or variceal hemorrhage as a single first decompensating event and to identify the events associated with poorer survival in these patients, two cohorts of patients with cirrhosis collected prospectively from a cooperative European-Latin American center (comprising data from 12 Italian centers, 4 Spanish centers. 1 German, 1 UK and 1 Argentinian center) were analyzed: The **total cohort included 2996** patients, of which 402 had decompensated prior to inclusion into the study, leaving 2226 patients to be analyzed:

1. Cohort 1 consisted of 1006 patients with compensated cirrhosis at inclusion into the study. In a follow-up period of 2252 days, **638 patients** developed a decompensating event while 368 remained compensated.
2. Cohort of **1588 patients** with decompensation at inclusion.

Of the 2226 patients (638 + 1588) with the first decompensation included in the analysis, 1058 (48%) developed ascites alone; 432 (19%) developed bleeding alone, 57 (3%), developed encephalopathy, 98 (4%) developed jaundice alone, 202 (9%) developed two complications at once (ascites + bleed), 133 (6% presented with any combination that included AKI/SBP, and 245 (11%) presented with any combo with HE or jaundice.

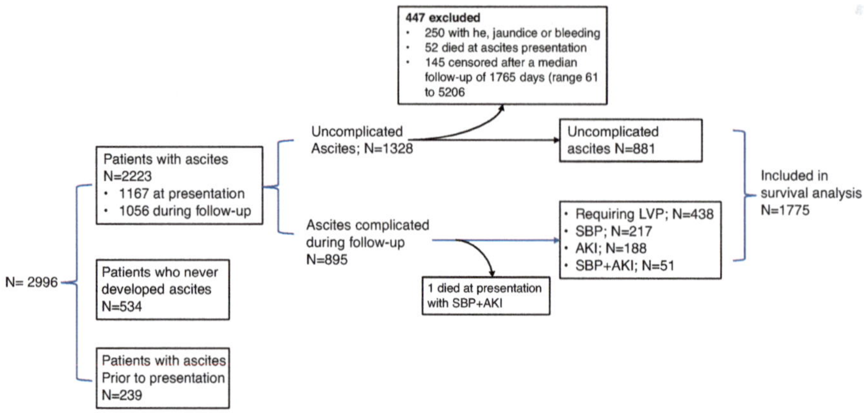

Fig. 47.1 Flow-diagram of inclusion of patients with ascites who presented or not specific complications of ascites in survival analysis. Over a total of 2223 patients who presented ascites during the study, 1775 were suitable for survival analysis

Outcomes After Presentation with Ascites

A shown in Fig. 47.1, of the total of **2996 patients** included in the two cohorts, 2223 presented with ascites (N = 1167) or developed ascites during the follow-up (N = 1056), 534 never developed ascites and 239 had already developed ascites prior to inclusion in the study. To assess survival following the presentation of ascites, two analyses are presented.

The first analysis was aimed at investigating the impact of ascites complications on survival. In this analysis, survival was separately assessed for patients who never presented complications of ascites and for those who required large-volume paracentesis (LVP) or developed spontaneous bacterial peritonitis (SBP) or acute kidney injury (AKI) or both SBP + AKI. Time zero for survival analysis was respectively the time of development of ascites for patients who did not experience complications of ascites and the time of occurrence of each specific complication for the other groups of patients. Of the 2233 patients with ascites, 1328 had uncomplicated ascites (no LVP requirement or SBP and/or AKI), 438 had required LVP, 217 had SBP, 188, had AKI, and 52 had both AKI and SBP. Figure 47.2 shows 1- and 2-year cumulative proportions of surviving patients for each of these patient groups.

The second analysis was performed to explore the impact of further decompensating events, (respectively bleeding, hepatic encephalopathy, and jaundice) and their interplay with the specific complications of ascites (LVP, SBP, AKI). As shown in the inclusion flow diagram (Fig. 47.3), a total of 1392 patients were included in this analysis. Of them, 304 remained free of other complications after the development of ascites (no bleeding or hepatic encephalopathy or jaundice, neither required LKVP or developed SBP or AKI); 219 developed specific complications of ascites (LVP, SBP, or AKI); 272 developed a single second event (either bleeding, or

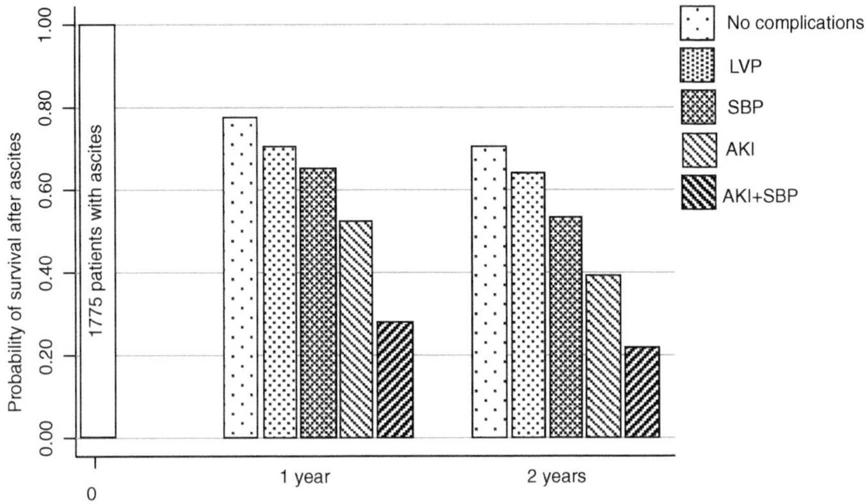

Number at risk

No complications	881	627	482
LVP	438	267	203
SBP	217	117	78
AKI	188	88	55
AKI+SBP	51	11	7

Fig. 47.2 1 and 2-year cumulative probability of survival after presentation of ascites or of complications of ascites. Time zero for survival analysis was when each complication occurred. LVP need for large volume paracentesis, SBP spontaneous bacterial peritonitis, AKI acute kidney injury

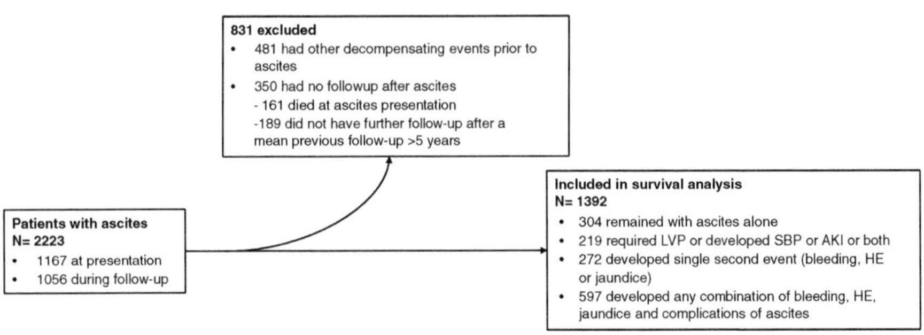

Fig. 47.3 Flow-diagram of inclusion of patients with ascites who presented either new decompensating events or specific complications of ascites, in survival analysis. Over a total of 2223 patients with ascites, 1392 were suitable for this analysis

encephalopathy or jaundice); and 597 developed any combination of events (either complicating ascites or not).

Figure 47.4 shows how, compared to patients who remained with ascites alone, those who developed one or two additional decompensating events (VH, HE, or jaundice) had significantly poorer survival.

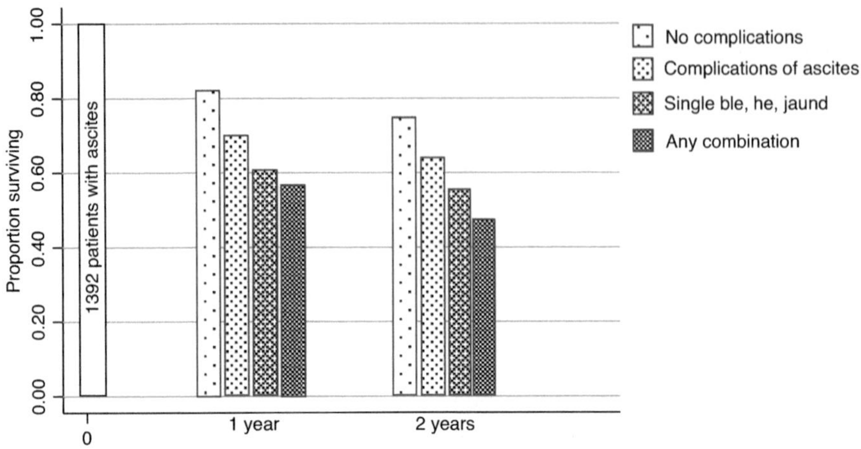

Number at risk			
No complications	304	230	186
Complications of ascites	219	137	103
Single ble, he, jaund	272	148	101
Any combination	597	299	196

Fig. 47.4 1- and 2-year cumulative probability of survival after presentation with ascites according to whether other subsequent decompensating events or specific complications of ascites. Time zero for survival analysis was when each new event or complication of ascites occurred

Outcomes After Presentation With Variceal Bleeding

The second most common decompensating event is variceal hemorrhage, which is associated with a significantly increased risk of further decompensation and death compared to compensated patients [11].

To investigate the impact of further decompensation after variceal bleeding on survival, we performed a survival analysis of patients surviving a variceal bleeding according to whether they developed or not further decompensating events after bleeding. The analysis has been performed in the same European-:Latin American study. Of the 2226 patients (see above) 1005 presented a variceal bleeding, 214 died at bleeding, 107 had already experienced other decompensating events, and information on further decompensation was missing in 1 patient, leaving a total of 683 patients suitable for this analysis. Of them, 624 (91%) had a second decompensating event (ascites, encephalopathy, or jaundice), occurring at the time of bleeding in 338, within 42 days of the acute episode in 157, and > 42 days after bleeding in 129 while 59 had variceal bleeding without further decompensating events. Overall, 371 patients rebled after a mean (± of 1027 ± 1180 days): 14 in the group who did not experience further decompensating events, 158 in the group who presented with bleeding together with decompensating events, 137 in patients who presented further decompensation within 6 weeks of index bleeding and 62 in those with further decompensation later on.

Mortality in these patients is increased if there is a recurrence of hemorrhage and/or if they develop a second decompensating event (ascites, encephalopathy, or jaundice). Remarkably, although survival was decreased in patients who developed a second event >42 days after presentation with acute bleed, survival was no different between patients with VH alone and those who developed a second event <42 days after presentation.

To explore the prognostic value of previous decompensating events in patients presenting with bleeding, we also used preliminary data from an individual patient data meta-analysis still in progress (Table 47.1). Patient-level data were obtained from 10 published studies, either RCTs or observational studies reporting on survival of patients surviving a variceal bleeding episode. Data from a total of 2631 patients were available for this analysis. Of them, 1824 had a previous decompensation and 807 had bleeding as the first decompensating event: 2-years survival was 70% vs. 58% ($p < 0.001$). Increased mortality after a previous decompensation was also confirmed in separate analyses according to the type of the previous decompensation whether this was ascites, encephalopathy, or jaundice. Information for previous ascites was available in 2509 patients: 1266 had never had ascites while1243 had a history or present ascites. The 2-year survival was 70% without and 55% with previous ascites ($p < 0001$). Similarly, survival was significantly higher without previous encephalopathy and without previous ascites (Fig. 47.5).

Although preliminary, overall, these findings from the multicenter European-Latino-American study and from the IPD metanalysis of long-term outcomes after variceal bleeding support a significant worsening of the outcome after further decompensation either following or preceding variceal bleeding.

Table 47.1 List of studies included in an ongoing individual patient data meta-anlysis of the outcome of patients with cirrhosis after variceal bleeding

Author	Reference	Design	*N* patients
Bucsics T	PLoS One 2018;13(1):e0189414	Observational	286
Garcia-pagan JC	Gut 2009;58:1144–50	RCT	158
Garcia-pagan JC	J Hepatol 2013;58:45–50	Observational	75
Garcia pagan JC	N Engl J Med 2010;362:2370–9	RCT	63
Sempere L	Rev Esp Enferm Dig 2009;101:236–48	Observational	201
Rout G	Dig Dis Sci 2019. https://doi.org/10.1007/s10620-019-05557-y	Observational	523
Rudler M	Aliment Pharmacol Ther 2014;40:1074–80	Observational	62
Thabut D	J Hepatol 2018;68:j73–81	Observational	964
Villanueva C	Hepatology 2017;65:1693–707	RCT	170
Lv Y	Lancet Gastroenterol Hepatol 2019;4:587–98	RCT	129

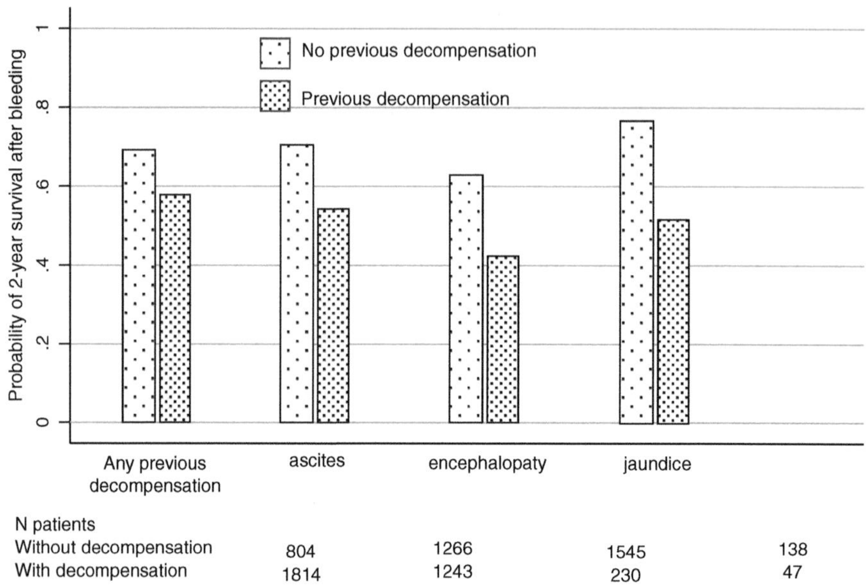

Fig. 47.5 Two-year probability of survival in an individual patient data meta-analysis of outcome after variceal bleeding in cirrhosis according to previous and type of decompensation. A total of 2631 patients was included

Recompensation

While it is now well-accepted that early to moderate fibrosis can regress and possibly even resolve, the concept of cirrhosis regression is not as clear. Perez-Tamayo in 1979 showed evidence of reversal of fibrosis and cirrhosis in both animal models and patients [13]. Later on, Wanless et al. presented serial biopsies from a patient with hepatitis B following antiviral treatment that showed apparent regression of hepatitis B with the shift from fully developed cirrhosis to incomplete septal cirrhosis. In this landmark paper, Wanless et al. reported also livers removed at transplantation having cirrhosis or incomplete septal cirrhosis with a complete description of histologic parameters that suggest progression or regression of fibrosis [14].

Thereafter reports have been published showing biopsy-proven regression of cirrhosis of various etiologies including alcohol-induced cirrhosis, hepatitis C (HCV), hepatitis B (HBV), autoimmune cirrhosis, and hemochromatosis after elimination/suppression of the etiological agents [7]. This has been more extensively analyzed in studies of HCV or HBV cirrhosis effectively treated with antiviral regimens. Both in HCV and HBV cirrhosis, regression to a non-cirrhotic stage has been shown to occur in more than 60% of patients with baseline biopsy showing advanced fibrosis/cirrhosis after 5–6 years of viral suppression [15–18].

Concordantly, HBV viral suppression using continuous lamivudine in patients with compensated advanced liver fibrosis/cirrhosis have resulted in lower development of decompensation and liver-related deaths compared to placebo [19].

Similarly, patients with HCV cirrhosis who achieve sustained virological response (SVR) have been shown to be less likely to develop decompensation and overall mortality and liver-related mortality compared to patients who do not achieve SVR [20]. In another study, SVR reduced further liver decompensation or mortality independent of the presence or absence of varices prior to treatment [21]. Notably, in this study patients with varices prior to antiviral therapy had a greater risk of developing decompensation/death compared to those without varices, indicating that clinically significant portal hypertension is relevant in determining decompensation even after the cure of etiological treatment. In fact, variceal bleeding after HCV elimination has occurred despite viral eradication and normalization of liver function tests in a patient with varices prior to treatment [22].

A number of studies evaluated the effects of direct-acting anti-HCV treatment on portal hypertension, assessed by the hepatic venous pressure gradient (HVPG) [23–25]. They have all shown a significant decrease in HVPG after HCV elimination. In the largest study [24], comprising 226 patients with Child A or B cirrhosis and baseline HVPG ≥10 mmHg who achieved SVR with DAAs, HVPG decreased significantly from 15 mmHg before treatment to 13 mmHg after SVR and it decreased by ≥10% (response) in 62% of the patients. Interestingly, baseline serum albumin below 3.5 g/dL was the only predictor of the lack of an HVPG response.

The impact of etiologic treatment on the clinical course of **already decompensated cirrhosis** is less predictable. After viral elimination, about one-third of patients with decompensated cirrhosis experience an improvement in MELD and Child-Pugh score. In a large cohort study of 707 patients who presented with decompensated HBV cirrhosis, a significant improvement in liver function was observed in treated vs. untreated patients, with 34% of treated patients de-listed for liver transplantation. Patients with early treatment (within 3 months of the decompensating event) had better clinical outcomes than those with delayed treatment. Survival was dependent on antiviral response, being significantly better in responders than in non-responders or untreated cases [26]. These results underscore the importance of promptly administering antiviral drugs to patients under consideration for liver transplantation. It may be speculated that there is a certain 'point-of-no-return' in patients with decompensated cirrhosis, in whom neither cure/suppression/removal of the primary etiological factor nor other supporting medical therapies can halt disease progression and liver failure cannot be prevented without liver transplantation. However, it must be emphasized that etiologic treatment is a cornerstone in the treatment of decompensated cirrhosis and is a requirement to allow for clinical re-compensation of patients with decompensated cirrhosis.

Removing patients from the liver transplant list ("delisting") constitutes evidence of improvement to a point where the patient is deemed to no longer require liver transplantation to survive. Although it has been observed with other etiologies, most of the evidence derives from studies in patients listed for liver transplants that receive antiviral therapy. After the introduction of treatment with direct antiviral agents for HCV, the number of patients with decompensated cirrhosis on the liver transplant wait-list decreased by 30% in the United States [27]. Similarly, data from the European Liver Transplant Registry show the proportion of liver transplants performed for HCV in Europe was constant (approximately 23%) from 2007 to 2014 and then decreased sharply (to 10.6%) in 2017, after the introduction of DAAs [28].

Table 47.2 shows data on recent series of patients having been "delisted" because of improvement in cirrhosis status. As can be observed, de-listing percentages range from 8% in patients with alcohol-related cirrhosis who become alcohol-abstinent [29, 30] to up to ~60% in patients with HCV [31] or HBV- [32] cirrhosis in whom the virus is eliminated with antiviral therapy. Criteria for delisting vary widely and are not very precise. In most series it is based on decreases in MELD score to <15 [29, 31, 33, 34]. MELD score is more indicative of liver/kidney function but does not reflect resolution of decompensating events. Some series also required patients to be free of ascites or encephalopathy or to be "compensated" [29, 33] but some allowed low doses of diuretics or "easy" control of decompensation [30, 34]. The only study with clearer criteria, did not rely on MELD score but on Child-Pugh score with all

Table 47.2 Data on recent series of patients having been "delisted" from the transplant list

First author (year)	Listed for transplant	*N* delisted	Etiology of cirrhosis	Reasons for improvement	Criteria for de-listing
Aravinthan (2017) [29]	935	77 (8%)	All etiologies (mostly alcohol)	"Spontaneous", specific therapy, TIPS	• Absence of ascites, hepatic hydrothorax, oedema after d/c diuretics • Absence HE, after d/c prophylactic tx • MELD <15
Pascasio (2017) [33]	49	11 (22%)	HCV	Antiviral therapy	• In general, patients had to be compensated and had a MELD ≤15 points
Perricone (2018) [34]	142	44 (31%)	HCV	Antiviral therapy	• Regression of HE and ascites (low doses of diuretics acceptable) • MELD <15
Nabatchikova (2021) [31]	45	26 (58%)	HCV	Antiviral therapy	"Durable" MELD <15 and a CTP score < 7
Pose (2021) [30]	420	36 (8%)	Alcohol	Alcohol abstinence	Absence or easy control of decompensations and a significant improvement in liver function
Kim (2021) [32]	311 Child B/C	193 (62%)	HBV	Antiviral therapy	Recovery to a child–Pugh A5

patients with HBV cirrhosis on the transplant list belonging to Child B/C and who were delisted after viral suppression once they had reverted to a Child-Pugh A5 [32]. The Child-Pugh score provides a better assessment of the stages of cirrhosis (Child A mostly compensated, Child B decompensated, Child C further decompensated) and includes both decompensating events and liver synthetic function tests, particularly serum albumin which, after HVPG, is the best predictor of decompensation [2, 35]. In fact, a recent study performed in patients with hepatitis C (75% of whom had cirrhosis) showed that pretreatment serum albumin was the most significant predictor of liver-related mortality independent of SVR [36]. Therefore, an increase in serum albumin to normal levels (without the use of intravenous albumin) after resolution of etiology is a strong indicator of cirrhosis recompensation.

Prior to the Baveno VII meeting, a survey among experts participating in the conference indicated that >80% considered recompensation in the following scenarios: patient with ascites in whom ascites disappears and no longer require diuretics and a patient with encephalopathy who is free of encephalopathy and no longer requires specific therapy. Furthermore, patients in whom ascites/variceal bleeding resolves after TIPS cannot be considered recompensated unless a resolution is associated with etiological therapy and improvement of liver function. In the patient with variceal bleeding, the majority considered 12 months as the timeframe that would define recompensation although the majority would not discontinue NSBB unless there was a resolution of CSPH.

Therefore, per Baveno consensus, definition of cirrhosis recompensation requires fulfillment of all the following three criteria:

1. Removal/suppression/cure of the primary etiology of cirrhosis (viral elimination for hepatitis C, sustained viral suppression for hepatitis B, sustained alcohol abstinence for alcohol-induced cirrhosis).
2. Resolution of ascites (off diuretics), encephalopathy (off lactulose/rifaximin), and absence of recurrent variceal hemorrhage (for at least 12 months).
3. Stable improvement of liver function tests (albumin, INR, bilirubin).

References

1. Bosch J, Groszmann RJ, Shah VH. Evolution in the understanding of the pathophysiological basis of portal hypertension: how changes in paradigm are leading to successful new treatments. J Hepatol. 2015;62:S121–30.
2. Ripoll C, Groszmann R, Garcia-Tsao G, Grace N, Burroughs A, Planas R, Escorsell A, et al. Hepatic venous pressure gradient predicts clinical decompensation in patients with compensated cirrhosis. Gastroenterology. 2007;133:481–8.
3. Abraldes JG, Trebicka J, Chalasani N, D'Amico G, Rockey DC, Shah VH, Bosch J, et al. Prioritization of therapeutic targets and trial design in cirrhotic portal hypertension. Hepatology. 2019;69:1287–99.
4. Turco L, Garcia-Tsao G, Magnani I, Bianchini M, Costetti M, Caporali C, Colopi S, et al. Cardiopulmonary hemodynamics and C-reactive protein as prognostic indicators in compensated and decompensated cirrhosis. J Hepatol. 2018;68(5):949–58.

5. Wiest R, Garcia-Tsao G. Bacterial translocation (BT) in cirrhosis. Hepatology. 2005;41:422–33.
6. Wiest R, Lawson M, Geuking M. Pathological bacterial translocation in liver cirrhosis. J Hepatol. 2014;60:197–209.
7. Costa D, Simbrunner B, Jachs M, Hartl L, Bauer D, Paternostro R, Schwabl P, et al. Systemic inflammation increases across distinct stages of advanced chronic liver disease and correlates with decompensation and mortality. J Hepatol. 2021;74:819–28.
8. Ruiz-del-Arbol L, Achecar L, Serradilla R, Rodriguez-Gandia MA, Rivero M, Garrido E, Natcher JJ. Diastolic dysfunction is a predictor of poor outcomes in patients with cirrhosis, portal hypertension and a normal creatinine. Hepatology. 2013;58(5):1732–41.
9. Bruno S, Saibeni S, Bagnardi V, Vandelli C, De LM, Felder M, Fracanzani AL, et al. Mortality risk according to different clinical characteristics of first episode of liver decompensation in cirrhotic patients: a nationwide, prospective, 3-year follow-up study in Italy. Am J Gastroenterol. 2013;108:1112–22.
10. Planas R, Montoliu S, Balleste B, Rivera M, Miquel M, Masnou H, Galeras JA, et al. Natural history of patients hospitalized for management of cirrhotic ascites. Clin Gastroenterol Hepatol. 2006;4:1385–94.
11. D'Amico G, Morabito A, D'Amico M, Pasta L, Malizia G, Rebora P, Valsecchi MG. Clinical states of cirrhosis and competing risks. J Hepatol. 2018;68:563–76.
12. D'Amico G, Pasta L, Morabito A, D'Amico M, Caltagirone M, Malizia G, Tine F, et al. Competing risks and prognostic stages of cirrhosis: a 25-year inception cohort study of 494 patients. Aliment Pharmacol Ther. 2014;39:1180–93.
13. Pérez-Tamayo R. Cirrhosis of the liver: a reversible disease? Pathol Annu. 1979;14(Pt 2):183–213.
14. Wanless IR, Nakashima E, Sherman M. Regression of human cirrhosis. Morphologic features and the genesis of incomplete septal cirrhosis. Arch Pathol Lab Med. 2000;124:1599–607.
15. Poynard T, McHutchison J, Manns M, Trepo C, Linday K, Goodman Z, Ling M-H, et al. Impact of pegylated interferon alfa-2b and ribavirin on liver fibrosis in patients with chronic hepatitis C. Gastroenterology. 2002;122:1303–13.
16. D'Ambrosio R, Aghemo A, Rumi MG, Ronchi G, Donato MF, Paradis V, Colombo M, et al. A morphometric and immunohistochemical study to assess the benefit of a sustained virological response in hepatitis C virus patients with cirrhosis. Hepatology. 2012;56:532–43.
17. Marcellin P, Gane E, Buti M, Afdhal N, Sievert W, Jacobson IM, Washington MK, et al. Regression of cirrhosis during treatment with tenofovir disoproxil fumarate for chronic hepatitis B: a 5-year open-label follow-up study. Lancet. 2013;381:468–75.
18. Sun Y, Zhou J, Wang L, Wu X, Chen Y, Piao H, Lu L, et al. New classification of liver biopsy assessment for fibrosis in chronic hepatitis B patients before and after treatment. Hepatology. 2017;65:1438–50.
19. Liaw YF, Sung JJ, Chow WC, Farrell G, Lee CZ, Yuen H, Tanwandee T, et al. Lamivudine for patients with chronic hepatitis B and advanced liver disease. N Engl J Med. 2004;351:1521–31.
20. Nahon P, Bourcier V, Layese R, Audureau E, Cagnot C, Marcellin P, Guyader D, et al. Eradication of hepatitis C virus infection in patients with cirrhosis reduces risk of liver and non-liver complications. Gastroenterology. 2017;152:142–56. e142
21. Di Marco V, Calvaruso V, Ferraro D, Bavetta MG, Cabibbo G, Conte E, Cammà C, et al. Effects of eradicating hepatitis C virus infection in patients with cirrhosis differ with stage of portal hypertension. Gastroenterology. 2016;151:130–9. e132
22. Sack J, Garcia-Tsao G. Variceal hemorrhage in a patient with hepatitis C virus cirrhosis in whom liver synthetic function had normalized after viral elimination. Hepatology. 2016;63:1733–5.
23. Afdhal N, Everson GT, Calleja JL, McCaughan GW, Bosch J, Brainard DM, McHutchison JG, et al. Effect of viral suppression on hepatic venous pressure gradient in hepatitis C with cirrhosis and portal hypertension. J Viral Hepat. 2017;24:823–31.
24. Lens S, Alvarado E, Marino Z, Londono MC, Llop E, Martinez J, Fortea JI, et al. Effects of all-oral anti-viral therapy on HVPG and systemic hemodynamics in patients with hepatitis C virus-associated cirrhosis. Gastroenterology. 2017;153(5):1273–1283.e1.

25. Mandorfer M, Kozbial K, Schwabl P, Freissmuth C, Schwarzer R, Stern R, Chromy D, et al. Sustained virologic response to interferon-free therapies ameliorates HCV-induced portal hypertension. J Hepatol. 2016;65:692–9.
26. Jang JW, Choi JY, Kim YS, Woo HY, Choi SK, Lee CH, Kim TY, et al. Long-term effect of antiviral therapy on disease course after decompensation in patients with hepatitis B virus-related cirrhosis. Hepatology. 2015;61:1809–20.
27. Flemming JA, Kim WR, Brosgart CL, Terrault NA. Reduction in liver transplant wait-listing in the era of direct-acting antiviral therapy. Hepatology. 2017;65:804–12.
28. Belli LS, Perricone G, Adam R, Cortesi PA, Strazzabosco M, Facchetti R, Karam V, et al. Impact of DAAs on liver transplantation: major effects on the evolution of indications and results. An ELITA study based on the ELTR registry. J Hepatol. 2018;69:810–7.
29. Aravinthan AD, Barbas AS, Doyle AC, Tazari M, Sapisochin G, Cattral MS, Ghanekar A, et al. Characteristics of liver transplant candidates delisted following recompensation and predictors of such delisting in alcohol-related liver disease: a case-control study. Transpl Int. 2017;30:1140–9.
30. Pose E, Torrents A, Reverter E, Perez-Campuzano V, Campos-Varela I, Avitabile E, Gratacós-Ginès J, et al. A notable proportion of liver transplant candidates with alcohol-related cirrhosis can be delisted because of clinical improvement. J Hepatol. 2021;75:275–83.
31. Nabatchikova EA, Abdurakhmanov DT, Rozina TP, Nikulkina EN, Tanaschuk EL, Moiseev SV. Delisting and clinical outcomes of liver transplant candidates after hepatitis C virus eradication: a long-term single-center experience. Clin Res Hepatol Gastroenterol. 2021;45:101714.
32. Kim TH, Um SH, Lee YS, Yim SY, Jung YK, Seo YS, Kim JH, et al. Determinants of re-compensation in patients with hepatitis B virus-related decompensated cirrhosis starting antiviral therapy. Aliment Pharmacol Ther. 2022;55(1):83–96.
33. Pascasio JM, Vinaixa C, Ferrer MT, Colmenero J, Rubin A, Castells L, Manzano ML, et al. Clinical outcomes of patients undergoing antiviral therapy while awaiting liver transplantation. J Hepatol. 2017;67:1168–76.
34. Perricone G, Duvoux C, Berenguer M, Cortesi PA, Vinaixa C, Facchetti R, Mazzarelli C, et al. Delisting HCV-infected liver transplant candidates who improved after viral eradication: outcome 2 years after delisting. Liver Int. 2018;38:2170–7.
35. Ripoll C, Bari K, Garcia-Tsao G. Serum albumin can identify patients with compensated cirrhosis with a good prognosis. J Clin Gastroenterol. 2015;49:613–9.
36. Calvaruso V, Petta S, Cacciola I, Cabibbo G, Cartabellotta F, Distefano M, Scifo G, et al. Liver and cardiovascular mortality after hepatitis C virus eradication by DAA: data from RESIST-HCV cohort. J Viral Hepat. 2021;28:1190–9.

Prevention of Variceal Bleeding and Rebleeding

48

Vincenzo La Mura, Laura Turco, Hélène Larrue, and Christophe Bureau

Abbreviations

ACLF	Acute on chronic liver failure
AKI/HRS	Acute kidney injury/hepato-renal syndrome
CI	Cardiac index
EVL	Endoscopic variceal ligation
HVPG	Hepatic venous pressure gradient
IPD	Individual patient data
ISMN	Isosorbide mono-nitrate
LVP	Large volume paracentesis
MAP	Mean arterial pressure
NSBB	Non-selective beta-blockers
OHE	Overt hepatic encephalopathy
PH	Portal hypertension
PPF	Primary prophylaxis failure
PTFE	Polytetrafluoroethylene

V. La Mura (✉)
Fondazione IRCCS Ca' Granda Ospedale Maggiore Policlinico, Angelo Bianchi Bonomi Hemophilia and Thrombosis Center, Milan, Italy

Università degli Studi di Milano, Department of Pathophysiology and Transplantation, Milan, Italy
e-mail: vincenzo.lamura@unimi.it

L. Turco
Division of Internal Medicine for the Treatment of Severe Organ Failure, IRCCS Azienda Ospedaliero, Universitaria di Bologna, Bologna, Italy

H. Larrue · C. Bureau
Servie d'Hépatologie Hôpital Rangueil CHU Toulouse cedex et Université Paul Sabatier, Toulouse, France
e-mail: larrue.h@chu-toulouse.fr; bureau.c@chu-toulouse.fr

R. de Franchis (ed.), *Portal Hypertension VII*,
https://doi.org/10.1007/978-3-031-08552-9_48

RCT	Randomized controlled trial
SAP	Systolic arterial pressure
SOC	Standard of care
TIPS	Transjugular intrahepatic portosystemic shunt

Introduction

Because this section is focused on the prevention of further decompensation, it will deal with two main topics: (a) prevention of first variceal hemorrhage in patients with ascites and (b) prevention of recurrent variceal bleeding in both compensated and decompensated cirrhosis. In both, a mainstay of therapy is the use of non-selective beta-blockers (NSBB). Nonselective beta-blockers, first introduced in secondary prophylaxis in 1981 [1], act by decreasing the cardiac output (beta 1 adrenergic blockade) and by reducing the splanchnic vasodilation (beta 2 adrenergic blockade) leading to decreased portal pressure which in turn decreases intraluminal variceal pressure [1].

Because safety issues have been raised with the use of NSBB in patients with decompensated cirrhosis, this specific issue as well as the timing for consideration of the placement of a Transjugular Intrahepatic Portosystemic Shunt (TIPS) will be addressed in this chapter.

Prevention of First Variceal Bleeding in Patients With Ascites

While in compensated cirrhosis, the objective is now to prevent decompensation, in patients with ascites (already decompensated), the goal is still to prevent variceal bleeding. Therefore, in patients who present with ascites, a screening endoscopy should be performed to investigate the presence of varices at a high risk of bleeding so that preventative measures can be implemented.

Primary prophylaxis is recommended with either NSBB or endoscopic variceal ligation (EVL) [2], although recent evidence suggests that NSBB (or carvedilol) is more effective than EVL in preventing first variceal bleeding [3] and is associated with lower rates of decompensation/further decompensation and death [4, 5]. Further evidence on the type of NSBB recommended in the setting of patients with ascites is provided in the section below.

Prevention of Recurrent Variceal Bleeding

Patients who have recovered from an episode of acute variceal bleeding are at high risk of rebleeding and death within the first year after the index episode (60% and 33%, respectively) [6]. Starting effective secondary prophylaxis to prevent the recurrence of variceal bleeding is therefore crucial in these patients. Combination of

NSBB and endoscopic variceal ligation (EVL) is the first-line therapy for the prevention of rebleeding [2, 7, 8] as consistently demonstrated by observational studies, randomized controlled trials, and meta-analyses comparing monotherapy (NSBB or EVL alone) vs. combination therapy (NSBB+EVL) [9, 10].

EVL is a local treatment aimed at the prevention of variceal rupture. Importantly, adding NSBB to EVL improves the efficacy of EVL alone and reduces mortality, whereas adding EVL to NSBB alone is associated with a nonsignificant decrease in rebleeding and mortality [10]. In a recent individual patient data meta-analysis [11], the authors analyzed the benefit of combination therapy, stratifying patients according to Child A (patients without any decompensation other than previous variceal bleeding) vs. Child B/C class of risk (patients with another decompensating event in addition to variceal bleeding). Interestingly, in Child A combination therapy was associated with lower rate of rebleeding without any effect on mortality. In Child B/C patients, combination therapy was associated with lower rate of rebleeding only in trials in which it was compared to EVL alone, indicating, once again, that NSBB is the key element to prevent rebleeding especially in advanced patients. Moreover, these patients also had a lower rate of death, showing that NSBB does not just decrease the incidence of rebleeding, but also improve survival [11] and are a key element in combination to EVL as the standard of care (SOC).

Type of NSBB Recommended in the Prevention of Variceal Bleeding/Rebleeding

Regarding the type of NSBB, most studies performed in the prevention of bleeding/rebleeding had used propranolol or propranolol plus nitrates. Carvedilol is a non-selective beta-blocker with additional anti-alpha1-adrenergic activity. It decreases portal pressure more effectively than traditional NSBB (propranolol or nadolol) [12–14], therefore it can be assumed that the combination of carvedilol + EVL would be preferred. Unfortunately, this potential positive effect on portal pressure is counterbalanced by a higher risk of arterial hypotension [14]. Furthermore, long-term treatment with carvedilol has been associated with the need for reduction of dose/discontinuation of the therapy and increase of diuretics [12], therefore the indication of carvedilol in patients with advanced stages of cirrhosis, particularly those with refractory ascites, has been questioned [15].

Few studies have explored the benefit/risk ratio of carvedilol in the setting of primary prophylaxis (prevention of the first bleed) in patients with ascites or the secondary prophylaxis (prevention of recurrent bleed) in any patient with cirrhosis independent of the presence or absence of ascites).

As mentioned above, a study comparing carvedilol vs. ligation for primary prophylaxis demonstrated the superiority of carvedilol, with 51% of patients included in the study having ascites [3].

Regarding secondary prophylaxis, a recent meta-analysis from Malandris et al. [16] compared carvedilol vs. one of the following comparators: EVL (three studies, including 112/230 patients in the carvedilol arm), NSBB plus ISMN (two studies, including 108/207 patients in the carvedilol arm), propranolol alone (two studies, including 33/61 patients in the carvedilol arm). Among these studies, only one compared carvedilol + EVL (21 patients) vs. the standard of care, propranolol + EVL (17 patients). Overall, carvedilol was as effective as any of the comparator treatments regarding rebleeding rate and mortality. Interestingly, in the subgroup analysis comparing carvedilol vs. propranolol, there was a tendency for lower mortality with carvedilol (OR: 0.39, 95%-CI: 0.15–1.03) thus, suggesting a favorable efficacy of carvedilol (vs. propranolol) in secondary prophylaxis. When integrating the available evidence on the benefits and safety of carvedilol, we recommend invariably NSSB or carvedilol for the prevention of the first hemorrhage in patients with ascites and the combination of NSBB or carvedilol + EVL for secondary prophylaxis.

Safety of NSBB/Carvedilol in Patients With Ascites

More than 10 years ago, in a landmark study by Serstè et al. [17], concerns were raised regarding the potentially harmful effect of NSBB in patients with refractory ascites. In this single-center observational study including 151 patients with refractory ascites, 77 of them on NSBB for the prevention of variceal bleeding, the authors found that patients on NSBB had a significantly lower rate of survival (19% vs. 64% at 1-year, $p < 0.0001$) compared to patients, not on NSBB. This study started an intense debate on the real safety of NSBB in decompensated patients based on the hypothesis of a therapeutic window for their use in clinical practice which would be open when patients develop varices but would be closed when patients develop refractory ascites due to the risk of renal hypoperfusion and, ultimately, AKI/HRS [17]. Several observational studies have been published since then with contrasting results. Indeed, some of them have produced confirmatory data on this hypothesis [18], others have limited the cautionary note just to patients with spontaneous bacterial peritonitis [19], others have refuted this observation even in the clinical setting of patients with refractory ascites [20].

All these data should be reconsidered in light of the important hemodynamic changes that occur in the pathophysiology of portal hypertension. Indeed, as liver cirrhosis progresses from a compensated to a decompensated stage there is a progressive decrease in the mean arterial pressure (MAP) together with changes in the cardiac index (CI), which is maximum in the early stage of diuretic-controlled ascites but decreases in the late stage of refractory ascites [21]. In this last clinical condition, the heart cannot further compensate for the progressive decrease of Mean Arterial Pressure (MAP) with detrimental effects on systemic hemodynamics, organ perfusion, and survival. As a matter of fact, patients with refractory ascites and a CI <3.2 L/min/m^2 have lower survival than patients with a CI above this threshold. The use of NSBB in this group of patients could further lower the CI and can decrease survival. In a study including 403 patients with cirrhosis (213 compensated and 190 decompensated), Alvarado-Tapias et al. demonstrated the deleterious short-term effect of NSBB in patients with decompensated cirrhosis and a low CI (<3 L/min/m^2), confirming that when the heart can no longer compensate the extreme systemic vasodilation, NSBB may further decrease the cardiac compensation and increase the risk of death [22]. Accordingly, Tellez et al. [23] measured the ejection intra-ventricular pressure difference (a noninvasive marker of systolic function) before and after NSBB in patients with ascites. The authors stratified patients according to the type of ascites (diuretic-controlled ascites vs. refractory ascites) and found no significant difference in systolic function before and after NSBB in patients with diuretic-controlled ascites, while there was a significant decrease in patients with refractory ascites. Importantly, in patients with refractory ascites, the reduced systolic function associated with NSBB had a severe impact on kidney function when the renal perfusion pressure dropped below 65 mmHg, a critical threshold for organ autoregulation of perfusion. More than half of the patients (11/20 patients, 55%) decreased the renal perfusion under that critical value, with the consequence that 4 out of these 11 patients had an increase in creatinine meeting the criteria for HRS-AKI. Analyzing additional data, kindly provided us by the authors, patients with refractory ascites who decreased the renal perfusion pressure below 65 mmHg had a significantly lower MAP (median 70 [65–73 IQR]) and Systolic Blood Pressure (SAP) (median 74 [69–78 IQR]) after NSBB compared to those who maintained a renal perfusion pressure above 65 mmHg (median MAP 90 [85–95 IQR]; median SBP 80 [75–85 IQR]). Based on this data, the critical threshold to safely use NSBB in patients with refractory ascites could be 65 mmHg for MAP which could be proposed as an additional criterion of safety together with the threshold of safety for systolic blood pressure over 90 mmHg traditionally recommended by experts, which is also supported by this data [2]. This observation is consistent with data from a posthoc analysis on the efficacy of satavaptan in patients with ascites, including 1198 patients stratified according to the type of ascites [20]. In this large series composed of patients with diuretic-responsive ascites ($n = 462$), patients with ascites managed with diuretics and occasional paracentesis ($n = 496$), and patients with refractory ascites ($n = 240$), none had a negative impact on survival associated with the use of NSBB, even if patients were suffering from refractory ascites. Only in patients with a MAP below 70 mmHg, NSBB suggest a risk for survival, confirming what was specifically addressed by Tellez et al. and commented above [23].

Attention should also be paid to the other events that can, per se, impact blood and renal pressure, like infections. In a retrospective study by Mandorfer et al., [19] including more than 600 patients with ascites managed by repeated paracenteses, patients on NSBB had a 25% lower risk of death [HR = 0.75; 95%CI: 0.581–0.968, $p = 0.027$] compared with those not on NSBB. However, in patients who developed spontaneous bacterial peritonitis, the use of NSBB was associated with a 58% higher risk of mortality [HR = 1.58; 95%CI: 1.098–2.274, $p = 0.014$]. The mechanism behind this observation is, again, the low renal perfusion, further reduced by the superimposed infection, which leads to HRS-AKI. The rate of HRS development was, in fact, significantly higher in patients with NSBB compared with those without NSBB treatment (24% vs. 11%, respectively; $p = 0.027$), with a higher risk of death in the NSBB group.

Even though, as mentioned above, carvedilol has a greater portal hypotensive effect and appears to be more useful in preventing variceal bleeding and rebleeding, its additional alpha-adrenergic blockade effect could lead to further falls in systemic/arterial blood pressure and renal perfusion pressure. Therefore patients, particularly those with ascites, should be monitored closely for the development of systemic hypotension.

Therefore, to safely use traditional NSBBs or carvedilol in patients with ascites, it is advisable to reduce the dose or discontinue the drug in case of systolic blood pressure persistently below 90 mmHg, mean arterial pressure below 65 mmHg, and/or if HRS-AKI is detected. Once blood pressure returns to baseline and/or HRS-AKI resolves, NSBB should be re-initiated or re-titrated. If the patient remains intolerant to NSBB, EVL is then recommended in order to prevent variceal hemorrhage, but TIPS could be a valid alternative, particularly in patients fulfilling the criteria of refractory/diuretic intolerant ascites or recurrent ascites [24].

Patients with Intolerance or Failure of NSBB

As of today, due to the recent extension of the indications for the use of NSBB in patients with compensated cirrhosis in order to prevent the first decompensation [25], an increasing number of patients with cirrhosis could experience their first episode of variceal hemorrhage while on NSBB, which means clinical failure of primary prophylaxis [26]. Furthermore, a nonnegligible proportion of patients may not tolerate chronic therapy with this class of drugs (up to 15% withdrawn in RCTs) or fail to achieve the adequate prophylactic dose [27]. As a consequence, there is a heterogeneous group of candidates for secondary prophylaxis that could not be fully protected by the SOC and no RCTs have been targeted in this subgroup of patients. Therefore, patients who are intolerant to NSBB frequently could receive EVL alone both for primary and secondary prophylaxis, or, in selected cases (see below), advanced lines of therapy (e.g., TIPS).

Role of TIPS

Primary Prophylaxis of Variceal Bleeding in Decompensated Patients

Importantly, **patients with decompensated cirrhosis should be evaluated for liver transplantation** as this represents a definitive treatment of the underlying liver disease (i.e., for decompensation) and is associated with favorable long-term prognosis, especially in patients with a high MELD and/or high Child-Pugh score and in patients with previous recurrent or chronic overt hepatic encephalopathy. No study has been yet performed to investigate TIPS in the setting of primary prophylaxis of bleeding in the absence of any other indication for a shunt creation. Thus, **TIPS cannot be recommended in such a setting. However,** considering the favorable results with

TIPS for recurrent/refractory ascites (lower rate of further decompensation including less PH-related bleeding and a survival benefit), TIPS is effective to prevent variceal bleeding (and should thus be used for primary prophylaxis) in patients with varices that have never bleed who also suffer from recurrent/refractory ascites.

Secondary Prophylaxis of Variceal Bleeding and Prevention of Further Decompensation

Previous and current guidelines recommend the association of NSBB and EVL in secondary prophylaxis of recurrent variceal bleeding. Several meta-analyses have shown that the combination of both NSBB and EVL was superior to either treatment alone in preventing rebleeding. In most recent meta-analyses, although there was no significant difference in survival, TIPS was found to be superior to the control group in decreasing the incidence of both rebleeding and bleeding-related death [28, 29]. Therefore, while TIPS has never been specifically assessed by RCT, neither in the context of the failure of a well conducted secondary prophylaxis study nor in the case of intolerance or contraindication to NSBB and/or EVL, TIPS remains the second-line therapy. At the time of rebleeding, it is crucial to check that the patient does not meet the high-risk criteria so as not to miss a pre-emptive TIPS and a survival benefit.

Furthermore, it is also important to point out that TIPS could also be indicated in secondary prophylaxis, particularly in those patients with recurrent ascites or portal vein thrombosis. When dealing with a patient with cirrhosis, we do not only call for a decreased risk of bleeding or rebleeding but also a decreased risk of any further decompensation, since any episode of further decompensation is associated with a worse prognosis [30].

On one side, we know, from controlled studies with TIPS in patients with recurrent ascites that the risk of other liver-related events is reduced. For example, in the RCT from Bureau et al. in patients with recurrent ascites [24], the number of days in the hospital, the occurrence of bleeding or rebleeding (see above), and the number of hernia-related complications were significantly lower in the TIPS group compared to the large-volume paracentesis group. Additionally, serum creatinine improves over time in patients treated with TIPS and importantly, plasma renin activity, a relevant prognostic marker, dramatically decreases over 1 year.

On the other side, looking at the results focusing on preemptive TIPS, we observe that the proportion of patients with new-onset or relapse of ascites is lower in the TIPS group when compared to standard treatment (NSBB+EVL). This result was consistently found in European studies [31, 32], Asian studies [33], and in the recent IPD meta-analysis [34].

One additional benefit could be the rate of recanalization of the portal vein in a patient with portal vein thrombosis. Two RCTs [35, 36], which aimed to assess the risk of PH rebleeding in patients with portal vein thrombosis, observed that the rate of recanalization was significantly higher in the TIPS group compared to the control group.

The RCTs mentioned above in patients with PVT and preemptive TIPS failed to show any difference regarding the risk of encephalopathy between TIPS and control groups. However, two RCTs in secondary prophylaxis show a higher rate of OHE in the TIPS group as compared to the control group

- the German study: 18% in the TIPS arm versus 8% in the control arm ($p = 0.05$), [37],
- the Holster study: 35% versus 14% at 1 year ($p = 0.035$) [34].

Finally, the benefit of TIPS on survival remains a controversial issue but a survival advantage has been shown:

- in the pivotal study of pre-emptive TIPS: 1-year survival 86% in TIPS group versus 61% in the control group ($p < 0.001$) [38],
- in the Asian RCT: 1-year survival 86% in the TIPS group versus 73% in the control group ($p = 0.046$), [33],
- and confirmed in the recent IPD meta-analysis: 1-year survival 79% % in the TIPS group versus 62% in the control group ($p < 0.001$) [39].

Prevention of Bleeding/Rebleeding: Is it Time for a Personalized Approach?

NSBB+EVL represents the SOC to reduce the risk of rebleeding but 20%–25% of patients can rebleed or even die during the first 6 weeks of therapy [40]. Furthermore, some patients can experience the first bleeding episode while on NSBB or can be intolerant to NSSB which can substantially reduce the efficacy of SOC on rebleeding and survival [11, 26]. Therefore, any recommendation about treatment should be based on the individualized prognosis by considering more advanced lines of therapy in patients who are likely not to benefit from SOC. TIPS and liver transplantation are potential alternatives to SOC to reduce the risk of further decompensation and death, however, the most appropriate time for their indication in secondary prophylaxis is still debated. To date, the multistage model of risk proposed for cirrhosis [30] and an HVPG-guided risk stratification [41] could bring a valuable prognostic information to guide the clinical decision based on the individual risk. The first model gives an additive prognostic effect to any single episode of decompensation, it is easy to apply in clinical practice and is noninvasive. However, no studies have tested this model to direct patients to more advanced lines of therapy in secondary prophylaxis. On the contrary, in a randomized controlled trial, Villanueva et al. compared 84 patients assigned to a stepwise increase of portal hypotensive drug therapy based on HVPG monitoring vs. 86 controls with better results on rebleeding and survival in the former group [41]. This study confirmed the importance of HVPG reduction to change the natural history of cirrhosis; however, it showed that three subsequent hemodynamic studies were needed in up to 44% of patients assigned to the HVPG-guided arm, which makes the applicability of such HVPG-based approach unfeasible and unaffordable. In an attempt to maintain the informative contribution of the clinical staging system and the HVPG measurement, recently La Mura et al. proposed a new prognostic algorithm limiting the measurement of the HVPG-response to NSBB to patients who had bled and had another decompensation (e.g., ascites or hepatic encephalopathy) and a baseline HVPG over 16 mmHg [42]. Interestingly, the algorithm had the same prognostic performance as the classic chronic HVPG-response, but, as an advantage, the number of hemodynamic studies needed for prognostication was halved. This algorithm was validated in an independent series confirming its accuracy in prognostication, however, its efficacy in the clinical decision-making process needs to be tested and validated before drawing definitive conclusions.

Summary and Conclusions

All patients after the first episode of decompensation are candidates for any strategy of treatment able to reduce the risk of further decompensation. Particularly in this chapter, we have covered the main issues regarding the prevention of the first bleeding episode in patients with ascites and the prevention of rebleeding after the first episode of variceal bleeding. In both situations, the use of traditional NSBB (propranolol or nadolol) or carvedilol is the mainstay of medical therapy because of their lowering effect on portal pressure. TIPS has been discussed as the most appropriate advanced line of treatment before transplantation just in highly selected cases, patients with recurrent ascites or patients after failure of rebleeding prophylaxis. As of today, there is no evidence to support systematically a different approach, therefore all patients after the first episode of decompensation are treated in the same way by combining drugs+/−endoscopy or TIPS depending on the clinical stage of the disease. It is necessary that future studies take into account the vast heterogeneity of candidates to the prevention of further decompensation. When integrating the available evidence on the patients included in the clinical studies commented on in this chapter, the subgroup of patients who should be prioritized for the inclusion in new clinical studies are certainly those intolerants to NSBB/carvedilol, those who bleed despite adequate medical therapy to prevent the first bleeding episode, those with ascites/hepatic encephalopathy who bleed and do not reduce HVPG under therapy. For some of these areas of study, making an adequately sized randomized controlled trial is really problematic. However, well-designed collaborative studies of effectiveness are mandatory to warrant new algorithms of treatment based on the individual risk of developing further decompensation.

References

1. Lebrec D, Poynard T, Hillon P, et al. Propranolol for prevention of recurrent gastrointestinal bleeding in patients with cirrhosis: a controlled study. N Engl J Med. 1981;305:1371–4.
2. Angeli P, Bernardi M, Villanueva C, et al. EASL clinical practice guidelines for the management of patients with decompensated cirrhosis. J Hepatol. 2018;69:406–60.
3. Tripathi D, Ferguson JW, Kochar N, et al. Randomized controlled trial of carvedilol versus variceal band ligation for the prevention of the first variceal bleed. Hepatology. 2009;50:825–33.

4. Turco L, Villanueva C, La Mura V, et al. Lowering portal pressure improves outcomes of patients with cirrhosis, with or without ascites: a meta-analysis. Clin Gastroenterol Hepatol. 2020;18:313–27.
5. McDowell HR, Chuah CS, Tripathi D, et al. Carvedilol is associated with improved survival in patients with cirrhosis: a long-term follow-up study. Aliment Pharmacol Ther. 2021;53:531–9.
6. Bosch J, García-Pagán JC. Prevention of variceal rebleeding. Lancet. 2003;361:952–4.
7. De Franchis R, Abraldes JG, Bajaj J, et al. Expanding consensus in portal hypertension report of the Baveno VI consensus workshop: stratifying risk and individualizing care for portal hypertension. J Hepatol. 2015;63:743–52.
8. Garcia-Tsao G, Abraldes JG, Berzigotti A, et al. Portal hypertensive bleeding in cirrhosis: risk stratification, diagnosis, and management: 2016 practice guidance by the American association for the study of liver diseases. Hepatology. 2017;65:310–35.
9. Gonzalez R, Zamora J, Gomez-Camarero J, et al. Meta-analysis: combination endoscopic and drug therapy to prevent variceal rebleeding in cirrhosis. Ann Intern Med. 2008;149:109–22.
10. Puente A, Hernández-Gea V, Graupera I, et al. Drugs plus ligation to prevent rebleeding in cirrhosis: an updated systematic review. Liver Int. 2014;34:823–33.
11. Albillos A, Zamora J, Martínez J, et al. Stratifying risk in the prevention of recurrent variceal hemorrhage: results of an individual patient meta-analysis. Hepatology. 2017;66:1219–31.
12. Bañares R, Moitinho E, Piqueras B, et al. Carvedilol, a new nonselective beta-blocker with intrinsic anti- Alpha1-adrenergic activity, has a greater portal hypotensive effect than propranolol in patients with cirrhosis. Hepatology. 1999;30:79–83.
13. Reiberger T, Ulbrich G, Ferlitsch A, et al. Carvedilol for primary prophylaxis of variceal bleeding in cirrhotic patients with haemodynamic non-response to propranolol. Gut. 2013;62:1634–41.
14. Sinagra E, Perricone G, D'Amico M, et al. Systematic review with meta-analysis: the haemodynamic effects of carvedilol compared with propranolol for portal hypertension in cirrhosis. Aliment Pharmacol Ther. 2014;39:557–68.
15. Reiberger T, Mandorfer M. Beta adrenergic blockade and decompensated cirrhosis. J Hepatol. 2017;66:849–59.
16. Malandris K, Paschos P, Katsoula A, et al. Carvedilol for prevention of variceal bleeding: a systematic review and meta-analysis. Ann Gastroenterol. 2019;32:287–97.
17. Sersté T, Melot C, Francoz C, et al. Deleterious effects of beta-blockers on survival in patients with cirrhosis and refractory ascites. Hepatology. 2010;52:1017–22.
18. Kalambokis GN, Christodoulou D, Baltayiannis G, et al. Propranolol use beyond 6 months increases mortality in patients with child-Pugh C cirrhosis and ascites. Hepatology. 2016;64:1806–8.
19. Mandorfer M, Bota S, Schwabl P, et al. Nonselective β blockers increase risk for hepatorenal syndrome and death in patients with cirrhosis and spontaneous bacterial peritonitis. Gastroenterology. 2014;146:1680–90.
20. Bossen L, Krag A, Vilstrup H, et al. Nonselective β-blockers do not affect mortality in cirrhosis patients with ascites: post hoc analysis of three randomized controlled trials with 1198 patients. Hepatology. 2016;63:1968–76.
21. Turco L, Garcia-Tsao G, Magnani I, et al. Cardiopulmonary hemodynamics and C-reactive protein as prognostic indicators in compensated and decompensated cirrhosis. J Hepatol. 2018;68:949–58.
22. Alvarado-Tapias E, Ardevol A, Garcia-Guix M, et al. Short-term hemodynamic effects of β-blockers influence survival of patients with decompensated cirrhosis. J Hepatol. 2020;73:829–41.
23. Téllez L, Ibáñez-Samaniego L, Pérez del Villar C, et al. Non-selective beta-blockers impair global circulatory homeostasis and renal function in cirrhotic patients with refractory ascites. J Hepatol. 2020;73:1404–14.
24. Bureau C, Thabut D, Oberti F, et al. Transjugular intrahepatic portosystemic shunts with covered stents increase transplant-free survival of patients with cirrhosis and recurrent ascites. Gastroenterology. 2017;152:157–63.

25. Villanueva C, Albillos A, Genescà J, et al. β blockers to prevent decompensation of cirrhosis in patients with clinically significant portal hypertension (PREDESCI): a randomised, double-blind, placebo-controlled, multicentre trial. Lancet. 2019;393:1597–608.
26. De Souza AR, La Mura V, Reverter E, et al. Patients whose first episode of bleeding occurs while taking a β-blocker have high long-term risks of rebleeding and death. Clin Gastroenterol Hepatol. 2012;10:670–6.
27. Tandon P, Saez R, Berzigotti A, et al. A specialized, nurse-run titration clinic: a feasible option for optimizing beta-blockade in non-clinical trial patients. Am J Gastroenterol. 2010;105:1917–21.
28. Miao Z, Lu J, Yan J, et al. Comparison of therapies for secondary prophylaxis of esophageal variceal bleeding in cirrhosis: a network meta-analysis of randomized controlled trials. Clin Ther. 2020;42:1246–75.
29. Jing L, Zhang Q, Chang Z, et al. Nonsurgical secondary prophylaxis of esophageal Variceal bleeding in cirrhotic patients: a systematic review and network meta-analysis. J Clin Gastroenterol. 2021;55:159–68.
30. D'Amico G, Pasta L, Morabito A, et al. Competing risks and prognostic stages of cirrhosis: a 25-year inception cohort study of 494 patients. Aliment Pharmacol Ther. 2014;39:1180–93.
31. Garcia-Pagán JC, Di Pascoli M, Caca K, et al. Use of early-TIPS for high-risk variceal bleeding: results of a post-RCT surveillance study. J Hepatol. 2013;58:45–50.
32. Hernández-Gea V, Procopet B, Giráldez Á, et al. Preemptive-TIPS improves outcome in high-risk variceal bleeding: an observational study. Hepatology. 2019;69:282–93.
33. Lv Y, Yang Z, Liu L, et al. Early TIPS with covered stents versus standard treatment for acute variceal bleeding in patients with advanced cirrhosis: a randomised controlled trial. Lancet Gastroenterol Hepatol. 2019;4:587–98.
34. Holster IL, Tjwa ETTL, Moelker A, et al. Covered transjugular intrahepatic portosystemic shunt versus endoscopic therapy + β-blocker for prevention of variceal rebleeding. Hepatology. 2016;63:581–9.
35. Lv Y, Qi X, He C, et al. Covered TIPS versus endoscopic band ligation plus propranolol for the prevention of variceal rebleeding in cirrhotic patients with portal vein thrombosis: a randomised controlled trial. Gut. 2018;67:2156–68.
36. Luo X, Wang Z, Tsauo J, et al. Advanced cirrhosis combined with portal vein thrombosis: a randomized trial of TIPS versus endoscopic band ligation plus propranolol for the prevention of recurrent esophageal Variceal bleeding. Radiology. 2015;276:286–93.
37. Sauerbruch T, Mengel M, Dollinger M, et al. Prevention of Rebleeding from esophageal Varices in patients with cirrhosis receiving small-diameter stents versus Hemodynamically controlled medical therapy. Gastroenterology. 2015;149:660–8.
38. García-Pagán JC, Caca K, Bureau C, et al. Early use of TIPS in patients with cirrhosis and variceal bleeding. N Engl J Med. 2010;362:2370–9.
39. Nicoară-Farcău O, Han G, Rudler M, et al. Effects of early placement of Transjugular Portosystemic shunts in patients with high-risk acute Variceal bleeding: a meta-analysis of individual patient data. Gastroenterology. 2021;160:193–205.
40. Villanueva C, Balanzó J, Novella MT, et al. Nadolol plus isosorbide mononitrate compared with sclerotherapy for the prevention of variceal rebleeding. N Engl J Med. 1996;334:1624–9.
41. Augustin S, Muntaner L, Altamirano JT, et al. Predicting early mortality after acute variceal hemorrhage based on classification and regression tree analysis. Clin Gastroenterol Hepatol. 2009;7:1347–54.
42. La Mura V, Garcia-Guix M, Berzigotti A, et al. A prognostic strategy based on stage of cirrhosis and HVPG to improve risk stratification after Variceal bleeding. Hepatology. 2020;72:1353–65.

Prevention of Further Decompensation in Patients With Ascites

49

Salvatore Piano, Thomas Reiberger, Hélène Larrue, and Christophe Bureau

Ascites and Further Decompensation After Ascites as the Index Decompensating Event

In cirrhotic patients, ascites is the most common first decompensating event, with about two-thirds of cases [1, 2]. The occurrence of ascites is associated with poor quality of life and a high risk of hospital readmissions, but importantly also with poor survival [3, 4]. Specifically, the probability of survival after the development of ascites is 85% and 57% at 1 year and 5 years, respectively [5]. In terms of the most eminent short-term risk after the development of ascites, patients frequently develop *further decompensating events*, that can be either non-ascites-related PH-driven events (such as variceal bleeding, hepatic encephalopathy) or ascites-related events (such as progression to recurrent/refractory ascites, spontaneous bacterial peritonitis [SBP] or hepatorenal syndrome [HRS]). The pathophysiology underlying ascites development and the main drivers of further decompensation after ascites are summarized in Fig. 49.1. Importantly, there are three distinct pathophysiologic mechanisms, which are strictly related and considered crucial for ascites development and its progression to further decompensation: (1) portal hypertension, (2) splanchnic arterial vasodilation (3) systemic inflammation [6].

S. Piano (✉)
Unit of Internal Medicine and Hepatology, Department of Medicine—DIMED, University and Hospital of Padova, Padova, Italy

T. Reiberger
Division of Gastroenterology and Hepatology, Department of Medicine III, Medical University of Vienna, Vienna, Austria
e-mail: Thomas.reiberger@meduniwien.ac.at

H. Larrue · C. Bureau
Servie d'Hépatologie Hôpital Rangueil, CHU Toulouse, Toulouse, France
e-mail: larrue.h@chu-toulouse.fr; bureau.c@chu-toulouse.fr

R. de Franchis (ed.), *Portal Hypertension VII*,
https://doi.org/10.1007/978-3-031-08552-9_49

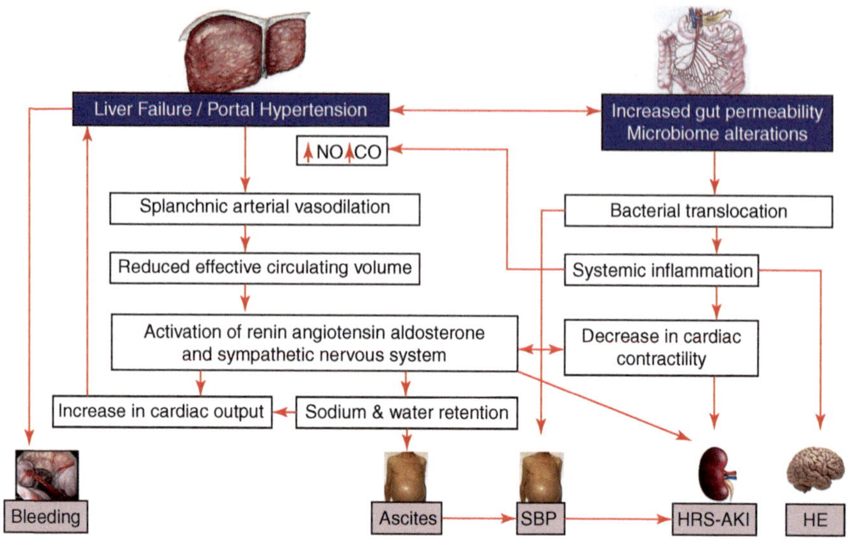

Fig. 49.1 Pathophysiology of ascites and further decompensation in patients with cirrhosis. NO nitric oxide, CO carbon monoxide, HE hepatic encephalopathy, AKI-HRS acute kidney injury hepatorenal syndrome

Portal Hypertension (PH) is the main driver for the occurrence of decompensation in patients with compensated cirrhosis [7] and causes splanchnic congestion and gut barrier abnormalities [8], which favor splanchnic arterial vasodilation and increase intestinal permeability to gut microbes, respectively [9]. Splanchnic arterial vasodilation occurs due to the release of vasodilators (such as nitric oxide, carbon monoxide) and is responsible for a decrease in effective circulating volume, decrease in systemic blood pressure, activation of endogenous vasoconstrictors systems (renin angiotensin aldosterone system (RAAS), sympathetic Nervous System (SNS) and non-osmotic secretion of Antidiuretic Hormone (ADH, formerly known as Arginine Vasopressin, AVP)), which causes avid sodium and water retention in the kidney. Moreover, the activation of the SNS induces an adrenergic drive for increased cardiac output which aims to counteract the reduced systemic vascular resistance in an attempt to maintain adequate systemic blood pressure. The hyperdynamic circulation and the increase in splanchnic blood flow contributes to the increase in portal pressure (i.e., driven by splanchnic hyperperfusion), while concomitantly the exaggerated SNS activity also reduces gut motility favoring intestinal bacterial overgrowth and dysbiosis. Finally, the combination of increased intestinal permeability, gut bacterial overgrowth, and changes in the microbiome facilitates the occurrence of a pathological translocation of bacteria or bacterial fragments (i.e., pathogen-associated molecular patterns, PAMPs) from the gut to the systemic circulation and mesenteric lymph nodes [10]. Bacterial translocation causes the activation of the immune system (macrophages, dendritic cells) and the release of inflammatory cytokines, which causes oxidative stress, and the release of vasodilators with a further worsening of splanchnic vasodilation [10]. Moreover, in advanced cirrhosis, systemic inflammation is also responsible for a decrease in cardiac contractility [11]. The reduction of cardiac contractility and cardiac output further aggravates circulatory dysfunction and is associated with the occurrence of further decompensating events such as refractory ascites [12] and HRS-AKI [13, 14]. SBP represents the most severe clinical manifestation of bacterial translocation and is associated with a high risk of HRS-AKI and mortality.

Bacterial translocation is not the sole cause of inflammation in patients with cirrhosis, because liver damage per se can cause the release of Danger-Associated Molecular Patterns (DAMPs) from hepatocytes, which can trigger an inflammatory response. Consequently, systemic inflammation is present across all stages of ACLD, however, it seems to be most pronounced in decompensated cirrhosis where it drives *further decompensation* [15].

Clinical events that cause acute inflammatory responses, such as bacterial infections, heavy drinking, or HBV flares can precipitate the occurrence of further decompensating events in patients with cirrhosis and ascites.

Interestingly, all of the aforementioned pathophysiological mechanisms constitute potential targets for treatment aiming at prevention of *further decompensation.*

Classification of Ascites

Ascites can be classified as uncomplicated and complicated. Uncomplicated ascites are ascites that are not infected and which are not associated with the development of acute kidney injury/hepatorenal syndrome (HRS-AK) [16].

In addition, ascites can be quantified according to its amount in the peritoneal cavity into three progressive grades: (i) mild ascites only detectable by ultrasound; (ii) moderate ascites with symmetrical distension of the abdomen and (iii) large/tense ascites. The natural history of mild ascites is not well defined, and it does not qualify to define decompensated cirrhosis. However, recent data suggest that mild ascites are already associated with some degree of systemic inflammation and importantly, with lower survival as compared to patients with compensated cirrhosis and without any ascites [17]. Thus, the prognostic value of mild ascites should be further investigated, and patients with mild ascites should also be more closely monitored.

In Baveno-VII, further decompensating events examined, that are specifically related to ascites, were SBP, HRS-AKI, and recurrent/refractory ascites—with all of them representing hallmarks of complicated ascites. Ascites are considered refractory when it cannot be mobilized or its early (within 1 month) recurrence cannot be prevented by medical (diuretic) treatment [18]. Recurrent ascites are ascites that recur in a quantity that requires large-volume paracentesis (LVP) at least three times within a year. Importantly, the definition of recurrent ascites does not need an observation period of 1 year but can also be established when three LVPs are required within a shorter time frame. However, it also has to be emphasized that, if repeated

LVPs are required around an acute decompensation event (e.g., variceal bleeding) that may have only 'short-term' impact on renal function and systemic circulation, this may not be 'true' recurrent ascites, and clinical judgement for determining the 'reversibility' of severe/tense ascites is always required. Still, as TIPS represents a very effective treatment for recurrent ascites, and also prevents other further decompensation events (e.g., variceal bleeding) and even improves transplant-free survival [19], the establishment of a diagnosis of recurrent ascites (i.e., the indication for TIPS) should not require a full year of observations (and of repeated LVPs).

Changing Paradigm in Treating Ascites: From Ascites Control to Prevention of Further Decompensation

Traditionally, treatment of ascites in patients with cirrhosis has been focused on managing ascites by removing abdominal fluid by paracentesis and counteracting sodium and water retention [16]. Although this strategy is effective, at least temporarily, in controlling ascites, it does not provide any survival benefit and the treatment of patients with decompensated cirrhosis should instead be aimed at preventing further decompensation and improving survival. The underlying mechanisms that drive the development of further decompensation in cirrhosis include ongoing/active/progressive etiology of liver disease, the severity of PH (HVPG), abnormalities in the gut-liver axis, pronounced systemic inflammation, and splanchnic arterial vasodilation. Targeting these disease-driving mechanisms can prevent the occurrence of further decompensation and improve survival. Worth noting, all patients with cirrhosis and ascites should be referred to liver transplant centers for the assessment of indication and eligibility for liver transplantation.

Ascites Control by Diuretics and Paracentesis

In all patients with the first occurrence of moderate/tense ascites, a diagnostic paracentesis should be performed. In fact, ascitic fluid analysis can establish if the accumulation of ascites is a consequence of PH (serum albumin ascites gradient $\geq$1.1 g/dL), the presence/absence of SBP (polymorphonuclear cell count >250 cells/μL), and the risk of developing SBP during follow up (total ascitic fluid protein content<1.5 g/dL) [20, 21]. Moreover, a diagnostic paracentesis should be performed to rule-out SBP in all patients with ascites hospitalized for an episode of further decompensation (e.g., jaundice, hepatic encephalopathy, variceal bleeding, recurrent ascites, AKI) and/or experiencing a clinical deterioration [20, 21].

In patients with tense ascites, a large volume paracentesis (LVP) is recommended, since it can rapidly control ascites [22]. LVP (in particular if >5 L of ascitic fluid are removed) can cause Post-Paracentesis circulatory dysfunction (PPCD), a syndrome characterized by the hyperactivation of the endogenous vasoconstrictor system and associated with early recurrence of ascites, refractory ascites, HRS-AKI and poor survival [23]. Albumin infusion (8 g/L of ascites removed) prevents

PPCD. After paracentesis, all patients should receive diuretic treatment and a moderately low sodium diet to prevent the recurrence of ascites [20]. Regarding diuretic treatment, anti-aldosterone drugs are the treatment of choice [24], while loop diuretics (furosemide) can be combined with anti-aldosterone drugs in patients with recurrent ascites [25]. Diuretic treatment counteracts sodium and water retention but has small effects on the pathophysiological mechanisms driving further decompensation of cirrhosis.

Etiologic Treatment

Etiologic treatment is crucial in the management of patients with cirrhosis since it has been demonstrated to prevent the occurrence of decompensation and improve survival. However, the impact of etiologic treatment on the clinical course of already decompensated cirrhosis is less predictable. After achieving SVR, about one-third of patients with decompensated cirrhosis experience an improvement in MELD and Child-Pugh score. However, patients with HCV-related decompensated cirrhosis have a lower probability of getting a resolution of clinically significant portal hypertension and remain at higher risk of developing further decompensating events and liver-related death [26–28]. Importantly, decompensated HCV patients do not achieve $a \geq 10\%$ decrease in HVPG after SVR and remained at higher risk for further decompensation [29]. Also, the reported effects and benefits of abstinence in patients with alcohol-related cirrhosis who are already decompensated are heterogeneous, with some studies showing an improvement in survival in those maintaining abstinence [30] while others did not [31]. It may be speculated that there is a certain 'point-of-no-return' in patients with decompensated cirrhosis, in whom neither cure/suppression/removal of the primary etiological factor nor other supporting medical therapies can halt disease progression and liver failure cannot be prevented without liver transplantation. However, it must be emphasized that etiologic treatment is a cornerstone in the treatment of decompensated cirrhosis and is a requirement to allow for clinical recompensation of patients with decompensated cirrhosis.

Betablocker Therapy to Ameliorate PH

Treatment of portal hypertension by traditional non-selective beta-blockers (NSBBs) or carvedilol is the standard of care for primary bleeding prophylaxis (also in patients with decompensated cirrhosis/ascites) and secondary bleeding prophylaxis, meaning that NSBBs and carvedilol are clearly able to prevent bleeding-related further decompensation. This aspect in the prevention of further decompensation is described in detail in another chapter of this Baveno-VII proceeding book. However, non-hemodynamic effects of NSBB therapy that likely contribute to a reduced risk of further decompensation need to be explored, and thus, NSBBs—at this state of knowledge—cannot be recommended for the prevention of non-bleeding related further decompensation.

Antibiotics

As previously mentioned, bacterial infections are a common trigger of further decompensation in patients with cirrhosis [32]. Indeed, they are associated with a high risk of acute kidney injury [33], hepatic encephalopathy [34], rebleeding from varices [35], and are associated with poor survival [36]. Overall, bacterial infections increase mortality fourfold in patients with cirrhosis and are very frequent, being diagnosed in almost 40% of patients hospitalized for decompensated cirrhosis [37]. The most common infections are SBP, urinary tract infections, pneumonia, bloodstream infections, and skin and soft tissue infections, which constitute more than 80% of all the infections [37–39]. Infections can be subtle in patients with cirrhosis and the onset of decompensation can be the sole sign of infection. This is the reason why infections should be ruled out in all patients with first or new decompensating events and/or when clinical deterioration occurs. The minimal work-up for infections should include diagnostic paracentesis, chest X-ray, cultures of blood, ascites and urine, and skin examination. In patients with infections, antibiotic treatment should be timely administered. Principles guiding the choice of antibiotic treatment include the severity of infections, local epidemiology, and risk factors for multidrug-resistant bacteria (e.g., nosocomial infections, previous treatment with antibiotics, etc.) [40]. Fungal infections should be suspected in patients not responding to antibiotic treatment and in those with specific risk factors, namely with prior bacterial infections, diabetes, AKI, carrying a central venous catheter, being on parenteral nutrition, or having received a gastrointestinal endoscopy in the last month [41, 42].

Patients with ascites and a low ascitic fluid protein concentration have a high risk of developing spontaneous bacterial peritonitis. In those having ascites protein levels <15 g/L and advanced liver failure (Child-Pugh score ≥ 9 points with serum bilirubin level ≥ 3 mg/dL) or impaired renal function (SCr level ≥ 1.2 mg/dL, blood urea nitrogen level ≥ 25 mg/dL, or serum sodium level ≤ 130 mEq/L), norfloxacin (400 mg/day) significantly reduced the incidence of SBP versus placebo at 12 months (7% vs. 61%) [43]. Moreover, norfloxacin also reduced the probability of developing HRS at 12 months (28% vs. 41%) and improved survival at 3 months (94% vs. 62%), although the difference in survival at 12 months was not significantly different between the two groups (60% vs. 48%). More recently, in the NORFLOCIR trial, norfloxacin did neither reduce the incidence of SBP nor improve survival vs. placebo in patients with cirrhosis and Child-Pugh class C [44]. However, in a post hoc analysis, in the subgroup of patients with ascites protein levels <15 g/L, mortality was significantly lower in the norfloxacin group than in the placebo group. In patients surviving an episode of SBP, recurrence is very frequent and norfloxacin prophylaxis significantly reduces SBP recurrence [45].

Rifaximin is a poorly absorbable antibiotic that is currently indicated in preventing the recurrence of hepatic encephalopathy [46]. In patients undergoing TIPS placement, rifaximin treatment (started 14 days before TIPS) significantly reduced the incidence of hepatic encephalopathy at 168 days vs. placebo (34% vs. 53%) [47]. In a small randomized controlled trial in patients with hepatic encephalopathy, rifaximin reduced gut-derived inflammation and was associated with a reduced incidence of infections [48]. Rifaximin could be a promising alternative to norfloxacin for prophylaxis of SBP [49], however, further data are needed to confirm this finding.

TIPS

Generally, TIPS allows for a significant reduction of portal pressure and restoration of portal blood flow, resulting in a significant decrease in the risk of recurrence of tense ascites and rebleeding.

In patients with recurrent/refractory ascites, older RCTs using uncovered stents [50–54] clearly showed that TIPS is more effective than LVP in preventing the recurrence of ascites. However, patients treated with TIPS were consistently found to have an increased risk of encephalopathy and the benefit in terms of survival differed across the study. Several meta-analyses on the comparison of TIPS vs. LVPs in patients with ascites have also been published. The first [55, 56] found that TIPS was more effective in preventing the recurrence of ascites, but there was an increased risk of encephalopathy and survival was unchanged compared to LVP. However, in the sole IPD meta-analysis published by Salerno et al. [57], the actuarial rate of transplant-free survival was better in patients allocated to the TIPS arm than the rate in the LVP group (63% and 52% at 1 year, respectively). The most recent RCT [19] (not included in the meta-analysis), which compared TIPS using PTFE covered stents to LVP + albumin in patients with recurrent ascites, showed a significantly better transplant-free survival at 1 year (93% in the TIPS group versus 52% in the LVP + A group). This suggests that in a subset of selected patients with recurrent ascites, TIPS is not only able to decrease the risk of tense ascites, but it also improves survival. In the C Bureau et al. study, 72% of patients included had no history of PH-related bleeding. 6/33 (18%) PH-related bleeding was observed within 1 year in the LVP group compared to 0/29 (0%) in the TIPS group ($p = 0.01$). Moreover, the number of hernia-related complications was significantly lower in the TIPS group compared to the large-volume paracentesis group, serum creatinine improved over time in patients treated with TIPS and importantly, plasma renin activity dramatically decreased over 1 year.

One recent retrospective study from India [58] has explored the use of TIPS at an 'early stage' of decompensated cirrhosis. Three groups of patients were compared: (i) one treated with TIPS with well-validated indications (standard TIPS group: recurrent ascites/hydrothorax $n = 23$; prevention of rebleeding $n = 7$), (ii) the second group (ascites/hydrothorax $n = 20$; variceal bleeding $n = 15$) with medical/endoscopic treatment alone, and (iii) the third one (Anticipant group) treated with TIPS after the first episode of ascites ($n = 15$) or first episode of bleeding ($n = 12$). The authors demonstrated improved 1-year survival in the Anticipant group when compared to the TIPS group (81% vs. 40% $p < 0.001$). Additionally, they observed that a higher rate of patients in the standard medical treatment group developed varices (26% vs. 0%), had new-onset or recurrence of ascites (49% vs. 11%), and required hospital admission (46% vs. 19%) at 1 year compared to Anticipant group but without difference regarding 1-year survival (81% and 82%). No definitive conclusion can be drawn since this is a small, nonrandomized, retrospective study with limited confidence regarding the comparability of the different groups. However, this would deserve further investigation, probably using the smallest diameter-controlled expansion stent as possible, with potential sequential 'fine-tuning' interventions in order to optimize the effect and/or limiting the risk of portosystemic 'over-shunting.' The use of NSBBs and rifaximin may further help to optimize patient outcomes.

Furthermore, one main concern is the risk of OHE triggered by the shunt procedure. The most recent studies did not observe a significant difference between the TIPS arm and the control arm:

- in patients with recurrent ascites: the 1-year rate was 35% in both groups in the French RCT [19] and 43% vs. 32%, in the TIPS and control groups, respectively in a large observational Austrian study [59],
- in secondary prophylaxis in patients with PVT: 16% vs. 18% in the TIPS and control groups, respectively in the Luo study [60]; and 25% vs. 16% in the TIPS and control groups, respectively, in the Lv study [61],
- and pre-emptive TIPS studies: 19% vs. 10% in the TIPS and control groups, respectively in the pivotal study [62]; 35% and 36% in the TIPS and control groups, respectively in the Lv study [63], and 35% vs. 26% in the TIPS and control groups, respectively in the IPD meta-analysis [64].

In order to answer the question if TIPS is able to prevent further decompensation (i.e., defined by the development of PH-related-bleeding, tense ascites, or OHE), an individual patient data meta-analysis was conducted for the Baveno VII conference. Controlled studies, randomized or not, comparing TIPS (only covered stents) versus Standard Medical Treatment (SMT) were considered for inclusion. Out of 14 eligible studies, 12 had available data and therefore were included. The main endpoint was the 1-year incidence of further decompensation defined as the onset of PH-related bleeding or tense ascites or OHE after the index decompensation. Secondary endpoints were 1-year survival and the occurrence of each decompensation considered separately.

Preliminary results showed that TIPS increased the 1-year probability of remaining free of:

- further decompensation: 55% vs. 40% as compared to SMT ($p < 0.0001$),
- bleeding: 91% vs. 69% as compared to SMT ($p < 0.0001$),
- tense ascites: 89% vs. 78% as compared to the SMT ($p < 0.0001$),

The 1-year rate of OHE was increased in the TIPS group: 35% vs. 29% compared to SMT ($p = 0.004$), but a higher survival was observed in the TIPS group compared to SMT: 78% vs. 73% ($p = 0.0003$).

Considering the cumulative results obtained from trials using covered TIPS vs. standard medical therapy in patients with recurrent/refractory ascites, recurrent bleeding, or high risk of rebleeding, we can conclude that TIPS decreases the risk of further decompensation. Consequently, TIPS should be considered in all these situations. It is crucial to apply this recommendation to patients who meet the criteria of indications used in RCTs. Importantly, more research is still needed regarding the optimal assessment of the risk related to the shunt procedure and for post-TIPS complications in order to select the best candidates for TIPS.

Albumin

Human Albumin (HA) infusion counteracts the reduction of the effective circulating volume and may exert anti-inflammatory, antioxidant and immune-modulating effects [65]. HA is currently recommended in patients receiving large-volume paracentesis to prevent PPCD [23]. Moreover, albumin is recommended in patients with SBP, because it can prevent the occurrence of HRS-AKI [66]. Albumin is also used in combination with vasoconstrictors for the treatment of HRS-AKI, because it enhances the effects of vasoconstrictors [67]. In patients hospitalized for decompensated cirrhosis, a 14-day treatment with HA aiming at normalizing albumin concentration is not useful, maybe detrimental, and should be avoided [68]. Conversely, the data on the long-term use of HA in patients with uncomplicated ascites are controversial. In the ANSWER study, 440 patients with cirrhosis and ascites were randomized to receive HA (40 g twice a week for 2 weeks and 40 g per week thereafter) or standard of care [69]. Long-term HA administration was associated with improved survival at 18 months and lower incidence of refractory ascites, HRS-AKI, hepatic encephalopathy, SBP, and other infections vs. standard of care. In the MACHT study, 196 patients with ascites on the liver transplant waiting list were randomized to receive HA (40 g every 15 days) plus midodrine (15–30 mg/day) or placebo [70]. HA and midodrine did neither prevent complications of cirrhosis nor improve survival. The discrepancy between the two studies can be related to differences in patient population (sicker in MACHT), different doses of albumin used (higher in ANSWER), different duration of treatment (shorter in MACHT) effects of midodrine or open-label design in ANSWER trial, with more frequent access to healthcare services in the treatment group. Another open-label trial [71] and a case-control study [72] showed better survival and better control of ascites. Therefore, although long-term HA treatment may prevent further decompensation in certain patients with ascites, further data are needed to confirm those findings.

Conclusions

In summary, patients with decompensated cirrhosis should be considered for liver transplantation. The cornerstones in the prevention of further decompensation include the treatment of the primary etiologic factor, accurate preventive measures against SBP and HRS-AKI, the adequate use of covered TIPS in selected patients, and the early identification and treatment of risk factors and triggers of further decompensation (Fig. 49.2).

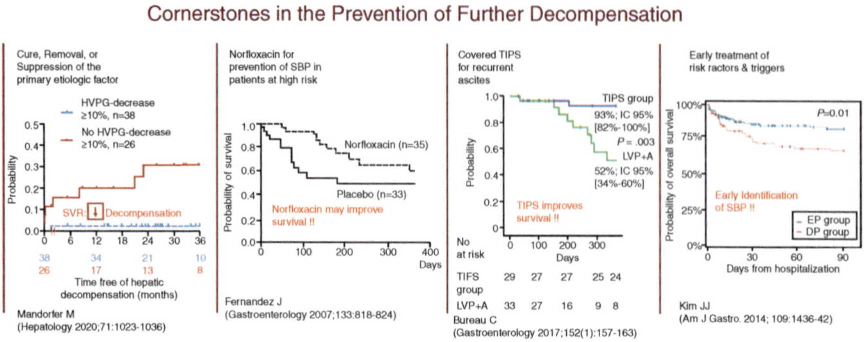

Fig. 49.2 Cornerstones in the prevention of further decompensation of cirrhosis. SVR sustained virological response, TIPS transjugular intrahepatic portosystemic shunt, SBP spontaneous bacterial peritonitis

In patients with ascites, bleeding prophylaxis should be performed using NSBBs or carvedilol, however, in case of persistently low blood pressure (systolic blood pressure < 90 mmHg or mean arterial pressure < 65 mmHg) and/or HRS-AKI, NSBBs or carvedilol should be dose-reduced or discontinued. Once blood pressure returns to baseline and/or HRS-AKI resolves, NSBB can be re-initiated or re-titrated.

Management of patients with ascites should always consider the treatment of the underlying etiology of liver disease but requires most often the combination of diuretics and LVP. If complications occur in patients with ascites, i.e., SBP and/or HRS-AKI adequate antibiotic therapy and albumin administration should be used as indicated. The role of antibiotic therapy and albumin in the prevention of further decompensation remains to be further defined. Importantly, in patients developing recurrent ascites—defined by the requirement of ≥3 large-volume paracenteses within 1 year (or a shorter time period)—TIPS should be considered for prevention of further decompensation since it improves transplant-free survival.

References

1. Ginés P, Quintero E, Arroyo V, Terés J, Bruguera M, Rimola A, et al. Compensated cirrhosis: natural history and prognostic factors. Hepatology. 1987;7:122–8.
2. D'Amico G, Pasta L, Morabito A, D'Amico M, Caltagirone M, Malizia G, et al. Competing risks and prognostic stages of cirrhosis: a 25-year inception cohort study of 494 patients. Aliment Pharmacol Ther. 2014;39:1180–93.
3. Solà E, Watson H, Graupera I, Turón F, Barreto R, Rodríguez E, et al. Factors related to quality of life in patients with cirrhosis and ascites: relevance of serum sodium concentration and leg edema. J Hepatol. 2012;57:1199–206.
4. Volk ML, Tocco RS, Bazick J, Rakoski MO, Lok AS. Hospital readmissions among patients with decompensated cirrhosis. Am J Gastroenterol. 2012;107:247–52.
5. Planas R, Montoliu S, Ballesté B, Rivera M, Miquel M, Masnou H, et al. Natural history of patients hospitalized for management of cirrhotic ascites. Clin Gastroenterol Hepatol. 2006;4:1385–94.

6. Bernardi M, Moreau R, Angeli P, Schnabl B, Arroyo V. Mechanisms of decompensation and organ failure in cirrhosis: from peripheral arterial vasodilation to systemic inflammation hypothesis. J Hepatol. 2015;63:1272–84.
7. Ripoll C, Groszmann R, Garcia-Tsao G, Grace N, Burroughs A, Planas R, et al. Hepatic venous pressure gradient predicts clinical decompensation in patients with compensated cirrhosis. Gastroenterology. 2007;133:481–8.
8. Reiberger T, Ferlitsch A, Payer BA, Mandorfer M, Heinisch BB, Hayden H, et al. Non-selective betablocker therapy decreases intestinal permeability and serum levels of LBP and IL-6 in patients with cirrhosis. J Hepatol. 2013;58:911–21.
9. Engelmann C, Clària J, Szabo G, Bosch J, Bernardi M. Pathophysiology of decompensated cirrhosis: portal hypertension, circulatory dysfunction, inflammation, metabolism and mitochondrial dysfunction. J Hepatol. 2021;75(Suppl 1):S49–66. https://doi.org/10.1016/j.jhep.2021.01.002.
10. Wiest R, Lawson M, Geuking M. Pathological bacterial translocation in liver cirrhosis. J Hepatol. 2014;60:197–209.
11. Yang Y-Y, Liu H, Nam SW, Kunos G, Lee SS. Mechanisms of TNFalpha-induced cardiac dysfunction in cholestatic bile duct-ligated mice: Interaction between TNFalpha; and endocannabinoids. J Hepatol. 2010;53:298–306.
12. Turco L, Garcia-Tsao G, Magnani I, Bianchini M, Costetti M, Caporali C, et al. Cardiopulmonary hemodynamics and C-reactive protein as prognostic indicators in compensated and decompensated cirrhosis. J Hepatol. 2018;68:949–58.
13. Ruiz-del-Arbol L, Monescillo A, Arocena C, Valer P, Ginès P, Moreira V, et al. Circulatory function and hepatorenal syndrome in cirrhosis. Hepatology. 2005;42:439–47.
14. Krag A, Bendtsen F, Henriksen JH, Møller S. Low cardiac output predicts development of hepatorenal syndrome and survival in patients with cirrhosis and ascites. Gut. 2009;59:105–10.
15. Costa D, Simbrunner B, Jachs M, Hartl L, Bauer D, Paternostro R, et al. Systemic inflammation increases across distinct stages of advanced chronic liver disease and correlates with decompensation and mortality. J Hepatol. 2021;74:819–28.
16. Moore KP, Wong F, Gines P, Bernardi M, Ochs A, Salerno F, et al. The management of ascites in cirrhosis: report on the consensus conference of the international ascites Club. Hepatology. 2003;38:258–66.
17. Tonon M, Piano S, Gambino CG, Romano A, Pilutti C, Incicco S, et al. Outcomes and mortality of grade 1 ascites and recurrent ascites in patients with cirrhosis. Clin Gastroenterol Hepatol. 2021;19:358–366.e8.
18. Arroyo V, Ginès P, Gerbes AL, Dudley FJ, Gentilini P, Laffi G, et al. Definition and diagnostic criteria of refractory ascites and hepatorenal syndrome in cirrhosis. Hepatology. 1996;23:164–76.
19. Bureau C, Thabut D, Oberti F, Dharancy S, Carbonell N, Bouvier A, et al. Transjugular intrahepatic portosystemic shunts with covered stents increase transplant-free survival of patients with cirrhosis and recurrent ascites. Gastroenterology. 2017;152:157–63.
20. Angeli P, Bernardi M, Villanueva C, Francoz C, Mookerjee RP, Trebicka J, et al. EASL clinical practice guidelines for the management of patients with decompensated cirrhosis. J Hepatol. 2018;69:406–60.
21. Biggins SW, Angeli P, Garcia-Tsao G, Ginès P, Ling SC, Nadim MK, et al. Diagnosis, evaluation, and management of ascites, spontaneous bacterial peritonitis and hepatorenal syndrome: 2021 practice guidance by the American Association for the Study of Liver Diseases. Hepatology. 2021;74:1014–48.
22. Ginés P, Arroyo V, Quintero E, Planas R, Bory F, Cabrera J, et al. Comparison of paracentesis and diuretics in the treatment of cirrhotics with tense ascites. Results of a randomized study. Gastroenterology. 1987;93:234–41.
23. Ginès A, Fernández-Esparrach G, Monescillo A, Vila C, Domènech E, Abecasis R, et al. Randomized trial comparing albumin, dextran 70, and polygeline in cirrhotic patients with ascites treated by paracentesis. Gastroenterology. 1996;111:1002–10.

24. Pérez-Ayuso RM, Arroyo V, Planas R, Gaya J, Bory F, Rimola A, et al. Randomized comparative study of efficacy of furosemide versus spironolactone in nonazotemic cirrhosis with ascites. Relationship between the diuretic response and the activity of the renin-aldosterone system. Gastroenterology. 1983;84:961–8.
25. Angeli P, Fasolato S, Mazza E, Okolicsanyi L, Maresio G, Velo E, et al. Combined versus sequential diuretic treatment of ascites in non-azotaemic patients with cirrhosis: results of an open randomised clinical trial. Gut. 2010;59:98–104.
26. Krassenburg LAP, Maan R, Ramji A, Manns MP, Cornberg M, Wedemeyer H, et al. Clinical outcomes following DAA therapy in patients with HCV-related cirrhosis depend on disease severity. J Hepatol. 2021;74:1053–63.
27. Lens S, Baiges A, Alvarado-Tapias E, LLop E, Martinez J, Fortea JI, et al. Clinical outcome and hemodynamic changes following HCV eradication with oral antiviral therapy in patients with clinically significant portal hypertension. J Hepatol. 2020;73:1415–24.
28. D'Ambrosio R, Degasperi E, Anolli MP, Fanetti I, Borghi M, Soffredini R, et al. Incidence of liver and non-liver-related outcomes in patients with HCV-cirrhosis after SVR. J Hepatol. 2022;76(2):302–10. https://doi.org/10.1016/j.jhep.2021.09.013.
29. Mandorfer M, Kozbial K, Schwabl P, Chromy D, Semmler G, Stättermayer AF, et al. Changes in hepatic venous pressure gradient predict hepatic Decompensation in patients who achieved sustained Virologic response to interferon-free therapy. Hepatology. 2020;71:1023–36.
30. Alvarez MA, Cirera I, Solà R, Bargalló A, Morillas RM, Planas R. Long-term clinical course of decompensated alcoholic cirrhosis: a prospective study of 165 patients. J Clin Gastroenterol. 2011;45:906–11.
31. Lackner C, Spindelboeck W, Haybaeck J, Douschan P, Rainer F, Terracciano L, et al. Histological parameters and alcohol abstinence determine long-term prognosis in patients with alcoholic liver disease. J Hepatol. 2017;66:610–8.
32. Piano S, Angeli P. Bacterial infections in cirrhosis as a cause or consequence of Decompensation? Clin Liver Dis. 2021;25:357–72.
33. Huelin P, Piano S, Solà E, Stanco M, Solé C, Moreira R, et al. Validation of a staging system for acute kidney injury in patients with cirrhosis and association with acute on chronic liver failure. Clin Gastroenterol Hepatol. 2017;15:438–45.
34. Merli M, Lucidi C, Pentassuglio I, Giannelli V, Giusto M, Di Gregorio V, et al. Increased risk of cognitive impairment in cirrhotic patients with bacterial infections. J Hepatol. 2013;59:243–50.
35. Chavez-Tapia NC, Barrientos-Gutierrez T, Tellez-Avila FI, Soares-Weiser K, Uribe M. Antibiotic prophylaxis for cirrhotic patients with upper gastrointestinal bleeding. Cochrane Database Syst Rev. 2010;2010:CD002907. https://doi.org/10.1002/14651858.CD002907.pub2.
36. Dionigi E, Garcovich M, Borzio M, Leandro G, Majumdar A, Tsami A, et al. Bacterial infections change natural history of cirrhosis irrespective of liver disease severity. Am J Gastroenterol. 2017;112:588–96.
37. Fernández J, Prado V, Trebicka J, Amoros A, Gustot T, Wiest R, et al. Multidrug-resistant bacterial infections in patients with decompensated cirrhosis and with acute-on-chronic liver failure in Europe. J Hepatol. 2019;70(3):398–411. https://doi.org/10.1016/j.jhep.2018.10.027.
38. Piano S, Singh V, Caraceni P, Maiwall R, Alessandria C, Fernandez J, et al. Epidemiology and effects of bacterial infections in patients with cirrhosis worldwide. Gastroenterology. 2019;156:1368–1380.e10.
39. Bajaj JS, O'Leary JG, Reddy KR, Wong F, Biggins SW, Patton H, et al. Survival in infection-related acute-on-chronic liver failure is defined by extrahepatic organ failures. Hepatology. 2014;60:250–6.
40. Piano S, Brocca A, Mareso S, Angeli P. Infections complicating cirrhosis. Liver Int. 2018;38:126–33.
41. Bajaj JS, Reddy RK, Tandon P, Wong F, Kamath PS, Biggins SW, et al. Prediction of fungal infection development and their impact on survival using the NACSELD cohort. Am J Gastroenterol. 2018;113(4):556–63.

42. Bartoletti M, Rinaldi M, Pasquini Z, Scudeller L, Piano S, Giacobbe DR, et al. Risk factors for candidaemia in hospitalized patients with liver cirrhosis: a multicentre case-control-control study. Clin Microbiol Infect. 2021;27:276–82.
43. Fernández J, Navasa M, Planas R, Montoliu S, Monfort D, Soriano G, et al. Primary prophylaxis of spontaneous bacterial peritonitis delays hepatorenal syndrome and improves survival in cirrhosis. Gastroenterology. 2007;133:818–24.
44. Moreau R, Elkrief L, Bureau C, Perarnau J-M, Thévenot T, Saliba F, et al. Effects of long-term norfloxacin therapy in patients with advanced cirrhosis. Gastroenterology. 2018;155:1816–1827.e9.
45. Ginès P, Rimola A, Planas R, Vargas V, Marco F, Almela M, et al. Norfloxacin prevents spontaneous bacterial peritonitis recurrence in cirrhosis: results of a double-blind, placebo-controlled trial. Hepatology. 1990;12:716–24.
46. Bass NM, Mullen KD, Sanyal A, Poordad F, Neff G, Leevy CB, et al. Rifaximin treatment in hepatic encephalopathy. N Engl J Med. 2010;362:1071–81.
47. Bureau C, Thabut D, Jezequel C, Archambeaud I, D'Alteroche L, Dharancy S, et al. The use of rifaximin in the prevention of overt hepatic encephalopathy after transjugular intrahepatic portosystemic shunt: a randomized controlled trial. Ann Intern Med. 2021;174:633–40.
48. Patel V, Lee S, McPhail M, Da Silva K, Guilly S, Zamalloa A, et al. Rifaximin reduces gut-derived inflammation and mucin degradation in cirrhosis and encephalopathy: RIFSYS randomised controlled trial. J Hepatol. 2021;76(2):332–42. https://doi.org/10.1016/j.jhep.2021.09.010.
49. Assem M, Elsabaawy M, Abdelrashed M, Elemam S, Khodeer S, Hamed W, et al. Efficacy and safety of alternating norfloxacin and rifaximin as primary prophylaxis for spontaneous bacterial peritonitis in cirrhotic ascites: a prospective randomized open-label comparative multicenter study. Hepatol Int. 2016;10:377–85.
50. Lebrec D, Giuily N, Hadengue A, Vilgrain V, Moreau R, Poynard T, et al. Transjugular intrahepatic portosystemic shunts: comparison with paracentesis in patients with cirrhosis and refractory ascites: a randomized trial. French Group of Clinicians and a Group of Biologists. J Hepatol. 1996;25:135–44.
51. Rössle M, Ochs A, Gülberg V, Siegerstetter V, Holl J, Deibert P, et al. A comparison of paracentesis and transjugular intrahepatic portosystemic shunting in patients with ascites. N Engl J Med. 2000;342:1701–7.
52. Ginès P, Uriz J, Calahorra B, Garcia-Tsao G, Kamath PS, Del Arbol LR, et al. Transjugular intrahepatic portosystemic shunting versus paracentesis plus albumin for refractory ascites in cirrhosis. Gastroenterology. 2002;123:1839–47.
53. Salerno F, Merli M, Riggio O, Cazzaniga M, Valeriano V, Pozzi M, et al. Randomized controlled study of TIPS versus paracentesis plus albumin in cirrhosis with severe ascites. Hepatology. 2004;40:629–35.
54. Narahara Y, Kanazawa H, Fukuda T, Matsushita Y, Harimoto H, Kidokoro H, et al. Transjugular intrahepatic portosystemic shunt versus paracentesis plus albumin in patients with refractory ascites who have good hepatic and renal function: a prospective randomized trial. J Gastroenterol. 2011;46:78–85.
55. Albillos A, Bañares R, González M, Catalina M-V, Molinero L-M. A meta-analysis of transjugular intrahepatic portosystemic shunt versus paracentesis for refractory ascites. J Hepatol. 2005;43:990–6.
56. D'Amico G, Luca A, Morabito A, Miraglia R, D'Amico M. Uncovered transjugular intrahepatic portosystemic shunt for refractory ascites: a meta-analysis. Gastroenterology. 2005;129:1282–93. https://doi.org/10.1053/j.gastro.2005.07.031.
57. Salerno F, Cammà C, Enea M, Rössle M, Wong F. Transjugular intrahepatic portosystemic shunt for refractory ascites: a meta-analysis of individual patient data. Gastroenterology. 2007;133:825–34.
58. Rajesh S, Philips CA, Betgeri SS, George T, Ahamed R, Mohanan M, et al. Transjugular intrahepatic portosystemic shunt (TIPS) placement at index portal hypertensive decompensation

(anticipant TIPS) in cirrhosis and the role of early intervention in variceal bleeding and ascites. Indian J Gastroenterol. 2021;40:361–72.
59. Bucsics T, Hoffman S, Grünberger J, Schoder M, Matzek W, Stadlmann A, et al. ePTFE-TIPS vs repetitive LVP plus albumin for the treatment of refractory ascites in patients with cirrhosis. Liver Int. 2018;38:1036–44.
60. Luo X, Wang Z, Tsauo J, Zhou B, Zhang H, Li X. Advanced cirrhosis combined with portal vein thrombosis: a randomized trial of TIPS versus endoscopic band ligation plus propranolol for the prevention of recurrent esophageal variceal bleeding. Radiology. 2015;276:286–93.
61. Lv Y, Qi X, He C, Wang Z, Yin Z, Niu J, et al. Covered TIPS versus endoscopic band ligation plus propranolol for the prevention of variceal rebleeding in cirrhotic patients with portal vein thrombosis: a randomised controlled trial. Gut. 2018;67:2156–68.
62. García-Pagán JC, Caca K, Bureau C, Laleman W, Appenrodt B, Luca A, et al. Early use of TIPS in patients with cirrhosis and variceal bleeding. N Engl J Med. 2010;362:2370–9.
63. Lv Y, Yang Z, Liu L, Li K, He C, Wang Z, et al. Early TIPS with covered stents versus standard treatment for acute variceal bleeding in patients with advanced cirrhosis: a randomised controlled trial. Lancet Gastroenterol Hepatol. 2019;4:587–98.
64. Nicoară-Farcău O, Han G, Rudler M, Angrisani D, Monescillo A, Torres F, et al. Effects of early placement of transjugular portosystemic shunts in patients with high-risk acute variceal bleeding: a meta-analysis of individual patient data. Gastroenterology. 2021;160:193–205.e10.
65. Bernardi M, Angeli P, Claria J, Moreau R, Gines P, Jalan R, et al. Albumin in decompensated cirrhosis: new concepts and perspectives. Gut. 2020;69:1127–38. https://doi.org/10.1136/gutjnl-2019-318843.
66. Sort P, Navasa M, Arroyo V, Aldeguer X, Planas R, Ruiz-del-Arbol L, et al. Effect of intravenous albumin on renal impairment and mortality in patients with cirrhosis and spontaneous bacterial peritonitis. N Engl J Med. 1999;341:403–9.
67. Ortega R, Ginès P, Uriz J, Cárdenas A, Calahorra B, De Las HD, et al. Terlipressin therapy with and without albumin for patients with hepatorenal syndrome: results of a prospective, nonrandomized study. Hepatology. 2002;36:941–8.
68. China L, Freemantle N, Forrest E, Kallis Y, Ryder SD, Wright G, et al. A randomized trial of albumin infusions in hospitalized patients with cirrhosis. N Engl J Med. 2021;384:808–17.
69. Caraceni P, Riggio O, Angeli P, Alessandria C, Neri S, Foschi FG, et al. Long-term albumin administration in decompensated cirrhosis (ANSWER): an open-label randomised trial. Lancet. 2018;391:2417–29.
70. Solà E, Solé C, Simón-Talero M, Martín-Llahí M, Castellote J, Garcia-Martínez R, et al. Midodrine and albumin for prevention of complications in patients with cirrhosis awaiting liver transplantation. A randomized placebo-controlled trial. J Hepatol. 2018;69:1250–9.
71. Romanelli R-G, La Villa G, Barletta G, Vizzutti F, Lanini F, Arena U, et al. Long-term albumin infusion improves survival in patients with cirrhosis and ascites: an unblinded randomized trial. World J Gastroenterol. 2006;12:1403–7.
72. Di Pascoli M, Fasolato S, Piano S, Bolognesi M, Angeli P. Long-term administration of human albumin improves survival in patients with cirrhosis and refractory ascites. Liver Int. 2019;39:98–105.

The Impact of Sarcopenia, Frailty, and Malnutrition on Further Decompensation

50

Sarah Wang and Puneeta Tandon

Abbreviations

LT	Liver transplant
HE	Hepatic encephalopathy
MAMC	Mid-arm muscle circumference
CT	Computed tomography
MRI	Magnetic resonance imaging
DEXA	Dual x-ray absorptiometry
US	Ultrasound
BIA	Bioelectrical impedance analysis
SMI	Skeletal muscle index
CFS	Clinical Frailty Scale
ADL	Activities of daily living
KPS	Karnofsky Performance Status
FFC	Fried Frailty Criteria
HGS	Handgrip strength
SPPB	Short physical performance battery
LFI	Liver frailty index
RFH-NPT	ROYAL Free Hospital-Nutritional Prioritizing Tool
TPMT	Transversal psoas muscle thickness
FFMA	Fat-free muscle area
ECOG	Eastern Cooperative Oncology Group
MELD	Model for end-stage liver disease
HPVG	Hepatic venous portal gradient
CP	Child-Pugh
ACLF	Acute on chronic liver failure

S. Wang · P. Tandon (✉)
Division of Gastroenterology (Liver Unit), University of Alberta, Edmonton, AB, Canada
e-mail: swang@ualberta.ca; ptandon@ualberta.ca

R. de Franchis (ed.), *Portal Hypertension VII*,
https://doi.org/10.1007/978-3-031-08552-9_50

RA	Refractory ascites
TIPS	Transjugular intrahepatic portosystemic shunt
RCT	Randomized control trial
ESPEN	European Society for Clinical Nutrition and Metabolism

Introduction

Sarcopenia, frailty, and malnutrition are common in patients with cirrhosis [1, 2]. Their influence on cirrhosis outcomes is becoming increasingly evident, with associations to mortality on the liver transplant (LT) waitlist [3–5], longer and more frequent hospitalizations [6–9], decreased quality of life [10], and worse survival post LT [11, 12]. Due to this growing evidence, major guidelines now recommend that clinicians assess sarcopenia, frailty, and malnutrition using validated standardized tests in the routine care of patients with cirrhosis [13–15].

While sarcopenia, frailty, and malnutrition are commonly viewed as complications of advanced cirrhosis, biologically it is plausible that these phenotypes may also play a role in disease progression. The strongest pathophysiological rationale for this process is hepatic encephalopathy (HE). Hyperammonemia, for example, can exhibit myotoxic effects, leading to loss of muscle mass. The loss of muscle mass in turn can predispose patients to HE due to decreased ammonia metabolism by skeletal muscle [16, 17]. Whether the treatment or prevention of sarcopenia, frailty, and malnutrition can reverse or delay further decompensation is an exciting area awaiting further research.

Definitions and Assessment of Sarcopenia, Frailty, and Malnutrition

While the phenotypes of sarcopenia, frailty, and malnutrition can be distinct from one another, they are closely interrelated, and often co-exist [14]. There is no single "gold-standard" measure for identifying each of these concepts. Existing studies have used a variety of tools and a range of cut-off measures across distinct populations, with consideration for separate cut-off values based on ethnicity [18–20].

Sarcopenia in the geriatric literature is classically defined as the loss of muscle mass and function [21]. In cirrhosis, however, the majority of studies have based the definition on the loss of muscle mass alone [14]. A variety of tools have been used to evaluate sarcopenia in cirrhosis (Table 50.1). These include anthropometric measures, such as the mid-arm muscle circumference (MAMC) [12, 16], CT-or MRI-based measures of muscle mass [5, 22, 23], dual x-ray absorptiometry (DEXA) [24, 25], thigh ultrasound (US) [26, 27], and bioelectrical impedance analysis (BIA) [28–30]. Among these tools, the CT-based Skeletal Muscle Index (SMI) at L3 is one of the most commonly evaluated, and well-validated measures [1]. Other research has supported the transversal psoas muscle thickness as a prognostically relevant

Table 50.1 Measures of sarcopenia in cirrhosis

Tool	Description	Equipment
Cross sectional imaging [5, 22, 23, 59]	Various techniques measuring muscle mass, including the total abdominal skeletal muscle area at L3 vertebrae normalized to height (SMI), the transversal diameter of the psoas muscle (TPMT), and total fat-free mass of the erector spinae muscle Aarea (FFMA)	CT or MRI
Ultrasound [26, 27]	Measurements of iliopsoas or thigh muscle groups	US
Phase angle [28–30, 60]	Metric calculated from BIA, reflecting tissue resistance and reactance	Impedance analyzer
Anthropometric [16]	Bedside measurements of MAMC, TSF	N/A
DEXA [24, 25]	Totally body composition as determined by DEXA scan	DEXA scanner

BIA bioelectrical impedance analysis, *DEXA* dual-energy x-ray absorptiometry, *FFMA* fat-free muscle area, *MAMC* mid-arm muscle circumference, *SMI* skeletal muscle index, *TPMT* transversal psoas muscle thickness, *TSF* triceps skin fold

measure of sarcopenia that can be evaluated without additional image processing [31, 32]. North American data using the psoas has not correlated as well with mortality when compared to the L3 SMI [33]. Beyond muscle quantity, muscle quality is also becoming increasingly recognized as a prognostic factor [34]. Myosteatosis reflects the degree of fat accumulation in muscle tissue and has itself been associated with mortality and HE [35, 36].

Frailty is the clinical syndrome of decreased physiologic reserve and increased vulnerability to health stressors [37, 38]. Operationally in cirrhosis, frailty has been classically defined by measures of impaired muscle contractile function [14]. Related measures of reduced aerobic capacity and even physical disability have also been considered under the frailty construct [14, 39]. Measurement tools [14, 39] range from global scores (without objective measurement) such as the Clinical Frailty Scale (CFS) [6, 40], activities of daily living (ADLs) [8, 41], and Karnofsky Performance Status (KPS) [42–44], to more objective measures, including the Fried Frailty Criteria (FFC) [9, 45, 46], Handgrip Strength (HGS) [47, 48], Short Gait Speed [7], 6 Minute Walk Test [49], Short Physical Performance Battery (SPPB) [45, 50], Liver Frailty Index (LFI) [4, 51, 52], and cardiopulmonary exercise testing [53, 54] (Table 50.2).

Malnutrition as a construct has perhaps been the most heterogeneous in how it has been evaluated in cirrhosis. With regard to screening tools for malnutrition, the Royal Free Hospital-Nutritional Prioritizing Tool (RFH-NPT) has been shown to predict clinical complications and transplant-free survival, and is recommended by existing guidelines [13, 14, 55, 56]. With regard to assessment tools, a recent meta-analysis showed that there were 32 distinct definitions used to assess malnutrition across 47 studies [57]. Conceptually, malnutrition is defined as the syndrome of nutrient imbalance that leads to measurable changes in body composition, tissue function, and/or clinical outcomes [58]. Practically, malnutrition manifests in frailty and sarcopenia, and as such, is often measured by overlapping metrics with those

Table 50.2 Measures of frailty in cirrhosis

Tool	Description
Clinical frailty scale (CFS) [6]	Rapid, subjective scale divided into nine categories, ranging from very fit to terminally ill.
Activities of daily living [8]	Dependence or independence for six tasks deemed essential for home living.
Karnofsky performance score (KPS) [43, 44]	Global assessment of functional status ranging from 100 (normal), 50 (needs considerable assistance or frequent medical care) to 0 (dead).
Eastern Cooperative Oncology Group (ECOG) scale [61, 62]	Functional limitations scale ranging from 0 to 5, with 0 = no limitations, 2 = able to self-care, up and about more than 50% of the time; 4 = completely disabled.
Fried Frailty Criteria (FFC) [9, 45]	Five components: Grip strength, gait speed, exhaustion, weight loss, and physical activity.
Handgrip strength (HGS) [63–65]	Gripping the dynamometer using the dominant hand, an average of three times
Short gait speed [7]	Time to walk 5 m at usual walking speed.
6-minute walk test [49]	Distance walked at usual speed for 6 min.
Short physical performance battery [45, 50]	Three components: Balance testing, gait speed, chair stands.
Liver frailty index [3, 4]	Three components: Handgrip strength, chair stands, and balance testing.
Cardiopulmonary exercise testing [53, 54]	Noninvasive measures of gas exchange at rest and in response to exercise.

two phenotypes [14]. Thus, for the purpose of examining the role of malnutrition in further decompensation, it has been encompassed under the review of sarcopenia and frailty.

Pathophysiology of Frailty, Sarcopenia, and Malnutrition in Cirrhosis and Cirrhosis Progression

In cirrhosis, malnutrition can develop due to both inadequate or excessive oral intake, altered macronutrient metabolism, and malabsorption syndromes [14, 56, 66]. Increased protein catabolism, decreased protein synthesis, and increased gluconeogenesis with insulin resistance all lead to an accelerated state of starvation [2, 56, 66]. Such changes can result in decreased protein anabolism even in the presence of appropriate nutritional and physical activity stimuli [1].

Cirrhosis itself can further contribute to sarcopenia. Hyperammonemia, inflammation, endotoxemia, and hormonal changes such as low testosterone and insulin resistance are all proposed mediators of sarcopenia in cirrhosis [1]. Ascites can contribute to poor oral intake and protein loss through the paracentesis. Furthermore, the etiology of cirrhosis may play a role [14]. Alcohol has been found to be a mechanistic contributor to sarcopenia [67]. Inflammation and insulin resistance may play a role in the development of sarcopenia in nonalcoholic steatohepatitis [68]. In cholestatic liver diseases, bile acids may induce skeletal muscle atrophy [69].

The pathogenic pathways in which frailty, sarcopenia, and malnutrition may contribute to further decompensation in cirrhosis are not well understood. Micronutrient deficiencies have been linked to decompensation, such as zinc and magnesium deficiency in association with cognitive dysfunction [14]. Decreased ammonia detoxification due to muscle dysfunction, along with decreased branched-chain amino acids, have been implicated in the development of HE [70]. Sarcopenia and frailty can also predispose to infections [36], a known predictor of poor outcomes in many chronic diseases including cirrhosis [71–73]. Furthermore, evidence suggests that frailty may be associated with systemic inflammation, a plausible link to further decompensation that invites future research in patients with cirrhosis [74–76].

Sarcopenia, Frailty, and Mortality

Across a range of measures, multiple studies have associated sarcopenia as an independent predictor of mortality in cirrhosis. The L3-SMI is one of the most commonly used measures, with sarcopenia increasing the risk of mortality in patients being evaluated for LT [5, 35, 77] and those listed for LT [22, 78, 79]. Additional measures, including the psoas muscle mass as evaluated by varying modalities [26, 59, 80], phase angle in BIA [60, 81], and DEXA body composition scans [25], have all been associated with mortality, predominantly in patients with decompensated cirrhosis. Furthermore, studies have associated the L3-SMI as having superior prognostic value to the Subjective Global Assessment, highlighting the importance of objective measures of sarcopenia [78, 82]

The predictive nature of sarcopenia may vary based on the clinical context and the population of patients that are included. While the MELD-Sarcopenia score, developed in one cohort of patients undergoing LT assessment, was a better predictor of mortality over the MELD score alone [5], in another cohort of patients awaiting LT, the MELD-Sarcopenia score did not improve prognostication beyond MELD alone [79]. These varying results may be due to differences in the cohorts (e.g., higher rates of transplantation and lower rates of death in the waitlisted group compared to the LT assessment group). Importantly, in both cohorts, the MELD-sarcopenia score was found to be more predictive in patients with MELD <15. This is consistent with other studies demonstrating that sarcopenia was predictive of mortality in patients with compensated or early decompensated cirrhosis and non-clinically significant portal hypertension [77], but was of no prognostic value in patients with acute on chronic liver failure, a population with the most severe liver dysfunction [83]. Another study demonstrated that there was no correlation between sarcopenia as measured by the TPMT in patients with advanced chronic liver disease and the hepatic venous pressure gradient (HVPG) [32]. Taken together, the prognostic relevance of sarcopenia may be more pronounced at earlier stages of cirrhosis.

In the prospective FrAILT cohort, the MELD-Frailty index score was found to better predict waitlist mortality above and beyond the MELD-Na alone, with frailty adding nine points to the MELD-Na score [3]. Similarly, the CFS, KPS, SBBP, FFC, and 6-minute walk-test were all instruments that predicted mortality in outpatients with cirrhosis [6, 40, 44, 45, 49, 50]. Among those hospitalized, ADLs and KPS were found to predict post-hospitalization mortality [8, 43]. Furthermore, changes in frailty on the waitlist have translated into worse prognosis, highlighting the dynamic nature of frailty and the need for longitudinal assessments [84]. Frailty has been identified as an independent predictor of mortality across compensated and decompensated cirrhosis, suggesting that this construct may be more robust across the spectrum of cirrhosis [52].

Sarcopenia, Frailty, and Further Decompensation

The evidence for the effects of sarcopenia on further decompensation is limited and predominantly based on the retrospective and cross-sectional studies. In keeping with the hypothesis of the role of skeletal muscle in ammonia detoxification, several studies have demonstrated an association between sarcopenia and increased incidence of HE [16, 28, 48, 85, 86]. This has been further confirmed in a prospective cohort whereby sarcopenia and myosteatosis as assessed by CT were independently associated with the development of minimal and overt, HE, with two-thirds of these patients having Child-Pugh B or C disease [87]. In another study using psoas muscle diameter to height ratio, sarcopenia was found to predict hospitalizations for decompensation events [26]. Furthermore, impaired muscle function in a cohort of patients undergoing transplant evaluation was linked to 6-month decompensation-free survival, defined by acute renal failure, new-onset or worsening ascites, and HE [88]. In contrast, in a study of patients who had undergone routine clinical MRI, Beer et al. demonstrated that TPMT/height ratio was not predictive of first or further decompensation, as defined by variceal bleeding, spontaneous bacterial peritonitis, and new or worsening ascites or HE [80]. These differences across studies may be related to the type of patients studied (transplant candidates vs. routine outpatients), study design limitations (retrospective, cross-sectional, lack of competing risk analyses), and the variations in sarcopenia measurement used. Further research is thus needed to delineate the predictive impact of sarcopenia in the process of further decompensation.

The impact of frailty on further decompensation has similarly undergone limited evaluation. Cross-sectional research has associated frailty with a higher prevalence of decompensation [42]. The well-known Alvares-da-Silva study from 2005, which evaluated 50 patients and associated abnormal handgrip strength with decompensation-related complications, was performed in predominantly compensated patients (88% Child-Pugh A), so conclusions for further decompensation are limited [64]. Most recently, in a retrospective multicenter cohort of 822 patients, the LFI was found to be independently associated with cirrhosis progression, both in compensated and decompensated patients, as determined by the D'Amico cirrhosis stages [52]. In the decompensated group, frailty, defined by an LFI >4.5, was independently associated with a hazard ratio of 2.6 for cirrhosis progression or death. This relationship continued to be significant in sensitivity analysis where death was removed as an outcome. These promising findings suggest that frailty may not be merely a consequence of advanced liver disease but may possibly play a role in further decompensation leading to worse outcomes.

Sarcopenia and Transjugular Intrahepatic Portosystemic Shunts (TIPS)

The impact of muscle loss on outcomes in patients undergoing TIPS is another area in sarcopenia that has garnered interest (Table 50.3). Here discordant results have been identified both in the proportion of patients who reverse their sarcopenia

Table 50.3 The effects of sarcopenia in patients undergoing TIPS

Author, year	Tool	*n*	% CP B/C	Design	% for RA	Study Conclusions
Artru, 2020 [90]	TPMT/height	179		Retro	53	Sarcopenia is independently associated with 6-month mortality. 94% were still sarcopenic at month 6
Benmassaoud, 2020 [92]	L3-SMI	107	100	Retro	100	Sarcopenia did not predict mortality or de novo HE
Cai, 2021 [93]	Psoas density	251		Retro	5.2	Sarcopenia is independently predictive of hepatic encephalopathy
Gioia, 2019 [91]	L3-SMI	35	52	Retro	54	Sarcopenia improved in 63%. Less episodes of overt HE if sarcopenia improved
Nardelli, 2017 [94]	SMI	47	76	Prosp	50	MELD and sarcopenia were independent predictors of HE post-TIPS
Praktiknjo, 2018 [23]	FFMA	116		Retro	53	Sarcopenia associated with risk of ACLF and death after TIPS; and persistent decompensation
Praktiknjo, 2019 [95]	TPMT/height	186	78	Retro	52	Sarcopenic vs. non-sarcopenic patients had higher rates of mortality, ascites, overt HE, and ACLF at 1 year after TIPS.
Tsien, 2013 [89]	Various CT measurements	57		Retro	72	Sarcopenia improved by 72%. Failure to reverse sarcopenia associated with increased mortality (44% vs. 10%)

ACLF acute on chronic liver failure, *CP* Child-Pugh, *FFMA* fat-free muscle area, *HE* hepatic encephalopathy, *Prosp* prospective, *RA* refractory ascites, *Retro* retrospective, *SMI* skeletal muscle index, *TPMT* transversal psoas muscle thickness, *TIPS* transjugular intrahepatic portosystemic shunt

post-TIPS (from 6% to 72%) [89–91] and in the post-TIPS clinical outcomes that are associated with pre-TIPS sarcopenia [23, 92–94]. Some studies have shown that sarcopenia predicts the risk of post TIPS HE [93, 94], slower to resolve ascites [23], Acute on Chronic Liver Failure (ACLF), and mortality [95], with the degree of improvement in sarcopenia linked to decreased HE and mortality [89, 91]. In contrast, a recent study found that in patients with refractory ascites who had a TIPS placed, sarcopenia did not predict new HE or mortality [92]. At this point, further studies are needed in order to determine how best to optimize the selection and management of patients with sarcopenia undergoing TIPS. Importantly, in the absence of other classical indications, sarcopenia in itself is not an indication of TIPS.

Effects of Treatment of Sarcopenia, Frailty, Malnutrition on Further Decompensation

The current mainstay of treatment of sarcopenia, frailty, and malnutrition is adequate nutrition and physical activity. Although variable across guidelines, the nutrition prescription most commonly suggests protein intake of 1.2–1.5 g/day, the daily caloric intake of at least 35 kcal/g in non-obese patients, and a tailored hypocaloric approach in obese patients [1]. Prolonged fasting should be avoided, and regular nutritional intake should occur every 3–4 h while awake [1]. Both aerobic and resistance training are recommended, with suggested intervals and prescreening exercise safety recommendations detailed in recent reviews [1, 14]. Most notably, given the acute rise in the portal pressure that can occur with exercise [96], adequate primary or secondary variceal prophylaxis must be in place before starting a moderate-intensity exercise program [97, 98].

Well-designed prospective Randomized Control Trials (RCTs) are required to determine the impact of nutrition and exercise therapy on the course of decompensation. In HE, a small study showed that periodic nutritional consultation with a dietician (recommending 30–35 kcal/kg of calories, and 1.0–1.5 g/kg of protein per day), led to improvement of minimal HE, albeit with no control group [99]. An RCT by Maharshi et al. subsequently showed that nutritional therapy following the European Society for Clinical Nutrition and Metabolism (ESPEN) guidelines improved minimal HE based on the psychometric HE scores [100]. Furthermore, Gioia et al. demonstrated that an improvement in body composition post-TIPS was associated with less HE episodes [91]. In ascites, conceptually, optimal management can improve sarcopenia and malnutrition by improving appetite, and oral intake, and possibly impacting the resting energy expenditure [101, 102]. An uncontrolled study by Vidot et al. ($n = 14$) in patients with malnutrition and ascites provided patients with nutritional support through continuous tube feeding. Study results noted an improvement in the degree of ascites and the need for paracentesis [103]. These findings, while intriguing, need to be confirmed by larger randomized controlled trials.

The impact of exercise on decompensation in cirrhosis remains largely unknown. The pathophysiologic mechanisms of portal hypertension and inflammation as drivers of decompensation may potentially be mitigated by exercise. Although acute exercise is associated with an increase in the HVPG and thereby necessitates adequate prophylaxis to reduce the risk of variceal hemorrhage [97, 98], in the chronic setting, two studies have associated exercise with a reduction in the HVPG [104, 105]. Outside of the cirrhosis literature, exercise has also been shown to decrease systemic sinflammatory markers [106]. Thus, beyond the improvement of muscle mass and function, the beneficial effects of exercise in cirrhosis await further research.

Future Directions

While our understanding of sarcopenia, frailty, and malnutrition and its impact on the natural history of cirrhosis is growing, it is still far from complete. The question of causality between these constructs and cirrhosis decompensation, including the impact of nutrition and exercise therapy, requires further exploration using well-designed pathophysiological studies and adequately powered RCTs with carefully selected patient populations. TIPS is emerging as an increasingly important therapy for portal hypertension in patients with cirrhosis, and future work will also be awaited that determines how better to risk-stratify and optimize the outcomes of patients with sarcopenia who are undergoing this procedure.

References

1. Tandon P, Montano-Loza AJ, Lai JC, Dasarathy S, Merli M. Sarcopenia and frailty in decompensated cirrhosis. J Hepatol. 2021;75(Suppl 1):S147–62.
2. Cheung K, Lee SS, Raman M. Prevalence and mechanisms of malnutrition in patients with advanced liver disease, and nutrition management strategies. Clin Gastroenterol Hepatol. 2012;10(2):117–25.
3. Lai JC, Covinsky KE, Dodge JL, Boscardin WJ, Segev DL, Roberts JP, et al. Development of a novel frailty index to predict mortality in patients with end-stage liver disease. Hepatology. 2017;66(2):564–74.
4. Lai JC, Rahimi RS, Verna EC, Kappus MR, Dunn MA, McAdams-DeMarco M, et al. Frailty associated with waitlist mortality independent of ascites and hepatic encephalopathy in a multicenter study. Gastroenterology. 2019;156(6):1675–82.
5. Montano-Loza AJ, Duarte-Rojo A, Meza-Junco J, Baracos VE, Sawyer MB, Pang JXQ, et al. Inclusion of sarcopenia within MELD (MELD-sarcopenia) and the prediction of mortality in patients with cirrhosis. Clin Transl Gastroenterol. 2015;6:e102.
6. Tandon P, Tangri N, Thomas L, Zenith L, Shaikh T, Carbonneau M, et al. A rapid bedside screen to predict unplanned hospitalization and death in outpatients with cirrhosis: a prospective evaluation of the clinical frailty scale. Am J Gastroenterol. 2016;111(12):1759–67.
7. Dunn MA, Josbeno DA, Tevar AD, Rachakonda V, Ganesh SR, Schmotzer AR, et al. Frailty as tested by gait speed is an independent risk factor for cirrhosis complications that require hospitalization. Am J Gastroenterol. 2016;111(12):1768–75.
8. Tapper EB, Finkelstein D, Mittleman MA, Piatkowski G, Lai M. Standard assessments of frailty are validated predictors of mortality in hospitalized patients with cirrhosis. Hepatology. 2015;62(2):584–90.

9. Sinclair M, Poltavskiy E, Dodge JL, Lai JC. Frailty is independently associated with increased hospitalisation days in patients on the liver transplant waitlist. World J Gastroenterol. 2017;23(5):899–905.
10. Tapper EB, Baki J, Parikh ND, Lok AS. Frailty, psychoactive medications, and cognitive dysfunction are associated with poor patient-reported outcomes in cirrhosis. Hepatology. 2019;69(4):1676–85.
11. Bhanji RA, Takahashi N, Moynagh MR, Narayanan P, Angirekula M, Mara KC, et al. The evolution and impact of sarcopenia pre- and post-liver transplantation. Aliment Pharmacol Ther. 2019;49(6):807–13.
12. Kalafateli M, Mantzoukis K, Choi Yau Y, Mohammad AO, Arora S, Rodrigues S, et al. Malnutrition and sarcopenia predict post-liver transplantation outcomes independently of the model for end-stage Liver disease score. J Cachexia Sarcopenia Muscle. 2017;8(1):113–21.
13. European Association for the Study of the Liver. EASL Clinical Practice Guidelines on nutrition in chronic liver disease. J Hepatol. 2019;70(1):172–93.
14. Lai JC, Tandon P, Bernal W, Tapper EB, Ekong U, Dasarathy S, et al. Malnutrition, frailty, and sarcopenia in patients with cirrhosis: 2021 practice guidance by the American Association for the Study of Liver Diseases. Hepatology. 2021;74(3):1611–44.
15. Bischoff SC, Bernal W, Dasarathy S, Merli M, Plank LD, Schütz T, et al. ESPEN practical guideline: Clinical nutrition in liver disease. Clin Nutr. 2020;39(12):3533–62.
16. Merli M, Giusto M, Lucidi C, Giannelli V, Pentassuglio I, Di Gregorio V, et al. Muscle depletion increases the risk of overt and minimal hepatic encephalopathy: results of a prospective study. Metab Brain Dis. 2013;28(2):281–4.
17. Chen H-W, Dunn MA. Muscle at risk: the multiple impacts of ammonia on sarcopenia and frailty in cirrhosis. Clin Transl Gastroenterol. 2016;7:e170.
18. Zeng X, Shi Z-W, Yu J-J, Wang L-F, Luo Y-Y, Jin S-M, et al. Sarcopenia as a prognostic predictor of liver cirrhosis: a multicentre study in China. J Cachexia Sarcopenia Muscle. 2021;12(6):1948–58. https://doi.org/10.1002/jcsm.12797.
19. Nishikawa H, Shiraki M, Hiramatsu A, Moriya K, Hino K, Nishiguchi S. Japan Society of Hepatology guidelines for sarcopenia in liver disease (1st edition): recommendation from the working group for creation of sarcopenia assessment criteria. Hepatol Res. 2016;46(10):951–63.
20. Chen L-K, Woo J, Assantachai P, Auyeung T-W, Chou M-Y, Iijima K, et al. Asian Working Group for Sarcopenia: 2019 consensus update on sarcopenia diagnosis and treatment. J Am Med Dir Assoc. 2020;21(3):300–7.e2.
21. Cruz-Jentoft AJ, Bahat G, Bauer J, Boirie Y, Bruyère O, Cederholm T, et al. Sarcopenia: revised European consensus on definition and diagnosis. Age Ageing. 2019;48(4):601.
22. Carey EJ, Lai JC, Wang CW, Dasarathy S, Lobach I, Montano-Loza AJ, et al. A multicenter study to define sarcopenia in patients with end-stage liver disease. Liver Transpl. 2017;23(5):625–33.
23. Praktiknjo M, Book M, Luetkens J, Pohlmann A, Meyer C, Thomas D, et al. Fat-free muscle mass in magnetic resonance imaging predicts acute-on-chronic liver failure and survival in decompensated cirrhosis. Hepatology. 2018;67(3):1014–26.
24. Giusto M, Lattanzi B, Albanese C, Galtieri A, Farcomeni A, Giannelli V, et al. Sarcopenia in liver cirrhosis: the role of computed tomography scan for the assessment of muscle mass compared with dual-energy X-ray absorptiometry and anthropometry. Eur J Gastroenterol Hepatol. 2015;27(3):328–34.
25. Sinclair M, Hoermann R, Peterson A, Testro A, Angus PW, Hey P, et al. Use of dual X-ray absorptiometry in men with advanced cirrhosis to predict sarcopenia-associated mortality risk. Liver Int. 2019;39(6):1089–97.
26. Hari A, Berzigotti A, Štabuc B, Caglevič N. Muscle psoas indices measured by ultrasound in cirrhosis—preliminary evaluation of sarcopenia assessment and prediction of liver decompensation and mortality. Dig Liver Dis. 2019;51(11):1502–7.
27. Kobayashi K, Maruyama H, Kiyono S, Ogasawara S, Suzuki E, Ooka Y, et al. Application of transcutaneous ultrasonography for the diagnosis of muscle mass loss in patients with liver cirrhosis. J Gastroenterol. 2018;53(5):652–9.

28. Ruiz-Margáin A, Macías-Rodríguez RU, Ampuero J, Cubero FJ, Chi-Cervera L, Ríos-Torres SL, et al. Low phase angle is associated with the development of hepatic encephalopathy in patients with cirrhosis. World J Gastroenterol. 2016;22(45):10064–70.
29. Pirlich M, Schütz T, Spachos T, Ertl S, Weiss ML, Lochs H, et al. Bioelectrical impedance analysis is a useful bedside technique to assess malnutrition in cirrhotic patients with and without ascites. Hepatology. 2000;32(6):1208–15.
30. Belarmino G, Gonzalez MC, Torrinhas RS, Sala P, Andraus W, D'Albuquerque LAC, et al. Phase angle obtained by bioelectrical impedance analysis independently predicts mortality in patients with cirrhosis. World J Hepatol. 2017;9(7):401–8.
31. Paternostro R, Lampichler K, Bardach C, Asenbaum U, Landler C, Bauer D, et al. The value of different CT-based methods for diagnosing low muscle mass and predicting mortality in patients with cirrhosis. Liver Int. 2019;39(12):2374–85.
32. Paternostro R, Bardach C, Hofer BS, Scheiner B, Schwabl P, Asenbaum U, et al. Prognostic impact of sarcopenia in cirrhotic patients stratified by different severity of portal hypertension. Liver Int. 2021;41(4):799–809.
33. Ebadi M, Wang CW, Lai JC, Dasarathy S, Kappus MR, Dunn MA, et al. Poor performance of psoas muscle index for identification of patients with higher waitlist mortality risk in cirrhosis. J Cachexia Sarcopenia Muscle. 2018;9(6):1053–62.
34. Ebadi M, Bhanji RA, Tandon P, Mazurak V, Baracos VE, Montano-Loza AJ. Review article: prognostic significance of body composition abnormalities in patients with cirrhosis. Aliment Pharmacol Ther. 2020;52(4):600–18.
35. Bhanji RA, Moctezuma-Velazquez C, Duarte-Rojo A, Ebadi M, Ghosh S, Rose C, et al. Myosteatosis and sarcopenia are associated with hepatic encephalopathy in patients with cirrhosis. Hepatol Int. 2018;12(4):377–86.
36. Montano-Loza AJ, Angulo P, Meza-Junco J, Prado CMM, Sawyer MB, Beaumont C, et al. Sarcopenic obesity and myosteatosis are associated with higher mortality in patients with cirrhosis. J Cachexia Sarcopenia Muscle. 2016;7(2):126–35.
37. Xue Q-L. The frailty syndrome: definition and natural history. Clin Geriatr Med. 2011;27(1):1–15.
38. Fried LP, Tangen CM, Walston J, Newman AB, Hirsch C, Gottdiener J, et al. Frailty in older adults: evidence for a phenotype. J Gerontol A Biol Sci Med Sci. 2001;56(3):M146–56.
39. Williams FR, Milliken D, Lai JC, Armstrong MJ. Assessment of the frail patient with end-stage liver disease: a practical overview of sarcopenia, physical function, and disability. Hepatol Commun. 2021;5(6):923–37.
40. Kremer WM, Nagel M, Reuter M, Hilscher M, Michel M, Kaps L, et al. Validation of the clinical frailty scale for the prediction of mortality in patients with liver cirrhosis. Clin Transl Gastroenterol. 2020;11:e00211. https://doi.org/10.14309/ctg.0000000000000211.
41. Lai JC, Dodge JL, McCulloch CE, Covinsky KE, Singer JP. Frailty and the burden of concurrent and incident disability in patients with cirrhosis: a prospective cohort study. Hepatol Commun. 2020;4(1):126–33.
42. McCabe P, Hirode G, Wong R. Functional status at liver transplant Waitlisting correlates with greater odds of encephalopathy, ascites, and spontaneous bacterial peritonitis. J Clin Exp Hepatol. 2020;10(5):413–20.
43. Tandon P, Reddy KR, O'Leary JG, Garcia-Tsao G, Abraldes JG, Wong F, et al. A Karnofsky performance status-based score predicts death after hospital discharge in patients with cirrhosis. Hepatology. 2017;65(1):217–24.
44. Orman ES, Ghabril M, Chalasani N. Poor performance status is associated with increased mortality in patients with cirrhosis. Clin Gastroenterol Hepatol. 2016;14(8):1189–95.e1.
45. Lai JC, Feng S, Terrault NA, Lizaola B, Hayssen H, Covinsky K. Frailty predicts waitlist mortality in liver transplant candidates. Am J Transplant. 2014;14(8):1870–9.
46. Román E, Parramón M, Flavià M, Gely C, Poca M, Gallego A, et al. Frailty in outpatients with cirrhosis: a prospective observational study. Liver Int. 2021;41(2):357–68. https://doi.org/10.1111/liv.14694.
47. Tapper EB, Derstine B, Baki J, Su GL. Bedside measures of frailty and cognitive function correlate with sarcopenia in patients with cirrhosis. Dig Dis Sci. 2019;64(12):3652–9.

48. Miwa T, Hanai T, Nishimura K, Maeda T, Ogiso Y, Imai K, et al. Handgrip strength stratifies the risk of covert and overt hepatic encephalopathy in patients with cirrhosis. JPEN J Parenter Enteral Nutr. 2022;46(4):858–66. https://doi.org/10.1002/jpen.2222.
49. Carey EJ, Steidley DE, Aqel BA, Byrne TJ, Mekeel KL, Rakela J, et al. Six-minute walk distance predicts mortality in liver transplant candidates. Liver Transpl. 2010;16(12):1373–8.
50. Essam Behiry M, Mogawer S, Yamany A, Rakha M, Awad R, Emad N, et al. Ability of the short physical performance battery frailty index to predict mortality and hospital readmission in patients with liver cirrhosis. Int J Hepatol. 2019;2019:8092865.
51. Lai JC, Covinsky KE, McCulloch CE, Feng S. The liver frailty index improves mortality prediction of the subjective clinician assessment in patients with cirrhosis. Am J Gastroenterol. 2018;113(2):235–42.
52. Wang S, Whitlock R, Xu C, Taneja S, Singh S, Abraldes JG, et al. Frailty is associated with increased risk of cirrhosis disease progression and death. Hepatology. 2022;75(3):600–9. https://doi.org/10.1002/hep.32157.
53. Prentis JM, Manas DMD, Trenell MI, Hudson M, Jones DJ, Snowden CP. Submaximal cardiopulmonary exercise testing predicts 90-day survival after liver transplantation. Liver Transpl. 2012;18(2):152–9.
54. Bernal W, Martin-Mateos R, Lipcsey M, Tallis C, Woodsford K, McPhail MJ, et al. Aerobic capacity during cardiopulmonary exercise testing and survival with and without liver transplantation for patients with chronic liver disease. Liver Transpl. 2014;20(1):54–62.
55. Borhofen SM, Gerner C, Lehmann J, Fimmers R, Görtzen J, Hey B, et al. The Royal Free Hospital-Nutritional Prioritizing Tool is an independent predictor of deterioration of liver function and survival in cirrhosis. Dig Dis Sci. 2016;61(6):1735–43.
56. Tandon P, Raman M, Mourtzakis M, Merli M. A practical approach to nutritional screening and assessment in cirrhosis. Hepatology. 2017;65(3):1044–57.
57. Ney M, Li S, Vandermeer B, Gramlich L, Ismond KP, Raman M, et al. Systematic review with meta-analysis: nutritional screening and assessment tools in cirrhosis. Liver Int. 2020;40(3):664–73.
58. Lochs H, Allison SP, Meier R, Pirlich M, Kondrup J, Schneider S, et al. Introductory to the ESPEN guidelines on enteral nutrition: terminology, definitions and general topics. Clin Nutr. 2006;25(2):180–6.
59. Durand F, Buyse S, Francoz C, Laouénan C, Bruno O, Belghiti J, et al. Prognostic value of muscle atrophy in cirrhosis using psoas muscle thickness on computed tomography. J Hepatol. 2014;60(6):1151–7.
60. Ruiz-Margáin A, Xie JJ, Román-Calleja BM, Pauly M, White MG, Chapa-Ibargüengoitia M, et al. Phase angle from bioelectrical impedance for the assessment of sarcopenia in cirrhosis with or without ascites. Clin Gastroenterol Hepatol. 2021;19(9):1941–9.e2.
61. Vouche M, Habib A, Ward TJ, Kim E, Kulik L, Ganger D, et al. Unresectable solitary hepatocellular carcinoma not amenable to radiofrequency ablation: multicenter radiology-pathology correlation and survival of radiation segmentectomy. Hepatology. 2014;60(1):192–201.
62. Hsu C-Y, Lee Y-H, Hsia C-Y, Huang Y-H, Su C-W, Lin H-C, et al. Performance status in patients with hepatocellular carcinoma: determinants, prognostic impact, and ability to improve the Barcelona Clinic Liver Cancer system. Hepatology. 2013;57(1):112–9.
63. Sinclair M, Chapman B, Hoermann R, Angus PW, Testro A, Scodellaro T, et al. Handgrip strength adds more prognostic value to the model for end-stage Liver disease score than imaging-based measures of muscle mass in men with cirrhosis. Liver Transpl. 2019;25(10):1480–7.
64. Alvares-da-Silva MR, Reverbel da Silveira T. Comparison between handgrip strength, subjective global assessment, and prognostic nutritional index in assessing malnutrition and predicting clinical outcome in cirrhotic outpatients. Nutrition. 2005;21(2):113–7.
65. Hanai T, Shiraki M, Imai K, Suetsugu A, Takai K, Moriwaki H, et al. Reduced handgrip strength is predictive of poor survival among patients with liver cirrhosis: a sex-stratified analysis. Hepatol Res. 2019;49(12):1414–26.

66. Bunchorntavakul C, Reddy KR. Review article: malnutrition/sarcopenia and frailty in patients with cirrhosis. Aliment Pharmacol Ther. 2020;51(1):64–77.
67. Thapaliya S, Runkana A, McMullen MR, Nagy LE, McDonald C, Naga Prasad SV, et al. Alcohol-induced autophagy contributes to loss in skeletal muscle mass. Autophagy. 2014;10(4):677–90.
68. Bhanji RA, Narayanan P, Allen AM, Malhi H, Watt KD. Sarcopenia in hiding: the risk and consequence of underestimating muscle dysfunction in nonalcoholic steatohepatitis. Hepatology. 2017;66(6):2055–65.
69. Abrigo J, Gonzalez F, Aguirre F, Tacchi F, Gonzalez A, Meza MP, et al. Cholic acid and deoxycholic acid induce skeletal muscle atrophy through a mechanism dependent on TGR5 receptor. J Cell Physiol. 2021;236(1):260–72.
70. Ooi PH, Gilmour SM, Yap J, Mager DR. Effects of branched chain amino acid supplementation on patient care outcomes in adults and children with liver cirrhosis: a systematic review. Clin Nutr ESPEN. 2018;28:41–51.
71. Arvaniti V, D'Amico G, Fede G, Manousou P, Tsochatzis E, Pleguezuelo M, et al. Infections in patients with cirrhosis increase mortality four-fold and should be used in determining prognosis. Gastroenterology. 2010;139(4):1246–56, 1256.e1–5
72. Kochar B, Cai W, Cagan A, Ananthakrishnan AN. Pretreatment frailty is independently associated with increased risk of infections after immunosuppression in patients with inflammatory bowel diseases. Gastroenterology. 2020;158(8):2104–11.e2.
73. Villanueva C, Albillos A, Genescà J, Garcia-Pagan JC, Brujats A, Calleja JL, et al. Bacterial infections adversely influence the risk of decompensation and survival in compensated cirrhosis. J Hepatol. 2021;75(3):589–99.
74. Soysal P, Stubbs B, Lucato P, Luchini C, Solmi M, Peluso R, et al. Corrigendum to "inflammation and frailty in the elderly: a systematic review and meta-analysis" [Ageing Res Rev. 31 (2016) 1–8]. Ageing Res Rev. 2017;35:364–5.
75. Franceschi C, Garagnani P, Parini P, Giuliani C, Santoro A. Inflammaging: a new immune-metabolic viewpoint for age-related diseases. Nat Rev Endocrinol. 2018;14(10):576–90.
76. Bhanji RA, Montano-Loza AJ, Watt KD. Sarcopenia in cirrhosis: looking beyond the skeletal muscle loss to see the systemic disease. Hepatology. 2019;70(6):2193–203.
77. Kang SH, Jeong WK, Baik SK, Cha SH, Kim MY. Impact of sarcopenia on prognostic value of cirrhosis: going beyond the hepatic venous pressure gradient and MELD score. J Cachexia Sarcopenia Muscle. 2018;9(5):860–70.
78. Tandon P, Ney M, Irwin I, Ma MM, Gramlich L, Bain VG, et al. Severe muscle depletion in patients on the liver transplant wait list: its prevalence and independent prognostic value. Liver Transpl. 2012;18(10):1209–16.
79. van Vugt JLA, Alferink LJM, Buettner S, Gaspersz MP, Bot D, Darwish Murad S, et al. A model including sarcopenia surpasses the MELD score in predicting waiting list mortality in cirrhotic liver transplant candidates: a competing risk analysis in a national cohort. J Hepatol. 2018;68(4):707–14.
80. Beer L, Bastati N, Ba-Ssalamah A, Pötter-Lang S, Lampichler K, Bican Y, et al. MRI-defined sarcopenia predicts mortality in patients with chronic liver disease. Liver Int. 2020;40(11):2797–807.
81. Saueressig C, Glasenapp JH, Luft VC, Alves FD, Ferreira PK, Hammes TO, et al. Phase angle is an independent predictor of 6-month mortality in patients with decompensated cirrhosis: a prospective cohort study. Nutr Clin Pract. 2020;35(6):1061–9.
82. Moctezuma-Velazquez C, Ebadi M, Bhanji RA, Stirnimann G, Tandon P, Montano-Loza AJ. Limited performance of subjective global assessment compared to computed tomography-determined sarcopenia in predicting adverse clinical outcomes in patients with cirrhosis. Clin Nutr. 2019;38(6):2696–703.
83. Li T, Xu M, Kong M, Song W, Duan Z, Chen Y. Use of skeletal muscle index as a predictor of short-term mortality in patients with acute-on-chronic liver failure. Sci Rep. 2021;11(1):12593.

84. Lai JC, Dodge JL, Kappus MR, Dunn MA, Volk ML, Duarte-Rojo A, et al. Changes in frailty are associated with waitlist mortality in patients with cirrhosis. J Hepatol. 2020;73(3):575–81. https://doi.org/10.1016/j.jhep.2020.03.029.
85. Badran H, Elsabaawy MM, Ragab A, Aly RA, Alsebaey A, Sabry A. Baseline sarcopenia is associated with lack of response to therapy, liver decompensation and high mortality in hepatocellular carcinoma patients. Asian Pac J Cancer Prev. 2020;21(11):3285–90.
86. Hanai T, Shiraki M, Watanabe S, Kochi T, Imai K, Suetsugu A, et al. Sarcopenia predicts minimal hepatic encephalopathy in patients with liver cirrhosis. Hepatol Res. 2017;47(13):1359–67.
87. Nardelli S, Lattanzi B, Merli M, Farcomeni A, Gioia S, Ridola L, et al. Muscle alterations are associated with minimal and overt hepatic encephalopathy in patients with liver cirrhosis. Hepatology. 2019;70(5):1704–13.
88. Oey RC, Aarts P, Erler NS, Metselaar HJ, Lakenman PLM, Riemslag Baas-van der Ree S, et al. Identification and prognostic impact of malnutrition in a population screened for liver transplantation. Clin Nutr ESPEN. 2020;36:36–44.
89. Tsien C, Shah SN, McCullough AJ, Dasarathy S. Reversal of sarcopenia predicts survival after a transjugular intrahepatic portosystemic stent. Eur J Gastroenterol Hepatol. 2013;25(1):85–93.
90. Artru F, Miquet X, Azahaf M, Labreuche J, Ntandja Wandji LC, Sergent G, et al. Consequences of TIPSS placement on the body composition of patients with cirrhosis and severe portal hypertension: a large retrospective CT-based surveillance. Aliment Pharmacol Ther. 2020;52(9):1516–26.
91. Gioia S, Ridola L, Cristofaro L, Merli M, Faccioli J, Riggio O, et al. The improvement in body composition including subcutaneous and visceral fat reduces ammonia and hepatic encephalopathy after transjugular intrahepatic portosystemic shunt. Liver Int. 2021;41(12):2965–73. https://doi.org/10.1111/liv.15060.
92. Benmassaoud A, Roccarina D, Arico F, Leandro G, Yu B, Cheng F, et al. Sarcopenia does not worsen survival in patients with cirrhosis undergoing transjugular intrahepatic portosystemic shunt for refractory ascites. Am J Gastroenterol. 2020;115(11):1911–4.
93. Cai W, Lin H, Qi R, Lin X, Zhao Y, Chen W, et al. Psoas muscle density predicts occurrences of hepatic encephalopathy in patients receiving transjugular intrahepatic portosystemic shunts within 1 year. Cardiovasc Intervent Radiol. 2022;45(1):93–101. https://doi.org/10.1007/s00270-021-02961-8.
94. Nardelli S, Lattanzi B, Torrisi S, Greco F, Farcomeni A, Gioia S, et al. Sarcopenia is risk factor for development of hepatic encephalopathy after transjugular intrahepatic portosystemic shunt placement. Clin Gastroenterol Hepatol. 2017;15(6):934–6.
95. Praktiknjo M, Clees C, Pigliacelli A, Fischer S, Jansen C, Lehmann J, et al. Sarcopenia is associated with development of acute-on-chronic liver failure in decompensated liver cirrhosis receiving transjugular intrahepatic portosystemic shunt. Clin Transl Gastroenterol. 2019;10(4):e00025.
96. García-Pagàn JC, Santos C, Barberá JA, Luca A, Roca J, Rodriguez-Roisin R, et al. Physical exercise increases portal pressure in patients with cirrhosis and portal hypertension. Gastroenterology. 1996;111(5):1300–6.
97. Tandon P, Ismond KP, Riess K, Duarte-Rojo A, Al-Judaibi B, Dunn MA, et al. Exercise in cirrhosis: translating evidence and experience to practice. J Hepatol. 2018;69(5):1164–77.
98. Bandi JC, García-Pagán JC, Escorsell A, François E, Moitinho E, Rodés J, et al. Effects of propranolol on the hepatic hemodynamic response to physical exercise in patients with cirrhosis. Hepatology. 1998;28(3):677–82.
99. Kato A, Tanaka H, Kawaguchi T, Kanazawa H, Iwasa M, Sakaida I, et al. Nutritional management contributes to improvement in minimal hepatic encephalopathy and quality of life in patients with liver cirrhosis: a preliminary, prospective, open-label study. Hepatol Res. 2013;43(5):452–8.

100. Maharshi S, Sharma BC, Sachdeva S, Srivastava S, Sharma P. Efficacy of nutritional therapy for patients with cirrhosis and minimal hepatic encephalopathy in a randomized trial. Clin Gastroenterol Hepatol. 2016;14(3):454–60.e3; quiz e33
101. Aqel BA, Scolapio JS, Dickson RC, Burton DD, Bouras EP. Contribution of ascites to impaired gastric function and nutritional intake in patients with cirrhosis and ascites. Clin Gastroenterol Hepatol. 2005;3(11):1095–100.
102. Dolz C, Raurich JM, Ibáñez J, Obrador A, Marsé P, Gayá J. Ascites increases the resting energy expenditure in liver cirrhosis. Gastroenterology. 1991;100(3):738–44.
103. Vidot H, Bowen DG, Carey S, McCaughan GW, Allman-Farinelli M, Shackel NA. Aggressive nutrition intervention reduces ascites and frequency of paracentesis in malnourished patients with cirrhosis and ascites. JGH Open. 2017;1(3):92–7.
104. Macías-Rodríguez RU, Ilarraza-Lomelí H, Ruiz-Margáin A, Ponce-de-León-Rosales S, Vargas-Vorácková F, García-Flores O, et al. Changes in hepatic venous pressure gradient induced by physical exercise in cirrhosis: results of a pilot randomized open clinical trial. Clin Transl Gastroenterol. 2016;7(7):e180.
105. Berzigotti A, Albillos A, Villanueva C, Genescá J, Ardevol A, Augustín S, et al. Effects of an intensive lifestyle intervention program on portal hypertension in patients with cirrhosis and obesity: the SportDiet study. Hepatology. 2017;65(4):1293–305.
106. Kelley GA, Kelley KS. Effects of aerobic exercise on C-reactive protein, body composition, and maximum oxygen consumption in adults: a meta-analysis of randomized controlled trials. Metabolism. 2006;55(11):1500–7.

Preventing Further Decompensation: Consensus Statements of Panel 7

51

Guadalupe Garcia-Tsao, Thomas Reiberger, Christophe Bureau, Gennaro D'Amico, Vincenzo La Mura, Salvatore Piano, Puneeta Tandon, and Laura Turco

G. Garcia-Tsao (✉)
Digestive Diseases Section, Yale University, New Haven, CT, USA

Digestive Diseases Section, VA-CT Healthcare System, West Haven, CT, USA
e-mail: Guadalupe.garcia-tsao@yale.edu

T. Reiberger
Division of Gastroenterology and Hepatology, Department of Medicine III, Medical University of Vienna, Vienna, Austria
e-mail: Thomas.reiberger@meduniwien.ac.at

C. Bureau
Service d'Hépato-Gastroentérologie, CHU Toulouse, Toulouse, France
e-mail: bureau.c@chu-toulouse.fr

G. D'Amico
Gastroenterology Unit, Ospedale V Cervello, Palermo, Italy

V. La Mura
Fondazione IRCCS Ca' Granda, Ospedale Maggiore Policlinico, Dipartimento di Scienze Biomediche per la Salute and CRC AM e A Migliavacca per lo Studio delle Malattie del Fegato, University of Milan, Milan, Italy
e-mail: Vincenzo.lamura@unimi.it

S. Piano
Unit of Internal Medicine and Hepatology, Department of Medicine, University of Padova, Padua, Italy
e-mail: Salvatore.piano@unipd.it

P. Tandon
Division of Gastroenterology, University of Alberta, Edmonton, AB, Canada
e-mail: ptandon@ualberta.ca

L. Turco
Division of Internal Medicine for the Treatment of Severe Organ Failure, IRCCS Azienda Ospedaliero-Universitaria di Bologna, Bologna, Italy

R. de Franchis (ed.), *Portal Hypertension VII*,
https://doi.org/10.1007/978-3-031-08552-9_51

Definition of "Further Decompensation"

7.1 Further decompensation in cirrhosis represents a prognostic stage associated with even higher mortality than that associated with the first decompensation. Specific events that define further decompensation are any of the following (B1) (New):

(a) Development of a second portal hypertension-driven decompensating event (ascites, variceal hemorrhage, or hepatic encephalopathy) and/or jaundice.

(b) Development of recurrent variceal bleeding, recurrent ascites (requirement of ≥3 large-volume paracenteses within 1 year), recurrent encephalopathy, development of spontaneous bacterial peritonitis (SBP) and/or hepatorenal syndrome (HRS-AKI).

(c) In patients presenting with bleeding alone, development of ascites, encephalopathy, or jaundice after recovery from bleeding but not if these events occur around the time of bleeding.

Preventing Further Decompensation in Patients with Ascites

7.2 Patients with decompensated cirrhosis should be considered for liver transplantation. (A1) (New)

7.3 Patients with ascites who are not on non-selective beta-blockers (NSBB, i.e., propranolol or nadolol) or carvedilol should undergo a screening endoscopy. (B1) (New)

7.4 TIPS should be considered in patients with recurrent ascites (requirement of ≥3 large-volume paracenteses within 1 year) irrespective of the presence or absence of varices or history of variceal hemorrhage. (A1) (New)

7.5 In patients with ascites and low-risk varices (small [<5 mm], no red signs, not Child C), NSBB or carvedilol may be used to prevent first variceal hemorrhage. (B2) (Changed)

7.6 In patients with ascites and high-risk varices (large varices [≥5 mm]), or red spot signs, or Child C), prevention of first variceal hemorrhage is indicated, with NSBB or carvedilol being preferred over endoscopic variceal ligation (EVL). (B1) (Changed)

7.7 In patients with ascites, NSBBs or carvedilol should be dose-reduced or discontinued in case of persistently low blood pressure (systolic blood pressure < 90 mmHg or mean arterial pressure < 65 mmHg) and/or HRS-AKI (B1). Once blood pressure returns to baseline and/or HRS-AKI resolves, NSBB can be re-initiated or re-titrated. (B1). If a patient remains intolerant to NSBB, EVL is then recommended to prevent variceal hemorrhage (Changed)

Preventing Recurrent Variceal Hemorrhage (Secondary Prophylaxis)

7.8 First-line therapy for the prevention of recurrent variceal hemorrhage is the combination of NSBB or carvedilol and EVL. (A1) (Changed)

7.9 TIPS is the treatment of choice in patients who rebleed despite NSBB or carvedilol and EVL. (B1) (Unchanged)

7.10 In patients who cannot get/tolerate EVL or carvedilol or NSBB, any of these therapies can be maintained alone (A1) and TIPS should be considered in patients with recurrent ascites. (B1) (Changed)

7.11 In patients who bled despite adherence to NSBB or carvedilol as primary prophylaxis, the combination of NSBB or carvedilol and EVL is recommended, and TIPS should be considered in those with recurrent ascites. (B1) (New)

Preventing Recurrent Bleeding from Portal Hypertensive Gastropathy (PHG)

7.12 PHG and portal hypertension-associated gastric or small intestinal polypoid lesions (PHP) have to be distinguished from gastric antral vascular ectasia (GAVE) because treatments are different. (B1) (Changed)

7.13 NSBB are first-line therapy in preventing recurrent bleeding from PHG. (A1) (Unchanged)

7.14 Endoscopic therapy (e.g., argon plasma coagulation or hemospray) may be used to treat recurrent bleeding from PHG. (D1) (New)

7.15 TIPS should be considered for transfusion-dependent PHG despite NSBB or carvedilol and endoscopic therapy. (C1) (Changed)

Role of Infections in Decompensated Cirrhosis

7.16 Bacterial infections are common in patients with decompensated cirrhosis and may cause further decompensation. (A1) (New)

7.17 In all patients hospitalized with decompensation, bacterial infections should be ruled out. The minimal work-up for infections should include diagnostic paracentesis, chest X-ray, cultures of blood, ascites and urine, and skin examination. (A1) (New)

7.18 Patients with bacterial infections should be promptly treated with antibiotics. The empirical antibiotic treatment should be tailored to local epidemiology, risk factors for multidrug-resistant bacteria, and severity of infection. (A1). If no response to antibiotics is observed, consider viral and fungal infections. (C1) (Changed)

The Role of Sarcopenia and Frailty in Further Decompensation

7.19 Frailty, malnutrition, and sarcopenia have an impact on survival in patients with decompensated cirrhosis. They should be evaluated with available standardized tools. (B1) (New)

7.20 All patients with decompensated cirrhosis should receive nutrition consultation and be advised regarding the benefits of regular exercise. (B1) (New)

7.21 While sarcopenia improves in some patients after TIPS, pre-procedural sarcopenia has also been associated with poor outcomes (e.g., encephalopathy, slower resolution of ascites) and higher mortality. Therefore, sarcopenia by itself should not be an indication tor TIPS. (C2) (New)

Definition of Cirrhosis Recompensation

7.22 The concept of recompensation implies that there is at least partial regression of the structural and functional changes of cirrhosis after the removal of the etiology of cirrhosis. (A1) (New)

7.23 Clinically, the definition of "recompensation" is based on expert consensus and requires fulfillment of <u>all</u> of the following criteria: (C2) (New)
   (d) Removal/suppression/cure of the primary etiology of cirrhosis (viral elimination for hepatitis C, sustained viral suppression for hepatitis B, sustained alcohol abstinence for alcohol-induced cirrhosis).
   (e) Resolution of ascites (off diuretics), encephalopathy (off lactulose/rifaximin), and absence of recurrent variceal hemorrhage (for at least 12 months).
   (f) Stable improvement of liver function tests (albumin, INR, bilirubin).

7.24 Because clinically significant portal hypertension (CSPH) may persist despite recompensation, NSBB should not be discontinued unless CSPH resolves. (B1) (New)

7.25 Resolution of ascites (while on diuretics or after TIPS) and/or lack of recurrent variceal hemorrhage (while on NSBB or carvedilol + EVL or after TIPS) without removal/suppression/cure of the primary etiologic factor and without improvement in liver synthetic function is not evidence of recompensation. (B1) (New)

Research Agenda

Further Decompensation and re-Compensation

- Investigate the effect of time to further decompensation on prognosis.
- Obtain data to support the suggested concept of cirrhosis recompensation, particularly regarding the timeframe necessary to consider a patient truly recompensated.
- Association between re-compensation and resolution of CSPH.

- Impact of etiological therapy other than alcohol abstinence and antiviral therapy on re-compensation.

NSBB and Further Decompensation

- Prospective studies should assess if NSBB treatment prevents further (non-rebleeding) decompensation in decompensated patients.
- Prospective studies should assess if HVPG-guided (NSBB/carvedilol) therapy is more efficient to prevent further decompensation over non-HVPG guided strategies.
- Optimal blood pressure cut-offs (mean arterial pressure/systolic arterial pressure) to define the safe use of NSBB/carvedilol therapy and whether dose reduction (vs. discontinuation) is safe.
- The impact of NSBB discontinuation on the natural history of decompensated cirrhosis.
- Assessment of the benefit of carvedilol over traditional NSBBs in secondary prophylaxis of variceal hemorrhage.

TIPS and Further Decompensation

- The benefit of TIPS in secondary prophylaxis in patients with NSBB intolerance/NSBB non-response should be assessed in patients with ascites not meeting "strict" criteria for recurrent ascites.
- Establish whether TIPS placement past the 72-h preemptive TIPS window is still beneficial.
- Hemodynamic and non-hemodynamic effects of NSBB in patients after TIPS.

Sarcopenia, Frailty and Nutrition, and Further Decompensation

- Impact of nutritional interventions on the natural history of decompensation.
- Impact of therapies targeting sarcopenia and frailty on the natural history of decompensation.
- Define the role of sarcopenia in the selection of patients for TIPS.

Part X

Vascular Liver Disorders 1: Splanchnic Vein Thrombosis

Portal Vein Thrombosis: Anticoagulation Vs. Interventional Radiology

52

Fanny Turon, Anna Baiges, Marta Barrufet, and Patricia Bermudez

Introduction

Portal vein thrombosis (PVT) is defined by the presence of a thrombus within the main portal vein trunk and/or in the intrahepatic portal branches, which can sometimes extend towards the splenic vein or the superior mesenteric vein. In this chapter, we will focus on the management of non-tumoral non-cirrhotic portal vein thrombosis, a distinctive entity that requires a specific approach different from cirrhotic PVT and from tumoral PVT. Non-tumoral non-cirrhotic PVT (from now on referred to as PVT) can present either acutely – with or without symptoms – or can be identified at a chronic stage where it can lead to portal hypertension complications. Differentiation of acute from chronic PVT is crucial because the two entities are managed differently.

Diagnosis

The diagnosis of both acute and chronic PVT is reached through non-invasive imaging tests [1, 2]. Usually, the diagnosis of symptomatic acute PVT will be triggered by symptoms such as abdominal pain, diarrhea, and/or ischemic colitis. On the contrary, clinically silent chronic PVT is usually first identified incidentally through

F. Turon (✉) · A. Baiges
Barcelona Hepatic Hemodynamic Laboratory, Liver Unit, Hospital Clínic, IDIBAPS, CIBERehd, Health Care Provider of the European Reference Network on Rare Liver Disorders (ERN Liver), Barcelona, Spain
e-mail: fturon@clinic.cat; abaigesa@clinic.cat

M. Barrufet · P. Bermudez
Department of Radiology, Hospital Clínic de Barcelona, Barcelona, Spain

Interventional Radiology Unit, CDI, Hospital Clinic, Barcelona, Spain
e-mail: BARRUFET@clinic.cat; pbermude@clinic.cat

R. de Franchis (ed.), *Portal Hypertension VII*,
https://doi.org/10.1007/978-3-031-08552-9_52

ultrasound showing hyperechoic material in the vessel lumen with distension of the portal vein and its tributaries. Doppler imaging will show the absence of flow in all or part of the lumen and is used to estimate flow rate. The presence of multiple small vessels where the portal vein would be expected to be is characteristic of cavernomatous transformation. However, the splenic and mesenteric veins can be difficult to visualize with ultrasound. Therefore, for the assessment of thrombus extension within the splanchnic system, CT or MRI angiography are more sensitive techniques than Doppler US [2, 3].

Determination of the Presence/Absence of Cirrhosis

Defining the presence or absence of cirrhosis is fundamental due to prognostic and long-term management implications. It is important to remark that contour nodularity may be present in chronic non-cirrhotic PVT and that imaging findings alone should not induce the assumption of cirrhosis [4]. Transient elastography, hepatic venous pressure gradient, and biopsy can reliably help to rule out cirrhosis in patients with portal vein thrombosis.

Etiology

Etiological factors for PVT can be divided into local and systemic factors. The most frequent local risk factors in non-cirrhotic PVT are intra-abdominal surgery, infections, or abdominal inflammation. Abdominal infection/inflammation may be present in approximately 30% of cases, as activation of coagulation is an important response in the host's defense against infections as it prevents the dissemination of microorganisms [5]. In pediatric age, previous umbilical catheterization is the most prevalent local factor [6, 7]. Regarding systemic etiological factors for PVT, it is possible to identify an underlying prothrombotic factor in up to 60% of patients [8, 9] and multiple etiologic factors coexist in over 15% of patients. Inherited deficiencies of natural inhibitors of the coagulation system, increased levels of coagulation factors, and genetic mutations of coagulant factors are associated with an increased risk of PVT. Also, patients with myeloproliferative neoplasms (MPNs) – chronic clonal hematopoietic stem cell disorders characterized by an overproduction of granulocytes, erythrocytes, and/or platelets – are at high risk of arterial and venous thrombotic complications [10, 11], being the most frequent etiological cause of non-cirrhotic PVT (30%–40% of cases). Identification of other inherited deficiencies of natural coagulant inhibitors such as deficiencies of antithrombin, protein C, or protein S can be difficult to assess in PVT patients [12] given that in this population there can be a non-specific decrease in the liver synthesis of both coagulation factors and inhibitors, even when liver dysfunction is not obvious [13]. A comprehensive evaluation of deficiencies of anticoagulant inhibitors is the key to finding the etiological factor of PVT development and should be included in the diagnostic work-up. Summing up, diagnosing the underlying etiological factor for the development of PVT is important since it may have therapeutic and prognostic implications (need of long-term anticoagulation).

Acute Portal Vein Thrombosis

Prompt diagnosis and management of acute symptomatic portal vein thrombosis is essential. Failure to detect and efficiently treat thrombosis can result in mesenteric ischemia, sepsis, and potentially death. Importantly, even if the patient successfully overcomes an acute PVT, the physician must be aware that it can evolve to chronic PVT with cavernomatous transformation and with ulterior portal hypertension complications. Thus, the aim of treatment in the setting of acute PVT is to prevent ischemic complications, achieve thrombus recanalization, and prevent progression to chronic PVT.

The severity of acute PVT presentation depends mainly on the extent of thrombosis. At diagnosis, acute PVT evaluation with CT scan or magnetic resonance imaging is warranted to evaluate the extent of the thrombosis, assess the etiology, and provide information about associated complications such us bowel ischemia. Extensive acute thrombosis involving the superior mesenteric vein is more likely to produce bowel ischemia and infarction, which is associated with higher morbidity and mortality.

Management

Anticoagulation

Resolution of acute PVT without any treatment is rare [8, 14]. Anticoagulation is the standard and first-line treatment for acute non-tumoral PVT and should be initiated immediately after diagnosis. The early initiation of therapy for acute PVT is essential to prevent progression of the thrombus and restore adequate venous outflow to prevent bowel infarction and the sequelae of portal hypertension [8, 15]. Indeed, studies reporting the efficacy of anticoagulation have shown that timing of anticoagulation is associated with the rate of recanalization. Specifically, when anticoagulation is initiated immediately after diagnosis complete recanalization of the portal vein occurs in about 50% of cases [8, 14–16], while when there is a delay in anticoagulation initiation the likelihood of recanalization significantly drops [8]. Importantly, the presence of ascites and splenic vein thrombosis at diagnostic has been reported as factors independently associated to absence of recanalization [8].

The recommended duration of anticoagulation is at least 6 months [8, 14] in all patients with acute non-cirrhotic PVT, although anticoagulation therapy should be maintained indefinitely in patients with high-risk procoagulant status or history of intestinal infarction. Prospective multicentric data including 95 patients with acute PVT treated with anticoagulants showed no improvement in recanalization rates beyond the sixth months of treatment if thrombosis affected the portal vein trunk. On the contrary, when the thrombosis involved the superior mesenteric vein or splenic vein, recanalization was documented to occur after 6 months. In all reported series, major complications of anticoagulation are relatively uncommon, occurring in less than 5% of patients [8, 15].

Low-molecular weight heparin (or unfractionated heparin if invasive procedures are pending) is the preferred initial treatment for acute PVT, which can be followed by oral anticoagulation therapy with either vitamin K antagonists or DOACS once abdominal pain is relieved, and no invasive procedures are planned. The use of DOACS in the setting of PVT is supported by recent observational studies that report successful and safe use of DOAC in non-cirrhotic acute PVT [17, 18].

Thrombolysis and Interventional Vascular Procedures

Interventional vascular radiology with or without pharmacological thrombolysis (local or systemic) has been proposed as an emergent approach, on the one side to re-establish physiological splanchnic venous outflow, prevent the cavernomatous transformation, and avoid complications of portal hypertension and, on the other, be used as an adjunct to anticoagulation in cases of acute PVT and imminent signs of bowel ischemia.

Several endovascular techniques have been described for treating acute PVT via direct or indirect access into the portal system, including catheter-delivered fibrinolytic therapy and mechanical/suction thrombectomy with or without fibrinolytic therapy:

- Direct percutaneous access can be achieved through a transhepatic or transsplenic approach, although it involves an increased risk of bleeding, and it may require embolization of the access tract. The portal system can also be directly accessed through creation of a transjugular intrahepatic portosystemic shunt (TIPS).
- Indirect access to the portal system includes infusion of thrombolytics into the superior mesenteric artery (SMA), exposing portal venous thrombus to fibrinolytic agents via the small bowel capillaries.
- Mechanical thrombectomy involves restoring flow within the main portal vein using a balloon thrombectomy, rheolytic thrombectomy, or thromboaspiration. Successful use of mechanical thrombectomy-assisted thrombolysis has been previously described in the literature, although no prospective data are available to compare different techniques and outcomes.
- Thrombolytic therapy: If there is no contraindication, passive infusion of fibrinolytic medications directly into the acute thrombus can be considered.

However, available data are scarce, only case reports and small series have been published and, due to the rareness and severity of this condition, no randomized controlled trials are available to compare the interventional vascular procedures with the standard treatment with anticoagulation. Varying success rates of recanalization with significant procedure-related complications and associated morbimortality have been reported (Table 52.1).

Table 52.1 Main studies approaching interventional recanalization in acute non-cirrhotic PVT

	Hollingshead	Smalberg	Wang	Liu	Cao	Rosenqvuist	Klinger	Wolter	Benmassaoud	Rössle	Li	Oguslu
	2005	2008	2009	2009	2013	2016	2017	2018	2019	2020	2021	2021
N	20	4	2	32	12	4	17	11	22	39	23	9
Follow-up, months (mean)	NA	NA	40	NA	NA	NA	28	24	17	19	12	23
Intestinal ischemia	20	NA	2	NA	NA	4	10	11 abd pain	22	NA	23 abd pain	9
TIPS placement	0	1	0	26	0	3	8	7	11	10	23	0
Recanal												
complete	3	1	2	26	10	0	9	7	11	22	6	8
Partial	12	1	0	6	1	4	7	2	8	9	15	1
Failure	5	2	0	0	1	0	1	2	3	8	2	0
Gut resection	0	NA	0	NA	NA	1	2	NA	1	4	NA	
Complications	60% 12 major (bleeding, sepsis)	50% 2 major bleeding	0	3% 1 abdominal abscess and MOF	8% 1 resp. distress	75% 2 liver hematoma 1 hemothorax	17% 2 HIT 1 hepatic artery pseudo-neurism	18% 1 HE 1 MOF	40% 8 minor bleeding 1 hepatic hematoma and HE 1 neck hematoma	3 intra-peritoneal hemorrhage 4 cutaneous bleeding 4 liver hematoma	95% 2 hepatic hematoma 18 hemoglobinuria 2 HE	11% 1 DVT and pulmonary embolism
Deaths	1	0	0	1	1	1	0	0	0	1	0	1
Re-thrombosis		NA	0	3	5	3 TIPS occlusion	2 PVT 3 TIPS occlusion 1 portal cavernoma	3 TIPS occlusion 2 cavernoma	3	NA	NA	1

NA not available, *HIT* heparin-induced thrombocytopenia, *MOF* multiorganic failure, *HE* hepatic encephalopaty, *DVT* deep vein thrombosis

Thrombolysis and interventional procedures should be considered if there is symptomatic progression of the thrombus despite medical therapy and when there are features of imminent bowel infarction [19]. Recently, small case series have reported favorable results combining transjugular thrombolysis with TIPS placement (as summarized in Table 52.1).

Interestingly, a stepwise management has been proposed in a recent report including 22 patients [19] with acute PVT and evidence of imminent intestinal ischemia with ongoing symptoms despite a systemic anticoagulation. An initial approach using low dose of systemic alteplase was done, and in patients with ongoing abdominal pain without radiological improvement 48–72 h thereafter, it was followed by TIPS with local thrombolysis (alteplase local infusion) and mechanical thrombolysis. With this stepwise protocol, symptoms resolved in 91% patients and recanalization was achieved in 86%, while only one patient required resection for intestinal ischemia. Major complications occurred in two patients (9%) with no deaths.

Recently, Rössle et al. [20] reported results from an observational study comparing 30 patients with acute PVT who received standard medical treatment vs. 35 patients who received interventional treatment, showing that the interventional group had improved partial and complete recanalization rates but at the cost of increased bleeding complications. Indeed, four patients in each group required bowel resection due to intestinal gangrene.

Unfortunately, these studies have several concerns that hinder their applicability. The selection of patients included is heterogeneous and the treatment and technical approaches are very different among the different studies. Also, one of the main limitations is the lack of a clear definition of absence of response to anticoagulation and, therefore, lack of definition of the optimal time at which the patient would benefit from additional radiological intervention.

However, and despite no strong recommendations can be made, in selected patients with increased risk of intestinal infarction despite anticoagulation (persistent severe abdominal pain, bloody diarrhea, lactic acidosis, bowel loop distention …) early radiological intervention can be considered. Desirably, it requires a multidisciplinary approach in referral centers.

Chronic Portal Vein Thrombosis

PVT is defined as chronic when at least 6 months have passed since the acute event. Careful comparison with prior imaging tests can be useful to determine chronicity given that asymptomatic acute PVT may be unnoticed and undiagnosed. In the absence of recanalization, cavernomatous transformation can occur. Although portal cavernoma is usually a sign of chronicity, it is important to remark that cavernomatous transformation has been reported in as few as 6 days [21] and, thus, it is important not to assume that all cavernomas are chronic PVT. Determining the chronicity of portal vein thrombosis is important because of clinical management implications. A detailed clinical history looking for possible inciting events (surgery, abdominal infection) or possible symptoms that could mark the moment of acute PVT (abdominal pain, new-onset ascites, etc.) can be helpful to establish the timeline.

Management

The management of chronic PVT involves issues related to both portal hypertension and to avoid thrombosis recurrence or progression.

Portal Hypertension Complications

In the presence of portal cavernoma, the most relevant and common complications are related to the development of portal hypertension, while usually liver function is well preserved. These complications are treated analogously to patients with cirrhotic portal hypertension.

- **Esophageal and Gastric Varices (and/or Ectopic Varices).** The probability of developing esophageal and gastric varices requiring primary prophylaxis has been shown to be 13%, 40%, and 50% at 1, 3, and 5 years, respectively [22] and, as in cirrhosis, the risk of variceal bleeding is higher in medium or large varices and in the presence of red signs. Due to the low prevalence of PVT, specific studies aiming at determining adequate strategies for endoscopic screening and management of varices are scarce and currently it is accepted to follow the same recommendations regarding varices screening validated for patients with cirrhosis and portal hypertension.
- **Ascites and Hepatic Encephalopathy.** Ascites, overt encephalopathy, and bacterial infections are rarer than in cirrhosis. In the acute setting, ascites affects up to 40% of patients and is usually transient, while in the chronic setting, although it is rarer than in cirrhosis, it can be a complicating feature of chronic non-cirrhotic PVT due to pre-sinusoidal portal hypertension. Treatment of ascites and hepatic encephalopathy in chronic PVT follows the same recommendations used in cirrhosis.
- **Portal Biliopathy.** Patients with portal cavernoma frequently present with radiological manifestations of portal cavernoma cholangiopathy, such as extrinsic indentation on the bile duct. Magnetic resonance cholangiography is the gold-standard examination for diagnosing portal biliopathy. It is usually asymptomatic but occasionally (5%–30%) can present with cholangitis, obstructive jaundice, or cholecystitis. Specific treatment of portal biliopathy should be only considered in the presence of symptoms (pruritus, cholangitis) and should be endoscopically—based in the presence of severe stenosis and bile stones in the main bile duct. Importantly, the high risk of bleeding from bile duct varices has to be also taken into account before indicating this procedure. Treatment with ursodesoxycholic acid has also proved useful to improve jaundice and diminish the probability of cholangitis.

Anticoagulation

As above mentioned, in the setting of acute PVT, anticoagulation achieves complete recanalization in around 50% of patients, while half of them evolve to cavernoma and chronic PVT. Once PVT evolves to portal cavernoma, anticoagulation is less likely to be effective in achieving recanalization, but it can still prove to be beneficial. Indeed, anticoagulation therapy has been shown to reduce the rate of thrombus progression and of recurrent thrombosis [14, 16, 23]. Promising data from a recent randomized controlled trial show improved recurrent thrombosis-free survival rates in patients with non-cirrhotic PVT without risk factors for thrombosis treated with rivaroxaban 15 mg/day compared with patients not receiving anticoagulation, without increasing the risk of bleeding [24]. Anticoagulation has been also associated with improved survival [23]. Regarding the bleeding risk, several studies have shown that anticoagulation is safe and does not increase the risk of variceal bleeding. Indeed, even in cases where variceal bleeding occurs, anticoagulation does not play a role in the severity of the episode [25, 26].

In current guidelines long-term anticoagulation therapy in patients with chronic PVT, as long as we do not have new data available, is recommended only in those individuals with major underlying thrombophilia risk factors and/or previous history of thrombotic events and/or previous history of intestinal infarction.

Portal Vein Recanalization

The management of variceal bleeding and ascites—the main portal hypertension complications affecting patients with PVT—is based on the recommendations used in patients with cirrhosis. However, despite this approach, patients can present with recurrent complications. Second-line treatments for non-responders to standard treatment are not well defined. TIPS is an effective therapy to reduce portal hypertension and recanalize the portal vein obtaining sustained patency, but when portal cavernoma is present TIPS placement can be challenging and sometimes not feasible. New portal vein recanalization (PVR) techniques have been developed and have emerged as a successful approach in symptomatic patients despite conservative treatment (non-selective beta-blockers, variceal band ligation and/or anticoagulation therapy) [27–36]. Transhepatic and trans-splenic portal vein recanalization—TIPS (PVR-TIPS) was used initially in cirrhotic patients as a tool to increase patient's candidacy for transplant and decrease post-transplant morbidity [37, 38], but its use has been extended to non-cirrhotic non-tumoral PVT, eliminating portal hypertension-related complications also in this population. To evaluate the feasibility of PVR, it must be noted that preoperative imaging with CT scan or MRI lacks diagnostic accuracy and that feasibility is better assessed with pre-procedure portography. In non-cirrhotic PVT, the need of TIPS placement has to be assessed individually as the benefit of shunting the liver in patients without intrahepatic portal hypertension is debatable. However, in patients with extensive intrahepatic obstruction, TIPS can be required to ensure adequate outflow. Similarly, in patients with systematic recurrent portal vein obstruction, TIPS should be considered to improve outcome.

In Table 52.2, we summarize the main studies approaching PVR-TIPS in chronic PVT and we detail its outcomes and main conclusions. Knight et al. [28] reported the largest cohort published until now, including 39 non-cirrhotic patients with extrahepatic portal vein thrombosis. Most patients (77%) presented an underlying thrombophilia while in the remaining patients PVTs were associated to a local factor or were idiopathic. All patients presented with cavernomatous transformation, involving in most cases both the superior mesenteric vein and the splenic vein. In all

Table 52.2 Main studies approaching portal vein recanalization-transjugular intrahepatic portosystemic shunt (PVR-TIPS) in chronic PVT

	Bilbao B 2004	M Senzolo 2006	Fanelli 2011	Luo X 2014	Kallini 2016	Klinger 2018	Enterazi 2020	Kobe A 2021	Knight 2021	Abud A 2021
Technique	Transsplenic transhepatic	Transjugular	Transjugular	Standard TIPS	Transjugular transsplenic	Transjugular	Mesenteric access	Transsplenic	Transjugular Transhepatic Transsplenic Mesenteric	Transjugular
N	6	15	12	15	5	17	3	10	39	1
Follow-up months	23.3	18.1	17.4	45.2	8	22.8	NA	19.3	36	4
Indication	Refractary PH complication	Refractary PH complication	Refractary PH complication	Refractary PH complication	Refractary PH complication	Refractary PH complication	Refractary PH compli-cation	Refractary PH complication.	Refractary PH complic.	Refractary PH complic.
TIPS placement	Yes	Yes	Yes	Yes	Yes	Yes	Yes	90%	Yes	Yes
Recanalization	100%	83.3%	83.3%	73.3%	100%	76.5%	100%	80%	81%	100
Major complications	0	0	25%	13%	NA	11.8%	100%	10%	10%	0
Deaths	0	6%	0	13%	0	6%	0	10%	0	0
Re-thrombosis	33%	30%	25%	13%	0	23%	0	3.3%	30%	0

Refractory PH complications: Refractory portal hypertension complications

cases, the indication of PVR-TIPS was the development of portal-hypertension related complications, with variceal bleeding being the most frequent (61.5%). Several approaches were used in this cohort, including trans-splenic access, trans-hepatic access, and trans-mesenteric vein access. Only 31% of patients could undergo a purely transjugular access procedure. Importantly, this study shows that at 36 months of follow-up 63% of patients were free of primary TIPS thrombosis and that this could be improved to 81% by doing a close follow-up and performing additional interventions such as angioplasty or re-stenting when needed. The rate of severe adverse events was low.

Summing up, the evidence published shows that PVR-TIPS is a challenging but life-changing procedure. PVR-TIPS is currently indicated in patients that have already developed refractory complications of portal hypertension. More evidence assessing the risk/benefit of this procedure will be needed before expanding the indication of PVR-TIPS to asymptomatic compensated patients with portal cavernoma.

References

1. Haddad MC, Clark DC, Sharif HS, al Shahed M, Aideyan O, Sammak BM. MR, CT, and ultrasonography of splanchnic venous thrombosis. Gastrointest Radiol. 1992;17(1):34–40. http://www.ncbi.nlm.nih.gov/pubmed/1544556
2. Kreft B, Strunk H, Flacke S, Wolff M, Conrad R, Gieseke J, et al. Detection of thrombosis in the portal venous system: comparison of contrast-enhanced MR angiography with intraarterial digital subtraction angiography. Radiology. 2000;216(1):86–92. http://www.ncbi.nlm.nih.gov/pubmed/10887231
3. Ueno N, Sasaki A, Tomiyama T, Tano S, Kimura K. Color Doppler ultrasonography in the diagnosis of cavernous transformation of the portal vein. J Clin Ultrasound. 1997;25(5):227–33. http://www.ncbi.nlm.nih.gov/pubmed/9314103
4. Vilgrain V, Condat B, Bureau C, Hakimé A, Plessier A, Cazals-Hatem D, et al. Atrophy-hypertrophy complex in patients with cavernous transformation of the portal vein: CT evaluation. Radiology. 2006;241(1):149–55. http://www.ncbi.nlm.nih.gov/pubmed/16908681
5. Hernández-Gea V, De Gottardi A, Leebeek FWG, Rautou P-E, Salem R, Garcia-Pagan JC. Current knowledge in pathophysiology and management of Budd-Chiari syndrome and non-cirrhotic non-tumoral splanchnic vein thrombosis. J Hepatol. 2019;71(1):175–99. http://www.ncbi.nlm.nih.gov/pubmed/30822449
6. El-Karaksy H, El-Raziky M. Splanchnic vein thrombosis in the mediterranean area in children. Mediterr J Hematol Infect Dis. 2011;3(1):e2011027. http://www.ncbi.nlm.nih.gov/pubmed/21869913
7. Sarin SK, Agarwal SR. Extrahepatic portal vein obstruction. Semin Liver Dis. 2002;22(1):43–58. http://www.ncbi.nlm.nih.gov/pubmed/11928078
8. Plessier A, Darwish-Murad S, Hernandez-Guerra M, Consigny Y, Fabris F, Trebicka J, et al. Acute portal vein thrombosis unrelated to cirrhosis: a prospective multicenter follow-up study. Hepatology. 2010;51(1):210–8. http://www.ncbi.nlm.nih.gov/pubmed/19821530
9. Denninger MH, Chaït Y, Casadevall N, Hillaire S, Guillin MC, Bezeaud A, et al. Cause of portal or hepatic venous thrombosis in adults: the role of multiple concurrent factors. Hepatology. 2000;31(3):587–91. http://www.ncbi.nlm.nih.gov/pubmed/10706547
10. Marchetti M, Castoldi E, Spronk HMH, van Oerle R, Balducci D, Barbui T, et al. Thrombin generation and activated protein C resistance in patients with essential thrombocythemia and polycythemia vera. Blood. 2008;112(10):4061–8. http://www.ncbi.nlm.nih.gov/pubmed/18768782

11. Chait Y, Condat B, Cazals-Hatem D, Rufat P, Atmani S, Chaoui D, et al. Relevance of the criteria commonly used to diagnose myeloproliferative disorder in patients with splanchnic vein thrombosis. Br J Haematol. 2005;129(4):553–60. http://www.ncbi.nlm.nih.gov/pubmed/15877740
12. Janssen HL, Meinardi JR, Vleggaar FP, van Uum SH, Haagsma EB, van Der Meer FJ, et al. Factor V Leiden mutation, prothrombin gene mutation, and deficiencies in coagulation inhibitors associated with Budd-Chiari syndrome and portal vein thrombosis: results of a case-control study. Blood. 2000;96(7):2364–8.
13. Klinger C, Riecken B, Schmidt A, De Gottardi A, Meier B, Bosch J, et al. Transjugular local thrombolysis with/without TIPS in patients with acute non-cirrhotic, non-malignant portal vein thrombosis. Dig Liver Dis. 2017;49(12):1345–52. http://www.ncbi.nlm.nih.gov/pubmed/28733177
14. Condat B, Pessione F, Helene Denninger M, Hillaire S, Valla D. Recent portal or mesenteric venous thrombosis: increased recognition and frequent recanalization on anticoagulant therapy. Hepatology. 2000;32(3):466–70. http://www.ncbi.nlm.nih.gov/pubmed/10960436
15. Turnes J, García-Pagán JC, González M, Aracil C, Calleja JL, Ripoll C, et al. Portal hypertension-related complications after acute portal vein thrombosis: impact of early anticoagulation. Clin Gastroenterol Hepatol. 2008;6(12):1412–7. http://www.ncbi.nlm.nih.gov/pubmed/19081529
16. Amitrano L, Guardascione MA, Scaglione M, Pezzullo L, Sangiuliano N, Armellino MF, et al. Prognostic factors in noncirrhotic patients with splanchnic vein thromboses. Am J Gastroenterol. 2007;102(11):2464–70.
17. De Gottardi A, Trebicka J, Klinger C, Plessier A, Seijo S, Terziroli B, et al. Antithrombotic treatment with direct-acting oral anticoagulants in patients with splanchnic vein thrombosis and cirrhosis. Liver Int. 2017;37(5):694–9.
18. Nery F, Valadares D, Morais S, Gomes MT, De Gottardi A. Efficacy and safety of direct-acting oral anticoagulants use in acute portal vein thrombosis unrelated to cirrhosis. Gastroenterology Res. 2017;10(2):141–3. http://www.ncbi.nlm.nih.gov/pubmed/28496539
19. Benmassaoud A, AlRubaiy L, Yu D, Chowdary P, Sekhar M, Parikh P, et al. A stepwise thrombolysis regimen in the management of acute portal vein thrombosis in patients with evidence of intestinal ischaemia. Aliment Pharmacol Ther. 2019;50(9):1049–58. http://www.ncbi.nlm.nih.gov/pubmed/31489698
20. Rössle M, Bettinger D, Trebicka J, Klinger C, Praktiknjo M, Sturm L, et al. A prospective, multicentre study in acute non-cirrhotic, non-malignant portal vein thrombosis: comparison of medical and interventional treatment. Aliment Pharmacol Ther. 2020;52(2):329–39. http://www.ncbi.nlm.nih.gov/pubmed/32506456
21. De Gaetano AM, Rinaldi P, Barbaro B, Mirk P, Di Stasi C, Gui B, et al. Intrahepatic portosystemic venous shunts: color Doppler sonography. Abdom Imaging. 2007;32(4):463–9. http://www.ncbi.nlm.nih.gov/pubmed/17334878
22. Noronha Ferreira C, Seijo S, Plessier A, Silva-Junior G, Turon F, Rautou P-E, et al. Natural history and management of esophagogastric varices in chronic noncirrhotic, nontumoral portal vein thrombosis. Hepatology. 2016;63(5):1640–50. http://www.ncbi.nlm.nih.gov/pubmed/26799606
23. Orr DW, Harrison PM, Devlin J, Karani JB, Kane PA, Heaton ND, et al. Chronic mesenteric venous thrombosis: evaluation and determinants of survival during long-term follow-up. Clin Gastroenterol Hepatol. 2007;5(1):80–6. http://www.ncbi.nlm.nih.gov/pubmed/17142105
24. Plessier A, Goria O, Al E. Prophylaxis of recurrent thrombosis by rivaroxaban in patients with non-cirrhotic portal vein thrombosis a multicentre randomized controlled study testing rivaroxaban vs no anticoagulation. J Hepatol. 2021;75(Suppl 2):S191–865.
25. Condat B, Pessione F, Hillaire S, Denninger MH, Guillin MC, Poliquin M, et al. Current outcome of portal vein thrombosis in adults: risk and benefit of anticoagulant therapy. Gastroenterology. 2001;120(2):490–7. http://www.ncbi.nlm.nih.gov/pubmed/11159889
26. Spaander MCW, Hoekstra J, Hansen BE, Van Buuren HR, Leebeek FWG, Janssen HLA. Anticoagulant therapy in patients with non-cirrhotic portal vein thrombosis: effect on

new thrombotic events and gastrointestinal bleeding. J Thromb Haemost. 2013;11(3):452–9. http://www.ncbi.nlm.nih.gov/pubmed/23289370
27. Abud A, Maddur H, Salem R. Management of symptomatic portal cavernoma cholangiopathy with transplenic portal vein recanalization and transjugular intrahepatic portosystemic shunt. Hepatology. 2021;73(1):456–9. http://www.ncbi.nlm.nih.gov/pubmed/32500579
28. Knight GM, Clark J, Boike JR, Maddur H, Ganger DR, Talwar A, et al. TIPS for adults without cirrhosis with chronic mesenteric venous thrombosis and EHPVO refractory to standard-of-care therapy. Hepatology. 2021;74(5):2735–44. http://www.ncbi.nlm.nih.gov/pubmed/34021505
29. Kobe A, Puippe G, Müllhaupt B, Pfammatter T. Recanalization of chronic noncirrhotic, nonmalignant splanchnic thromboses is feasible: a transsplenic assisted patient-tailored approach. J Vasc Interv Radiol. 2021;32(9):1377–85. http://www.ncbi.nlm.nih.gov/pubmed/34462082
30. Entezari P, Riaz A, Thornburg B, Salem R. Percutaneous ultrasound-guided superior and inferior mesenteric vein access for portal vein recanalization-transjugular intrahepatic portosystemic shunt: a case series. Cardiovasc Intervent Radiol. 2021;44(3):496–9. http://www.ncbi.nlm.nih.gov/pubmed/33230650
31. Klinger C, Riecken B, Schmidt A, De Gottardi A, Meier B, Bosch J, et al. Transjugular portal vein recanalization with creation of intrahepatic portosystemic shunt (PVR-TIPS) in patients with chronic non-cirrhotic, non-malignant portal vein thrombosis. Z Gastroenterol. 2018;56(3):221–37. http://www.ncbi.nlm.nih.gov/pubmed/29113006
32. Kallini JR, Gabr A, Kulik L, Ganger D, Lewandowski R, Thornburg B, et al. Noncirrhotic complete obliterative portal vein thrombosis: Novel management using trans-splenic transjugular intrahepatic portosystemic shunt with portal vein recanalization. Hepatology. 2016;63(4):1387–90. http://www.ncbi.nlm.nih.gov/pubmed/2670923
33. Luo X, Nie L, Zhou B, Yao D, Ma H, Jiang M, et al. Transjugular intrahepatic portosystemic shunt for the treatment of portal hypertension in noncirrhotic patients with portal cavernoma. Gastroenterol Res Pract. 2014;2014:659726. http://www.ncbi.nlm.nih.gov/pubmed/24868203
34. Fanelli F, Angeloni S, Salvatori FM, Marzano C, Boatta E, Merli M, et al. Transjugular intrahepatic portosystemic shunt with expanded-polytetrafuoroethylene-covered stents in non-cirrhotic patients with portal cavernoma. Dig Liver Dis. 2011;43(1):78–84. http://www.ncbi.nlm.nih.gov/pubmed/20637712
35. Senzolo M, Tibbals J, Cholongitas E, Triantos CK, Burroughs AK, Patch D. Transjugular intrahepatic portosystemic shunt for portal vein thrombosis with and without cavernous transformation. Aliment Pharmacol Ther. 2006;23(6):767–75. http://www.ncbi.nlm.nih.gov/pubmed/16556179
36. Bilbao JI, Elorz M, Vivas I, Martínez-Cuesta A, Bastarrika G, Benito A. Transjugular intrahepatic portosystemic shunt (TIPS) in the treatment of venous symptomatic chronic portal thrombosis in non-cirrhotic patients. Cardiovasc Intervent Radiol. 2004;27(5):474–80. http://www.ncbi.nlm.nih.gov/pubmed/15383850
37. Salem R, Vouche M, Baker T, Herrero JI, Caicedo JC, Fryer J, et al. Pretransplant portal vein recanalization-transjugular intrahepatic portosystemic shunt in patients with complete obliterative portal vein thrombosis. Transplantation. 2015;99(11):2347–55.
38. Thornburg B, Desai K, Hickey R, Hohlastos E, Kulik L, Ganger D, et al. Pretransplantation portal vein recanalization and transjugular intrahepatic portosystemic shunt creation for chronic portal vein thrombosis: final analysis of a 61-patient cohort. J Vasc Interv Radiol. 2017;28(12):1714–1721.e2. https://linkinghub.elsevier.com/retrieve/pii/S1051044317307546

53 Staging of Portal Vein Thrombosis: Recurrent Thrombosis and Prognostic Factors for Recurrence in Non-Cirrhotic Non-Tumoral Portal Vein Thrombosis (PVT)

A. Plessier and A. Shukla

Abbreviations

DVT	Deep vein thrombosis
LT	Liver transplantation
PVT	Portal vein thrombosis
PV	Portal vein
SV	Splenic
SMV	Superior mesenteric vein

Introduction

Portal vein thrombosis (PVT) is defined as the formation of a complete or partial non-tumoral thrombus within the main portal vein (PV) or left or right branches, which may also extend into the splenic (SV) and/or superior mesenteric vein

A. Plessier (✉)
Université de Paris cité, AP-HP, Hôpital Beaujon, Service d'Hépatologie, DMU DIGEST, Centre de Référence des Maladies Vasculaires du Foie, FILFOIE, Valdig an EASL consortium, ERN RARE-LIVER, Centre de recherche sur l'inflammation, Inserm, UMR 1149, Paris, France
e-mail: aurelie.plessier@aphp.fr

A. Shukla
Department of Gastroenterology, Seth GS Medical College & KEM Hospital, Mumbai, India

R. de Franchis (ed.), *Portal Hypertension VII*,
https://doi.org/10.1007/978-3-031-08552-9_53

(SMV) [1, 2]. PVT may occur as a complication of cirrhosis or hepatocellular carcinoma. Tumoral invasion of the portal vein may also occur in HCC patients. These clinical entities are different with respect to pathogenesis, treatment, and prognosis and will not be discussed in this manuscript. We will discuss only non-cirrhotic non-malignant portal vein thrombosis. Non-cirrhotic portal vein thrombosis is diagnosed either at a recent stage (the so-called recent or acute) or at a chronic stage. These two stages of the same disease share similar causes [3]. Portal cavernoma defines the presence of porto-portal collaterals as a sequel of portal vein obstruction. Portal cavernoma is generally used to refer to chronic portal vein thrombosis, but it should be noted that it can be present, early after PVT occurrence. The prevalence of non-cirrhotic, non-malignant portal vein thrombosis is about 2.5‰ in the general population. The natural course of PVT probably differs according to the underlying condition (cirrhosis vs. non -cirrhosis, cancer, and other major or low-risk factors for thrombosis) [4]. A standardized documentation of initial site, extent, degree of luminal obstruction, and chronicity of clot formation is required as a starting point for course evaluation.

Staging/Classification of Portal Vein Thrombosis

There are multiple staging systems of PVT, based broadly on anatomical location, degree of obstruction, the presence or absence of collaterals, functional implications, or combinations of these. The initial classifications, mainly described in cirrhosis, were based on anatomical involvement (Table 53.1) [5–10]. Yerdel's classification initially described for surgical management at the time of liver transplantation is commonly used in the setting of liver transplantation (LT) [5]. Long-term data in the context of liver transplantation are lacking. Other grading systems also considered collaterals and cavernoma formation. The third category involved additionally functional variables like duration of clot, the presence of symptoms, underlying liver disease, and degree of portal hypertension. The following classifications are described according to anatomic (Table 53.1) [5–10] or functional classifications (Table 53.2) [2, 11–13].

The last classification recommended in AASLD guidelines aims at having a simple serial assessment of splanchnic veins obstruction in order to assess relevant correlation with or without therapy [2].

Table 53.1 Anatomic classifications of portal vein thrombosis [5–10]

Author and year	Classification	New input
Stieber 1991	**Type A:** Segment of main PV **Type B:** Main PV and SMV **Type C**: Main PV and SV or main PV, SMV, SV, and IMV	
Nonami 1992	**Grade 1:** Thrombosis of intrahepatic portal vein branches **Grade 2:** Thrombosis of the right or left portal branch or at the bifurcation **Grade 3**: Partial obstruction of the portal vein trunk **Grade 4**: Complete obstruction of the portal vein trunk	Considered PV branch involvement
Gayowski 1996	**Grade 1:** Mural thrombosis of main portal trunk (± extension below) with residual flow **Grade 2:** Complete thrombosis of the main portal trunk not extending to the confluence **Grade 3:** Complete thrombosis of the main portal trunk extending to the confluence **Grade 4:** Complete thrombosis of the main portal trunk with extension below the confluence	Considered extension below confluence
Yerdel 2000	**Grade 1**: Minimally or partially thrombosed portal vein (PV), in which the thrombus is mild or, at the most, confined to <50% of the vessel lumen with or without minimal extension into the superior mesenteric vein (SMV) **Grade 2**: >50% occlusion of the PV, including total occlusions, with or without minimal extension into the SMV **Grade 3**: Complete thrombosis of both PV and proximal SMV. Distal SMV is open **Grade 4:** Complete thrombosis of the PV and proximal as well as distal SMV	Lumen obstruction minimal vs. complete
Jamieson 2000	**Type 1:** Thrombosis confined to the portal vein beyond the confluence of the SV and SMV **Type 2:** Extension of thrombus into the SMV but with patent mesenteric vessels **Type 3:** Diffuse thrombosis of splanchnic venous system but with large collaterals **Type 4:** Extensive splanchnic venous thrombosis but with only thin collaterals	
Bhangui 2019	Simple vs. complex based on anatomical factors and presence of collaterals **Simple**: Yerdel grade 1–3; having a partial or complete involvement restricted to the portal vein and/or the very distal part of the SV and/or the SM **Complex:** Yerdel grade 4 and Jamieson and Charco grades 3 and 4	Adds notion: – simple vs. complex – Collaterality – stratifies strategy for portal inflow reconstruction

Table 53.2 Anatomo-functional classifications of portal vein thrombosis [2, 11–13]

Anatomic and functional aspects of PVT	Includes non-cirrhotic PVT	
Bauer 2006	Clot burden in mesenteric, portal, and splenic veins and graded according to occlusion **Grade I**: Less than 25% **Grade II:** 26%–50% **Grade III:** 51%–75% **Grade IV:** 76%–100% It is also stratified according to clot location and cavernous transformation	For TIPS insertion
Ma 2014	1. Onset of symptoms – Abdominal distension or pain, nausea, vomiting, diarrhea, anorexia and fever 2. Duration of clot – PVT was defined as acute if the symptoms are within <60 days without evidence of portal cavernoma formation and portal hypertension complications 3. Presence of complications of portal hypertension including splenomegaly, gastroesophageal variceal bleeding, and ascites	Based on radiological contrast-enhanced computed tomography [CT]
Sarin 2016	**Site of PVT** – (type 1, 2a, 2b, 3) type 1: Only trunk type 2: Only branch: 2a, one branch; 2b, both branches type 3: Trunk and branches **Degree of portal venous system occlusion** O: Occlusive: No flow visible in PV lumen on imaging/ Doppler study NO: Nonocclusive: Flow visible in PV lumen through imaging/Doppler study **Duration and presentation (R, C)** **R: Recent** (first time detected in previously patent PV, presence of hyperdense thrombus on imaging, absent or limited collateral circulation, dilated PV at the site of occlusion) Asymptomatic: (as) Symptomatic: (S), acute PVT features (with or without ABI) **C: Chronic** (no hyperdense thrombus; previously diagnosed PVT on follow-up, portal cavernoma and clinical features of PHT) Asymptomatic Symptomatic: Features of portal hypertension (with or without PHT) **Extent of PV system occlusion (S, M, SM)** splenic vein, mesenteric vein or both **Type and presence of underlying liver disease**: Cirrhotic, noncirrhotic liver disease, post-liver transplant, HCC, local malignancies, and associated conditions	

Table 53.2 (continued)

Anatomic and functional aspects of PVT	Includes non-cirrhotic PVT	
Northup 2021	**Time course** Recent (<6 months) Chronic (>6 months) **Percent occlusion of main PV** Completely occlusive no persistent lumen Partially occlusive clot >50% of lumen Minimally occlusive clot <50% of lumen Cavernous transformation gross Porto-portal collaterals without original PV seen **Response to treatment or interval change** Progressive: Increases in size or progresses to more complete occlusion Stable: No appreciable change in size or occlusion Regressive: Decreases in size or degree of occlusion	Includes clinical outcome correlation with treatment

Recurrent Thrombosis and Prognostic Factors for Recurrence in Non-Cirrhotic Non-Tumoral PVT

Recurrent Deep Vein Thrombosis (DVT)

Studies assessing the impact of permanent anticoagulation vs. 3-month anticoagulation in deep vein thrombosis are based on estimating a balance between reduction in recurrent DVT and increase in major bleeding. An estimation of both indicators will then determine recommendations for permanent anticoagulation. Newly diagnosed DVT occur in around 5 per 10,000 of the whole population per year (i.e., 1 per 2000) of whom 2 per 10,000 are idiopathic. Trauma is associated with a 13-fold increase and malignancy with a fivefold increase in risk and oral contraceptives or hormone replacement therapy is associated with a risk around 2 ± 4-fold compared to non-use. Inherited or acquired thrombophilia may increase risk very substantially. In a prospective follow-up study of 738 consecutive DVT patients, the 5-year cumulative incidence of recurrent venous thromboembolic events is 21.5% (95% confidence interval [CI], 17.7%–25.4%) after a first DVT and 27.9% (95% CI, 19.7%–36.1%) after a second DVT [14]. The annual incidence of recurrent thrombosis gradually increases up to 30% after 8-year follow-up [15, 16]. Several scores and models are currently used to predict recurrence in DVT [17–20]. Moreover, various factors found to be predictive of the recurrence for DVT and pulmonary embolism are enumerated in Table 53.3.

The risk of DVT recurrence is higher in the following:

- Idiopathic events: 10% per year the first 2 years with an odds ratio of 2.4.
- In male subjects: before the age of 60, with an odds ratio of 2–4.
- In patients with persistently elevated D-dimer level (odds ratio of 2.3, compared with normal level) [21].

Table 53.3 Predictive factors of the recurrence for DVT and pulmonary embolism

Factor	Description	Salient features
D-dimer	In patients with unprovoked VTE, stopping anticoagulants after at least 3 months, a negative D-dimer result has 3.5% annual risk for recurrent disease, whereas a positive D-dimer result has 8.9% annual risk for recurrence [27]	D-dimer is assessed at least 4 weeks after stopping anticoagulation. Randomized controlled trials and meta-analyses have given consistent results [28]
Factor VIII	High levels have a RR of 5.43 of recurrence in patients with idiopathic VTE. The RR increases with the increase in plasma level of factor VIII [29–31]	Prolonging anticoagulation (if high factor VIII levels) reduces risk of recurrence [32]
D-dimer + Prothrombin fragment 1& 2 (1 + 2 + residual venous obstruction	Abnormally high 1 + 2 levels are associated with increased risk of VTE recurrence (OR 2.4) similar to high d-dimer (OR 3.1) and combined elevated 1 + 2 & D-dimer (OR 4.3) [33]	Residual venous obstruction and presence of thrombophilic disorders are not associated with VTE recurrence [33]
Thrombin generation (TG)	Endogenous thrombin potential >960 nM-min (ETP; adjusted HR in the presence and absence of thrombomodulin 3.27 and 2.37, respectively), peak thrombin >193 nM-min (HR 4.36 and 2.56) and lag time < 14.5 min (HR 3.11 and 2.91) [34]	Large overlap in thrombin generation is seen between patients with and without recurrences [35] Advice to individual patients on anticoagulation based on TG is unreliable
Clot phenotype	Dense fiber networks displaying reduced plasmin-induced lysability, low clot permeability, and increased clot lysis time [36]	Mechanisms still not elucidated. Clots from recurrent VTE patients are less elastic and viscous [37]
Fibrinogen	High fibrinogen (>4.00 g/L) level in the presence of low HDL-C (<1.08 mmol/L) only at baseline predicted recurrent cerebral venous thrombosis (HR4.69) [38]	This was the only predictor found in this study
Factor IX (FIX) levels	Recurrence of VTE is 23% at 3 years in those with elevated FIX. Adjusted RR is 1.08 for every 10 IU/dL rise in the FIX level [39]	The RR for recurrence is 1.6
Genetic markers	Gene encoding for acyl-CoA Synthetase family member 2(ACSF2), is under expressed in those with recurrent DVT and across both genders [40]	This was a proof-of-concept study and needs to be validated
Complements	Role of complements in triggering thrombosis and re-thrombosis is being increasingly recognized. Plasmin, rather than thrombin efficiently, generates C5a in the process of venous clot formation [41]	Activity in the highest quintile of classical compliment pathway can trigger VTE [42]

- During the first 2 years after discontinuation of treatment.
- The presence of cancer and of impaired coagulation inhibition increase the risk for recurrent venous thromboembolism [15].

When a transient risk factor is identified, the risk for recurrent venous thromboembolism seems to be decreased [15, 16].

Therefore, provoking factors for DVT, localization of DVT, past personal or familial history of DVT, male sex and d-dimer, thrombophilia, and residual thrombus balanced to the bleeding risk are indicators that will influence the decision to continue or stop anticoagulation after 3 months [22–26].

Recurrence of Non-Cirrhotic PVT

Prothrombotic disorders are found in 60% to 75% of non-cirrhotic PVT [1, 2, 43–45]. A hereditary or acquired prothrombotic condition (Table 53.1) is found in approximately 10%–50% of cases and an exogenous risk factor for venous thrombosis (hormonal, local cause, …) is present in approximately 6%–30% of cases. Several factors are associated in the same patient in 10% of cases, which is more frequent than expected by chance alone. In expert recommendations for deep vein thrombosis, major risk associated causes are myeloproliferative neoplasms, antiphospholipid syndrome or homozygous or composite heterozygous G20210A factor II and G1691A factor V mutations, personal or first degree unprovoked family history of venous thrombosis. The benefit of permanent anticoagulation in patients with low-risk factors for thrombosis is unknown. Patients with recent or acute PVT receive anticoagulant therapy for 6 months in the absence of severe bleeding contraindication [44, 46]. Currently, permanent anticoagulation is recommended in patients harboring major risk factors or past history of gut resection for mesenteric infarction [1, 2, 47].

Recurrence risk is also balanced by the duration of anticoagulation. In a multicenter prospective registry of splanchnic vein thrombosis (SVT), 138 out of 177 patients with incidentally detected splanchnic vein thrombosis had PVT, while 322 of 420 patients with clinically suspected splanchnic vein thrombosis had PVT [48]. Twenty of the 177 patients with incidentally detected splanchnic vein thrombosis had recurrent splanchnic thrombosis. The incidence of thrombosis on treatment was calculated to be 3.9 events (1.6–9.5) per 100 patient-years; and off-treatment was 11.9 events (5.0–28.7) per 100 patient-years after treatment discontinuation and 11.5 events (6.2–21.3) per 100 patient-years in patients who never received treatment. Twenty-one of the 420 patients with clinically suspected splanchnic vein thrombosis had recurrent splanchnic thrombosis. Each month of anticoagulant treatment was associated with lower rates of thrombotic events (HR 0.85, 95% CI 0.76–0.96, $p = 0.0084$).

The rationale for permanent anticoagulation in major risk factor PVT relies on multiple myeloproliferative neoplasm studies, although all uncontrolled and retrospective:

- In a European cohort of 604 patients with splanchnic vein thrombosis, including 28% of cirrhotic patients, a thrombotic event occurred in 7.3 per 100 patient-years. In 465 patients who had anticoagulation the rate was 5.6 per 100 patients-years. The authors observed a high recurrent thrombotic rate of 10.5 per 100 patient-years (6.8–16.3) after anticoagulation discontinuation, whereas the recurrence rate while on VKA was still 3.9 per 100 patient-years. Common risk factors for SVT in this population were MPN in 20%, cancer in 22%, and cirrhosis in 18%.The case fatality rate of thrombotic events was 13.2% (95%CI, 6.60%–24.15%). Male sex, solid cancer, myeloproliferative neoplasms, and unprovoked SVT were associated with an increased risk of vascular events [49].
- In a European cohort of 181 patients with myeloproliferative neoplasms and splanchnic vein thrombosis, the recurrent thrombosis incidence rate was 4.2 per 100 pts-years. It was reduced to 3.9 per 100 pts-years in patients treated with VKA:, whereas in the small fraction (15%) not receiving VKA it was as high as 7.2 per 100 pts-years [50].
- In a cohort of 44 PVT/MPN patients with a long-term follow-up (median 5.8 years (0.4–21 years), recurrent thrombosis occurred in 12 patients (24%). It was the cause of death in 18% of the patients [51].
- In the retrospective study by Condat et al., which included 136 patients with acute or chronic portal venous thrombosis followed for 4 years, 84 patients received anticoagulant treatment, 30 of whom received treatment for a limited time only. Fifty-two patients did not receive anticoagulant treatment. Thirty-six thrombotic events were reported in 26 patients, including 18 lower limb venous thromboses, 5 pulmonary embolisms, 8 mesenteric venous infarctions, and 1 splenic infarction. In multivariate analysis, only the absence of thrombophilia and the presence of anticoagulant treatment were significantly associated with survival without new thrombotic events. A history of thrombosis, age, gender, and age of thrombosis were not predictive of thrombosis recurrence. With documented thrombophilia, the risk of recurrence in the portal system and of mesenteric or splenic infarction was 0.82 and 5.19/100 BP with and without anticoagulants, respectively (RR 6.3, 95% CI 1.3–30.4; $p = 0.01$). Only two patients had a mesenteric infarction on anticoagulants. Overall, this study suggests a benefit of anticoagulant treatment on the risk of thrombosis in patients with thrombophilia without increasing the risk of bleeding (Table 53.4) [52].
- In a small Irish cohort of 14 SVT patients, recurrence of SVT occurred mostly in the setting of interventional liver procedures. Recurrent thrombosis outside of the splanchnic venous system occurred in 28.5% of patients, predominantly off therapeutic anticoagulation [53].

Table 53.4 Recurrence of thrombotic events in two studies

Recurrent thrombosis incidence per 100 patient-years	Ageno $N = 604$ [49]	Condat $N = 136$[a] [52]
Total	7.3	5.5
With anticoagulation	5.6	3.8/0.82[a]
Anticoagulation discontinuation	10.5	6.3/5.19[a]

[a] With documented thrombophilia

- Similarly, patients with active cancer were at increased risk for VTE recurrences (HR: 3.06; 95%CI: 1.14–8.17).

Data on PVT recurrence risk in patients with low-risk factors for recurrence were up to now limited. A recent unpublished multicenter, open-label-randomized controlled trial assessed the effects of rivaroxaban 15 mg/day, on the risk of recurrent venous thrombosis in chronic PVT patients with no major risk factors for thrombosis. In 111 included patients, the incidence of thrombosis was 0 per 100 person-years (PY) in the rivaroxaban group, and 19.71 per 100 PY (95% CI [7.49–31.92]) in the no-anticoagulation group (log-rank p-value = 0.0008). After a median follow-up of 30.3 months (95%CI [29.8–35.9]), major bleeding occurred in two patients receiving rivaroxaban, and in one without anticoagulant. D-dimers concentration were identified as specific biomarkers to predict the risk of recurrence with a negative predictive value of 94%. There were no deaths [54]. Data from the RIETE registry included 521 patients with SVT, of which, 212 (41%) had symptomatic and the rest incidental thrombosis [55]. Seven patients developed SVT recurrence on anticoagulation therapy. On multivariable analysis, patients with incidental SVT had a slightly higher risk for symptomatic VTE recurrences although the difference did not reach statistical significance (adjusted hazard ratio [HR]: 2.04; 95%CI: 0.71–5.88).

Conclusion

Homogeneous staging in cirrhotic and non-cirrhotic PVT including time course, extension in diameter, and length and outcome over time is needed to assess natural history and response to therapy. Currently, acute PVT therapeutics include treatment of a cause and anticoagulation for 6 months. Long-term anticoagulation may be needed when recurrent thrombotic risk exceeds bleeding risk and should be periodically reassessed. Recurrent thrombosis seems high in patients with major prothrombotic risk factors and even higher in low-risk/idiopathic PVT, and anticoagulation with rivaroxaban 15 mg/day reduces the risk in the last population. When a transient risk factor (local or estrogen containing pill) is identified, the risk for recurrent venous thromboembolism seems to be decreased.

References

1. European Association for the Study of the Liver. EASL Clinical Practice Guidelines: vascular diseases of the liver. J Hepatol. 2016;64:179–202. https://doi.org/10.1016/j.jhep.2015.07.040.
2. Northup PG, Garcia-Pagan JC, Garcia-Tsao G, Intagliata NM, Superina RA, Roberts LN, et al. Vascular liver disorders, portal vein thrombosis, and procedural bleeding in patients with liver disease: 2020 practice guidance by the American Association for the Study of Liver Diseases. Hepatology. 2021;73:366–413. https://doi.org/10.1002/hep.31646.
3. Condat B, Pessione F, Helene Denninger M, Hillaire S, Valla D. Recent portal or mesenteric venous thrombosis: increased recognition and frequent recanalization on anticoagulant therapy. Hepatology. 2000;32:466–70. https://doi.org/10.1053/jhep.2000.16597.

4. Plessier A, Rautou P-E, Valla D-C. Management of hepatic vascular diseases. J Hepatol. 2012;56(Suppl 1):S25–38. https://doi.org/10.1016/S0168-8278(12)60004-X.
5. Yerdel MA, Gunson B, Mirza D, Karayalçin K, Olliff S, Buckels J, et al. Portal vein thrombosis in adults undergoing liver transplantation: risk factors, screening, management, and outcome. Transplantation. 2000;69:1873–81. https://doi.org/10.1097/00007890-200005150-00023.
6. Jamieson NV. Changing perspectives in portal vein thrombosis and liver transplantation. Transplantation. 2000;69:1772–4. https://doi.org/10.1097/00007890-200005150-00006.
7. Bhangui P, Lim C, Levesque E, Salloum C, Lahat E, Feray C, et al. Novel classification of non-malignant portal vein thrombosis: a guide to surgical decision-making during liver transplantation. J Hepatol. 2019;71:1038–50. https://doi.org/10.1016/j.jhep.2019.08.012.
8. Gayowski TJ, Marino IR, Doyle HR, Echeverri L, Mieles L, Todo S, et al. A high incidence of native portal vein thrombosis in veterans undergoing liver transplantation. J Surg Res. 1996;60:333–8. https://doi.org/10.1006/jsre.1996.0053.
9. Nonami T, Yokoyama I, Iwatsuki S, Starzl TE. The incidence of portal vein thrombosis at liver transplantation. Hepatology. 1992;16:1195–8.
10. Stieber AC, Zetti G, Todo S, Tzakis AG, Fung JJ, Marino I, et al. The spectrum of portal vein thrombosis in liver transplantation. Ann Surg. 1991;213:199–206. https://doi.org/10.1097/00000658-199103000-00003.
11. Bauer J, Johnson S, Durham J, Ludkowski M, Trotter J, Bak T, et al. The role of TIPS for portal vein patency in liver transplant patients with portal vein thrombosis. Liver Transpl. 2006;12:1544–51. https://doi.org/10.1002/lt.20869.
12. Ma J, Yan Z, Luo J, Liu Q, Wang J, Qiu S. Rational classification of portal vein thrombosis and its clinical significance. PLoS One. 2014;9:e112501. https://doi.org/10.1371/journal.pone.0112501.
13. Sarin SK, Philips CA, Kamath PS, Choudhury A, Maruyama H, Nery FG, et al. Toward a comprehensive new classification of portal vein thrombosis in patients with cirrhosis. Gastroenterology. 2016;151:574–577.e3. https://doi.org/10.1053/j.gastro.2016.08.033.
14. Hansson PO, Sörbo J, Eriksson H. Recurrent venous thromboembolism after deep vein thrombosis: incidence and risk factors. Arch Intern Med. 2000;160:769–74. https://doi.org/10.1001/archinte.160.6.769.
15. Prandoni P, Lensing AW, Cogo A, Cuppini S, Villalta S, Carta M, et al. The long-term clinical course of acute deep venous thrombosis. Ann Intern Med. 1996;125:1–7. https://doi.org/10.7326/0003-4819-125-1-199607010-00001.
16. Baglin T, Luddington R, Brown K, Baglin C. Incidence of recurrent venous thromboembolism in relation to clinical and thrombophilic risk factors: prospective cohort study. Lancet. 2003;362:523–6. https://doi.org/10.1016/S0140-6736(03)14111-6.
17. Timp JF, Braekkan SK, Lijfering WM, van Hylckama VA, Hansen J-B, Rosendaal FR, et al. Prediction of recurrent venous thrombosis in all patients with a first venous thrombotic event: the Leiden thrombosis recurrence risk prediction model (L-TRRiP). PLoS Med. 2019;16:e1002883. https://doi.org/10.1371/journal.pmed.1002883.
18. Eichinger S, Heinze G, Jandeck LM, Kyrle PA. Risk assessment of recurrence in patients with unprovoked deep vein thrombosis or pulmonary embolism: the Vienna prediction model. Circulation. 2010;121:1630–6. https://doi.org/10.1161/CIRCULATIONAHA.109.925214.
19. Tosetto A, Iorio A, Marcucci M, Baglin T, Cushman M, Eichinger S, et al. Predicting disease recurrence in patients with previous unprovoked venous thromboembolism: a proposed prediction score (DASH). J Thromb Haemost. 2012;10:1019–25. https://doi.org/10.1111/j.1538-7836.2012.04735.x.
20. Rodger MA, Kahn SR, Wells PS, Anderson DA, Chagnon I, Le Gal G, et al. Identifying unprovoked thromboembolism patients at low risk for recurrence who can discontinue anticoagulant therapy. CMAJ. 2008;179:417–26. https://doi.org/10.1503/cmaj.080493.
21. Avnery O, Martin M, Bura-Riviere A, Barillari G, Mazzolai L, Mahé I, et al. D-dimer levels and risk of recurrence following provoked venous thromboembolism: findings from the RIETE registry. J Intern Med. 2020;287:32–41. https://doi.org/10.1111/joim.12969.

22. Douketis J, Tosetto A, Marcucci M, Baglin T, Cushman M, Eichinger S, et al. Patient-level meta-analysis: effect of measurement timing, threshold, and patient age on ability of D-dimer testing to assess recurrence risk after unprovoked venous thromboembolism. Ann Intern Med. 2010;153:523–31. https://doi.org/10.7326/0003-4819-153-8-201010190-00009.
23. Adeboyeje G, Sylwestrzak G, Barron JJ, White J, Rosenberg A, Abarca J, et al. Major bleeding risk during anticoagulation with Warfarin, Dabigatran, Apixaban, or Rivaroxaban in patients with nonvalvular atrial fibrillation. J Manag Care Spec Pharm. 2017;23:968–78. https://doi.org/10.18553/jmcp.2017.23.9.968.
24. Marchiori A, Mosena L, Prins MH, Prandoni P. The risk of recurrent venous thromboembolism among heterozygous carriers of factor V Leiden or prothrombin G20210A mutation. A systematic review of prospective studies. Haematologica. 2007;92:1107–14. https://doi.org/10.3324/haematol.10234.
25. Bounameaux H, Perrier A. Duration of anticoagulation therapy for venous thromboembolism. Hematology Am Soc Hematol Educ Program. 2008:252–8. https://doi.org/10.1182/asheducation-2008.1.252.
26. Kearon C, Akl EA, Comerota AJ, Prandoni P, Bounameaux H, Goldhaber SZ, et al. Antithrombotic therapy for VTE disease: antithrombotic therapy and prevention of thrombosis, 9th ed: American College of Chest Physicians Evidence-Based Clinical Practice Guidelines. Chest. 2012;141:e419S–96S. https://doi.org/10.1378/chest.11-2301.
27. Verhovsek M, Douketis JD, Yi Q, Shrivastava S, Tait RC, Baglin T, et al. Systematic review: D-dimer to predict recurrent disease after stopping anticoagulant therapy for unprovoked venous thromboembolism. Ann Intern Med. 2008;149:481–90., W94. https://doi.org/10.7326/0003-4819-149-7-200810070-00008.
28. Palareti G, Cosmi B, Legnani C, Tosetto A, Brusi C, Iorio A, et al. D-dimer testing to determine the duration of anticoagulation therapy. N Engl J Med. 2006;355:1780–9. https://doi.org/10.1056/NEJMoa054444.
29. Cristina L, Benilde C, Michela C, Mirella F, Giuliana G, Gualtiero P. High plasma levels of factor VIII and risk of recurrence of venous thromboembolism. Br J Haematol. 2004;124:504–10. https://doi.org/10.1046/j.1365-2141.2003.04795.x.
30. Kyrle PA, Minar E, Hirschl M, Bialonczyk C, Stain M, Schneider B, et al. High plasma levels of factor VIII and the risk of recurrent venous thromboembolism. N Engl J Med. 2000;343:457–62. https://doi.org/10.1056/NEJM200008173430702.
31. Timp JF, Lijfering WM, Flinterman LE, van Hylckama VA, le Cessie S, Rosendaal FR, et al. Predictive value of factor VIII levels for recurrent venous thrombosis: results from the MEGA follow-up study. J Thromb Haemost. 2015;13:1823–32. https://doi.org/10.1111/jth.13113.
32. Eischer L, Gartner V, Schulman S, Kyrle PA, Eichinger S, AUREC-FVIII Investigators. 6 versus 30 months anticoagulation for recurrent venous thrombosis in patients with high factor VIII. Ann Hematol. 2009;88:485–90. https://doi.org/10.1007/s00277-008-0626-1.
33. Poli D, Antonucci E, Ciuti G, Abbate R, Prisco D. Combination of D-dimer, F1+2 and residual vein obstruction as predictors of VTE recurrence in patients with first VTE episode after OAT withdrawal. J Thromb Haemost. 2008;6:708–10. https://doi.org/10.1111/j.1538-7836.2008.02900.x.
34. Tripodi A, Legnani C, Chantarangkul V, Cosmi B, Palareti G, Mannucci PM. High thrombin generation measured in the presence of thrombomodulin is associated with an increased risk of recurrent venous thromboembolism. J Thromb Haemost. 2008;6:1327–33. https://doi.org/10.1111/j.1538-7836.2008.03018.x.
35. Chaireti R, Jennersjö C, Lindahl TL. Is thrombin generation at the time of an acute thromboembolic episode a predictor of recurrence? The LInköping study on thrombosis (LIST)—a 7-year follow-up. Thromb Res. 2013;131:135–9. https://doi.org/10.1016/j.thromres.2012.11.015.
36. Undas A. Prothrombotic fibrin clot phenotype in patients with deep vein thrombosis and pulmonary embolism: a new risk factor for recurrence. Biomed Res Int. 2017;2017:8196256. https://doi.org/10.1155/2017/8196256.

37. Baker SR, Zabczyk M, Macrae FL, Duval C, Undas A, Ariëns RAS. Recurrent venous thromboembolism patients form clots with lower elastic modulus than those formed by patients with non-recurrent disease. J Thromb Haemost. 2019;17:618–26. https://doi.org/10.1111/jth.14402.
38. Ma X, Ji X-M, Fu P, Ding Y-C, Xue Q, Huang Y. Coexistence of high fibrinogen and low high-density lipoprotein cholesterol levels predicts recurrent cerebral venous thrombosis. Chin Med J. 2015;128:1732–7. https://doi.org/10.4103/0366-6999.159345.
39. Weltermann A, Eichinger S, Bialonczyk C, Minar E, Hirschl M, Quehenberger P, et al. The risk of recurrent venous thromboembolism among patients with high factor IX levels. J Thromb Haemost. 2003;1:28–32. https://doi.org/10.1046/j.1538-7836.2003.00038.x.
40. Montes R, Guruceaga E, González-Porras JR, Reverter JC, Marco P, Pina E, et al. Identification of new markers of recurrence in patients with unprovoked deep vein thrombosis by gene expression profiling: the retro study. Eur J Haematol. 2016;97:128–36. https://doi.org/10.1111/ejh.12692.
41. Foley JH, Walton BL, Aleman MM, O'Byrne AM, Lei V, Harrasser M, et al. Complement activation in arterial and venous thrombosis is mediated by plasmin. EBioMedicine. 2016;5:175–82. https://doi.org/10.1016/j.ebiom.2016.02.011.
42. Høiland II, Liang RA, Hindberg K, Latysheva N, Brekke O-L, Mollnes TE, et al. Associations between complement pathways activity, mannose-binding lectin, and odds of unprovoked venous thromboembolism. Thromb Res. 2018;169:50–6. https://doi.org/10.1016/j.thromres.2018.06.019.
43. Denninger MH, Chaït Y, Casadevall N, Hillaire S, Guillin MC, Bezeaud A, et al. Cause of portal or hepatic venous thrombosis in adults: the role of multiple concurrent factors. Hepatology. 2000;31:587–91. https://doi.org/10.1002/hep.510310307.
44. Plessier A, Darwish-Murad S, Hernandez-Guerra M, Consigny Y, Fabris F, Trebicka J, et al. Acute portal vein thrombosis unrelated to cirrhosis: a prospective multicenter follow-up study. Hepatology. 2010;51:210–8. https://doi.org/10.1002/hep.23259.
45. Janssen HL, Meinardi JR, Vleggaar FP, van Uum SH, Haagsma EB, van Der Meer FJ, et al. Factor V Leiden mutation, prothrombin gene mutation, and deficiencies in coagulation inhibitors associated with Budd-Chiari syndrome and portal vein thrombosis: results of a case-control study. Blood. 2000;96:2364–8.
46. Spaander MCW, Hoekstra J, Hansen BE, Van Buuren HR, Leebeek FWG, Janssen HLA. Anticoagulant therapy in patients with non-cirrhotic portal vein thrombosis: effect on new thrombotic events and gastrointestinal bleeding. J Thromb Haemost. 2013;11:452–9. https://doi.org/10.1111/jth.12121.
47. de Franchis R, Baveno VI Faculty. Expanding consensus in portal hypertension: report of the Baveno VI consensus workshop: stratifying risk and individualizing care for portal hypertension. J Hepatol. 2015;63:743–52. https://doi.org/10.1016/j.jhep.2015.05.022.
48. Riva N, Ageno W, Schulman S, Beyer-Westendorf J, Duce R, Malato A, et al. Clinical history and antithrombotic treatment of incidentally detected splanchnic vein thrombosis: a multicentre, international prospective registry. Lancet Haematol. 2016;3:e267–75. https://doi.org/10.1016/S2352-3026(16)30020-5.
49. Ageno W, Riva N, Schulman S, Beyer-Westendorf J, Bang SM, Senzolo M, et al. Long-term clinical outcomes of splanchnic vein thrombosis: results of an international registry. JAMA Intern Med. 2015;175:1474–80. https://doi.org/10.1001/jamainternmed.2015.3184.
50. De Stefano V, Vannucchi AM, Ruggeri M, Cervantes F, Alvarez-Larrán A, Iurlo A, et al. Splanchnic vein thrombosis in myeloproliferative neoplasms: risk factors for recurrences in a cohort of 181 patients. Blood Cancer J. 2016;6:e493. https://doi.org/10.1038/bcj.2016.103.
51. Hoekstra J, Bresser EL, Smalberg JH, Spaander MCW, Leebeek FWG, Janssen HLA. Long-term follow-up of patients with portal vein thrombosis and myeloproliferative neoplasms. J Thromb Haemost. 2011;9:2208–14. https://doi.org/10.1111/j.1538-7836.2011.04484.x.
52. Condat B, Pessione F, Hillaire S, Denninger MH, Guillin MC, Poliquin M, et al. Current outcome of portal vein thrombosis in adults: risk and benefit of anticoagulant therapy. Gastroenterology. 2001;120:490–7. https://doi.org/10.1053/gast.2001.21209.

53. Greenfield G, McMullin MF. Splanchnic venous thrombosis in JAK2 V617F mutation positive myeloproliferative neoplasms—long term follow-up of a regional case series. Thrombosis J. 2018;16:33. https://doi.org/10.1186/s12959-018-0187-z.
54. Plessier OG, Cervoni JP, Ollivier I, Bureau C, Poujol-Robert A, Minello A, Houssel-Debry P, Rautou PE, Payancé A, Scoazec G, Bruno O, Corbic M, Durand F, Vilgrain V, Paradis V, Boudaoud L, de Raucourt E, Roy C, Gault N, Valla DA. Rivaroxaban prevents thrombosis recurrence in non-cirrhotic portal vein thrombosis: a multicentric randomized controlled trial. J Hepatol. 2021;(Suppl 2):S191–865.
55. Tufano A, Ageno W, Di Micco P, Niglio A, Rosa V, Ballaz A, et al. Outcomes during anticoagulation in patients with symptomatic vs. incidental splanchnic vein thrombosis. Thromb Res. 2018;164:69–74. https://doi.org/10.1016/j.thromres.2018.02.143.

Myeloproliferative Neoplasms and Splanchnic Vein Thrombosis

54

Lina Benajiba and Jean-Jacques Kiladjian

Introduction

Splanchnic vein thromboses (SVT), including portal vein thrombosis (PVT) and thrombosis of the hepatic veins causing Budd–Chiari syndrome (BCS), are severe vascular events [1]. The pathogenesis of primary SVT is mostly related to the presence of underlying prothrombotic conditions like inherited or acquired thrombophilia, paroxysmal nocturnal hemoglobinuria, or myeloproliferative neoplasms (MPN). Among these, MPN represents the most frequent cause of BCS and PVT, found in approximately 30% to 40% of the patients [2]. All subtypes of MPNs have been reported to cause SVT, although the most frequent appears to be polycythemia vera (PV), followed by essential thrombocythemia (ET), while myelofibrosis (MF) is rarely involved. Portal hypertension is a virtually constant feature in patients with SVT and the subsequent hypersplenism and hemodilution make the diagnosis of MPN more difficult due to changes in usual key diagnostic features such as high blood cell counts and splenomegaly. MPN diagnosis has therefore been for a long time based on bone marrow biopsy findings and endogenous erythroid colony (EEC) formation assessment, both of which have limitations [3–5]. The discovery of the JAK2V617F mutation, followed by mutations in *MPL* (the thrombopoietin receptor) and *calreticulin* (CALR) genes as driver mutations causing MPN, clearly improved the diagnostic algorithm for these diseases [6]. Several studies have also shown that screening for these mutations improves the accuracy of MPN diagnosis in patients with SVT and it is now clearly indicated and should be carried out as a standard practice in the etiological work-up of patients presenting with SVT. However, proper diagnosis and classification of MPN cannot rely only on the presence of a driver mutation and still requires a series of additional criteria

L. Benajiba · J.-J. Kiladjian (✉)
Centre d'Investigations Cliniques, INSERM, CIC1427, Université Paris Cité, Hôpital Saint-Louis, AP-HP, Paris, France
e-mail: lina.benajiba@aphp.fr; jean-jacques.kiladjian@aphp.fr

R. de Franchis (ed.), *Portal Hypertension VII*,
https://doi.org/10.1007/978-3-031-08552-9_54

according to the World Health Organization (WHO) classification [7]. Such classification is of importance since the risk of recurrent thrombosis, the incidence of long-term hematological complications including evolution to myelofibrosis (MF), or acute myeloid leukemia (AML) are different according to MPN subtype. For example, these long-term risks seem lower in essential thrombocythemia (ET) compared to polycythemia vera (PV). In addition, treatment of these MPN may be different, PV requiring the reduction of the red cell mass, while control of platelet count in high risk ET seems sufficient as a marker of treatment efficacy [8].

Should we Still Use the Good Old Diagnostic Tools?

The in vitro growth of erythroid colonies in the absence of erythropoietin, known as EECs, was considered for 30 years as a diagnostic marker of PV, although it was also observed in some ET and MF patients, and used in the context of SVT [5]. The availability of EEC culture was limited because culture methods were difficult to standardize, expensive, and laborious. However, EEC was considered as an important diagnostic criterion in the WHO classification, and this diagnostic tool remained a minor criterion in the WHO diagnostic classification until 2008. The emergence and development of molecular markers in the 2005–2015 period made the EEC test questionable since JAK2 mutations were associated to EEC formation. Because of its cost, the JAK2V617F molecular marker detection standardization, and the shift from functional to molecular tests, EEC formation as a diagnostic marker disappeared from the diagnostic criteria in the WHO 2016 classification [7].

Bone marrow biopsy (BMB) remains a major tool to diagnose MPN in the latest 2016 WHO MPN classification, in part due to the description of new entities such as masked PV (mPV) and prefibrotic MF (pMF). Both mimic ET and patients often present with isolated thrombocytosis [9]. In mPV, the differential diagnosis can be based either by the identification of an absolute erythrocytosis by measuring the red cell mass [10] or by careful analysis of bone marrow histology showing differences in the cellularity and hematopoietic progenitors distribution. Indeed, this new entity may be particularly relevant in patients with PV and SVT since their polycythemia as evaluated by their hemoglobin or hematocrit values is often “masked” by portal hypertension and hypersplenism. Besides identification and classification of MPN subtypes, BMB can also provide important prognostic information. For example, identification of some degree of fibrosis at diagnosis of PV or a diagnosis of pMF is associated with a higher risk of poor outcome and evolution to overt MF [11].

Molecular Markers

Altogether, it is now well established that all patients with SVT should be investigated for MPN, first by testing for JAK2V617F mutation in peripheral blood, as well as for other prothrombotic factors and systemic diseases [12]. In that perspective, close collaboration between hepatologists, hematologists and specialists in internal medicine is recommended, underlining the multidisciplinary approach required for optimal management of these diseases with complex consequences. JAK2V617F mutation is by far the most frequent mutation found in patients with SVT and an underlying MPN [13]. In BCS, 80% of patients with MPN are JAK2V617F-positive and the prevalence of JAK2V617F mutation is around 40% [2]. In addition, screening for this mutation may allow a diagnosis of MPN in 17% of patients without classical MPN features in peripheral blood. In PVT, 86% of patients with MPN are JAK2V617F-positive and the prevalence of JAK2V617F mutation is slightly lower, estimated around 30% [2]. In this context also, screening for the mutation allows for a diagnosis of otherwise overlooked MPN in 15% of patients. Less frequently, a mutation in the exon 12 of JAK2 may cause polycythemia, but JAK2-exon12 mutated PV were very rarely found in the diagnostic workup of patients with SVT, two recent reports suggesting that the prevalence of this mutation is around 1% [14, 15].

For an MPN diagnosis, hematologists usually search for one of the two other driver mutations, in MPL and CALR, in patients who are JAK2V617F-negative [13, 16]. Indeed, these mutations activating the thrombopoietin signaling are only observed in ET and MF: CALR mutations are detected in about 20%–25%, while MPL mutations are found in less than 5% of ET and MF patients, respectively. The diagnostic performance of the screening for MPL and CALR mutations as a marker of MPN in patients with SVT is low, and they were found in approximately 1% of patients only in large published series [2, 16].

Since the discovery of JAK2V617F mutation, a series of additional non-driver somatic mutations targeting genes involved in the regulation of various intracellular pathways such as epigenetics, mRNA splicing, signalization, and transcription have been identified in patients with MPN [17]. These mutations may be present in addition to, or independently from, the three driver mutations and they may be a helpful marker of clonality (i.e., a diagnostic criterion for MPN) in a subgroup of ET and MF patients lacking any of the driver mutations, commonly referred to as "triple-negative" patients [18]. Of note, most of these non-driver mutations can also be found in other hematological myeloid malignancies like acute myeloid leukemia and myelodysplastic syndromes. Furthermore, they have also been reported in healthy individuals, the frequency of which is increasing with age, and many studies now look at the clinical significance of these clonal hematopoiesis of indeterminate potential (CHIP) or age-related clonal hematopoiesis (ARCH) [19]. In all, although finding one or several of these additional non-driver mutations can be a helpful tool to diagnose MPN, their clinical and pathophysiological impact remains sometimes debatable. We and others have recently assessed the role of screening of these mutations using targeted next-generation sequencing (NGS) in the diagnosis of SVT. Magaz and colleagues suggested that this approach may improve the diagnostic of MPN in patients with SVT [14]. In a cohort of 80 patients (including "triple-negative" patients), they found that NGS was able to detect genetic variants in non-driver genes in 37.8% of patients with idiopathic NC-SVT. In our hands, molecular profiling of MPN-SVT patients with NGS identified a series of mutations associated with poorer event-free and overall survival [15]. In particular, molecular profile could predict the risk of evolution of the MPN to myelofibrosis or acute leukemia, and therefore potentially help in selecting the optimal cytoreductive therapy for these patients. Altogether, these two studies strongly suggest that NGS analysis should now be proposed in the diagnostic work-up of all patients with SVT because it can provide crucial information for both the diagnosis and the prognosis of an underlying MPN [20].

MPN Diagnosis in the Context of SVT

During the Baveno VII consensus conference, experts agreed on the etiological work-up for MPN in primary thrombosis of the portal venous system or hepatic venous outflow tract. First of all, in all adult patients, MPN should be searched for by testing for JAK2V617F mutation in peripheral blood (grade A1). In patients with undetectable JAK2V617F mutation, consider additional investigations for MPN, including somatic calreticulin and JAK2-exon12 mutations, and next-generation sequencing (grade A1). In patients with primary thrombosis of the splanchnic veins without any biomarker of MPN, bone marrow biopsy is recommended for a diagnosis of MPN, irrespective of peripheral blood cell counts (grade B2).

Again, one should stress the fact that MPN diagnosis is not obvious in the particular context of SVT and requires some degree of hematological expertise [18]. Building a multidisciplinary specific care path for these patients is of utmost importance to ensure optimal management (including proper diagnosis and treatment strategy) and ultimately the best possible outcome. In this regard, referring the patient to a dedicated hematologist seems appropriate in all cases. When the patient is found to be JAK2V617F-positive, the disease still has to be better diagnosed with a precise MPN subtype and specific treatment and follow-up strategies must be discussed [7, 8]. In this regard, decision of BMB and NGS analysis must be done in a specialized center for an adequate interpretation. If the patient is JAKV617F-negative, further molecular analyses and BMB should also be discussed in an experienced center.

Treatment of MPN in the Context of SVT

Obviously, anticoagulation remains the cornerstone of management and MPN-directed cytoreductive therapy has to be discussed in addition to this central treatment [21]. There are no randomized prospective trials of MPN-specific treatment in patients with SVT. However, in MPN patients, a history of thrombosis is a clear indication for cytoreductive therapy in international guidelines [8]. The main objectives of such treatment are to decrease the risks of recurrence of thrombosis and liver complications [22]. Another long-term objective is to diminish the risk of hematological evolution to more aggressive diseases like acute leukemia. In patients with elevated blood counts, starting cytoreductive therapy is simple, and the target of therapy is usually to normalize these counts. However, due to hypersplenism and hemodilution, most of patients with PV and increased red cell mass have hemoglobin and hematocrit levels within the normal range. A randomized trial has shown that PV patients maintaining a hematocrit below 45% had lower risk of vascular events [23], and this threshold must at least be used for patients with SVT. Whether a lower threshold should be applied in this specific population is an important question, and this should be evaluated in a prospective study.

Main potential cytoreductive drugs that may be used to treat MPN in patients with SVT include hydroxyurea, interferon alpha, and Ruxolitinib. Hydroxyurea is an antimetabolite approved for the treatment of PV and ET, has a long history of efficacy and safety in MPN, and remains the most frequently used cytoreductive therapy worldwide. One limitation could be its absence of disease-modifying activity that may not prevent the risk of long-term hematological transformation [24], an issue of particular relevance in MPN patients with SVT due to the high proportion of younger patients compared to "standard" MPN population. On the other hand, interferon alpha has been regularly shown to target the MPN mutant cells [25–28] that could result in a lower risk of hematological transformation, in particular, for patients at higher molecular risk identified by NGS analyses [15]. More recently, Ruxolitinib, a JAK inhibitor, has been shown to be safe and useful to reduce spleen size and disease-related symptoms in a phase 2 study in patients with MPN associated with SVT [29].

Conclusion

More than 30 years after the identification of MPN as the main cause of primary SVT, diagnostic and therapeutic methods are slowly evolving. Erythroid colony formation in vitro has been abandoned while bone marrow biopsy still has an important role for the diagnosis and prognosis of MPN. In parallel, identification of many molecular markers beyond JAK2V617F mutation offers new diagnostic and prognostic tools of particular importance in patients with MPN and SVT. Altogether, we believe that close collaboration and discussion between hepatologists and hematologists skilled in the very particular presentation and evolution of patients with MPN and SVT is key to improve the outcome of our patients. After all, there is only one letter distinguishing these two specialties, so it should not be so difficult!

References

1. Plessier A, Rautou PE, Valla DC. Management of hepatic vascular diseases. J Hepatol. 2012;56(Suppl 1):S25–38.
2. Smalberg JH, Arends LR, Valla DC, Kiladjian JJ, Janssen HL, Leebeek FW. Myeloproliferative neoplasms in Budd-Chiari syndrome and portal vein thrombosis: a meta-analysis. Blood. 2012;120(25):4921–8.
3. Chait Y, Condat B, Cazals-Hatem D, Rufat P, Atmani S, Chaoui D, et al. Relevance of the criteria commonly used to diagnose myeloproliferative disorder in patients with splanchnic vein thrombosis. Br J Haematol. 2005;129(4):553–60.

4. De Stefano V, Teofili L, Leone G, Michiels JJ. Spontaneous erythroid colony formation as the clue to an underlying myeloproliferative disorder in patients with Budd-Chiari syndrome or portal vein thrombosis. Semin Thromb Hemost. 1997;23(5):411–8.
5. Valla D, Casadevall N, Lacombe C, Varet B, Goldwasser E, Franco D, et al. Primary myeloproliferative disorder and hepatic vein thrombosis. A prospective study of erythroid colony formation in vitro in 20 patients with Budd-Chiari syndrome. Ann Intern Med. 1985;103(3):329–34.
6. Vainchenker W, Delhommeau F, Constantinescu SN, Bernard OA. New mutations and pathogenesis of myeloproliferative neoplasms. Blood. 2011;118(7):1723–35.
7. Arber DA, Orazi A, Hasserjian R, Thiele J, Borowitz MJ, Le Beau MM, et al. The 2016 revision to the World Health Organization classification of myeloid neoplasms and acute leukemia. Blood. 2016;127(20):2391–405.
8. Barbui T, Tefferi A, Vannucchi AM, Passamonti F, Silver RT, Hoffman R, et al. Philadelphia chromosome-negative classical myeloproliferative neoplasms: revised management recommendations from European LeukemiaNet. Leukemia. 2018;32(5):1057–69.
9. Barbui T, Thiele J, Carobbio A, Guglielmelli P, Rambaldi A, Vannucchi AM, et al. Discriminating between essential thrombocythemia and masked polycythemia vera in JAK2 mutated patients. Am J Hematol. 2014;89(6):588–90.
10. Maslah N, Soret J, Dosquet C, Vercellino L, Belkhodja C, Schlageter MH, et al. Masked polycythemia vera: analysis of a single center cohort of 2480 red cell masses. Haematologica. 2020;105(3):e95–e7.
11. Barbui T, Thiele J, Carobbio A, Passamonti F, Rumi E, Randi ML, et al. Disease characteristics and clinical outcome in young adults with essential thrombocythemia versus early/prefibrotic primary myelofibrosis. Blood. 2012;120(3):569–71.
12. Valla DC. Budd-Chiari syndrome/hepatic venous outflow tract obstruction. Hepatol Int. 2018;12(Suppl 1):168–80.
13. Kiladjian JJ, Cervantes F, Leebeek FW, Marzac C, Cassinat B, Chevret S, et al. The impact of JAK2 and MPL mutations on diagnosis and prognosis of splanchnic vein thrombosis: a report on 241 cases. Blood. 2008;111(10):4922–9.
14. Magaz M, Alvarez-Larrán A, Colomer D, López-Guerra M, García-Criado M, Mezzano G, et al. Next-generation sequencing in the diagnosis of non-cirrhotic splanchnic vein thrombosis. J Hepatol. 2021;74(1):89–95.
15. Debureaux PE, Cassinat B, Soret-Dulphy J, Mora B, Verger E, Maslah N, et al. Molecular profiling and risk classification of patients with myeloproliferative neoplasms and splanchnic vein thromboses. Blood Adv. 2020;4(15):3708–15.
16. Poisson J, Plessier A, Kiladjian JJ, Turon F, Cassinat B, Andreoli A, et al. Selective testing for calreticulin gene mutations in patients with splanchnic vein thrombosis: a prospective cohort study. J Hepatol. 2017;67(3):501–7.
17. Vainchenker W, Kralovics R. Genetic basis and molecular pathophysiology of classical myeloproliferative neoplasms. Blood. 2017;129(6):667–79.
18. Cassinat B, Giraudier S, Kiladjian JJ. How much does 2016 WHO classification of myeloproliferative neoplasms affect the clinic? Expert Rev Hematol. 2019;12(7):473–6.
19. Cassinat B, Harrison C, Kiladjian JJ. Clonal hematopoiesis and atherosclerosis. N Engl J Med. 2017;377(14):1400–1.
20. Kiladjian JJ, Debureaux PE, Plessier A, Soret-Dulphy J, Valla D, Cassinat B, et al. Benefits of molecular profiling with next-generation sequencing for the diagnosis and prognosis of myeloproliferative neoplasms in splanchnic vein thrombosis. J Hepatol. 2021;74(1):251–2.
21. Stein BL, Martin K. From Budd-Chiari syndrome to acquired von Willebrand syndrome: thrombosis and bleeding complications in the myeloproliferative neoplasms. Blood. 2019;134(22):1902–11.
22. Soret J, Debray D, Fontbrune FS, Kiladjian JJ, Saadoun D, Latour RP, et al. Risk factors for vascular liver diseases: vascular liver diseases: position papers from the francophone network for vascular liver diseases, the French Association for the Study of the liver (AFEF), and ERN-rare liver. Clin Res Hepatol Gastroenterol. 2020;44(4):410–9.

23. Marchioli R, Finazzi G, Specchia G, Cacciola R, Cavazzina R, Cilloni D, et al. Cardiovascular events and intensity of treatment in polycythemia vera. N Engl J Med. 2013;368(1):22–33.
24. Kiladjian JJ, Chevret S, Dosquet C, Chomienne C, Rain JD. Treatment of polycythemia vera with hydroxyurea and pipobroman: final results of a randomized trial initiated in 1980. J Clin Oncol. 2011;29(29):3907–13.
25. Kiladjian JJ, Cassinat B, Chevret S, Turlure P, Cambier N, Roussel M, et al. Pegylated interferon-alfa-2a induces complete hematologic and molecular responses with low toxicity in polycythemia vera. Blood. 2008;112(8):3065–72.
26. Verger E, Cassinat B, Chauveau A, Dosquet C, Giraudier S, Schlageter MH, et al. Clinical and molecular response to interferon alpha therapy in essential thrombocythemia patients with CALR mutations. Blood. 2015;126(24):2585–91.
27. Verger E, Soret-Dulphy J, Maslah N, Roy L, Rey J, Ghrieb Z, et al. Ropeginterferon alpha-2b targets JAK2V617F-positive polycythemia vera cells in vitro and in vivo. Blood Cancer J. 2018;8(10):94.
28. Gisslinger H, Klade C, Georgiev P, Krochmalczyk D, Gercheva-Kyuchukova L, Egyed M, et al. Ropeginterferon alfa-2b versus standard therapy for polycythaemia vera (PROUD-PV and CONTINUATION-PV): a randomised, non-inferiority, phase 3 trial and its extension study. Lancet Haematol. 2020;7(3):e196–208.
29. Pieri L, Paoli C, Arena U, Marra F, Mori F, Zucchini M, et al. Safety and efficacy of ruxolitinib in splanchnic vein thrombosis associated with myeloproliferative neoplasms. Am J Hematol. 2017;92(2):187–95.

Splanchnic Vein Thrombosis: Consensus Statements of Panel 8

55

Shiv K. Sarin, Dominique-Charles Valla, Anna Baiges, Marta Barrufet, Lina Benajiba, Jean Jacques Kiladjian, Aurelie Plessier, Massimo Primignani, and Fanny Turon

S. K. Sarin (✉)
Institute of Liver Biliary Sciences, New Delhi, India
e-mail: sksarin@ilbs.in

D.-C. Valla
Service d'hépatologie, Hôpital Beaujon, Clichy, France
e-mail: dominique.valla@aphp.fr

A. Baiges · F. Turon
Barcelona Hepatic Hemodynamic Laboratory, Liver Unit, Hospital Clínic, IDIBAPS, CIBERehd, Health Care Provider of the European Reference Network on Rare Liver Disorders (ERN Liver), Barcelona, Spain
e-mail: abaigesa@clinic.cat; fturon@clinic.cat

M. Barrufet
Interventional Radiology Unit, CDI, Hospital Clinic, Barcelona, Spain
e-mail: BARRUFET@clinic.cat

L. Benajiba · J. J. Kiladjian
Centre d'Investigations Cliniques, Hôpital Saint-Louis (AP-HP), Paris, France
e-mail: lina.benajiba@aphp.fr; jean-jacques.kiladjian@aphp.fr

A. Plessier
Université de Paris, AP-HP, Hôpital Beaujon, Service d'Hépatologie, DMU DIGEST, Centre de Référence des Maladies Vasculaires du Foie, FILFOIE, ERN RARE-LIVER, Centre de recherche sur l'inflammation, Inserm, UMR 1149, Paris, France
e-mail: aurelie.plessier@aphp.fr

M. Primignani
First Gastroenterology Unit, IRCCS Ca' Granda Policlinico Foundation, Milan, Italy
e-mail: massimo.primignani@policlinico.mi.it

R. de Franchis (ed.), *Portal Hypertension VII*,
https://doi.org/10.1007/978-3-031-08552-9_55

Aetiological Work-Up in Primary Thrombosis of the Portal Venous System or Hepatic Venous Outflow Tract

8.1 For patients with primary thrombosis of the splanchnic veins in the absence of cirrhosis, close collaboration with subspecialists is recommended for complete work-up for prothrombotic factors and systemic diseases (A,1). (Changed)

8.2 Various combinations of risk factors for thrombosis can be present so that identification of one risk factor does not deter from a complete work-up (A,1). (New)

8.3 In all adult patients, myeloproliferative neoplasia (MPN) should be searched for by testing for V617F JAK2 mutation in peripheral blood (A,1). (Unchanged)

8.4 In patients with undetectable JAK2 V617F mutation, consider additional investigations for MPN, including somatic calreticulin and JAK2-exon12 mutations, and next-generation sequencing (A,1). (Changed)

8.5 In all adult patients with primary thrombosis of the splanchnic veins without MPN driver mutation, bone marrow biopsy should be discussed in collaboration with haematologists to rule out MPN, irrespective of blood cell counts. Bone marrow biopsy should be considered particularly in patients without major risk factors for thrombosis (B,2). (Changed)

Budd–Chiari Syndrome—Definition

8.6 Budd–Chiari syndrome (BCS) is the consequence of an obstruction to the hepatic venous outflow. Obstruction can be located from the level of the small hepatic veins to the level of the entrance of the inferior vena cava into the right atrium (A,1). (Unchanged)

8.7 BCS is the preferred designation for any primary hepatic venous outflow tract obstruction (HVOTO) (D,1). (New)

8.8 BCS is considered secondary when the mechanism for venous obstruction is extrinsic compression, for example, by a benign or malignant tumour. BCS is considered primary otherwise (A,1). (Changed)

Budd–Chiari Syndrome—Diagnosis

8.9 BCS presentation and manifestations are extremely diverse so that the diagnosis must be considered in any patient with acute, acute-on-chronic, or chronic liver disease (A,1). (Changed)

8.10 BCS is diagnosed by the demonstration of an obstruction of the venous lumen, or by the presence of hepatic vein collaterals together with the absence of patent hepatic veins (A,1). (Unchanged)

8.11 Liver biopsy should not be performed to diagnose BCS when vascular imaging demonstrates obstruction of the hepatic venous outflow tract (B,1). (Unchanged)

8.12 Liver biopsy is necessary to diagnose BCS if obstruction of the small hepatic veins is not seen on imaging (B,1). (Changed)

8.13 In patients with BCS, hepatic nodules are frequent and most often benign. However, HCC may occur and therefore patients should be monitored with periodic imaging and alpha-fetoprotein measurements. (B,1). (Changed)

8.14 A 6-month interval can be proposed for periodic imaging (C,1). (New)

8.15 It is still unclear which ultrasonography or magnetic resonance imaging should be used for periodical imaging screening (C,1). (New)

8.16 Patients developing nodules should be referred to centres experienced in managing BCS (D,1). (Unchanged)

8.17 Characterization of the nodule may first include magnetic resonance imaging using hepatobiliary contrast agents (C,1). Biopsy of the lesion is indicated for a definitive diagnosis of hepatocellular carcinoma (C,1). (New)

Budd–Chiari Syndrome—Management

8.18 Management of BCS should be undertaken using a stepwise approach including anticoagulation, angioplasty/stent/thrombectomy/thrombolysis, TIPS and orthotopic liver transplantation, at experienced centres (B,1). (Unchanged)

8.19 Long-term anticoagulation should be given to all patients with primary BCS (B,1). (Changed)

8.20 Because of the increased risk of heparin-induced thrombocytopenia, the use of unfractionated heparin is generally not recommended and may only be reserved for special situations (e.g. glomerular filtration rate < 30 mL/min, pending invasive procedures) (D,2). (New)

8.21 Stenoses that are amenable to percutaneous angioplasty/stenting (short length stenoses) should be actively looked for and treated accordingly (B,1). (Unchanged)

8.22 TIPS insertion should be attempted by operators with specific experience in BCS when angioplasty/stenting/thrombectomy/thrombolysis is not feasible, and when the patient does not improve on medical therapy including anticoagulants (B,1). (Unchanged)

8.23 Consider improvement as a combination of several of the following outcomes: decreasing rate of ascites formation, decreasing serum bilirubin, serum creatinine and INR when elevated (or increasing factor V in patients receiving vitamin K antagonists) (D,1). (New)

8.24 BCS-TIPS Prognostic Index score can be used to predict outcomes in patients in whom TIPS insertion is considered (B,1). (Changed)

8.25 Liver transplantation should be considered in patients with uncontrolled clinical manifestation despite a stepwise approach, or in patients with high BCS-TIPS Prognostic Index score (>7) before TIPS placement (C,1). (Changed)

8.26 In patients with BCS presenting as acute liver failure, urgent liver transplantation should be considered. Emergency TIPS should be performed, if possible, independently of listing for liver transplantation (C,1). (New)

Portal Vein Thrombosis and Portal Cavernoma in the Absence of Cirrhosis—Definition

8.27 Portal vein thrombosis is characterized by the presence of a thrombus in the portal vein trunk or its branches. Portal cavernoma is a network of porto-portal collaterals that develops as a consequence of prior portal vein obstruction (D,1). Obstruction leading to cavernoma is mostly related to thrombosis in adults, but less likely so in children and young adults (B,1). (Changed)

8.28 Portal vein thrombosis should be distinguished by imaging tools from the extravascular compression of the venous lumen by a neighbouring space-occupying formation (D,1). (New)

8.29 Cirrhosis and/or malignancy should be ruled out and other underlying liver diseases (e.g. PSVD or other chronic liver diseases) should be investigated (D,1). (Changed)

Portal Vein Thrombosis and Portal Cavernoma in the Absence of Cirrhosis—Diagnosis

8.30 For diagnosis of portal vein thrombosis or cavernoma, Doppler ultrasound, CT or MR angiography should demonstrate solid intraluminal material not showing enhancement after injection of vascular contrast agents; or a network of porto-portal collaterals, respectively (B,1). If diagnosed by Doppler ultrasound, confirmation with contrast-enhanced CT or MR angiography is needed (D,1). (Changed)

8.31 A standardized documentation (as proposed in Table 55.1) of the initial site, extent degree of luminal obstruction and chronicity of clot formation is

Table 55.1 Recommended standardized nomenclature for the description of portal vein thrombosis and portal cavernoma in both the clinical and research setting [18]

Feature	Definition
Time course	
Recent	PVT presumed to be present for <6 months
Chronic	PVT present or persistent for >6 months
Percent occlusion of main PV	
Completely occlusive	No persistent lumen
Partially occlusive	Clot obstructing >50% of original vessel lumen
Minimally occlusive	Clot obstructing <50% of original vessel lumen
Cavernous transformation	Gross Porto-portal collaterals without original PV seen
Response to treatment or interval change	
Progressive	Thrombus increases in size or progresses to more complete occlusion
Stable	No appreciable change in size or occlusion
Regressive	Thrombus decreases in size or degree of occlusion

required to allow subsequent evaluation of the spontaneous course and/or response to treatment (D,1). (New)

8.32 Portal vein thrombosis and portal cavernoma in adults are frequently associated with one or more risk factors for thrombosis, which may be occult at presentation and should be investigated (B,1). (Unchanged)

8.33 In patients with portal vein thrombosis following abdominal surgery or pancreatitis, invasive procedures (e.g. bone marrow biopsy and liver biopsy) should be discussed on an individual basis considering the expected low diagnostic yield in such populations and the risk of morbidity associated with these procedures (C,2). (New)

8.34 If the liver is dysmorphic on imaging or liver tests are persistently abnormal, liver biopsy and HVPG measurement are recommended to rule out cirrhosis or PSVD (B,1). Liver stiffness by TE may be useful to exclude cirrhosis, although precise cut-offs cannot be proposed yet (C,2). (Changed)

Portal Vein Thrombosis and Portal Cavernoma in the Absence of Cirrhosis—Management

8.35 In the absence of cirrhosis, recent portal vein thrombosis rarely resolves spontaneously. Therefore, anticoagulation should be started at a therapeutic dosage immediately at diagnosis (B,1). (Changed)

8.36 Because of the increased risk of heparin-induced thrombocytopenia, the use of unfractionated heparin is not generally recommended and may only be reserved for special situations (e.g. glomerular filtration rate < 30 mL/min, pending invasive procedures) (D,2). (New)

8.37 As a primary treatment option for recent portal vein thrombosis in the absence of cirrhosis, start with low molecular weight heparin and switch to vitamin K antagonists when possible (B,1) (Changed). DOACS can be considered as the primary option in selected cases in the absence of the so-called 'triple positive' anti-phospholipid syndrome, although data are limited (C,2). (New)

8.38 Anticoagulation should be given for at least 6 months in all patients with recent portal vein thrombosis in the absence of cirrhosis (B,1). (Unchanged)

Recent Portal Vein Thrombosis in the Absence of Cirrhosis—Management

8.39 After 6 months, long-term anticoagulation is recommended in patients with the permanent underlying prothrombotic state (B1) and should also be considered in patients without an underlying prothrombotic state (B,2). (New)

8.40 If anticoagulation is discontinued, D-dimers <500 ng/mL 1 month after discontinuation may be used to predict a low risk of recurrence (C,2). (New)

8.41 In patients without cirrhosis who do not develop complications of recent portal vein thrombosis, despite the absence of portal vein recanalization, interventions other than anticoagulation are not required (B,2). (Changed)

8.42 A follow-up contrast-enhanced CT scan should be performed 6 months after recent portal vein thrombosis (C,1). (New)

8.43 Because of the risk of recurrence of splanchnic vein thrombosis, patients need to be followed up, irrespective of anticoagulation discontinuation (C,1). (New)

8.44 The risk of intestinal infarction and organ failure is increased in patients with recent portal vein thrombosis and (i) persistent severe abdominal pain despite anticoagulation therapy, (ii) bloody diarrhoea, (iii) lactic acidosis, (iv) bowel loop distention, or (v) occlusion of second-order radicles of the superior mesenteric vein. Therefore, a multidisciplinary approach with early image-guided intervention, thrombolysis and surgical intervention should be considered in referral centres (C,2). (New)

Past Portal Vein Thrombosis or Cavernoma in the Absence of Cirrhosis—Management

8.45 In patients with past portal vein thrombosis or cavernoma, including those with incomplete resolution of recent portal vein thrombosis at 6 months, long-term anticoagulation is recommended in patients with a permanent underlying prothrombotic state (B,1) and should also be considered in patients without an underlying prothrombotic state (B,2). (New)

8.46 No data are available to recommend or discourage anticoagulation in childhood-onset past portal vein thrombosis or cavernoma in the absence of an underlying prothrombotic state (C,1). (New)

8.47 In patients with past portal vein thrombosis or cavernoma not yet receiving anticoagulants, anticoagulation should be started after adequate portal hypertensive bleeding prophylaxis has been initiated in patients with high-risk varices (C,2). (Changed)

8.48 Mesenteric-left portal vein bypass (Meso-Rex operation) should be considered in all children with complications of portal cavernoma, and these patients should be referred to centres with experience in treating this condition (B,1). (Unchanged)

8.49 Patients with refractory complications of portal vein thrombosis or cavernoma should be referred to expert centres to consider percutaneous recanalization of the portal vein or other vascular interventional procedures (C,1). (New)

Treatment of Portal Hypertension in EHPVO

8.50 There is insufficient data on whether beta-blockers or endoscopic therapy could be preferred for primary prophylaxis of portal hypertension-related bleeding in patients with past portal vein thrombosis or cavernoma. Guidelines for cirrhosis should be applied (C,2). (Changed)

8.51 Oesophageal variceal band ligation can be performed safely without withdrawing vitamin K antagonists (C,2). (New)

8.52 All patients in whom thrombosis have not been recanalized should be screened for gastroesophageal varices within 6 months of the acute episode. In the absence of varices, endoscopy should be repeated at 12 months and 2 years thereafter (B,1). (Unchanged)

8.53 In patients with acute portal hypertension-related bleeding, recommendations for patients with cirrhosis may be applied (D,1). (Changed)

8.54 Based on the recommendations for cirrhosis, combination of non-selective beta-blockers and band ligation is recommended for secondary prophylaxis (D,1). (New)

Research Agenda

Budd–Chiari Syndrome

- Risk factors for hepatocellular carcinoma in patients with BCS
- Non-invasive diagnosis of hepatocellular carcinoma in patients with BCS
- Short-term (8 days) evolution criteria predicting a good mid-long-term outcome (i.e. criteria for 'treatment response') in patients with BCS

Portal Vein Thrombosis Without Cirrhosis

- Predictors of development, progression and spontaneous resolution of PVT
- Influence of beta-blockers on the natural history of PVT
- Effect of early recanalization using interventional radiology or TIPS vs. fibrinolytic agents and/or anticoagulants in patients with recent PVT
- Efficacy of anticoagulation in children/young adults with PVT on recanalization and on prevention of progression of PVT
- Pathophysiology and management of cytopenia in patients with non-cirrhotic portal hypertension

Part XI

Vascular Liver Disorders 2: Other Issues in Vascular Liver Disorders

Porto-Sinusoidal Vascular Disorder 56

Andrea De Gottardi and Valérie Paradis

Introduction

Idiopathic non-cirrhotic portal hypertension includes a heterogeneous group of vascular liver diseases that may lead to portal hypertension in the absence of parenchymal cirrhotic nodules [1]. It corresponds to a variety of histopathologic entities that have been referred to as hepatoportal sclerosis, non-cirrhotic portal fibrosis, nodular regenerative hyperplasia, or incomplete septal fibrosis/cirrhosis [2]. Until very recently, there were no conclusive diagnostic methods or characteristic histopathologic findings available for diagnosing idiopathic non-cirrhotic portal hypertension, which was thus made after excluding all other possible causes of liver disease. The pathophysiology of idiopathic non-cirrhotic portal hypertension is still poorly understood, and therapy is restricted to the manifestations of portal hypertension. Idiopathic non-cirrhotic portal hypertension has gained increased attention over the last two decades in parallel to the increased use of immunosuppressive drugs for autoimmune and hematological disorders, and to the increased prevalence of treated Human Immunodeficiency Virus (HIV) infection, all conditions etiologically linked to idiopathic non-cirrhotic portal hypertension [3, 4]. Increased awareness and widespread use of liver elastography for fibrosis assessment have permitted diagnosis in patients in whom prominent features of portal hypertension contrast with low liver stiffness [5]. In some patients with inconspicuous clinical features of portal hypertension, the diagnosis is made after detecting specific liver lesions at biopsy. Patients with extrahepatic splanchnic venous thrombosis may have idiopathic non-cirrhotic portal

A. De Gottardi (✉)
Department of Gastroenterology and Hepatology, Ente Ospedaliero Cantonale, Lugano, Switzerland

Faculty of Biomedical Sciences, Università della Svizzera Italiana, Lugano, Switzerland

V. Paradis
Department of d'Anatomie Pathologique, Hôpital Beaujon, Clichy, France
e-mail: valerie.paradis@aphp.fr

R. de Franchis (ed.), *Portal Hypertension VII*,
https://doi.org/10.1007/978-3-031-08552-9_56

hypertension as an underlying condition. Last, the previous definition, based on ruling out causes for cirrhosis, has excluded from specific attention patients with non-cirrhotic portal hypertension when concomitant causes for the liver disease were present (e.g., hepatitis C, alcohol consumption, or metabolic syndrome).

The complexity and unclear pathogenesis of the entity so-called idiopathic non-cirrhotic portal hypertension prompted the Vascular Liver Diseases Interest Group (VALDIG) to organize a multidisciplinary conference in February 2017. Experts in vascular liver diseases assembled to discuss the definition and terminology of portal vascular lesions, as well as the pathogenesis, causes, diagnostic workup, and treatment of idiopathic non-cirrhotic portal hypertension. The term *porto-sinusoidal vascular disorder* was proposed as a denomination for an entity incorporating various vascular liver diseases based on clear criteria [6]. This chapter aims to clarify the new denomination of porto-sinusoidal vascular disorder, as well as to provide a comprehensive view of its pathophysiology, diagnosis, and treatment.

Definition

According to the previous definition, idiopathic non-cirrhotic portal hypertension was characterized by direct and/or indirect signs of portal hypertension, including a mild increase in hepatic venous pressure gradient, esophageal varices, non-malignant ascites, splenomegaly or hypersplenism, portosystemic collaterals, and the absence of cirrhosis on liver biopsy. Additionally, all other causes of chronic liver disease leading to cirrhotic and non-cirrhotic portal hypertension (sarcoidosis, schistosomiasis) and portal or hepatic vein thrombosis had to be excluded. This previous definition of idiopathic non-cirrhotic portal hypertension included the histopathologic entities previously known as obliterative portal venopathy, hepatoportal sclerosis, nodular regenerative hyperplasia, non-cirrhotic portal fibrosis, and incomplete septal cirrhosis.

However, several key limitations to this definition were addressed at the 2017 VALDIG conference. First, this definition may be too restrictive because, in early stages of the disease, lesions can be present while significant portal hypertension has not yet developed or will not develop. Corresponding cases would be erroneously excluded. Second, the previous definition excludes any thrombosis of hepatic or portal venous systems; therefore, patients who develop portal vein thrombosis as a complication of their underlying intrahepatic vascular liver disease would similarly be erroneously excluded. Last, the previous definition did not allow for the presence of concomitant liver diseases, although it is well known that some common diseases such as viral hepatitis, HIV infection, or alcoholic or non-alcoholic fatty liver disease can concur with vascular liver disease.

The term *porto-sinusoidal vascular* disorder (PSVD) was developed to group several conditions that, despite diverse pathophysiology, are characterized by lesions involving sinusoids and small-sized portal veins. This new denomination encompasses the whole spectrum of the disease spanning idiopathic non-cirrhotic portal hypertension, obliterative portal venopathy, incomplete septal cirrhosis, and nodular regenerative hyperplasia [6]. The main components of this definition include the absence of cirrhosis in a liver biopsy and the detection of histological findings with or without portal hypertension (Fig. 56.1).

In contrast with the criteria of the previous definition, the presence of causes for liver disease (i.e., alcohol misuse, metabolic syndrome, or viral hepatitis) does not exclude PSVD, *if* the liver biopsy shows specific findings indicative of PSVD. In such overlapping cases, the relative contribution of PSVD and parenchymal liver disease to the development or degree of severity of portal hypertension remains an open question.

Similar to the previous definition, conditions affecting the hepatic veins or specific diseases that have been well characterized as causing microvascular disease such as sarcoidosis or congenital hepatic fibrosis are excluded. Sinusoidal obstruction syndrome, which occurs mainly after hematopoietic stem cell transplantation, is characterized by specific criteria and is not included in PSVD. Although extrahepatic portal vein thrombosis can cause, per se, non-cirrhotic portal hypertension, it does not constitute exclusion criteria if liver biopsy shows specific findings indicative of PSVD. This is justified by its most frequent secondary occurrence in PSVD patients.

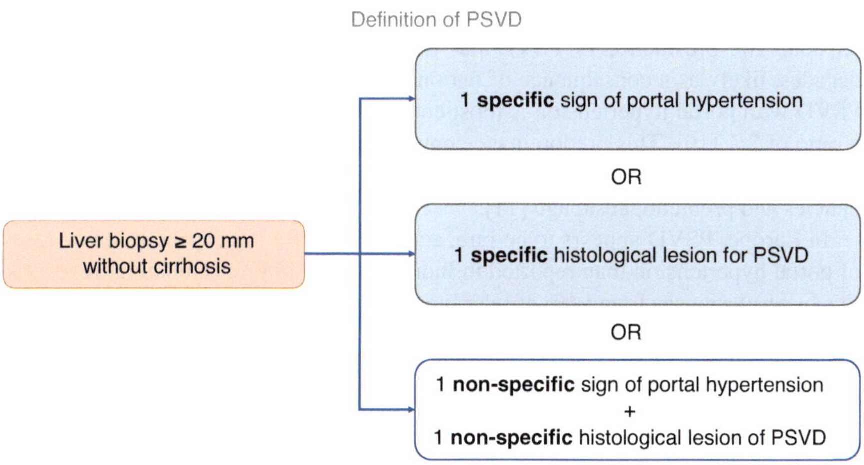

	Features of portal hypertension	Histological lesions suggestive of PSVD assessed by an expert pathologist
Specific	Gastric, esophageal or ectopic varices Portal hypertensive bleeding Porto-systemic collaterals at imaging	Obliterative portal venopathy(thickening of vessel wall, occlusion of the lumen, vanishing of portal veins) Nodular regenerative hyperplasia Incomplete septal fibrosis (also called incomplete septal cirrhosis); this latter feature can only be assessed on liver explants and not on liver biopsies
Not specific	Ascites Platelet count < 150'000/mm^3 Spleen size ≥ 13 cm in the largest axis	Portal tract abnormalities (multiplication, increased number of arteries, periportal vascular channels, aberrant vessels) Architectural disturbance: irregular distribution of the portal tracts and central veins Non-zonal sinusoidal dilatation Mild perisinusoidal fibrosis

Fig. 56.1 Definition of porto-sinusoidal vascular disorder (PSVD)

Overall, this new denomination is intended to clarify and improve diagnosis. From a research perspective, this inclusive definition is expected to facilitate studies on this condition, by providing uniform criteria. On the other hand, it may be argued that these new criteria and terms may be overly simplistic and decrease precision to define a complex disease, and eventually introduce bias for studies that otherwise would not be concentrated on the same disease. It will therefore be important to gather further experience with this denomination and refine the definition accordingly.

Epidemiology

The overall prevalence of PSVD worldwide is unknown. Prevalence and denominations have differed according to geographic areas. Therefore, the new denomination was elaborated in order to be applicable independent of location.

In India, the corresponding condition was known as non-cirrhotic portal fibrosis. The prevalence in this area, although decreasing, is still high, accounting in some studies for 34% of all cases of portal hypertension [7]. Socioeconomic status and sanitary/hygiene conditions have been suggested to be associated with its development. Males aged 30–49 years have been predominantly affected [8].

In Japan, the corresponding condition was denominated idiopathic portal hypertension. The prevalence of PSVD has dramatically shrunk during the last four decades, likely as a consequence of national health services policies [9]. In Japan, PSVD with portal hypertension is most common in women aged 40–59 years with a ratio of 2:1 [10]. This predominance could be related to the autoimmune disease being more common in women than in men and to hormonal factors related to pregnancies and premenopausal age [11].

In Europe, PSVD appears to be rare, accounting for a lower proportion of cases of portal hypertension than reported in India or Japan. In France, nodular regenerative hyperplasia was found 4% of liver biopsies performed for various reasons [12]. The condition was found to predominantly affect men in France and UK (3:1) [13, 14]. In the USA and Canada, this prevalence was 3% to 7%; men aged 60–69 years were predominantly affected [15–17].

Associated Conditions

PVSD is associated with rare conditions in about 50% of patients. These varied conditions can be categorized as drug exposure, immunological, coagulation disorders, infectious, and congenital or familial defects [4, 18, 19]. Several of these conditions can be simultaneously present in occasional patients.

Drug Exposure

Older age and cumulative exposure to didanosine and stavudine were shown to be independent predictors for the development of nodular regenerative hyperplasia in patients with HIV infection [20]. The overall prevalence of HIV infection in PSVD patients was 4% in a Dutch study [4]. Mallet et al. reported a significantly lower protein S activity in patients with HIV infection and nodular regenerative hyperplasia than in controls, but the unifying factor of PSVD in HIV patients was previous drug exposure [21]. Didanosine and stavudine currently being no longer used, if their responsibility is real, a decrease in the prevalence of PSVD among HIV-infected patients is to be expected over the next decades.

PSVD has also been related to prior exposure to immunosuppressive or antineoplastic agents (in particular azathioprine and oxaliplatin) as well as to numerous other drugs [22].

Immunological Disorders

Immune disorders, including acquired and congenital immune deficiencies and autoimmune diseases, have been detected in 10% of PSVD patients [23]. Conversely, PSVD has been found in up to 84% of patients with common variable immune deficiency [24], hyper-IgM syndrome, primary antibody-deficiency syndromes such as Bruton's disease [25], and Felty's syndrome [26].

In patients with inflammatory bowel disease, the prevalence of PSVD was reported to be 6% [27]. However, it is difficult to decipher whether PSVD is mainly linked to the underlying inflammatory bowel disease or to azathioprine exposure. Adult celiac disease has also been associated with PSVD [13]. It has been proposed that the sinusoidal changes found in patients with conditions of disordered immunity are related to intrasinusoidal cytotoxic T lymphocytes and granulomas, causing portal vein or sinusoidal endothelitis. This concept is in line with overexpression of lymphocyte activation genes in blood samples from PSVD patients [28, 29].

Thrombophilia

An increasing amount of evidence suggests that microthrombosis and platelet aggregation contribute to the development of PSVD [10, 30]. In fact, thickening or occlusion and obliteration of portal vein venules detected at liver biopsy is generally regarded as indicating the previous thrombosis. Moreover, prothrombotic conditions such as protein C deficiency have been associated with a higher risk of PSVD [31]. Portal vein thrombosis is relatively common in these patients further pointing at a procoagulant tendency in these patients. Future studies should elucidate the prevalence and impact of prothrombotic risk factors in PSVD.

Infections

Results from epidemiological studies have shown a relationship between low hygienic living conditions and PSVD, which has been interpreted as supporting a role for infections. Such a mechanism, however, could not be reproduced in experimental models [32]. Intra-abdominal infections may serve as a trigger for PSVD through recurrent small to medium portal branch occlusion [33, 34].

Hereditary and Genetic Disorders

PSVD has been linked to genetic disorders such as Adams–Oliver syndrome, Turner's syndrome, familial obliterative portal venopathy, and cystic fibrosis [35–37].

In hereditary studies, the familial aggregation has been found regarding PSVD and HLA-DR3 [38] and mutations in the telomerase gene complex [39]. Interestingly, a link between didanosine exposure in HIV patients and PSVD has been associated with certain single nucleotide polymorphisms of genes involved in the purine metabolic pathway [40]. Whole exome sequencing in families affected with PSVD led to the discovery of various mutations, but independent validation in other cohorts is still lacking [41, 42].

Clinical Presentation

PSVD With Portal Hypertension

In higher income countries, patients with portal hypertension and PSVD are mainly middle-aged men and are usually asymptomatic. In most cases, liver synthesis function is maintained, and in about 80%, there is a slight increase (< double upper limit of normal) in liver biochemistry values, alanine aminotransferase or alkaline phosphatase. Some are detected through noninvasive methods with thrombocytopenia generally around 100 G/L, splenomegaly, or an irregular liver aspect on ultrasound. Most patients have serum albumin and bilirubin levels within the normal range and prothrombin time slightly decreased, which aids in distinguishing them from those with cirrhosis. On the other hand, some patients develop complications of portal hypertension, mostly variceal bleeding, which is the initial manifestation in around 20%–40%, whereas ascites and encephalopathy are uncommon presenting manifestations in comparison.

The natural history of patients with idiopathic non-cirrhotic portal hypertension is characterized by the presence of large varices, which were found at initial presentation in two-thirds of the patients with PSVD and portal hypertension or developed in 20% of patients within an average of 10 years of diagnosis [19]. Over time, PSVD patients with portal hypertension can develop ascites in 20%–50% of cases with a precipitant factor identified in the majority of cases and usually transient [18, 19]. Within 5 years of diagnosis, portal vein thrombosis develops in around a third of patients, but is completely obstructive (i.e., occupying more than 80% of the vessel lumen) in only a third of these [4, 18, 19]. The risk of thrombosis is increased in patients with a history of bleeding and with associated conditions, namely HIV infection.

As far as longer term prognosis is concerned, one study with 69 patients showed minimal changes in markers of liver function in these patients, suggesting that PSVD is stable [18]. Patients can develop porto-pulmonary hypertension, hepatopulmonary syndrome, and liver regenerative nodules, but the precise risk factors leading to these complications are currently unidentified. Concerning mortality, the presence of ascites, age, and associated diseases are known risk factors [4]. Previous published series have demonstrated that the mortality can reach 15%–20% after an 8-year follow-up period [4, 18, 19, 43]. The referral rate for liver transplantation (5%–37%) appears to be very variable depending on the assessment of the risk of progression to end-stage liver disease.

PSVD Without Portal Hypertension

Altered liver tests in the absence of an identified cause and without any signs of portal hypertension (splenomegaly, gastro-esophageal varices, portosystemic collaterals, ascites, or hepatic encephalopathy) could represent a pre-clinical stage of PSVD [18, 44], which may be followed by the development of manifest signs of portal hypertension [18]. Indeed, it appears that the prevalence of PSVD without portal hypertension is higher than previously thought (19% of cases with cryptogenic liver disease). Moreover, Cazals-Hatem et al. hypothesized that the presence of slightly impaired liver function tests, a higher rate of prothrombotic conditions, and immune diseases were likely to contribute to the progression to portal hypertension [18]. The diagnosis is established on specific findings at liver biopsy performed in asymptomatic patients with slight changes in liver biochemistry.

Given the lack of longitudinal studies analyzing patients with PSVD without portal hypertension, there is currently insufficient data to clarify the natural history and risk factors of this form of the disease.

Histopathological Findings

The diagnosis of PSVD may be incidental and can be considered in a variety of settings, including altered liver biochemistry of unknown cause, signs of portal hypertension without liver dysfunction, abnormal ultrasound findings, or portal hypertension with a low liver stiffness level. From a histological point of view, there is an important diversity of lesions among patients that remains currently unexplained.

Nodular regenerative hyperplasia, obliterative portal venopathy, and incomplete septal cirrhosis are specific enough to be regarded as diagnostic for PSVD even in the absence of other clinical, laboratory, or imaging alterations (Fig. 56.2).

Obliterative portal venopathy and hepatoportal sclerosis/phlebosclerosis are characterized by incomplete or complete stenosis of mainly medium- and small-sized intrahepatic portal vein branches with or without thickening of the wall. Moreover, scarring and stenosis of small portal vein branches along with an increased number of small vascular channels within the portal tracts and incomplete thin fibrous septa have been described. The portal vein branch is not always obliterated or absent (venopenia), as it can still be visible although with a narrowed lumen [15]. Although portal venous changes are common, they can be difficult to detect at biopsy due to a heterogeneous distribution, supporting adequate tissue sampling (>20 mm, >7 portal tracts).

Nodular regenerative hyperplasia is defined by diffuse or focal nodular regeneration, dilated sinusoids in areas of hepatocellular cell atrophy, and no or very mild

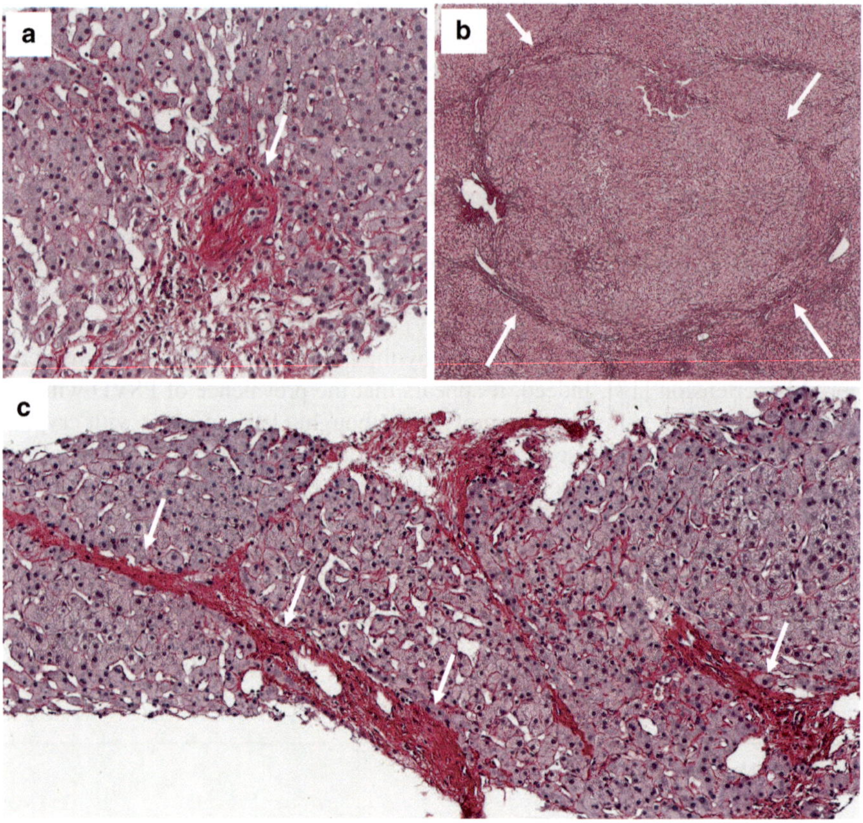

Fig. 56.2 Specific histological features of porto-sinusoidal vascular disorder. (**a**) Obliterative venopathy, section from a liver biopsy (red sirius staining) showing a portal tract slightly sclerotic (white arrow) devoid of normal portal vein. (**b**) Nodular regenerative hyperplasia, section from a liver biopsy (reticulin staining) showing nodular organization of the liver parenchyma within the lobule (nodule surrounded by atrophic cell plates (white arrows) without fibrosis); (**c**) Incomplete septal fibrosis, section from a liver biopsy (red sirius staining) showing portal tracts enlarged and prolonged by incomplete thin septa (white arrows) in a non-cirrhotic parenchyma

perisinusoidal fibrosis. These nodules are generally lighter in tone and less well defined compared to cirrhosis. The lobules are distorted and replaced by nodules of hyperplastic hepatocytes arranged in thicker plates. These are surrounded at the periphery by compressed atrophic cell plates and a condensed reticulin network but without significant fibrosis. Portal tract remnants (small portal tracts with inconspicuous or sometimes absent portal vein branches) can be found. Reticulin staining is required for diagnosis, although the diagnosis is generally demanding and requires expert and experienced histopathologists [45].

Diffuse and poorly demarcated nodules and slender fibrous septa which span into the parenchyma without connection with other portal tracts or venules illustrate incomplete septal fibrosis. Isolated collagen bundles within the parenchyma are associated with disturbed vascular relationships and can be linked with incomplete septal fibrosis. In fact, these lesions were described in a period when cirrhosis was thought to have an irreversible progressive course, without the prospect of regression. Recent advances in chronic liver disease have clearly shown that hepatic architecture is in constant remodeling as a response to tissue damage and repair. Actually, incomplete septal fibrosis may derive, in some cases, from cirrhosis that had regressed [17]. Vascular lesions related to cirrhosis may still be evident for many years after fibrosis regression and may explain the persistence of portal hypertension. It is still unclear how such vascular changes and subsequent portal hypertension evolve over time.

In addition, these specific findings are frequently associated with other changes including fine perforated septa, isolated thick collagen fibers, thin periportal fibrous spikes, portal tract remnants, herniated portal vein directly abutting periportal parenchyma, periportal abnormal vessels defined as thin-walled vessels in the paraportal area, foci of sinusoidal dilation or peliosis, and prominent arteries or artery multiplication (Fig. 56.3) [46–48].

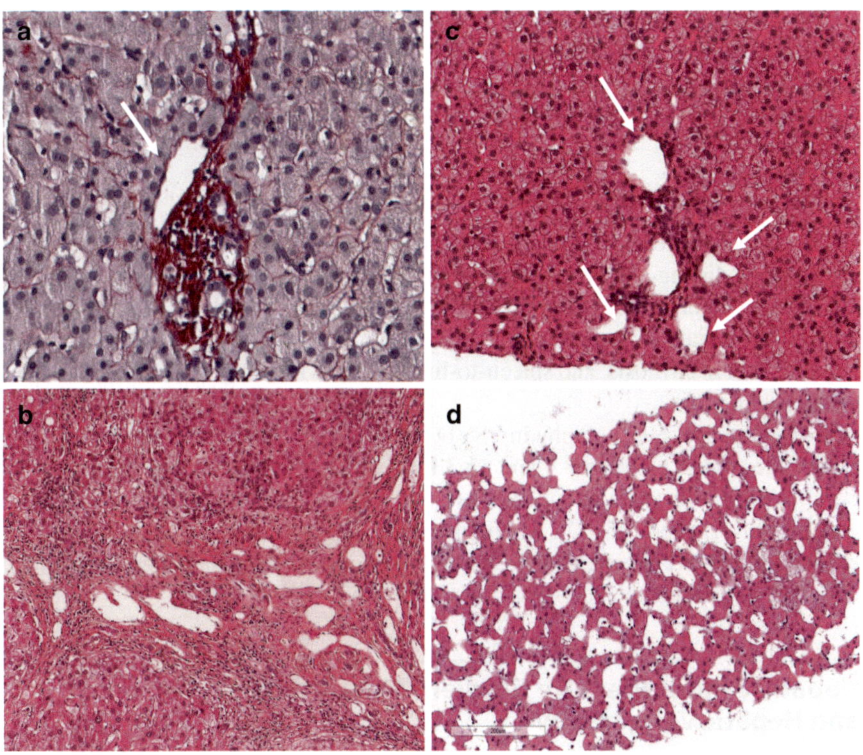

Fig. 56.3 Nonspecific histological features of porto-sinusoidal vascular disorder. (**a**) Herniated portal vein (white arrow), section from a liver biopsy (red sirius staining). (**b**) Hypervascularized portal tract, section from a liver biopsy (hematein and eosin staining) showing multiple thin-walled vascular spaces within the portal tract. (**c**) Periportal abnormal vessels section from a liver biopsy (hematein & eosin staining) showing multiple thin-walled vascular spaces (white arrows) in close contact to the portal tract. (**d**) Sinusoidal dilatation, section from a liver biopsy (hematein and eosin staining) showing liver cell plates outlined by dilated sinusoids

Imaging

Patients with PSVD and portal hypertension present signs indicative of the latter including splenomegaly and porto-systemic collaterals. In patients without portal hypertension, particular imaging features may be present although not specific for portal hypertension. Such features include an increased hepatic artery diameter, although also common in patients with cirrhosis. Hypertrophy of segments IV and I and atrophy of the remaining segments aids in distinguishing PSVD or portal vein thrombosis from cirrhosis where, by contrast, there is atrophy of segment IV and hypertrophy of segment I. The increase in size of segments IV and I and atrophy of the surrounding segments is related to an impaired flow in the portal vein leading to hypertrophy of the central part of the liver and atrophy of the periphery [49, 50]. Furthermore, in comparison with patients with cirrhosis, patients with PSVD have more frequently a reduced caliber, an occlusive thrombosis, or a lack of visibility of intrahepatic portal vein branches and focal nodular hyperplasia-like nodules [50].

The presence of thrombosis of portal veins is currently not an exclusion criterion for a diagnosis of non-cirrhotic portal hypertension because patients with PSVD may develop secondary portal vein thrombosis.

Elastography

In the last decade, the widespread implementation of liver and spleen elastography has aided in distinguishing among patients with clinically evident portal hypertension those with or without liver cirrhosis. Actually, patients with PSVD usually have liver stiffness values much lower than the cutoffs for clinically significant portal hypertension in cirrhosis, and spleen-to-liver stiffness ratio higher than in other liver diseases [51, 52].

Current data of elastography in PSVD are relatively limited. Reported liver stiffness values range between 8.4 and 11.3 kPa, which is higher than in patients with portal vein thrombosis without PVSD (6.4–8.4 kPa) and significantly lower than in patients with cirrhosis [5, 52]. Although elastography promises to be most useful in evaluating for PSVD patients with portal hypertension, the data being still limited, liver biopsy remains the basis for diagnosis.

Hepatic Venous Pressure Gradient Measurement and Hepatic Venography

Hepatic venous pressure gradient measurement allows for documenting PSVD in patients with signs of obvious portal hypertension. A majority of patients have a portal pressure gradient below 10 mmHg, the cutoff for the so-called clinically significant portal hypertension, despite signs of obvious portal hypertension. Additionally, in patients with PSVD, the hepatic venography performed during the portal pressure assessment commonly shows large hepatic veno-venous communications, a finding thus far incompletely understood [52].

Additional Considerations

Focal Liver Lesions

Hepatocellular nodules can develop in PSVD patients, as they do in other vascular liver diseases, although less commonly than in patients with Budd–Chiari syndrome [53]. These nodules are generally benign, being for the most part focal nodular hyperplasia-like, and rarely hepatocellular adenomas. These nodules are considered to develop because of a distorted local blood perfusion combining enhanced arterialization and decreased portal venous perfusion, in addition to other probable hormonal and gender-related factors. Development or progression to hepatocellular carcinoma appears to be very rare [54, 55].

Pregnancy

Pregnancy, per se, is not a recognized risk factor for PSVD. From a practical aspect, pregnancy desire should be addressed routinely in patients with PSVD, as about 15% of patients with PSVD are women of childbearing age, rendering reproductive issues particularly relevant [56–58]. It is paramount that liver disease remains stable before considering pregnancy.

Three small retrospective series including 40 women reported variceal bleeding in 15% of cases. Terlipressin is contraindicated during pregnancy. Low-molecular weight heparin use was associated with post-partum genital bleeding; no deaths were observed. Nevertheless, in patients with PSVD with previous portal vein thrombosis, low-molecular weight heparin can be safely used, and a 24-h interruption is recommended before delivery, ideally vaginal whenever possible [56, 58–61]. Ten to 25% of the pregnancies in these series resulted in fetal loss.

Primary and secondary prophylaxis for variceal bleeding should by routinely started, following the rules recommended for patients with liver cirrhosis. There is currently no evidence supporting primary prophylaxis of thrombosis with anticoagulation in pregnant women with PSVD.

Non-Hepatic Abdominal Surgery

A retrospective VALDIG study, including 47 patients with PSVD and portal hypertension, reported portal hypertension-related complications in 30% of patients within 3-month post-op; these were more common in those with extrahepatic comorbidities. In patients with preserved renal function, 6-month survival was very good [62]. No information is available regarding patients with PSVD without portal hypertension.

Management

Anticoagulation

One of the most common denominators of PSVD is thickening, narrowing, or obliteration of intrahepatic portal venules. Such a narrowing is thought to produce an ischemic atrophy of the hepatocytes, as seen in nodular regenerative hyperplasia. Among explanted livers, portal venules were found to be obliterated in 100%, and large portal veins in 67% [18]. Portal vein thrombosis occurs in 13%–45% of PSVD patients during follow-up [4, 18, 43, 63]. Whether the increased risk of splanchnic thrombosis is related with local venule endothelial factors or mechanical causes related to blood stasis and portal hypertension, or a combination of all these factors, remains unknown [64]. Lastly, patients with PSVD commonly have underlying disorders associated with an increased risk of thrombosis (0%–18%) [4, 14, 18, 19].

Moreover, in patients with non-cirrhotic portal hypertension secondary to portal vein thrombosis, recanalization occurs in less than half of those treated with early anticoagulation [19, 65]. Such poor outcomes might be avoided with prophylactic anticoagulation in PSVD patients at risk for portal vein thrombosis. Randomized trials are required to assess the benefit risk ratio of prophylactic anticoagulation in patients with PSVD. Anticoagulation therapy is currently recommended for patients with high-risk prothrombotic disorders or those developing portal vein thromboses [66].

Treatment of Portal Hypertension

The incidence and risk factors for progression of portal hypertension in PSVD patients are still unclear so that no preventive therapy can currently be recommended [18, 67].

In patients with PSVD and portal hypertension, current practice guidelines propose treating varices following the recommendations elaborated for patients with cirrhosis [66]. The effectiveness of this approach has been demonstrated [19, 68]. The cornerstone of therapy is beta-blockers, either carvedilol or propranolol, and endoscopic variceal ligation.

In situations where drug exposure or associated conditions exist, drug cessation or disease directed therapy could theoretically improve the outcomes of PSVD, although uncertain. The optimal strategy and interval to screen for portal hypertension signs such as varices is currently undefined.

Transjugular intrahepatic portosystemic shunt (TIPS) can be an effective treatment option in patients with PSVD and complications of portal hypertension such as variceal bleeding and refractory ascites. A multicenter study of 41 patients with PSVD and portal hypertension described a comparable outcome to that of patients with cirrhosis and similar liver function. Normal kidney function and the absence of severe extrahepatic comorbidities were prognostic factors for a better outcome [69].

Based on limited data, the overall outcome of PVSD patients with portal hypertension treated with abdominal shunt surgery appears to be favorable [62, 70]. Portosystemic shunting or splenectomy have been mainly reported in adults or children from India and Turkey [32, 62, 70–73]. In most patients, porto-systemic surgical shunts and splenectomies were performed in patients with either complications related to portal hypertension or symptoms caused be splenomegaly [32]. Although shunt surgery was effective in reducing portal hypertension and no operative mortality was described, delayed morbidity was frequent, occurring in 20%–50% of patients [70–73]. Variceal rebleeding (10%), ascites, and hepatic encephalopathy (up to 18% of cases) were the most frequently reported complications [71, 72]. This data highlights an important rate of complications in PSVD with portal hypertension treated surgically with shunt. In cases of severe hypersplenism, partial splenic embolization and splenectomy have been performed, but given the risks related to it, it must be only considered in rare, individual cases with symptomatic hypersplenism [69, 74].

Scarce reported data have demonstrated that survival of PSVD patients after liver transplantation is favorable [58]. Post-transplant (recurrent) PSVD has been reported, although its incidence is unclear [2].

Current and Future Perspectives in Translational and Clinical Research

Translational

There are currently some animal models that reproduce human PSVD. Among these are models replicating nodular regenerative hyperplasia and venous occlusion. Vascular embolization animal models using microspheres of dextran and serum bovine albumin, surgical models after splenic extraction and models with direct injection of bacteria into the portal vein did not accurately create PSVD [75–77]. Genetic models, namely *NOTCH1* knockout mice replicated all of the histological findings of nodular regenerative hyperplasia and portal hypertension [78, 79]. JAK1 (IL-6–JAK–STAT pathway) mutated mice induce a phenotype similar to autoimmune disease with histological signs of nodular regenerative hyperplasia [80]. Rats fed selenium-rich diet also generated nodular regenerative hyperplasia with portal hypertension [81]. Although promising, until now, no specific therapies have been tested in animal models.

Clinical

The establishment of a new terminology to combine vascular liver diseases affecting the porto-sinusoidal area and its wide dissemination will allow for a better understanding of the epidemiology of the disease. Furthermore, cohort studies with patients with PSVD will advance knowledge of this condition and possibly help answer fundamental questions. It remains to be clarified why some patients develop portal hypertension, while others remain asymptomatic. It is also unknown which is

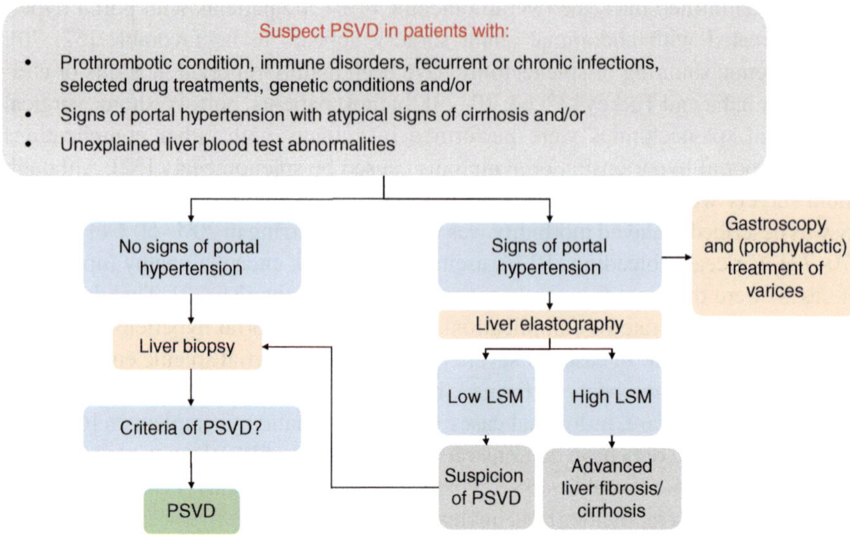

Fig. 56.4 Diagnostic flowchart of porto-sinusoidal vascular disorder with and without portal hypertension. PSVD porto-sinusoidal vascular disorder, HIV human immunodeficiency virus, NRH nodular regenerative hyperplasia, EGD esophagogastroduodenoscopy, EVL endoscopic variceal ligation, LSM liver stiffness measurement, SSM spleen stiffness measurement, PH portal hypertension

the best method to diagnose clinically significant portal hypertension considering that portal pressure gradient is not accurate in these patients, and at what intervals should they be screened. Additionally, it is obscure how PSVD should be defined histologically in patients with PSVD and concomitant liver diseases of other etiologies and what its relative impact is.

Lastly, the inclusion of patients under the terminology of PSVD will also facilitate the development of multicenter clinical trials testing the effect of directed therapy such as anticoagulation.

Conclusions

The establishment of the new terminology aims to cover a heterogeneous group of conditions under clearly defined diagnostic criteria. Liver biopsy remains fundamental for diagnosis. Figure 56.4 suggests a pragmatic approach in the management of patients with a suspicion of PSVD. The implementation of the term *porto-sinusoidal vascular disorder* is key to facilitate multicenter, collaborative cohort studies to address the critical questions regarding this entity.

References

1. European Association for the Study of the Liver. EASL clinical practice guidelines: vascular diseases of the liver. J Hepatol. 2016;64(1):179–202.

2. Schouten JN, Garcia-Pagan JC, Valla DC, Janssen HL. Idiopathic noncirrhotic portal hypertension. Hepatology. 2011;54(3):1071–81.
3. Chang PE, Miquel R, Blanco JL, Laguno M, Bruguera M, Abraldes JG, et al. Idiopathic portal hypertension in patients with HIV infection treated with highly active antiretroviral therapy. Am J Gastroenterol. 2009;104(7):1707–14.
4. Schouten JN, Van der Ende ME, Koeter T, Rossing HH, Komuta M, Verheij J, et al. Risk factors and outcome of HIV-associated idiopathic noncirrhotic portal hypertension. Aliment Pharmacol Ther. 2012;36(9):875–85.
5. Vuppalanchi R, Mathur K, Pyko M, Samala N, Chalasani N. Liver stiffness measurements in patients with noncirrhotic portal hypertension-the devil is in the details. Hepatology. 2018;68(6):2438–40.
6. De Gottardi A, Rautou PE, Schouten J, Rubbia-Brandt L, Leebeek F, Trebicka J, et al. Porto-sinusoidal vascular disease: proposal and description of a novel entity. Lancet Gastroenterol Hepatol. 2019;4(5):399–411.
7. Dhiman RK, Chawla Y, Vasishta RK, Kakkar N, Dilawari JB, Trehan MS, et al. Non-cirrhotic portal fibrosis (idiopathic portal hypertension): experience with 151 patients and a review of the literature. J Gastroenterol Hepatol. 2002;17(1):6–16.
8. Mukta V, Panicker LC, Sivamani K, Goel A, Basu D, Dhanapathi H. Non-cirrhotic portal fibrosis at a tertiary care Centre in South India. Trop Dr. 2017;47(1):26–30.
9. Imai FKK, Komaba M. Interim report on IPH survey. In: Futagawa S, editor. Report of the Research Committee on Aberrant Portal Hemodynamics 1992. Tokyo: Ministry of Health and Welfare; 1993. p. 107–10.
10. Murai Y, Ohfuji S, Fukushima W, Tamakoshi A, Yamaguchi S, Hashizume M, et al. Prognostic factors in patients with idiopathic portal hypertension: two Japanese nationwide epidemiological surveys in 1999 and 2005. Hepatol Res. 2012;42(12):1211–20.
11. Saito K, Nakanuma Y, Takegoshi K, Ohta G, Obata Y, Okuda K, et al. Non-specific immunological abnormalities and association of autoimmune diseases in idiopathic portal hypertension. A study by questionnaire. Hepato-Gastroenterology. 1993;40(2):163–6.
12. Barge S, Grando V, Nault JC, Broudin C, Beaugrand M, Ganne-Carrie N, et al. Prevalence and clinical significance of nodular regenerative hyperplasia in liver biopsies. Liver Int. 2016;36(7):1059–66.
13. Eapen CE, Nightingale P, Hubscher SG, Lane PJ, Plant T, Velissaris D, et al. Non-cirrhotic intrahepatic portal hypertension: associated gut diseases and prognostic factors. Dig Dis Sci. 2011;56(1):227–35.
14. Hillaire S, Bonte E, Denninger MH, Casadevall N, Cadranel JF, Lebrec D, et al. Idiopathic non-cirrhotic intrahepatic portal hypertension in the west: a re-evaluation in 28 patients. Gut. 2002;51(2):275–80.
15. Mikkelsen WP, Edmondson HA, Peters RL, Redeker AG, Reynolds TB. Extra- and intrahepatic portal hypertension without cirrhosis (hepatoportal sclerosis). Ann Surg. 1965;162(4):602–20.
16. Villeneuve JP, Huet PM, Joly JG, Marleau D, Cote J, Legare A, et al. Idiopathic portal hypertension. Am J Med. 1976;61(4):459–64.
17. Wanless IR, Nakashima E, Sherman M. Regression of human cirrhosis. Morphologic features and the genesis of incomplete septal cirrhosis. Arch Pathol Lab Med. 2000;124(11):1599–607.
18. Cazals-Hatem D, Hillaire S, Rudler M, Plessier A, Paradis V, Condat B, et al. Obliterative portal venopathy: portal hypertension is not always present at diagnosis. J Hepatol. 2011;54(3):455–61.
19. Siramolpiwat S, Seijo S, Miquel R, Berzigotti A, Garcia-Criado A, Darnell A, et al. Idiopathic portal hypertension: natural history and long-term outcome. Hepatology. 2014;59(6):2276–85.
20. Cotte L, Benet T, Billioud C, Miailhes P, Scoazec JY, Ferry T, et al. The role of nucleoside and nucleotide analogues in nodular regenerative hyperplasia in HIV-infected patients: a case control study. J Hepatol. 2011;54(3):489–96.
21. Mallet VO, Varthaman A, Lasne D, Viard JP, Gouya H, Borgel D, et al. Acquired protein S deficiency leads to obliterative portal venopathy and to compensatory nodular regenerative hyperplasia in HIV-infected patients. AIDS. 2009;23(12):1511–8.

22. Ghabril M, Vuppalanchi R. Drug-induced nodular regenerative hyperplasia. Semin Liver Dis. 2014;34(2):240–5.
23. Zuo C, Chumbalkar V, Ells PF, Bonville DJ, Lee H. Prevalence of histological features of idiopathic noncirrhotic portal hypertension in general population: a retrospective study of incidental liver biopsies. Hepatol Int. 2017;11(5):452–60.
24. Pulvirenti F, Pentassuglio I, Milito C, Valente M, De Santis A, Conti V, et al. Idiopathic non cirrhotic portal hypertension and spleno-portal axis abnormalities in patients with severe primary antibody deficiencies. J Immunol Res. 2014;2014:672458.
25. Malamut G, Ziol M, Suarez F, Beaugrand M, Viallard JF, Lascaux AS, et al. Nodular regenerative hyperplasia: the main liver disease in patients with primary hypogammaglobulinemia and hepatic abnormalities. J Hepatol. 2008;48(1):74–82.
26. Stock H, Kadry Z, Smith JP. Surgical management of portal hypertension in Felty's syndrome: a case report and literature review. J Hepatol. 2009;50(4):831–5.
27. Seinen ML, van Asseldonk DP, de Boer NK, Bouma G, van Nieuwkerk CM, Mulder CJ, et al. Nodular regenerative hyperplasia of the liver in patients with IBD treated with allopurinol-Thiopurine combination therapy. Inflamm Bowel Dis. 2017;23(3):448–52.
28. Kotani K, Kawabe J, Morikawa H, Akahoshi T, Hashizume M, Shiomi S. Comprehensive screening of gene function and networks by DNA microarray analysis in Japanese patients with idiopathic portal hypertension. Mediat Inflamm. 2015;2015:349215.
29. Yamaguchi N, Tokushige K, Haruta I, Yamauchi K, Hayashi N. Analysis of adhesion molecules in patients with idiopathic portal hypertension. J Gastroenterol Hepatol. 1999;14(4):364–9.
30. Holmes E, Wijeyesekera A, Taylor-Robinson SD, Nicholson JK. The promise of metabolic phenotyping in gastroenterology and hepatology. Nat Rev Gastroenterol Hepatol. 2015;12(8):458–71.
31. Nayak NC. Idiopathic portal hypertension (noncirrhotic portal fibrosis), thrombosis in portal venous system and protein C deficiency. Hepatology. 1989;10(5):902.
32. Khanna R, Sarin SK. Non-cirrhotic portal hypertension—diagnosis and management. J Hepatol. 2014;60(2):421–41.
33. Omanwar S, Rizvi MR, Kathayat R, Sharma BK, Pandey GK, Alam MA, et al. A rabbit model of non-cirrhotic portal hypertension by repeated injections of E.coli through indwelling cannulation of the gastrosplenic vein. Hepatobiliary Pancreat Dis Int. 2004;3(3):417–22.
34. Sarin SK, Kumar A. Noncirrhotic portal hypertension. Clin Liver Dis. 2006;10(3):627–51. x
35. Besmond C, Valla D, Hubert L, Poirier K, Grosse B, Guettier C, et al. Mutations in the novel gene FOPV are associated with familial autosomal dominant and non-familial obliterative portal venopathy. Liver Int. 2018;38(2):358–64.
36. Girard M, Amiel J, Fabre M, Pariente D, Lyonnet S, Jacquemin E. Adams-Oliver syndrome and hepatoportal sclerosis: occasional association or common mechanism? Am J Med Genet A. 2005;135(2):186–9.
37. Roulot D, Degott C, Chazouilleres O, Oberti F, Cales P, Carbonell N, et al. Vascular involvement of the liver in Turner's syndrome. Hepatology. 2004;39(1):239–47.
38. Sarin SK, Mehra NK, Agarwal A, Malhotra V, Anand BS, Taneja V. Familial aggregation in noncirrhotic portal fibrosis: a report of four families. Am J Gastroenterol. 1987;82(11):1130–3.
39. Calado RT, Regal JA, Kleiner DE, Schrump DS, Peterson NR, Pons V, et al. A spectrum of severe familial liver disorders associate with telomerase mutations. PLoS One. 2009;4(11):e7926.
40. Vispo E, Cevik M, Rockstroh JK, Barreiro P, Nelson M, Scourfield A, et al. Genetic determinants of idiopathic noncirrhotic portal hypertension in HIV-infected patients. Clin Infect Dis. 2013;56(8):1117–22.
41. Koot BG, Alders M, Verheij J, Beuers U, Cobben JM. A de novo mutation in KCNN3 associated with autosomal dominant idiopathic non-cirrhotic portal hypertension. J Hepatol. 2016;64(4):974–7.
42. Vilarinho S, Sari S, Yilmaz G, Stiegler AL, Boggon TJ, Jain D, et al. Recurrent recessive mutation in deoxyguanosine kinase causes idiopathic noncirrhotic portal hypertension. Hepatology. 2016;63(6):1977–86.

43. Hollande C, Mallet V, Darbeda S, Vallet-Pichard A, Fontaine H, Verkarre V, et al. Impact of Obliterative portal Venopathy associated with human immunodeficiency virus. Medicine. 2016;95(11):e3081.
44. Guido M, Sarcognato S, Sonzogni A, Luca MG, Senzolo M, Fagiuoli S, et al. Obliterative portal venopathy without portal hypertension: an underestimated condition. Liver Int. 2016;36(3):454–60.
45. Jharap B, van Asseldonk DP, de Boer NK, Bedossa P, Diebold J, Jonker AM, et al. Diagnosing nodular regenerative hyperplasia of the liver is thwarted by low Interobserver agreement. PLoS One. 2015;10(6):e0120299.
46. Guido M, Sarcognato S, Sacchi D, Colloredo G. Pathology of idiopathic non-cirrhotic portal hypertension. Virchows Arch. 2018;473(1):23–31.
47. Nakanuma Y, Hoso M, Sasaki M, Terada T, Katayanagi K, Nonomura A, et al. Histopathology of the liver in non-cirrhotic portal hypertension of unknown aetiology. Histopathology. 1996;28(3):195–204.
48. Guido M, Alves VAF, Balabaud C, Bathal PS, Bioulac-Sage P, Colombari R, et al. Histology of portal vascular changes associated with idiopathic non-cirrhotic portal hypertension: nomenclature and definition. Histopathology. 2019;74(2):219–26.
49. Arora A, Sarin SK. Multimodality imaging of primary extrahepatic portal vein obstruction (EHPVO): what every radiologist should know. Br J Radiol. 2015;88(1052):20150008.
50. Glatard AS, Hillaire S, d'Assignies G, Cazals-Hatem D, Plessier A, Valla DC, et al. Obliterative portal venopathy: findings at CT imaging. Radiology. 2012;263(3):741–50.
51. Berzigotti A. Non-invasive evaluation of portal hypertension using ultrasound elastography. J Hepatol. 2017;67(2):399–411.
52. Seijo S, Reverter E, Miquel R, Berzigotti A, Abraldes JG, Bosch J, et al. Role of hepatic vein catheterisation and transient elastography in the diagnosis of idiopathic portal hypertension. Dig Liver Dis. 2012;44(10):855–60.
53. Moucari R, Rautou PE, Cazals-Hatem D, Geara A, Bureau C, Consigny Y, et al. Hepatocellular carcinoma in Budd-Chiari syndrome: characteristics and risk factors. Gut. 2008;57(6):828–35.
54. Montenovo MI, Jalikis FG, Yeh M, Reyes JD. Progression of hepatic adenoma to carcinoma in the setting of Hepatoportal sclerosis in HIV patient: case report and review of the literature. Case Rep Hepatol. 2016;2016:1732069.
55. Sempoux C, Paradis V, Komuta M, Wee A, Calderaro J, Balabaud C, et al. Hepatocellular nodules expressing markers of hepatocellular adenomas in Budd-Chiari syndrome and other rare hepatic vascular disorders. J Hepatol. 2015;63(5):1173–80.
56. Andrade F, Shukla A, Bureau C, Senzolo M, D'Alteroche L, Heurgue A, et al. Pregnancy in idiopathic non-cirrhotic portal hypertension: a multicentric study on maternal and fetal management and outcome. J Hepatol. 2018;69(6):1242–9.
57. Bissonnette J, Durand F, de Raucourt E, Ceccaldi PF, Plessier A, Valla D, et al. Pregnancy and vascular liver disease. J Clin Exp Hepatol. 2015;5(1):41–50.
58. Krasinskas AM, Eghtesad B, Kamath PS, Demetris AJ, Abraham SC. Liver transplantation for severe intrahepatic noncirrhotic portal hypertension. Liver Transpl. 2005;11(6):627–34. discussion 10–1
59. Aggarwal N, Sawhney H, Vasishta K, Dhiman RK, Chawla Y. Non-cirrhotic portal hypertension in pregnancy. Int J Gynaecol Obstet. 2001;72(1):1–7.
60. Henriksson P, Westerlund E, Wallen H, Brandt L, Hovatta O, Ekbom A. Incidence of pulmonary and venous thromboembolism in pregnancies after in vitro fertilisation: cross sectional study. BMJ. 2013;346:e8632.
61. Sumana G, Dadhwal V, Deka D, Mittal S. Non-cirrhotic portal hypertension and pregnancy outcome. J Obstet Gynaecol Res. 2008;34(5):801–4.
62. Elkrief L, Ferrusquia-Acosta J, Payance A, Moga L, Tellez L, Praktiknjo M, et al. Abdominal surgery in patients with idiopathic noncirrhotic portal hypertension: a multicenter retrospective study. Hepatology. 2019;70(3):911–24.

63. Matsutani S, Maruyama H, Akiike T, Kobayashi S, Yoshizumi H, Okugawa H, et al. Study of portal vein thrombosis in patients with idiopathic portal hypertension in Japan. Liver Int. 2005;25(5):978–83.
64. Rajekar H, Vasishta RK, Chawla YK, Dhiman RK. Noncirrhotic portal hypertension. J Clin Exp Hepatol. 2011;1(2):94–108.
65. Plessier A, Darwish-Murad S, Hernandez-Guerra M, Consigny Y, Fabris F, Trebicka J, et al. Acute portal vein thrombosis unrelated to cirrhosis: a prospective multicenter follow-up study. Hepatology. 2010;51(1):210–8.
66. de Franchis R, Baveno VIF. Expanding consensus in portal hypertension: report of the Baveno VI consensus workshop: stratifying risk and individualizing care for portal hypertension. J Hepatol. 2015;63(3):743–52.
67. Elkrief L, Rautou PE. Idiopathic non-cirrhotic portal hypertension: the tip of the obliterative portal venopathies iceberg? Liver Int. 2016;36(3):325–7.
68. Sarin SK, Gupta N, Jha SK, Agrawal A, Mishra SR, Sharma BC, et al. Equal efficacy of endoscopic variceal ligation and propranolol in preventing variceal bleeding in patients with noncirrhotic portal hypertension. Gastroenterology. 2010;139(4):1238–45.
69. Bissonnette J, Garcia-Pagan JC, Albillos A, Turon F, Ferreira C, Tellez L, et al. Role of the transjugular intrahepatic portosystemic shunt in the management of severe complications of portal hypertension in idiopathic noncirrhotic portal hypertension. Hepatology. 2016;64(1):224–31.
70. Sharma BC, Singh RP, Chawla YK, Narasimhan KL, Rao KL, Mitra SK, et al. Effect of shunt surgery on spleen size, portal pressure and oesophageal varices in patients with non-cirrhotic portal hypertension. J Gastroenterol Hepatol. 1997;12(8):582–4.
71. Karagul S, Yagci MA, Tardu A, Ertugrul I, Kirmizi S, Sumer F, et al. Portosystemic shunt surgery in patients with idiopathic noncirrhotic portal hypertension. Ann Transplant. 2016;21:317–20.
72. Mitra SK, Rao KL, Narasimhan KL, Dilawari JB, Batra YK, Chawla Y, et al. Side-to-side lienorenal shunt without splenectomy in noncirrhotic portal hypertension in children. J Pediatr Surg. 1993;28(3):398–401; discussion −2
73. Pal S, Radhakrishna P, Sahni P, Pande GK, Nundy S, Chattopadhyay TK. Prophylactic surgery in non-cirrhotic portal fibrosis:is it worthwhile? Indian J Gastroenterol. 2005;24(6):239–42.
74. Hirota S, Ichikawa S, Matsumoto S, Motohara T, Fukuda T, Yoshikawa T. Interventional radiologic treatment for idiopathic portal hypertension. Cardiovasc Intervent Radiol. 1999;22(4):311–4.
75. Albini B, Ito S, Brentjens J, Andres G. Splenomegaly and immune complex splenitis in rabbits with experimentally induced chronic serum sickness: immunopathological findings. J Reticuloendothel Soc. 1983;34(6):485–500.
76. Kathayat R, Pandey GK, Malhotra V, Omanwar S, Sharma BK, Sarin SK. Rabbit model of non-cirrhotic portal fibrosis with repeated immunosensitization by rabbit splenic extract. J Gastroenterol Hepatol. 2002;17(12):1312–6.
77. Komeichi H, Katsuta Y, Aramaki T, Okumura H. A new experimental animal model of portal hypertension. Intrahepatic portal obstruction by injecting DEAE-cross-linked dextran microspheres into the portal vein in the rabbit. Nihon Ika Daigaku zasshi. 1991;58(3):273–84.
78. Croquelois A, Blindenbacher A, Terracciano L, Wang X, Langer I, Radtke F, et al. Inducible inactivation of Notch1 causes nodular regenerative hyperplasia in mice. Hepatology. 2005;41(3):487–96.
79. Dubuisson L, Boussarie L, Bedin CA, Balabaud C, Bioulac-Sage P. Transformation of sinusoids into capillaries in a rat model of selenium-induced nodular regenerative hyperplasia: an immunolight and immunoelectron microscopic study. Hepatology. 1995;21(3):805–14.
80. McEntee MF, Wright KN, Wanless I, DeVovo R, Schneider JF, Shull R. Noncirrhotic portal hypertension and nodular regenerative hyperplasia of the liver in dogs with mucopolysaccharidosis type I. Hepatology. 1998;28(2):385–90.
81. Nielsen H, Binder V, Daugharty H, Svehag SE. Circulating immune complexes in ulcerative colitis. I. Correlation to disease activity. Clin Exp Immunol. 1978;31(1):72–80.

57 Anticoagulation in Splanchnic Vein Thrombosis With and Without Underlying Liver Disease

Marco Senzolo and Alberto Zanetto

Introduction

The occurrence of thrombosis in the liver venous vessels is determined by an alteration in the physiological equilibrium that regulates the balance between coagulation and anticoagulation. As hypercoagulability plays a major role in the pathophysiology of splanchnic and hepatic vein thrombosis, anticoagulation is a cornerstone in the treatment of these patients, including those with cirrhosis. This chapter reviews the current knowledge regarding anticoagulation in splanchnic and hepatic vein thrombosis.

Budd–Chiari Syndrome

The use of anticoagulation became systematic from the mid-80s with the comprehension of thrombophilic conditions associated with Budd–Chiari syndrome (BCS) [1]. Early studies showed that Factor V Leyden was present in up to 20% of patients [2]. Later studies confirmed that JAK2 V617F mutation was found in 60% of cases (24/41) [3]. The widespread adoption of anticoagulants led to the improvement in patient's survival [1]. Yet, preliminary reports showed that only a minority of anticoagulated patients (approximately 30%) achieved resolution/stabilization of thrombosis [4, 5]. Different radiological and surgical treatments were proposed and combined with anticoagulation, and this finally led to the stepwise algorithm which is currently recommended [6, 7].

According to this algorithm, anticoagulation must be initiated as early as possible after diagnosis. All patients should be treated, including those without an underlying prothrombotic disorder and those who are asymptomatic [8]. The aim of

M. Senzolo (✉) · A. Zanetto
Gastroenterology and Multivisceral Transplant Unit, Department of Surgery, Oncology, and Gastroenterology, Padua University Hospital, Padua, Italy

R. de Franchis (ed.), *Portal Hypertension VII*,
https://doi.org/10.1007/978-3-031-08552-9_57

anticoagulation is to restore hepatic vein outflow or, at least, prevent progression of thrombosis, which can be achieved in approximately 15%–25% of patients, particularly in those with mild/moderate disease [9, 10]. In patients without ongoing improvement and in those with progression of thrombosis despite anticoagulation, additional interventions will be considered. Consensus on how to define lack of response to anticoagulation has not yet been reached. Some criteria were proposed to define treatment failure and identify the optimal timing at which to move a patient along the algorithm (Table 57.1); however, they lack prospective validation [6].

Anticoagulation should start with low-molecular weight heparin (LMWH) and then be switched to long-term vitamin K antagonists (VKA), with the goal of maintaining the INR > 2.5 [8]. Intravenous heparin should be avoided because of the association between BCS and myeloproliferative disease (MPD), which increases the risk of heparin-induced thrombocytopenia (HIT) (10% in a previous study including 58 JAK2V617F-positive patients) [11]. Increased level of platelet activation markers has been reported in heparin-treated patients, which could explain the predisposition for HIT due to formation of PF4/heparin complexes [12]. In the same way, HT has been reported in patients with polycythemia vera (PV) and essential thrombocythemia (TE). Randi et al. [13] reported that among 29 patients with MPD treated with heparin, 5 (17%) developed a new clinically significant thrombotic complication between 11 and 55 days after start of heparin therapy. Among these five patients (two with PV and three with TE), 40% had unequivocal HT.

Lifelong anticoagulation is recommended to reduce the risk of thrombosis recurrence [8]. De Stefano and colleagues retrospectively studied 181 patients who presented a first episode of splanchnic vein thrombosis (67 with PV, 67 with TE, and 47 with primary myelofibrosis) [14]. BCS and portal vein thrombosis (PVT) were diagnosed in 31 (17%) and 109 (60%) patients, respectively, while isolated thrombosis of mesenteric or splenic veins was detected in 18 (10%) and 23 (13%) cases,

Table 57.1 Evaluation of response to treatment in patients with BCS

	Ongoing treatments response (2 weeks)	Complete treatment response
Ascites	Yes	No clinically detectable
Creatinine	Normal or improving	Normal
Na^+	Normal of improving	Normal
Balance (water-Na^+)	Negative	
NaCl intake	Moderate	Moderate
Factor V	Improving	Above 40% of normal reference
Direct bilirubin	Improving	Below 15 mmol/L
Portal hypertensive-related bleeding	–	No
Spontaneous bacterial infections	–	No
Body mass index	–	>20 kg/m^2

Adapted from Plessier et al. [6]

respectively. After this index event, patients were followed for 735 patient years and experienced 31 recurrences corresponding to an incidence rate of 4.2 per 100 patient-years. Factors associated with a higher risk of recurrence were BCS (OR: 3.03), history of thrombosis (OR: 3.62), splenomegaly (OR: 2.66), and leukocytosis (OR: 2.8). VKA were prescribed in 85% of patients and the recurrence rate was 3.9 per 100 patient-years in treated patients, whereas in the small fraction (15%) not receiving VKA more recurrences (7.2 per 100 patient-years) were reported. In patients with prothrombotic disorders such as MPD, paroxysmal nocturnal hemoglobinuria, and Behcet's syndrome, specific treatment of these conditions is recommended to reduce the risk of thrombosis progression or recurrence [15–17].

BCS-associated MPDs are not absolute contraindications for LT [18–20]. However, LT does not cure most of the BCS-associated prothrombotic disorders and BCS may recur after LT. In a previous study of 36 patients with BCS who underwent LT, approximately 1/3 developed liver-related thrombotic complications and 10 of them needed re-transplantation [21]. Therefore, it is mandatory to start early anticoagulation after LT and maintain it life-long. The current approach is to use a combination including anticoagulation (VKAs), aspirin, and hydroxyurea (anti-proliferative drug) [22, 23]. In a previous series of patients with BCS who underwent LT, those with MPD (n = 12) treated with this combination showed a low risk of recurrence (8%), and thrombosis was not associated with graft dysfunction or reduced survival [23]. In recipients at risk of bleeding or who experience bleeding while on anticoagulation, it may be reasonable to avoid or stop VKAs and keep them on aspirin and anti-proliferative treatment. An exception for lifelong anticoagulation after LT could be considered in those recipients whose prothrombotic disorder is corrected by LT.

Patients with BCS on anticoagulation may be at risk of bleeding. In a prospective study including 94 patients with BCS, after a median follow-up of 43 months, 47 patients had 92 major bleeding episodes. Forty episodes were related to invasive procedures for BCS, while the origin of the other 52 episodes was gastrointestinal in 26 (including 15 related to portal hypertension) and genital in 10; 26 were spontaneous and 26 provoked. Excess anticoagulation was identified in 13 (27%) out of 49 documented episodes. These results indicate that invasive procedures and portal hypertension are the major drivers for bleeding in BCS, while excess anticoagulation plays a secondary role [24].

Few data on direct oral anticoagulants (DOACs) to treat BCS are available [25, 26]. DOACs may be considered in patients with BCS and normal liver function. However, DOACs are not registered for this indication, and if used, one should pay attention to potential comorbidities associated with BCS (i.e.: renal failure).

Portal Vein Thrombosis in Absence of Cirrhosis

Recent Portal Vein Thrombosis

Recent non-cirrhotic portal vein thrombosis (NCPVT) may present with varying degrees of severity [27]. Most patients with NCPVT only have modest and non-specific symptoms; however, recent NCPVT can be rarely associated with significant morbidity and mortality. Mesenteric venous infarction is the most feared acute complication of NCPVT and may lead to perforation, peritonitis, and multiorgan failure. Early anticoagulation has lowered the incidence of this complication that is uncommon in anticoagulated patients (2% in a large multicenter study) [28]. In patients at risk for infarction but who do not show signs of peritonitis, interventional radiology may be an option [29]. These are patients with worsening abdominal pain despite 48 to 72 h of anticoagulation and in whom one would presume a lack of efficacy of anticoagulation. Anticoagulation, however, must be continued even after a successful radiological procedure. Previous data on the use of local thrombolysis have suggested a 60% rate of major bleeding [30]. A more recent series including 22 patients with NCPVT treated with a stepwise protocol (low-dose systemic alteplase followed by local clot dissolution therapy through a TIPS in patients with ongoing abdominal pain and no evidence of radiological improvement after 48–72 h of systemic thrombolysis) demonstrated a good rate of portal vein recanalization with no episode of intracranial bleeding [31]. Patients who show or develop peritoneal signs have to be evaluated for surgery. In patients who undergo intestinal resection, anticoagulation appears to improve the outcome [32].

The goal of treatment of NCPVT is to recanalize the obstructed veins, which will prevent intestinal infarction and portal hypertension. In a retrospective study, 50% of patients not achieving recanalization developed gastroesophageal varices, with a 2-year risk of variceal hemorrhage of 12%, and 16% developed ascites [33]. Severe portal biliopathy was observed in one/third of patients with NCPVT within 1 year after diagnosis [34, 35].

As spontaneous recanalization is very rare in untreated patients [33, 36], all NCPVT should be anticoagulated for at least 6 months. LMWH should be started early after diagnosis and then switched to VKA [8]. Unfractionated heparin is not recommended due to the high risk of HIT (20% in patients with MPD) [13]. It may be considered in special conditions such as glomerular filtration rate < 30 mL/min or pending procedures.

Two retrospective studies included consecutive patients with NCPVT treated with UFH or LMWH and switched to VKAs [33, 37]. If initiated immediately, anticoagulation led to recanalization in 40%–50% of patients. Bleeding complications were rare and mostly mild [33, 37]. By contrast, delays in anticoagulation (>1 week) were associated with a lower chance to achieve recanalization (20% vs. 60%) [33]. Among baseline factors, extension of thrombosis was associated with lower response to anticoagulation in one study [36] but not in another [33].

In 2010, Plessier et al. reported the first large, prospective multicenter study examining safety and efficacy of anticoagulation in 95 patients with NCPVT [28]. Patients received early of LMWH followed by VKAs targeting an INR of 2–3. Only 2% of patients developed intestinal infarction. Full recanalization was obtained in 38% of patients after 6 months of anticoagulation. Failure to achieve recanalization was independently related to ascites (HR 3.8) and splenic vein thrombosis (HR 3.5). A year after the diagnosis of NCPVT, 40% of the patients had permanent occlusion of portal vein and portal cavernoma. Bleeding was observed in 9% of the patients.

After a median follow-up of 8 months, mortality rate was 2% and was not related to bleeding or NCPVT [28].

DOACs have been sporadically used for the treatment of recent NCPVT [38, 39]. Recently, a retrospective study examining 330 patients with recent NCPVT treated with VKAs ($n = 108$), LMWH ($n = 70$), DOACs ($n = 93$), fondaparinux ($n = 2$), and no anticoagulation ($n = 57$) was published [40]. DOACs were associated with higher rate of complete thrombus resolution (HR 2.91) and lower rate of major bleeding (HR 0.20) compared with warfarin [40]. In another retrospective study looking at thrombosis recurrence and risk of bleeding in 26 patients with NCPVT treated with DOACs vs. 23 patients with NCPVT treated with enoxaparin, there was no difference in recurrent thrombosis and major bleeding between the two groups. Unfortunately, data regarding response to anticoagulation were not presented [38].

In patients with underlying thrombophilia, it is recommended to prolong anticoagulation independently of thrombosis resolution [8]. In a retrospective multicenter study including 109 patients with MPDs and NCPVT ($n = 63$) or BCS ($n = 46$), cytoreductive therapy was associated with a lower risk of liver-related events and vascular complications [15]. In NCPVT patients with paroxysmal nocturnal hemoglobinuria, eculizumab is indicated to achieve recanalization and prevent thrombosis recurrence [41].

Extrahepatic Portal Vein Obstruction (Chronic Portal Vein Thrombosis)

The aim of long-term anticoagulant therapy in extrahepatic portal vein obstruction (EHPVO) is to prevent thrombosis extension/recurrence and associated complications. Once prophylaxis for gastrointestinal bleeding has been implemented, we suggest considering permanent anticoagulation in patients with *any* prothrombotic condition, past history of intestinal ischemia, or in those with recurrent thrombosis during follow-up. This extends the previous recommendation by the European Association for the Study of the Liver [8], which considered permanent anticoagulation only in the presence of a *strong* prothrombotic condition, past history of intestinal ischemia or recurrent thrombosis on follow-up. No specific recommendation can be formulated for patients without persistent provoking factors.

The evidence regarding anticoagulation in patients with EHPVO is based on three-old retrospective studies [37, 42, 43] and two recent prospective cohorts [44, 45], which all suggest that long-term anticoagulation lowers the risk of thrombosis extension and recurrence. The impact of anticoagulation on patient's survival is not as clear as anticoagulation was independently associated with improved survival in one study [46] but not in other two [42, 43].

Condat et al. showed that anticoagulation in EHPVO was associated with a lower risk of mesenteric venous infarction (rate 0.82 and 5.2 per 100 patient-years in 84 treated vs. 52 non-treated patients, respectively) [42]. Similar results by Amitrano showed that the risk of recurrent thrombosis in treated patients was lower than in untreated patients (0% [0/10] vs. 46% [39/85]) [37].

In a recent international study including patients with splanchnic vein thrombosis ($n = 604$), the risk of recurrent thrombosis during a 2-year follow-up nearly doubled the risk of bleeding and risk of death related to thrombotic events was more than double than that related to bleeding [44]. Although the population was heterogenous due to inclusion of both patients with and without cirrhosis, the multivariate analysis in the subgroup of patients without cirrhosis ($n = 437$) showed that MPD and unprovoked SVT were independently associated with increased risk of vascular events (OR 9.02; $p = 0.01$ and OR 3.74; $p = 0.02$, respectively). On the other hand, anticoagulation reduced the risk of thrombosis (OR 0.88; $p < 0.001$) and was not associated with increased risk of major bleeding.

Current guidelines suggest that no anticoagulation should be given to patients with incidentally detected SVT [47]. By comparing patients with incidentally diagnosed SVT ($n = 177$) vs. those with clinically suspected PVT ($n = 420$) from the same multicenter registry, Riva et al. [45] demonstrated that the two conditions have similar clinical course and prognosis, which indirectly suggests that the same strategy should be considered in both groups. Unfortunately, this study did not exclude patients with cirrhosis and the higher number of cirrhotic patients in the incidentally detected group compared to the clinically suspected group (82/177 [46%] vs. 84/416 [20%], respectively; $p < 0.0001$) hinders the strength of the results.

The most frequent complication in patients with EHPVO is gastrointestinal bleeding related to portal hypertension [37, 42, 46]. Two retrospective studies show that long-term anticoagulation is not associated with increased risk of bleeding provided that prophylaxis is implemented prior to start of anticoagulation [37, 42]. Conversely, in another cohort in which a strategy for prophylaxis was not evaluated, anticoagulation was associated with a higher risk of bleeding (OR 2.0; $p = 0.01$) [43]. Yet, the severity of bleeding was not increased in treated vs. non-treated patients and recurrent thrombosis, but neither bleeding nor anticoagulation was associated with increased mortality (OR 3.1; $p = 0.02$) [43].

Current recommendation is therefore to start anticoagulation once prophylaxis for gastrointestinal bleeding is implemented [8]. However, preliminary data suggest that endoscopic variceal ligation (EVL) can be performed in patients with EHPVO without withdrawal of anticoagulation. No difference was found in a recent bicentric study between risk of post-EVL bleeding between patients who underwent EVL while on VKA vs. patients in whom anticoagulation was withdrawn (9 bleeding out of 121 session in 31 patients receiving VKA vs. 6 episodes out of 130 session in 13 patients not receiving VKA [48], similarly to patients with cirrhosis [49]).

Regarding DOACs, patients with SVT were excluded from phase III clinical trials and no large prospective study has yet investigated the safety of DOACs in the setting of EHPVO. The VALDIG consortium reported on the outcome of 38 patients with PVT without liver disease treated with different DOACs. After a median follow-up of approximately 26 months, 20% of patients (5/26) experienced side effects: 3 bleeding (1 VH, 1 due to gastric ulcer, and 1 prolonged menstrual bleeding) and 2 thrombosis (1 progressive PVT and 1 thrombosis of mesocaval shunt).

DOACs were stopped in four cases [50]. An independent study from North America compared three groups of patients: 63 patients with VTE in atypical locations (of whom 26 with SVT treated with rivaroxaban and apixaban), 23 patients with VTE (of whom 22 with SVT treated with enoxaparin), and 352 patients with VTE in typical locations treated with DOACs. Rates of recurrent thrombosis and major bleeding in patients receiving DOACs were comparable to those receiving traditional anticoagulation [38]. While awaiting prospective controlled studies, these preliminary data would suggest that DOACs may be considered for the treatment of patients with SVT.

Portal Vein Thrombosis in Cirrhosis

Portal vein thrombosis (PVT) in cirrhosis is a dynamic process and spontaneous recanalization may occur. Rate of recanalization ranges according to thrombosis and patient characteristics. In compensated patients with non-occlusive PVT, it is as high as 70% [51]. In decompensated patients with occlusive PVT, it is much lower (2% in a study including 42 patients with ascites) [52]. In decompensated patients or in those awaiting transplantation, partial PVT progresses in between 50% and 70% of patients at 2 years follow-up [52–54].

The influence of PVT on the risk of hepatic decompensation is still a matter of debate [55]. In compensated patients, partial PVT is not predictive of decompensation [51]. In decompensated patients, the impact of PVT on further decompensation is not as clear [56]. PVT may be asymptomatic and incidentally diagnosed at follow-up ultrasound; however, it may also be associated with risk of early treatment failure after VH [57], longer time to achieve variceal eradication, and higher risk of variceal relapse after eradication [58]. In addition, complete PVT has been related to a significantly higher post-liver transplant mortality [59].

Different treatment strategies have been adopted in cirrhosis complicated by PVT (Table 57.2) [53, 54, 56, 60–70, 73–75, 77]. There is no agreement on which drug should be used and the evidence is based on case series which show that recanalization rate varies between ~50% and ~75% (Table 57.2) [78]. In a recent meta-analysis including eight studies and 353 patients, a higher proportion of patients treated with anticoagulation had PVT recanalization compared with patients who were not anticoagulated (71% vs. 42%, respectively; $p < 0.0001$) [79]. Early start of anticoagulation (<6 months after PVT diagnosis) is the most important factor for predicting a response to therapy [8]. Involvement of mesenteric veins and/or the severity of cirrhosis have been reported as negative predictive factors for response [63, 75] but with conflicting results [56]. When anticoagulation is withdrawn, recurrence of thrombosis is frequent [73, 74]. Prolongation of anticoagulation treatment after recanalization may reduce the risk of re-thrombosis [53].

Cirrhosis is frequently associated with alterations of hemostasis [80–83] and there is still hesitation to anticoagulate these patients due to concerns of bleeding. However, current evidence does not show increased adverse outcomes in

Table 57.2 Published cohorts evaluating the efficacy of anticoagulant therapy for the treatment of portal vein thrombosis in patients with cirrhosis

Study	Patients (*n*)	Anticoagulation T`ype	Anticoagulation Duration (months)	PVT Characteristics Total/ partial	PVT Characteristics Extension to SMV/SV	PVT Outcome No/partial/ complete recanalization
Retrospective						
Werner, 2013 [60]	28	VKA	12	NA	15	5/12/11
Chung, 2014 [61]	14	VKA	3.7	NA	3	3/5/6
Naeshiro, 2014 [62]	26	Daparanoid ± AT	0.5	NA	11	6/16/4
Chen, 2016 [63]	30[a]	VKA	7.6	NA	20	7/NA/NA
La Mura, 2018 [64]	63	VKA	23	15/48	32	19/13/31
Artaza, 2018 [65]	32	LMWH [29] VKA [3]	12	7/25	16	9/6/17
Scheiner, 2018 [66]	10	VKA	12	6/6	NA	2/1/7
Hayashi, 2019 [67]	52	Daparanoid ± AT	0.5	6/46	9	NA
Noronha-Ferreira, 2019 [68]	37	LMWH [15] VKA [22]	NA	NA	14	NA/18
Rodriguez-Castro, 2019 [69]	65[b]	LMWH	12	18/47	20	18/15/28
Pettinari, 2019 [70]	81	LMWH [56] Fondaparinux [15] VKA [10]	13.4	8/51[c]	29	35/15/31
Senzolo, 2021 [71]	124	FPX [41] LMWH [72]	8/12	43/81	14/41	21/18/61
Prospective						
Francoz, 2005 [54]	19	VKA	8.1	18/1	NA	11/0/8
Amitrano, 2010 [73]	28	LMWH	6	5/23	20	5/14/9
Delgado, 2012 [74]	55	LWMH [47] VKA [8]	7	14/41	27	22/8/25
Senzolo, 2012 [53]	33	LMWH	6	11/24	14	12/9/12
Cui, 2015 [75]	65	LMWH	6	11/54	NA	51/8/6

Table 57.2 (continued)

Study	Patients (*n*)	Anticoagulation		PVT		
				Characteristics		Outcome
		Type	Duration (months)	Total/ partial	Extension to SMV/SV	No/partial/ complete recanalization
Senzolo, 2018 [56]	92	UFH [6] LMWH [76] VKA [32]	6.5	NA	76	45/47[d]
Kwon, 2018 [77]	91	LMWH	5.7	14/77	38	32/36/20[e]

LMWH low-molecular weight heparin, *VKA* vitamin K antagonist, *UFH* unfractioned heparin, *AT* antithrombin, *NA* not available

[a]Although 30 patients were treated, only 22 had follow-up. Partial or total recanalization was seen in 15 patients, but differentiation between partial and total was not reported

[b]Anticoagulant treatment was suspended in four patients

[c]Include 22 patients with intra-hepatic thrombosis

[d]Includes both partial and complete recanalization

[e]Two patients were lost to follow-up

anticoagulated patients (Table 57.3) [53, 54, 56, 60–70, 73–75, 77]. A platelet count <50 × 10^9/L and the use of VKA were the only factors related to bleeding in a previous study including 55 patients with cirrhosis and PVT [74]. Therefore, in patients with platelet counts <50 × 10^9/L, LMWH may represent the best choice [84].

One concern regarding LMWH in cirrhosis is the reduction of plasmatic antithrombin, owing to the fact that LMWH requires antithrombin to exert its action. Unfortunately, anti-Xa assay (a test that is used to monitor LMWH activity) cannot be used in cirrhosis. Indeed, AT-deficient plasma, such that observed in cirrhosis, yields false anti-Xa determination due to decreased accuracy of classical anti-Xa assays [85]. In patients treated with VKA, a close monitoring of the INR is important to determine the therapeutic range and dosing of these drugs may be challenging due to preexisting elevations of INR. In a cohort study evaluating 29,000 INR measurements, liver disease and alcohol abuse were independently correlated with excessive anticoagulation [86].

Fondaparinux (FPX) has a linear pharmacokinetic profile with longer half-life, which allows once-daily administration, and does not bind to plasma proteins, which makes the development of immune thrombocytopenia unlikely [87, 88]. In a recent retrospective study comparing 124 patients with cirrhosis and PVT, 41 (33%) treated with FPX and 83 (67%) with LMWH, we found that FPX was associated with a higher probability of recanalization at 36 months. Interestingly, FPX remained effective also when used at lower dose due to coexisting thrombocytopenia. Yet, a higher bleeding rate observed in FPX-treated patients suggests caution in the use of FPX in cirrhosis [71].

DOACs that inhibit thrombin (dabigatran) or activated factor X (rivaroxaban, apixaban, edoxaban) are appealing alternatives. Advantages of DOACs include quick onset of action, oral administration, and no need for routine monitoring drug

Table 57.3 Bleeding episodes in patients with cirrhosis and portal vein thrombosis treated with anticoagulation

Study	Patients (n)	Child class (A/B/C)	Anticoagulation (type)	Adverse events associated with anticoagulation
Retrospective				
Werner, 2013 [60]	28	NA	VKA	1 vaginal bleeding
Chung, 2014 [61]	14	6/8/0	VKA	None
Naeshiro, 2014 [62]	26	13/8/5	Daparanoid ± AT	None
Chen, 2016 [63]	30	6/7/15	VKA	4 hematemesis; 1 epistaxis; 1 gingival bleeding
La Mura, 2018 [64]	63	28/35[a]	VKA	Major [8]: 5 GI bleeding; 1 intracranial bleeding; 1 hematoma; 1 hematuria; Minor [13]: 8 epistaxis; 1 lower GI bleeding; 1 anemia; 2 hematoma
Artaza, 2018 [65]	32	18/12/2	LMWH [29] VKA [3]	2 EV bleeding; 1 intracranial bleeding
Scheiner, 2018 [66]	10	NA	VKA	None
Hayashi, 2019 [67]	52	13/25/14	Daparanoid ± AT	None
Noronha-Ferreira, 2019 [68]	37	12/16/9	LMWH [15] VKA [22]	None
Rodriguez-Castro, 2019 [69]	65	27/23/15	LMWH	Major [2]: 1 intracranial bleeding; 1 congestive gastropathy; Minor [2]: 1 epistaxis; 1 hematuria
Pettinari, 2019 [70]	81	43/33/5	LMWH [56] FPX [15] VKA [10]	4 variceal bleeding, 6 hemorrhoidal bleeding; 2 GAVE; 4 post-traumatic.
Senzolo, 2021 [71]	124		FPX [41] LMWH [72]	22 episodes of bleeding, of whom 7 GI non-PH related; 3 VH; 1 post-paracentesis, and 1 intra-cranial
Prospective				
Francoz, 2005 [54]	19	2/13/4	VKA	1 EV bleeding after EVL
Amitrano, 2010 [73]	28	14/14[a]	LMWH	2 anemia secondary to portal hypertensive gastropathy requiring iron transfusion
Delgado, 2012 [74]	55	25/21/9	LWMH [47] VKA [8]	1 lower GI bleeding; 1 obscure gastrointestinal bleeding; 1 oral bleeding after dental extraction; 1 vaginal bleeding; 1 surgical wound hemorrhage

Table 57.3 (continued)

Study	Patients (n)	Child class (A/B/C)	Anticoagulation (type)	Adverse events associated with anticoagulation
Senzolo, 2012 [53]	33	11/16/8	LMWH	1 intracranial bleeding; 1 epistaxis; 1 variceal bleed; 1 hematuria
Cui, 2015 [75]	65	Pugh 7 (IQR 6–8)	LMWH	3 injections sites; 5 epistaxis; 2 hematuria
Senzolo, 2018 [56]	92	Pugh 7 (IQR 5–8)	UFH [6] LMWH [76] VKA [32]	Major [19]: 10 EV bleeding; 3 GI; 3 intracranial/intraspinal bleeding
Kwon, 2018 [77]	91	45/42/4	LMWH	Major [4]: 1 EV bleeding; 2 intracranial/intraspinal bleeding; 1 not specified; Minor [7]: not described

LMWH low-molecular weight heparin, *VKA* vitamin K antagonist, *UFH* unfractioned heparin, *FPX* fondaparinux, *NA* not available, *EV* esophageal varices, *EVL* esophageal variceal ligation, *GI* gastrointestinal, *GAVE* gastric antral vascular ectasia, *AT* antithrombin
[a] Includes both Child B and C

levels or effect. However, each agent has distinctive properties requiring particular attention to dosing, absorption, and clearance. Careful consideration of use of DOAC in patients with renal impairment is needed as DOAC are all dependent on renal clearance to varying degrees. Another issue to be aware, particularly for dabigatran and edoxaban, is the potential drug–drug interaction with NSBBs, statins, and PPIs [89, 90]. The volume of distribution of DOACs may vary according to body mass index and is different in underweight vs. obese patients. These conditions frequently occur in cirrhosis [91] and should be considered in prescribing or confirming DOACs in these patients [92]. In fact, a theoretical risk of excessive anticoagulation exists when using DOAC in cirrhosis and serum levels of DOACs are not sufficient for monitoring. However, Potze et al. demonstrated a decreased in vitro effect of rivaroxaban in patients with cirrhosis by using TG. In contrast, an increased response to dabigatran was found. Interestingly, the enhanced effect of dabigatran on TG was proportional to the severity of cirrhosis [93]. The same group examined the in vitro anticoagulant potency of apixaban [94]. Twenty-five ng/mL of apixaban or 50 ng/mL of rivaroxaban were added to plasma samples of 11 healthy individuals and 14 patients with cirrhosis (Child B and C). While a fixed dose of the drugs decreased total TG in healthy volunteers by 55 ± 6% (rivaroxaban) and 51 ± 4% (apixaban), the mean decrease in TG in cirrhosis was lower (30 ± 9% for rivaroxaban, $p < 0.0001$; 32 ± 10% for apixaban, $p < 0.0001$).

Although they are currently used off-label, some preliminary experiences have been reported (Table 57.4) [50, 95–98, 100]. The first retrospective case-series was described by Intagliata et al. [95] who looked at safety of DOACs in 20 patients with cirrhosis (Child A and B) treated with apixaban ($n = 11$) or rivaroxaban ($n = 9$). Indication for anticoagulation included PVT (approximately 2/3 of patients), atrial fibrillation, and non-splanchnic thrombosis. Median duration of AC was 270 days

Table 57.4 Published cohorts of patients with cirrhosis treated with direct oral anticoagulants

Study	Patients (n)	Patient characteristics	Anticoagulation			Adverse events
			Type	Duration (months)	Indication	
Intagliata, 2016 [95]	20	Child A [9] Child B [11] MELD 12 [10–15]	Rivaroxaban 10 mg or 20 mg [9] Apixaban 2.5 mg or 5 mg [11]	9	SVT [12] Non-splanchnic [4] Atrial fibrillation [4]	Major [1]: intracranial bleeding Moderate [1]: GI bleeding Mild [2]: GI bleeding and vaginal bleeding
De Gottardi, 2017 [50]	36	Child-Pugh 6 (range 5–8)	Rivaroxaban 15 mg [30] Apixaban 5 mg [4] Dabigatran 110 mg or 220 mg [2]	7	PVT [22] Concomitant BCS [5] Cardiac arrythmia [5] DVT [4]	Major [1]: lower GI Minor [4]: portal hypertensive bleedings, lower GI, epistaxis; post band ligation
Hum, 2017 [96]	27	Child A [11] Child B [12] Child C [4]	Rivaroxaban 15 mg with or without an initial 20 mg dose [17] Apixaban 5 mg, with or without an initial 10 mg dose [10]	7	Atrial fibrillation [15] DVT [12] PVT [4]	Major [1]; moderate [4]; minor [3]; 5/8 bleeding were GI-related)
Nagaoki, 2018 [97]	20	Child A [15] Child B [5]	Danaparoid for 2 weeks switched to edoxaban 60 mg [4] or 30 mg [16]	6	PVT	2 lower GI and 1 small bowel bleeding
Hanafi, 2019 [98]	40	HCV-cirrhosis Child A–B MELD 11 ± 1.4	LMWH for 3 days switched to rivaroxaban 10 mg	~3–7	PVT	None

Table 57.4 (continued)

Study	Patients (n)	Patient characteristics	Anticoagulation Type	Duration (months)	Indication	Adverse events
Semmler, 2021 [99]	104	Child A [53] Child B [44] Child C [7]	Edoxaban in 59 patients Apixaban in 16 patients Rivaroxaban in 21 patients Dabigatran in 2 Sequential treatment in 6 61 (58.7% full dose) and 43 (41.3% reduced dose)	10.5	PVT [74] BCS [9] AF [15] AF and PVT [7] DVT/PE [6] Others [2]	6 procedure related bleedings (1 major and 5 minor) 33 spontaneous bleeding events

MELD model for end-stage liver disease, *PVT* portal vein thrombosis, *SVT* splanchnic vein thrombosis, *DVT* deep vein thrombosis, *PE* pulmonary embolism, *BCS* Budd–Chiari syndrome, *GI* gastro-intestinal
[a] Dose unknown

and DOACs were given at full dose (10 mg apixaban or 20 mg rivaroxaban) in 15 patients and at lower dosages in 5 patients (5 mg apixaban or 10 mg rivaroxaban daily). Major bleeding occurred in 5% of patients: 1 non-fatal intracranial bleeding, 2 GI, and 1 vaginal bleeding. Similar results were found in a European cohort of 36 patients with Child A and B cirrhosis (22 with PVT) treated with rivaroxaban (83%), dabigatran (11%), and apixaban (6%). After a median follow-up of 15 months, 1 major bleeding (lower GI) and 4 minor bleedings (portal hypertensive gastropathy, lower GI, post bend ligation) were reported [50]. A recent, multicenter retrospective study evaluated safety of DOACs in 123 patients with cirrhosis/vascular liver diseases treated with DOACs vs. 58 controls treated with LMWH/VKA [99]. Rate of major bleeding was higher in Child B/C patients vs. Child A; however, no significant difference was found between DOAC and LMWH/VKA groups (28.6% vs. 19.0%, respectively; $p = 0.162$) [99].

A recent Egyptian RCT evaluated safety and efficacy of DOACs in patients with cirrhosis and PVT [98]. The investigators randomized 80 patients with Child A or B HCV-related cirrhosis complicated by acute PVT to rivaroxaban (low dosage, 10 mg daily) vs. VKA (monitored by INR). Interestingly, despite the relatively low dosage of rivaroxaban, recanalization of portal vein was obtained in 85% of cases vs. 45% of patients treated with VKA. Not only rivaroxaban was effective in treating PVT, but it was also safe, and rate of bleeding was higher in warfarin (43%) vs. rivaroxaban (no episode of bleeding).

Few cases of probable rivaroxaban-induced liver injury have been reported [101]. Most patients showed transient elevation of serum transaminases and bilirubin which spontaneously resolved after rivaroxaban discontinuation. However, this underlines the need for further data on pharmacovigilance in this setting.

In conclusion, although preliminary and mostly retrospective, these data suggest that DOACs in cirrhosis may be efficient and safe in patients belonging to Child A class. Data regarding Child B are few and DOACs should be used with caution in these patients. In Child C, DOACs should be considered on a case-by-case basis. There is no evidence to suggest any particular DOAC for patients with Child A cirrhosis. Based on pharmacodynamics properties and a favorable drug-induced liver injury profile, it appears that apixaban may be the preferred DOAC in Child B patients. Further prospective studies are awaited to assess the risk/benefit ratio of individual DOACs in cirrhosis, particularly in decompensated patients.

References

1. Zanetto A, Pellone M, Senzolo M. Milestones in the discovery of Budd-Chiari syndrome. Liver Int. 2019;39:1180–5.
2. Mahmoud AE, Elias E, Beauchamp N, Wilde JT. Prevalence of the factor V Leiden mutation in hepatic and portal vein thrombosis. Gut. 1997;40:798–800.
3. Patel RK, Lea NC, Heneghan MA, Westwood NB, Milojkovic D, Thanigaikumar M, Yallop D, et al. Prevalence of the activating JAK2 tyrosine kinase mutation V617F in the Budd-Chiari syndrome. Gastroenterology. 2006;130:2031–8.
4. McCarthy PM, van Heerden JA, Adson MA, Schafer LW, Wiesner RH. The Budd-Chiari syndrome. Medical and surgical management of 30 patients. Arch Surg. 1985;120:657–62.
5. Wang ZG, Jones RS. Budd-Chiari syndrome. Curr Probl Surg. 1996;33:83–211.
6. Plessier A, Sibert A, Consigny Y, Hakime A, Zappa M, Denninger MH, Condat B, et al. Aiming at minimal invasiveness as a therapeutic strategy for Budd-Chiari syndrome. Hepatology. 2006;44:1308–16.
7. Seijo S, Plessier A, Hoekstra J, Dell'era A, Mandair D, Rifai K, Trebicka J, et al. Good long-term outcome of Budd-Chiari syndrome with a step-wise management. Hepatology. 2013;57:1962–8.
8. European Association for the Study of the Liver. EASL Clinical Practice Guidelines: vascular diseases of the liver. J Hepatol. 2016;64:179–202.
9. Hadengue A, Poliquin M, Vilgrain V, Belghiti J, Degott C, Erlinger S, Benhamou JP. The changing scene of hepatic vein thrombosis: recognition of asymptomatic cases. Gastroenterology. 1994;106:1042–7.
10. Zeitoun G, Escolano S, Hadengue A, Azar N, El Younsi M, Mallet A, Boudet MJ, et al. Outcome of Budd-Chiari syndrome: a multivariate analysis of factors related to survival including surgical portosystemic shunting. Hepatology. 1999;30:84–9.
11. Castelli R, Gallipoli P, Schiavon R, Teatini T, Deliliers GL, Bergamaschini L. High prevalence of heparin induced thrombocytopenia with thrombosis among patients with essential thrombocytemia carrying V617F mutation. J Thromb Thrombolysis. 2018;45:106–13.
12. Bellucci S, Ignatova E, Jaillet N, Boffa MC. Platelet hyperactivation in patients with essential thrombocythemia is not associated with vascular endothelial cell damage as judged by the level of plasma thrombomodulin, protein S, PAI-1, t-PA and vWF. Thromb Haemost. 1993;70:736–42.
13. Randi ML, Tezza F, Scapin M, Duner E, Scarparo P, Scandellari R, Fabris F. Heparin-induced thrombocytopenia in patients with Philadelphia-negative myeloproliferative disorders and unusual splanchnic or cerebral vein thrombosis. Acta Haematol. 2010;123:140–5.

14. De Stefano V, Vannucchi AM, Ruggeri M, Cervantes F, Alvarez-Larran A, Iurlo A, Randi ML, et al. Splanchnic vein thrombosis in myeloproliferative neoplasms: risk factors for recurrences in a cohort of 181 patients. Blood Cancer J. 2016;6:e493.
15. Chagneau C, Roy L, Guilhot J, Gloria O, Ollivier-Hourmand I, Bureau C. Impact o cytoreductive therapy on the outcome of patients with myeloproliferative neoplasms and hepatosplanchnic vein thrombosis. J Hepatol. 2013;58:857A.
16. Desbois AC, Rautou PE, Biard L, Belmatoug N, Wechsler B, Resche-Rigon M, Zarrouk V, et al. Behcet's disease in Budd-Chiari syndrome. Orphanet J Rare Dis. 2014;9:104.
17. Alvarez-Larran A, Pereira A, Magaz M, Hernandez-Boluda JC, Garrote M, Cuevas B, Ferrer-Marin F, et al. Natural history of polycythemia vera and essential thrombocythemia presenting with splanchnic vein thrombosis. Ann Hematol. 2020;99:791–8.
18. Dogrul AB, Yankol Y, Mecit N, Kanmaz T, Acarli K, Kalayoglu M. Orthotopic liver transplant for Budd-Chiari syndrome: an analysis of 14 cases. Exp Clin Transplant. 2016;14:641–5.
19. Mentha G, Giostra E, Majno PE, Bechstein WO, Neuhaus P, O'Grady J, Praseedom RK, et al. Liver transplantation for Budd-Chiari syndrome: a European study on 248 patients from 51 centres. J Hepatol. 2006;44:520–8.
20. Potthoff A, Attia D, Pischke S, Mederacke I, Beutel G, Rifai K, Deterding K, et al. Long-term outcome of liver transplant patients with Budd-Chiari syndrome secondary to myeloproliferative neoplasms. Liver Int. 2015;35:2042–9.
21. Westbrook RH, Lea NC, Mohamedali AM, Smith AE, Orr DW, Roberts LN, Heaton ND, et al. Prevalence and clinical outcomes of the 46/1 haplotype, Janus kinase 2 mutations, and ten-eleven translocation 2 mutations in Budd-Chiari syndrome and their impact on thrombotic complications post liver transplantation. Liver Transpl. 2012;18:819–27.
22. Chinnakotla S, Klintmalm GB, Kim P, Tomiyama K, Klintmalm E, Davis GL, Trotter JF, et al. Long-term follow-up of liver transplantation for Budd-Chiari syndrome with antithrombotic therapy based on the etiology. Transplantation. 2011;92:341–5.
23. Melear JM, Goldstein RM, Levy MF, Molmenti EP, Cooper B, Netto GJ, Klintmalm GB, et al. Hematologic aspects of liver transplantation for Budd-Chiari syndrome with special reference to myeloproliferative disorders. Transplantation. 2002;74:1090–5.
24. Rautou PE, Douarin L, Denninger MH, Escolano S, Lebrec D, Moreau R, Vidaud M, et al. Bleeding in patients with Budd-Chiari syndrome. J Hepatol. 2011;54:56–63.
25. Husova L. Use of idarucizumab in clinical practice: a case report. Vnitr Lek. 2019;65:377–8.
26. Sharma S, Kumar R, Rout G, Gamanagatti SR, Shalimar. Dabigatran as an oral anticoagulant in patients with Budd-Chiari syndrome post-percutaneous endovascular intervention. J Gastroenterol Hepatol. 2020;35(4):654–62.
27. Singal AK, Kamath PS, Tefferi A. Mesenteric venous thrombosis. Mayo Clin Proc. 2013;88:285–94.
28. Plessier A, Darwish-Murad S, Hernandez-Guerra M, Consigny Y, Fabris F, Trebicka J, Heller J, et al. Acute portal vein thrombosis unrelated to cirrhosis: a prospective multicenter follow-up study. Hepatology. 2010;51:210–8.
29. Kumar S, Sarr MG, Kamath PS. Mesenteric venous thrombosis. N Engl J Med. 2001;345:1683–8.
30. Hollingshead M, Burke CT, Mauro MA, Weeks SM, Dixon RG, Jaques PF. Transcatheter thrombolytic therapy for acute mesenteric and portal vein thrombosis. J Vasc Interv Radiol. 2005;16:651–61.
31. Benmassaoud A, AlRubaiy L, Yu D, Chowdary P, Sekhar M, Parikh P, Finkel J, et al. A stepwise thrombolysis regimen in the management of acute portal vein thrombosis in patients with evidence of intestinal ischaemia. Aliment Pharmacol Ther. 2019;50:1049–58.
32. Brunaud L, Antunes L, Collinet-Adler S, Marchal F, Ayav A, Bresler L, Boissel P. Acute mesenteric venous thrombosis: case for nonoperative management. J Vasc Surg. 2001;34:673–9.
33. Turnes J, Garcia-Pagan JC, Gonzalez M, Aracil C, Calleja JL, Ripoll C, Abraldes JG, et al. Portal hypertension-related complications after acute portal vein thrombosis: impact of early anticoagulation. Clin Gastroenterol Hepatol. 2008;6:1412–7.

34. Franceschet I, Zanetto A, Ferrarese A, Burra P, Senzolo M. Therapeutic approaches for portal biliopathy: a systematic review. World J Gastroenterol. 2016;22:9909–20.
35. Llop E, de Juan C, Seijo S, Garcia-Criado A, Abraldes JG, Bosch J, Garcia-Pagan JC. Portal cholangiopathy: radiological classification and natural history. Gut. 2011;60:853–60.
36. Condat B, Pessione F, Helene Denninger M, Hillaire S, Valla D. Recent portal or mesenteric venous thrombosis: increased recognition and frequent recanalization on anticoagulant therapy. Hepatology. 2000;32:466–70.
37. Amitrano L, Guardascione MA, Scaglione M, Pezzullo L, Sangiuliano N, Armellino MF, Manguso F, et al. Prognostic factors in noncirrhotic patients with splanchnic vein thromboses. Am J Gastroenterol. 2007;102:2464–70.
38. Janczak DT, Mimier MK, McBane RD, Kamath PS, Simmons BS, Bott-Kitslaar DM, Lenz CJ, et al. Rivaroxaban and apixaban for initial treatment of acute venous thromboembolism of atypical location. Mayo Clin Proc. 2018;93:40–7.
39. Nery F, Valadares D, Morais S, Gomes MT, De Gottardi A. Efficacy and safety of direct-acting oral anticoagulants use in acute portal vein thrombosis unrelated to cirrhosis. Gastroenterology Res. 2017;10:141–3.
40. Naymagon L, Tremblay D, Zubizarreta N, Moshier E, Troy K, Schiano T, Mascarenhas J. The efficacy and safety of direct oral anticoagulants in noncirrhotic portal vein thrombosis. Blood Adv. 2020;4:655–66.
41. van Bijnen ST, van Rijn RS, Koljenovic S, te Boekhorst P, de Witte T, Muus P. Possible high risk of thrombotic events in patients with paroxysmal nocturnal haemoglobinuria after discontinuation of eculizumab. Br J Haematol. 2012;157:762–3.
42. Condat B, Pessione F, Hillaire S, Denninger MH, Guillin MC, Poliquin M, Hadengue A, et al. Current outcome of portal vein thrombosis in adults: risk and benefit of anticoagulant therapy. Gastroenterology. 2001;120:490–7.
43. Spaander MC, Hoekstra J, Hansen BE, Van Buuren HR, Leebeek FW, Janssen HL. Anticoagulant therapy in patients with non-cirrhotic portal vein thrombosis: effect on new thrombotic events and gastrointestinal bleeding. J Thromb Haemost. 2013;11:452–9.
44. Ageno W, Riva N, Schulman S, Beyer-Westendorf J, Bang SM, Senzolo M, Grandone E, et al. Long-term clinical outcomes of splanchnic vein thrombosis: results of an international registry. JAMA Intern Med. 2015;175:1474–80.
45. Riva N, Ageno W, Schulman S, Beyer-Westendorf J, Duce R, Malato A, Santoro R, et al. Clinical history and antithrombotic treatment of incidentally detected splanchnic vein thrombosis: a multicentre, international prospective registry. Lancet Haematol. 2016;3:e267–75.
46. Orr DW, Harrison PM, Devlin J, Karani JB, Kane PA, Heaton ND, O'Grady JG, et al. Chronic mesenteric venous thrombosis: evaluation and determinants of survival during long-term follow-up. Clin Gastroenterol Hepatol. 2007;5:80–6.
47. Kearon C, Akl EA, Comerota AJ, Prandoni P, Bounameaux H, Goldhaber SZ, Nelson ME, et al. Antithrombotic therapy for VTE disease: Antithrombotic Therapy and Prevention of Thrombosis, 9th ed: American College of Chest Physicians Evidence-Based Clinical Practice Guidelines. Chest. 2012;141:e419S–96S.
48. Guillaume M, Christol C, Plessier A, Corbic M, Peron JM, Sommet A, Rautou PE, et al. Bleeding risk of variceal band ligation in extrahepatic portal vein obstruction is not increased by oral anticoagulation. Eur J Gastroenterol Hepatol. 2018;30:563–8.
49. Ponthus S, Spahr L, Casini A, Berney T, Frossard JL, Majno P, Elkrief L. Safety of variceal band ligation in patients with cirrhosis and portal vein thrombosis treated with anticoagulant therapy: a retrospective study. Eur J Gastroenterol Hepatol. 2020;32:395–400.
50. De Gottardi A, Trebicka J, Klinger C, Plessier A, Seijo S, Terziroli B, Magenta L, et al. Antithrombotic treatment with direct-acting oral anticoagulants in patients with splanchnic vein thrombosis and cirrhosis. Liver Int. 2017;37:694–9.
51. Nery F, Chevret S, Condat B, de Raucourt E, Boudaoud L, Rautou PE, Plessier A, et al. Causes and consequences of portal vein thrombosis in 1,243 patients with cirrhosis: results of a longitudinal study. Hepatology. 2015;61:660–7.

52. Luca A, Caruso S, Milazzo M, Marrone G, Mamone G, Crino F, Maruzzelli L, et al. Natural course of extrahepatic nonmalignant partial portal vein thrombosis in patients with cirrhosis. Radiology. 2012;265:124–32.
53. Senzolo M, Sartori TM, Rossetto V, Burra P, Cillo U, Boccagni P, Gasparini D, et al. Prospective evaluation of anticoagulation and transjugular intrahepatic portosystemic shunt for the management of portal vein thrombosis in cirrhosis. Liver Int. 2012;32:919–27.
54. Francoz C, Belghiti J, Vilgrain V, Sommacale D, Paradis V, Condat B, Denninger MH, et al. Splanchnic vein thrombosis in candidates for liver transplantation: usefulness of screening and anticoagulation. Gut. 2005;54:691–7.
55. Zanetto A, Garcia-Tsao G. Some answers and more questions about portal vein thrombosis in patients with decompensated cirrhosis. Clin Gastroenterol Hepatol. 2020;18(11):2432–4.
56. Senzolo M, Riva N, Dentali F, Rodriguez-Castro K, Sartori MT, Bang SM, Martinelli I, et al. Long-term outcome of splanchnic vein thrombosis in cirrhosis. Clin Transl Gastroenterol. 2018;9:176.
57. D'Amico G, De Franchis R, Cooperative SG. Upper digestive bleeding in cirrhosis. Post-therapeutic outcome and prognostic indicators. Hepatology. 2003;38:599–612.
58. Zanetto A, Shalaby S, Feltracco P, Gambato M, Germani G, Russo FP, Burra P, et al. Recent advances in the management of acute variceal hemorrhage. J Clin Med. 2021;10(17):3818.
59. Zanetto A, Rodriguez-Kastro KI, Germani G, Ferrarese A, Cillo U, Burra P, Senzolo M. Mortality in liver transplant recipients with portal vein thrombosis—an updated meta-analysis. Transpl Int. 2018;31:1318–29.
60. Werner KT, Sando S, Carey EJ, Vargas HE, Byrne TJ, Douglas DD, Harrison ME, et al. Portal vein thrombosis in patients with end stage liver disease awaiting liver transplantation: outcome of anticoagulation. Dig Dis Sci. 2013;58:1776–80.
61. Chung JW, Kim GH, Lee JH, Ok KS, Jang ES, Jeong SH, Kim JW. Safety, efficacy, and response predictors of anticoagulation for the treatment of nonmalignant portal-vein thrombosis in patients with cirrhosis: a propensity score matching analysis. Clin Mol Hepatol. 2014;20:384–91.
62. Naeshiro N, Aikata H, Hyogo H, Kan H, Fujino H, Kobayashi T, Fukuhara T, et al. Efficacy and safety of the anticoagulant drug, danaparoid sodium, in the treatment of portal vein thrombosis in patients with liver cirrhosis. Hepatol Res. 2015;45:656–62.
63. Chen H, Liu L, Qi X, He C, Wu F, Fan D, Han G. Efficacy and safety of anticoagulation in more advanced portal vein thrombosis in patients with liver cirrhosis. Eur J Gastroenterol Hepatol. 2016;28:82–9.
64. La Mura V, Braham S, Tosetti G, Branchi F, Bitto N, Moia M, Fracanzani AL, et al. Harmful and beneficial effects of anticoagulants in patients with cirrhosis and portal vein thrombosis. Clin Gastroenterol Hepatol. 2018;16:1146–52. e1144
65. Artaza T, Lopes M, Romero M, Gomez AZ, de la Cruz G, Sanchez JJ, Gonzalez C, et al. Efficacy and safety of anticoagulation in non-malignant portal vein thrombosis in patients with liver cirrhosis. Gastroenterol Hepatol. 2018;41:611–7.
66. Scheiner B, Stammet PR, Pokorny S, Bucsics T, Schwabl P, Brichta A, Thaler J, et al. Anticoagulation in non-malignant portal vein thrombosis is safe and improves hepatic function. Wien Klin Wochenschr. 2018;130:446–55.
67. Hayashi T, Takatori H, Horii R, Nio K, Terashima T, Iida N, Kitahara M, et al. Danaparoid sodium-based anticoagulation therapy for portal vein thrombosis in cirrhosis patients. BMC Gastroenterol. 2019;19:217.
68. Noronha Ferreira C, Reis D, Cortez-Pinto H, Tato Marinho R, Goncalves A, Palma S, Leite I, et al. Anticoagulation in cirrhosis and portal vein thrombosis is safe and improves prognosis in advanced cirrhosis. Dig Dis Sci. 2019;64:2671–83.
69. Rodriguez-Castro KI, Vitale A, Fadin M, Shalaby S, Zerbinati P, Sartori MT, Landi S, et al. A prediction model for successful anticoagulation in cirrhotic portal vein thrombosis. Eur J Gastroenterol Hepatol. 2019;31:34–42.
70. Pettinari I, Vukotic R, Stefanescu H, Pecorelli A, Morelli M, Grigoras C, Sparchez Z, et al. Clinical impact and safety of anticoagulants for portal vein thrombosis in cirrhosis. Am J Gastroenterol. 2019;114:258–66.

71. Senzolo M, Piano S, Shalaby S, Tonon M, Tonello S, Zanetto A, Sacerdoti D, et al. Comparison of fondaparinux and low-molecular-weight heparin in the treatment of portal vein thrombosis in cirrhosis. Am J Med. 2021;134:1278–85. e1272
72. Tripodi A, Primignani M, Braham S, Chantarangkul V, Clerici M, Moia M, Peyvandi F. Coagulation parameters in patients with cirrhosis and portal vein thrombosis treated sequentially with low molecular weight heparin and vitamin K antagonists. Dig Liver Dis. 2016;48:1208–13.
73. Amitrano L, Guardascione MA, Menchise A, Martino R, Scaglione M, Giovine S, Romano L, et al. Safety and efficacy of anticoagulation therapy with low molecular weight heparin for portal vein thrombosis in patients with liver cirrhosis. J Clin Gastroenterol. 2010;44:448–51.
74. Delgado MG, Seijo S, Yepes I, Achecar L, Catalina MV, Garcia-Criado A, Abraldes JG, et al. Efficacy and safety of anticoagulation on patients with cirrhosis and portal vein thrombosis. Clin Gastroenterol Hepatol. 2012;10:776–83.
75. Cui SB, Shu RH, Yan SP, Wu H, Chen Y, Wang L, Zhu Q. Efficacy and safety of anticoagulation therapy with different doses of enoxaparin for portal vein thrombosis in cirrhotic patients with hepatitis B. Eur J Gastroenterol Hepatol. 2015;27:914–9.
76. Russo FP, Zanetto A, Campello E, Bulato C, Shalaby S, Spiezia L, Gavasso S, et al. Reversal of hypercoagulability in patients with HCV-related cirrhosis after treatment with direct-acting antivirals. Liver Int. 2018;38:2210–8.
77. Kwon J, Koh Y, Yu SJ, Yoon JH. Low-molecular-weight heparin treatment for portal vein thrombosis in liver cirrhosis: efficacy and the risk of hemorrhagic complications. Thromb Res. 2018;163:71–6.
78. Northup PG, Garcia-Pagan JC, Garcia-Tsao G, Intagliata NM, Superina RA, Roberts LN, Lisman T, et al. Vascular liver disorders, portal vein thrombosis, and procedural bleeding in patients with liver disease: 2020 practice guidance by the American Association for the Study of Liver Diseases. Hepatology. 2021;73:366–413.
79. Loffredo L, Pastori D, Farcomeni A, Violi F. Effects of anticoagulants in patients with cirrhosis and portal vein thrombosis: a systematic review and meta-analysis. Gastroenterology. 2017;153:480–7. e481
80. Zanetto A, Campello E, Bulato C, Gavasso S, Saggiorato G, Shalaby S, Spiezia L, et al. More pronounced hypercoagulable state and hypofibrinolysis in patients with cirrhosis with versus without HCC. Hepatol Commun. 2021;5(12):1987–2000.
81. Campello E, Zanetto A, Bulato C, Maggiolo S, Spiezia L, Russo FP, Gavasso S, et al. Coagulopathy is not predictive of bleeding in patients with acute decompensation of cirrhosis and acute-on-chronic liver failure. Liver Int. 2021;41:2455–66.
82. Zanetto A, Rinder HM, Senzolo M, Simioni P, Garcia-Tsao G. Reduced clot stability by thromboelastography as a potential indicator of procedure-related bleeding in decompensated cirrhosis. Hepatol Commun. 2021;5:272–82.
83. Zanetto A, Rinder HM, Campello E, Saggiorato G, Deng Y, Ciarleglio M, Wilson FP, et al. Acute kidney injury in decompensated cirrhosis is associated with both hypo-coagulable and hyper-coagulable features. Hepatology. 2020;72:1327–40.
84. Under the auspices of the Italian Association for the Study of Liver Diseases (AISF) and the Italian Society of Internal Medicine (SIMI). Hemostatic balance in patients with liver cirrhosis: report of a consensus conference. Dig Liver Dis. 2016;48:455–67.
85. Lehman CM, Rettmann JA, Wilson LW, Markewitz BA. Comparative performance of three anti-factor Xa heparin assays in patients in a medical intensive care unit receiving intravenous, unfractionated heparin. Am J Clin Pathol. 2006;126:416–21.
86. Brigden ML, Kay C, Le A, Graydon C, McLeod B. Audit of the frequency and clinical response to excessive oral anticoagulation in an out-patient population. Am J Hematol. 1998;59:22–7.
87. Schindewolf M, Steindl J, Beyer-Westendorf J, Schellong S, Dohmen PM, Brachmann J, Madlener K, et al. Use of fondaparinux off-label or approved anticoagulants for management of heparin-induced thrombocytopenia. J Am Coll Cardiol. 2017;70:2636–48.

88. Bauer KA, Hawkins DW, Peters PC, Petitou M, Herbert JM, van Boeckel CA, Meuleman DG. Fondaparinux, a synthetic pentasaccharide: the first in a new class of antithrombotic agents—the selective factor Xa inhibitors. Cardiovasc Drug Rev. 2002;20:37–52.
89. Burnett AE, Mahan CE, Vazquez SR, Oertel LB, Garcia DA, Ansell J. Guidance for the practical management of the direct oral anticoagulants (DOACs) in VTE treatment. J Thromb Thrombolysis. 2016;41:206–32.
90. Steffel J, Verhamme P, Potpara TS, Albaladejo P, Antz M, Desteghe L, Haeusler KG, et al. The 2018 European Heart Rhythm Association Practical Guide on the use of non-vitamin K antagonist oral anticoagulants in patients with atrial fibrillation. Eur Heart J. 2018;39:1330–93.
91. Sciarrone SS, Zanetto A, Russo FP, Germani G, Gambato M, Battistella S, Pellone M, et al. Malnourished cirrhotic patient: what should we do? Minerva Gastroenterol (Torino). 2021;67:11–22.
92. Chen A, Stecker E, Warden BA. Direct oral anticoagulant use: a practical guide to common clinical challenges. J Am Heart Assoc. 2020;9:e017559.
93. Potze W, Arshad F, Adelmeijer J, Blokzijl H, van den Berg AP, Meijers JC, Porte RJ, et al. Differential in vitro inhibition of thrombin generation by anticoagulant drugs in plasma from patients with cirrhosis. PLoS One. 2014;9:e88390.
94. Potze W, Adelmeijer J, Lisman T. Decreased in vitro anticoagulant potency of rivaroxaban and Apixaban in plasma from patients with cirrhosis. Hepatology. 2015;61:1435–6.
95. Intagliata NM, Henry ZH, Maitland H, Shah NL, Argo CK, Northup PG, Caldwell SH. Direct oral anticoagulants in cirrhosis patients pose similar risks of bleeding when compared to traditional anticoagulation. Dig Dis Sci. 2016;61:1721–7.
96. Hum J, Shatzel JJ, Jou JH, Deloughery TG. The efficacy and safety of direct oral anticoagulants vs traditional anticoagulants in cirrhosis. Eur J Haematol. 2017;98:393–7.
97. Nagaoki Y, Aikata H, Daijyo K, Teraoka Y, Shinohara F, Nakamura Y, Hatooka M, et al. Efficacy and safety of edoxaban for treatment of portal vein thrombosis following danaparoid sodium in patients with liver cirrhosis. Hepatol Res. 2018;48:51–8.
98. Hanafy AS, Abd-Elsalam S, Dawoud MM. Randomized controlled trial of rivaroxaban versus warfarin in the management of acute non-neoplastic portal vein thrombosis. Vasc Pharmacol. 2019;113:86–91.
99. Semmler G, Pomej K, Bauer DJM, Balcar L, Simbrunner B, Binter T, Hartl L, et al. Safety of direct oral anticoagulants in patients with advanced liver disease. Liver Int. 2021;41:2159–70.
100. Weinberg EM, Palecki J, Reddy KR. Direct-acting oral anticoagulants (DOACs) in cirrhosis and cirrhosis-associated portal vein thrombosis. Semin Liver Dis. 2019;39:195–208.
101. Russmann S, Niedrig DF, Budmiger M, Schmidt C, Stieger B, Hurlimann S, Kullak-Ublick GA. Rivaroxaban postmarketing risk of liver injury. J Hepatol. 2014;61:293–300.

Medical Approach to Fontan Patients

58

Luis Téllez, Antonio Guerrero, and Agustín Albillos

Introduction

Fontan surgery is the final stage of the surgical treatment of several congenital cardiac malformations with univentricular physiology. These malformations are characterized by mixing deoxygenated and oxygenated blood in a single ventricle, resulting from the elimination of the double circulation (pulmonary and systemic) [1]. The main objective of Fontan surgery is to restore the double vascular circuit, avoiding cyanosis. However, the new circulatory system created chronically elevates systemic venous pressure and eventually decreases cardiac output [2]. These chronic changes lead to a hemodynamic breakdown and long-term complications involving the lungs, kidneys, brain, gut, and liver [3].

Fontan-associated liver disease (FALD) encompasses the broad spectrum of structural and functional liver alterations present in this population [4, 5]. FALD is almost universally present after Fontan surgery and can progress through several steps in some cases and reach the final stage, when the main complications of portal hypertension and hepatocellular carcinoma (HCC) may occur. This chapter focuses on the pathophysiology and staging of FALD and proposes a comprehensive medical approach in this population.

L. Téllez (✉) · A. Guerrero · A. Albillos
Department of Gastroenterology and Hepatology, Hospital Universitario Ramón y Cajal, Instituto Ramón y Cajal de Investigación Sanitaria (IRYCIS), Centro de Investigación Biomédica en Red (CIBEREHD), Universidad de Alcalá, Madrid, Spain
e-mail: luis.tellez@salud.madrid.org; antonio.guerrero@salud.madrid.org; agustin.albillos@uah.es

R. de Franchis (ed.), *Portal Hypertension VII*,
https://doi.org/10.1007/978-3-031-08552-9_58

Pathophysiology and Natural History

The Fontan procedure was first described in 1971 and has mainly been employed in patients with several congenital heart malformations when the biventricular repair is not feasible [6]. In simple terms, the Fontan technique creates a connection between the systemic venous return from both the inferior and superior vena cava and the pulmonary arteries, avoiding the right ventricle, which will passively transmit the blood to the single ventricular chamber. Two major variants of Fontan surgery have been described: the atrio-pulmonary connection and the bicavo-pulmonary connection (Fig. 58.1a). Regardless of the type of surgery performed, the immediate consequence of Fontan circulation is the development of a pressure gradient as the mechanism that passively drives blood from the vena cavae to the pulmonary vasculature and finally to the left atrium (Fig. 58.1b). The resulting pulmonary artery input impedance hinders venous return through the pulmonary bed and leads to chronic venous congestion. In the long term, the reduced pulmonary wall strain and adverse vessel remodeling in the non-pulsatile pulmonary circulation result in increased intimal fibrosis, disrupted endothelial integrity, and loss of vascular smooth muscle cells [7]. These changes lead to progressively increased pulmonary vascular resistance, systemic venous collaterals, and cyanosis [2]. Hence, the volume load into the single ventricle markedly falls, resulting in decreased cardiac output. "Fontan failure" is used for this hemodynamic breakdown, which is clinically characterized by multisystem organ dysfunction. These hemodynamic changes lead to hepatic damage mediated by three mechanisms mainly related to the vascular supply and drainage of the liver:

- **Liver congestion.** The high pressure transmitted by the hepatic veins to the sinusoids leads to sinusoidal dilatation, hyperfiltration, and perisinusoidal edema. In turn, sinusoidal shear stress promotes the mechanical activation of hepatic stellate cells, resulting in liver fibrosis. These changes hamper the diffusion of oxygen and nutrients and foster hepatocellular necrosis and atrophy, which are more marked in the centrilobular zone.
- **Hypoxia and hepatic ischemia.** Several hemodynamic insults cumulatively develop from birth (neonatal hemodynamic instability and cardiac surgeries). In the long term, a high proportion of Fontan patients can develop severe systolic and diastolic dysfunction with reduced cardiac output that is markedly resistant to metabolic demands. The aberrant circulation can also facilitate the formation of right-to-left shunts that chronically further aggravate the hypoxia [8].
- **Prothrombotic state.** Thromboembolic events are frequent in the Fontan population due to the anatomical and functional characteristics of the Fontan circulation and to a state of thrombophilia [9]. The hypercoagulability state is characterized by low serum levels of antithrombin III, alpha2-antiplasmin, thrombomodulin, and C and S proteins and a high concentration of thrombin–antithrombin complex, similar to that observed in cirrhosis [10]. In consequence, liver microthrombosis has been proposed as another profibrogenic mechanism in FALD [11].

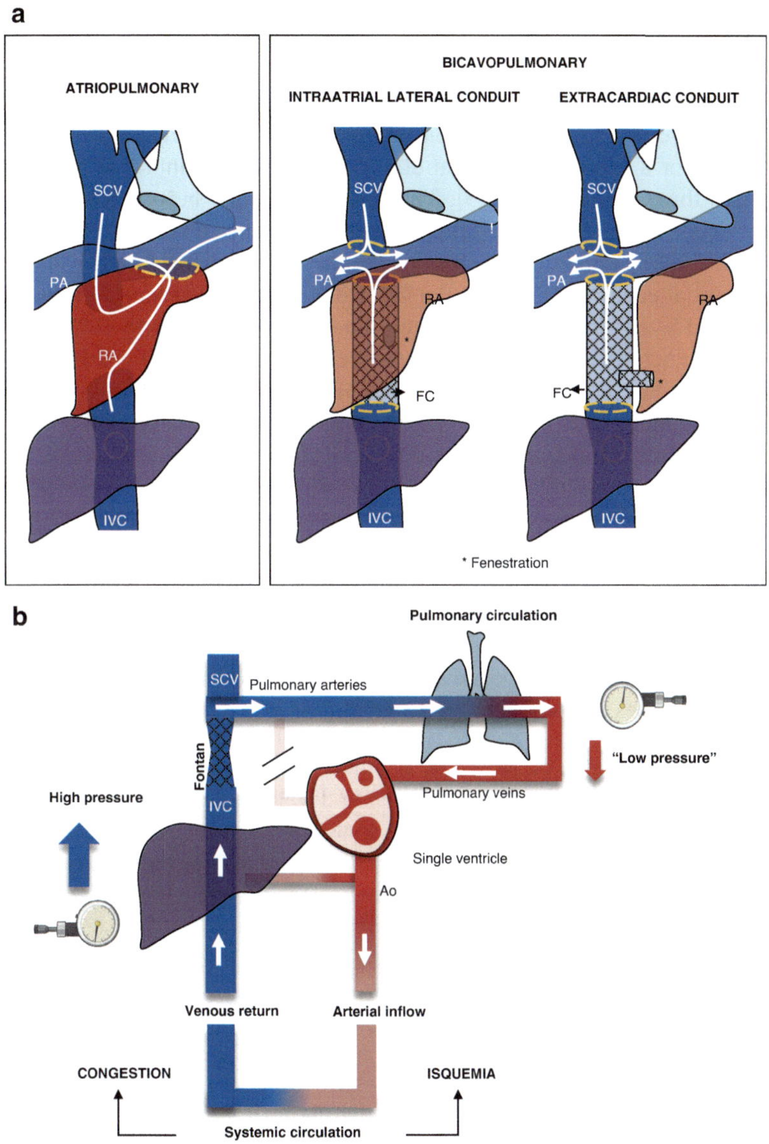

Fig. 58.1 (**a**) Major variants of Fontan surgery. Atrio-pulmonary Fontan (left): the superior and inferior cavae drain to the right atrium, which is connected to the pulmonary artery. Intra-atrial lateral conduit (middle): the superior vena cava directly drains to the right pulmonary artery, and the inferior vena cava is connected through an intra-atrial tunnel to the right pulmonary artery. Extracardiac conduit (right): the superior vena cava is connected directly to the right pulmonary artery and the inferior vena cava is connected to the right pulmonary artery through an extracardiac conduit. In both modalities, a fenestration can be left open between the tunnel/conduit and the left atrium to decrease the central venous pressure and maintain higher cardiac output at the expense of mild cyanosis (fenestrated Fontan). (**b**) The new cardiopulmonary system passively drives blood from the vena cavae to the pulmonary vasculature and finally to the left atrium. Systemic venous congestion and chronic systemic ischemia lead to end-organ damage, especially in the liver. (**b**) Fontan circulation. FC Fontan conduit, IVC inferior vena cava, PA pulmonary artery, PV pulmonary vein, RA right atrium, SVC superior vena cava

Natural History of Fontan-Associated Liver Disease

Liver damage is almost universal after Fontan surgery and generally develops slowly and silently over time. Consequently, the first manifestation of FALD frequently overlaps the dysfunction of another organ, such as protein-losing enteropathy or a drop-in functional class [12]. As shown in Table 58.1, FALD comprises three main stages [13]. The **first stage** starts before Fontan surgery and continues to childhood. Patients are usually asymptomatic and exhibit mild hyperbilirubinemia and elevated gamma-glutamyl transpeptidase (GGT) due to perisinusoidal edema and canalicular bile congestion. The **second stage** occurs around 10–15 years after Fontan surgery and is characterized by perisinusoidal fibrosis, regenerative nodules, and hepatocellular necrosis. The **third stage** is generally reached in adulthood and is clinically indistinguishable from other forms of end-stage liver disease: patients might show hypoalbuminemia, a prolonged coagulation time, significant thrombocytopenia, and an increased risk of liver-related complications, such as ascites, variceal bleeding, encephalopathy, and HCC.

The transition rate from one stage to the next is not uniform, and several variables listed in Table 58.2 have been associated with an increased risk of severe liver fibrosis. Among those risk factors of advanced liver damage, the most important is the time since Fontan surgery, which is a surrogate of the failure of the Fontan circulation. Indeed, since the risk of advanced FALD is low in the first 5 years after surgery, it increases by ninefold after 15 years [14].

Table 58.1 Natural history of Fontan-associated liver disease

	Stage 1	Stage 2	Stage 3
Timing	Before surgery to 10 years after surgery	10–20 years after surgery[a]	>20 years after surgery[a]
Clinical manifestations	Asymptomatic Painful hepatomegaly Hepatojugular reflux	Asymptomatic Painful hepatomegaly Hepatojugular reflux	Ascites Esophageal varices HCC
Laboratory tests	↑GGT ↑Bilirubin (indirect)	↑↑↑GGT ↑↑Bilirubin (indirect) ↑AST-ALT	↓Albumin ↓Platelets ↓Prothrombin
Histology	Sinusoidal dilatation Hepatocellular necrosis Mild perisinusoidal fibrosis	Sinusoidal dilatation Hepatocellular necrosis Moderate fibrosis Regenerative nodules	Bridging centro-portal fibrosis Regenerative nodules "Cardiac cirrhosis"

ALT alanine aminotransferase, *AST* aspartate aminotransferase, *GGT* gamma-glutamyl transpeptidase, *HCC* hepatocellular carcinoma

[a]Or before if Fontan failure has occurred

Table 58.2 Risk factors for advanced Fontan-associated liver disease

Risk factors for advanced liver disease after Fontan surgery
Fontan duration
Decreased ventricular function
Tachyarrhythmia
Sinus node dysfunction
Fontan conduit thrombosis or stenosis
Hypoplastic left ventricular syndrome
Atrio-pulmonary type surgery
Chronic use of amiodarone and other hepatotoxic drugs
Chronic hepatitis C virus infection
Protein-losing enteropathy and systemic inflammation

Diagnosis and Staging of Fontan-Associated Liver Disease

Liver Biopsy

Liver biopsy is the gold standard for grading the severity of liver fibrosis in FALD and excluding other etiologies. The typical histological pattern of FALD is a moderate to severe sinusoidal dilatation without significant periportal inflammation. Liver fibrosis is almost universally present after surgery. In the early stages, fibrosis appears predominantly in the perisinusoidal area instead of the centrilobular pattern observed in other forms of congestive hepatopathy [15–17]. In the long term, FALD typically combines portal and sinusoidal fibrosis. Thus, the classic scoring systems (i.e., Ishak, Knodell, or METAVIR) used to grade inflammatory-mediated liver diseases with primarily portal fibrosis seem inadequate in FALD. Lately, the Congestive Hepatic Fibrosis Score, used in other congestive hepatopathies, has been proposed in this population [17, 18]. The scale awards centrilobular fibrosis a score from 0 to 4: 0 to 2 are considered low-grade fibrosis, and 3 to 4 are high-grade fibrosis. However, the patchy nature of fibrosis in FALD may be underestimated in biopsies, as suggested by a study comparing the liver histology before liver transplantation and in explant specimens [19]. The usefulness of liver biopsies in predicting clinical outcomes is summarized in Table 58.3.

Liver Function Tests and Serological Biomarkers

Mild increases in serum GGT and alkaline phosphatase, as markers of bile canalicular congestion, are commonly seen in patients with FALD. However, the interpretation of other important liver function parameters (i.e., bilirubin, albumin, or coagulation time) is extremely difficult. Low serum albumin levels may be secondary to protein-losing enteropathy or nephropathy, malnutrition, or a chronic wasting state [12, 20]. Hyperbilirubinemia in this setting can result from liver injury, hemolysis, toxicity, ischemic damage, or canalicular congestion [21]. The model for end-stage liver disease excluding the international normalized ratio (MELD-XI) has

Table 58.3 Liver biopsy in Fontan-associated liver disease

Author, year	Patients included	Age (median)	Via	Score	(%) severe liver fibrosis	Clinical outcomes
Cho, 2021 [74]	22 (mixed)	14.7	TJ	CHFS	27.3%	Fibrosis was universally present, but not related to central pressure or time from Fontan surgery
Rathgeber, 2020 [75]	14 (children)	11.7	TJ	Scheuer	0%	Fibrosis was universally present in pediatric patients
Téllez, 2021[a]	143 (adults)	27.6	TJ, PC	CHFS	55.2%	Fibrosis was related to central and wedged hepatic venous pressures
Munsterman, 2019 [23]	38 (adults)	27.0	PC	CHFS	57.9%	
Silva-Sepúlveda, 2019 [29]	49 (mixed)	17.8	TJ	CHFS, Ishak	NR	Fibrosis was related to central pressure
Schachter, 2018 [76]	14 (adults)	26.4	PC	CHFS	35.7%	No relationship with clinical events
Wu, 2017 [77]	68 (mixed)		TJ, PC, A			No relationship with liver function or exercise capacity
Goldberg, 2017 [16]	67 (mixed)	17.3	PC	%CD	23.0%	Fontan complications are not related to fibrosis
Surrey, 2016 [17]	74 (mixed)	17.7	TJ, PC	CHFS	39.2%	No relationship with clinical events
Pundi, 2016 [37]	24 (mixed)	–	NR	–	~50%	35%—5-year survival after cirrhosis diagnosis
Evans, 2016 [22]	70 (mixed)	16.0	TJ	0–8	NR	Fibrosis was related to MELD-XI
Wu, 2015 [35]	68 (mixed)	23.2	TJ, PC		73%	Liver fibrosis did not predict 3-year transplant-free survival

Table 58.3 (continued)

Author, year	Patients included	Age (median)	Via	Score	(%) severe liver fibrosis	Clinical outcomes
Schwartz, 2013 [78]	13 (mixed)	19.1	TJ	Scheuer	69.0%	Fibrosis was inversely related to platelets count
Kendall, 2008 [15]	18 (mixed)	–	TJ	–	–	–
Kiessewetter, 2007 [44]	12 (mixed)	24.6	TJ	METAVIR	58%	Fibrosis was related to hepatic and central pressures and esophageal varices
Ghaferi, 2004 [79]	9 (children)	–	Autopsy	–	–	Time from Fontan surgery was related to fibrosis

TJ Transjugular, *PC* Percuteneous, *CHFS* Congestive Hepatic Fibrosis Score
[a]Unpublished data. Available as abstract PO-7-3-YI (Bleeding, thrombosis, and vascular liver diseases-EASL monothematic conference. Geneva. 1–10 October 2021)

been designed to overcome the main limitation of MELD in this population, namely an increased international normalized ratio due to warfarin therapy. MELD-XI has demonstrated a statistical correlation with liver fibrosis and poor prognosis after cardiac transplantation, but specific MELD-XI cutoffs for predicting clinical outcomes have not been identified [22]. Progressive thrombocytopenia is frequently found in FALD and reflects hypersplenism due to portal hypertension or increased systemic venous pressure.

Most serological tests used to estimate liver fibrosis in other chronic liver diseases have been evaluated in FALD. However, the studies generally lacked histology as the gold standard and did not consider the intrinsic limitations of these noninvasive parameters [14]. A small single-center study comparing patients with histologically proven mild and severe liver fibrosis did not find differences in ELF score, APRI, and FIB-4 values between the two groups [23].

Abdominal Imaging

Liver ultrasound mainly reflects hepatic venous outflow obstruction. The typical findings include heterogeneous echogenicity, hepatomegaly, enlarged caudate lobe, dilated hepatic veins, nodular liver surface, and small-sized, peripheral, and hyperechoic nodules [24, 25]. Doppler imaging may reveal high resistance and pulsatility indices in the celiac trunk and decreased portal flow velocity [40]. Loss of the three-phase Doppler pattern of the hepatic veins is universal in the bicavo-pulmonary variant (because of the absence of the atrial beat). Still, the presence of a monophasic pattern could imply more advanced liver fibrosis [26].

Through three-phasic (dynamic) evaluation, the liver parenchyma typically shows a patchy, poor, and irregular enhancement in the periphery compared with the hilar area, due to systemic congestion [25, 27]. Cross-sectional imaging also provides extrahepatic information such as the presence of portosystemic shunts, splenomegaly, ascites, and esophageal and gastric varices. However, ultrasonography, computed tomography (CT), and magnetic resonance imaging (MRI) may not accurately predict the extent of liver fibrosis when compared with biopsy specimens [26, 27].

Elastography

Hepatic congestion increases liver stiffness, which is seen immediately after Fontan surgery [28]. This hampers the applicability of elastography to FALD, with contradictory results in the literature [23, 29]. Liver stiffness and fibrosis increase in parallel to the time from surgery, which suggests that, in addition to congestion, fibrosis progressively contributes to stiffness [30]. New liver stiffness cutoffs should be identified and validated in this particular population to allow clinicians to confirm and, more importantly, rule out severe liver fibrosis. Furthermore, it is essential to highlight that liver stiffness can also predict non-liver-related events (i.e., protein-losing enteropathy) and a poor cardio-dynamic state in Fontan patients (i.e., higher central venous and end-diastolic pressures and a lower cardiac index and ejection fraction) [12, 31]. Consequently, liver elastography may be a suitable noninvasive method for monitoring Fontan patients over time [32].

Hepatic Hemodynamics

The hepatic venous pressure gradient (HVPG), as the difference between wedged and free hepatic venous pressures, is considered the best prognostic marker in patients with chronic liver disease and sinusoidal portal hypertension. The HVPG also facilitates the differential diagnosis of portal hypertension: high hepatic pressures (free and wedged) with a normal HVPG suggest a posthepatic origin, the most frequent finding in advanced FALD [33]. HVPG is not elevated in FALD, even in patients with cirrhosis, and does not correlate with liver disease severity or clinical outcomes [34].

Hepatic Complications of Fontan-Associated Liver Disease

Ascites

The prevalence of ascites in Fontan patients ranges from 4% to 17% [17, 35]. However, in FALD, several causes of ascites can coexist. It can be present in patients with severe systemic venous hypertension (due to failure of the Fontan circulation), hypoalbuminemia (secondary to protein-losing enteropathy and nephropathy, malnutrition, or liver insufficiency), and sinusoidal portal hypertension (in case of severe liver fibrosis).

Ascites are usually manageable via optimizing cardiac function, nutrition, loop diuretics, and anti-aldosterone drugs [36]. Large-volume paracentesis could be a rescue treatment. Transjugular intrahepatic portosystemic shunt (TIPS) does not appear to be a good option in Fontan patients with refractory ascites, and no case reports have been published.

Esophageal Varices

The prevalence of esophageal varices in FALD is highly variable and ranges from 9.4% in children to 43% in adults [30, 37–39]. The presence of esophageal varices along with other manifestations of portal hypertension increased the risk of death, heart transplantation, and HCC [31]. In the Fontan circulation, the increased systemic venous pressure is directly transmitted through the azygos system to the esophageal venous plexus, contributing to variceal development in the upper (downhill varices) and lower esophagus [40]. Unlike classic portal hypertension secondary to cirrhosis, the portal hypertension model in FALD is characteristically hypodynamic due to low cardiac output [1]. Consequently, the esophageal varices are usually small with a low risk of bleeding, and the use of primary prophylaxis with nonselective beta-blockers is thus debatable. Moreover, given the high rate of sinus node dysfunction and low cardiac output in the Fontan population, beta-blockade should be employed cautiously. In the absence of specific studies on FALD, acute episodes of variceal bleeding should be managed with vasoactive drugs and endoscopic band ligation according to Baveno VII recommendations. In cases of refractory bleeding, TIPS insertion could be an option, although the acute increase in preload may precipitate cardiac failure.

Hepatic Encephalopathy

In FALD, hepatic encephalopathy is a scarcely documented event that can be explained by preserved liver function, even in patients with cirrhosis [4, 37, 41]. However, the actual incidence of this complication may be underestimated by the retrospective nature of most studies and the presence of other neurological complications in FALD (i.e., anoxic encephalopathy, cardioembolic stroke, and cognitive impairment).

Hepatic Nodules and Hepatocellular Carcinoma

Liver nodules are detected in 50% of patients with FALD [24–27, 35, 42–45]. They are usually small (<2 cm), multiple, and located in the periphery of the liver [25].

Most of them correspond to focal nodular hyperplasia (FNH) [42, 46], a polyclonal lesion resulting from a hyperplastic response to increased blood flow induced by a focal vascular abnormality [47]. The origin of these nodules has been associated with perfusion defects secondary to Fontan circulation: (1) Hepatic congestion and reduced cardiac output cause a decrease in portal blood flow predominantly in the periphery of the liver, resulting in focal ischemia and (2) Extinction of the parenchyma secondary to ischemia triggers an arterial vasodilatory response that stimulates the proliferation of healthy hepatocytes resulting in nodular regenerative hyperplasia [48, 49].

Patients with FALD have an increased risk of HCC, which grows over time. The incidences reported at 10, 20, and 30 years after surgery are 1%, 3%, and 13%, respectively [50, 51]. A systematic review confirmed that HCC generally develops 10 years after Fontan surgery (median patient age, 28 years) [52]. Most cases of HCC in FALD are diagnosed at advanced stages when curative treatments are not an option. Consequently, the survival rate reported is as low as 1 year after diagnosis, which highlights the urgent need for surveillance programs in this population.

HCC in FALD almost universally displays the typical radiological features of this condition accepted in other forms of cirrhosis (i.e., hyperenhancement in the arterial phase with portal washout). However, in FALD, non-neoplastic nodules can mimic this radiological behavior because the delayed washout would probably be due to contrast retained by the surrounding congested parenchyma [25, 53]. Accordingly, the HCC diagnostic criteria applied for cirrhosis are helpful to establish the diagnostic suspicion but are not suitable for a definitive diagnosis, and a confirmatory biopsy is always required. As in Budd-Chiari syndrome, serum alpha-fetoprotein (AFP) is a valuable complementary tool for diagnosing HCC because this biomarker is above the upper limit of normal in nearly 80% of Fontan patients diagnosed with HCC [54]. In contrast, in a large prospective series with a high prevalence of non-neoplastic nodules, no patient with Fontan surgery without HCC showed elevated AFP [25]. Accordingly, elevated AFP levels in patients with FALD should always be suspicious of HCC. On the other hand, hyper-enhancing nodules without delayed washout, stability during follow-up, and normal AFP should be interpreted as benign. The management of HCC in FALD should follow clinical practice guidelines [55], although the practitioners have to tailor them to the specific considerations secondary to the unique features of Fontan patients [56].

Medical Management of Fontan-Associated Liver Disease

Strategies of Prevention and Amelioration of Liver Damage

There is no specific treatment for FALD; thus, we should focus our efforts on preventing liver deterioration and the early identification of liver-related complications. During early childhood, even before Fontan surgery, perioperative ischemia and hypoxemia should be avoided. Bicavo-pulmonary variants of surgery avoiding small-diameter conduits are preferred to elude the "atrio-hepatic reflux" that appear with atrio-pulmonary techniques [31]. During childhood and adolescence, early identification of cardiac (i.e., arrhythmias) and noncardiac (i.e., protein-losing enteropathy) complications may provide early treatment, even in subclinical stages. Adolescents should be instructed to avoid obesity and overweight, limit alcohol intake and smoking, and perform exercise [57, 58]. Exposure to hepatotoxic drugs, especially amiodarone, should be limited, and screening for hepatitis B/D and C infection is recommended. In early adulthood, correct risk stratification of advanced FALD is mandatory, in conjunction with an exhaustive cardiac evaluation if heart problems are suspected. Antifibrotic therapies, such as ACE inhibitors and anticoagulants, should be properly evaluated in the future and cannot be routinely recommended.

Risk Stratification of Advanced FALD and Screening Strategies

FALD progression parallels the breakdown of the Fontan circulation. We lack the appropriate tools to stratify the severity of liver damage in this particular circulation [59–62]. Since advanced liver damage is rare within the first 10 years, periodic liver assessment in early childhood is probably not needed unless there are signs of Fontan failure. Some authors have suggested performing a liver biopsy to stage fibrosis in all patients 10 years after surgery [4, 63, 64]. However, the lack of a specific treatment based on liver histology calls into question the value of protocolized liver biopsies in clinical practice. Otherwise, liver biopsy is advisable for heart transplant candidates, liver disease of unknown etiology, and during hepatic or cardiac catheterization. In FALD, noninvasive diagnostic methods, such as serological markers, abdominal imaging, and liver elastography, are not well validated. However, the combination of several diagnostic methods and a comprehensive clinical evaluation can alert us of advanced FALD and Fontan failure. Patients with suspected advanced FALD require closer monitoring and therapeutic interventions to improve hepatic outflow and ameliorate hepatic congestion, as proposed in Fig. 58.2. This complex evaluation and treatment are only possible in highly experienced centers with multidisciplinary teams, including cardiologists with expertise in congenital cardiac diseases and hemodynamics, abdominal and cardiac surgeons, radiologists, hepatologists, nutritionists, and advanced practice nurses [65].

HCC screening programs are recommended for those patients who underwent surgery more than 10 years earlier. Although the accuracy of ultrasound for the early diagnosis of HCC is not well established in FALD, the available evidence suggests that abdominal ultrasound adequately detects most cases of HCC [52, 64]. Accordingly, biannual abdominal ultrasound and AFP determination may be useful to detect incident liver nodules. Contrast-enhanced cross-sectional imaging (CT/MRI) is mandatory to characterize liver nodules and should be strongly considered for baseline assessment. A helpful tip would be to offer hepatic MRI when cardiac MRI is performed [52].

Following current clinical guidelines, screening for esophageal varices is recommended in all patients with chronic liver disease and suspected portal hypertension

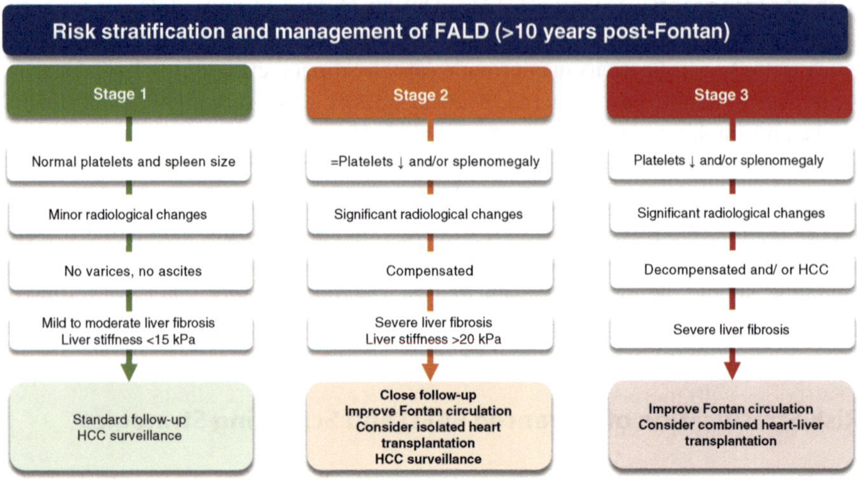

Fig. 58.2 Risk stratification and management of Fontan-associated liver disease

for starting primary prophylaxis with nonselective beta-blockers or band ligation. However, in Fontan patients, primary prophylaxis cannot be consistently recommended because the esophageal varices are usually small and have a low risk of bleeding. However, esophageal varix presence identifies those cases with worse prognoses [31]. Accordingly, screening for esophageal varices is advisable for staging purposes, but a strong recommendation about timing for upper endoscopy indication cannot be made.

Heart and Liver Transplantation

Heart transplantation is a curative option for patients with failure of the Fontan circulation. Still, the transplantation rate has been very low because of reported poor outcomes even in the latest decades [66]. One of the major concerns in adults with severe FALD is the risk of liver failure or decompensation after heart transplantation, raising a dilemma concerning which patient subgroups require an isolated heart or combined liver-heart transplantation [67, 68]. This controversy remains unresolved due to the availability of scarce evidence. First, no differences in survival have been found between cirrhotic and non-cirrhotic patients who undergo heart transplantation [69]. Therefore, severe fibrosis and compensated cirrhosis should not be absolute contraindications for isolated heart transplantation. However, predicting which patients will show stabilization, worsening, or regression of their liver damage after isolated cardiac transplantation is a matter of debate. For decades, cirrhosis has been considered an irreversible disease; however, current knowledge indicates that cirrhosis is potentially reversible after removing the causative factor [70]. In cardiac cirrhosis, experimental models and small case series have found that the liver disease may also improve and even fully resolve if cardiac function is restored [71, 72]. Some authors consider combined heart–liver transplantation the most radical curative option because it simultaneously eliminates both the liver and heart diseases, preventing future complications. Based on this hypothesis, D'Souza et al. reported a case series of seven patients with Fontan circulation who received a double transplant. The criterion for heart-liver transplantation was the presence of bridging fibrosis or cirrhosis in a pretransplant biopsy. All patients survived after a median follow-up of 4.6 years [73]. This approach could be too aggressive because the degree of liver damage in compensated patients is probably insufficient to justify liver transplantation, especially given the current shortage of organs and the potential improvement in liver function and fibrosis after heart transplantation. Combined heart–liver transplantation should probably be reserved for those cases with previous severe liver-related complications. While we await stronger evidence, institutions with more experience in this field recommend an individualized analysis of each case in a multidisciplinary committee.

References

1. Téllez L, Rodríguez-Santiago E, Albillos A. Fontan-associated liver disease: a review. Ann Hepatol. 2018;17:192–204.
2. Hebson CL, McCabe NM, Elder RW, Mahle WT, McConnell M, Kogon BE, et al. Hemodynamic phenotype of the failing Fontan in an adult population. Am J Cardiol. 2013;112:1943–7.
3. Alsaied T, Bokma JP, Engel ME, Kuijpers JM, Hanke SP, Zuhlke L, et al. Predicting long-term mortality after Fontan procedures: a risk score base don 6707 patients from 28 studies. Congenit Heart Dis. 2017;12:393–8.
4. Rychik J, Veldtman G, Rand E, Russo P, Rome JJ, Krok K, et al. The precarious state of the liver after a Fontan operation: summary of a multidisciplinary symposium. Pediatr Cardiol. 2012;33:1001–12.
5. Sessa A, Allaire M, Lebray P, Medmoun M, Tiritilli A, Laria P, et al. From congestive hepatopathy to hepatocellular carcinoma, how can we improve patient management. JHEP Rep. 2021;2:100249.
6. Fontan F, Baudet E. Surgical repair of tricuspid atresia. Thorax. 1971;26:240–8.
7. Gewillig M, Brown SC. The Fontan circulation after 45 years: update in physiology. Heart Br Card Soc. 2016;102:1081–6.
8. Heinemann M, Breuer J, Steger V, Steil E, Sieverding L, Ziemer G. Incidence and impact of systemic venous collateral development after Glenn and Fontan procedures. Thorac Cardiovasc Surg. 2001;49:172–8.
9. Faircloth JM, Roe I, Alsaied T, Palumbo JS, Vinks A, Veldtman GR. Intermediate term thrombotic risk in contemporary total cavo-pulmonary connection for single ventricle circulations. J Thromb Thrombolysis. 2017;44:275–80.
10. Tomkiewicz-Pajak L, Hoffman P, Trojnarska O, Lipczyńska M, Podolec P, Undas A. Abnormalities in blood coagulation, fibrinolysis, and platelet activation in adult patients after the Fontan procedure. J Thorac Cardiovasc Surg. 2014;147:1284–90.
11. Simonetto DA, Yang H, Yin M, de Assuncao TM, Kwon JH, Hilscher M, et al. Chronic passive venous congestion drives hepatic fibrogenesis via sinusoidal thrombosis and mechanical forces. Hepatology. 2015;61:648–59.
12. Rodriguez de Santiago E, Téllez L, Garrido-Lestache Rodríguez-Monte E, Garrido-Gómez E, Aguilera-Castro L, Álvarez-Fuente M, Del Cerro MJ, et al. Fontan protein-losing enteropathy is associated with advanced liver disease and proinflammatory intestinal and systemic state. Liver Int. 2020;40:638–45.

13. Téllez L, Rodríguez de Santiago E, Albillos A. Fontan-associated liver disease. Rev Esp Cardiol (Engl Ed). 2018;71:192–202.
14. Baek JS, Bae EJ, Ko JS, Kim GB, Kwon BS, Lee SY, et al. Late hepatic complications after Fontan operation; non-invasive markers of hepatic fibrosis and risk factors. Heart Br Card Soc. 2010;96:1750–5.
15. Kendall TJ, Stedman B, Hacking N, Haw M, Vettukattill JJ, Salmon AP, et al. Hepatic fibrosis and cirrhosis in the Fontan circulation: a detailed morphological study. J Clin Pathol. 2008;61:504–8.
16. Goldberg DJ, Surrey LF, Glatz AC, Dodds K, O'Byrne ML, Lin HC, et al. Hepatic fibrosis is universal following Fontan operation, and severity is associated with time from surgery: a liver biopsy and hemodynamic study. J Am Heart Assoc. 2017;6:e004809.
17. Surrey LF, Russo P, Rychik J, Goldberg DJ, Dodds K, O'Byrne ML, et al. Prevalence and characterization of fibrosis in surveillance liver biopsies of patients with Fontan circulation. Hum Pathol. 2016;57:106–15.
18. Bosch DE, Koro K, Richards E, Hoch BL, Jalikis F, Koch LK, et al. Validation of a congestive hepatic fibrosis scoring system. Am J Surg Pathol. 2019;49:766–72.
19. Vaikunth SS, Higgings JP, Concepcion W, Haeffele C, Wright GE, Chen S, et al. Does liver biopsy accurately measure fibrosis in Fontan-associated liver disease? A comparison of liver biopsy pre-combined heart and liver transplant and liver explant post-trasplant. Clin Transplant. 2020;34:e14120.
20. Opotowsky AR, Baraona FR, Mc Causland FR, Loukas B, Landzberg E, Landzberg MJ, et al. Estimated glomerular filtration rate and urine biomarkers in patients with single-ventricle Fontan circulation. Heart. 2017;103:434–42.
21. Fujishiro J, Sugiyama M, Ishimaru T, Watanabe M, Sato K, Hoshino N, et al. Direct hyperbilirubinemia in infants with congenital heart disease. Pediatr Int. 2018;60:179–82.
22. Evans WN, Acherman RJ, Ciccolo ML, Carrillo SA, Galindo A, Rothman A, et al. MELD-XI scores correlate with post-Fontan hepatic biopsy fibrosis scores. Pediatr Cardiol. 2016;37:1274–7.
23. Munsterman ID, Duijnhouwer AL, Kendall TJ, Bronkhorst CM, Ronot M, van Wettere M, et al. The clinical spectrum of Fontan-associated liver disease: results from a prospective multimodality screening cohort. Eur Heart J. 2019;40:1057–68.
24. Bae JM, Jeon TY, Kim JS, Kim S, Hwang SM, Yoo S-Y, et al. Fontan-associated liver disease: Spectrum of US findings. Eur J Radiol. 2016;85:850–6.
25. Téllez L, Rodríguez de Santiago E, Mínguez B, Payancé A, Clemente A, Baiges A, et al. Prevalence, features and predictive factors of liver nodules in Fontan surgery patients: the VALDIG Fonliver prospective cohort. J Hepatol. 2020;72:702–10.
26. Wallihan DB, Podberesky DJ. Hepatic pathology after Fontan palliation: spectrum of imaging findings. Pediatr Radiol. 2013;43:330–8.
27. Bulut OP, Romero R, Mahle WT, McConnell M, Braithwaite K, Shehata BM, et al. Magnetic resonance imaging identifies unsuspected liver abnormalities in patients after the Fontan procedure. J Pediatr. 2013;163:201–6.
28. Deorsola L, Aidala E, Cascarano MT, Valori A, Agnoletti G, Pace NC. Liver stiffness modifications shortly after total cavopulmonary connection. Interact Cardiovasc Thorac Surg. 2016;23:513–8.
29. Silva-Sepulveda JA, Fonseca Y, Vodkin I, Vaughn G, Newbury R, Vavinskaya V, et al. Evaluation of Fontan liver disease: correlation of transjugular liver biopsy with magnetic resonance and hemodynamics. Congenit Heart Dis. 2019;14:600–8.
30. Agnoletti G, Ferraro G, Bordese R, Marini D, Gala S, Bergamasco L, et al. Fontan circulation causes early, severe liver damage. Should we offer patients a tailored strategy? Int J Cardiol. 2016;209:60–5.
31. Elder RW, McCabe NM, Hebson C, Veledar E, Romero R, Ford RM, et al. Features of portal hypertension are associated with major adverse events in Fontan patients: the VAST study. Int J Cardiol. 2013;168:3764–9.

32. Egbe A, Miranda WR, Connolly HM, Khan AR, Al-Otaibi M, Venkatesh SK, et al. Temporal changes in liver stiffness after Fontan operation: results of serial magnetic resonance elastography. Int J Cardiol. 2018;258:299–304.
33. Mori M, Hebson C, Shioda K, Elder RW, Kogon BE, Rodriguez FH, et al. Catheter-measured hemodynamics of adult Fontan circulation: associations with adverse event and end-organ dysfunctions. Congenit Heart Dis. 2016;11:589–97.
34. Egbe AC, Miranda WR, Veldtman GR, Graham RP, Kamath PS. Hepatic venous pressure gradient in Fontan physiology has limited diagnostic and prognostic significance. CJC Open. 2020;2:360–54.
35. Wu FM, Jonas MM, Opotowsky AR, Harmon A, Raza R, Ukomadu C, et al. Portal and centrilobular hepatic fibrosis in Fontan circulation and clinical outcomes. J Heart Lung Transplant. 2015;34:883–91.
36. Hilscher MB, Johnson JN, Cetta F, Driscoll DJ, Poterucha JJ, Sanchez E, et al. Surveillance for liver complications after the Fontan procedure. Congenit Heart Dis. 2017;12:124–32.
37. Pundi K, Pundi KN, Kamath PS, Cetta F, Li Z, Poterucha JT, et al. Liver disease in patients after the Fontan operation. Am J Cardiol. 2016;117:456–60.
38. Tellez L, Rodriguez de Santiago E, Garrido E, Baiges A, La Mura V, Garrido-Lestache E, et al. High risk of esophageal varices in patients with Fontan surgery: The VALDIG Fonliver study. J Hepatol. 2018;68:S715. (Abstract)
39. Isoura Y, Yamamoto A, Cho Y, Ehara E, Jogo A, Suzuki T, et al. Platelet count and abdominal dynamic CT are useful in predicting and screening for gastroesophageal varices after Fontan surgery. PLoS One. 2021;16:e0257441.
40. Machogu E, Balistrieri G, Hehir D, Quintero D. A case of tracheal varices in an adolescent patient with cyanotic heart disease. Ann Am Thorac Soc. 2013;10:26–30.
41. Koteda Y, Suda K, Kishimoto S, Iemura M. Portal-systemic encephalopathy after Fontan-type operation in patient with polysplenia syndrome. Eur J Cardio-Thorac Surg. 2009;35:1083–5.
42. Bryant T, Ahmad Z, Milward-Sadler H, Burney K, Stedman B, Kendall T, et al. Arterialised hepatic nodules in the Fontan circulation: hepático-cardiac interactions. Int J Cardiol. 2011;151:268–72.
43. Poterucha JT, Johnson JN, Qureshi MY, O'Leary PW, Kamath PS, Lennon RJ, et al. Magnetic resonance elastography: a novel technique for the detection of hepatic fibrosis and hepatocellular carcinoma after the Fontan operation. Mayo Clin Proc. 2015;90:882–94.
44. Kiesewetter CH, Sheron N, Vettukattill JJ, Hacking N, Stedman B, Millward-Sadler H, et al. Hepatic changes in the failing Fontan circulation. Heart. 2007;93:579–84.
45. Horvat N, Rocha MS, Chagas AL, Oliveira BC, Pacheco MP, Binotto MA, et al. Multimodality screening of hepatic nodules in patients with congenital heart disease after Fontan procedure: role of ultrasound, ARFI Elastography, CT, and MRI. AJR Am J Roentgenol. 2018;211:1212–20.
46. Engelhardt EM, Trout AT, Sheridan RM, Veldtman GR, Dillman JR. Focal liver lesions following Fontan palliation of single ventricle physiology: a radiology-pathology case series. Congenit Heart Dis. 2019;14:380–8.
47. Sempoux C, Balabaud C, Paradis V, Bioulac-Sage P. Hepatocellular nodules in vascular liver diseases. Virchows Arch. 2018;473:33–44.
48. Cazals-Hatem D, Vilgrain V, Genin P, Denninger MH, Durand F, Belghiti J, et al. Arterial and portal circulation and parenchymal changes in Budd-Chiari syndrome: a study in 17 explanted livers. Hepatology. 2003;37:510–9.
49. Van Wettere M, Purcell Y, Bruno O, Payancé A, Plessier A, Rautou PE, et al. Low specificity of washout to diagnose hepatocellular carcinoma in nodules showing arterial hyperenhancement in patients with Budd-Chiari syndrome. J Hepatol. 2019;70:1123–32.
50. Sagawa T, Kogiso T, Sugiyama H, Hashimoto E, Yamamoto M, Tokushige K. Hepatol Res. 2020;50:853–62.
51. Nandwana SB, Olaiya B, Cox K, Sahu, Mittal P. Abdominal imaging surveillance in adult patients after Fontan procedure: risk of chronic liver disease and hepatocellular carcinoma. Curr Probl Diagn Radiol. 2018;47:19–22.

52. Rodríguez de Santiago E, Téllez L, Guerrero A, Albillos A. Hepatocellular carcinoma after Fontan surgery: a systematic review. Hepatol Res. 2021;51:116–34.
53. Wells ML, Hough DM, Fidler JL, Kamath PS, Poterucha JT, Venkatesh SK. Benign nodules in post-Fontan livers can show imaging features considered diagnostic for hepatocellular carcinoma. Abdom Radiol (NY). 2017;42:2623–31.
54. European Association for the Study of the Liver. EASL Clinical Practice Guidelines: management of hepatocellular carcinoma. J Hepatol. 2018;69:182–236.
55. Kwon S, Scovel L, Yeh M, Dorsey D, Dembo G, Krieger EV, et al. Surgical management of hepatocellular carcinoma after Fontan procedure. J Gastrointest Oncol. 2015;6:E55–60.
56. Angelico R, Lisignoli V, Monti L, Pariante R, Grimaldi C, Saffioti MC, et al. Laparoscopic liver resection for hepatocellular carcinoma in Fontan-associated chronic liver disease. The first case report. Int J Surg Case Rep. 2019;59:144–7.
57. Scheffers LE, Em Vd Berg L, Ismailova G, Dulfer K, Takkenberg JJ, Helbing WA. Physical exercise training in patients with Fontan circulation: a systematic review. Eur J Prev Cardiol. 2021;28(11):1269–78.
58. Weinreb SJ, Dodds KM, Burstein DS, Huang J, Rand EB, Mancilla E, et al. End-organ function and exercise performance in patients with Fontan circulation: what characterizes the high performers? J Am Heart Assoc. 2020;9:e016850.
59. Warnes CA, Williams RG, Bashore TM, Child JS, Connolly HM, Dearani JA, et al. ACC/AHA 2008 guidelines for the management of adults with congenital heart disease: executive summary: a report of the American College of Cardiology/ American Heart Association task force on practice guidelines (writing committee to develop guidelines for the management of adults with congenital heart disease). Circulation. 2008;118:2395–451.
60. Baumgartner H, Bonhoeffer P, De Groot NMS, de Haan F, Deanfield JE, Galie N, et al. ESC guidelines for the grown-up congenital heart disease (new version 2010). Eur Heart J. 2010;31:2915–57.
61. Daniels CJ, Bradley EA, Landberg MJ, Aboulhosn J, Beekman RH, Book W, et al. Fontan-associated liver disease: Proceedings from the American College of Cardiology Stakeholders Meeting, October 1 to 2, 2015, Washington DC. J Am Coll Cardiol. 2017;70:3173–94.
62. Rychik J, Atz AM, Celermaier DS, Deal BJ, Gatzoulis MA, Gewillig MH, et al. Evaluation and management of the child and adult with Fontan circulation: a scientific statement from the American Heart Association. Circulation. 2019; https://doi.org/10.1161/CIR.0000000000000696.
63. Emamaullee J, Zaidi AN, Schiano T, Kahn J, Valentino PL, Hofer RE, et al. Fontan-associated liver disease: screening, management, and transplat considerations. Circulation. 2020;142:591–604.
64. Téllez L, Rodríguez de Santiago E, Albillos A. Fontan-associated liver disease: pathophysiology, staging, and management. Semin Liver Dis. 2021;41:538–50.
65. Di Maria MV, Barrett C, Rafferty C, Wolfe K, Kelly SL, Liptzin DR, et al. Initiating a Fontan multidisciplinary clinic: decreasing care variability improving surveillance, and subsequent treatment of Fontan survivors. Congenit Heart Dis. 2019;14:5990–9.
66. Tabarsi N, Guan M, Simmonds J, Toma M, Kiess M, Tsang V, et al. Meta-analysis of the effectiveness of heart transplantation in patients with a failing Fontan. Am J Cardiol. 2017;119:1269–74.
67. D'Souza BA, Fuller S, Gleason LP, Hornsby N, Wald J, Krok K, et al. Single-center outcomes of combined heart and liver transplantation in the failing Fontan. Clin Transplant. 2017;31 https://doi.org/10.1111/ctr.12892.
68. Zhao K, Wang R, Kamoun M, Olthoff K, Hoteit M, Rame E, et al. Liver allograft provides protection against cardiac allograft rejection in combined heart and liver transplantation. Am J Transplant. 2019;19:114.
69. Simpson KE, Esmaeeli A, Khanna G, White F, Turnmelle Y, Eghtesady P, et al. Liver cirrhosis in Fontan patients does not affect 1-year post-heart transplant mortality or markers of liver function. J Heart Lung Transplant. 2014;33:170–7.

70. Ellis EL, Mann DA. Clinical evidence for the regression of liver fibrosis. J Hepatol. 2012;56:1171–80.
71. Crespo-Leiro MG, Robles O, Paniagua MJ, Marzoa R, Naya C, Flores X, et al. Reversal of cardiac cirrhosis following orthotopic heart transplantation. Am J Transplant. 2008;8:1336–9.
72. Bouchardy J, Meyer P, Yerly P, Blanche C, Hullin R, Giostra E, et al. Regression of advanced liver fibrosis after heart transplantation in a patient with prior Fontan surgery for complex congenital heart disease. Circ Heart Fail. 2018;11:e003754.
73. D'Souza BA, Fuller S, Gleason LP, Hornsby N, Wald J, Krok K, et al. Single-center outcomes of combined heart and liver transplantation in the failing Fontan. Clin Transpl. 2017;31 https://doi.org/10.1111/ctr.12892.
74. Cho Y, Kabata D, Ehara E, Yamamoto A, Mizuochi T, Mushiake S, et al. Assessing liver stiffness with conventional cut-off values overestimates liver fibrosis staging in patients who received the Fontan procedure. Hepatol Res. 2021;51:593–602.
75. Rathgeber SL, Guttman OR, Lee AF, Voss C, Hemphill NM, Schreiber RA, et al. Fontan-associated liver disease: spectrum of disease in children and adolescents. J Am Heart Assoc. 2020;9:e012529.
76. Schachter JL, Patel M, Horton SR, Mike Devane A, Ewing A, Abrams GA. FibroSURE and elastography poorly predict the severity of liver fibrosis in Fontan-associated liver disease. Congenit Heart Dis. 2018;13:764–70.
77. Wu FM, Earing MG, Aboulhosn JA, Johncilla ME, Singh MN, Odze RD, et al. Predictive value of biomarkers of hepatic fibrosis in adult Fontan patients. J Heart Lung Transplant. 2017;36:211–9.
78. Schwartz MC, Sullivan LM, Glatz AC, Rand E, Russo P, et al. Portal and sinusoidal fibrosis are common on liver biopsy after Fontan surgery. Pediatr Cardiol. 2013;34:135–42.
79. Ghaferi AA, Hutchins GM. Progression of liver pathology in patients undergoing the Fontan procedure: chronic passive congestion, cardiac cirrhosis, hepatic adenoma, and hepatocellular carcinoma. J Thorac Cardiovasc Surg. 2005;129:1348–52.

Other Issues in Vascular Liver Disorders: Consensus Statements of Panel 9

59

Andrea De Gottardi, Pierre-Emmanuel Rautou, Sarwa Darwish Murad, Valerie Paradis, Marco Senzolo, and Luis Téllez

Use of Anticoagulants in Non-Cirrhotic Vascular Liver Diseases

9.1 Low-molecular weight heparin and vitamin K antagonists are widely accepted and used in primary thrombosis of the portal venous system or hepatic venous outflow tract (A,1). (Unchanged)

9.2 There are no major concerns with safety of DOACs in patients with non-cirrhotic vascular liver diseases, as long as liver function is preserved. DOACs

A. De Gottardi
Gastroenterology and Hepatology, Ente Ospedaliero Cantonale, Lugano, Switzerland

Faculty of Biomedical Sciences, Università della Svizzera Italiana, Lugano, Switzerland

P.-E. Rautou (✉)
Service d'Hépatologie, and Inserm, UMR1149, Hôpital Beaujon, Clichy, France
e-mail: pierre-emmanuel.rautou@inserm.fr

S. D. Murad
Department of Gastroenterology and Hepatology, Erasmus MC University Medical Center Rotterdam, Rotterdam, The Netherlands
e-mail: s.darwishmurad@erasmusmc.nl

V. Paradis
Pathology Unit, Hôpital Beaujon, Assistance Publique-Hôpitaux de Paris, Clichy, France
e-mail: valerie.paradis@aphp.fr

M. Senzolo
Multivisceral Transplant Unit, Department of Surgery, Oncology and Gastroenterology, Padua University Hospital, Padua, Italy

L. Téllez
Servicio de Gastroenterología y Hepatología, Hospital Universitario Ramón y Cajal, IRYCIS, CIBERehd, Universidad de Alcalá, Madrid, Spain
e-mail: luis.tellez@salud.madrid.org

R. de Franchis (ed.), *Portal Hypertension VII*,
https://doi.org/10.1007/978-3-031-08552-9_59

should be used with caution in patients with impaired liver function (equivalent to Child-Pugh class B), as well as in patients with creatinine clearance below 30 mL/min. The use of DOACs in patients with severe liver dysfunction (equivalent to Child-Pugh C) is not recommended outside study protocols (C,2). (New)

Anticoagulation and Portal Vein Thrombosis (PVT) in Cirrhosis

9.3 Screening for PVT is recommended in all patients who are potential liver transplant candidates, at the time of screening for hepatocellular carcinoma (D,2). (Changed)

9.4 Occurrence of PVT in the presence of HCC does not directly imply vascular malignant invasion, but further imaging is recommended (CT scan and/or magnetic resonance imaging and/or contrast-enhanced ultrasonography) (D,2). (Changed)

9.5 Anticoagulation is recommended in patients with cirrhosis and (i) recent (<6 months) completely or partially occlusive (>50%) thrombosis of the portal vein trunk with or without extension to the SMV, (ii) symptomatic portal vein thrombosis, independently of the extension, or (iii) portal vein thrombosis in potential candidates for liver transplantation, independently of the degree of occlusion and extension (C,2). (New)

9.6 In potential liver transplant candidates, the goal of anticoagulation is to prevent re-thrombosis or progression of thrombosis to facilitate adequate portal anastomosis in LT and reduce post-transplant morbidity and mortality (C,1). (Changed)

9.7 Anticoagulation should be considered in patients with cirrhosis and minimally occlusive (<50%) thrombosis of the portal vein trunk that (i) progresses on short-term follow-up (1–3 months) or (ii) compromises the superior mesenteric vein. (C,2) (New)

9.8 Anticoagulation should be (i) maintained until portal vein recanalization or for a minimum of 6 months, (ii) continued after recanalization in patients awaiting liver transplantation, and (iii) considered to be continued on anticoagulation after recanalization in all others, while balancing benefits in preventing recurrence and increasing survival and the risk of bleeding (C,1). (New)

9.9 Patients with low platelet count (e.g., <50 G/L) are not only at higher risk of PVT, but also of bleeding complications on anticoagulation and require a case-by-case assessment (C,2). (Changed)

9.10 TIPS is recommended in patients with thrombosis of the portal vein trunk without recanalization on anticoagulation, especially in patients listed for liver transplantation (C,2). (New)

9.11 Anticoagulation is preferably initiated with LMWH and maintained with either LMWH, VKA, or DOAC. Advantages of LMWH are that its use is based on solid data. VKA carry challenges with regard to INR monitoring in patients with cirrhosis. Advantages of DOACs are that they are easier to use but less data are available (C,1). (Changed)

9.12 Currently available data suggest that there are no major safety concerns for DOACs in patients with Child-Pugh class A cirrhosis. Due to the possibility of accumulation, DOACs should be used with caution in Child-Pugh class B patients, as well as in patient with creatinine clearance below 30 mL/min. The use of DOACs in Child-Pugh class C patients is not recommended outside study protocols (B,2). (New)

9.13 DOACs likely have different safety-efficacy profiles in patients with cirrhosis, although at the moment no recommendation can be made in favor of a specific DOAC in this setting (D,2). (New)

Porto-Sinusoidal Vascular Disorder (PSVD)

9.14 Porto-sinusoidal vascular disorder (PSVD) is a broad clinicopathological entity encompassing non-cirrhotic portal fibrosis, idiopathic portal hypertension or non-cirrhotic intrahepatic portal hypertension, and various overlapping histological patterns including nodular regenerative hyperplasia, obliterative portal venopathy, hepatoportal sclerosis, and incomplete septal cirrhosis (B,1).(New)

9.15 Absence of portal hypertension does not rule out PSVD. Presence of common causes of liver disease (e.g., viral hepatitis, excessive alcohol consumption, metabolic syndrome, etc.) does not rule out PSVD, and both can coexist. Presence of portal vein thrombosis does not rule out PSVD, and both can coexist (B,1). (New)

9.16 PSVD should be considered in the following situations: (i) signs of portal hypertension contrasting with atypical features for cirrhosis (e.g., HVPG <10 mmHg; liver stiffness measurement <10 kPa; smooth liver surface and no atrophy of segment IV; hepatic vein-to-vein communications; although none of these features is considered pathognomonic for PSVD) or (ii) liver blood test abnormalities or portal hypertension in a patient with a condition known to be associated with PSVD (Table 59.1); or (iii) unexplained liver blood test abnormalities even without signs of portal hypertension (B,1). (New)

Diagnosis of PSVD

9.17 Porto-sinusoidal vascular disorder can be observed in the absence of clinical, laboratory, or imaging features of portal hypertension (B,1). (New)

9.18 A liver biopsy specimen of adequate size (>20 mm) and of minimal fragmentation—or otherwise considered adequate for interpretation by an expert pathologist—is required for the diagnosis of PSVD (C,1). (New)

Table 59.1 Conditions associated with PSVD

Type	Specific condition
Blood diseases	Aplastic anemia Myeloproliferative disorder Hodgkin's lymphoma Multiple myeloma, Waldenstrom
Prothrombotic conditions	Protein C/S deficiency Factor II or V gene mutation Antiphospholipid syndrome ADAMTS13 deficiency
Immunological disorders	Common variable immune deficiency (significant hypogammaglobulinemia of unknown cause, failure to produce specific antibodies after immunizations and susceptibility to bacterial infections) Autoimmune hepatitis Systemic lupus erythematosus Scleroderma Rheumatoid arthritis HIV Celiac disease Inflammatory bowel disease
Infectious	Repeated gastrointestinal infections
Drug-induced	Didanosine Azathioprine 6-Thioguanine Oxaliplatin
Genetic	Turner's syndrome Adams–Oliver syndrome TERT mutations Cystic fibrosis Familial cases

Adapted from Ref. [1]

9.19 Diagnosis of PSVD requires the exclusion of cirrhosis and of other causes of portal hypertension (B,1), together with one of the following three criteria (C,2): (i) at least one feature specific for portal hypertension; (ii) at least one histologic lesion specific for PSVD; or (iii) at least one feature not specific for portal hypertension together with at least one histologic lesion compatible although not specific for PSVD (Table 59.2). (New)

Table 59.2 Criteria in the definition of PSVD

	Feature of portal hypertension	Histological lesions suggestive of PSVD assessed by an expert pathologist
Specific	– Gastric, esophageal, or ectopic varices – Portal hypertensive bleeding – Porto-systemic collaterals at imaging	– Obliterative portal venopathy (thickening of vessel wall, occlusion of the lumen, vanishing of portal veins) – Nodular regenerative hyperplasia – Incomplete septal fibrosis (also called incomplete septal cirrhosis); this latter feature can only be assessed on liver explants and not on liver biopsies
Not specific	– Ascites – Platelet count $<150{,}000/\text{mm}^3$ – Spleen size ≥ 13 cm in the largest axis	– Portal tract abnormalities (multiplication, dilatation of arteries, periportal vascular channels, aberrant vessels) – Architectural disturbance: Irregular distribution of the portal tracts and central veins – Non-zonal sinusoidal dilatation – Mild perisinusoidal fibrosis

Adapted from Ref. [1]

Management of PSVD

9.20 Once the diagnosis of PSVD is made, patients should be screened for associated immunological diseases, prothrombotic or genetic disorders and exposure to drugs/toxins (D,2). (New) (Table 59.1).

9.21 Endoscopic screening for gastroesophageal varices is required at diagnosis of PSVD (C,1). (New)

9.22 The noninvasive Baveno VII criteria for screening of esophageal varices used in patients with cirrhosis cannot be applied to patients with PSVD (B,1). (New)

9.23 During follow-up, the frequency of endoscopic screening for varices has not yet been defined. Management according to cirrhosis guidelines is recommended, expect for stopping rules (D,2). (New)

9.24 There is insufficient data on which therapy should be preferred for portal hypertension prophylaxis in PSVD. Management according to cirrhosis guidelines is recommended (D,2). (New)

9.25 A contrast enhanced CT scan is suggested at diagnosis of PSVD in order to assess the anatomy/patency of the portal venous system and potential porto-systemic collaterals (D,2). (New)

9.26 Screening for portal vein thrombosis in patients with PSVD: there is no data on the best screening method and interval (D,2) (New). Doppler ultrasound every 6 months is suggested in patients with PSVD and features of portal hypertension (C,1) (New). In case of abdominal pain, Doppler ultrasound or cross-sectional imaging should be performed to rule out splanchnic vein thrombosis (B,1). (New)

9.27 No recommendation can be made for anticoagulation therapy to prevent the development of portal vein thrombosis in PSVD (D,2). (New)
9.28 In those patients developing PVT, anticoagulant therapy should be started according to recommendations for non-cirrhotic PVT (C,1). (New)
9.29 TIPS can be considered to treat severe complications of portal hypertension. Underlying/associated conditions, which negatively impact post-TIPS outcome, must be taken into account in making individual decision for TIPS insertion (C,2). (New)
9.30 Liver transplantation is an option in selected PSVD patients with severe or refractory complications of portal hypertension or with advanced liver dysfunction. Indication should be discussed in expert centers (D,2). (New)

Research Agenda

Anticoagulation in PVT in Cirrhosis

- Assessment of the safety and efficacy of each single DOAC in patients with cirrhosis
- Identification of indicators associated with a favorable outcome in patients with cirrhosis and PVT treated with anticoagulants
- Stopping rules of long-term anticoagulant treatment in patients with cirrhosis and PVT
- Advantages and disadvantages of prophylactic vs. full-dose anticoagulation in patients with cirrhosis and PVT
- Definition of response to treatment in patients with cirrhosis and PVT

PSVD

- Natural history of PSVD without portal hypertension
- Improvement of noninvasive methods to screen for PSVD (e.g., cross-sectional imaging, spleen stiffness measurement)
- Prophylaxis for PVT in patients with PSVD and signs of portal hypertension
- Incidence and predictors of development of PVT in patients with PSVD and efficacy of anticoagulation in this setting

Reference

1. De Gottardi A, Rautou PE, Schouten J, Rubbia-Brandt L, Leebeek F, Trebicka J, Murad SD, Vilgrain V, Hernandez-Gea V, Nery F, Plessier A, Berzigotti A, Bioulac-Sage P, Primignani M, Semela D, Elkrief L, Bedossa P, Valla D, Garcia-Pagan JC, VALDIG group. Porto-sinusoidal vascular disease: proposal and description of a novel entity. Lancet Gastroenterol Hepatol. 2019;4(5):399–411. https://doi.org/10.1016/S2468-1253(19)30047-0. PMID:30957754.

GPSR Compliance

The European Union's (EU) General Product Safety Regulation (GPSR) is a set of rules that requires consumer products to be safe and our obligations to ensure this.

If you have any concerns about our products, you can contact us on ProductSafety@springernature.com

In case Publisher is established outside the EU, the EU authorized representative is:

Springer Nature Customer Service Center GmbH
Europaplatz 3
69115 Heidelberg, Germany

Batch number: 10732023

Printed by Printforce, the Netherlands